AF333095

The Chemistry of the Morita–Baylis–Hillman Reaction

The Chemistry of the Morita–Baylis–Hillman Reaction

Min Shi
State Key Laboratory of Organometallic Chemistry, Shanghai Institute of Organic Chemistry, Chinese Academy of Science, Shanghai, P. R. China

Fei-Jun Wang
School of Chemistry and Molecular Engineering, East China University of Science and Technology, Shanghai, P. R. China

Mei-Xin Zhao
School of Chemistry and Molecular Engineering, East China University of Science and Technology, Shanghai, P. R. China

Yin Wei
State Key Laboratory of Organometallic Chemistry, Shanghai Institute of Organic Chemistry, Chinese Academy of Science, Shanghai, P. R. China

RSC Publishing

RSC Catalysis Series No. 8

ISBN: 978-1-84973-129-4
ISSN: 1757-6725

A catalogue record for this book is available from the British Library

Published by The Royal Society of Chemistry,
Thomas Graham House, Science Park, Milton Road,
Cambridge CB4 0WF, UK

Registered Charity Number 207890

For further information see our web site at www.rsc.org

Printed and bound in Great Britain by CPI Antony Rowe, Chippenham and Eastbourne

Preface

The Morita–Baylis–Hillman (MBH) reaction has received remarkable and increasing interest since it combines two important requirements, namely, atom economy and generation of functional groups. Indeed, the last decade has seen an exponential growth of the MBH reaction and its applications. In fact, the MBH reaction has already become one of the most powerful carbon–carbon bond-forming methods and is widely used in organic synthesis. Since the 1990s, more and more research groups have initiated work on different facets of this reaction, involving the scope of the substrates, novel catalysts (especially chiral catalysts), understanding of the mechanism and various synthetic applications of MBH adducts. As practitioners in the field of the Morita–Baylis–Hillman reaction, we have long felt the need for a reference book that is intended to give a better understanding of the chemistry of Morita–Baylis–Hillman reaction for many synthetic organic chemists, although several major reviews and a monograph in *Organic Reactions* (John Wiley & Sons, 1997) have been published. This book provides a more complete overview of the chemistry of the Morita–Baylis–Hillman reaction and is divided into the following chapters. The origin and growth of the Morita–Baylis–Hillman reaction, the reactant classes and reaction conditions are described in Chapter 1, including the corresponding catalytic mechanisms. Achiral or chiral catalytic systems to promote the Morita–Baylis–Hillman reaction are discussed in Chapter 2. Chapter 3 is devoted mainly to illustrating various basic transformations of functional groups in the Morita–Baylis–Hillman reaction adducts. The use of Morita–Baylis–Hillman adducts, or derivatives, as starting materials to construct compounds having a carbocyclic or a heterocyclic framework is reviewed in Chapter 4. Finally, Chapter 5 records applications of the Morita–Baylis–Hillman reaction in the total synthesis of natural products.

Some brilliant work may be left out owing to the inevitable publication deadline; thus we apologize sincerely and hope that in the event of a second

RSC Catalysis Series No. 8
The Chemistry of the Morita–Baylis–Hillman Reaction
By Min Shi, Fei-Jun Wang, Mei-Xin Zhao and Yin Wei
© Min Shi, Fei-Jun Wang, Mei-Xin Zhao and Yin Wei 2011
Published by the Royal Society of Chemistry, www.rsc.org

volume we can rectify these omissions. We also apologize in advance for any errors in this book, and welcome constructive comments from our readers in order to correct such errors in future editions. Feedback, further hints and tips would be most welcome.

We hope that this book – besides being of interest to chemists in academia and industry that require an introduction, an update or part of a coherent review to the field of Morita–Baylis–Hillman chemistry – will also stimulate the interest of undergraduate and graduate level students.

We are very grateful to the students in our group for their assistance with this book. We wish to acknowledge valuable suggestions and corrections given by Dr. Yin Wei, who read the whole manuscript. Our appreciation is also extended to the Royal Society of Chemistry, in particular to Dr. Merlin Fox, for his support and numerous discussions during the preparation of this book.

Contents

RSC Catalysis Series No. 8
The Chemistry of the Morita–Baylis–Hillman Reaction
By Min Shi, Fei-Jun Wang, Mei-Xin Zhao and Yin Wei
© Min Shi, Fei-Jun Wang, Mei-Xin Zhao and Yin Wei 2011
Published by the Royal Society of Chemistry, www.rsc.org

Chapter 5 Application of Morita–Baylis–Hillman Reaction for the Synthesis of Natural Products   **485**
Fei-Jun Wang, Yin Wei and Min Shi

CHAPTER 1

Morita–Baylis–Hillman Reaction

MEI-XIN ZHAO, YIN WEI AND MIN SHI

1.1 Introduction

The formation of carbon–carbon bonds is one of the most fundamental reactions in organic chemistry and, therefore, has been and remains an important challenge and a fascinating area in organic synthesis. Numerous reactions for the formation of carbon–carbon bonds have been discovered and exploited. Recent progress in organic chemistry has clearly established that the development of a reaction is dependent on two main criteria: atom economy and selectivity (chemo-, regio- and stereo-). Among the carbon–carbon bond-forming reactions, the Morita–Baylis–Hillman (MBH) reaction has become one of the most useful and popular routes, with enormous synthetic utility, promise and potential. The origin of Morita–Baylis–Hillman reaction dates back to 1968 to a pioneering report presented by Morita (phosphine catalyzed reaction)[1] and, subsequently, Baylis and Hillman described a similar amine-catalyzed reaction in 1972.[2] Although this reaction is promising and fascinating, unfortunately, it was ignored by organic chemists for almost a decade after its discovery. At the beginning of the 1980s, organic chemists such as Drewes, Hoffmann, Perlmutter, Basavaiah started looking at this reaction and exploring various aspects of it.[3] Especially, since the mid-1990s, particularly in last decade, this reaction and its applications have received remarkable growing interest, and the exponential growth of this reaction and its importance are evidenced by numerous research papers and several major reviews.[4] The reasons for the rapid growth of MBH reaction can be attributed to its several advantages:

1. the starting materials are commercially available and the reaction is suitable for large-scale production;
2. atom-economic nature;

RSC Catalysis Series No. 8
The Chemistry of the Morita–Baylis–Hillman Reaction
By Min Shi, Fei-Jun Wang, Mei-Xin Zhao and Yin Wei
© Min Shi, Fei-Jun Wang, Mei-Xin Zhao and Yin Wei 2011
Published by the Royal Society of Chemistry, www.rsc.org

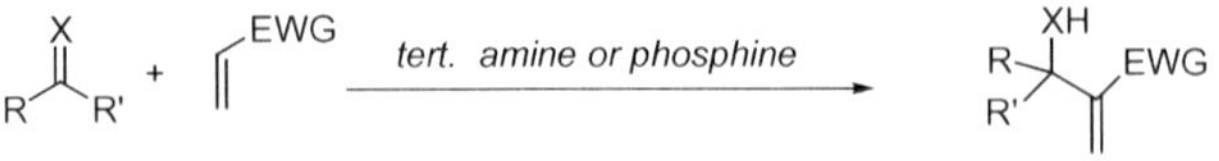

R = aryl, alkyl, heteroaryl, *etc.*; R' = H, CO$_2$R", alkyl, *etc.*
X = O, NCO$_2$R", NSO$_2$Ar, *etc.*
EWG = COR", CHO, CN, CO$_2$R", PO(OEt)$_2$, SO$_2$Ph, SO$_3$Ph, SOPh, *etc.*

Scheme 1.1

3. MBH adducts are flexible and multifunctional;
4. usually involves a nucleophilic organocatalytic system without the heavy-metal pollution;
5. mild reaction conditions.

The Morita–Baylis–Hillman (MBH) reaction can be broadly defined as a condensation of an electron-deficient alkene and an aldehyde catalyzed by tertiary amine or phosphine. Instead of aldehydes, imines can also participate in the reaction if they are appropriately activated, and in this case the process is commonly referred to as the aza-Morita–Baylis–Hillman (aza-MBH) reaction (Scheme 1.1). These operationally simple and atom-economic reactions afford α-methylene-β-hydroxy-carbonyl or α-methylene-β-amino-carbonyl derivatives, which consist of a contiguous assembly of three different functionalities.

Although several major reviews have discussed MBH/aza-MBH reaction and their applications in synthesis, it was difficult to completely overview the chemistry of MBH reaction due to a boom of research results in recent years. We hope that this book will satisfy the expectations of readers who are interested in the development of the field and looking for complete and up-to-date information on the chemistry of MBH reaction.

1.2   Mechanism of Morita–Baylis–Hillman Reaction

1.2.1   Amine-catalyzed Mechanism

Indisputably, a thorough understanding of reaction mechanism can lead to a better design of ligand or catalytic system. Whilst the elementary steps of the MBH reaction were postulated in the earliest publications,[1] the fine details of the reaction, in particular those controlling asymmetric induction, have been highlighted only recently, and remain at the core of mechanistic discussions. In 1983, Hoffmann first proposed a mechanism for the MBH reaction,[5] which was refined by kinetic studies by Hill and Isaacs[6] and others.[7] Their proposed mechanism is described in Scheme 1.2. The first reaction step I involves 1, 4-addition of the catalytic tertiary amine **1** to the activated alkene **2** (α,β-unsaturated carbonyl compounds, nitriles, *etc.*) to generate the *zwitterionic* aza-enolate (**3**). In step II, **3** forms intermediate **5** by adding to aldehyde **4** *via* an aldolic addition reaction. Step III involves intramolecular proton shift within **5** to form **6**, which subsequently generates the final MBH adduct and releases the catalyst **1** *via* E2 or E1cb elimination in step IV. Owing to the low kinetic

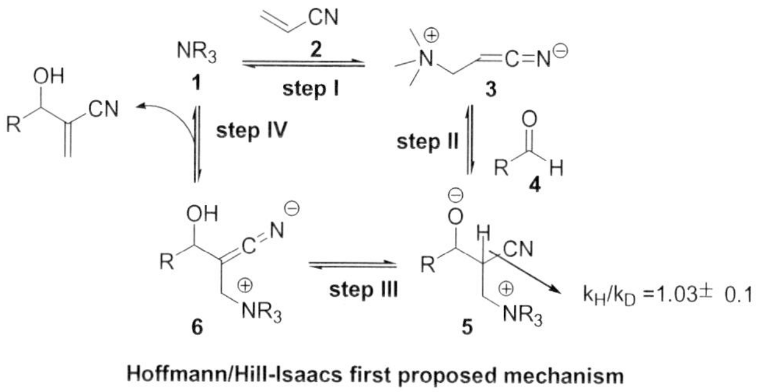

Hoffmann/Hill-Isaacs first proposed mechanism

Scheme 1.2

isotopic effect (KIE = 1.03±0.1, using acrylonitrile as electron-deficient alkene and acetaldehyde as carbon electrophile for the MBH reaction) measured by Hill and Isaacs and the dipole increase by charge separation, step II was initially considered as the MBH rate-determining step (RDS, Scheme 1.2).

Many observations in the MBH reactions could be explained by the above mechanism; however, it failed in some critical cases.[8] First, the mechanism did not provide any clue as to why stereocontrol is so difficult in MBH reactions. Privileged nucleophilic chiral catalysts,[9] which in the past have usually allowed good results in related asymmetric transformations, afforded only modest asymmetric induction. This fact pointed out that the basic factors governing the reactivity and selectivity of catalysts in MBH reaction were not fully understood. Other observations, such as the rate acceleration by the build-up of product (*i.e.*, the autocatalytic effect)[10] and also the formation of a considerable amount of "unusual" dioxanone byproduct, such as **11** (Scheme 1.3)[11] in the MBH reaction of aryl aldehydes with acrylates, warned of the limits of the discussed mechanism.

Recently, McQuade *et al.*[12] and Aggarwal *et al.*[13] have re-evaluated the MBH mechanism using both kinetics and theoretical studies, focusing on the proton-transfer step. According to McQuade, the MBH reaction is second order relative to the aldehyde and shows a significant kinetic isotopic effect (KIE: $k_H/k_D = 5.2 \pm 0.6$ in DMSO). Interestingly, regardless of the solvents (DMF, MeCN, THF, CHCl$_3$) the KIE were found to be greater than 2, indicating the relevance of proton abstraction in the rate-determining step. Based on these new data, McQuade *et al.* proposed a new mechanism for the proton-transfer step (Scheme 1.3), suggesting the proton transfer step as the RDS. Soon after, based on their kinetic studies, Aggarwal also proposed that the proton transfer step is the rate-determining step but only at its beginning ($\leq 20\%$ of conversion), then step II is the RDS when the product concentration builds up and proton transfer becomes increasingly efficient. Apparently, the MBH adducts **10** may act as a proton donor and therefore can assist the proton-transfer step *via* a six-membered intermediate (Scheme 1.3). This model also explains the autocatalytic effect of the product.

In addition, the Aggarwal proposed model shed light on the asymmetric catalysis of the MBH reaction. It suggested that all four diastereomers of the

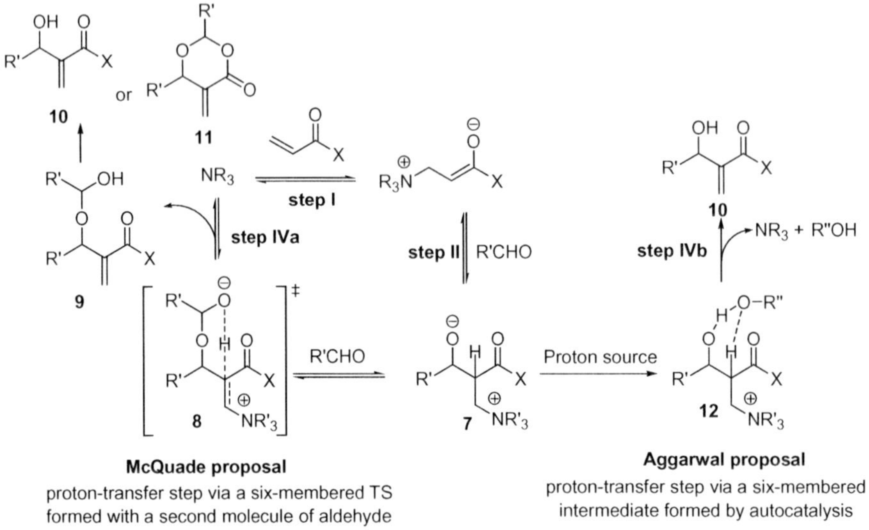

McQuade proposal
proton-transfer step via a six-membered TS
formed with a second molecule of aldehyde

Aggarwal proposal
proton-transfer step via a six-membered
intermediate formed by autocatalysis

Scheme 1.3

intermediate alkoxide are formed in the reaction, but only one has the hydrogen-bond donor suitably positioned to allow fast proton transfer, while the other diastereomers revert back to starting materials. These mechanistic studies directed attention to the proton-donor ability of the catalyst. If either the Brønsted acid or the Lewis base could be appropriately positioned on a chiral molecule, the Lewis base would react with the substrate (Michael addition), while the acid in an asymmetric environment would allow the chiral proton transfer. The Brønsted acid remains hydrogen-bonded to the resulting enolate in the enolate-addition step to the aldehyde,[14] and finally ensures efficient proton transfer in the rate-determining proton abstraction step. The action of the Brønsted co-catalysts, which are often employed in MBH reaction, is not limited to a role in proton transfer step. It rather promotes conjugate addition by binding to the zwitterionic enolate, and stabilizing these intermediates.

This new kinetic evidence has stimulated further theoretical studies on the MBH mechanism, conducted initially by Xu[15] and Sunoj.[16] Recently, Aggarwal performed an extensive theoretical study, which supported their own kinetic observations and those of McQuade about the proton transfer step.[17] They proposed that the proton-transfer step can proceed *via* two pathways: (i) addition of a second molecular of aldehyde to form a hemiacetal alkoxide (hemi1) followed by rate-limiting proton transfer as proposed by McQuade (non-alcohol-catalyzed pathway) and (ii) an alcohol that acts as a shuttle to transfer a proton from the α-position to the alkoxide of int2 (Scheme 1.4).

To investigate the MBH mechanism, Coelho and Eberlin *et al.* have also used electrospray ionization mass spectrometry [ESI-MS/MS)] to characterize key MBH reaction intermediates.[8] Using ESI-MS-(/MS), they also shed light on the co-catalytic role of ionic liquids in MBH reactions.[18] Stimulated by the

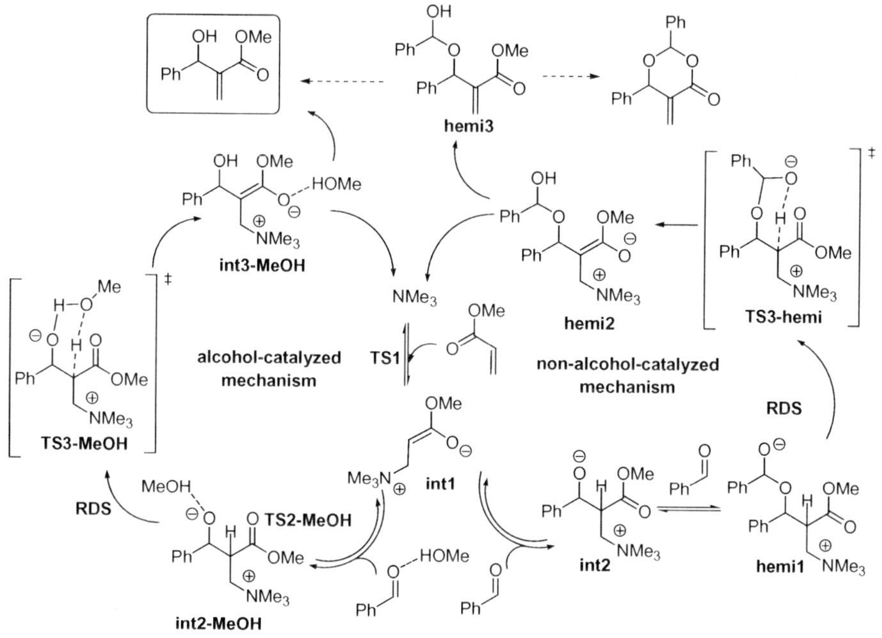

Scheme 1.4

mechanistically important new propositions about the proton-transfer step of the MBH reaction just discussed, Coelho *et al.* have performed complementary investigations on the MBH reaction mechanism *via* ESI-MS(/MS). New key intermediates for the rate-determining step of the MBH reaction have been successfully interpreted and structurally characterized (Figure 1.1), providing the first structural evidence supporting the mechanistic propositions made by McQuade *et al.* and Aggarwal *et al.*, based on kinetic experiments and theoretical calculations, for the dualistic nature of the proton-transfer step of the MBH mechanism.[19]

1.2.2 Phosphine-catalyzed Mechanism

The most likely mechanism of the MBH reaction catalyzed by tertiary phosphines is identical to that of the amine-catalyzed reaction except that the initially formed zwitterion **13** can isomerize to phosphorus ylide **14**, which can then undergo a Wittig reaction to give olefins **15** (Scheme 1.5). The latter process may require elevated temperatures, since it is not observed in reactions such as the (aza)-MBH reaction involving the more reactive α,β-unsaturated ketones under mild conditions.

This proposed mechanism has been supported by theoretical studies on the aza-MBH reaction between acrolein and mesyl imine catalyzed by trimethylamine and trimethylphosphine.[20] The relative energies of the crucial transition states for the PMe$_3$-catalyzed reaction have been found to be lower

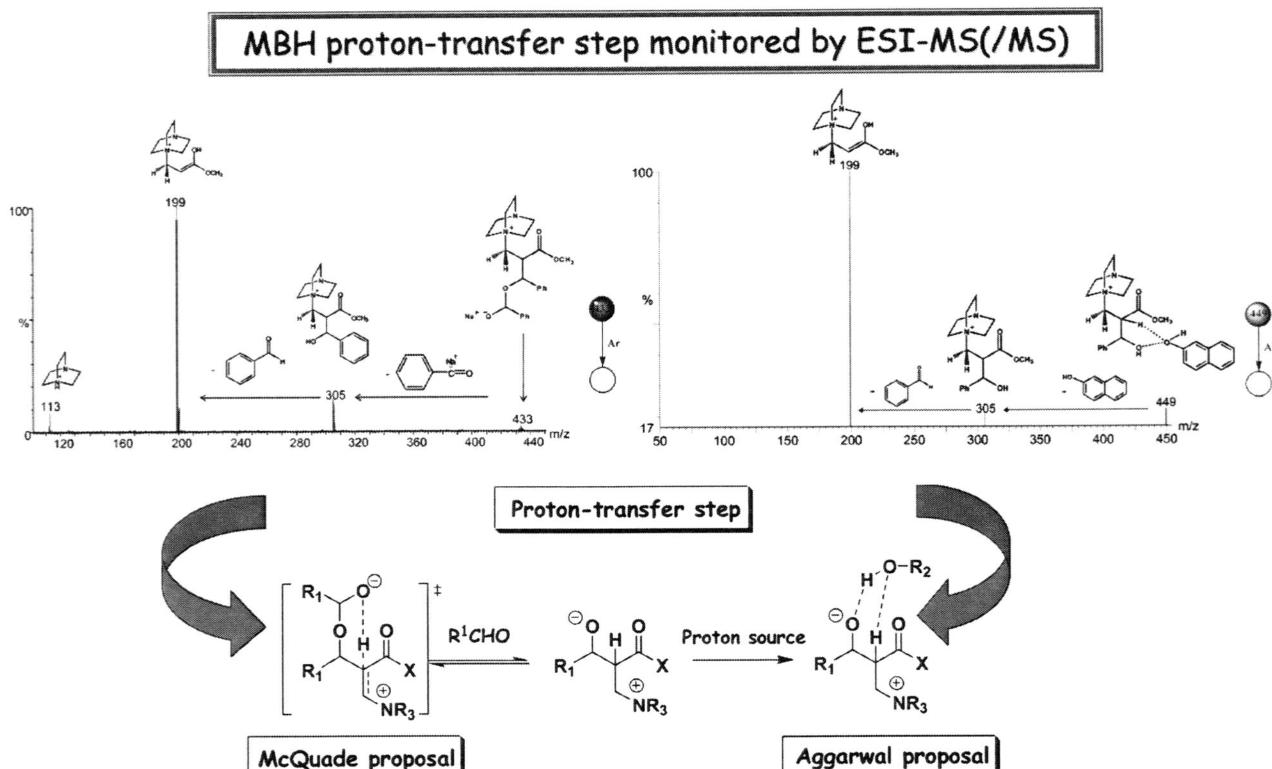

Figure 1.1 Key intermediates for the rate-determining step of the MBH reaction.

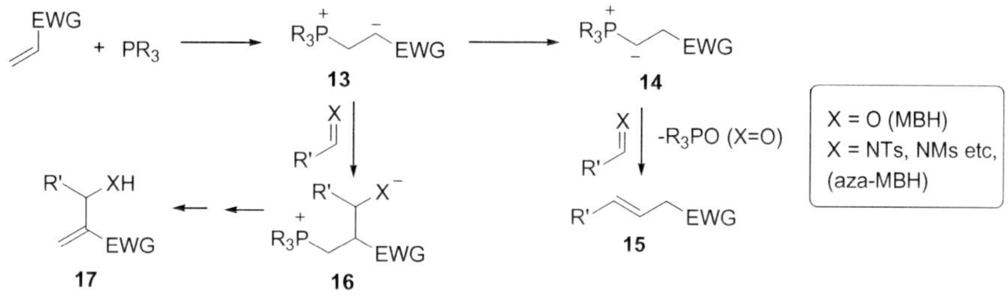

Scheme 1.5

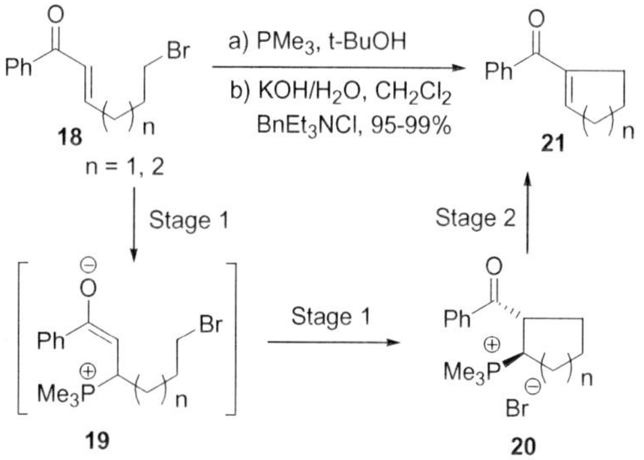

Scheme 1.6

than those of the corresponding NMe_3-catalyzed reaction. The kinetic advantage of the PMe_3-catalyzed reaction is also evident in the proton transfer step, where the energies of the transition states are much lower than those of the corresponding NMe_3-catalyzed reaction. These predictions are consistent with the available experimental reports in which faster reaction rates are in general noticed for the phosphine-catalyzed aza-MBH reaction.[21]

In addition, various phosphonium salts, the key intermediate in MBH reaction, have been synthesized and characterized to verify the phosphine-catalyzed MBH mechanism. Krafft *et al.* first isolated phosphonium salts **20** from a MBH alkylation and determined its structure by X-ray crystallography (Scheme 1.6).[22] The phosphonium salts **20** exhibit unprecedented *trans* geometry of the phosphonium salt and acyl group under kinetically controlled conditions, and lack the previously accepted electrostatic stabilization of the zwitterionic intermediate, suggesting that this electrostatic interaction is not the overriding electronic influence defining the stereochemical outcome of the cyclization. Moreover, these results also suggested that the oxygen–phosphorus electrostatic interaction in the transition state, long considered to be a key component in the traditional MBH reaction, is not a requirement for successful MBH alkylation.

Subseqently, Kwon *et al.* described the synthesis of stable phosphonium enolate zwitterions **22**, which have been proposed as intermediates in MBH

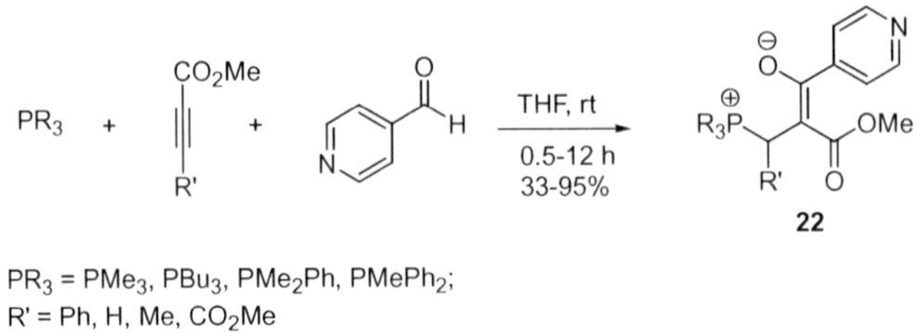
PR3
CO2Me
R'
N
O
H
THF, rt
0.5-12 h
33-95%
R3P
OMe
R'
O
22
PR3 = PMe3, PBu3, PMe2Ph, PMePh2;
R' = Ph, H, Me, CO2Me

Scheme 1.7

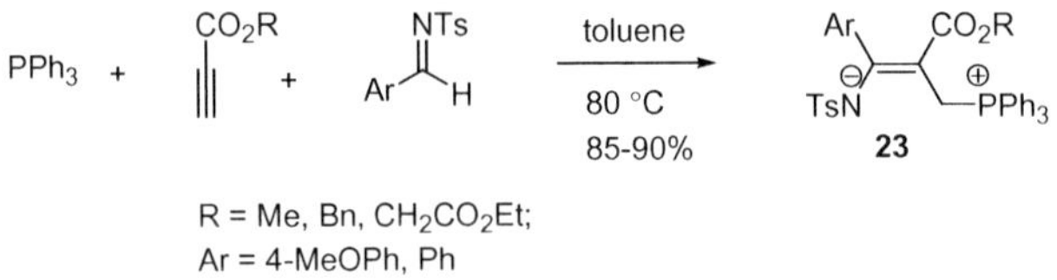
PPh3
CO2R
NTs
Ar
H
toluene
80 °C
85-90%
Ar
CO2R
TsN
PPh3
23
R = Me, Bn, CH2CO2Et;
Ar = 4-MeOPh, Ph

Scheme 1.8

reactions, through novel three-component coupling reactions of tertiary phosphines, alkynoates and aldehydes (Scheme 1.7). Notably, the reaction of $PMePh_2$ with methyl phenylpropiolate did produce a zwitterion that was observable in solution (NMR spectroscopy) but not isolable; however, no detectable zwitterion was observed in the case of PPh_3. These results are consistent with the hypothesis that electron-releasing alkyl substituents on the phosphonium center play a critical role in stabilizing phosphonium enolate zwitterions. Moreover, according to X-ray crystallography, such phosphonium enolate zwitterions (**22**) established the tetravalent nature of their phosphorus atoms unequivocally, which stands in contrast to those of the well-established pentavalent 1,2-λ^5-oxaphospholenes, and might explain the instability and high reactivity of phosphonium enolate zwitterions in MBH-type reactions.[23]

Most recently, Tong *et al.* have isolated a stable phosphonium–enamine zwitterion (**23**), which has long been postulated as one of the key intermediates in the aza-MBH reaction, from the PPh_3-catalyzed reaction between propiolate and *N*-tosylimine (Scheme 1.8).[24] Slightly different form Kwon's elegant work, PPh_3 worked well in this reaction probably because of the introduction of *N*-tosylimine as the electrophile. Thus, it was believed that either electron-releasing alkyl substituents on the phosphonium centre or electron-withdrawing groups on the anion centre play an important role in stabilizing the phosphonium zwitterions.

1.3 Activated Olefins

During the past 40 years, the Morita–Baylis–Hillman reaction has seen exponential growth in terms of three components, that is, the activated olefins, electrophiles and catalysts. Both substrate compatibility problems and selectivity issues have improved considerably, though they are not yet solved completely, and the range of olefin reagents has been extended. With the exception of the unusually high reactivity of phenyl vinylsulfonate, the reactivity of activated olefins increases with the electronegativity of the activating group, *i.e.* phenyl vinyl sulfoxide $\approx$ acrylamides < phenyl vinyl sulfone < acrylic esters $\approx$ ethyl vinylphosphonate < acrylonitrile < α,β-unsaturated ketones < α-crolein $\approx$ phenyl vinylsulfonate, as would be expected based on the mechanism of the MBH reaction. The nature of the catalyst does not appear to influence the reactivity order.[25]

1.3.1 Acrylates

Guided by initial studies on the addition of acetaldehyde with ethyl acrylate and acrylonitrile developed by Baylis and Hillman,[2] acrylates[3a-b,7b,11b,26] were first studied as activated olefins to react with aldehydes for the MBH reaction. To date, acrylates have constituted by far the largest group of activated olefins employed in the MBH reaction, probably due to the versatility of the ester group in further reactions.

In a series of additions of benzaldehyde to alkyl acrylates,[7b] it clearly appeared that the reaction rate decreased with steric bulk and with chain length of the alcohol, probably due it impeding the approach of the reagents (Scheme 1.9). The latter effect may also be steric in nature if the chain folded back on itself, or it could be a consequence of a less polar reaction medium since the reactions were carried out with a 30% excess of acrylate and without a solvent.

Esters with electro-withdrawing group at the β-position relative to the oxygen favored the MBH reaction. However, electronic effects are not the only factors since, for example, 2-fluoroethyl and 2-chloroethyl acrylates react

R	time (days)	yield (%)
Me	6	89
Me[a]	33	73
Et	7	79
n-Bu	4	85
i-Bu	16	85
t-Bu	28	65
n-Hexyl	9	82
n-Octyl	12	78
n-Decyl	14	75
2-Adamentyl[a]	62	40

a. dioxane as solvent

Scheme 1.9

rapidly, whereas 2-bromoethyl acrylate fail to react. Steric hindrance, as described above, also plays an important role, which could be responsible for the extremely low reactivity of acrylates containing a long-chain alkyl halide due to cluster aggregation around the reation sites. However, why such aggregation did not take place with 6-thiocyanohexyl acrylate is unclear (Scheme 1.10).

Aromatic esters of acrylic acid react more rapidly than aliphatic ones,[7b,11b] and there is no simple correlation of substituent σ values with rate (Scheme 1.11). Although it is not easy *to* interpret the effect of an electron-withdrawing group in the benzene ring, generally speaking it disfavors the MBH reaction, probably due to a decrease of the nucleophilicity of intermediate zwitterion. With strong electron-withdrawing groups, such as 4-trifluoromethylphenol

R	time (days)	yield (%)
CH_2CH_2F	3	81
CH_2CH_2Cl	3	61
CH_2CH_2Br	2	--
CH_2CH_2OH	2	20
CH_2CH_2OMe	4	89
$CH_2CH_2N(CH_3)_2$	8	82
$(CH_2)_6Cl$	15	--
$(CH_2)_6Br$	15	--
$(CH_2)_6SCN$	6	68
$(CH_2)_2SCN$	2	66

Scheme 1.10

R	σ	time (days)	yield (%)
$4\text{-}NMe_2$	-0.63	84	62
4-MeO	-0.28	8	54
4-Me	-0.14	36	55
2-Me	-0.06	36	54
H	0	5	55
4-F	0.15	72	39
4-Cl	0.24	72	42
$4\text{-}CO_2Me$	0.44	20	37
$3\text{-}CF_3$	0.46	24	22

Scheme 1.11

($\sigma = 0.53$), 3-cyanophenol ($\sigma = 0.62$), 4-cyanophenol ($\sigma = 0.70$) and 4-nitrophenol ($\sigma = 0.81$), only traces of expected adducts are observed after a much longer time.

Aliphatic aldehydes also react with aryl acrylates more rapidly than with alkyl acrylates, but yield instead the cyclic acetals **25**, arising from reaction of the initial MBH adduct **24** with a second molecule of aldehyde, exclusively or in a mixture with the normal adducts **24** (Scheme 1.12).[11b] More recently, α-naphthyl acrylates have also been shown to have a significant rate accelaration for the DABCO-catalyzed MBH reactions. Either the normal MBH adduct **26** or 1,3-dioxan-4-ones **27** could be obtained by controlling the substrate ratios and reaction time, respectively (Scheme 1.13).[11b,26]

Formation of cyclic acetals was also observed in the MBH reaction of aliphatic aldehydes with pantolactone acrylate **28**.[11a,27] The more stable *cis* isomers **29** are formed predominantly, and mixed products **30** can be isolated by sequential addition of two different aldehydes (Scheme 1.14).[11a] However, benzaldehyde failed to give the cyclic adduct on reaction with the pantolactone ester under the same conditions – electronic effects rather than steric hindrance

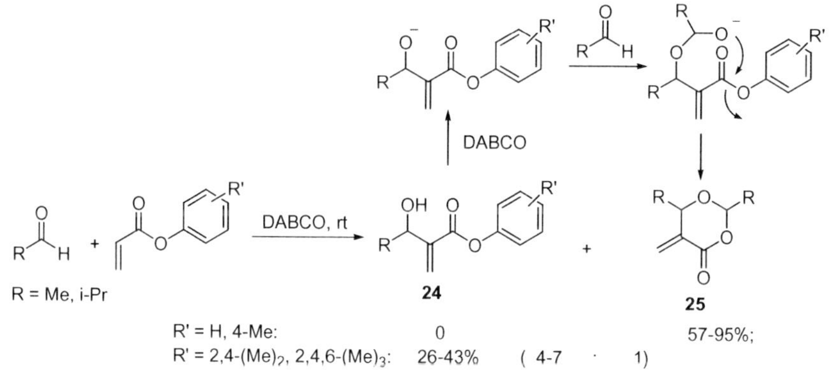

Scheme 1.12

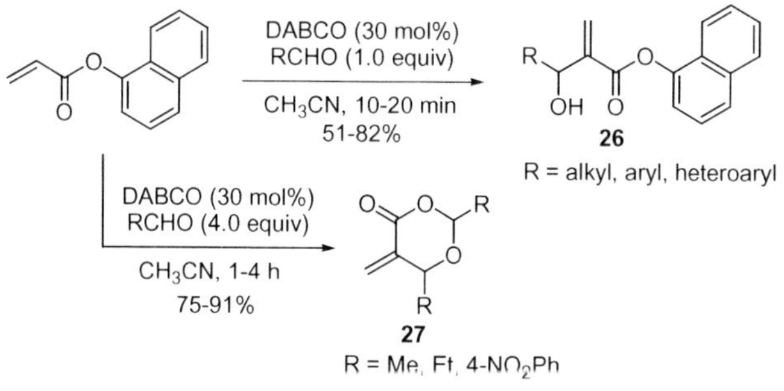

Scheme 1.13

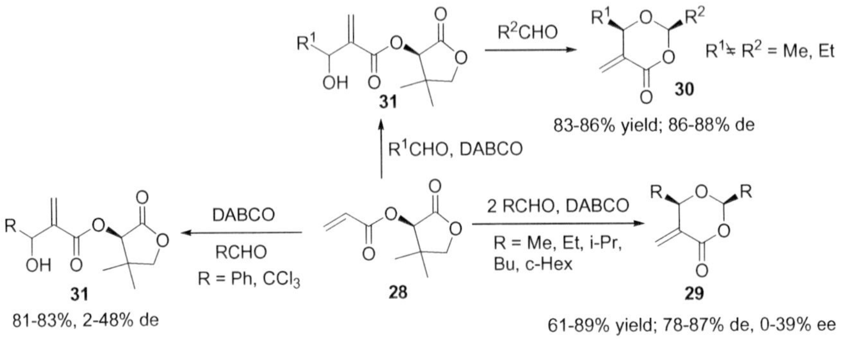

Scheme 1.14

were thought to prevent cyclization. Trichloroacetaldehyde did not cyclize either, which can be rationalized on a similar basis (Scheme 1.14). Acrylate esters such as the lactate and mandelate also show enhanced reactivity relative to aliphatic acrylates (24 h), but slower than for the pantolactone ester.[11a]

Drewes *et al.* found that a significant reduction in half lives ($t_{1/2}$) could be achieved if the substrate (activated olefins, such as methyl acrylate and methyl vinyl ketone) was reacted with highly electrophilic aldehydes and 3-hydroxyquinuclidine (3-HQD) was used as catalyst instead of DABCO, presumably due to the promotation by hydrogen bonding.[25,28a] However, in the reactions of aliphatic and aromatic aldehydes with a series of ω-hydroxyalkyl acrylates **32**, only small rate enhancements were observed even though addition of small amounts of an alcohol increased the rates of MBH reactions (Scheme 1.15).[28b]

Polymer-bound acrylic ester reacts with aldehydes in a MBH reaction to form 3-hydroxy-2-methylidenepropionic acids[29] or with aldehydes and sulfonamides in a three-component reaction to form 2-methylidene-3-[(arylsulfonyl)amino]propionic acids (aza-MBH adduct),[30] which can be used as a template for multiple core structure libraries (MCSL), such as aryl ketone, pyrazolones, allylic amines, isoxazole-based combinatorial libraries, *etc* (Chapter 2).[31]

To accomplish the highly diastereoselective MBH reaction and obtain the chiral MBH adducts, several chiral auxially contained acrylates, such as bornyl acrylate esters **33**,[32] sugar-derived acrylates **34**[33] and L-menthyl acrylates **35**,[34] have been developed and reacted with aldehydes catalyzed by DABCO or trimethylamine, leading to moderate to good diastereoselectivities (Scheme 1.16). Additionally, Krishna *et al.* have developed the strategy of double asymmetric induction by the coupling of chiral aldehydes with chiral sugar-containing acrylate **34** to obtain the corresponding adducts with high *syn* diastereoselectivities (de $>90\%$) in moderate to good yields (Scheme 1.16).[33b]

1.3.2 Acrylonitriles

Comparison of the data of the MBH reaction of aromatic and aliphatic aldehydes with methyl acrylate and acrylonitrile recorded in the literature showed that acrylonitrile appears to be somewhat more reactive towards aldehydes

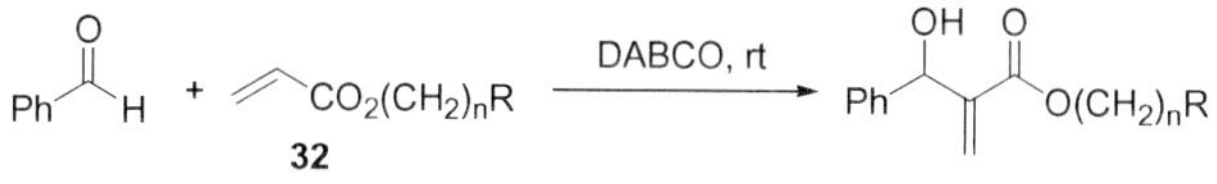

Scheme 1.15

n	R	time (days)	yield (%)
2	H	7	79
2	OH	5	50
3	OH	3	66
4	OH	3	85
6	OH	4	80
10	OH	6	78

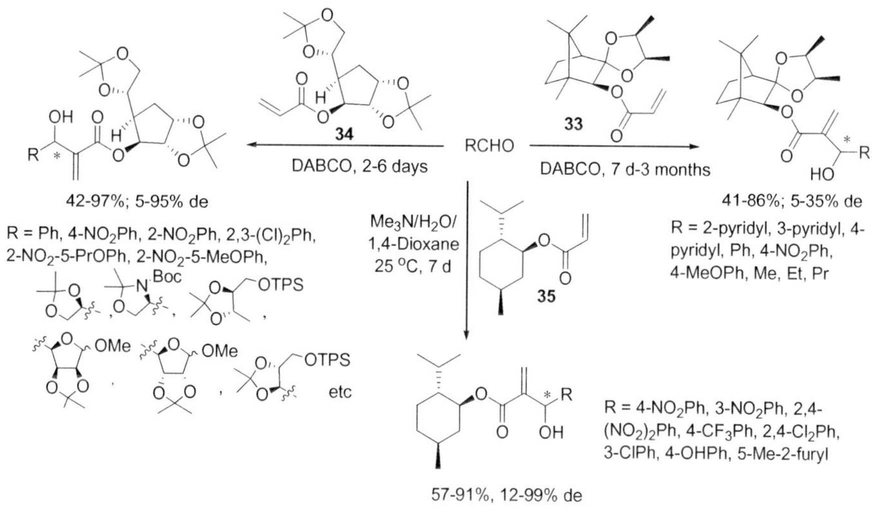

Scheme 1.16

than alkyl acrylates (Scheme 1.17).[35] In some cases, phosphine catalysts, especially an acid–base complex catalyst (tributylphosphine and triethyl-aluminium) appeared to be well suited for these reactions (Scheme 1.18).[10,36]

1.3.3　Allenes, Acetylenes and Dienes

More recently, allenes, acetylenes and activated dienes have been added as olefin equivalents, leading to products of more complex skeletons. Though the first report on allenic compounds involved in a MBH reaction dates back to the DABCO-catalyzed and butyllithium-promoted aldol condensation of allenic ester with aldehydes reported by Tsuboi *et al*,[37] there are very few reports of MBH reactions involving allenoates because they easily undergo cycloaddition reactions under different Lewis bases. For instance, five-membered pyrrolidine

R	EWG	time	yield (%)
Me	CO_2Me	7 days	88
Me	CN	40 h	76
Ph	CO_2Me	4 days	92
Ph	CN	40 h	79
MeCH=CH-	CO_2Me	20 days	31
MeCH=CH-	CN	2 days	59

Scheme 1.17

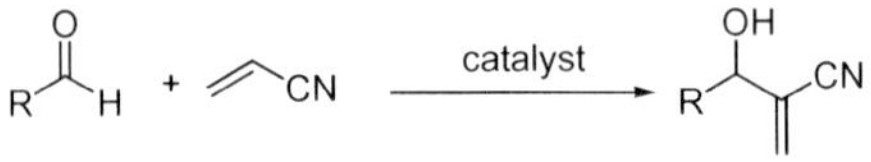

R = n-Bu, DABCO, no reaction;

    n-Bu$_3$P, Et$_3$Al, C$_6$H$_{14}$, CH$_2$Cl$_2$, sealed tube, 80 °C, 22 h, 90%;

R = n-nonyl, DABCO, rt, 7 days, 84% yield;

    n-Bu$_3$P, Et$_3$Al, CH$_2$Cl$_2$, sealed tube, 80 °C, 22 h, 74%;

R = 3-NO$_2$Ph, DABCO, rt, 24 h, 22%;

    (C$_6$H$_{11}$)$_3$P, dioxane, 30 °C, 6 h, 74%

Scheme 1.18

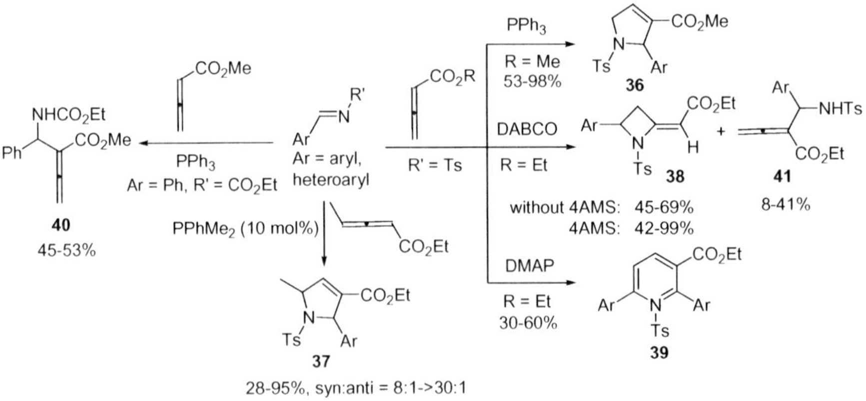

Scheme 1.19

derivatives **36** and **37**,[38] azetidine **38** or dihydropyridine derivatives **39**[38d,39] were obtained from the reactions of *N*-tosylated imines and 2,3-butadienoates or 2,3-pentadienoates catalyzed by phosphine, DABCO or DMAP, respectively (Scheme 1.19). Since *β*-substituted acrylates normally do not undergo the

MBH reaction, it was assumed that a change of mechanism involving direct formation of the anion by abstraction of the α proton had taken place when a stoichiometric amount of butyllithium was employed.

The normal aza-MBH adducts **40** could be formed in moderate yields in the PPh$_3$-catalyzed reaction between methyl 2,3-butadienoate and *N*-(ethoxy-carbonyl)benzaldimine,[38b] while only trace normal aza-MBH adducts **41** were afforded for DABCO-catalyzed reactions of *N*-tosylated imines with ethyl 2,3-butadienoate (Scheme 1.19).[38d] These results show that the reactivities of both imines and catalysts influence the final products using the same starting materials.

Recently, we reported the different reactivity patterns shown by nitrogen- and phosphorus-containing Lewis bases as catalysts in the reactions of *N*-Boc-imines **42** with ethyl 2,3-butadienoate – they differ from the previous observations in the normal aza-MBH reactions of other imines and are beyond the scope of the aza-MBH reactions.[40] The normal aza-MBH products **43** were obtained in good to excellent yields with DABCO as catalyst, whereas novel rearrangement product **44** could be formed in moderate yields by using PPh$_3$ as catalyst (Scheme 1.20).

At almost the same time, Miller *et al.* published a communication on pyridylalanine (Pal)-peptide **45** catalyzed enantioselective allenoate additions to *N*-acylimines **46** with moderate to good yields with high enantioselectivities (Scheme 1.21).[41]

As an analogue of allenic ester, allenic ketone **48**, such as 3-methylpenta-3,4-dien-2-one and 3-benzylpenta-3,4-dien-2-one, have also been used as activated olefins for the MBH reaction. The reactivities of both allenic ketones and

Scheme 1.20

Scheme 1.21

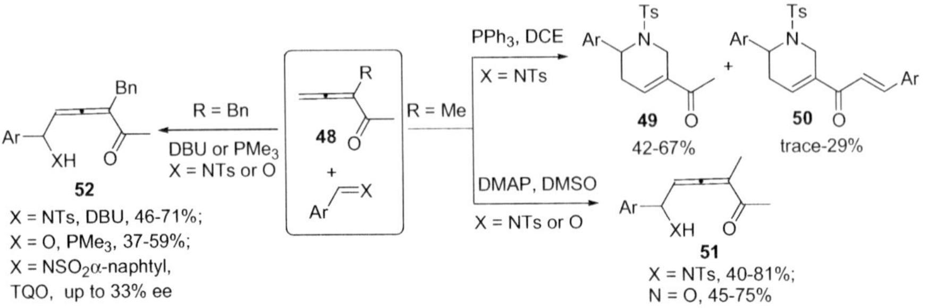

Scheme 1.22

catalysts, rather than imines or aldehydes, influence the final products. For 3-methylpenta-3,4-dien-2-one, tetrahydropyridine derivatives **49** and **50** were obtained from *N*-tosyl aldimines under the catalysis of tributylphosphine, while DMAP led to the acyclic adducts **51** as a pair of diastereoisomeric mixtures in a 1:1 ratio from *N*-tosyl aldimines and aryl aldehydes (Scheme 1.22).[39b] Even 3-benzylpenta-3,4-dien-2-one, with its greater steric hindrance, can undergo a MBH reaction with sulfonyl aldimines or aryl aldehydes to afford the corresponding acyclic adducts **52** in moderate yields in DMSO under, respectively, DBU or PMe$_3$ catalysis (Scheme 1.22). In comparison with 3-methylpenta-3,4-dien-2-one, 3-benzylpenta-3,4-dien-2-one provided better diastereoselectivities for the MBH reaction with *N*-arylmethylidene-1-naphthalene-sulfonamides catalyzed by the chiral catalyst cinchona alkaloid derivative TQO.[42]

Nemoto first developed the reaction of propiolates and aldehydes mediated by DABCO, which is analogous to the MBH reaction, obtaining novel β-functionalized MBH adducts **53** as major product in the case of aromatic aldehydes, whereas alkyne products **55** were afforded exclusively with aliphatic aldehydes (Scheme 1.23).[43] Subsequently, Xue *et al.* have reported that the three-component reaction of aldehydes and acetylenic ketones with 1,3-dicarbonyl compounds can be carried out under MBH reaction conditions to afford multi-carbonyl compounds **56** in moderate to high yields (Scheme 1.23).[44]

Back *et al.* have investigated the 3-hydroxyquinuclidine (3-HQD) catalyzed aza-MBH reaction of *N*-(benzenesulfonyl)imines with conjugated dienes **57**, where the *p*-toluenesulfonyl, ester, keto and cyano moieties were employed as the activating group, to afford the corresponding adducts **58** in moderate to high yields (Scheme 1.24).[45] Moreover, cyclization of the (*E*)-isomers of the products **58a** and **58b**, activated by sulfonyl and ester groups, respectively, was affected by base-catalyzed intramolecular conjugate addition of the sulfonamide group to the δ-position of the activated diene moiety (Scheme 1.24). Although the (*Z*)-isomers proved unreactive, their photoisomerization to the required (*E*)-configuration was readily achieved *in situ*, thus providing a convenient route to functionalized piperidine derivatives **59a,b** (Scheme 1.24).[45a,b] In addition, we have also reported the aza-MBH reaction of phenyl 2,4-pentadienoate with several imines.[46]

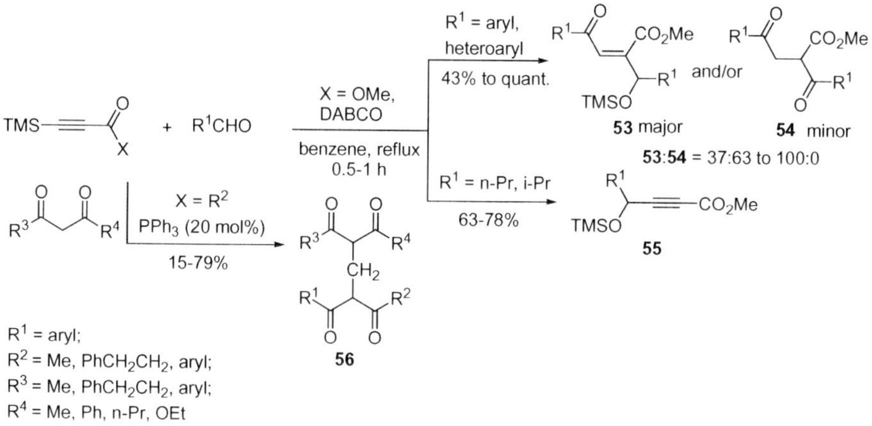

Scheme 1.23

1.3.4 Acrylamides

In an earlier report, the MBH reaction of aldehydes with acrylamides was thought to be inert under atmospheric pressure and at ambient temperature, and even the very reactive 2-pyridinecarboxaldehyde failed to react with either acrylamide or *N,N*-dimethylacrylamide.[47] With the aid of physical methods for accelerating MBH reaction rate, the amine-catalyzed reaction of acetaldehyde or acetone with acrylamide under 5 kbar pressure has since been reported to give the MBH product in 83% and 5% yield, respectively.[48] Similarly, microwave irradiation of a mixture of 3,4,5-trimethoxybenzaldehyde, acrylamide and DABCO in methanol for 25 min produces the MBH product in 40% yield.[49] The poor reactivities of acrylamides are not surprising since they are less electrophilic than most of the other activated olefins discussed in this chapter. However, camphor-derived Oppolzer's sultam **60**, which carries a second electron-withdrawing group on nitrogen, reacts readily with aliphatic aldehydes to give the cyclic products **61** with very high enantiomeric purity.[50] These cyclic products can easily be opened to give optically pure MBH esters **62** and **63** (Scheme 1.25). Other amides of this type, such as acryloyl imide **64**, can react with ethyl glyoxalate smoothly in the presence of DABCO to give mixtures of the corresponding MBH adducts **65** or **65′** enriched in either isomer, depending on the absence or presence of LiClO$_4$ in the reaction mixture (Scheme 1.26).[51]

Moreover, *N*-aryl acrylamide **66** have also been found to be a more activated Michael acceptor than acrylamides and *N*-alkyl acrylamides for MBH reaction. It is suggested that the delocalization of the lone pair of electron on the nitrogen of *N*-aryl acrylamide towards the aryl group increases the electron withdrawing capability of the carbonyl group, thus making *N*-aryl acrylamide a more reactive Michael acceptor.[52] *N*-Aryl acrylamides **66** could undergo a MBH addition reaction with activated aromatic aldehydes with DMF as

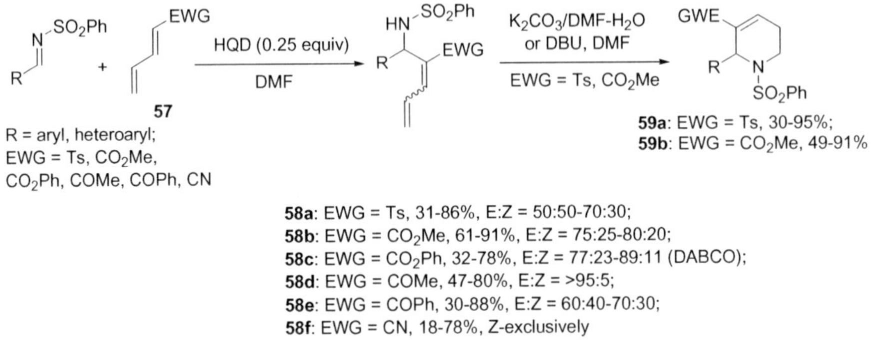

SO2Ph
N
R
+
EWG
57
R = aryl, heteroaryl;
EWG = Ts, CO2Me,
CO2Ph, COMe, COPh, CN
HQD (0.25 equiv)
DMF
HN SO2Ph
R
EWG
K2CO3/DMF-H2O
or DBU, DMF
EWG = Ts, CO2Me
GWE
R N
SO2Ph
59a: EWG = Ts, 30-95%;
59b: EWG = CO2Me, 49-91%
58a: EWG = Ts, 31-86%, E:Z = 50:50-70:30;
58b: EWG = CO2Me, 61-91%, E:Z = 75:25-80:20;
58c: EWG = CO2Ph, 32-78%, E:Z = 77:23-89:11 (DABCO);
58d: EWG = COMe, 47-80%, E:Z = >95:5;
58e: EWG = COPh, 30-88%, E:Z = 60:40-70:30;
58f: EWG = CN, 18-78%, Z-exclusively

Scheme 1.24

R H
+
N
S=O
O
60
DABCO (10 mol%)
CH2Cl2, 0 °C
33-98%, >99% ee
61
MeOH, CSA
R = Et
MeO2C
62 OH
85%
MeOH, Et3N
R = Et, i-Bu,
AcOCH2
MeO2C
R
63 OH
75-76%
R = Me, Et, n-Pr, i-Pr, PhCH2CH2, AcOCH2, i-Bu, BPSOCH2

Scheme 1.25

EtO2C H
+
N N
Ph
64
DABCO (50 mol%)
additive, solvent
OH O O
EtO2C
N N
Ph
65
+
OH O O
EtO2C
N N
Ph
65'
DMSO, rt, no LiClO4: 68% yield, dr = 85:15
DCM, 0 °C, LiClO4 (1 equiv): 67% yield, dr = 10:90

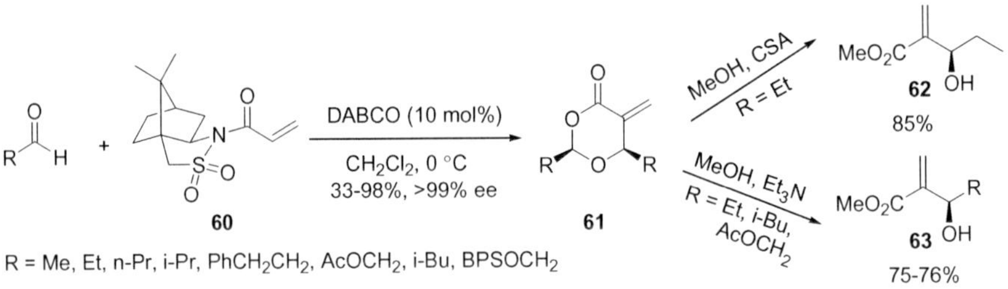

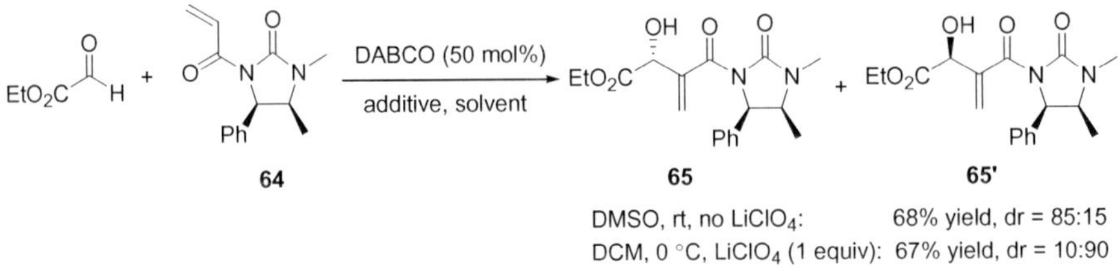

Scheme 1.26

solvent and DABCO as catalyst, giving a series of *N*-aryl acrylamide derivatives **67** and **68** (Scheme 1.27).[52a]

Furthermore, on the basis of convenient strategies to accelerate the MBH reaction by using a protic solvent and co-catalyst system, Hu and Yu[53] have found that acrylamide undergoes MBH coupling with aromatic aldehydes at ambient temperature in a 1:1 dioxane–H_2O solvent mixture in the presence of a stoichiometric amount of DABCO to give the corresponding MBH adducts in good yields. Water was thought to favor the formation of the zwitterionic intermediates and thus promoted the reaction. However, less electrophilic

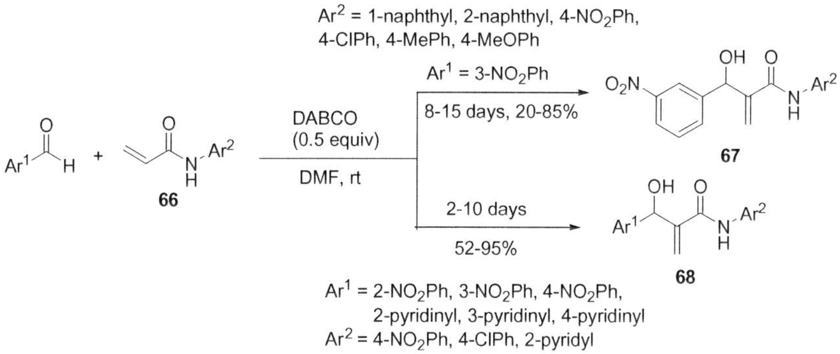

Ar^2 = 1-naphthyl, 2-naphthyl, 4-NO$_2$Ph,
4-ClPh, 4-MePh, 4-MeOPh

Ar^1 = 3-NO$_2$Ph

DABCO (0.5 equiv)

DMF, rt

8-15 days, 20-85%

2-10 days

52-95%

66

67

68

Ar^1 = 2-NO$_2$Ph, 3-NO$_2$Ph, 4-NO$_2$Ph,
2-pyridinyl, 3-pyridinyl, 4-pyridinyl
Ar^2 = 4-NO$_2$Ph, 4-ClPh, 2-pyridyl

Scheme 1.27

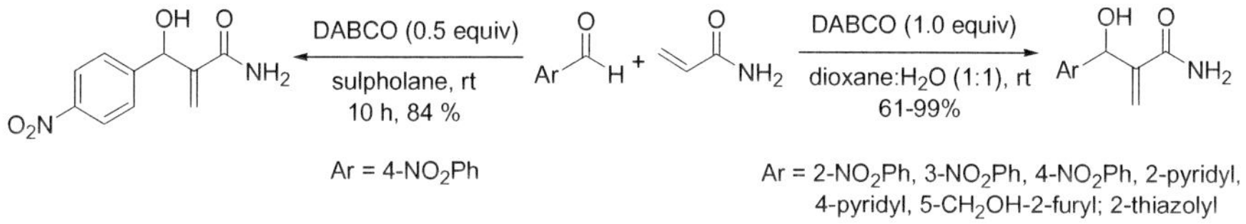

DABCO (0.5 equiv)
sulpholane, rt
10 h, 84 %

Ar = 4-NO$_2$Ph

DABCO (1.0 equiv)
dioxane:H$_2$O (1:1), rt
61-99%

Ar = 2-NO$_2$Ph, 3-NO$_2$Ph, 4-NO$_2$Ph, 2-pyridyl,
4-pyridyl, 5-CH$_2$OH-2-furyl; 2-thiazolyl

Scheme 1.28

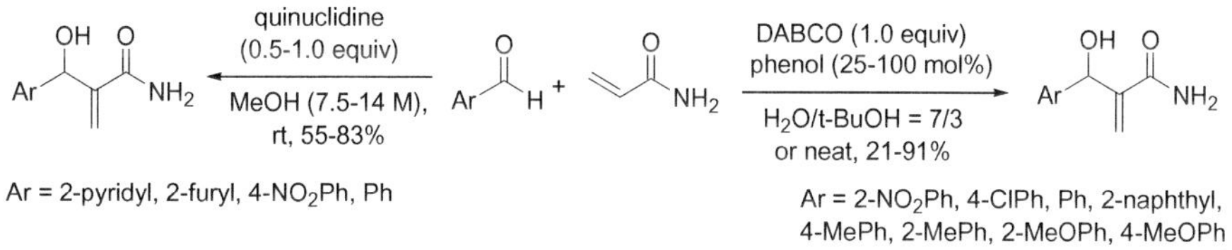

quinuclidine
(0.5-1.0 equiv)
MeOH (7.5-14 M),
rt, 55-83%

Ar = 2-pyridyl, 2-furyl, 4-NO$_2$Ph, Ph

DABCO (1.0 equiv)
phenol (25-100 mol%)
H$_2$O/t-BuOH = 7/3
or neat, 21-91%

Ar = 2-NO$_2$Ph, 4-ClPh, Ph, 2-naphthyl,
4-MePh, 2-MePh, 2-MeOPh, 4-MeOPh

Scheme 1.29

aldehydes (*i.e.* 4-fluorobenzaldehyde and 5-methylfuraldehyde) were unreactive (Scheme 1.28). By using sulfolane as solvent, an isolated example of the reaction between *p*-nitrobenzaldehyde and acrylamide has been reported by Krishna *et al.*,[54] affording the corresponding adduct in good yield without the need of a co-catalyst.

Subsequently, Aggarwal *et al.* reported the use of methanolic quinuclidine to promote the efficient reaction between acrylamide and activated aromatic aldehydes, including previously unreactive partners (benzaldehyde with acrylamide), to afford the corresponding adduct in 55% yield (Scheme 1.29).[55] In addition, Connon *et al.* have found that by the use of elevated temperature and the introduction of phenol as an additive, DABCO can effect clean, selective MBH reactions between acrylamide and aldehydes in alcoholic/aqueous media in which more basic nucleophilic catalysts, DMAP, DBU and quinuclidine,

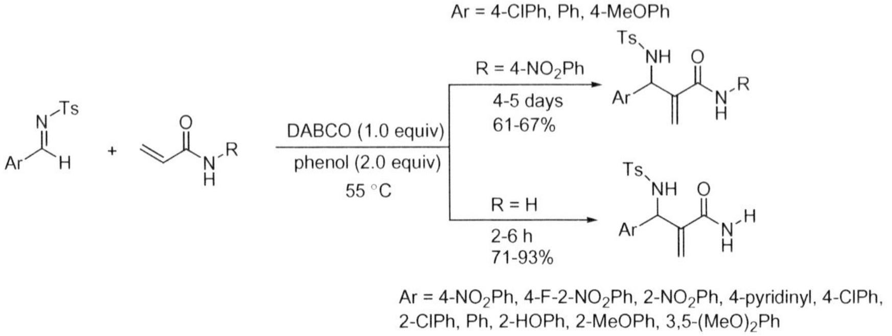

Scheme 1.30

promote MBH reaction preferentially.[56] Optimization of the reaction conditions has allowed acrylamide to be reacted even with deactivated aromatic aldehydes that were previously beyond the scope of the MBH reaction with acrylamide, affording products in moderate to excellent yields (Scheme 1.29).

Stimulated by Connon's report, Guo *et al.* have developed a highly efficient aza-MBH reaction between acrylamides or *N*-arylacrylamide with *N*-tosylated imines, using phenol as additive and DABCO as catalyst in the absence of solvent, to give the corresponding adducts in good yields (Scheme 1.30).[52b]

1.3.5   Acrolein

Acrolein rarely features as the activated olefins in the MBH reaction with aldehydes, clearly because of its propensity to form oligomers or polymers under the basic catalysts employed. However, this hurdle has been surmounted by selection of appropriate reaction conditions. The DABCO-catalyzed addition of acetaldehyde and propionaldehyde with acrolein proceeded in good yields under a low catalyst concentration, perhaps to minimize the polymerization of acrolein.[10] It is also the exclusive path in the attempted addition of 2-pyridinecarboxaldehyde[47] and α-diketones[57] to acrolein. In addition, successful MBH reactions of acrolein with aldehydes have been reported under high pressure, which affected a dramatic acceleration in the rate of MBH reaction.[48b] More reactive electrophiles, including halo ketones,[58] fluoro-carbonyls[59] and activated imines,[60] all reacted very rapidly with acrolein.

1.3.6   α,β-Unsaturated Ketones

The additions of aldehydes or imines to alkyl vinyl ketones proceed well with tertiary amine,[3f,35,61] tertiary phosphine,[62] and rhodium or ruthenium complex as catalysts (Scheme 1.31).[62c]

However, during our own investigation into the simple reaction of aryl aldehydes with methyl vinyl ketone (MVK) or ethyl vinyl ketone (EVK)

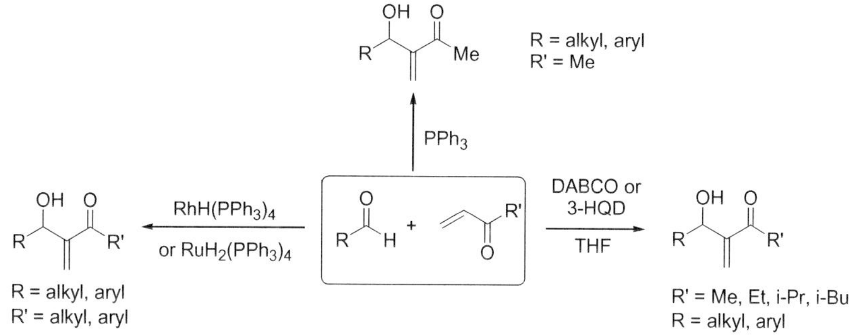

R = alkyl, aryl
R' = Me

PPh₃

RhH(PPh₃)₄ or RuH₂(PPh₃)₄

R = alkyl, aryl
R' = alkyl, aryl

DABCO or 3-HQD

THF

R' = Me, Et, i-Pr, i-Bu
R = alkyl, aryl

Scheme 1.31

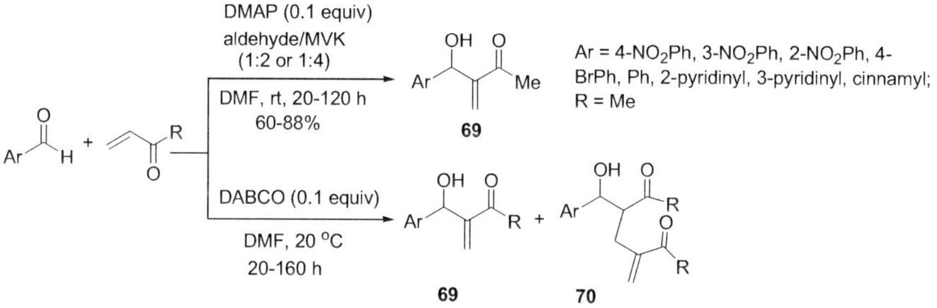

DMAP (0.1 equiv)
aldehyde/MVK (1:2 or 1:4)
DMF, rt, 20-120 h
60-88%

69

Ar = 4-NO₂Ph, 3-NO₂Ph, 2-NO₂Ph, 4-BrPh, Ph, 2-pyridinyl, 3-pyridinyl, cinnamyl; R = Me

DABCO (0.1 equiv)
DMF, 20 °C
20-160 h

69   **70**

aldehyde/MVK = 1:2	50-83%	0-25%	R = Me; Ar = 4-NO₂Ph, 3-NO₂Ph, 2-NO₂Ph, 4-BrPh
aldehyde/MVK = 1:4	50-62%	24-53% (*syn/anti* = 3:2 to 1:3)	R = Me; Ar = 4-NO₂Ph, 3-NO₂Ph, 4-BrPh, 2-pyridyl, cinnamyl
aldehyde/MVK = 1:8	41%	55% (*syn/anti* = 2:3)	R = Me; Ar = 4-NO₂Ph
aldehyde/EVK = 1:4	41-83%	trace	R = Et; Ar = 4-NO₂Ph, 3-NO₂Ph, 2-NO₂Ph, 2-pyridyl, 3-pyridyl, cinnamyl

Scheme 1.32

catalyzed by DABCO, we found that the reaction products are not as simple as those reported previously.[63] Using DMAP as catalyst, normal MBH adducts **69** were formed exclusively. In the DABCO-catalyzed MBH reaction of aryl aldehydes and MVK, besides the normal MBH adducts **69**, the diadduct **70** can also be formed at the same time; the yield of **70** can reach 55% by increasing the amount of methyl vinyl ketone (aldehyde/MVK = 1:8) (Scheme 1.32), but for EVK, only the normal MBH adduct **69** was obtained.

In addition, in the aza-MBH reaction between MVK and *N*-Ts imines, a significant Lewis base effect on this reaction has been observed.[64] Though DMAP, DABCO and dppe were effective for the aza-MBH reaction, the weaker Lewis base PPh₃ was the best catalyst and gave the normal MBH

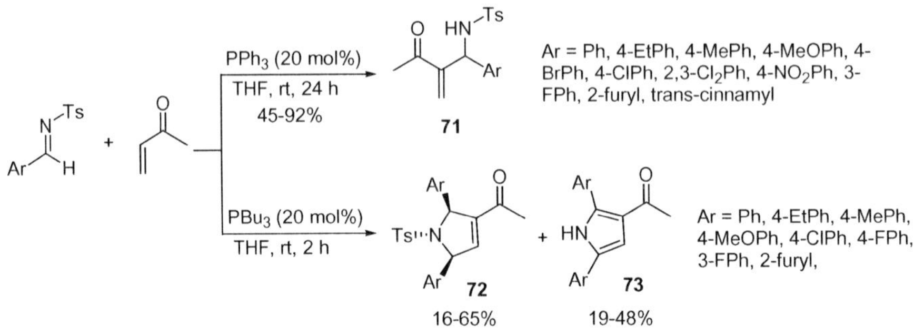

Scheme 1.33

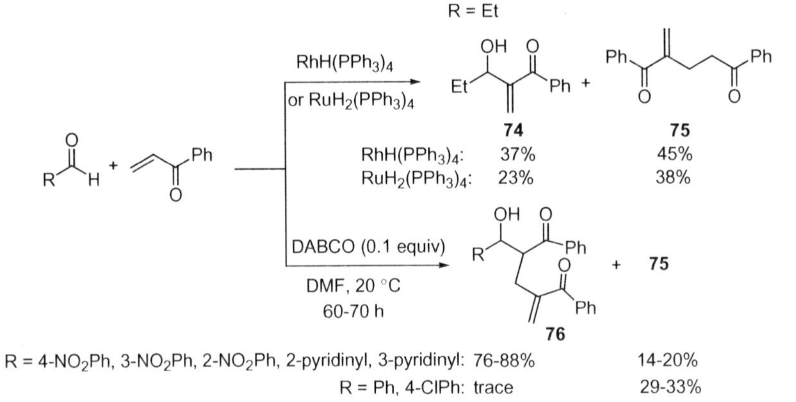

R = Et

Scheme 1.34

R = 4-NO$_2$Ph, 3-NO$_2$Ph, 2-NO$_2$Ph, 2-pyridinyl, 3-pyridinyl: 76-88%   14-20%

R = Ph, 4-ClPh: trace   29-33%

adducts **71** in good to very high yields (Scheme 1.33). When using the stronger Lewis base Bu$_3$P as catalyst, the abnormal MBH adducts **72** and **73** were formed under the same reaction conditions (Scheme 1.33).

For aryl vinyl ketones, probably due to their rapid dimerization in the presence of DABCO, there is only one example of an MBH reaction, between phenyl vinyl ketone (PVK) and propionaldehyde catalyzed by rhodium or ruthenium complexes,[62c] providing the corresponding adduct **74**, along with PVK dimer **75**. However, during our reinvestigation on the MBH reactions of aromatic aldehydes with PVK, we found that with electron-deficient aryl aldehydes, such as nitrobenzaldehydes or pyridyl aldehydes, the reaction proceeded smoothly to give exclusively diadduct **76** in good yields. With *p*-chlorobenzaldehyde or benzaldehyde, only trace amounts of the diadduct **76** were obtained and the PVK dimer **75** was formed almost exclusively (Scheme 1.34). Notably, increasing the amount of PVK did not improve the yields of diadduct **76** and the normal MBH adduct was not formed in all cases.[65]

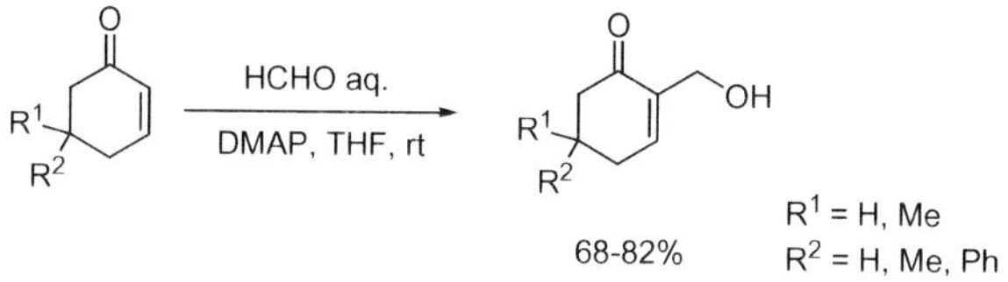

Scheme 1.35

Cyclic enones, which are less sterically encumbered than acyclic β-substituted activated olefins, have been widely employed in the MBH reaction in the past decade. During their synthesis of vitamin D3 metabolites, Uskokoviæ *et al.* demonstrated, for the first time, the tributylphosphine-catalyzed MBH reaction between 2-cyclopentenone and formaldehyde.[66] Later, Gaied *et al.* systematically investigated the coupling between substituted 2-cyclohexenones and formaldehyde, catalyzed by DMAP in THF, affording the corresponding adducts in good yields (Scheme 1.35).[67]

Since then, the MBH reaction of cyclic enones with various aldehydes has been widely studied. Various catalyst systems have been found to be viable catalysts in promoting the reactions of various aldehydes and cyclic enones, such as DMAP[68] and N,N,N',N'-tetramethyl-1,3-propanediamine (TMPDA)[69] in aqueous THF, triazole[70] and imidazoles[71] in alkaline solution, DBU[72] and methoxide anion[72b] in methanol, and tertiary phosphine.[73]

However, during our investigation on the aza-MBH reactions of cyclic enones with N-Ts imines, we found that the reaction is very complicated. The Lewis bases, solvents, substrates and the ring-size of the α,β-unsaturated cyclic ketone can all affect significantly the aza-MBH reaction rate and even the reaction product.[74] The aza-MBH reactions of N-Ts imines with 2-cyclohexenone or 2-cyclopentenone can be greatly accelerated in the presence of catalytic amounts of DMAP to give the normal MBH adducts **77** and **78** in good or excellent yields. Moreover, using PBu$_3$ as a Lewis base in the reaction of N-Ts imines with 2-cyclopentenone, affords the normal MBH adducts **77** in very high yields within 5–6 h; however, using PBu$_3$ or DBU as a Lewis base in the reaction of N-Ts imines with 2-cyclohexenone, besides the normal MBH adduct **78**, gives the abnormal MBH adduct **79** at the same time. In addition, the MBH reactions between N-Ts imines and 2-cycloheptenone or 2-cyclooctenone are very sluggish in the presence of a range of Lewis bases, and the abnormal MBH adducts **81** or **82** derived from aldol condensations are obtained in moderate yields together with the normal MBH adduct **80**. The MBH reaction between imines and 2-cyclooctenone in methanol gave **83**, derived from Michael additions of methanol to **82**, as the major products, along with traces of **82** (Scheme 1.36) (Chapter 2.3.1). From above results, we concluded that, in general, cyclopentenone undergoes an MBH reaction exclusively, while the MBH reaction and the aldol condensation reaction take place simultaneously for cyclohexenone or cycloheptenone; for large-sized α,β-unsaturated cyclic ketones, such as 2-cycloocten-1-one, only aldol condensation reactions occurred.

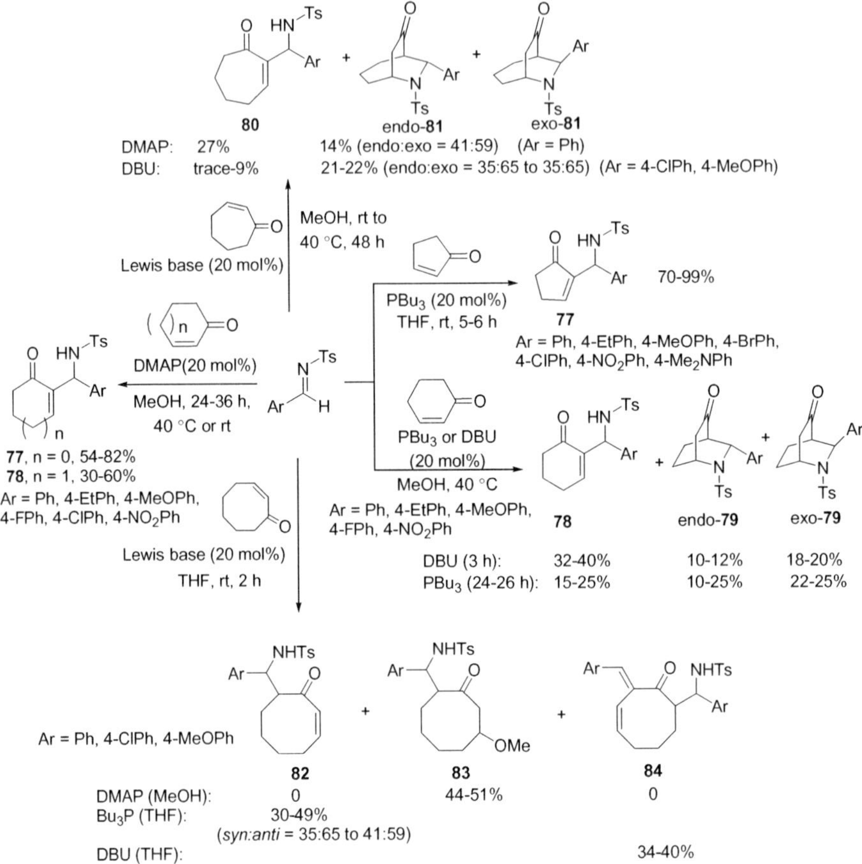

Scheme 1.36

1.3.7 β-Substituted Activated Olefins

According to the literature reported to date, successful examples of the MBH reaction of aldehydes with β-substituted activated olefins such as crotonate, crotononitrile or crotonaldehyde have only been achieved under high pressure[10,75] or by microwave irradiation[49] (Scheme 1.37). This is not surprising because the zwitterionic ammonium species, generated by Michael addition of the nitrogen nucleophilic Lewis base (DABCO) to the α,β-unsaturated enone according to the generally accepted mechanism of the MBH reaction, is difficult to form in high concentration in the reaction solution due to the steric bulkiness of β-substituted activated olefins.

We were encouraged by the fact that using *N*-tosyl imines instead of their aryl aldehydes means that MBH coupling with MVK or methyl acrylate can be accelerated to some extent to provide β-amino carbonyl compounds in higher yields, especially for those having electron-donating groups such as Et or MeO on the benzene ring, which usually showed low reactivities in the MBH

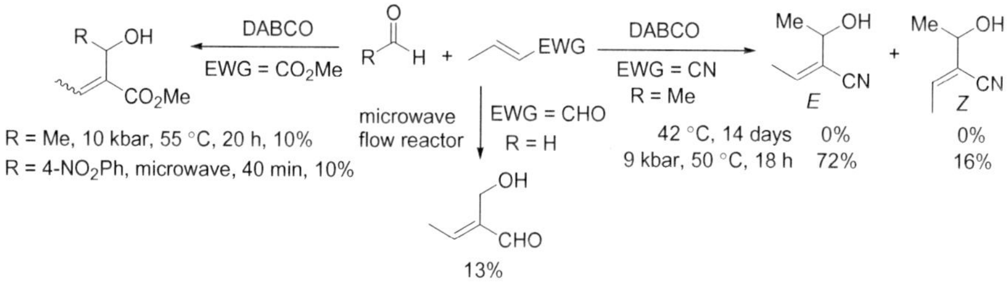

Scheme 1.37

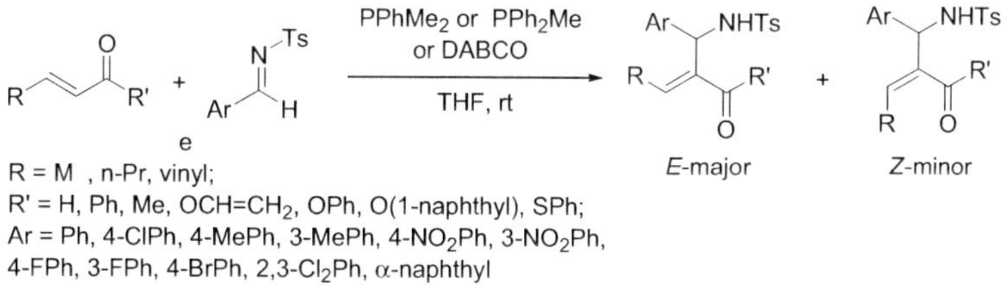

R = Me, 10 kbar, 55 °C, 20 h, 10%
R = 4-NO$_2$Ph, microwave, 40 min, 10%

EWG = CO$_2$Me

microwave flow reactor

EWG = CHO
R = H

13%

DABCO
EWG = CN
R = Me

42 °C, 14 days   0%   0%
9 kbar, 50 °C, 18 h   72%   16%

E   *Z*

Scheme 1.38

PPhMe$_2$ or PPh$_2$Me or DABCO

THF, rt

E-major   *Z*-minor

R = Me, n-Pr, vinyl;
R' = H, Ph, Me, OCH=CH$_2$, OPh, O(1-naphthyl), SPh;
Ar = Ph, 4-ClPh, 4-MePh, 3-MePh, 4-NO$_2$Ph, 3-NO$_2$Ph,
4-FPh, 3-FPh, 4-BrPh, 2,3-Cl$_2$Ph, α-naphthyl

reactions using amine or phosphine as catalyst.[76] Consequently, we demonstrated, for the first time, the aza-MBH reaction of α,β-substituted activated olefins such as crotonaldehyde, (*E*)-propenyl phenyl ketone, hex-2-enal or pent-3-en-2-one and β-substituted α,β-unsaturated esters with *N*-tosyl imines in the presence of tertiary phosphine or amine Lewis base promoter to give the corresponding adducts in major (*E*)-form in moderate to good yields under mild reaction conditions (Scheme 1.38).[46,77]

Subsequently, the MBH reaction of β-substituted crotonaldehyde with imines or nitroolefins catalyzed by the combination of imidazole or DABCO with proline was independently reported by Barbas[78] and Córdova *et al.*[79] to furnish highly enantiomerically enriched MBH-type products with β-substituted enal moieties with excellent (*E*)-selectivity (Chapter 2.6.1).

1.3.8 Nitroalkenes

Despite of the fact that Baylis and Hillman reported the synthesis of α-hydroxyethylated nitroethylene through the reaction between nitroethylene and acetaldehyde in the presence of DABCO, nitroalkenes employed as activated olefins in MBH reaction have not received much attention until recently. Prompted by the fact that nitroalkenes have shown superior Michael acceptor abilities, and that the first step in the MBH reaction is the Michael-type addition of the catalyst to substrate, Namboothiri *et al.* have published a series of papers on nitroalkenes involved the MBH reaction. The MBH reactions between nitroalkenes and various electrophiles such as formaldehyde,[80] activated carbonyl compounds,[81] imines,[82] alkenes[83] and azodicarboxylates[84] in the presence of imidazole or DMAP as catalyst have met with success in recent years. Except for both aromatic and aliphatic conjugated nitroalkenes, nitrodiene was also a suitable activated olefin to react with formaldehyde, trifluoropyruvate or glyoxylate, affording the corresponding MBH adduct in good yields.[81b] Additionally, Ballini has reported the MBH reaction using nitroalkenes as activated alkenes, ethyl-2-bromomethylacrylate as electrophilic acceptor and DBU as catalyst to give nitro dienes in good yields in very short reaction times.[85]

1.3.9 Vinyl Sulfones, Vinyl Sulfoxides and Vinyl Sulfonates

Only the additions of aldehydes to these substrates that contain a phenyl group as the activating substituent have been reported so far. Phenyl vinyl sulfone reacts with aldehydes at ambient pressure and temperature, but reaction times of a couple of days to weeks are common, especially with less reactive aldehydes (Scheme 1.39).[86] In view of the poor reactivity, the reaction conditions were further optimized. Among the various catalysts screened, DABCO proved to be the best; DBU was too basic and led to polymerization. Heating (130 °C, sealed tube) could speed up the reaction but yields were lower by 10–20%.[86b]

Since vinyl sulfoxides are poorer Michael acceptors than vinyl sulfones, the DABCO-catalyzed MBH reaction of vinyl sulfoxide with aldehydes required more forcing conditions to achieve a "satisfactory" yield, albeit with low diastereo-selectivity (Scheme 1.40).[87] No reaction was observed at ambient pressure.[86b]

Phenyl vinyl sulfonate is surprisingly reactive (Scheme 1.41),[88] but only one example of this class of compounds has been reported.

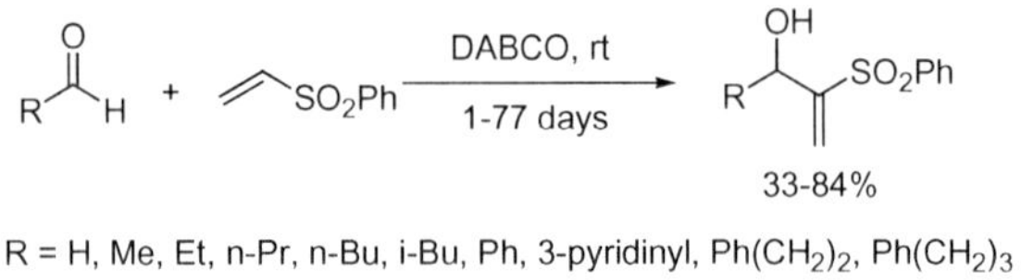

R = H, Me, Et, n-Pr, n-Bu, i-Bu, Ph, 3-pyridinyl, Ph(CH$_2$)$_2$, Ph(CH$_2$)$_3$

Scheme 1.39

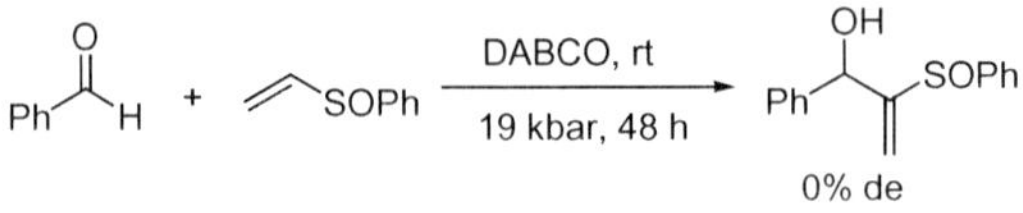

Scheme 1.40

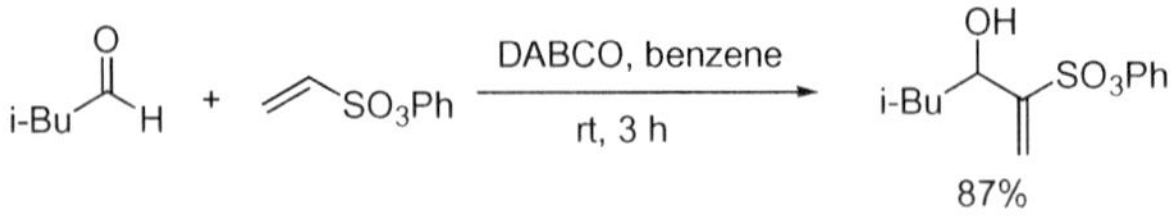

Scheme 1.41

1.3.10 Vinylphosphonates

The MBH reactions of diethyl vinylphosphonate dates back to 1990, and they present similar reactivities to alkyl acrylates.[89] It was found that diethyl vinylphosphonate can be coupled with various aliphatic aldehydes in the presence of DABCO to give moderate to high yields of the corresponding α-hydroxyalkyl phosphonates **85**. However, this approach was not effective for aqueous formaldehyde and polyoxymethylene, and the corresponding α-hydroxymethyl phosphonates **87** can only be obtained through a Wittig–Horner-type reaction from tetraethyl methylenediphosphonate **86** and formaldehyde (30% aq.) in the presence of a weak base K_2CO_3 (6–8 M) (Scheme 1.42).

1.3.11 Miscellaneous Activated Olefins

As well as the above-mentioned activated olefins, many novel activated alkenes (Figure 1.2) have been developed in recent years, including [α-(ethoxycarbonyl)vinyl]aluminum **88**,[90] chiral acryloylhydrazide **89**,[91] α,β-unsaturated thiol esters **90**,[92] uridine derivatives **91**,[93] 1-benzopyran-4(4*H*)-one derivatives

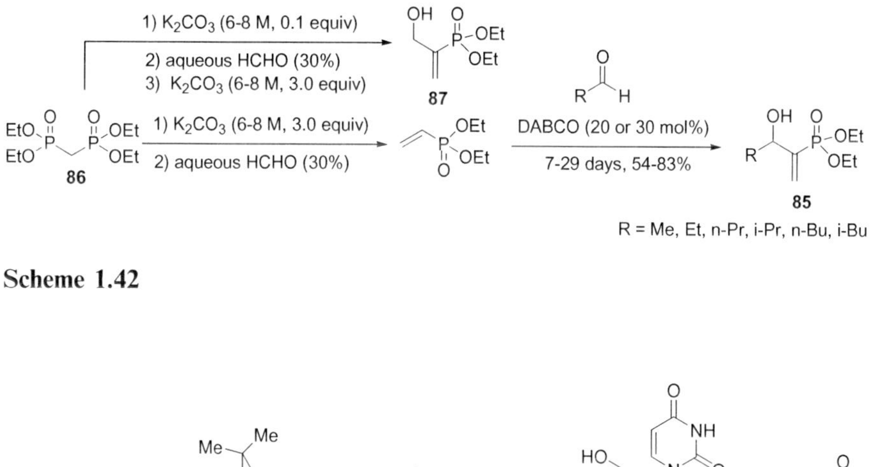

Scheme 1.42

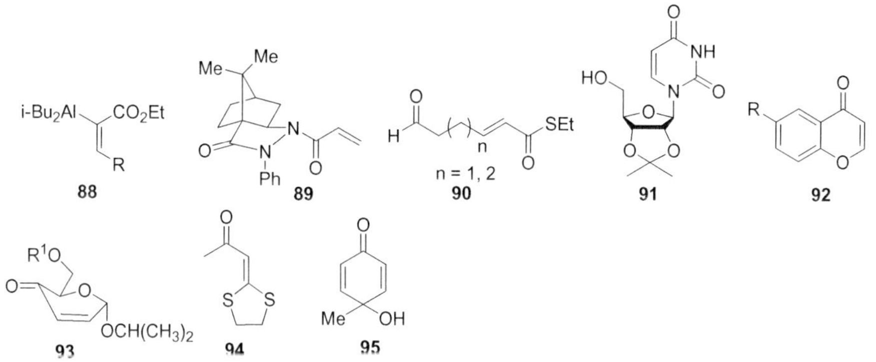

Figure 1.2 Examples of novel activated alkenes.

92,[94] sugar derived chiral-activated alkenes **93**,[95] α-oxo ketene dithioacetals **94**[96] and *p*-methylquinols **95**,[97] for the synthesis of diverse products *via* MBH reaction with aldehydes or other electrophiles in the presence of tertiary amine or TiCl$_4$ catalyst.

1.4   Electrophiles

1.4.1   Aldehydes

Aldehydes have been the primary source of electrophiles since the initial MBH reaction was explored. On the basis of both electronic and steric considerations, they are much more active than simple ketones. Formaldehyde can be employed as an aqueous solution (formalin), the polymer (paraformaldehyde), as a solution of the monomer in an organic solvent or as a hemiacetal (Scheme 1.43).[98] Similar results were obtained in couplings of aldehydes with acrylonitrile, methyl vinyl ketone and phenyl vinyl sulfone. However, aqueous formaldehyde can not react with diethyl vinylphosphonate.[89]

Other aliphatic aldehydes with chains of about six carbons or less appear to be only slightly less reactive than formaldehyde.[99] Longer chain[100] and especially α-branched aldehydes[101] react rather slowly (Scheme 1.44). With DABCO catalysis, pivalaldehyde (2,2-dimethylpropanal) does not react with methyl acrylate[102] and only very slowly with phenyl vinyl sulfone.[86b] Electron-withdrawing groups on the α carbon enhance reactivity (Scheme 1.44).[3b,103]

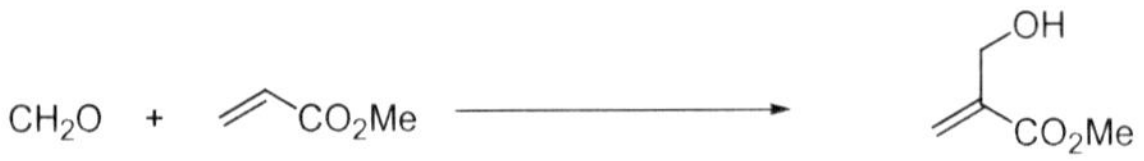

Formalin	DABCO, MeOH/H$_2$O, rt, 48 h	75%
Paraformaldehyde	Me$_3$N, H$_2$O, 60 °C, 3 h	80%
Monomer	DABCO, EtOH, rt, 72 h	59%
Cyclohexanol hemiacetal	DABCO, cyclohexanol, rt, 70 h	45%

Scheme 1.43

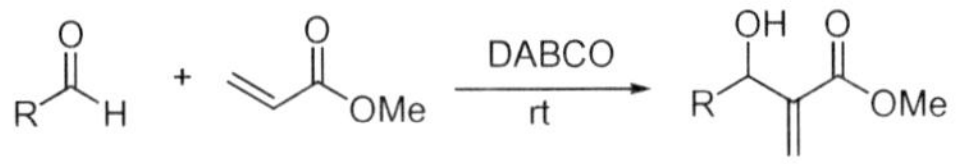

R	time	yield (%)
Me	7 days	88
n-Bu	9 days	72
i-Pr	13 weeks	68
t-Bu	--	--
Cl$_3$C	20 h	>55
CO$_2$Me	48 h	74

Scheme 1.44

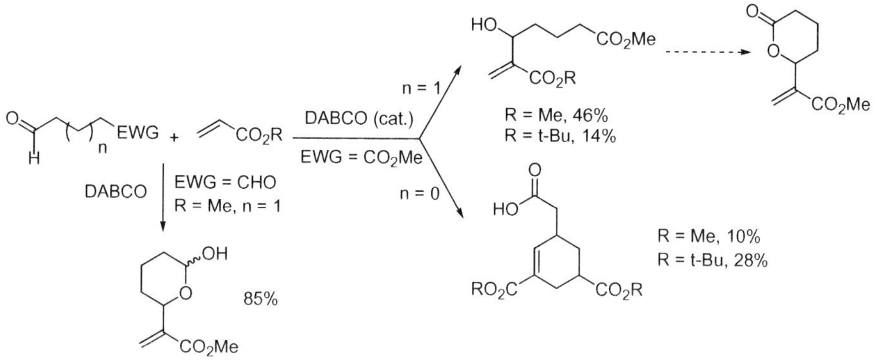

Scheme 1.45

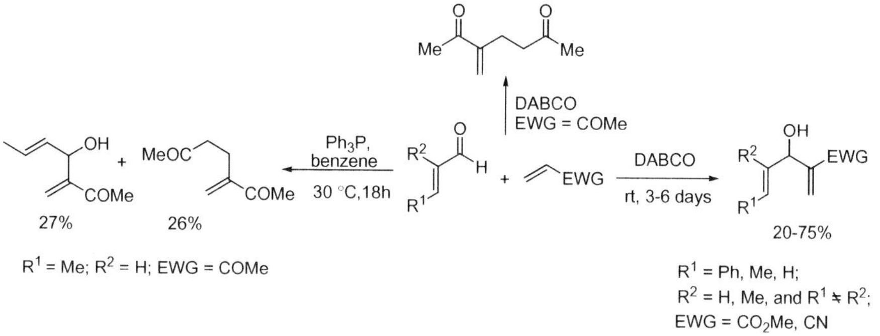

Scheme 1.46

Partial lactonization[104] or hemiketalization[105] occurred when the aldehyde contained an ester or carbonyl group at a favorable distance in the chain (Scheme 1.45).

Attempts to add acrolein to activated olefins led to polymerization, but methacrolein, crotonaldehyde and cinnamaldehyde react normally with acrylates and acrylonitrile (Scheme 1.46).[106] Upon treatment of these α,β-unsaturated aldehydes with methyl vinyl ketone in the presence of DABCO, the self-condensation product 3-methylenehepta-2,6-dione was obtained exclusively;[106] some MBH adduct was isolated with phosphine catalysis (Scheme 1.46).[62a] In addition, crotonaldehyde does not add to phenyl vinyl sulfone.[86b]

Acetylenic aldehydes **96** have been utilized as electrophiles in MBH reactions to give allyl propargyl alcohols **97** in moderate to good yields.[107] Aldehydes **98**, as a masked formybutadiene, react with activated olefins to give the corresponding MBH adducts **99** *via* a tandem sequence base-induced elimination–MBH reaction catalyzed by DABCO (Scheme 1.47).[108]

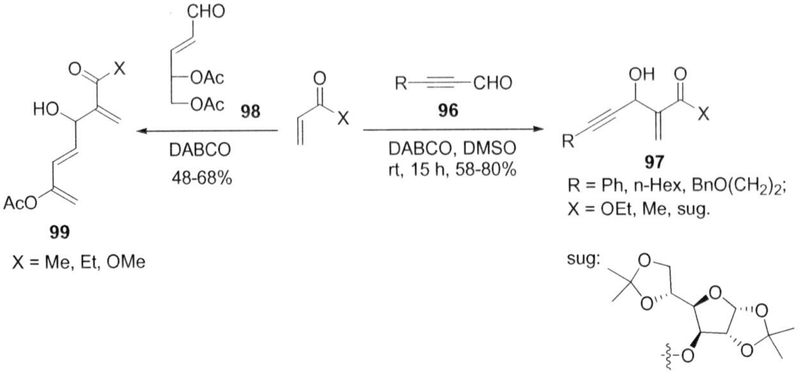

Scheme 1.47

R	Time	Yield (%)
H	2 days	92%
	40 h	18% [RhH(PPh$_3$)$_4$]
		33% [RuH$_2$(PPh$_3$)$_4$]
2-Cl	4 days	90%
4-NO$_2$	18 h	95%
4-MeO	20 days	90%

Scheme 1.48

With substituted benzaldehydes it is well known that substitution on the benzene ring has direct implications on the reactivity of the formyl group. Nitro- or trifluoromethyl-substituted benzaldehydes were some of the fastest reacting substrates due to their electron-withdrawing character, while the sluggish nature of methoxy- or dimethoxybenzaldehyde was attributed to their electron-donating nature.[105,109] In the presence of DABCO, aromatic aldehydes, especially those containing electron-withdrawing groups, react with activated olefins in high yields (Scheme 1.48), whereas lower yields have been obtained when aromatic aldehydes were used in combination with rhodium and ruthenium complexes.[62c]

The efficiency of the DABCO-catalyzed MBH reaction of activated olefins with aromatic aldehydes can be increased significantly by complexation of the arene to the electrophilic Cr(CO)$_3$ group, and the dependence of diastereo-selectivity on the nature of the ortho substituent was observed.[109a] Excellent diastereoselectivities have been achieved in reactions of tricarbonylchromium complexes of *o*-methoxybenzaldehyde with acrylonitrile and complexes of *o*-methoxybenzaldehyde and *o*-chlorobenzaldehyde with methyl acrylate; however, lower de values were obtained in the reactions of methyl acrylate with complexes of *o*-fluorobenzaldehyde and *o*-tolualdehyde (Scheme 1.49).

An attempt to use 2-hydroxybenzaldehyde for the DABCO-catalyzed reaction with methyl acrylate in chloroform led to diadduct **103** in 10% yield;[110] this product is also obtained on microwave irradiation in a shorter time and with higher yield.[49] However, on using methylene chloride instead of chloroform as the solvent, quaternary salt **104** was isolated, which for the first time proved that the elusive Michael adduct, postulated by all researchers in the area, did exist; the counterion chloride is presumably derived from the solvent (Scheme 1.50).[111]

In the reaction of aromatic aldehydes with acrylonitrile, complexes **105** containing two molecules of MBH adduct and one molecule of DABCO

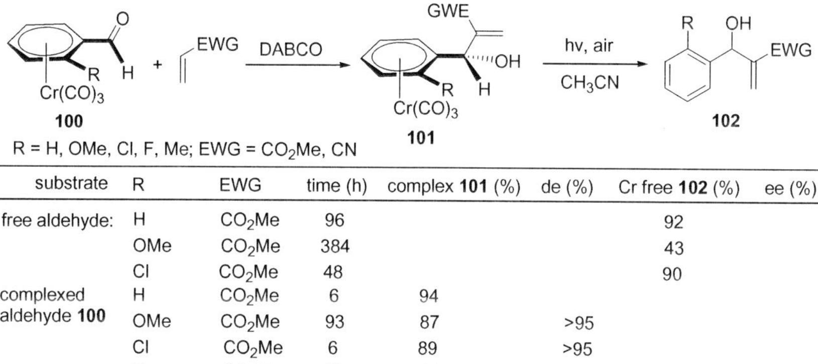

R = H, OMe, Cl, F, Me; EWG = CO$_2$Me, CN

substrate	R	EWG	time (h)	complex **101** (%)	de (%)	Cr free **102** (%)	ee (%)
free aldehyde:	H	CO$_2$Me	96			92	
	OMe	CO$_2$Me	384			43	
	Cl	CO$_2$Me	48			90	
complexed aldehyde **100**	H	CO$_2$Me	6	94			
	OMe	CO$_2$Me	93	87	>95		
	Cl	CO$_2$Me	6	89	>95		
	F	CO$_2$Me	7	92	84		
	Me	CO$_2$Me	48	90	68		
	S-(+)OMe	CO$_2$Me	93	85	>95	90	>98%
	S-(+)OMe	CN	11	88	>95	93	>98%

Scheme 1.49

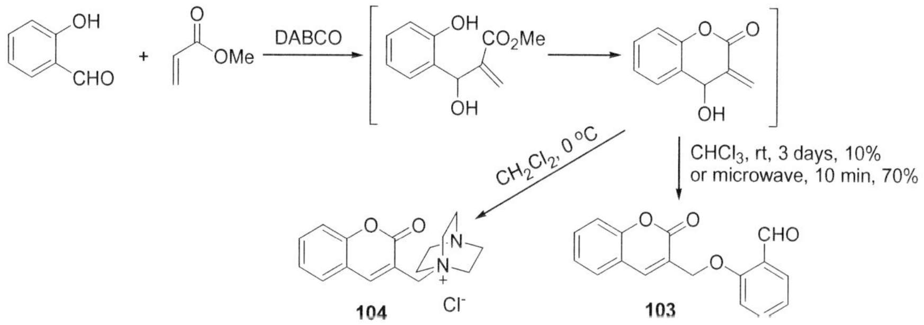

Scheme 1.50

have, sometimes, been isolated, and were readily decomposed by treatment with acid to give the DABCO salt and the corresponding MBH adduct **106** (Scheme 1.51).[112]

In the DABCO-catalyzed MBH reaction between aromatic or heteroaromatic dialdehydes **107** with methyl acrylate, mono-adduct **108** can be obtained selectively in excellent yields; the rate of the first condensation of a carbonyl group is considerably increased by the electron-withdrawing effect of the second carbonyl group (compare with the condensation rate of benzaldehyde), and of course the second condensation was considerably slowed. However, upon increasing the amount of used methyl acrylate, diadducts **109** and **110** were also obtained in moderate to good yields (Scheme 1.52).[105,113] With orthophthalaldehyde, cyclic hemiacetals **111** were the sole product (Scheme 1.52).[105]

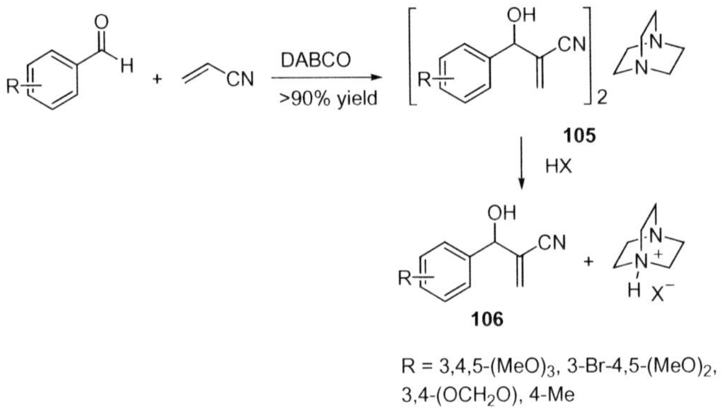

R = 3,4,5-(MeO)$_3$, 3-Br-4,5-(MeO)$_2$, 3,4-(OCH$_2$O), 4-Me

Scheme 1.51

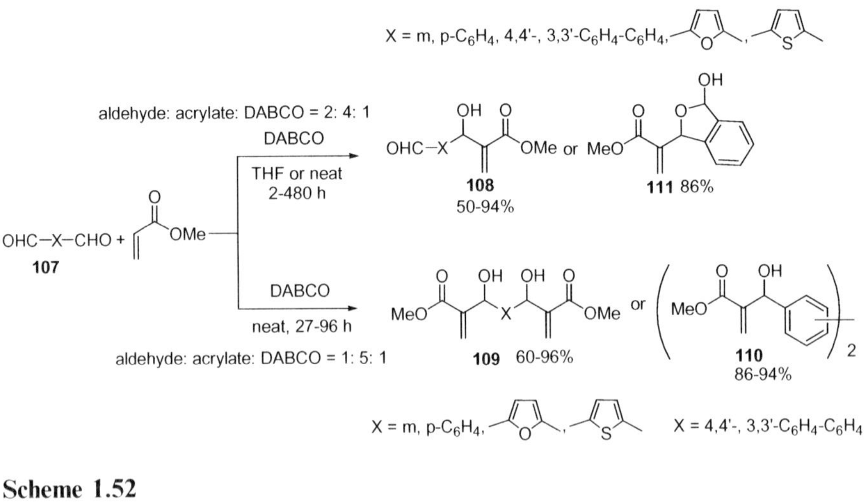

Scheme 1.52

For polynuclear aromatic aldehydes, the reactivity decreases in the order benzaldehyde > 1-naphthaldehyde > 9-anthraldehyde. The last does not add to methyl acrylate with DABCO catalysis at room temperature.[114]

Heteroaromatic aldehydes are excellent electrophiles in the MBH reaction because of their increased electrophilicity. The heteroatom also facilitates the proton transfers involved in the reaction (Scheme 1.53).[3b] In the presence of DABCO, pyridine-2-carboxaldehyde derivatives react rapidly with acrylates, methyl vinyl ketone and acrylonitrile to give excellent yields of the MBH adducts **112**, which can be further transformed into indolizines **114** by thermal cyclization of their acetate derivatives **113** (Scheme 1.54).[47] Notably, a small amount of indolizine **114** was formed along with normal MBH adduct **112** in the reaction of pyridine-2-carboxaldehyde with methyl vinyl ketone.[47] Similarly, substituted 2-chloronicotin aldehydes **115** are reactive and efficient in the MBH reaction of methyl acrylates, acrylonitrile and cyclic enones catalyzed by various tertiary amine catalysts, such as DABCO, DBU, imidazole and 1-methylimidazole, to provide novel MBH adducts **116** and **117** for biological activity screening (Scheme 1.55).[115]

2-Formylimidazole (**118**), in which the electrophilicity of the aldehyde functionality was decreased due to π-electron pair donation from N1 into the aromatic ring, formed 1,4-addition adduct **119** as well as a disubstituted product **120** as major products upon reaction with methyl acrylate and

Ar
H
O
+
OMe
DABCO (cat.)
rt
OH
O
OMe

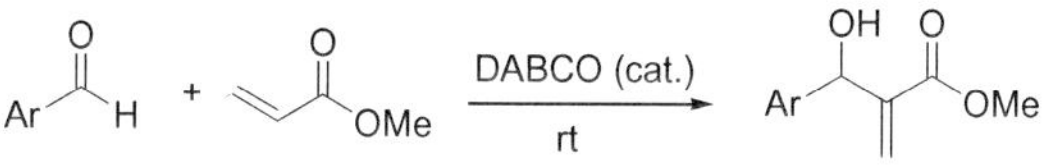

Ar	time	yield
Ph	6 days	39%
2-furyl	18 h	63%
3-pyridyl	4 h	>82%

Scheme 1.53

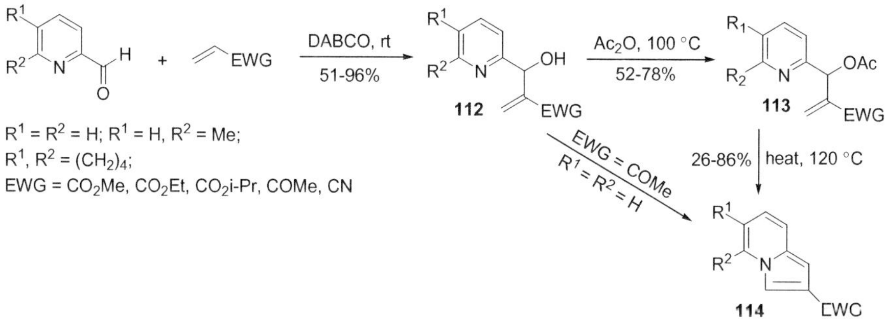

+ EWG
DABCO, rt
51-96%
112
Ac2O, 100 °C
52-78%
113
OAc
EWG = COMe
R1 = R2 = H
26-86%
heat, 120 °C
114
R1 = R2 = H; R1 = H, R2 = Me;
R1, R2 = (CH2)4;
EWG = CO2Me, CO2Et, CO2i-Pr, COMe, CN

Scheme 1.54

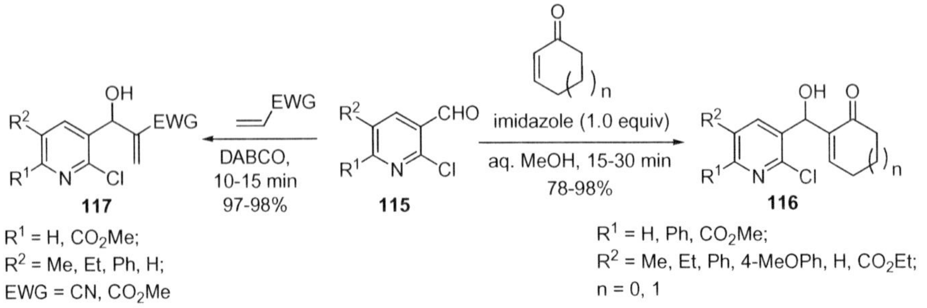

Scheme 1.55

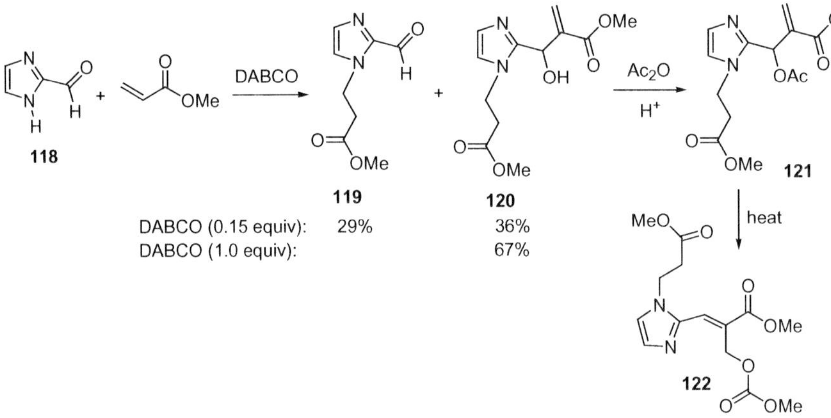

R¹ = H, CO₂Me;
R² = Me, Et, Ph, H;
EWG = CN, CO₂Me

R¹ = H, Ph, CO₂Me;
R² = Me, Et, Ph, 4-MeOPh, H, CO₂Et;
n = 0, 1

Scheme 1.56

DABCO (0.15 equiv): **119** 29% **120** 36%
DABCO (1.0 equiv): 67%

DABCO. The acetate of the latter product (**121**) rearranged on warming to afford a novel β-substituted acrylate ester **122** *via* an initial intramolecular attack by the imidazole nitrogen on allylic acetate, followed by S_N2 substitution (Scheme 1.56).[116]

The difference in reactivity of the formyl group present at different positions within a heterocyclic system for the MBH reaction has been developed. Within a heterocycle the formyl group present on the carbon atom adjacent to the heteroatom has better reactivity for the MBH reaction than a formyl group present on the carbon away from the heteroatom. A formyl group present at the 3- or 5-position in an isoxazole ring (**123** and **124**) undergoes a rapid MBH reaction as compared to the formyl group present at the 4-position (**125**); this has been attributed to a hypothetical assumption that the lone pair of electrons on heteroatoms present in proximity to the carbon atom bearing the formyl group helps in eliminating the tertiary base from the intermediate in the final step of the reaction (Figure 1.3).[117]

Moreover, the electronic or steric factors or the effect of the basicity of the heteroatom present near the carbon-bearing formyl group has been discussed.

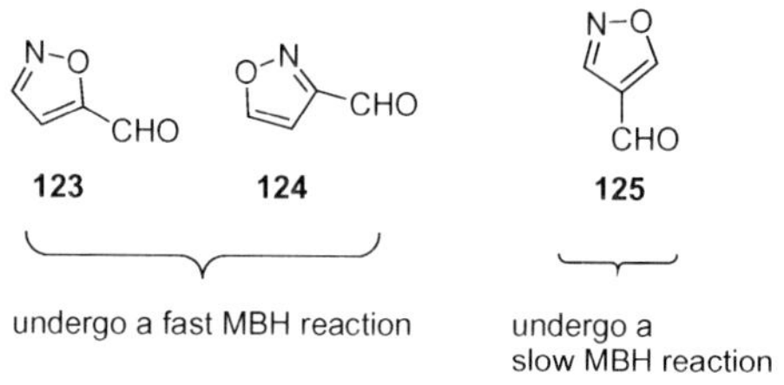

Figure 1.3 Reactivity of the formyl group in relation to its position on the isoxazole ring.

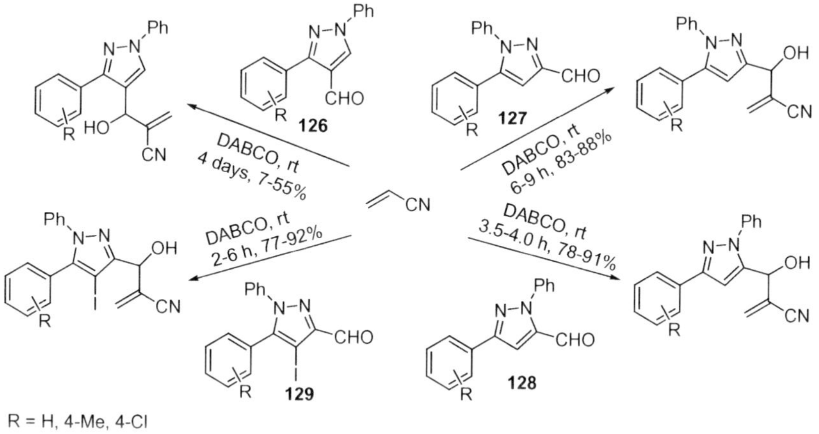

Scheme 1.57

The presence of an electron-withdrawing group such as a methoxy carbonyl group on the carbon adjacent to the one bearing the formyl group further increases its reactivity for this MBH reaction.[118] In addition, substituted 2-chloro-3-pyridinecarbaldehydes **115** have been shown to be a fast reacting substrate for MBH reaction, which is another powerful example.[115a] Although there are no direct comparisons of 2-chloropyridine-3-carbaldehydes with pyridine-3-carbaldehyde, it can be surmised from the literature that 2-chloro-pyridine-3-carbaldehydes have, perhaps, better reactivity than the 3-pyridinecarbaldehyde. This may be because the halide on the carbon adjacent to the one bearing the formyl group has some effect on reactivity for the MBH reaction.

Recently, Batra *et al.* have provided further insight into the fact mentioned above by investigating the addition of various substituted pyrazole aldehydes (**126–129**) with activated olefins. Scheme 1.57 gives representative comparative examples.[119]

Chromone-3-carbaldehydes **130** can undergo a MBH reaction with activated alkenes, such as acrylonitrile,[120] methyl acrylate and methyl vinyl ketone,[121] by

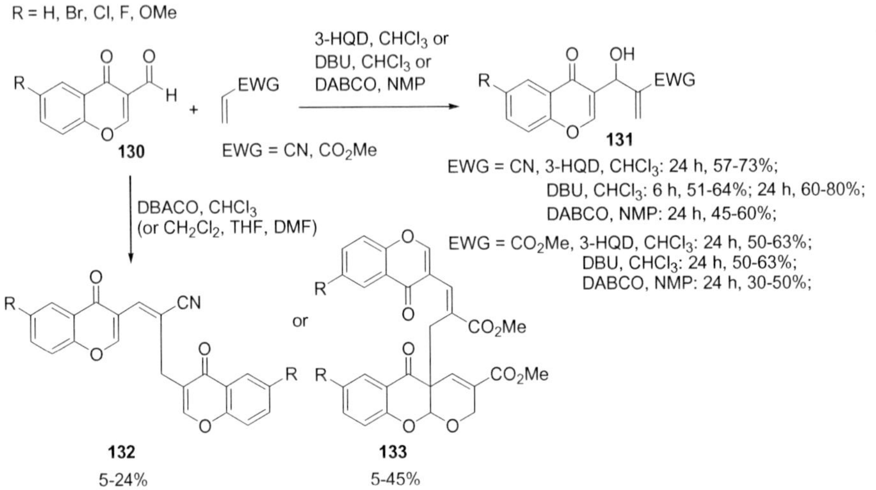

R = H, Br, Cl, F, OMe
3-HQD, CHCl3 or
DBU, CHCl3 or
DABCO, NMP
EWG
130
EWG = CN, CO2Me
131
EWG = CN, 3-HQD, CHCl3: 24 h, 57-73%;
DBU, CHCl3: 6 h, 51-64%; 24 h, 60-80%;
DABCO, NMP: 24 h, 45-60%;
EWG = CO2Me, 3-HQD, CHCl3: 24 h, 50-63%;
DBU, CHCl3: 24 h, 50-63%;
DABCO, NMP: 24 h, 30-50%;
DBACO, CHCl3
(or CH2Cl2, THF, DMF)
or
CN
CO2Me
CO2Me
132
5-24%
133
5-45%

Scheme 1.58

EWG
DABCO (10%)
rt, 1-4 days
R' = H
134
135
THF, rt, 1-4 h
EWG = CO2Et
50-82%
i-Bu2Al
CO2Et
134
136
RF = CF3, C3F7; R = H; EWG = CO2Et, 16-20%
RF = CF3; R = Me; EWG = CO2Et, 0%
RF = CF3, C3F7; R = H; EWG = CN, 0%
RF = C6F5; R = H; EWG = CO2Et, CN, 71-74%
RF = CF3, C3F7, C6F5, CCl3
R = H, Me, Ph, 2-thioph
R' = H, Ph

Scheme 1.59

using 3-HQD, DBU or DABCO as catalyst (Scheme 1.58). However, the choice of catalyst and solvent often had a significant effect on the efficiency and rate of MBH reactions.[122] Using 3-HQD or DBU as catalyst and $CHCl_3$ as solvent, the MBH adducts **131** appeared to be formed exclusively or as major products in moderated to good yields. Similar chemoselectivity appeared to occur when DABCO is used in the presence of 1-methyl-2-pyrrolidinone (NMP) as solvent, whereas, conversely, MBH dimers **132** and **133** were obtained by using $CHCl_3$ as solvent.

The simplest fluorinated aldehydes, fluoral, known to undergo polymerization instantaneously in the presence of amines, failed to undergo efficient MBH reaction with ethyl acrylate and acrylonitrile. However, pentafluorobenzaldehyde could provide good yields of MBH adduct **135**. Alternatively, unsubstituted and β-substituted [α-(ethoxycarbonyl)vinyl]aluminium **136** reacts with perfluoroalkyl and aryl aldehydes **134** to provide the fluorinated MBH adducts **135** in good to excellent yields (Scheme 1.59).[59b,90]

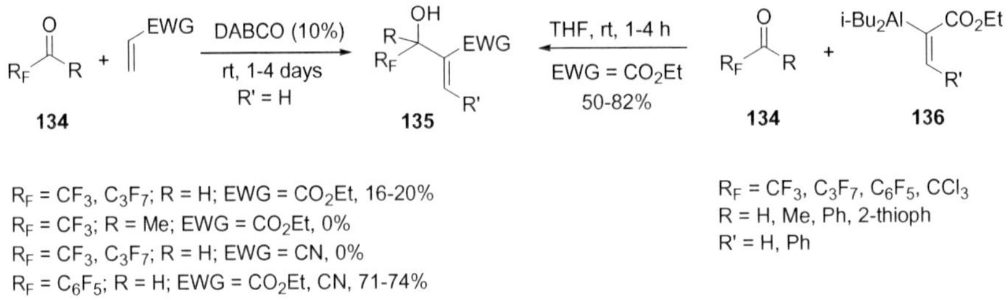

Moreover, a study has been made on the effect of fluorine substitution in the MBH reaction of various fluorocarbonyl partners with acrolein, methyl vinyl ketone, ethyl acrylate and acrylonitrile. Multifunctionalized fluorinated allyl alcohols were prepared from amine-sensitive aldehydes and olefins by balancing their reactivities.[59b] When the olefin is capable of reacting with itself in the presence of an amine (*e.g.* acrolein), the electrophile has to be very reactive as well (*e.g.* fluoral) to obtain a modest to good yield of MBH products. The reaction of a moderately reactive olefin (*e.g.* ethyl acrylate or acrylonitrile) and a very reactive electrophile (*e.g.* fluoral) resulted in self-reaction of the electrophile or very low yield of the allylic alcohol product.

Chiral α-amino aldehydes have been utilized in asymmetric MBH reactions[123] due to their ready availability in both enantiomeric forms from natural sources, as well as their pronounced versatility, owing to the presence of both the formyl group and suitably protected amino functionality in the same molecule. *N*-Protecting groups have a significant influence on the overall reactivity of the aldehyde and on the coupling diastereoselectivity.[124] Aldehydes having an electron-withdrawing *N*-protecting group were more reactive towards methyl acrylate in the presence of DABCO than their *N,N*-dibenzyl counterparts. *anti*-Diastereoselectivity was observed for the *N,N*-Bn$_2$ and the *N*-phthaloyl protecting groups, whereas the reversed stereoselectivity was found in the reaction of *N*-Boc-L-alaninal with methyl acrylate catalyzed by DABCO, probably due to the hydrogen bonds involved (Scheme 1.60).

In the presence of DABCO, the MBH reaction of *N*-protected α-amino aldehydes **137** with acrylamide proceeded rapidly and afforded *N*-acyl hemiaminal **138** in moderate to good yields, rather than the MBH adducts (Scheme 1.61).[125]

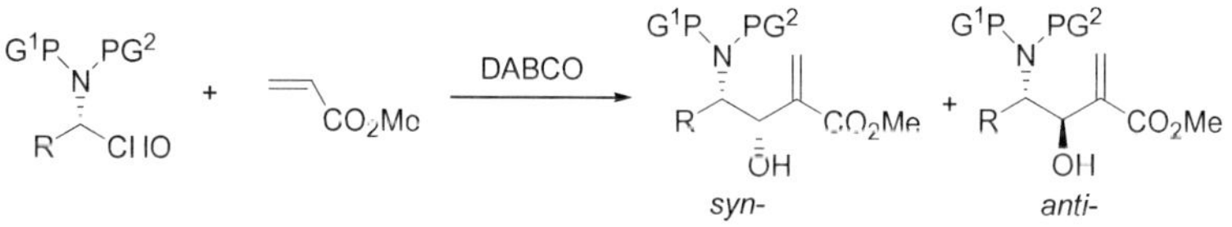

aldehyde	Time (days)	Yield (%)	*anti:syn*
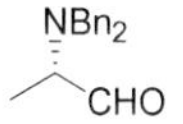 NBn$_2$ CHO	20	71	72:28
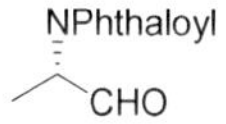 NPhthaloyl CHO	3.5	30	55:45
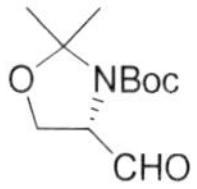 O NBoc CHO	7	85	86:14
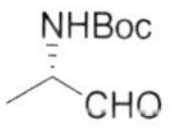 NHBoc CHO	1.5	76	29:71

Scheme 1.60

137 + acrylamide (with DABCO, 13-18 h, 29-85%) → **138**

Aldehydes: L-valinal, L-phenylalanal, L-Tryptophanal,
prolinal, Serinal (OBn), Glycinal, Lysinal (ε-Boc), Leucinal
P = Boc, Cbz;

Scheme 1.61

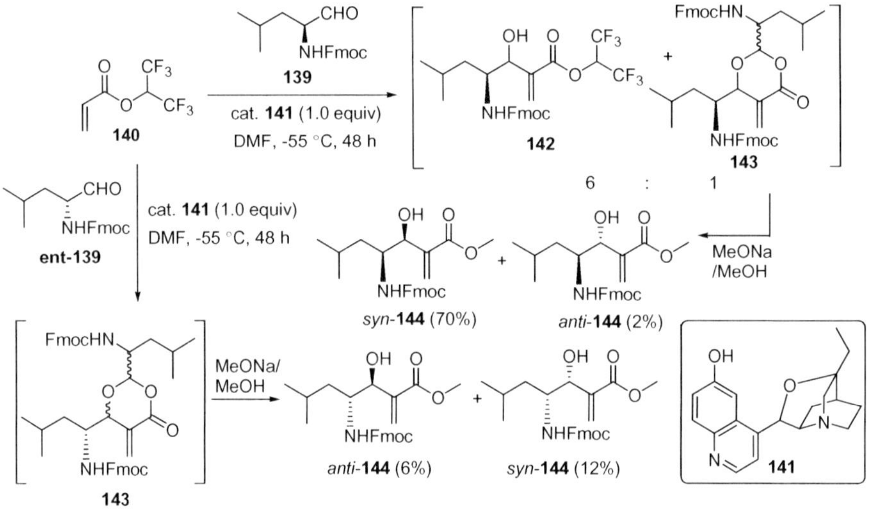

139 + **140** → **142** + **143** (cat. **141** (1.0 equiv), DMF, -55 °C, 48 h) 6 : 1

ent-139 (cat. **141** (1.0 equiv), DMF, -55 °C, 48 h) → **143**

143 → *syn*-**144** (70%) + *anti*-**144** (2%) (MeONa/MeOH)

143 → *anti*-**144** (6%) + *syn*-**144** (12%) (MeONa/MeOH)

141

Scheme 1.62

The reaction of *N*-Fmoc-L-leucinal (**139**) with 1,1,1,3,3,3-hexafluoroisopropyl acrylate (**140**) proceeded smoothly in the presence of a stoichiometric amount of cinchona alkaloid **141**, even at very low temperature, to give a 6:1 mixture of *syn*-ester **142** and dioxane derivative **143** (Scheme 1.62). However, the same reaction with N-Fmoc-D-leucinal (*ent*-**139**) turned out to be sluggish and only a mixture of dioxanones **143** was obtained in low yield. Thus, it can be concluded that the (*R*)-selectivity of the chiral amine catalyst matches well with L-configuration of the substrate, leading to high *syn*-selectivity.[126]

Sugar-derived aldehydes[127] and enals[128] have been used in diastereoselective MBH reactions to give the corresponding adducts in good yields with moderate to good diastereoselectivities. Notably, the yields and rate of reaction were slightly influenced by solvent, and the isopropylidene groups and their stereochemical disposition played a decisive role in diastereoselection, enhancing the de's, as observed in the case of L-sorbose, D-ribose-1-carboxaldehyde and D-mannose-1-carboxaldehyde.[127b]

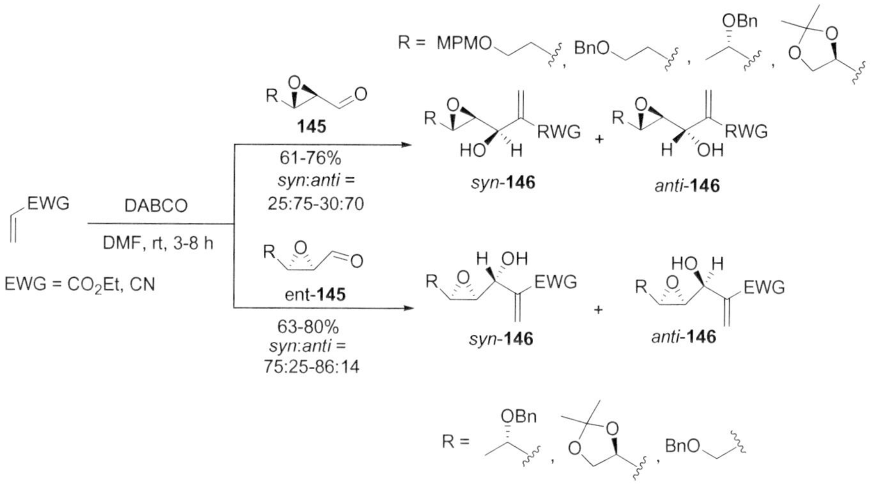

Scheme 1.63

Chiral 2,3-epoxy aldehydes **145** have been utilized for the first time as novel electrophiles in DABCO-catalyzed MBH reactions with activated alkenes, furnishing densely functionalized adducts **146** in good yields with moderate to good diastereoselectivities (Scheme 1.63).[129]

Other chiral aldehydes, containing small rings, such as *N*-tritylaziridine-2-(*S*)-carboxaldehyde (**147**),[130] 4-oxoazetidine-2-carbaldehydes **148**[131] and *trans*-(2*R*,3*R*)-cyclopropanecarbaldehydes **149**,[132] undergo a facile MBH reaction with various activated olefins in the presence of a catalytic amount of DABCO to furnish the corresponding adducts **150–152** in good yields and selectivities (Scheme 1.64). The ring conformation and substituents played a decisive role in the stereoselection of the product.

Inspired by the application of site isolation effects of crosslinked solid materials in so-called "wolf and lamb" reactions,[133] 4-nitrobenzaldehyde dimethyl acetal has, for the first time, as electrophile been reacted with methyl vinyl ketone *via* a sequential acid-catalyzed acetal hydrolysis followed by amine-catalyzed MBH reaction in the presence of star polymers containing core-confined PTSA analogues **153** and 4-(dialkylamino)pyridines catalysts **154** (Scheme 1.65).[134]

1.4.2 Ketones

Unreactive ketones do not undergo the MBH reaction under ambient conditions. Acetone reacts with *n*-butyl acrylate in the presence of DABCO at 120 °C, but the conversion is only 7% after 4–6 days.[135] Under high pressure, acetone, methyl ethyl ketone and cyclohexanone react with acrylonitrile,[10,48] but not with methyl acrylate.[136] Hindered ketones, such as diisopropyl ketone and aryl alkyl ketones, fail to react even under high pressure.[10] In contrast,

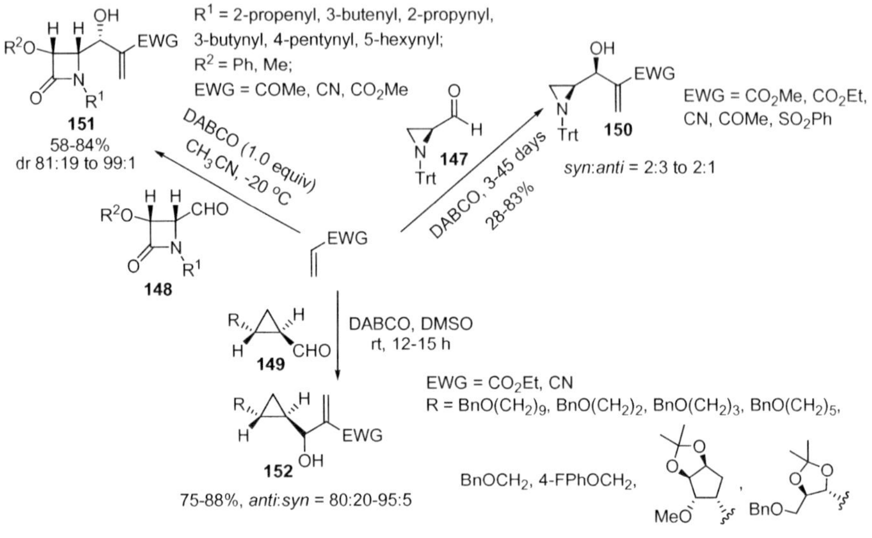

Scheme 1.64

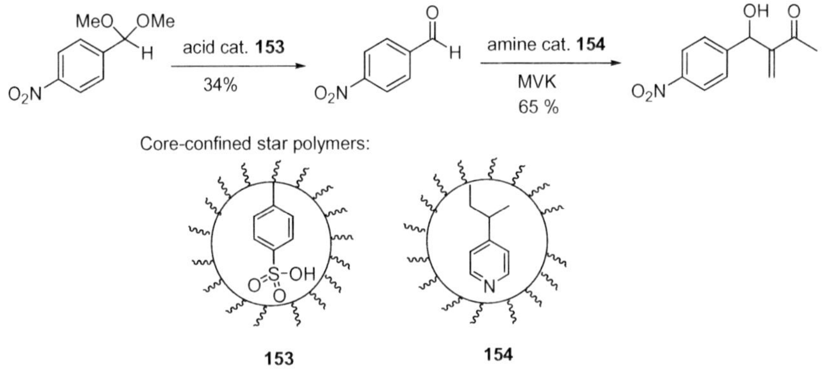

Scheme 1.65

halogenated ketones, even hindered ones, react smoothly with acrolein, acrylo-nitrile and ethyl acrylate.[58] Thus, the reluctance of ketones to undergo a MBH reaction appears to be for electronic rather than steric reasons.

Encouraged by the success with fluoral, fluoroketones have also been investigated systematically in the MBH reaction. Since 1,1,1-trifluoroacetone is known to trimerize in the presence of amines,[137] lower yields of MBH adducts were obtained at lower temperature when treated with acrolein and acrylo-nitrile. With methyl acrylate and MVK, however, only a polymeric material was isolated under the MBH reaction conditions.

To avoid the polymerization initiated by abstraction of the α-hydrogen atom, Ramchandran *et al.* have focused their attention on aryl trifluoromethyl ketones. By considering the match between the reactivities of the olefin and carbonyl partners, they succeeded in accomplishing the MBH reaction between moderately reactive electrophile (*e.g.* trifluoroacetophenone, trifluoroacetylthiophene, 3-trifluoroacetyl-indole, 2-chloro-2,2-difluoroacetophenone, 1,1,1-trifluoro-4-phenyl-3-butyn-2-one and 4,4,5,5,6,6,6-heptafluoro-1-phenyl-1-hexyn-3-one) with moderately reactive olefins (*e.g.* ethyl acrylate and acrylonitrile) in the presence of DABCO in high yields (Chapter 2.2.1).[59]

Alternatively, unsubstituted and β-substituted [α-(ethoxycarbonyl)vinyl]aluminum reacts with perfluoroalkyl ketones in the presence of DABCO to provide the fluorinated MBH adduct in good to excellent yields (Scheme 1.59).[90]

Recently, the MBH reaction of β-substituted α,β-unsaturated CF_3 ketones with acrylonitrile was also found to proceed in aqueous THF solution, with DABCO catalysis, to give the corresponding adducts in good yields. In the case of ketones containing EtO and Me_2N groups, only polymerization was observed and no target product was isolated (Chapter 2.2.1).[138]

1.4.3   α-Keto Esters, α-Keto Lactones, α-Keto Lactams and Diketones (α-Keto Carbonyl Compounds)

Although ketones are not generally considered to be reactive carbonyl partners in MBH reactions (except under high pressure), the enhanced reactivity of certain α-diketones towards aldol-type reactions[139] make them suitable partners for reaction with MBH-type vinyl carbanion equivalents. Indeed, α-keto esters have been found to possess the requisite reactivity and are very reactive electrophiles for the MBH reaction of acrylate, methyl vinyl ketone, acrylonitrile and acrolein in the presence of a tertiary amine, such as DABCO (Scheme 1.66).[58,140]

Chiral glyoxylic acid derivatives, *i.e.* (−)-8-phenylmenthyl glyoxylate, undergoes an MBH reaction with cyclic α,β-unsaturated ketones under the

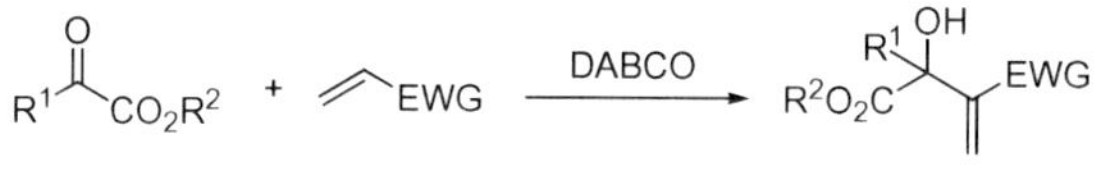

R¹	R²	EWG	conditions	Yield (%)
Ph	Et	CO_2Me	rt, 7 days	49
CO_2Et	Et	COMe	rt, 30 min	74
CF_3	Me	CHO	rt, 1 h	57
CF_3	Me	CN	rt, 2 h	30
CF_3	Me	CO_2Et	rt, 3 h	72
CF_3	Me	COMe	rt, 1 h	80

Scheme 1.66

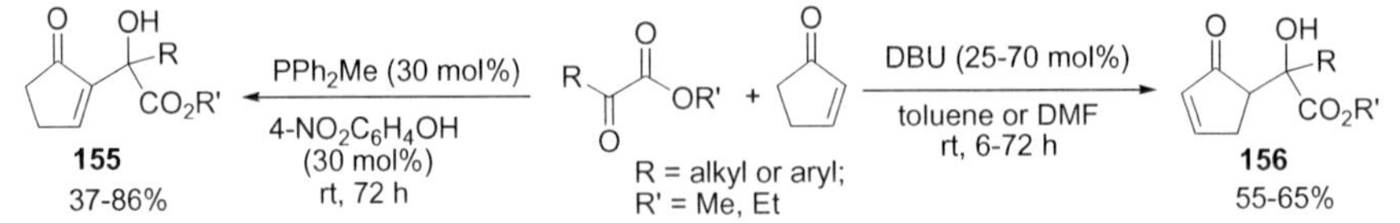

Scheme 1.67

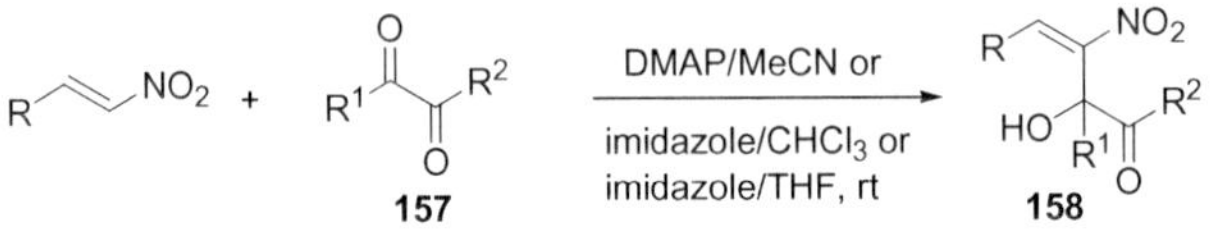

R = alkyl or aryl;
R' = Me, Et

155 37-86%   **156** 55-65%

PPh₂Me (30 mol%), 4-NO₂C₆H₄OH (30 mol%), rt, 72 h

DBU (25-70 mol%), toluene or DMF, rt, 6-72 h

Scheme 1.68

157 + → **158**

DMAP/MeCN or imidazole/CHCl₃ or imidazole/THF, rt

R = Aryl
R^1 = H, R^2 = OEt, DMAP: 5-30 min, 33-99%; imidazole: 2-24 h, 31-83%
R^1 = CF₃, R^2 = OMe, DMAP, 0.5-10 h, 40-80%;
R^1 = H, R^2 = Me, DMAP: 2 days, 26-35%; imidazole: 2 days, 20-36%;
R^{1}COCOR^2 = ninhydrin, DMAP: 15-45 min, 44-79%; imidazole: 5-24 h, 53-74%

catalytic influence of dimethyl sulfide in the presence of $TiCl_4$, providing the corresponding adducts with very high diastereoisomeric excess (over 95% de) and typical yields of 78% (Chapter 2.5.1, Scheme 2.169).[141]

In addition, during our studies on the MBH reaction of α-keto esters and cyclic enones, we found that the catalysts have a significant effect on the MBH adduct. The MBH reaction of α-keto esters with cyclopent-2-enone catalyzed by diphenylmethylphosphine provided the corresponding MBH adducts **155** in higher yields in the presence of *p*-nitrophenol, whereas the similar reaction of α-keto esters with cyclopent-2-enone catalyzed by tertiary-amine of DBU furnished the corresponding aldol adducts **156** with *syn*-configuration exclusively (Scheme 1.67).[142]

Recently, the MBH reactions of various conjugated nitroalkenes with activated non-enolizable carbonyl compounds **157**, such as glyoxylate, trifluoropyruvate and pyruvaldehyde, catalyzed by DMAP or imidazole have been developed. In most cases, the reactions catalyzed by DMAP in acetonitrile were faster and provided the desired MBH adducts **158** in higher yields than did the imidazole-catalyzed reactions (Scheme 1.68).[81a]

The MBH reactions of non-enolizable α-diketones precursors (Figure 1.4) with the activated olefins, *e.g.* acrolein, methyl acrylate and acrylonitrile, have been investigated systematically.[57] The reaction of 3,3,5,5-tetramethyl-cyclopentane-1,2-dione (**162**) with acrolein and acrylonitrile, but not methyl acrylate, afforded the mono-α-hydroxyalkylation products in high yields.[57] Other nonenolizable α-diketones, such as camphorquinone (**159**), homo-adamantane-2,3-dione (**160**) and bicyclo[3.3.2]decane-9,10-dione (**161**) reacted only with acrylonitrile, probably due to the hindered nature of the α-dicarbonyl compounds and the difference in steric demand between nitrile and ester.

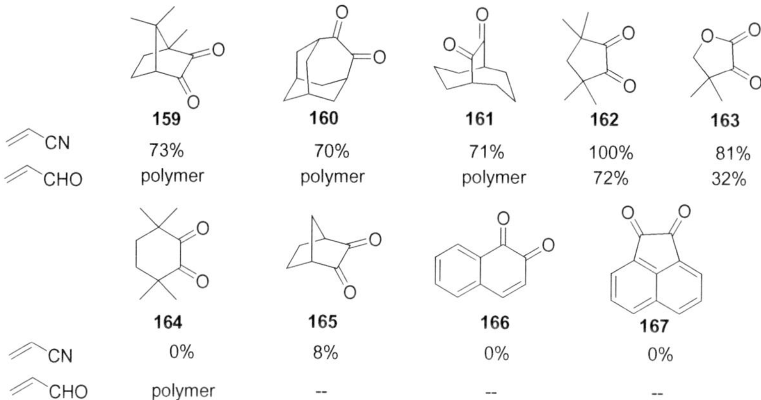

Figure 1.4 MBH reactions of non-enolizable α-diketones precursors with the activated olefins.

Unsurprisingly, the MBH reaction of these compounds with acrolein was unsuccessful, only the polymer of acrolein was obtained whatever the reaction conditions, even with ultrasonic radiation. In contrast to the readiness of **162** to undergo the MBH reaction with acrolein and acrylonirile, all attempts to conduct an MBH reaction with 3,3,6,6-tetramethylcycohexane-1,2-dione (**164**), under similar conditions, met with failure. This result may be attributed to both greater steric inhibition of nucleophilic attack at the carbonyl group of **164** and the lack of enhanced reactivity associated with "eclipsing" the two carbonyl groups (as in **162**). Norbornane-2,3-dione (**165**), an apparently logical candidate for testing, was reported to be unstable,[143] which undoubtedly contributed to the unsatisfactory results obtained on attempting the MBH rection with acrylonitrile (8% yield of adduct). None of the conjugated aryl diones 1,2-naphthoquinone (**166**), acenaphthenequinone (**167**) and **164** were sufficiently reactive to afford a satisfactory yield of MBH adduct with acrylonitrile and acrolein. The keto lactone **163** proved to be satisfactory substrate for the MBH reaction with acrylonitrile and acrolein, but failed to react with methyl acrylate, probably for steric reasons (Figure 1.4).[57]

Compared with chiral nonracemic α-amino carbonyl compounds – which are not suitable substrates for MBH reaction, mainly due to their racemization under normal conditions after prolonged exposure times to catalyst[124b] or due to poor diastereoselectivity[124a,126,130] – α-keto lactams, enantiopure 3-oxo-azetidin-2-ones **168**, readily react with various activated vinyl systems promoted by DABCO to afford the corresponding optically pure MBH adducts **169** without detectable epimerization (Scheme 1.69).[144] However, the Lewis acid-mediated reaction of electron-deficient alkynes with azetidine-2,3-diones **168** as an entry to β-halo MBH adducts was not very successful; the coupling product **170** was achieved with concomitant acetonide cleavage as a single (*E*)-isomer in low yield, in the presence of trimethylsilyl iodide under BF$_3$·OEt$_2$-induced catalysis (Scheme 1.69).

R = PMP, 2-propenyl, 3-butenyl, 4-pentenyl,
2-propynyl, 3-butynyl, 4-pentynyl;
EWG = COMe, CN, CO_2Me, SO_2Ph

Scheme 1.69

EWG = CO_2R (R = 1-naphthyl, 2-naphthyl, Ph, Bn Me), COMe, COEt, CN

Scheme 1.70

Ninhydrin and isatin derivatives, as the electrophilic component in the MBH reaction, readily react with acrylates,[145] acrylonitrile,[145] 1-benzopyran-4(4*H*)-one[94] and conjugated nitroalkenes[81a] to give good to excellent yields of the corresponding adducts (Chapter 2).

The chiral glyoxylate derivative *N*-glyoxyloyl-(2*R*)-bornane-10,2-sultam (**171**), whose diastereodifferentiation abilities had already been demonstrated in many types of reactions, was unstable under typical MBH conditions and no positive results were obtained by reacting with methyl acrylate, methyl vinyl ketone and 2-cyclohexenone in the presence of DABCO, Bu_3P, Bu_3P–Et_2AlCl or Me_2S–$TiCl_4$.[141] *N*-Glyoxyloyl-camphorpyrazolidinone **172** as another kind of chiral glyoxylate derivative, however, underwent smoothly the MBH reaction with various α,β-unsaturated carbonyls/nitrile to give the corresponding 2-hydroxy-3-methylene succinic acid derivative **173** in excellent diastereoselectivity (Scheme 1.70).[146]

1.4.4 Imines and Iminium Salts

As mentioned in Section 1.1, imines are also excellent electrophiles and could replace aldehydes to participate in an MBH reaction, the so-called aza-MBH reaction. The first example of imines involved in the aza-MBH reaction date back to 1984. Perlmutter and Teo reported that *N*-tosyl imines reacted with ethyl acrylate in the presence of DABCO as catalyst in a sealed tube to give adducts in moderate to good yields (Scheme 1.71).[3e] Since then, many kind of imines (**174–182**, Figure 1.5) bearing different activating groups have been developed and

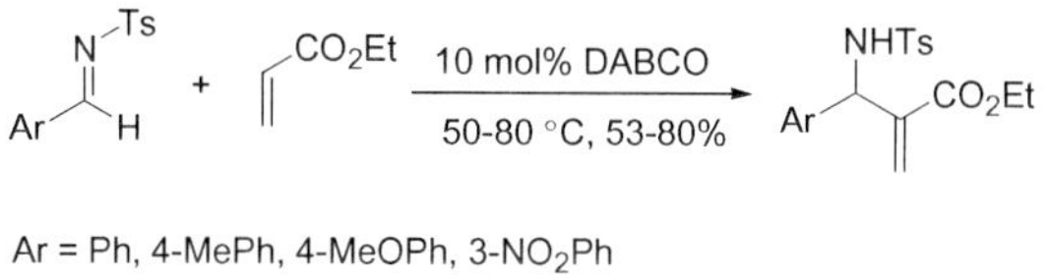

Ar = Ph, 4-MePh, 4-MeOPh, 3-NO$_2$Ph

Scheme 1.71

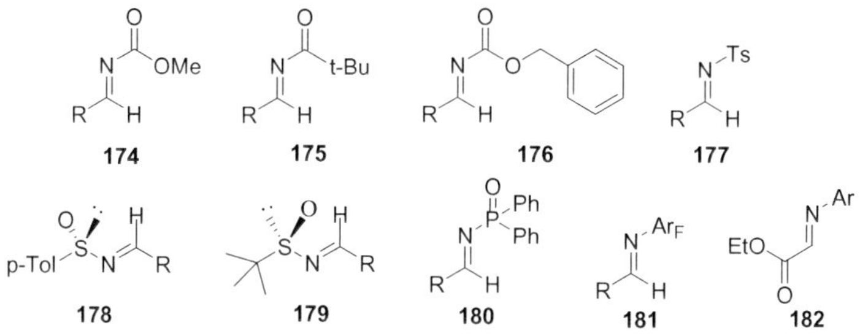

174   **175**   **176**   **177**

178   **179**   **180**   **181**   **182**

Figure 1.5   Imines **174–182** bearing different activating groups that have been developed and employed in the aza-MBH reaction.

employed in the aza-MBH reaction with various activated olefins. To date, the activating groups that have been employed include methoxycarbonyl,[147] *tert*-butylcarbonyl, carbobenzoxy, *p*-toluenesulfonyl,[64,76c,76e,82,148] enantiomerically pure *N*-*p*-toluenesulfinyl[149] and *tert*-butanesulfinyl,[149b] diphenylphosphinyl,[76a,150] per-(or poly)fluorophenyl,[151] aryl [*e.g.* ethyl (arylimino)acetate][152] and chiral thiophosphoryl,[153] *etc.* Among these imines, *p*-toluenesulfonyl imines **177** are often used because they can be conveniently prepared *in situ* from the aldehyde and amine. Hexafluoroacetone imines are the most reactive electrophiles among imines employed in the MBH reaction.[60] However, imines derived from an aliphatic aldehyde failed to react with activated olefins, probably due to enolization.[154]

When the imine **183** is part of a 1,3-diazabutadiene system, the initial adduct **184**, which can be observed by ^{19}F NMR spectroscopy, cyclizes to give a tetrahydropyrimidine **185** (Scheme 1.72).[155] When the diazabutadiene is part of an aromatic system, compound **186** could also undergo the similar reaction to furnish **187** (Scheme 1.72).[155]

Like the corresponding benzaldehyde derivatives,[109a] planar chiral benzaldehyde imine tricarbonylchromium complexes **188** readily react with activated olefins to give the corresponding adducts **189** with excellent diastereoselectivity.[156] After removing the metal by treatment with air and sunlight, chiral amines **190** have been obtained in high yields and enantiomeric excess (Scheme 1.73).

α-Amido sulfones **191**, derived from aromatic and heteroaromatic aldehydes, acting as an alternative equivalent of imine derivatives, which are inherently

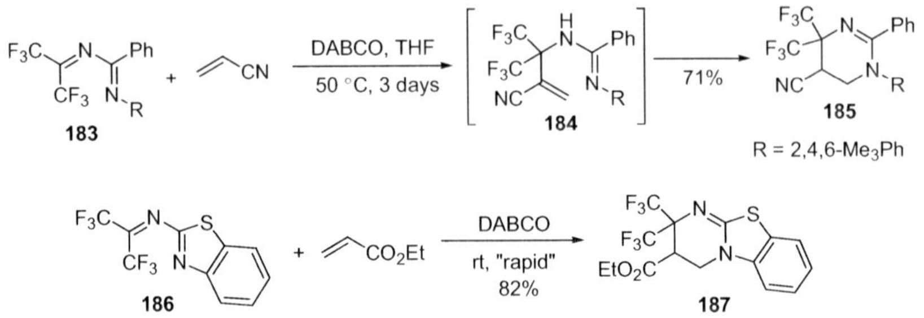

Scheme 1.72

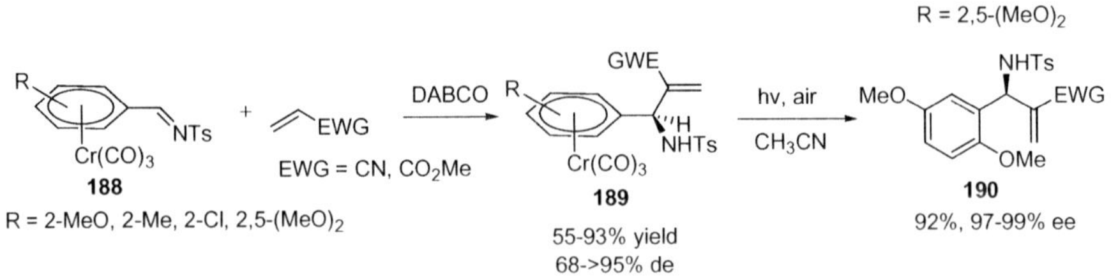

R = 2,5-(MeO)$_2$

EWG = CN, CO$_2$Me

188

R = 2-MeO, 2-Me, 2-Cl, 2,5-(MeO)$_2$

189

55-93% yield
68->95% de

190

92%, 97-99% ee

Scheme 1.73

highly reactive and highly sensitive to moisture, have been employed successfully in aza-MBH reactions with acrylates and acrylonitrile[157] and enantioselective aza-MBH reactions with α,β-unsaturated aldehydes (Scheme 1.74).[158] Moreover, aliphatic α-amidosulfones **195** as imine surrogates have also been developed for the enantioselective aza-MBH reaction with **196** by using the novel bifunctional catalyst 6′-deoxy-6′-acyl-amino-β-isocupreidine **197**, which served as a base to trigger *in situ* generation of *N*-sulfonylimine and also a chiral nucleophile to initiate the enantioselective aza-MBH reaction (Scheme 1.75).[159] Consequently, α-methylene-β-amino-β-alkyl carbonyl compounds **198**, which were difficult to access previously, can now be synthesized in excellent yields and enantioselectivities.

The iminium salt **199**, generated *in situ* from bis(dimethylamino)methane and acetyl chloride, was found to react with methyl vinyl ketone in the absence of a catalyst to give the α-aminoalkylation product **200** in high yield (Scheme 1.76).[160] The mechanism is similar to that of the MBH reaction, with chloride ion or a small amount of bis(dimethylamino)methane acting as the catalyst.

In situ prepared iminium salts **201** in a concentrated ethereal solution of lithium perchlorate are very effective electrophiles in aza-MBH reactions and react with methyl acrylate in the presence of a catalytic amount of a tertiary amine at ambient temperature to afford aza-MBH adducts **202**. The product undergoes conjugated addition with (trimethylsilyl)dialkylamines to give the diamines **203** in good yields (Scheme 1.77).[161]

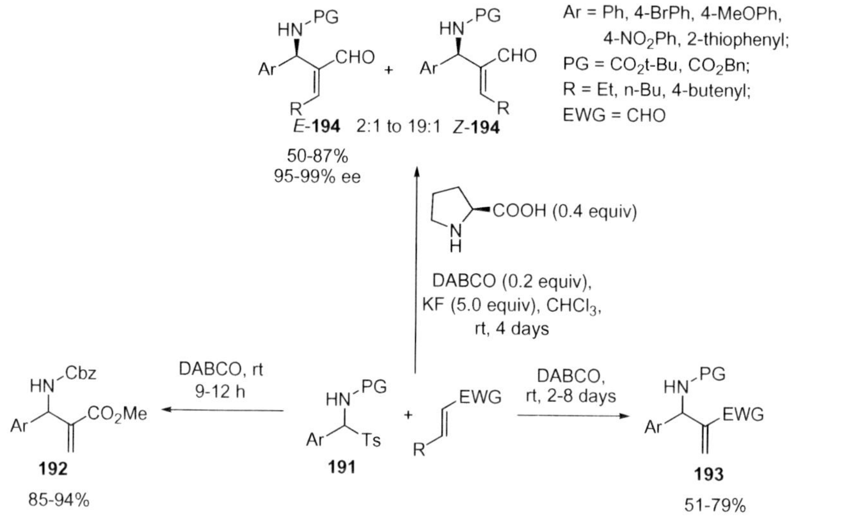

Ar = Ph, 4-BrPh, 4-MeOPh,
4-NO$_2$Ph, 2-thiophenyl;
PG = CO$_2$t-Bu, CO$_2$Bn;
R = Et, n-Bu, 4-butenyl;
EWG = CHO

E-**194** 2:1 to 19:1 *Z*-**194**

50-87%
95-99% ee

COOH (0.4 equiv)

DABCO (0.2 equiv),
KF (5.0 equiv), CHCl$_3$,
rt, 4 days

DABCO, rt
9-12 h

DABCO,
rt, 2-8 days

192

85-94%

191

193

51-79%

EWG = CO$_2$Me; R = H; PG = Cbz;
Ar = Ph, 4-ClPh, 4-MeOPh, 3,4-(MeO)$_2$Ph, 3,4,5-
(MeO)$_3$Ph, vanilyl, 4-NO$_2$Ph, 3-NO$_2$Ph, 4-tolyl, 4-
isopropylphenyl, 2-furyl, 2-thiophenyl, 2-naphthyl

EWG = CO$_2$Me, CO$_2$Et, CN; R = H;
PG = Boc, Cbz; Ar = Ph, 4-MeOPh, 4-
BrPh, 2-MeOPh, 2-ClPh, 2-furyl,
3-pyridyl, 1-naphthyl;

Scheme 1.74

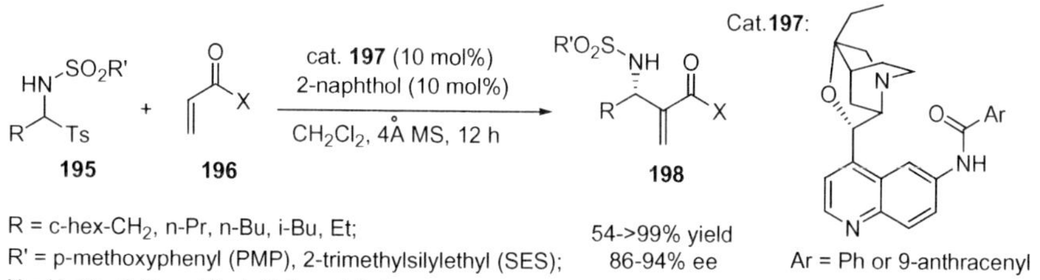

cat. **197** (10 mol%)
2-naphthol (10 mol%)

CH$_2$Cl$_2$, 4Å MS, 12 h

Cat.**197**:

195   **196**

198

54->99% yield
86-94% ee

Ar = Ph or 9-anthracenyl

R = c-hex-CH$_2$, n-Pr, n-Bu, i-Bu, Et;
R' = p-methoxyphenyl (PMP), 2-trimethylsilylethyl (SES);
X = H, Me, O-2-naphthyl, O-1-naphthyl

Scheme 1.75

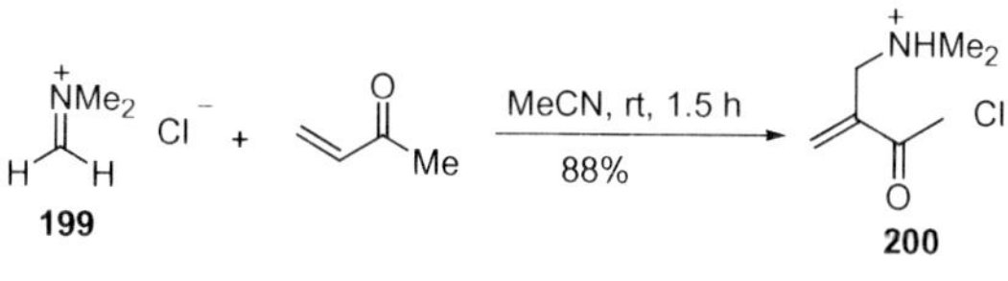

MeCN, rt, 1.5 h

88%

199

200

Scheme 1.76

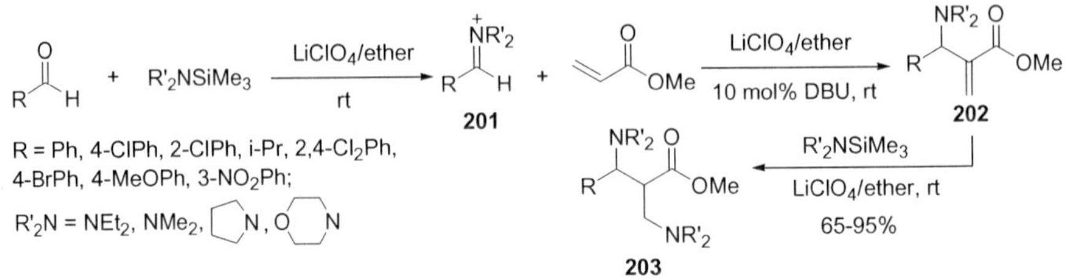

Scheme 1.77

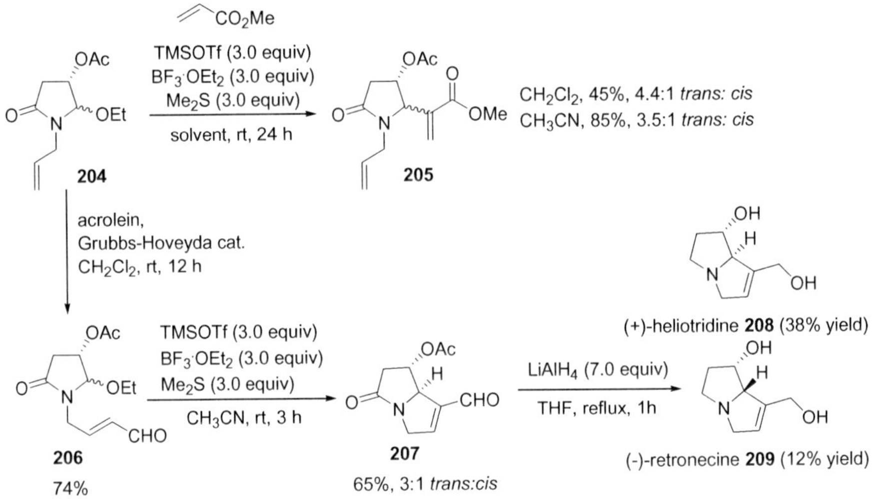

R = Ph, 4-ClPh, 2-ClPh, i-Pr, 2,4-Cl$_2$Ph,
4-BrPh, 4-MeOPh, 3-NO$_2$Ph;

R'$_2$N = NEt$_2$, NMe$_2$, [ ]N, O[ ]N

201

202

203

65-95%

Scheme 1.78

204

205

CH$_2$Cl$_2$, 45%, 4.4:1 *trans: cis*
CH$_3$CN, 85%, 3.5:1 *trans: cis*

acrolein,
Grubbs-Hoveyda cat.
CH$_2$Cl$_2$, rt, 12 h

206

74%

TMSOTf (3.0 equiv)
BF$_3$·OEt$_2$ (3.0 equiv)
Me$_2$S (3.0 equiv)

CH$_3$CN, rt, 3 h

207

65%, 3:1 *trans:cis*

LiAlH$_4$ (7.0 equiv)

THF, reflux, 1h

(+)-heliotridine **208** (38% yield)

(-)-retronecine **209** (12% yield)

In the presence of TMSOTf, BF$_3$.OEt$_2$ and dimethyl sulfide, the iminium ions **204** (masked as N,O-acetals) have been employed to couple with a very board range of readily available Michael acceptors, including acrolein and acrylates, in both an inter- and intramolecular MBH-type reaction to give densely functionalized heterocycles **207** (Scheme 1.78). The process has been rendered asymmetric and high enantioselectivity is obtained in reactions of iminium ions **210** (masked as N,O-acetals) and cyclic enones (Scheme 1.79). Finally, the usefulness of the methodology has been exemplified in a short synthesis of (+)-heliotridine **208** and (–)-retronecine **209** (Scheme 1.78).[162]

1.4.5   Other Electrophiles

A previous report demonstrated that esters, alkyl halides, vinyl ethers, acetic anhydride and epoxides did not react under MBH reaction conditions or give intractable mixtures;[10] however, many kinds of electrophiles have been

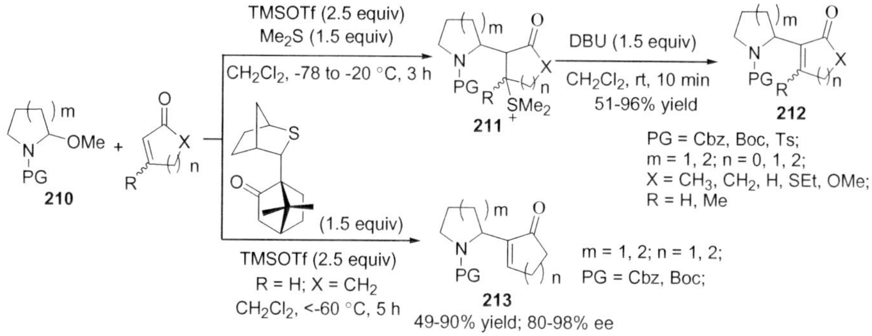

Scheme 1.79

developed more recently for the MBH reaction, such as allyl halide,[163] alkyl halide,[85,164] azodicarboxylates,[84,165] epoxides,[164,166] haloquinones,[167] aminomethylbenzotriazoles[168] and nitrostyrenes,[79b,83] to give satisfactory results in the presence of amine, phosphine or $TiCl_4$ catalyst.

1.5 Multicomponent One-pot Reaction

A three-component solution phase synthesis of substituted sulfonamides by reaction of an acrylic ester, an aldehyde and toluenesulfonamide under MBH reaction conditions has been reported. The reaction is catalyzed by $DABCO^{3e,169}$ or PPh_3.[148]

In 1989, Kahn *et al.* reported a three-component, one-pot procedure for the aza-MBH reaction.[148] However, β-substitution in the activated alkenes was not tolerated under the conditions (Scheme 1.80).

Subsequently, multicomponent reactions were less explored. In 1998, Richter and Jung[30] reported a three-component reaction of a polymer-bound acrylate, aldehydes and sulfonamides in the presence of DABCO as catalyst in dioxane. After cleavage from the polymer support, the desired Morita–Baylis–Hillman adducts **214** were obtained in good yields with moderate to high purities (Scheme 1.81).

Balan and Adolfsson have developed a three-component reaction with activated alkenes, aldehydes and tosylamines, affording aza-MBH adducts **215**, along with some normal MBH alcohols **216** as side products (Scheme 1.82). In the presence of DABCO or 3-HQD as catalyst, together with $La(OTf)_3$ and molecular sieves, the aza-MBH adducts **215** were obtained in moderate to high yields in propan-2-ol. Regarding the choice of base, DABCO was generally recommended since better chemoselectivities were obtained with this catalyst. For less electrophilic aldehydes, however, the use of 3-HQD resulted in higher yields of the aza-MBH adducts **215**.[170] Subsequently, the authors found that use of $Ti(OPr^i)_4$ as the Lewis acid co-catalyst was superior to that of $La(OTf)_3$, giving better chemoselectivity in the aza-MBH adducts **215** (Scheme 1.82).[171]

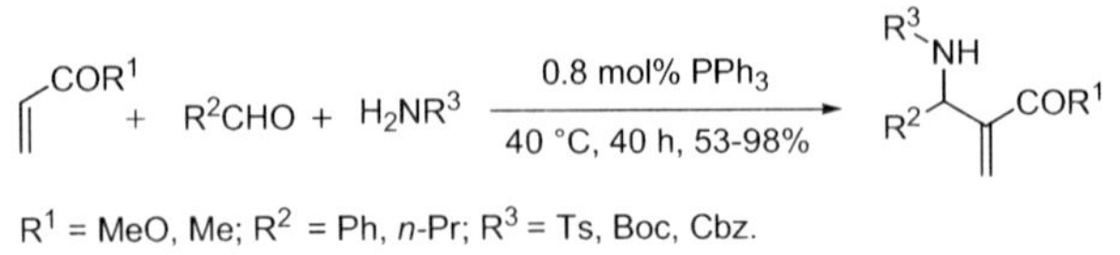
0.8 mol% PPh3
40 °C, 40 h, 53-98%
R1 = MeO, Me; R2 = Ph, n-Pr; R3 = Ts, Boc, Cbz.

Scheme 1.80

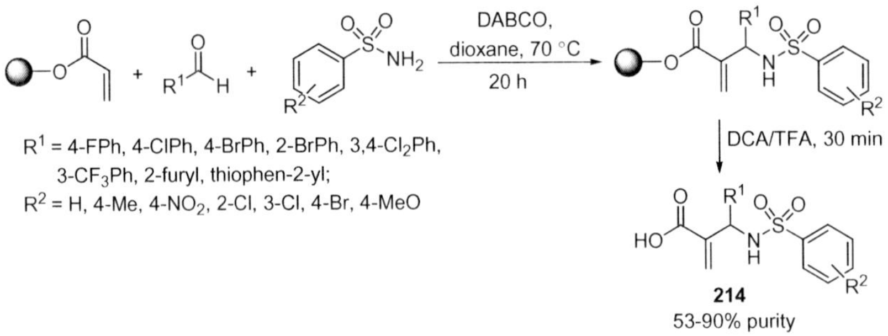
DABCO,
dioxane, 70 °C
20 h
DCA/TFA, 30 min
R1 = 4-FPh, 4-ClPh, 4-BrPh, 2-BrPh, 3,4-Cl2Ph,
3-CF3Ph, 2-furyl, thiophen-2-yl;
R2 = H, 4-Me, 4-NO2, 2-Cl, 3-Cl, 4-Br, 4-MeO
214
53-90% purity

Scheme 1.81

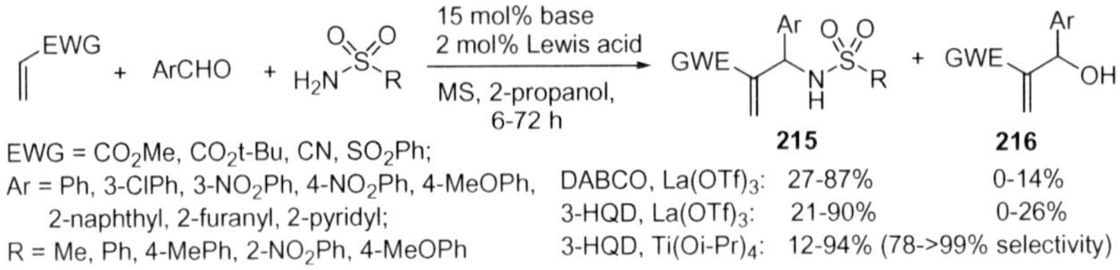
15 mol% base
2 mol% Lewis acid
MS, 2-propanol,
6-72 h
215
216
EWG = CO2Me, CO2t-Bu, CN, SO2Ph;
Ar = Ph, 3-ClPh, 3-NO2Ph, 4-NO2Ph, 4-MeOPh,
2-naphthyl, 2-furanyl, 2-pyridyl;
R = Me, Ph, 4-MePh, 2-NO2Ph, 4-MeOPh
DABCO, La(OTf)3: 27-87% 0-14%
3-HQD, La(OTf)3: 21-90% 0-26%
3-HQD, Ti(Oi-Pr)4: 12-94% (78->99% selectivity)

Scheme 1.82

Furthermore, we have found that one-pot three-component reactions of aryl aldehydes, diphenylphosphinamide and methyl vinyl ketone, in the presence of $TiCl_4$, PPh_3 and Et_3N, give the corresponding aza-MBH adducts **217** in moderate to good yields (Scheme 1.83).[172]

Subsequently, Balan and Adolfsson *et al.* employed chiral quinuclidine derivatives as catalysts in the asymmetric one-pot three-component aza-MBH reaction of aryl aldehydes, tosylamide and alkyl acrylates or acrylonitrile. A sterically non-hindered tricyclic derivative of quinidine (**141**) was the most efficient catalyst in transferring its chiral information. High conversions were ensured by using a catalytic amount of titanium isopropoxide and by the addition of molecular sieves (4 Å). The corresponding adducts **218** and **219** were obtained in good yields with up to 74% ee (Scheme 1.84).[173]

Lamaty *et al.* further presented the first examples of the *N*-supported aza-MBH reaction between PEG-SES amine **220**, prepared from a novel SES-type linker attached to an appropriate PEG polymer, acrylate and aldehydes

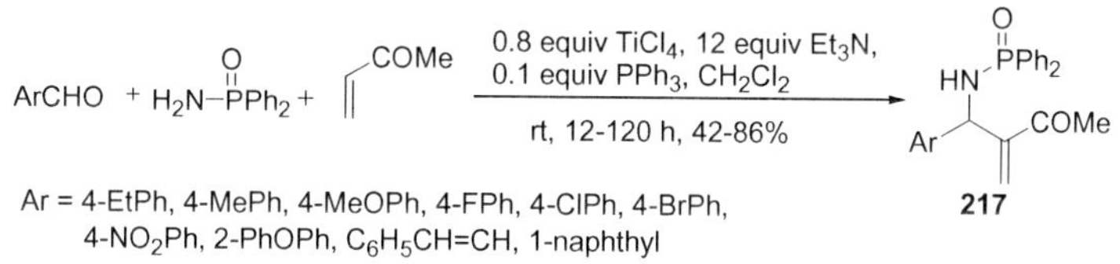

ArCHO
$H_2N-P(O)Ph_2$
COMe
0.8 equiv $TiCl_4$, 12 equiv Et_3N,
0.1 equiv PPh_3, CH_2Cl_2
rt, 12-120 h, 42-86%
HN
PPh_2
Ar
COMe
Ar = 4-EtPh, 4-MePh, 4-MeOPh, 4-FPh, 4-ClPh, 4-BrPh,
4-NO_2Ph, 2-PhOPh, $C_6H_5CH=CH$, 1-naphthyl
217

Scheme 1.83

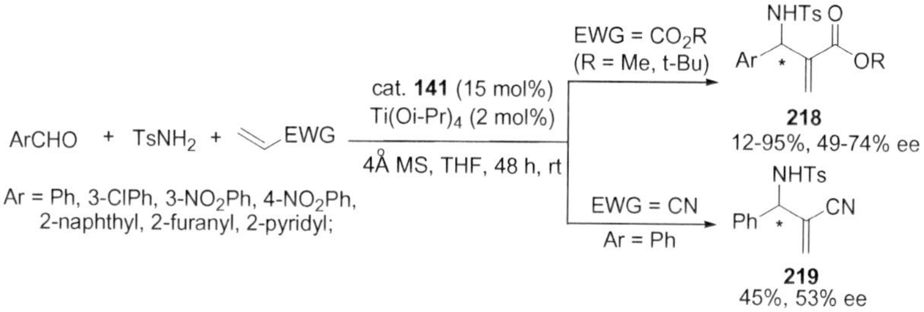

ArCHO + $TsNH_2$ + EWG
cat. 141 (15 mol%)
Ti(Oi-Pr)$_4$ (2 mol%)
4Å MS, THF, 48 h, rt
Ar = Ph, 3-ClPh, 3-NO_2Ph, 4-NO_2Ph,
2-naphthyl, 2-furanyl, 2-pyridyl;
EWG = CO_2R
(R = Me, t-Bu)
NHTs O
Ar
OR
218
12-95%, 49-74% ee
EWG = CN
Ar = Ph
NHTs
Ph
CN
219
45%, 53% ee

Scheme 1.84

catalyzed by DABCO[174] or 3-HQD[175] under thermal or microwave irradiation (Scheme 1.85). Notably, the absence of solvent significantly accelerates the reaction. A small library of PEG-supported aminoesters **221** was generated and further transformation (hydrogenation) was performed. Cleavage of **222** from the polymer support by the action of fluoride ions, followed by trapping with acetic anhydride, gave the acetylated aminoesters **223** in moderate yield.[174] This method has also been applied to the synthesis of diverse 2-substituted-3-methoxycarbonylpyrroles **224** (Scheme 1.85).[176]

Hashmi *et al.* have developed a short and high-yielding synthetic approach towards functionalized indoline derivatives **227** following a reaction sequence that consists of a three-component aza-MBH reaction from furfural catalyzed by La(OTf)$_3$ in combination with 3-HDQ, a simple sulfonamide propargylation and a gold-catalyzed cycloisomerization (Scheme 1.86).[177]

1.6 Intramolecular Morita–Baylis–Hillman Reaction

Although the MBH reaction, in general, has seen a high degree of growth with respect to all three essential components, the intramolecular version of this reaction has not been studied in depth. Intramolecular cyclization of $MeCO(CH_2)_2CH=CHCO_2Et$ (**228**) using (+)-CAMP produced cyclopentene **229** in 40% isolated yield with low enantioselectivity (Scheme 1.87).[178] CAMP was found to be superior to other phosphines, such as PBu_3, while DABCO and other amines were ineffective for the cyclization reaction. In the six-membered ring example the yield is only 17–23%.

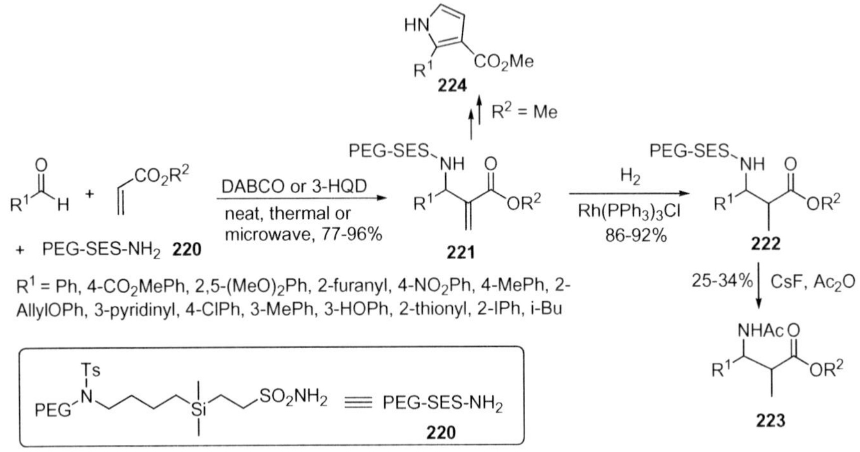

Scheme 1.85

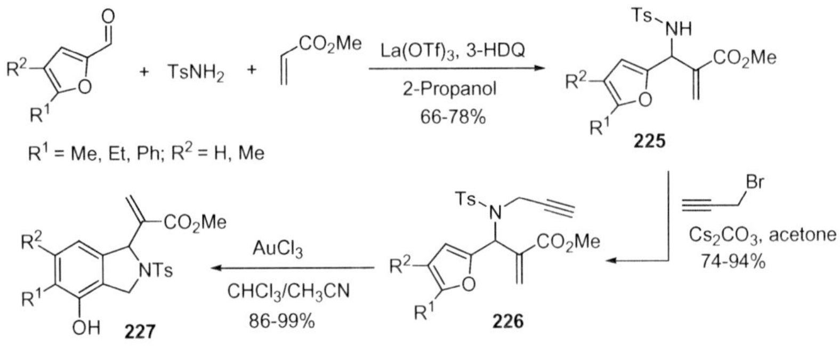

Scheme 1.86

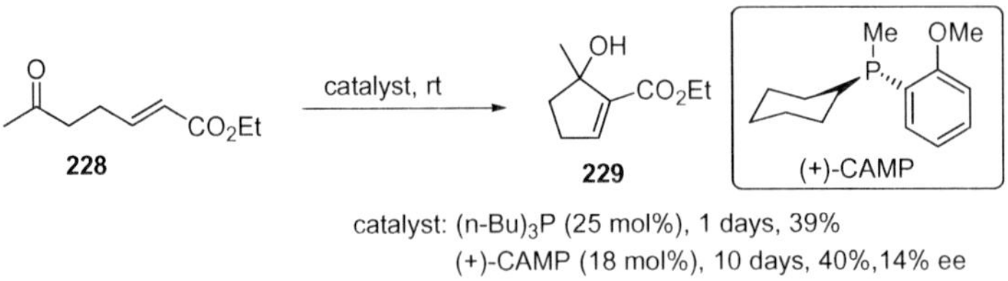

Scheme 1.87

Another example of intramolecular MBH reaction has been reported by Drewes *et al.* In the presence of DABCO, the acrylate ester of salicylaldehyde (**230**) afforded crystalline coumarin salt **231** (Scheme 1.88). The chloride ion in salt **231** originally comes from the solvent (CH_2Cl_2). Formation of this

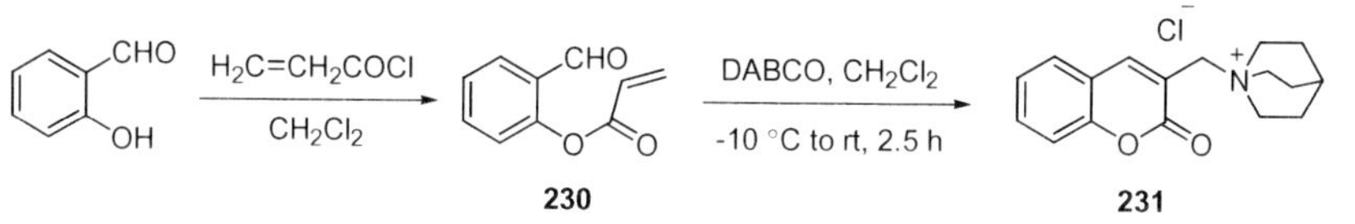

H₂C=CH₂COCl / **CH₂Cl₂**

DABCO, CH₂Cl₂ / -10 °C to rt, 2.5 h

230

231

Scheme 1.88

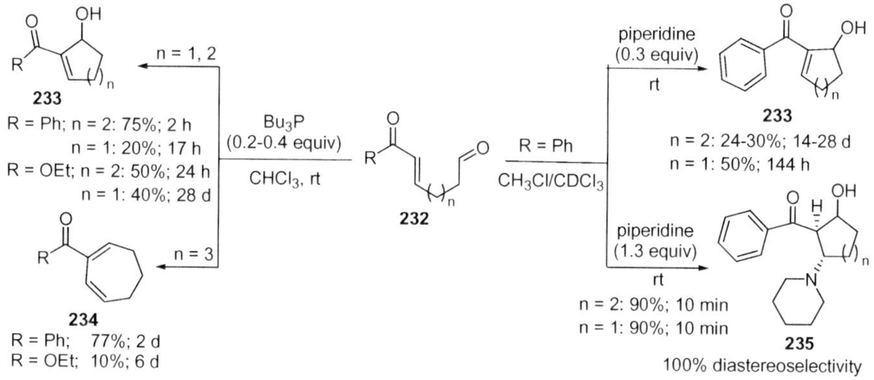

233
R = Ph; n = 2: 75%; 2 h
n = 1: 20%; 17 h
R = OEt; n = 2: 50%; 24 h
n = 1: 40%; 28 d

n = 1, 2

Bu₃P
(0.2-0.4 equiv)
CHCl₃, rt

232

R = Ph
CH₃Cl/CDCl₃

n = 3

234
R = Ph;   77%; 2 d
R = OEt; 10%; 6 d

piperidine
(0.3 equiv)
rt

233
n = 2: 24-30%; 14-28 d
n = 1: 50%; 144 h

piperidine
(1.3 equiv)
rt
n = 2: 90%; 10 min
n = 1: 90%; 10 min

235
100% diastereoselectivity

Scheme 1.89

derivative has been taken as evidence for the vital intermediate in the mechanism of the reaction.[111]

Murphy *et al.*[179] have systematically investigated the intramolecular Morita–Baylis–Hillman reaction using substrates containing the activated alkene and an electrophile, in the presence of amines or phosphines. Tributylphosphine gave directly intramolecular MBH adducts **233** in the case of five- and six-membered rings and cycloheptadienes **234** in the case of seven-membered rings. Stoichiometric amounts of piperidine provided aldol products **235** in the case of five- and six-membered rings, in high yields, when treated with corresponding phenyl alkenyl ketones, whereas a catalytic amount of piperidine afforded moderate yields of intramolecular MBH adducts **233** in these cases (Scheme 1.89).

Krische and co-workers[180] have developed an elegant Bu₃P-catalyzed cycloisomerization of bis-enones **236** to afford five- and six-membered rings. The effect of electronic (**236a** and **236b**) and steric factors (**236c**) on this cyclization was investigated (Scheme 1.90). In addition, this methodology was also extended to optically pure xylose-derived mono-enone mono-enoate **238** to provide the pentasubstituted cyclohexene **239** with high diastereoselectivity (Scheme 1.91).

At the same time, Roush and co-workers[181] reported phosphine-mediated intramolecular MBH reactions of di-activated 1,6-heptadienes **240** and 1,5-hexadienes **241** to provide an attractive synthesis of functionalized cyclohexene **242** and cyclopentene derivatives **243**, respectively. They also extended this

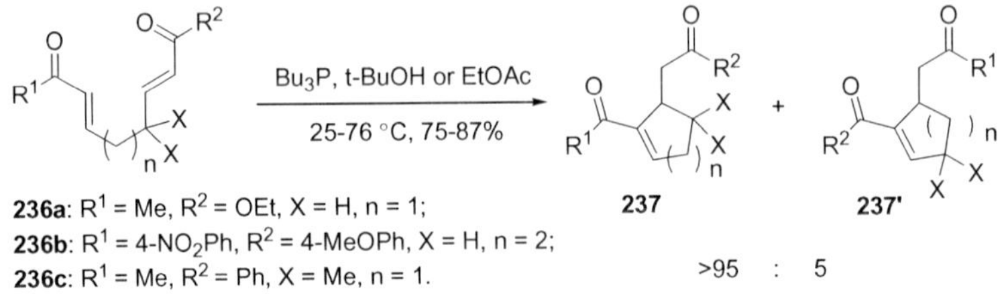

Bu3P, t-BuOH or EtOAc
25-76 °C, 75-87%
236a: R1 = Me, R2 = OEt, X = H, n = 1;
236b: R1 = 4-NO2Ph, R2 = 4-MeOPh, X = H, n = 2;
236c: R1 = Me, R2 = Ph, X = Me, n = 1.
237
237'
>95 : 5

Scheme 1.90

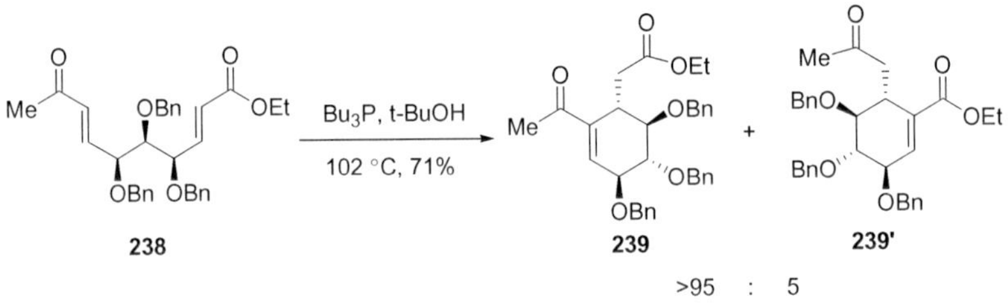

Bu3P, t-BuOH
102 °C, 71%
238
239
239'
>95 : 5

Scheme 1.91

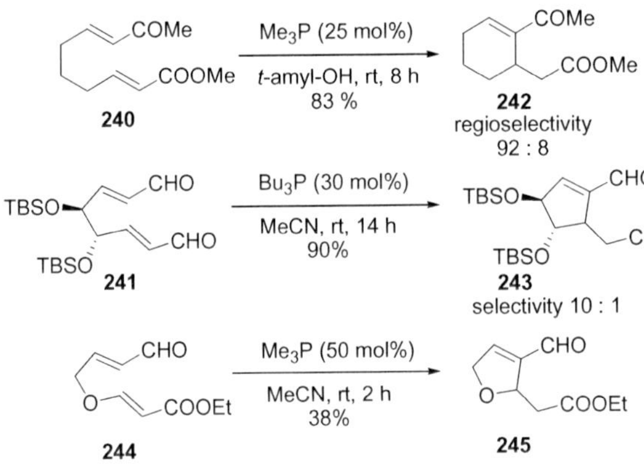

Me3P (25 mol%)
t-amyl-OH, rt, 8 h
83 %
240
242
regioselectivity
92 : 8
Bu3P (30 mol%)
MeCN, rt, 14 h
90%
241
243
selectivity 10 : 1
Me3P (50 mol%)
MeCN, rt, 2 h
38%
244
245

Scheme 1.92

methodology to the synthesis of substituted dihydrofuran derivatives **245** (Scheme 1.92). Subsequently, a concise synthesis of the spinosyn A tricyclic nucleus (**248**) has been developed by a route featuring a one-pot tandem intramolecular Diels–Alder reaction and intramolecular vinylogous Morita–Baylis–Hillman cyclization in which five stereocenters in tricycle **248** are set with excellent selectivity (Scheme 1.93).[182]

Keck and Welch have examined the intramolecular MBH reaction of α,β-unsaturated esters/thioesters containing an enolizable aldehyde group, under

Scheme 1.93

Scheme 1.94

various conditions.[92] In the case of thiol esters, cyclopentenol products **249** were formed in high yields when DMAP and DMAP–HCl in EtOH (at 78 °C for 1 h) or Me_3P in CH_2Cl_2 (at room temperature for 15 h) were employed. However, in the case of oxyesters, the desired cyclopentenol adducts **250** were obtained in low yields. Cyclohexenol products **251** were obtained in high yields when Me_3P was used as a reagent, whereas DMAP and DMAP–HCl provided **251** in low yields. A representative example for each case is described in Scheme 1.94.

Oshima and co-workers[183] have reported intramolecular Michael aldol cyclization of formyl α,β-enones under the influence of Lewis acids. Thus, reaction of **252** with $TiCl_4/Et_3N(CH_2Ph)Cl$ at 0 °C provides 2-benzoyl-3-chlorocyclohexanol (cyclo aldol adduct) **253**, whereas the treatment of **252** with Et_2AlI provides an intramolecular MBH adduct **254** (Scheme 1.95). A similar reaction of **255** with $TiCl_4/Et_3N(CH_2Ph)Cl$ provides intramolecular MBH adduct **256** in 33% yield along with intramolecular chloro aldol product **257** and dehydration product **258** (Scheme 1.96).

Krishna *et al.* have developed the first diastereoselective intramolecular MBH reaction of chiral substrates **259** and **260**, wherein both aldehyde and

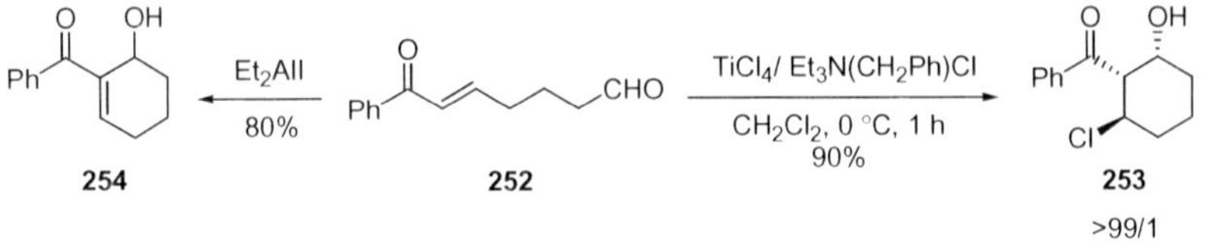

Scheme 1.95

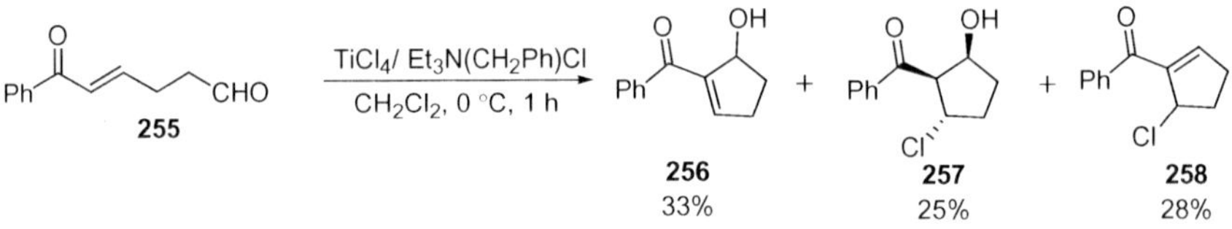

Scheme 1.96

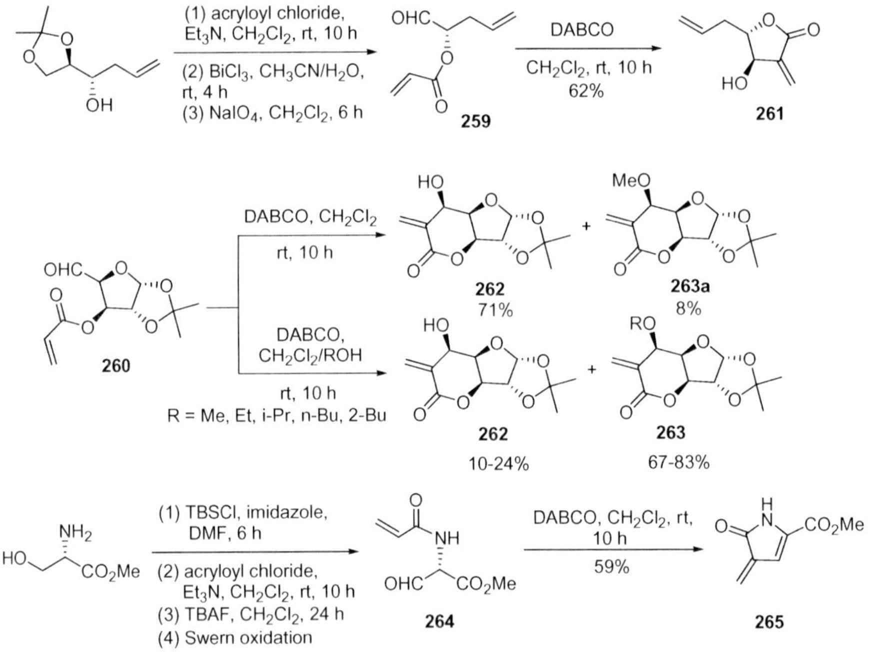

Scheme 1.97

activated olefin coexist as substituents, to afford α-methylene-β-hydro-xylactones **261** and **262**, respectively, in good yields exclusively as single isomers under the DABCO-catalyzed reaction conditions in CH_2Cl_2. They also observed the formation of alkoxylactones **263** by *in situ* derivatization of adducts under precise reaction conditions (Scheme 1.97).[184] Subsequently, they extended this study to chiral acrylamide aldehyde **264**, which was prepared from L-serine derivative in four steps, *viz.* (i) silylation, (ii) acryloylation, (iii) desilylation and (iv) Swern oxidation. However, lactam **265** was obtained as the product instead of the normal MBH adducts when **264** was subjected to an intramolecular MBH reaction with DABCO in CH_2Cl_2.

Koo *et al.* have developed an efficient method for the preparation of diverse ω-formyl-α,β-unsaturated carbonyl compounds **268** and **273** that relied on the $Pb(OAc)_4$-promoted oxidative ring cleavage of cyclic 1,2-diols **267** and **272**, respectively, which in turn can be readily obtained by the 1,2-addition of various nucleophiles to a α′-acetoxy-substituted conjugated cycloalkenones **266** and α-acetoxy cyclohexanone **271**. The authors also optimized the conditions for the intramolecular MBH reactions of **268** and **273**. The utility of this sequence is demonstrated by the syntheses of chromones **270** and the precursor (**274**) of the compound containing the 6,8-dioxabicyclo[3.2.1]octane ring (Scheme 1.98).[185]

In addition, we have found that the stereochemistry of the Michael acceptor also plays an important role in the efficiency of phosphine-mediated intramolecular MBH reactions. In all examined cases with PPh_3 or polymer-supported phosphine as the catalyst, cyclization substrates enones **275** possessing (*Z*)-alkenes afforded the desired product **276** in a higher yield than (*E*)-**275** under identical conditions (Scheme 1.99).[186] The reason for this difference in reactivity is most likely steric in nature, as substrates where the β-substituent is *cis* to the electron-withdrawing substituent are more accessible to reaction with the nucleophilic catalyst than their *trans* counterparts.

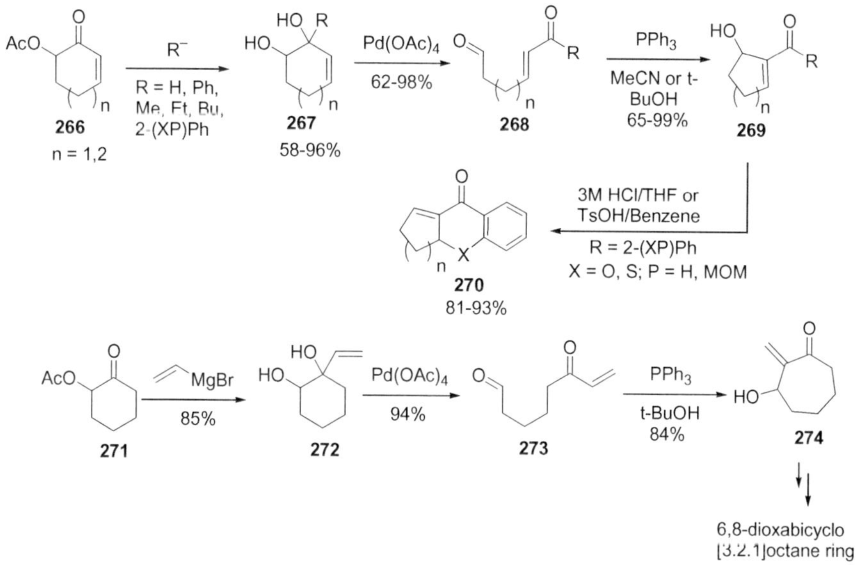

Scheme 1.98

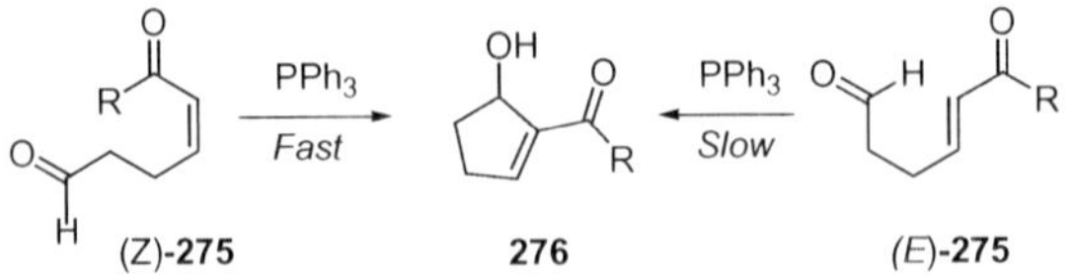

Scheme 1.99

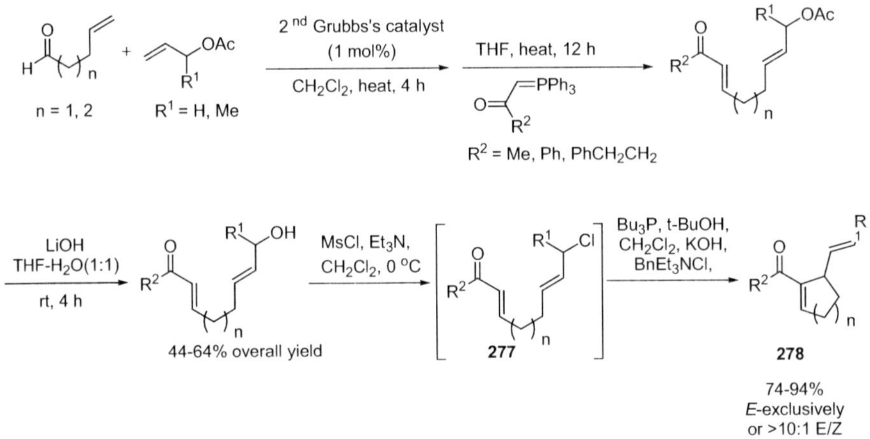

Scheme 1.100

As mentioned above, although some examples of intramolecular MBH reactions have been reported in the literature, this aspect is still in its infancy. Most known reports are based on the cyclizations of combinations of enone-enone, enone-acrylate, enone-aldehyde, unsaturated thioester-aldehyde, enone–allylic carbonate frameworks, *etc.* More recently, Krafft *et al.* have developed a novel, entirely organo-mediated intramolecular MBH reaction by using allyl chloride **277** as an alternative electrophile to afford densely functionalized cyclic enones **278**. This reaction tolerates modification of the enone and the use of primary and secondary allylic chlorides and generates both five- and six-membered rings in excellent yields. Both mono- and disubstituted alkenes are formed with excellent selectivity in the absence of a transition metal catalyst (Scheme 1.100).[187]

Subsequently, they were the first to demonstrate that saturated alkyl halides **279** as sp^3 hybridized electrophiles – never before used in an organo-mediated MBH reaction – facilitated the intramolecular α-alkylation of enones to afford five- and six-membered enone products **280** (Scheme 1.101).[188]

Later on, they also reported an interesting intramolecular MBH reaction of an enone-epoxide system. Opening of the MBH epoxide **281** afforded homologous aldol products **282** efficiently and embodies a C(sp^2)–C(sp^3) coupling with concomitant cyclization. Scheme 1.102 shows a representative example.[166]

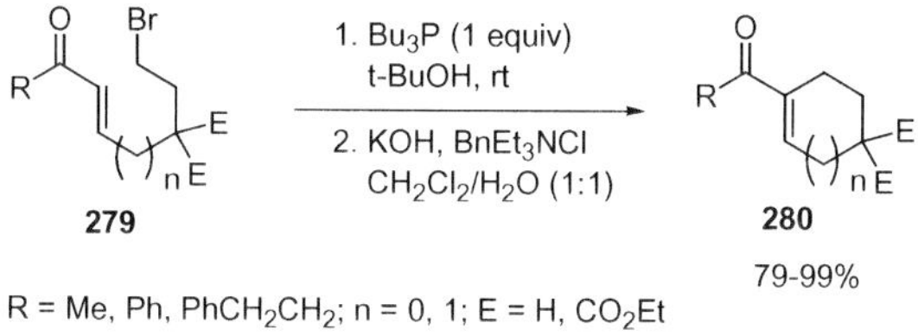

R = Me, Ph, PhCH$_2$CH$_2$; n = 0, 1; E = H, CO$_2$Et

Scheme 1.101

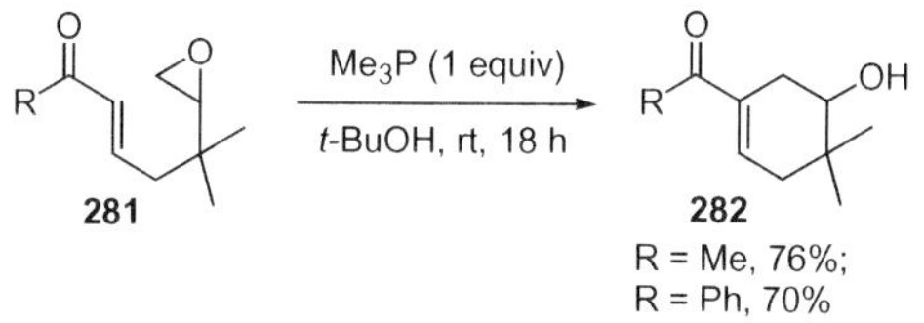

281

282
R = Me, 76%;
R = Ph, 70%

Scheme 1.102

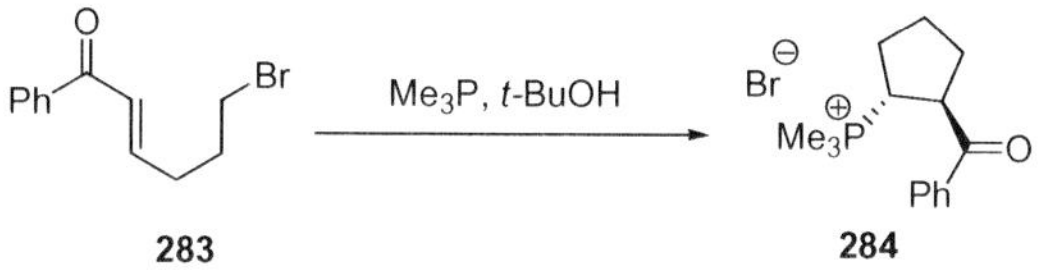

283   **284**

Scheme 1.103

In addition, during their work on the intramolecular MBH reaction of **283**,[22] Krafft and co-workers also isolated ketophosphonium salt **284** (Scheme 1.103), which, in fact, supports the proposed mechanistic pathway (as shown in Scheme 1.6) for the MBH reaction.

Recently, the fluorous phosphine P[(CH$_2$)$_3$R$_{f8}$]$_3$ [R$_{f8}$ = (CF$_2$)$_7$CF$_3$] **285** was found to be an efficient catalyst for intramolecular MBH reactions of enone-aldehyde or unsaturated thioester-aldehyde **286** in acetonitrile at 60–72 °C, affording the adducts **287** and **288** in good to high yields. Upon cooling, the catalyst precipitated from the organic solvent and was recycled. The products **287** and **288** were isolated from the supernatant (Scheme 1.104). They also extended significantly the catalyst systems, which were demonstrated to be recyclable by precipitation onto Teflon$^{®}$ tape, and on Gore-Rastex$^{®}$ fiber, for recovery. The latter fluoropolymer gives slightly better results, presumably due to a higher surface area.[189]

There have been only a few reports on asymmetric versions of the intra-molecular MBH reaction; Scheme 1.105 shows one representative example.[190] The intramolecular Morita–Baylis–Hillman reaction has been achieved with unprecedented levels of enantioselectivity by using a co-catalyst system invol-ving pipecolinic acid and *N*-methylimidazole; cyclic MBH products **289** were

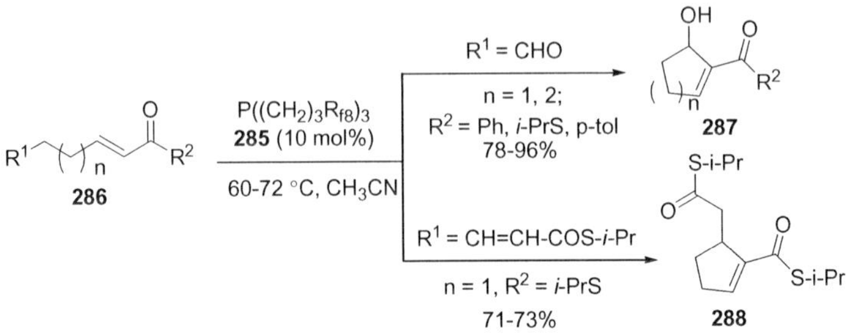

Scheme 1.104

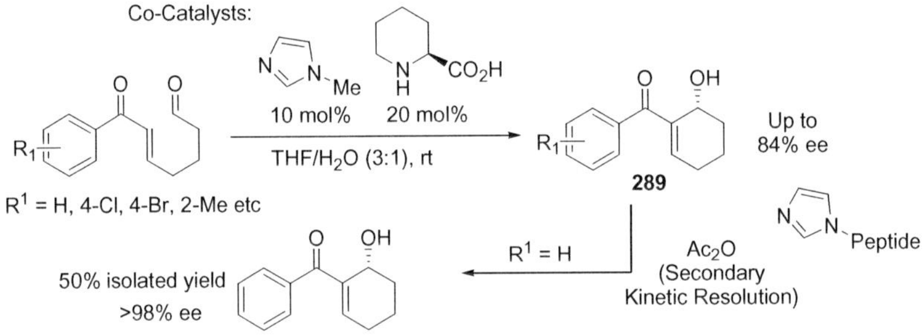

Scheme 1.105

obtained with up to 84% ee. In addition, if this reaction was carried out with a "kinetic resolution quench" involving acetic anhydride and an asymmetric acylation catalyst, an ee enhancement occurs to deliver products with >98% ee, with an isolated yield of 50%.

At the same time, Hong *et al.* also independently developed an efficient proline catalyzed enantioselective intramolecular Morita–Baylis–Hillman reaction of dialdehydes **290a**, providing the corresponding (*S*)-6-hydroxy-cyclohexene-1-carbaldehyde (**291**) in good yield with high enantioselectivity. Surprisingly, addition of imidazole to the mixture resulted in an inversion of selectivity. Temperature and solvent effects on the reactivity and enantio-selectivity were investigated and a possible mechanism for the two different selectivities has been proposed.[191] For **290b** and **290c**, the intramolecular Michael addition product **293** or the corresponding MBH adduct **292** was obtained in moderate yield with lower ee; no inversion of selectivity of **292** by addition of imidazole was observed (Scheme 1.106).

Recently, Gladysz *et al.* have developed an effective chiral rhenium-containing phosphine (**294**) catalyzed intramolecular MBH reaction of enone-aldehydes **295** in benzene, affording the corresponding adducts **296** in 88–99% yields with 38–74% ee (Scheme 1.107). Notably, the catalyst loadings can be much lower than those required with organophosphines, presumably due to the transition-metal-enhanced phosphorus nucleophilicities.[192]

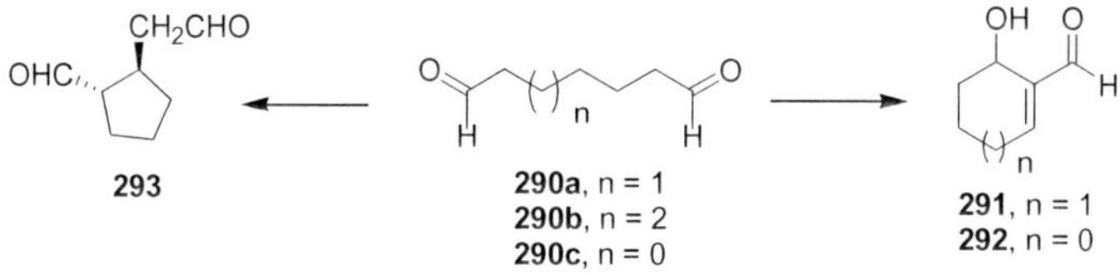

Scheme 1.106

	Cat.	Additive	Product (Configuration)
290a: n = 1	L-Proline	---	**291** (*S*)
	D-Proline	---	**291** (*R*)
	L-Proline	imidazole	**291** (*R*)
	D-Proline	imidazole	**291** (*S*)
290b: n = 2	L-Proline	---	**293** (*anti*)
290c: n = 0	L-Proline	---	**292** (*S*)
	D-Proline	---	**292** (*R*)
	L-Proline	imidazole	**292** (*S*)
	D-Proline	imidazole	**292** (*R*)

295　　n = 1, 2; R = Ph, S-*i*-Pr, *p*-Tol, Me

296　　88-99%, 38-74% ee

(S)-294

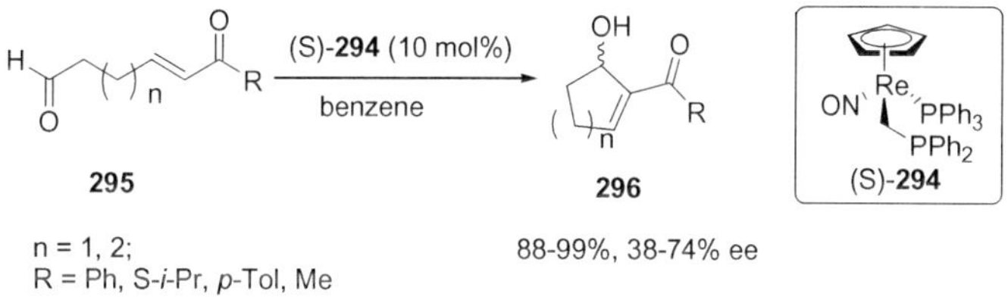

Scheme 1.107

1.7　Ionic Liquid as Reaction Media[193]

Although the MBH reaction is atom economic and allows the generation of a highly functionalized molecule in a single step, it suffers from slow reaction rates (often requiring days for completion), even under solvent-free conditions and in the presence of a large amount of base. Several attempts have been made to accelerate the MBH reaction. These include the use of ultrasound,[194] microwave irradiation,[49] high pressure,[10,195,48b] supercritical CO_2 (scCO_2),[196] polar solvent system[54,197] and the use of Lewis acids as co-catalysts (Chapter 2.6).

Since the rate enhancement of MBH reaction in protic solvents or by use of LBBA bifunctional catalysts and Lewis acids as co-catalysts is attributed to the involvement of hydrogen bonding, it seemed that imidazolium-based ionic liquids would be a good choice of solvent for the MBH reaction because the acidic proton at C(2) is known to act as a donor to hydrogen bond acceptors.[198]

Afonso *et al.* have reported that the ionic liquid 1-*n*-butyl-3-methylimidazolium hexafluorophosphate, [bmim][PF_6], can accelerate the MBH reaction between benzaldehyde and methyl acrylate in the presence of DABCO

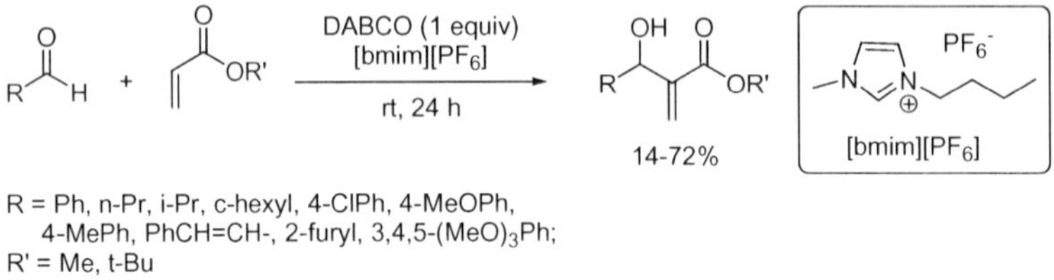

R = Ph, n-Pr, i-Pr, c-hexyl, 4-ClPh, 4-MeOPh,
4-MePh, PhCH=CH-, 2-furyl, 3,4,5-(MeO)$_3$Ph;
R' = Me, t-Bu

Scheme 1.108

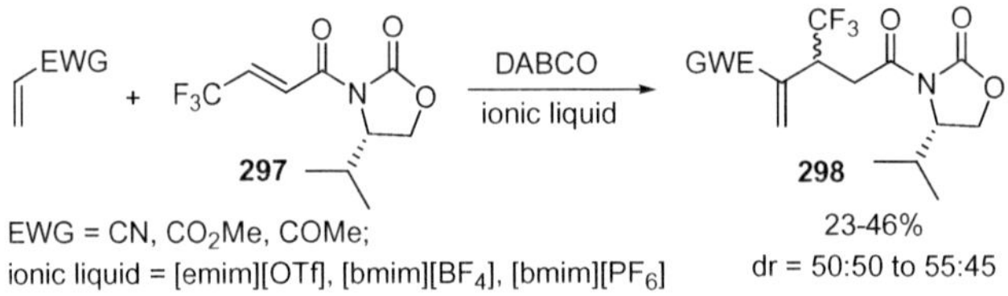

EWG = CN, CO$_2$Me, COMe;
ionic liquid = [emim][OTf], [bmim][BF$_4$], [bmim][PF$_6$]

23-46%
dr = 50:50 to 55:45

Scheme 1.109

(33 times faster than the reaction in CH$_3$CN, the reported conventional solvent), affording the desired products in moderate yields (Scheme 1.108). No significant improvements in reaction yield were observed by further addition of Lewis acids. Lithium perchlorate as an additive in this process increased the rate, but yields of the products were reduced.[199]

Kitazume *et al.* examined the first example of Michael additions *via* the MBH-type reaction that used 3-fluoromethylprop-2-enamide as a chiral auxiliary electrophile towards activated olefins in the DABCO–ionic liquid system.[200] The reaction of (4*S*)-3-[(*E*)-4,4,4-trifluorobut-2-enoyl]-4-isopropyl-2-oxazolidinone (**297**) with activated vinyl moiety proceeded smoothly at 80 °C to give the corresponding adducts **298** in moderate yields, albeit with low diastereoselectivity (Scheme 1.109).

Subsequently, Ko *et al.* investigated the MBH reaction of benzaldehyde with methyl acrylate in various bmim-based ion liquids and found that [bmim][PF$_6$] was the best, resulting in the maximum rate enhancement. A moderate acceleration in the reaction rates was observed if the above ionic liquid was used in combination with Lewis acid or H-bond donors. Notably, the rate-enhancing effects of the ionic liquid and Lewis acid or H-bond donors were not additive. A two-fold rate increase was achieved in mixtures of [bmim][PF$_6$], La(OTf)$_3$ and 2,2′,2-nitrilotris(ethanol) compared to the same reaction in MeCN.[201]

Prompted by these results, Aggarwal *et al.* reinvestigated their previous work on the MBH reaction in ionic liquids. It was found that under basic reaction conditions the aldehyde reacted with the imidazolium cation to form a side product (**299**), leading to the MBH adduct in low yield.[202] This finding demonstrated that ionic liquids are not always inert solvents – the acidic nature of the C(2) hydrogen of the imidazolium cation was responsible for this side reaction

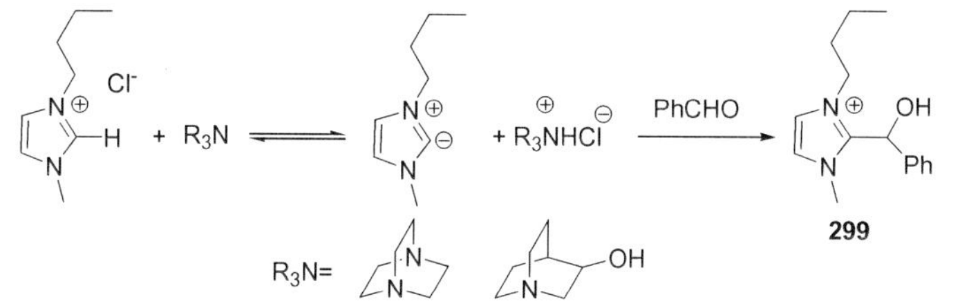

Scheme 1.110

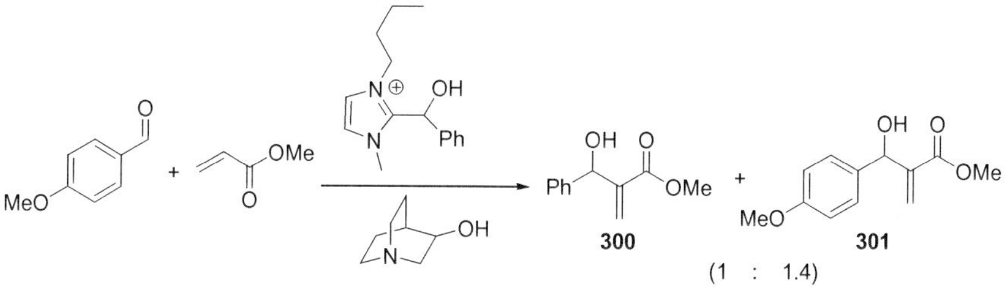

Scheme 1.111

(Scheme 1.110). Notably, since the formation of adduct **299** is reversible, the ionic liquid from one run of a MBH reaction can be recycled and used in a different MBH reaction to furnish a mixture of products (**300** and **301**, Scheme 1.111).

Since the problems associated with the use of imidazolium-based ionic liquids arise from the acidity of C(2) imidazolium cation, Hsu *et al.* have developed a novel 2-position substituted ionic liquid, [bdmim][PF$_6$], and applied it in the MBH reaction.[203] Unlike the commonly used [bmim][PF$_6$], which can react with electrophilic aldehydes under basic conditions, ionic liquid [bdmim][PF$_6$] is inert and the MBH reactions of various aldehydes and methyl acrylate proceed smoothly in better yields (Scheme 1.112).

Although substitution at the 2-position of the imidazolium cation was considered to prevent the side reaction in the MBH reaction, Handy and Okello have found that even the 2-methyl substituted imidazolium cation was not completely inert.[204] They found that the 2-methyl group underwent slow proton exchange even in the presence of a weak base such as triethylamine (Scheme 1.113). The acidic nature of this methyl group was further verified by analyzing the products obtained from attempted methylation of the imidazolium salt **302**. When **302** was treated with excess NaH and CH$_3$I, none of the expected product **304** was detected, instead product **303** was obtained (Scheme 1.114).[204]

Subsequently, Wilhelm *et al.* found that imidazolinium salts **305**, incorporating a phenyl ring at C2, were a suitable ionic liquid as a solvent for quinuclidine-catalyzed MBH reactions of aldehydes with acrylates, cyclohexenone or acrylamide. The corresponding adducts were obtained in moderate to good yields (Scheme 1.115).[205]

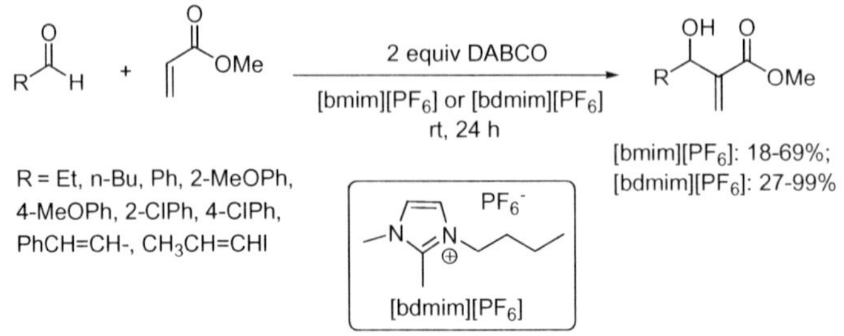

2 equiv DABCO

[bmim][PF$_6$] or [bdmim][PF$_6$]
rt, 24 h

R = Et, n-Bu, Ph, 2-MeOPh,
4-MeOPh, 2-ClPh, 4-ClPh,
PhCH=CH-, CH$_3$CH=CHI

[bmim][PF$_6$]: 18-69%;
[bdmim][PF$_6$]: 27-99%

PF$_6^-$

[bdmim][PF$_6$]

Scheme 1.112

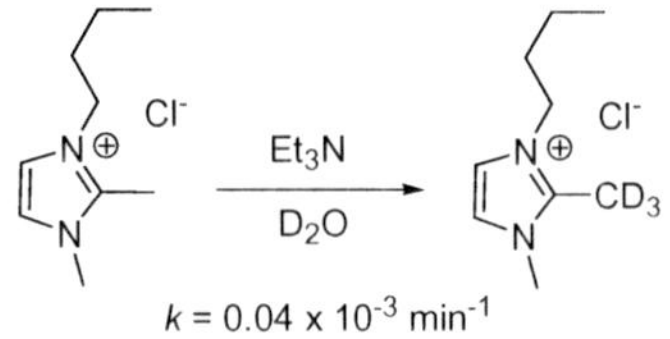

Et$_3$N

D$_2$O

k = 0.04 x 10^{-3} min^{-1}

Scheme 1.113

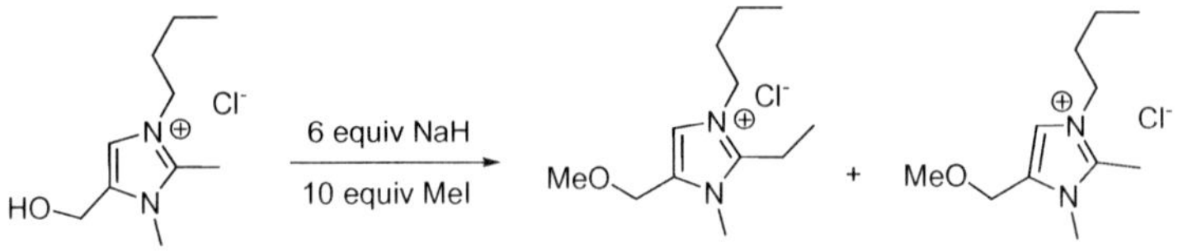

6 equiv NaH

10 equiv MeI

302

303

304

not detected

Scheme 1.114

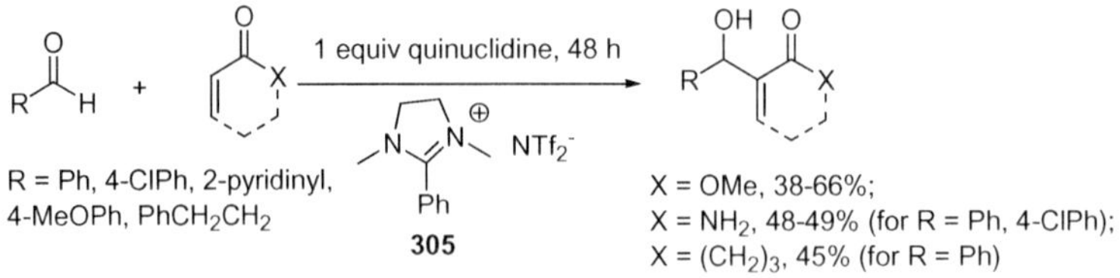

1 equiv quinuclidine, 48 h

R = Ph, 4-ClPh, 2-pyridinyl,
4-MeOPh, PhCH$_2$CH$_2$

NTf$_2^-$

Ph

305

X = OMe, 38-66%;
X = NH$_2$, 48-49% (for R = Ph, 4-ClPh);
X = (CH$_2$)$_3$, 45% (for R = Ph)

Scheme 1.115

Since MBH reactions carried out with common ionic liquids, such as [bmim][X], give lower to moderate yields, more recently, Tsai *et al.* have synthesized a new and highly active di-naphthalene imidazolium salt. They found that 1,3-bis[2-(naphthalene-2-yloxy)propyl]imidazolium bromide (**306**) promoted

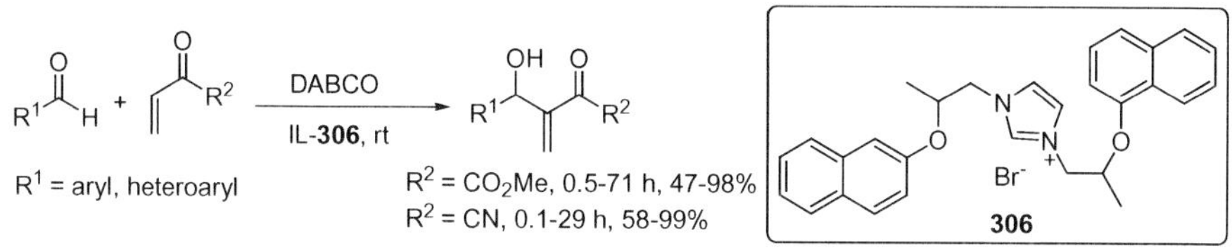

Scheme 1.116

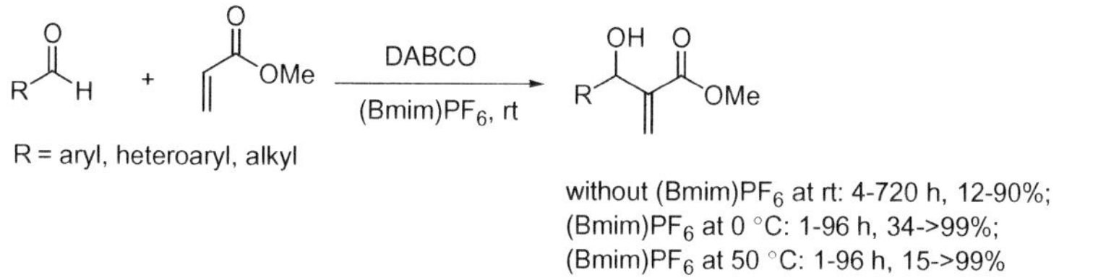

Scheme 1.117

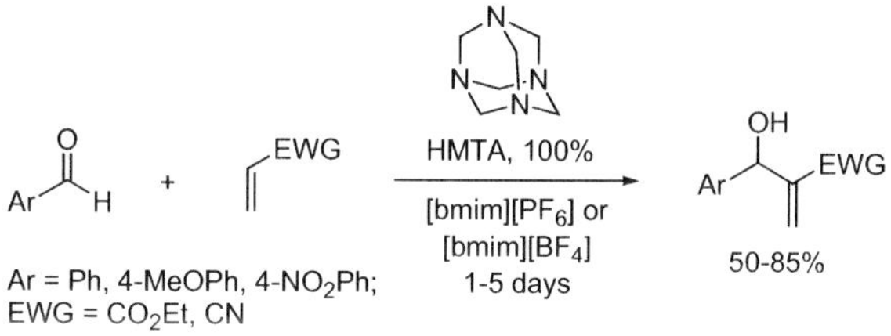

without (Bmim)PF$_6$ at rt: 4-720 h, 12-90%;
(Bmim)PF$_6$ at 0 °C: 1-96 h, 34->99%;
(Bmim)PF$_6$ at 50 °C: 1-96 h, 15->99%

Scheme 1.118

the MBH reaction of various aryl aldehyde compounds in the absence of solvents to yield MBH adducts in high yields and short reaction time (Scheme 1.116).[206]

Recently, Coelho has investigated systematically the effect of different catalytic conditions, including ultrasound, ionic liquids and temperature, independently or combined, on the reaction rate and yield of MBH reaction. A strong synergic effect, which significantly increases the reaction rates and yields, was observed when the reactions were performed using an imidazolic ionic liquid catalyst, [(Bmim)PF$_6$], at both 0 and 50 °C, compared with room temperature (Scheme 1.117).[194d]

Unlike DABCO, hexamethylenetetramine (HMTA) combined with bmim-type ionic liquids is an effective catalyst system for MBH reactions between aromatic aldehydes and acrylonitrile or methyl acrylate, affording the corresponding adducts in moderate to good yields (50–85%) in short reaction times (Scheme 1.118). Additionally, the ionic liquids can be recycled three times without any loss of activity.[207]

As well as imidazolium based ionic liquids, other kinds of ionic liquids have also been developed for the MBH reaction. Kumar *et al.* have found that ionic liquid chloroaluminates, consisting of AlCl$_3$ and *N*-1-butylpyridinium chloride

(BPC) or 1-ethyl-3-methyl-1*H*-imidazolium chloride (EMIC), as solvent can accelerate DABCO-catalyzed MBH reactions that are sluggish for other ionic liquids. In addition, the combination of EMIC with $AlCl_3$ is a more efficient chloroaluminate ionic liquid than the BPC with $AlCl_3$, as it offers higher yields in comparatively shorter time (Scheme 1.119). The chloroaluminates can be easily recycled and reused six times without loss of activity.[208]

Gong and Zhao *et al.* have independently presented the application of pyridinium-based ionic liquid as recyclable solvent for the DABCO-catalyzed MBH reaction. Compared with the commonly used imidazolium-based ionic liquids [bmim][PF6] and [bdmim][PF6], which evidently react with aldehydes under basic conditions, the pyridinium-based ionic liquid [EPy][BF4] and [BuPy][NO3] are inert and the MBH reactions of aldehydes with methyl acrylate, MVK or acrylonitrile proceed quickly in good yields.[209] Zhao *et al.* also examined hexamethyl-enetetramine (HMTA)-catalyzed MBH reactions by using the ionic liquid [EPy][BF4] as a reaction media; short reaction times and good to excellent yields was obtained (Scheme 1.120).[209b]

R	EWG	chloroaluminates	time (h)	yield (%)
Ph	CO_2Me	--	19	65
		BPC:AlCl3 (60%)	11	75
		EMIC:AlCl3 (60%)	8	80
	CN	--	9	82
		BPC:AlCl3 (60%)	5	93
		EMIC:AlCl3 (60%)	6	95
2-MeOPh	CO_2t-Bu	--	48	17
		BPC:AlCl3 (60%)	8	39
		EMIC:AlCl3 (60%)	7	39

Scheme 1.119

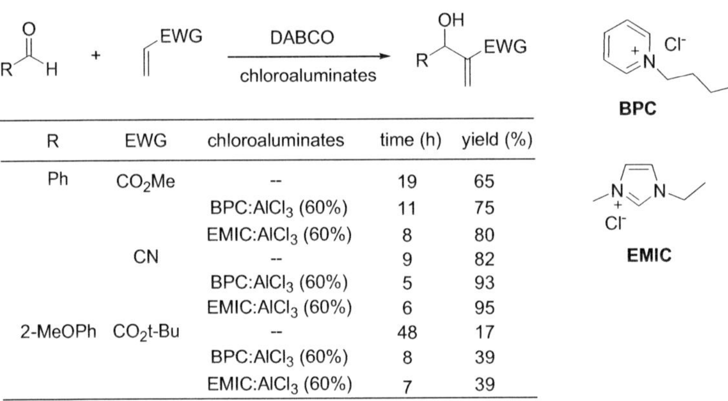

Scheme 1.120

Zhao *et al.* have synthesized novel ionic liquids **307** and **308**, which contain (3-chloro-2-hydroxypropyl)-functionalized pyridinium cations, by the ultrasound-assisted reaction of pyridine with acid (HCl and HBF$_4$) and 3-chloropropylene oxide at room temperature, in excellent yields and purity, in which the acid provided the corresponding anionic component of the ionic liquids. Furthermore, they examined the application of new ionic liquids as solvents in the MBH reaction and found that IL-BF$_4$ **308** showed a better effect in rate enhancement and an improved yield than IL-Cl **307** in some cases (Scheme 1.121).[210]

Following on from the pioneering strategy of applying chiral solvents in asymmetric synthesis, developed by Seebach and Oei,[211] chiral ILs have also been applied in the enantioselective MBH reaction. Vo-Thanh *et al.* first reported the asymmetric induction caused by a chiral IL **309** as the only source of chirality in an asymmetric reaction. In the DABCO-catalyzed MBH reactions of aldehydes and methyl acrylate, they obtained enantioselectivities up to 44% ee by using an IL **309** with chiral cations derived from (–)-*N*-methyl-ephedrine as reaction media (Scheme 1.122).[212] Importantly, the presence of the hydroxyl function on chiral ILs **309** is propitious for the transfer of chirality.

Subsequently, the novel bi-stereogenic chiral ionic liquids **310** and **311**, containing two chiral centers in the side chain bonded to the 2-position of the imidazolium cation and different anions, have been synthesized and applied to asymmetric MBH reactions of aldehydes with acrylates or cyclohexenone as reaction media. The corresponding adducts were obtained with fair enantio-selectivities (up to 25% ee), albeit in moderate to high yields (up to 97%), by using chiral ionic liquid **311** as solvent (Scheme 1.123).[213]

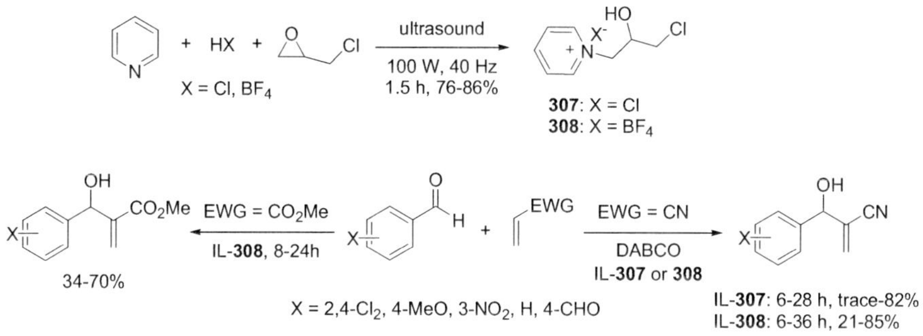

Scheme 1.121

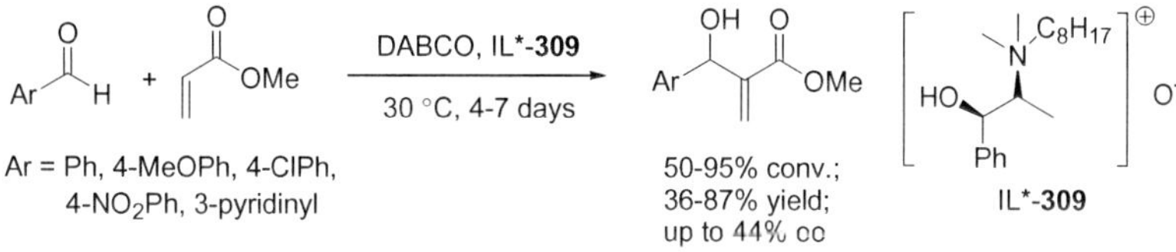

Scheme 1.122

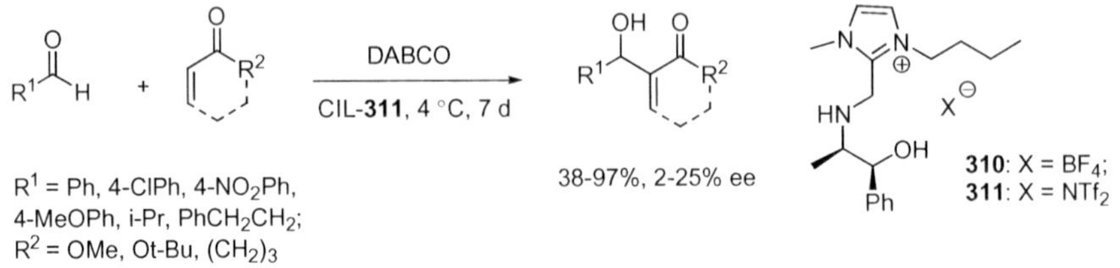

Scheme 1.123

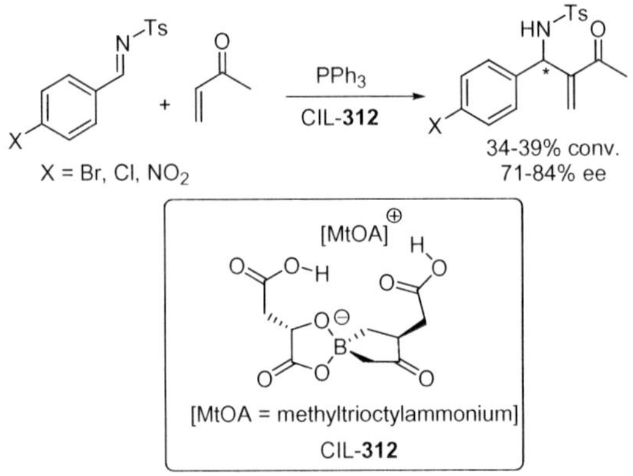

Scheme 1.124

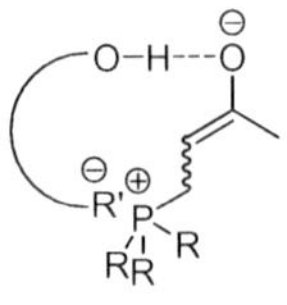

Figure 1.6 Possible bifunctional interaction of the zwitterionic intermediate of the aza-MBH reaction with the chiral anion of a CIL-**312**, which contains a hydrogen-bond donor.

Contrary to the view that applying chiral solvents in synthesis cannot result in appreciable enantioselectivities, Leitner *et al.* have reported the first example of an asymmetric reaction in which a chiral reaction medium induces a high level of enantioselectivity.[214] Using a specifically designed ionic liquid (**312**) with a chiral anion as the only source of chirality, up to 84% ee was obtained in the aza-MBH reaction of N-Ts arylaldimines with MVK (Scheme 1.124), which is comparable with values obtained with the best catalysts for the asymmetric aza-MBH reaction in conventional solvents (94% ee,[21] 83% ee[76d,215]). Possible bifunctional interaction of the zwitterionic intermediate of the aza-MBH reaction with the chiral anion of a CIL-**312** containing a hydrogen-bond donor was proposed (Figure 1.6).

References

1. (a) K. Morita, Z. Suzuki and H. Hirose, *Bull. Chem. Soc. Jpn.*, 1968, **41**, 2815; (b) K. Morita, *Japan Patent*, 6803364, 1968; *Chem. Abstr.*, 1968, **69**, 58828s.
2. A. B. Baylis and M. E. D. Hillman, German Patent 2155113, 1972; *Chem. Abstr.*, 1972, **77**, 34174q.
3. (a) S. E. Drewes and N. D. Emslie, *J. Chem. Soc., Perkin Tans. 1*, 1982, 2079; (b) H. M. R. Hoffmann and J. Rabe, *Angew. Chem. Int. Ed. Engl.*, 1983, **22**, 795; (c) H. M. R. Hoffmann and J. Rabe, *Helv. Chim. Acta*, 1984, **67**, 413; (d) H. M. R. Hoffmann and J. Rabe, *J. Org. Chem.*, 1985, **50**, 3849; (e) P. Perlmutter and C. C. Teo, *Tetrahedron Lett.*, 1984, **25**, 5951; (f) D. Basavaiah and V. V. L. Gowriswari, *Tetrahedron Lett.*, 1986, **27**, 2031.
4. (a) S. E. Drewes and G. H. P. Roos, *Tetrahedron*, 1988, **44**, 4653; (b) D. Basavaiah, P. Dharma Rao and R. Suguna Hyma, *Tetrahedron*, 1996, **52**, 8001; (c) E. Ciganek, in *Organic Reactions*, ed. L. A. Paquette, John Wiley & Sons, Inc., New York, 1997, **vol. 51**, p. 201; (d) D. Basavaiah, T. Satyanarayana and A. J. Rao, *Chem. Rev.*, 2003, **103**, 811; (e) V. Declerck, J. Martinez and F. Lamaty, *Chem. Rev.*, 2009, **109**, 1.
5. H. M. R. Hoffmann and J. Rabe, *Angew. Chem., Int. Ed. Engl.*, 1983, **22**, 796.
6. J. S. Hill and N. S. Isaacs, *J. Phys. Org. Chem.*, 1990, **3**, 285.
7. (a) P. T. Kaye and M. L. Bode, *Tetrahedron Lett.*, 1991, **32**, 5611; (b) Y. Fort, M.-C. Berthe and P. Caubère, *Tetrahedron*, 1992, **48**, 6371.
8. L. S. Santos, C. H. Pavam, W. P. Almeida, F. Coehlo and M. N. Eberlin, *Angew. Chem. Int. Ed.*, 2004, **43**, 4330.
9. The term was coined in analogy to pharmaceutical compound classes that are active against several different biological targets, see: T. P. Yoon and E. N. Jacobsen, *Science*, 2003, **299**, 1691.
10. J. S. Hill and N. S. Isaacs, *J. Chem. Res. (S)*, 1988, 330; *J. Chem. Res. (M)*, 1988, 2641.
11. (a) S. E. Drewes, N. D. Emslie, N. Karodia and A. A. Khan, *Chem. Ber.*, 1990, **123**, 1447; (b) P. Perlmutter, E. Puniani and G. Westman, *Tetrahedron Lett.*, 1996, **37**, 1715.
12. (a) K. E. Price, S. J. Broadwater, H. M. Jung and D. T. McQuade, *Org. Lett.*, 2005, **7**, 147; (b) K. E. Price, S. J. Broadwater, B. J. Walker and D. T. McQuade, *J. Org. Chem.*, 2005, **70**, 3980.
13. V. K. Aggarwal, S. Y. Fulford and G. C. Lloyd-Jones, *Angew. Chem., Int. Ed.*, 2005, **44**, 1706.
14. Y. Takemoto, *Org. Biomol. Chem.*, 2005, **3**, 4299.
15. J. J. Xu, *THEOCHEM*, 2006, **767**, 61.
16. D. Roy and R. B. Sunoj, *Org. Lett.*, 2007, **9**, 4873.
17. R. Robiette, V. K. Aggarwal and J. N. Harvey, *J. Am. Chem. Soc.*, 2007, **129**, 15513.
18. (a) G. W. Amarante, M. Benassi, A. A. Sabino, P. M. Esteves, F. Coelho and M. N. Eberlin, *Tetrahedron Lett.*, 2006, **47**, 8427; (b) L. S. Santos,

B. A. Silveira Neto, C. S. Consorti, C. H. Pavam, W. P. Almeida, F. Coelho, J. Dupont and M. N. Eberlin, *J. Phys. Org. Chem.*, 2006, **19**, 731.

19. G. W. Amarante, H. M. S. Milagre, B. G. Vaz, B. R. Vilachã Ferreira, M. N. Eberlin and F. Coelho, *J. Org. Chem.*, 2009, **74**, 3031.

20. D. Roy, C. Patel and R. B. Sunoj, *J. Org. Chem.*, 2009, **74**, 6936.

21. K. Matsui, S. Takizawa and H. Sasai, *J. Am. Chem. Soc.*, 2005, **127**, 3680.

22. M. E. Krafft, T. F. N. Haxell, K. A. Seibert and K. A. Abboud, *J. Am. Soc. Chem.*, 2006, **128**, 4174.

23. X.-F. Zhu, C. E. Henry and O. Kwon, *J. Am. Chem. Soc.*, 2007, **129**, 6722.

24. H. Liu, Q. Zhang, L. Wang and X. Tong, *Chem. Eur. J.*, 2010, **16**, 1968.

25. S. E. Drewes, S. D. Freese, N. D. Emslie and G. H. P. Roos, *Synth. Commun.*, 1988, **18**, 1565.

26. W.-D. Lee, K.-S. Yang and K. Chen, *Chem. Commun.*, 2001, 1612.

27. A. A. Khan, N. D. Emslie, S. E. Drewes, J. S. Field and N. Ramesar, *Chem. Ber.*, 1993, **126**, 1477.

28. (a) F. Ameer, S. E. Drewes, S. Freese and P. T. Kaye, *Synth. Commun.*, 1988, **18**, 495; (b) D. Basavaiah and P. K. S. Sarma, *Synth. Commun.*, 1990, **20**, 1611.

29. (a) H. Richter and G. Jung, *Mol. Divers.*, 1998, **3**, 191; (b) O. Rölfing, K. Prien, A. Thiel and H. Künzer, *Synlett*, 1997, 325.

30. H. Richter and G. Jung, *Tetrahedron Lett.*, 1998, **39**, 2729.

31. (a) B. A. Kulkarni and A. Ganesan, *J. Comb. Chem.*, 1999, **1**, 373; (b) H. Richter, T. Walk, A. Höltzel and G. Jung, *J. Org. Chem.*, 1999, **64**, 1362; (c) S. Batra, T. Srinivasan, S. K. Rastogi, B. Kundu, A. Patra, A. P. Bhaduri and M. Dixit, *Bioorg. Med. Chem. Lett.*, 2002, **12**, 1905.

32. M. D. Evans and P. T. Kaye, *Synth. Commun.*, 1999, **29**, 2137.

33. (a) P. R. Krishna, V. Kannan, A. Ilangovan and G. V. M. Sharma, *Tetrahedron: Asymmetry*, 2001, **12**, 829; (b) P. R. Krishna, R. Sachwani and V. Kannan, *Chem. Commun.*, 2004, 2580.

34. (a) K. He, Z. Zhou, G. Zhao and C. Tang, *Heteroatom Chem.*, 2006, **17**, 317; (b) K. He, Z.-H. Zhou, H. Y. Tang, G.-F. Zhao and C. Tang, *Chin. Chem. Lett.*, 2005, **16**, 1427.

35. H. Amri and J. Villieras, *Tetrahedron Lett.*, 1986, **27**, 4307.

36. (a) D. Basavaiah and V. V. L. Gowriswari, *Synth. Commun.*, 1987, **17**, 587; (b) T. Imagawa, K. Uemura, Z. Nagai and M. Kawanisi, *Synth. Commun.*, 1984, **14**, 1267; (c) Toyo Rayon Co. *Fr. Pat.* 1,506,132 (1967); *Chem. Abstr.*, 1969, **70**, 19613u.

37. (a) S. Tsuboi, S. Takatsuka and M. Utaka, *Chem. Lett.*, 1988, 2003; (b) S. Tsuboi, H. Kuroda, S. Takatsuka, T. Fukava, T. Sakai and M. Utaka, *J. Org. Chem.*, 1993, **58**, 5952.

38. (a) Z. Xu and X. Lu, *Tetrahedron Lett.*, 1997, **38**, 3461; (b) Z. Xu and X. Lu, *J. Org. Chem.*, 1998, **63**, 5031; (c) X. Lu, C. Zhang and Z. Xu, *Acc. Chem. Res.*, 2001, **34**, 535; (d) G.-L. Zhao and M. Shi, *J. Org. Chem.*, 2005, **70**, 9975.

39. (a) G.-L. Zhao, J.-W. Huang and M. Shi, *Org. Lett.*, 2003, **5**, 4737; (b) G.-L. Zhao and M. Shi, *Org. Biomol. Chem.*, 2005, **3**, 3686.

40. X.-Y. Guan, Y. Wei and M. Shi, *J. Org. Chem.*, 2009, **74**, 6343.

41. B. J. Cowen, L. B. Saunders and S. J. Miller, *J. Am. Chem. Soc.*, 2009, **131**, 6105.

42. M. Shi, Y.-W. Guo and H. B. Li, *Chin. J. Chem.*, 2007, **25**, 828.

43. Y. Matsuya, K. Hayashi and H. Nemoto, *J. Am. Chem. Soc.*, 2003, **125**, 646.

44. S. Xue, Q.-F. Zhou, L. Li and Q.-X. Guo, *Synlett*, 2005, 2990.

45. (a) T. G. Back, D. A. Rankic, J. M. Sorbetti and J. E. Wulff, *Org. Lett.*, 2005, **7**, 2377; (b) J. M. Sorbetti, K. N. Clary, D. A. Rankic, J. E. Wulff, M. Parvez and T. G. Back, *J. Org. Chem.*, 2007, **72**, 3326; (c) K. N. Clary and T. G. Back, *Synlett*, 2007, 2995.

46. Y.-L. Shi and M. Shi, *Tetrahedron*, 2006, **62**, 461.

47. (a) M. L. Bode and P. T. Kaye, *J. Chem. Soc., Perkin Trans. 1*, 1993, 1809; (b) M. L. Bode and P. T. Kaye, *J. Chem. Soc., Perkin Trans. 1*, 1990, 2612.

48. (a) N. S. Isaacs, J. S. Hill, Eur. Pat. Appl. EP 196, 708 (1986); *Chem.. Abstr.*, 1987, **106**, 19155h; (b) J. S. Hill and N. S. Isaacs, *Tetrahedron Lett.*, 1986, **27**, 5007.

49. (a) M. K. Kundu, S. B. Mukherjee, N. Balu, R. Padmakumar and S. V. Bhat, *Synlett*, 1994, 444; (b) C. R. Strauss, M. N. Galbraith, A. F. Faux, PCT Int. Appl. WO 91 18, 861 (1991); *Chem. Abstr.*, 1992, **116**, 128185v.

50. (a) L. J. Brzezinski, S. Rafel and J. W. Leahy, *J. Am. Chem. Soc.*, 1997, **119**, 4317; (b) L. J. Brzezinski, S. Rafel and J. W. Leahy, *Tetrahedron*, 1997, **53**, 16423.

51. E. Marcucci, G. Martelli, M. Orena and S. Rinaldi, *Monatsh. Chem.*, 2007, **138**, 27.

52. (a) W. Guo, W. Wu, N. Fan, Z. Wu and C. Xia, *Synth. Commun.*, 2005, **35**, 1239; (b) Z. Wu, G. Zhou, J. Zhou and W. Guo, *Synth. Commun.*, 2006, **36**, 2491.

53. C. Yu and L. Hu, *J. Org. Chem.*, 2002, **67**, 219.

54. P. R. Krishna, A. Manjuvani and G. V. M. Sharma, *Tetrahedron Lett.*, 2004, **45**, 1183.

55. V. K. Aggarwal, I. Emme and S. Y. Fulford, *J. Org. Chem.*, 2003, **68**, 692.

56. C. Faltin, E. M. Fleming and S. J. Connon, *J. Org. Chem.*, 2004, **69**, 6496.

57. G. M. Strunz, R. Bethell, G. Sampson and P. White, *Can. J. Chem.*, 1995, **73**, 1666.

58. A. S. Golubev, M. V. Galakhov, A. F. Kolomiets and A. V. Fokin, *Izv. Akad. Nauk Ser. Khim*, 1992, 2763; English transl. p. 2193.

59. (a) P. V. Ramchandran, M. V. Ram Reddy and M. T. Rudd, *Chem. Commun.*, 2001, 757; (b) M. V. Ram Reddy, M. T. Rudd and P. V. Ramchandran, *J. Org. Chem.*, 2002, **67**, 5382.

60. J. Cyrener and K. Burger, *Monatsh. Chem.*, 1994, **125**, 1279.

61. D. Basavaiah, T. K. Bharathi and V. V. L. Gowriswari, *Synth. Commun.*, 1987, **17**, 1893.

62. (a) T. Miyakoshi, H. Omichi and S. Saito, *Nippon Kagaku Kaishi*, 1980, 44; *Chem. Abstr.*, 1980, **92**, 146214u; (b) T. Miyakoshi and S. Saito, *Nippon Kagaku Kaishi*, 1983, 1623; (c) S. Sato, I. Matsuda and M. Shibata,

J. Organomet. Chem., 1989, **377**, 347; (d) S. I. Pereira, J. Adrio, A. M. S. Silva and J. C. Carretero, *J. Org. Chem.*, 2005, **70**, 10175.

63. (a) M. Shi, C.-Q. Li and J.-K. Jiang, *Chem. Commun.*, 2001, 833; (b) M. Shi, C.-Q. Li and J.-K. Jiang, *Tetrahedron*, 2003, **59**, 1181.

64. M. Shi and Y.-M. Xu, *Eur. J. Org. Chem.*, 2002, 696.

65. (a) M. Shi, C.-Q. Li and J.-K. Jiang, *Molecules*, 2002, **7**, 721; (b) M. Shi, C.-Q. Li and J.-K. Jiang, *Helv. Chim. Acta*, 2002, **85**, 1051.

66. M. M. Kabat, J. Kiegiel, N. Cohen, K. Toth, P. M. Wovkulich and M. R. Uskokoviæ, *J. Org. Chem.*, 1996, **61**, 118.

67. F. Rezgui and M. Moncef El Gaied, *Tetrahedron Lett.*, 1998, **39**, 5965.

68. K. Y. Lee, J. H. Gong and J. N. Kim, *Bull. Korean Chem. Soc.*, 2002, **23**, 659.

69. K. Y. Lee, S. GowriSankar and J. N. Kim, *Tetrahedron Lett.*, 2004, **45**, 5485.

70. S. Luo, X. Mi, P. G. Wang and J.-P. Cheng, *Tetrahedron Lett.*, 2004, **45**, 5171.

71. S. Luo, P. G. Wang and J.-P. Cheng, *J. Org. Chem.*, 2004, **69**, 555.

72. (a) V. K. Aggarwal and A. Mereu, *Chem. Commun.*, 1999, 2311; (b) S. Luo, X. Mi, H. Xu, P. G. Wang and J.-P. Cheng, *J. Org. Chem.*, 2004 **69**, 8413.

73. H. Ito, Y. Takenaka, S. Fukunishi and K. Iguchi, *Synthesis*, 2005, 3035.

74. (a) M. Shi and Y.-M. Xu, *Chem. Commun.*, 2001, 1876; (b) M. Shi, Y.-M. Xu, G.-L. Zhao and X.-F. Wu, *Eur. J. Org. Chem.*, 2002, **1**, 3666.

75. L. M. van Rozendaal, B. M. W. Voss and H. W. Scheeren, *Tetrahedron*, 1993, **49**, 6931.

76. For selected papers, see: refs. 64, 74 and; (a) M. Shi and G.-L. Zhao, *Tetrahedron Lett.*, 2002, **43**, 4499; (b) M. Shi and Y.-M. Xu, *Angew. Chem. Int. Ed.*, 2002, **41**, 4507; (c) M. Shi and Y.-M. Xu, *J. Org. Chem.*, 2003, **68**, 4784; (d) M. Shi and L. H. Chen, *Chem. Commun.*, 2003, 1310; and (e) M. Shi and Y.-M. Xu, *J. Org. Chem.*, 2004, **69**, 417

77. Y.-L. Shi, Y.-M. Xu and M. Shi, *Adv. Synth. Catal.*, 2004, **346**, 1220.

78. N. Utsumi, H. Zhang, F. Tanaka and C. F. Barbas III, *Angew. Chem. Int. Ed.*, 2007, **46**, 1878.

79. (a) J. Vesely, P. Dziedzic and A. Córdova, *Tetrahedron Lett.*, 2007, **48**, 6900; (b) J. Vesely, R. Rios and A. Córdova, *Tetrahedron Lett.*, 2008 **49**, 1137.

80. (a) N. Rastogi, I. N. N. Namboothiri and M. Cojocaru, *Tetrahedron Lett.*, 2004, **45**, 4745; (b) R. Mohan, N. Rastogi, I. N. N. Namboothiri, S. M. Mobin and D. Panda, *Bioorg. Med. Chem.*, 2006, **14**, 8073.

81. (a) I. Deb, M. Dadwal, S. M. Mobin and I. N. N. Namboothiri, *Org. Lett.*, 2006, **8**, 1201; (b) I. Deb, P. Shanbhag, S. M. Mobin and I. N. N. Namboothiri, *Eur. J. Org. Chem.*, 2009, 4091.

82. N. Rastogi, R. Mohan, D. Panda, S. M. Mobin and I. N. N. Namboothiri, *Org. Biomol. Chem.*, 2006, **4**, 3211.

83. M. Dadwal, R. Mohan, D. Panda, S. M. Mobin and I. N. N. Namboothiri, *Chem. Commun.*, 2006, **1**, 338.

84. M. Dadwal, S. M. Mobin and I. N. N. Namboothiri, *Org. Biomol. Chem.*, 2006, **4**, 2525.
85. R. Ballini, L. Barboni, G. Bosica, D. Fiorini, E. Mignini and A. Palmieri, *Tetrahedron*, 2004, **60**, 4995.
86. (a) P. Auvray, P. Knochel and J. F. Normant, *Tetrahedron Lett.*, 1986, **27**, 5095; (b) P. Auvray, P. Knochel and J. F. Normant, *Tetrahedron*, 1988, **44**, 6095; (c) H. M. R. Hoffmann, A. Weichert, A. M. Z. Slawin and D. J. Williams, *Tetrahedron*, 1990, **46**, 5591; (d) A. Weichert and H. M. R. Hoffmann, *J. Org. Chem.*, 1991, **56**, 4098.
87. D. Ando, C. Bevan, J. M. Brown and D. W. Price, *J. Chem. Soc., Chem. Commun.*, 1992, 592.
88. S. Z. Wang, K. Yamamoto, H. Yamada and T. Takahashi, *Tetrahedron*, 1992, **48**, 2333.
89. H. Amri, M. M. El Gaied and J. Villieras, *Synth. Commun.*, 1990, **20**, 659.
90. P. V. Ramachandran, M. V. R. Reddy, M. T. Rudd and J. R. de Alaniz, *Tetrahedron Lett.*, 1998, **39**, 8791.
91. K.-S. Yang and K. Chen, *Org. Lett.*, 2000, **2**, 729.
92. G. E. Keck and D. S. Welch, *Org. Lett.*, 2002, **4**, 3687.
93. H. Sajiki, A. Yamada, K. Yasunaga, T. Tsunoda, M. F. A. Amer and K. Hirota, *Nucl. Acids Res. (Supplement)*, 2002, 13.
94. D. Basavaiah and A. J. Rao, *Tetrahedron Lett.*, 2003, **44**, 4365.
95. (a) R. Sagar, C. S. Pant, R. Pathak and A. K. Shaw, *Tetrahedron*, 2004, **60**, 11399; (b) M. Saquib, M. K. Gupta, R. Sagar, Y. S. Prabhakar, A. K. Shaw, R. Kumar, P. R. Maulik, A. N. Gaikwad, S. Sinha, A. K. Srivastava, V. Chaturvedi, R. Srivastava and B. S. Srivastava, *J. Med. Chem.*, 2007, **50**, 2942.
96. S. Sun, Q. Zhang, Q. Liu, J. Kang, Y. Yin, D. Li and D. Dong, *Tetrahedron Lett.*, 2005, **46**, 6271.
97. M. C. Redondo, M. Ribagorda and M. C. Carreno, *Org. Lett.*, 2010, **12**, 568.
98. (a) R. Fiketscher, E. Hahn, A. Kud and A. Oftring, *Ger. Offen.* DE 3,444,098 (1986); *Chem. Abstr.*, 1986, **105**, 115538k; *US Pat.*, 4,654,432 (1987); (b) K. Nakagawa, M. Makino and Y. Kita, *Eur. Patent Appl.* EP 669,312 (1995); *Chem. Abstr.*, 1995, **123**, 285218v.
99. W. R. Roush and B. B. Brown, *J. Org. Chem.*, 1993, **58**, 2151.
100. D. Basavaiah and R. Saguna Hyma, *Tetrahedron*, 1996, **52**, 1253.
101. P. Perlmutter and M. Tabone, *J. Org. Chem.*, 1995, **60**, 6515.
102. Unpublished work carried out at Atochem, Paris, France, and cited in ref. 113a.
103. R. Fikentscher, E. Hahn, A. Kud and A. Oftring, *Ger. Offen.* DE 3,444,097 (1986); *Chem. Abstr.*, 1987, **107**, 7781s; *US Pat.* 4,652,669 (1987).
104. (a) W. Poly, D. Schomburg and H. M. R. Hoffmann, *J. Org. Chem.*, 1988, **83**, 3701; (b) P. Perlmutter and T. McCarthy, *Aust. J. Chem.*, 1993, **46**, 253.
105. Y. Fort, M.-C. Berthe and P. Caubère, *Synth. Commun.*, 1992, **22**, 1265.

106. F. R. van Heerden, J. J. Huyser and C. W. Holzapfel, *Synth. Commun.*, 1994, **24**, 2863.
107. P. R. Krishna, E. R. Sekhar and V. Kannan, *Tetrahedron Lett.*, 2003, **44**, 4973.
108. P. Areces, E. Carrasco, A. Mancha and J. Plumet, *Synthesis*, 2006, **1**, 946.
109. (a) E. P. Kündig, L. H. Xu, P. Romanens and G. Bernardinelli, *Tetrahedron Lett.*, 1993, **34**, 7049; (b) A. Foucaud and E. le Rouillé, *Synthesis*, 1990, **1**, 787; (c) A. Foucaud and F. El Guemmout, *Bull. Soc. Chim. Fr.*, 1989, **1**, 403.
110. M. L. Bode, R. B. English and P. T. Kaye, *S. Afr. J. Chem.*, 1992, **45**, 25.
111. S. E. Drewes, O. L. Njamela, N. D. Emslie, N. Ramesar and J. S. Field, *Synth. Commun.*, 1993, **23**, 2807.
112. A. Becker, *Fr. Demande* FR 2,602,507 (1988); *Chem. Abstr.*, 1989, **110**, 75549j; *US Pat.*, 4,789,743 (1988).
113. (a) M. C. Berthe, P. Caubère and Y. Fort, *Eur. Patent Appl.* EP 465,293 (1992); *Chem. Abstr.*, 1992, **116**, 152605c; *US Pat.*, 5,332,836 (1994); (b) P. Bauchat, N. Le Bras, L. Rigal and A. Foucaud, *Tetrahedron*, 1994, **50**, 7815.
114. E. Ciganek, *J. Org. Chem.*, 1995, **60**, 4635.
115. (a) P. Narender, U. Srinivas, B. Gangadasu, S. Biswas and V. Jayathirtha Rao, *Bioorg. Med. Chem. Lett.*, 2005, **15**, 5378; (b) P. Narender, B. Gangadasu, M. Ravinder, U. Srinivas, G. Y. S. K. Swamy, K. Ravikumar and V. Jayathirtha Rao, *Tetrahedron*, 2006, **62**, 954.
116. S. E. Drewes and M. B. Rohwer, *Synth. Commun.*, 1997, **27**, 415.
117. (a) A. Patra, S. Batra, B. Kundu, B. S. Joshi, R. Roy and A. P. Bhaduri, *Synthesis*, 2001, 276; (b) A. K. Roy and S. Batra, *Synthesis*, 2003, 1347; (c) A. K. Roy and S. Batra, *Synthesis*, 2003, 2325.
118. S. Batra and A. K. Roy, *Synthesis*, 2004, 2550.
119. S. Nag, V. Singh and S. Batra, *Arkivoc*, 2007, (xiv), 185.
120. P. T. Kaye, D. M. Molefe, A. T. Nchinda and L. V. Sabbagh, *J. Chem. Res.*, 2004, **4**, 303.
121. P. T. Kaye, A. T. Nchinda, L. V. Sabbagh and J. Bacsa, *J. Chem. Res. (S)*, 2003, 111; *J. Chem. Res. (M)*, 2003, 301.
122. D. M. Molefe and P. T. Kaye, *Synth. Commun.*, 2009, **39**, 3586.
123. For a review, see: D. Gryko, J. Chåko and J. Jurczak, *Chirality*, 2003, **15**, 514.
124. (a) T. Manickum and G. Roos, *Synth. Commun.*, 1991, **21**, 2269; (b) S. E. Drewes, A. A. Khan and K. Rowland, *Synth. Commun.*, 1993, **23**, 183.
125. J. C. Bussolari, K. Beers, P. Lalan, W. V. Murray, D. Gauthier and P. McDonnell, *Chem. Lett.*, 1998, 787.
126. Y. Iwabuchi, T. Sugihara, T. Esumi and S. Hatakeyama, *Tetrahedron Lett.*, 2001, **42**, 7867.
127. (a) P. Radha Krishna, V. Kannan, G. V. M. Sharma and R. M. H. V. Rao, *Synlett*, 2003, 888; (b) P. Radha Krishna, A. Manjuvani and V. Kannan, *Tetrahedron: Asymmetry*, 2005, **16**, 2691; (c) P. Radha Krishna and M. Narsingam, *J. Comb. Chem.*, 2007, **9**, 62; (d) P. Radha Krishna,

A. Manjuvani, M. Narsingam and G. Raju, *Eur. J. Org. Chem.*, 2010, 813.

128. R. Pathak, C. S. Pant, A. K. Shaw, A. P. Bhaduri, A. N. Gaikwad, S. Sinha, A. Srivastava, K. K. Srivastava, V. Chaturvedi, R. Srivastava and B. S. Srivastava, *Bioorg. Med. Chem.*, 2002, **10**, 3187.

129. P. Radha Krishna, K. R. Lopinti and V. Kannan, *Tetrahedron Lett.*, 2004, **45**, 7847.

130. S. K. Nayak, L. Thijs and B. Zwanenburg, *Tetrahedron Lett.*, 1999, **40**, 981.

131. B. Alcaide, P. Almendros and C. Aragoncillo, *J. Org. Chem.*, 2001, **66**, 1612.

132. P. R. Krishna, S. Kishore and P. S. Reddy, *Synlett*, 2009, 2605.

133. (a) F. Gelman, J. Blum and D. Avnir, *New J. Chem.*, 2003, **27**, 205; (b) F. Gelman, J. Blum and D. Avnir, *Angew. Chem.*, 2001, **113**, 3759; *Angew. Chem. Int. Ed.*, 2001, **40**, 3647; (c) F. Gelman, J. Blum and D. Avnir, *J. Am. Chem. Soc.*, 2000, **122**, 11999.

134. B. Helms, S. J. Guillaudeu, Y. Xie, M. McMurdo, C. J. Hawker and J. M. J. Fréchet, *Angew. Chem. Int. Ed.*, 2005, **44**, 6384.

135. C. R. Strauss, M. N. Galbraith and A. F. Faux, *PCT Int. Appl.* WO 91 18, 861 (1991); *Chem. Abstr.*, 1992, **116**, 128185v.

136. Unpublished work by N. S. Isaacs, and cited in ref. 10.

137. M. M. Dhingra and K. R. Tatta, *Org. Magn. Reson.*, 1977, **9**, 23.

138. V. G. Nenajdenko, S. V. Druzhinin and E. S. Balenkova, *Mendeleev Commun.*, 2006, **16**, 273.

139. G. M. Strunz and C. M. Yu, *Can. J. Chem.*, 1988, **66**, 1081.

140. (a) D. Basavaiah and V. V. L. Gowriswari, *Synth. Commun.*, 1989, **19**, 2461; (b) D. Basavaiah, T. K. Bharathi and V. V. L. Gowriswari, *Tetrahedron Lett.*, 1987, **28**, 4351.

141. T. Bauer and J. Tarasiuk, *Tetrahedron: Asymmetry*, 2001, **12**, 1741.

142. M. Shi and W. Zhang, *Tetrahedron*, 2005, **61**, 11887.

143. K. Alder, H. K. Shafer, H. Esser, H. Krieger and R. Reubke, *Justus Liebigs Ann. Chem.*, 1955, **593**, 23.

144. (a) B. Alcaide, P. Almendros and C. Aragoncillo, *Tetrahedron Lett.*, 1999, **40**, 7537; (b) B. Alcaide, P. Almendros, C. Aragoncillo and R. Rodríguez-Acebes, *J. Org. Chem.*, 2004, **69**, 826.

145. (a) S. J. Gardena and J. M. S. Skakle, *Tetrahedron Lett.*, 2002, **43**, 1969; (b) Y. M. Chung, Y. J. Im and J. N. Kim, *Bull. Korean Chem. Soc.*, 2002, **23**, 1651.

146. J.-F. Pan and K. Chen, *Tetrahedron Lett.*, 2004, **45**, 2541.

147. K. Yamamoto, M. Takagi and J. Tsuji, *Bull. Chem. Soc. Jpn.*, 1988, **61**, 319.

148. S. Bertenshaw and M. Kahn, *Tetrahedron Lett.*, 1989, **30**, 2731.

149. (a) M. Shi and Y.-M. Xu, *Tetrahedron: Asymmetry*, 2002, **13**, 1195; (b) V. K. Aggarwal, A. M. M. Castro, A. Mereu and H. Adams, *Tetrahedron Lett.*, 2002, **43**, 1577.

150. M. Shi and G.-L. Zhao, *Adv. Synth. Catal.*, 2004, **346**, 1205.

151. (a) Liu, J. Zhao, G. Jin, G. Zhao, S. Zhu and S. Wang, *Tetrahedron*, 2005, **61**, 3841; (b) X. Liu, Z. Chai, G. Zhao and S. Zhu, *J. Fluorine Chem.*, 2005, **126**, 1215.

152. J. Gao, G.-N. Ma, Q.-J. Li and M. Shi, *Tetrahedron Lett.*, 2006, **47**, 7685.

153. A. Lu, X. Xu, P. Gao, Z. Zhou, H. Song and C. Tang, *Tetrahedron: Asymmetry*, 2008, **19**, 1886.

154. E. M. Campi, A. Holmes, P. Perlmutter and C. C. Teo, *Aust. J. Chem.*, 1995, **48**, 1535.

155. J. Cyrener and K. Burger, *Monatsh. Chem.*, 1995, **126**, 319.

156. E. P. Kündig, L. H. Xu and B. Schnell, *Synlett*, 1994, 413.

157. (a) A. Gajda and T. Gajda, *J. Org. Chem.*, 2008, **73**, 8643; (b) B. Das, K. Damodar, N. Chowdhury, D. Saritha, B. Ravikanth and M. Krishnaiah, *Tetrahedron*, 2008, **64**, 9396.

158. S. Èíhalová, M. Remeš, I. Císaøová and J. Veselý, *Eur. J. Org. Chem.*, 2009, **1**, 6277.

159. N. Abermil, G. Masson and J. Zhu, *Adv. Synth. Catal.*, 2010, **352**, 656.

160. A. B. Koldovski, I. A. Milyutin and V. N. Kalinin, *Dokl. Akad. Nauk SSSR*, 1992, **324**, 1015; English transl. p. 119.

161. N. Azizi and M. R. Saidi, *Tetrahedron Lett.*, 2002, **43**, 4305.

162. (a) E. L. Myers, C. P. Butts and V. K. Aggarwal, *Chem. Commun.*, 2006, 4434; (b) E. L. Myers, J. G. de Vries and V. K. Aggarwal, *Angew. Chem.*, 2007, **119**, 1925; *Angew. Chem. Int. Ed.*, 2007, **46**, 1893.

163. D. Basavaiah, N. Kumaragurubaran and D. S. Sharada, *Tetrahedron Lett.*, 2001, **42**, 85.

164. I. Suarez del Villar, A. Gradillas, G. Dominguez and J. Perez-Castells, *Org. Lett.*, 2010, **12**, 2418.

165. (a) A. Kamimura, Y. Gunjigake, H. Mitsudera and S. Yokoyama, *Tetrahedron Lett.*, 1998, **39**, 7323; (b) M. Shi and G.-L. Zhao, *Tetrahedron*, 2004, **60**, 2083.

166. M. E. Krafft and J. A. Wright, *Chem. Commun.*, 2006, 2977.

167. (a) C. H. Lee and K.-J. Lee, *Synthesis*, 2004, 1941; (b) S. W. Lee, C. H. Lee and K.-J. Lee, *Bull. Korean Chem. Soc.*, 2006, **27**, 769.

168. A. R. Katritzky, M. S. Kim and K. Widyan, *Arkivoc*, 2008 (iii), 91.

169. E. M. Campi, A. Holmes, P. Perlmutter and C. C. Teo, *Aust. J. Chem.*, 1995, **48**, 1541.

170. D. Balan and H. Adolfsson, *J. Org. Chem.*, 2001, **66**, 6498.

171. D. Balan and H. Adolfsson, *J. Org. Chem.*, 2002, **67**, 2329.

172. M. Shi and G.-L. Zhao, *Tetrahedron Lett.*, 2002, **43**, 9171.

173. D. Balan and H. Adolfsson, *Tetrahedron Lett.*, 2003, **44**, 2521.

174. P. Ribière, C. Enjalbal, J.-L. Aubagnac, N. Yadav-Bhatnagar, J. Martinez and F. Lamaty, *J. Comb. Chem.*, 2004, **6**, 464.

175. P. Ribière, N. Yadav-Bhatnagar, J. Martinez and F. Lamaty, *QSAR Comb. Sci.*, 2004, **23**, 911.

176. V. Declerck, P. Ribière, J. Martinez and F. Lamaty, *J. Org. Chem.*, 2004, **69**, 8372.

177. A. S. K. Hashmi, S. Wagner and F. Rominger, *Aust. J. Chem.*, 2009, **62**, 657.

178. F. Roth, P. Gygax and G. Frater, *Tetrahedron Lett.*, 1992, **33**, 1045.

179. (a) G. P. Black, F. Dinon, S. Fratucello, P. J. Murphy, M. Nielsen, H. L. Williams and N. D. A. Walshe, *Tetrahedron Lett.*, 1997, **38**, 8561; (b) F. Dinon, E. Richards, P. J. Murphy, D. E. Hibbs, M. B. Hursthouse and K. M. A. Malic, *Tetrahedron Lett.*, 1999, **40**, 3279; (c) E. L. Richards, P. J. Murphy, F. Dinon, S. Fratucello, P. M. Brown, T. Gelbrich and M. B. Hursthouse, *Tetrahedron*, 2001, **57**, 7771.

180. L.-C. Wang, A. L. Luis, K. Agapiou, H.-Y. Jang and M. J. Krische, *J. Am. Chem. Soc.*, 2002, **124**, 2402.

181. S. A. Frank, D. J. Mergott and W. R. Roush, *J. Am. Chem. Soc.*, 2002, **124**, 2404.

182. D. J. Mergott, S. A. Frank and W. R. Roush, *Org. Lett.*, 2002, **4**, 3157.

183. K. Yagi, T. Turitani, H. Shinokubo and K. Oshima, *Org. Lett.*, 2002, **4**, 3111.

184. P. R. Krishna, V. Kannan and G. V. M. Sharma, *J. Org. Chem.*, 2004, **69**, 6467.

185. J. E. Yeo, X. Yang, H. J. Kim and S. Koo, *Chem. Commun.*, 2004, 236.

186. W.-D. Teng, R. Huang, C. K.-W. Kwong, M. Shi and P. H. Toy, *J. Org. Chem.*, 2006, **71**, 368.

187. M. E. Krafft and T. F. N. Haxell, *J. Am. Chem. Soc.*, 2005, **127**, 10168.

188. M. E. Krafft, K. A. Seibert, T. F. N. Haxell and C. Hirosawa, *Chem. Commun.*, 2005, 5772.

189. F. O. Seidela and J. A. Gladysz, *Adv. Synth. Catal.*, 2008, **350**, 2443.

190. C. E. Aroyan, M. M. Vasbinder and S. J. Miller, *Org. Lett.*, 2005, **7**, 3849.

191. S.-H. Chen, B.-C. Hong, C.-F. Su and S. Sarshar, *Tetrahedron Lett.*, 2005, **46**, 8899.

192. F. Seidel and J. A. Gladysz, *Synlett*, 2007, 986.

193. Review for ionic liquids reactivity, see: (a) C. A. M. Afonso, L. C. Branco, N. R. Candeias, P. M. P. Gois, N. M. T. Lourenc, N. M. M. Mateusb and J. N. Rosa, *Chem. Commun.*, 2007, 2669; and (b) S. Chowdhury, R. S. Mohanb and J. L. Scott, *Tetrahedron*, 2007, **64**, 2363

194. (a) G. H. P. Roos and P. Rampersadh, *Synth. Commun.*, 1993, **23**, 1261; (b) W. P. Almeida and F. Coehlo, *Tetrahedron Lett.*, 1998, **39**, 8609; (c) F. Coelho, W. P. Almeida, D. Veronese, C. R. Mateus, E. C. Silva Lopes, R. C. Rossi, G. P. C. Silveira and C. H. Pavam, *Tetrahedron*, 2002, **58**, 7437; (d) R. S. Porto, G. W. Amarante, M. Cavallaro, R. J. Poppi and F. Coelho, *Tetrahedron Lett.*, 2009, **50**, 1184.

195. (a) N. S. Isaacs, *Tetrahedron*, 1991, **47**, 8463; (b) R. J. W. Schuurman, A. V. den Linden, R. P. F. Grimbergen, R. J. M. Nolte and H. W. Scheeren, *Tetrahedron*, 1996, **52**, 8307; (c) I. E. Markó, P. R. Giles and N. J. Hindley, *Tetrahedron*, 1997, **53**, 1015; (d) Y. Hayashi, K. Okado, I. Ashimine and M. Shoji, *Tetrahedron Lett.*, 2002, **43**, 8683; (e) M. Shi and Y.-H. Liu, *Org. Biomol. Chem.*, 2006, **4**, 1468.

196. P. M. Rose, A. A. Clifford and C. M. Rayner, *Chem. Commun.*, 2002, 968.

197. (a) J. Cai, Z. Zhou, G. Zhao and C. Tang, *Org. Lett.*, 2002, **4**, 4723; (b) V. K. Aggarwal, A. Mereu, D. K. Dean and R. Williams, *J. Org. Chem.*, 2002, **67**, 510; (c) A. Kumar and S. S. Pawar, *Tetrahedron*, 2003, **59**, 5019; (d) Y. Song, H. Ke, N. Wang, L. Wang and G. Zou, *Tetrahedron*, 2009, **65**, 9086.

198. K. M. Dieter, C. J. Dymek Jr, N. E. Heimer, J. W. Rovang and J. S. Wilkes, *J. Am. Chem. Soc.*, 1988, **110**, 2722.

199. J. N. Rosa, C. A. M. Afonso and A. G. Santos, *Tetrahedron*, 2001, **57**, 4189.

200. T. Kitazume, K. Tamura, Z. Jiang, N. Miyake and I. Kawasaki, *J. Fluorine Chem.*, 2002, **115**, 49.

201. E. J. Kim and S. Y. Ko, *Helv. Chim. Acta*, 2003, **86**, 894.

202. V. K. Aggarwal, I. Emme and A. Mereu, *Chem. Commun.*, 2002, 1612.

203. J.-C. Hsu, Y.-H. Yen and Y.-H. Chu, *Tetrahedron Lett.*, 2004, **45**, 4673.

204. S. T. Handy and M. Okello, *J. Org. Chem.*, 2005, **70**, 1915.

205. V. Jurèík and R. Wilhelm, *Green Chem.*, 2005, **7**, 844.

206. Y.-S. Lin, C.-Y. Lin, C.-W. Liu and T. Y. R. Tsai, *Tetrahedron*, 2006, **62**, 872.

207. R. O. M. A. de Souza, P. H. Fregadolli, K. M. Gonçalves, L. C. Sequeira, V. L. P. Pereiraa, L. C. Filho, P. M. Esteves, M. L. A. A. Vasconcellos and O. A. C. Antunes, *Lett. Org. Chem.*, 2006, **3**, 936.

208. A. Kumar and S. S. Pawar, *J. Mol. Catal. A: Chem.*, 2004, **211**, 43.

209. (a) H. Gong, C. Cai, N. Yang, L. Yang, J. Zhang and Q.-H. Fan, *J. Mol. Catal. A: Chem.*, 2006, **260**, 236; (b) S.-H. Zhao, H.-R. Zhang, L.-H. Feng and Z.-B. Chen, *J. Mol. Catal. A: Chem.*, 2006, **260**, 251.

210. S. Zhao, E. Zhao, P. Shen, M. Zhao and J. Sun, *Ultrason. Sonochem.*, 2008, **15**, 955.

211. D. Seebach and H. A. Oei, *Angew. Chem.*, 1975, **87**, 629; *Angew. Chem. Int. Ed. Engl.*, 1975, **14**, 634.

212. B. Pégot, G. Vo-Thanh, D. Gori and A. Loupy, *Tetrahedron Lett.*, 2004, **45**, 6425.

213. S. Garre, E. Parker, B. Ni and A. D. Headley, *Org. Biomol. Chem.*, 2008, **6**, 3041.

214. R. Gausepohl, P. Buskens, J. Kleinen, A. Bruckmann, C. W. Lehmann, J. Klankermayer and W. Leitner, *Angew. Chem. Int. Ed.*, 2006, **45**, 3689.

215. M. Shi, L.-H. Chen and C.-Q. Li, *J. Am. Chem. Soc.*, 2005, **127**, 3790.

Catalytic Systems for the Morita–Baylis–Hillman Reaction

MEI-XIN ZHAO, YIN WEI AND MIN SHI

2.1 Introduction

As a versatile carbon–carbon bond-forming reaction, the Morita–Baylis–Hillman (MBH) reaction has attracted tremendous research activity. Considerable progress has been achieved in developing effective catalysts, and in establishing methodologies for asymmetric catalysis. Since this type of reaction is generally initiated by a Michael addition of a nucleophilic catalyst to substrate, tertiary amines and phosphines have been the most frequently employed catalysts for the MBH reaction. Moreover, chalcogenide/TiCl$_4$ has also been found to be an effective MBH catalyst in some cases. Recently, *in situ* generated halide ion has been demonstrated to behave like a nucleophile when TiCl$_4$ or Et$_2$AlI alone or combination with Bu$_4$NX, amine or alcohol was employed as a catalyst in certain MBH reactions, in which Lewis acids were considered to facilitate reactions by activating the substrates. As an important part of MBH catalyst system, co-catalysts and polymer-supported catalysts are also included in this chapter.

2.2 Amine-catalyzed System

2.2.1 Achiral Amine

Although the first report on an amine-catalyzed MBH reaction dates back to 1972,[1] in which Baylis and Hillman reported that the reaction of activated alkenes **1**, such as α,β-unsaturated esters, amides, nitriles and ketones, with various aldehydes, under the catalytic influence of a tertiary bicyclic amine,

RSC Catalysis Series No. 8
The Chemistry of the Morita–Baylis–Hillman Reaction
By Min Shi, Fei-Jun Wang, Mei-Xin Zhao and Yin Wei
© Min Shi, Fei-Jun Wang, Mei-Xin Zhao and Yin Wei 2011
Published by the Royal Society of Chemistry, www.rsc.org

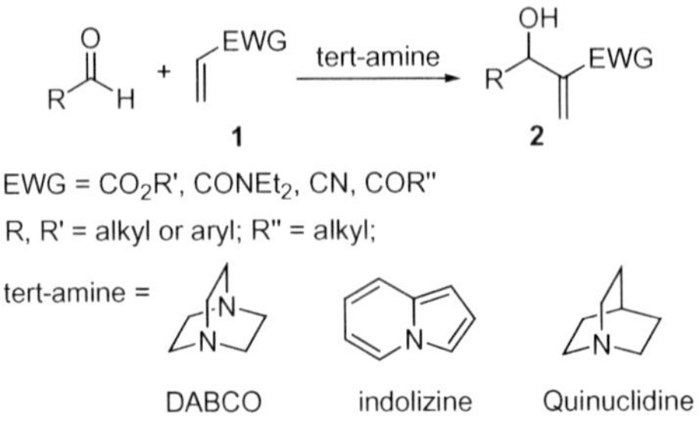

EWG = CO_2R', $CONEt_2$, CN, COR''

R, R' = alkyl or aryl; R'' = alkyl;

tert-amine =

DABCO   indolizine   Quinuclidine

Scheme 2.1

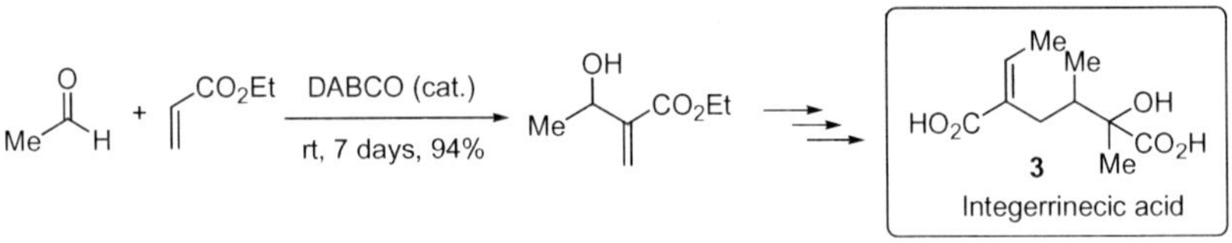

3

Integerrinecic acid

Scheme 2.2

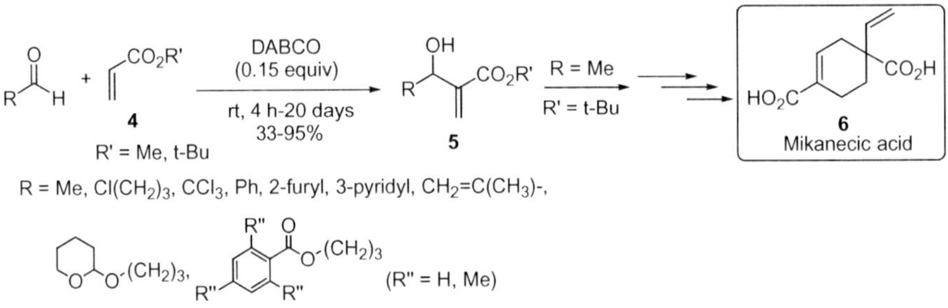

4

R' = Me, t-Bu

5

6

Mikanecic acid

R = Me, $Cl(CH_2)_3$, CCl_3, Ph, 2-furyl, 3-pyridyl, CH_2=$C(CH_3)$-,

O-$(CH_2)_3$,   (R'' = H, Me)

Scheme 2.3

such as 1,4-diazabicyclo[2.2.2]octane (DABCO), indolizine or quinuclidine, produced multifunctional molecules **2** (Scheme 2.1), this reaction did not receive proper attention from organic chemists for more than a decade.

In 1982, Drewes and Emslie first reported the reaction of ethyl acrylate with acetaldehyde in the presence of DABCO and successfully employed this adduct in the synthesis of integerrinecic acid (**3**) (Scheme 2.2).[2] Soon after that, Hoffmann *et al*. developed an interesting reaction between methyl/*tert*-butyl acrylates **4** and various aldehydes under the catalytic influence of DABCO to provide the corresponding α-hydroxyalkyl acrylate adducts **5** and elegantly applied one of these adducts to synthesis of racemic mikanecic acid (**6**) (Scheme 2.3).[3] Since then, a series of reports[4] have emerged and transformed the MBH reaction into a very useful and promising tool for the construction of carbon–carbon bonds in synthetic chemistry (Scheme 2.4).

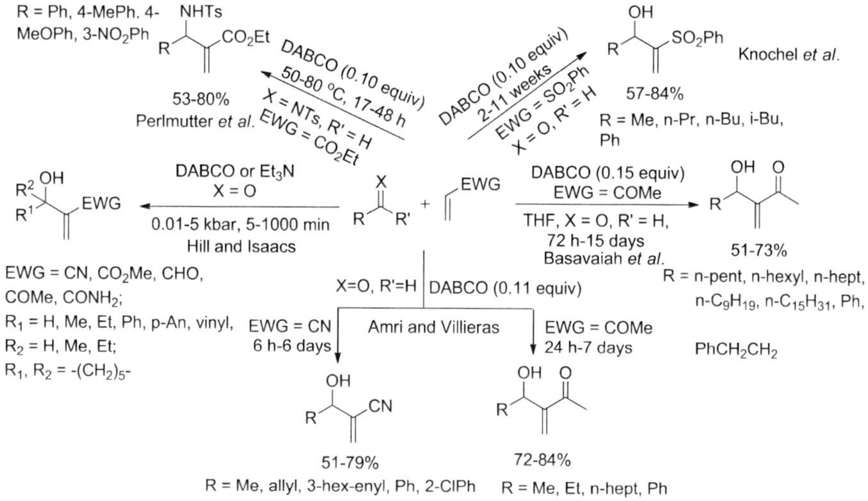

Scheme 2.4

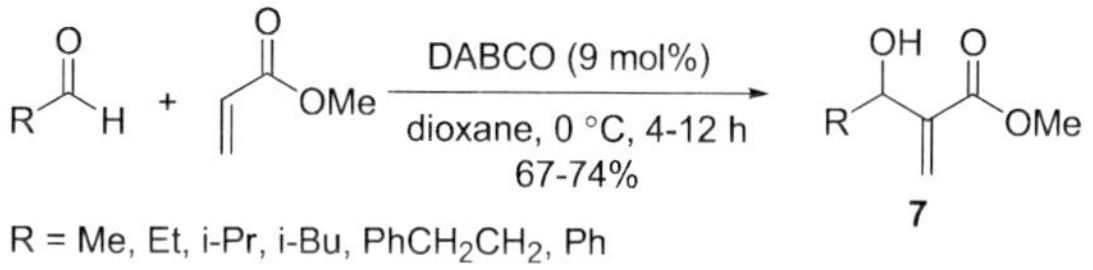

Scheme 2.5

Usually, the rate of MBH reaction increases upon warming the reaction mixture above room temperature, albeit the yield is not greater than that derived at room temperature. However, Rafel and Leahy observed significant rate acceleration when the reaction was conducted at 0 °C in dioxane. This apparent low-temperature acceleration can be extended to virtually any aldehydes, and even aromatic aldehydes undergo rapid conversion to furnish the corresponding MBH adducts **7** in short order (Scheme 2.5).[5]

Hu and co-workers[6] have developed a practical and efficient method by using stoichiometric **DABCO** and a dioxane–water medium to overcome problems commonly associated with the MBH reaction, such as low reaction yields and long reaction time, successfully converting various aliphatic and aromatic aldehydes into their corresponding MBH adducts **7** in shorter reaction times (Scheme 2.6). Less reactive activated alkene, acrylamide, could also undergo MBH reaction with reactive electrophiles under these conditions, to give the corresponding adducts **8** in moderate to good yields (Scheme 2.6).[7]

Later, Franck and Figadere[8] employed these conditions for the synthesis of racemic acaterin **10** *via* the coupling of γ-butyrolactone **9** with octanal (Scheme 2.7). However, their attempts to employ this strategy to obtain

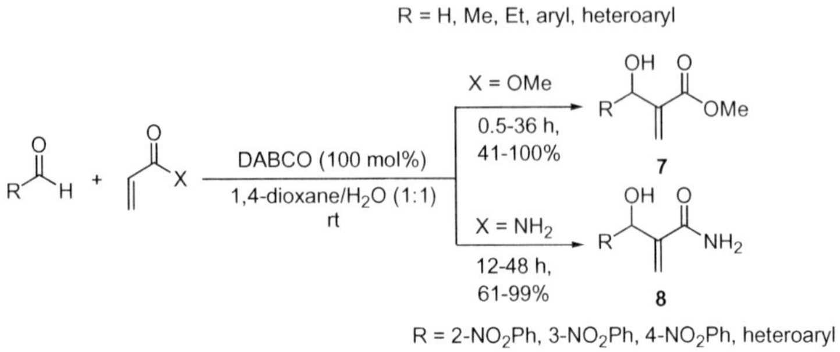

R = H, Me, Et, aryl, heteroaryl

Scheme 2.6

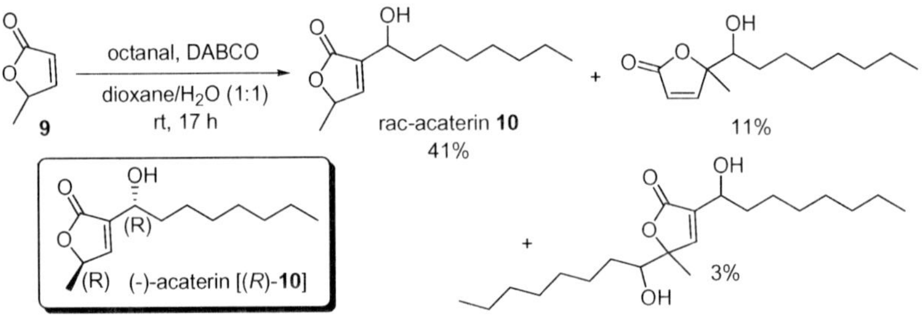

Scheme 2.7

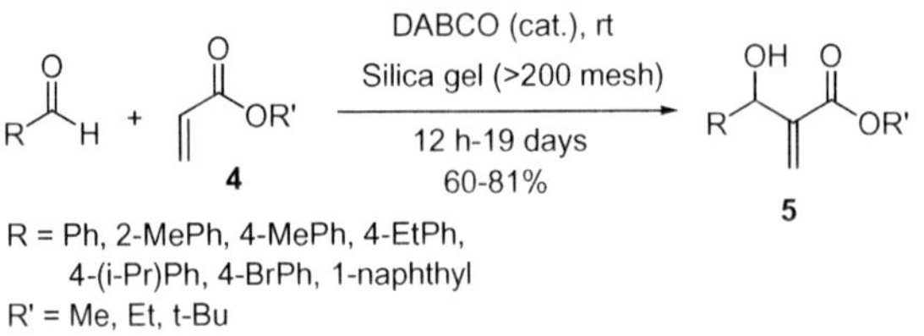

Scheme 2.8

enantiopure acaterin, (*R*)-**10** (a biologically active molecule), starting from (*S*)-γ-butyrolactone were unsuccessful due to the possible racemization.

In addition, Basavaiah *et al.* have observed a remarkable rate acceleration of this reaction in a silica-gel solid-phase medium.[9] They found that even the less reactive activated olefin, *tert*-butyl acrylate, could undergo the MBH reaction with various aromatic aldehydes under these conditions (Scheme 2.8).

During the synthesis of the lignan derivative hydroxy-β-piperonyl-γ-butyrolactone (**12**), Coelho *et al.*[10] successfully accomplished the MBH reaction between the less reactive piperonal and methyl acrylate under ultrasound conditions. The corresponding adduct **11** was subsequently transformed into the desired lignan derivative **12** (Scheme 2.9). More recently, they also

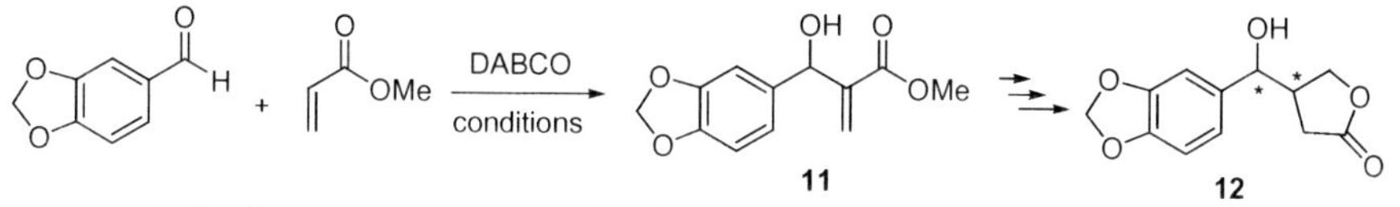

cat. DABCO

	time (days)	yield (%)
neat, rt	21	25
MeOH, dioxane, rt	20	30
MeOH, dioxane, rt ultrasound	8	40

(73% based on recovered aldehyde)

Scheme 2.9

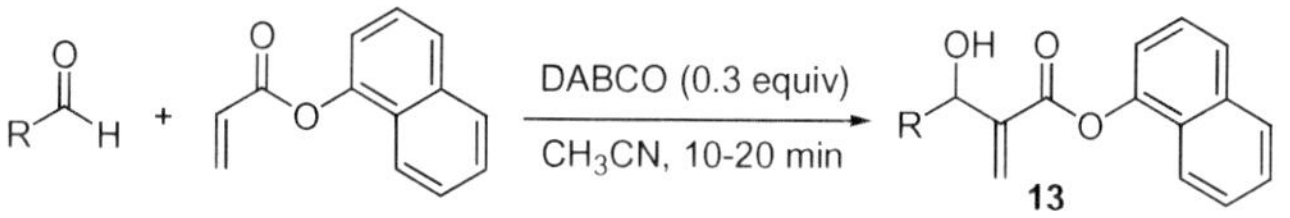

R = Me, Et, i-Bu, PhCH$_2$CH$_2$, i-Pr, *trans*-cinnamyl,
4-NO$_2$Ph, 4-FPh, 4-MePh, 3-pyridyl

51-82%

Scheme 2.10

systematically investigated the influence of ultrasound in accelerating the MBH reaction of various activated alkenes [methyl acrylate, MVK (methyl vinyl ketone) and acrylonitrile] with several aromatic and aliphatic aldehydes and found that DABCO was a more effective catalyst under ultrasound conditions than tri-*n*-butylphosphine. No effect on reaction rate was observed when the concentration of DABCO was increased.[11]

Chen *et al.* have hypothesized that substituted acrylates may provide stereo- and/or stereoelectronic effects that stabilize the oxy anion intermediate, which would shift the equilibrium forward and subsequently accelerate the following aldol reaction. Thus, they screened a range of substituted acrylates and found that an extremely rapid rate can be achieved by using α-naphthyl acrylate as the Michael acceptor for the MBH reaction.[12] The reaction of α-naphthyl acrylate with both aliphatic and aromatic aldehydes in the presence of DABCO provided the desired adducts **13** in reasonable yields within 20 min, which is one of the best rate acceleration systems for a wide range of aldehydes in the MBH reaction under atmospheric pressure (Scheme 2.10).

Wilcox *et al.* have demonstrated an application of the precipitation approach for MBH adduct isolation.[13] By using a diaryl alkene alcohol (**14**) as a precipitating auxiliary (based on the solubility switch of structural isomerization), the corresponding MBH acids **17** were obtained from the reaction of acrylate **15** and aldehydes in the presence of DABCO, followed by isomerization and cleavage, in moderate to good yield (Scheme 2.11).

Moreover, many new kinds of electrophiles have been developed for the DABCO-catalyzed MBH reaction to obtain various adducts for organic synthesis. Kamimura *et al.*[14] have employed azodicarboxylates **19** as electrophiles in an aza-MBH reaction with alkyl vinyl ketones (**18**) in the presence of

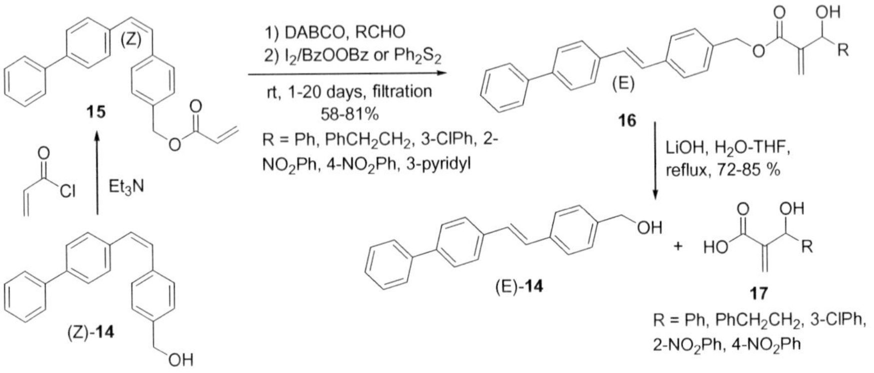

Scheme 2.11

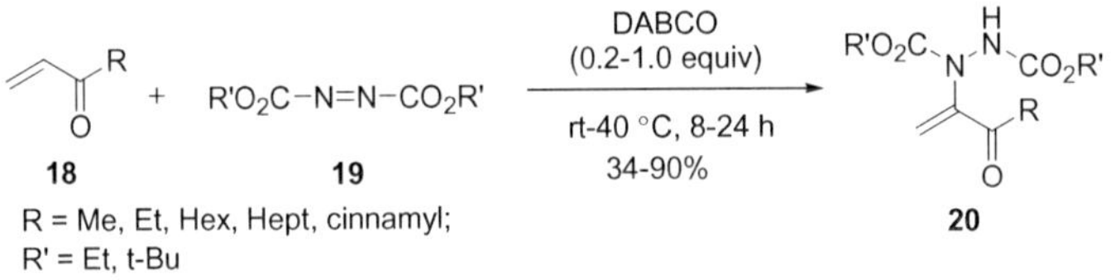

Scheme 2.12

DABCO as a catalyst (Scheme 2.12). Usually, alkyl vinyl ketones were reactive enough for the reaction in THF, whereas methyl acrylate gave no MBH adduct under the same conditions. With the divinyl ketone, the reaction exhibited good regioselectivity and occurred only for the terminal vinyl site, while a β-substituted olefin unit remained untouched. For less reactive di-*tert*-butyl azodicarboxylate, an equimolar amount of DABCO at 40 °C was required to enhance the reaction rate, affording the desired adduct **20** in moderate yields.

Later, our research group extended this methodology to aza-MBH reactions between DIAD or DEAD and acrylates or acrylonitrile (Scheme 2.13).[15] For various aryl acrylates, the reaction can proceed smoothly in the presence of DABCO in DMF and the corresponding aza-MBH adducts **22** were obtained in moderate to high yields. For other azo-compounds, for example, diphenyl azocarboxylate, 2,2′-azobisisobutyronitrile (AIBN) and azobenzene, no reaction occurred under the same conditions.

Kaye *et al.* have found that the DABCO-catalyzed MBH reaction between salicylaldehydes and methyl acrylate gives a mixture of various chromene and coumarin derivatives.[16] Subsequently, they[17] developed a simple one-pot methodology for the synthesis of 2*H*-1-chromenes **25** *via* the MBH reaction of 2-hydroxybenzaldehydes and 2-hydroxy-1-naphthaldehydes (**23**) with various activated alkenes (**24**) and subsequent regioselective cyclization (Scheme 2.14). In some cases competitive dimerization of the activated alkenes **24** was

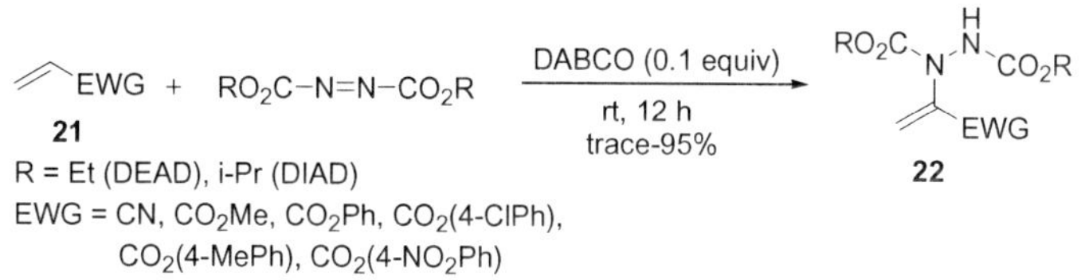
EWG + $RO_2C-N=N-CO_2R$
21
R = Et (DEAD), i-Pr (DIAD)
EWG = CN, CO_2Me, CO_2Ph, CO_2(4-ClPh), CO_2(4-MePh), CO_2(4-NO_2Ph)
DABCO (0.1 equiv)
rt, 12 h
trace-95%
22

Scheme 2.13

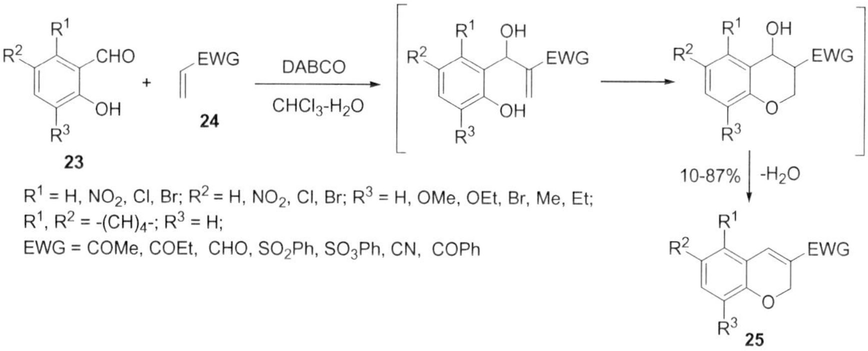
23
24
DABCO
$CHCl_3$-H_2O
R^1 = H, NO_2, Cl, Br; R^2 = H, NO_2, Cl, Br; R^3 = H, OMe, OEt, Br, Me, Et;
R^1, R^2 = -$(CH)_4$-; R^3 = H;
EWG = COMe, COEt, CHO, SO_2Ph, SO_3Ph, CN, COPh
10-87%
-H_2O
25

Scheme 2.14

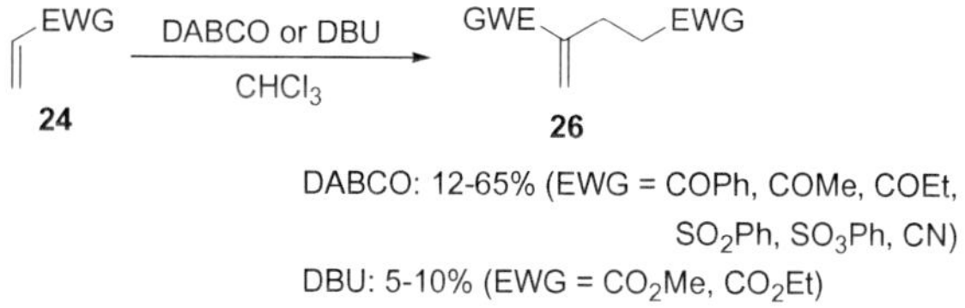
24
DABCO or DBU
$CHCl_3$
26
DABCO: 12-65% (EWG = COPh, COMe, COEt, SO_2Ph, SO_3Ph, CN)
DBU: 5-10% (EWG = CO_2Me, CO_2Et)

Scheme 2.15

observed, and direct dimerization in the presence of DABCO/DBU has been explored (Scheme 2.15).

Zwanenburg *et al.*[18] have demonstrated that (*S*)-*N*-tritylaziridine-2-carboxaldehyde (**27**) undergoes a facile MBH reaction with various activated olefins **28**, giving the MBH adducts **29** in good yields without racemization during the long exposure to DABCO. However, the diastereoselectivities in these reactions were found to be poor (Scheme 2.16).

Chiral acryloylhydrazide **30**, derived from novel camphor-based chiral auxiliary, react with aldehydes in the presence of DABCO to afford β-hydroxy-α-methylene carbonyl derivatives **31** and **32** with practical, high level diastereoselectivity (up to 98% de). Moreover, each diastereomer, **31** and **32**, can be prepared with high optical purity from the same chiral auxiliary by appropriate choice of reaction conditions (Scheme 2.17).[19]

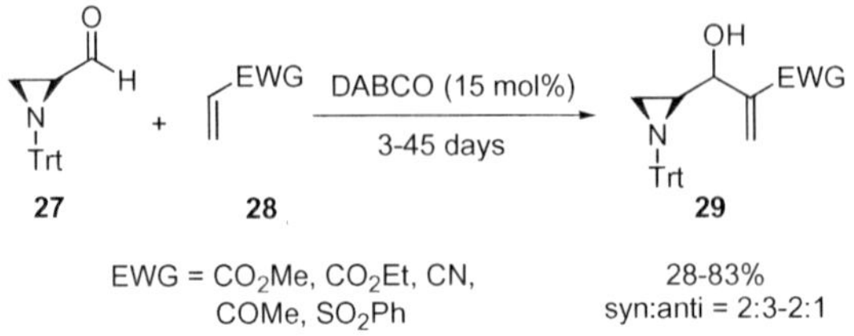

Scheme 2.16

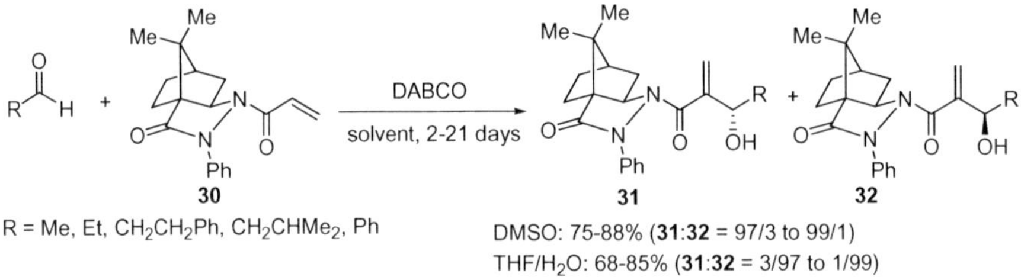

Scheme 2.17

In the presence of DABCO, the coupling reaction of $2',3'$-O-isopropylideneuridine derivatives **33** with aldehydes results in the corresponding adduct **35** in moderate to high yields. This reaction seems like an intramolecular base-catalyzed MBH reaction that involves, as a plausible reaction sequence, an initial nucleophilic attack at the 6-position of the uracil ring by the $5'$-hydroxy group that was activated by DABCO, which could give rise to an adduct **34**, followed by a ring-opening reaction to afford **35** (Scheme 2.18). However, uridine derivatives that possess no isopropylidene protective group or no $5'$-hydroxy group underwent almost no reaction under the same conditions.[20]

Batra *et al.* have identified 3-substituted 5-isoxazolecarboxaldehydes **36** as activated aldehydes in MBH reactions with various activated alkenes **37** in the presence of DABCO, affording the corresponding adducts **38** in moderate to excellent yields (Scheme 2.19).[21] Moreover, they have also developed a novel isoxazole-based scaffold for the generation of combinatorial libraries by using an MBH reaction of solid-phase-supported 5-isoxazolecarboxaldehyde **39** with ethyl acrylate[22] and 5-isoxazolecarboxaldehyde **36** with solid-phase-supported acrylate esters[23] as key step (Schemes 2.20 and 2.21, respectively). These libraries were further evaluated for their antithrombin activity *in vivo*.[23]

Isatin derivatives **45** (reactive cyclic α-keto amides) have been employed as electrophiles by Garden and Kim in MBH coupling with activated alkenes to give the corresponding adducts **46** and **47** in good to excellent yields (Scheme 2.22).[24] It is generally accepted that ketones only take part in the MBH reaction under relatively extreme conditions, with a few exceptions.

During their studies on the effect of fluorine substitution in the MBH reaction of various fluorocarbonyl partners with activated olefins, such as

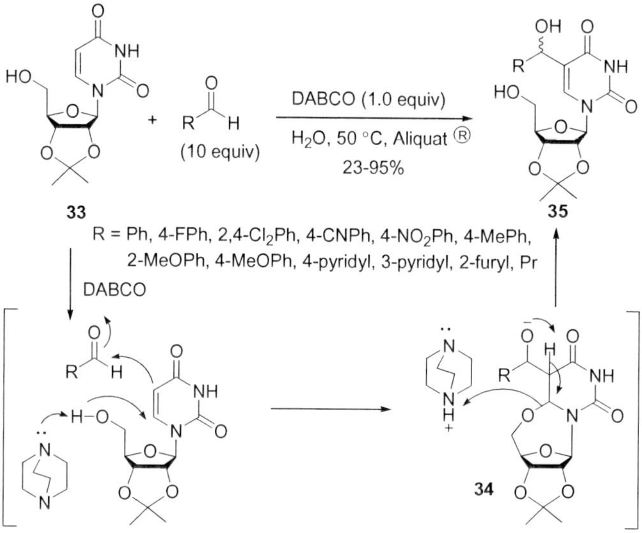

R = Ph, 4-FPh, 2,4-Cl$_2$Ph, 4-CNPh, 4-NO$_2$Ph, 4-MePh, 2-MeOPh, 4-MeOPh, 4-pyridyl, 3-pyridyl, 2-furyl, Pr

Scheme 2.18

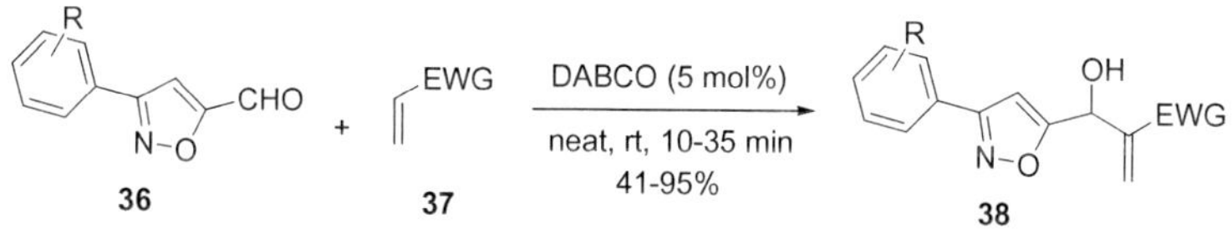

R = H, 4-Me, 4-OBn, 2-Cl, 3-NO$_2$;
EWG = CO$_2$Me, CO$_2$Et, CO$_2$t-Bu, CO$_2$Bu, CN, COMe

Scheme 2.19

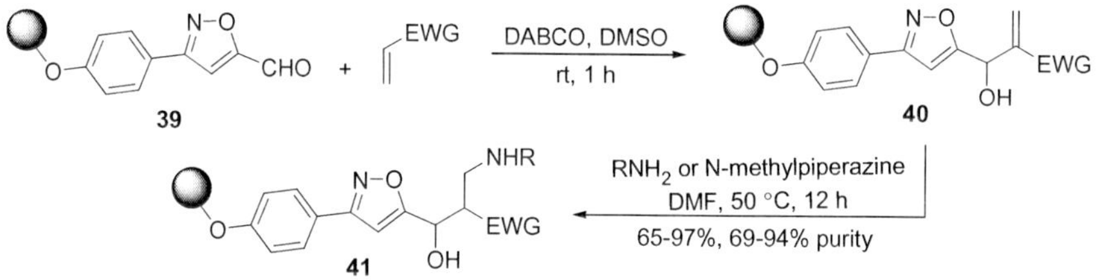

EWG = CO$_2$Et, CO$_2$t-Bu, CO$_2$Bu, CN;
R = CH$_2$CH$_2$NEt$_2$, CH$_2$Ph, CH$_2$CH$_2$-(1-pyrrolidinyl), 4-methyl-1-piperazinyl,
CH$_2$CH$_2$-(4-piperazinyl), CH$_2$-(2-furyl), CH$_2$-(4-piperidinyl), CH$_2$CH$_2$-(4-morpholinyl)

Scheme 2.20

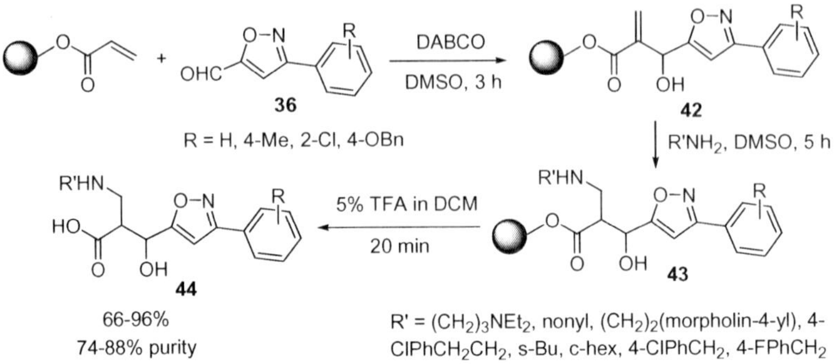

DABCO
DMSO, 3 h
36
R = H, 4-Me, 2-Cl, 4-OBn
42
R'NH2, DMSO, 5 h
5% TFA in DCM
20 min
44
43
66-96%
74-88% purity
R' = (CH2)3NEt2, nonyl, (CH2)2(morpholin-4-yl), 4-ClPhCH2CH2, s-Bu, c-hex, 4-ClPhCH2, 4-FPhCH2

Scheme 2.21

DABCO
(0.15-0.5 equiv)
THF or DMF, rt
90 min-21 days
50-94%
47
45
DABCO
EtOH or EtOH/THF
2-14 days, 70->95%
46
R1 = H, allyl, Bn, Ph, COMe, COEt, COPh;
R2 = H, 5-Br;
EWG = CO2Me, CO2Et, CN
R1 = Me, H, Bn;
R2 = H, 5-Br, 5-I, 5,7-Br2, 5-Br-7-NO2, 5-NO2;
EWG = CO2Et, CN

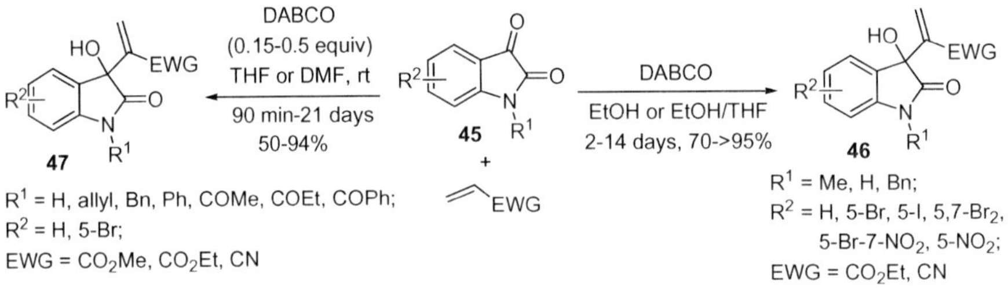

Scheme 2.22

acrolein, methyl vinyl ketone, ethyl acrylate and acrylonitrile, Ramchandran *et al.* have found that a match between the reactivities of the olefin **48** and carbonyl partners **49** is essential for obtaining reasonable yields of the products **50** and **51**. When the olefin (*e.g.* acrolein) can react with itself in the presence of an amine, the electrophile has to be very reactive as well (*e.g.* fluoral) to obtain MBH products in a modest to good yield; a moderately reactive electrophile (*e.g.* trifluoroacetophenone) provided a good yield of products with moderately reactive olefins (*e.g.* ethyl acrylate and acrylonitrile), whereas the reaction of a moderately reactive olefin (*e.g.* ethyl acrylate or acrylonitrile) and a very reactive electrophile (*e.g.* fluoral) results in either self-reaction of the electrophile or a very low yield of the allylic alcohol product (Scheme 2.23).[25]

Basavaiah *et al.*[26] were the first to demonstrate that application of allyl bromides/allyl chlorides (**52**) – derived from the corresponding MBH adducts (of methyl acrylate and MVK) – as electrophiles in the MBH reaction with acrylonitrile in the presence of DABCO affords 3-substituted functionalized 1,4-pentadienes **53** and **54** (Scheme 2.24). However, their attempts to couple acrylonitrile with methyl (2*Z*)-2-bromomethyl-hex-2-enoate, derived from the MBH adducts of corresponding aliphatic aldehydes, in the presence of DABCO were not successful.

Subsequently, they extended this strategy to allyl bromides (**55**) derived from alkyl 3-hydroxy-2-methylenepropanoates, thus developing a convenient, simple

Scheme 2.23

Scheme 2.24

one-pot methodology for the synthesis of 2,4-functionalized 1,4-pentadienes **56–59** *via* the MBH reaction between activated alkenes and **55** in the presence of DABCO or DBU (Scheme 2.25).[27]

Lee *et al.* found that the DABCO-catalyzed MBH reaction between methyl acrylate and methyl vinyl ketone with 2,3-dihalo-1,4-naphthoquinone **60** forms α-vinylnaphthoquinones in moderate to good yields.[28] The same authors subsequently reported additional examples of the use of DABCO attached enolate anion of methyl acrylate and MVK in substitution reactions of halo-quinones to form α-vinylquinone bonds, avoiding the use of organometallic reagents (Scheme 2.26).[29]

β-Substituted α,β-unsaturated trifluoromethyl ketones **67**, which usually behave as Michael acceptors, have been employed as primary electrophiles to react with acrylonitrile in the presence of DABCO, affording tertiary diallylic alcohols **68** *via* chemoselective 1,2-addition (Scheme 2.27).[30]

de Souza *et al.* have developed two different efficient reaction conditions depending on the used Michael acceptors. For acrylonitrile, *tert*-butanol–water (60:40) was the system of choice, while for methyl acrylate, DMSO–water (60:40) gave better results (Scheme 2.28). In comparison with other tertiary amines, such as DBU, DMAP, HMT, imidazole and triethylamine, DABCO was the catalyst of choice and quantitative proportions gave the best yields.[31]

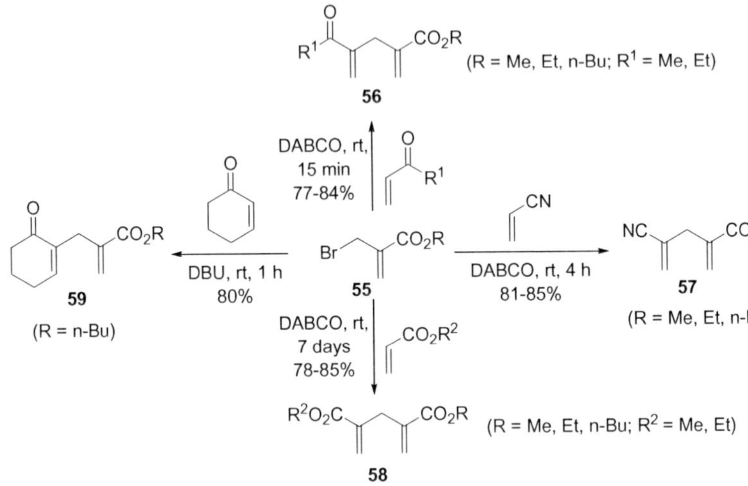

56 (R = Me, Et, n-Bu; R^1 = Me, Et)

DABCO, rt,
15 min
77-84%

DABCO, rt,
7 days
78-85%

DBU, rt, 1 h
80%

DABCO, rt, 4 h
81-85%

59
(R = n-Bu)

55

57
(R = Me, Et, n-Bu)

58 (R = Me, Et, n-Bu; R^2 = Me, Et)

Scheme 2.25

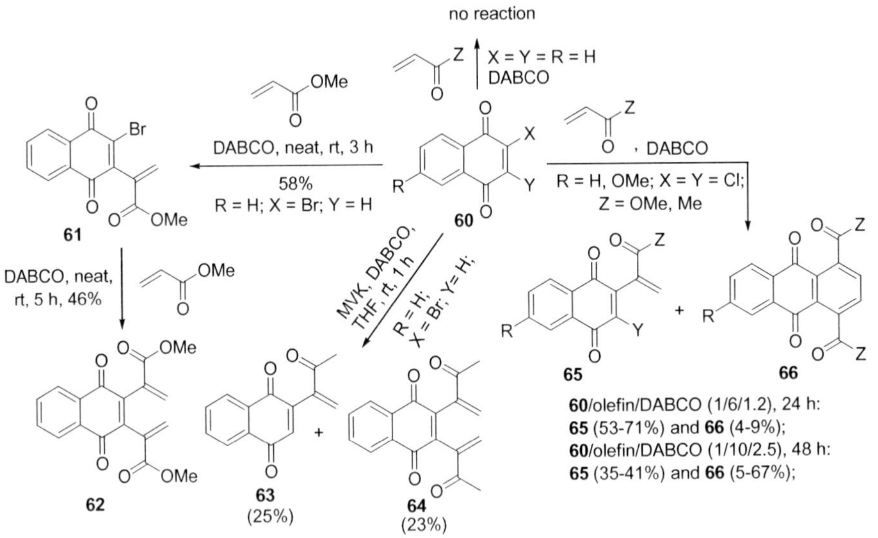

no reaction

X = Y = R = H
DABCO

DABCO, neat, rt, 3 h
58%
R = H; X = Br; Y = H

, DABCO
R = H, OMe; X = Y = Cl;
Z = OMe, Me

61

60

DABCO, neat,
rt, 5 h, 46%

MVK, DABCO,
THF, rt, 1 h
R = H; Y = H;
X = Br; Y = H;

62

63
(25%)

64
(23%)

65

66

60/olefin/DABCO (1/6/1.2), 24 h:
65 (53-71%) and **66** (4-9%);
60/olefin/DABCO (1/10/2.5), 48 h:
65 (35-41%) and **66** (5-67%);

Scheme 2.26

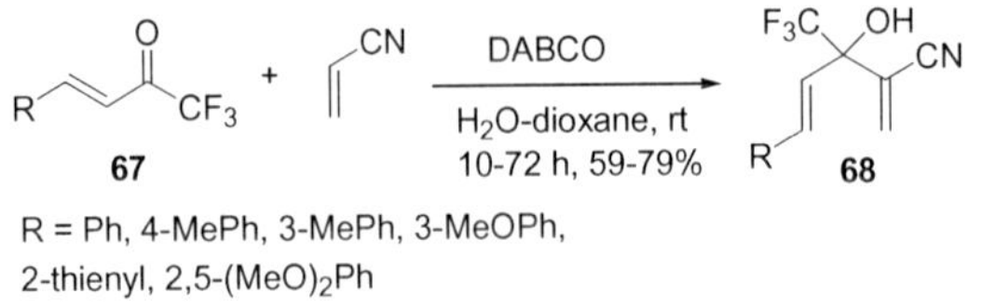

67

DABCO
H$_2$O-dioxane, rt
10-72 h, 59-79%

68

R = Ph, 4-MePh, 3-MePh, 3-MeOPh,
2-thienyl, 2,5-(MeO)$_2$Ph

Scheme 2.27

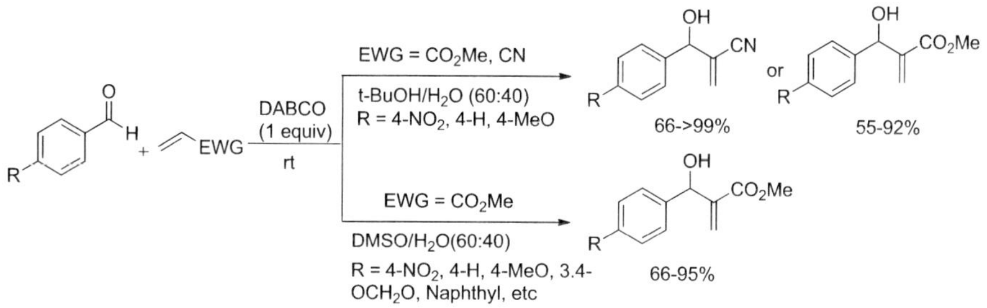

Scheme 2.28

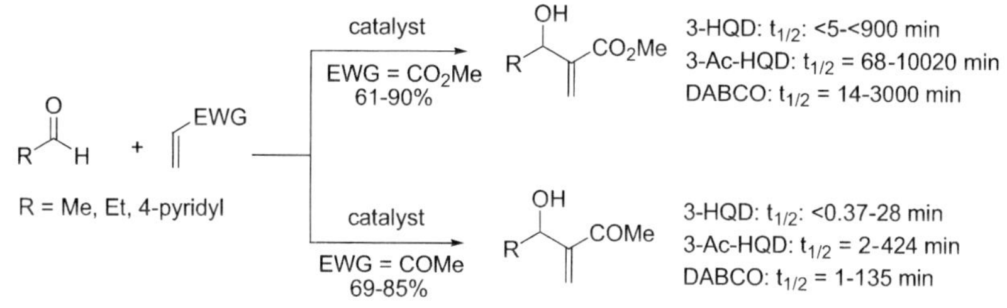

Scheme 2.29

Drewes *et al.* have found that the half-lives can be reduced significantly if the methyl acrylate is reacted with strong electrophilic aldehydes by using 3-hydroxyquinuclidine (3-HQD), instead of DABCO, as catalyst. It was speculated that the hydrogen-bonding stabilization of the catalyst–acrylate adduct by means of the hydroxyl group led to a rate-enhancement effect over DABCO.[32] Subsequently, they further confirmed this hypothesis with the fact that 3-hydroxyquinuclidine enhanced the rate of the MBH reaction whereas its acetylated derivative was a poor catalyst.[33] Moreover, a comparative study showed that methyl vinyl ketone was more reactive than methyl acrylate in the MBH reaction (Scheme 2.29).

In the presence of 3-HQD or DABCO, the MBH reaction between resin-bound acrylate and various aldehydes with different reactivities yielded allylic alcohols **69** and **70** in moderate to excellent yields by simply varying the base or the reaction time,[34] which led to a combinatorial synthesis of substituted racemic 3-hydroxypropionamides,[34a] amino alcohols[34b] or arylpropanolamines[34c] *via* the treatment with amines and the use of different cleavage methods (Scheme 2.30).

More recently, Hayashi *et al.*[35] have observed an efficient rate enhancement in the 3-HQD-catalyzed MBH reaction under high pressure induced by freezing water in a sealed autoclave. In most cases, moderate to good yields were obtained for the MBH reaction between various aldehydes with acrylates under 200 MPa pressure (Scheme 2.31).

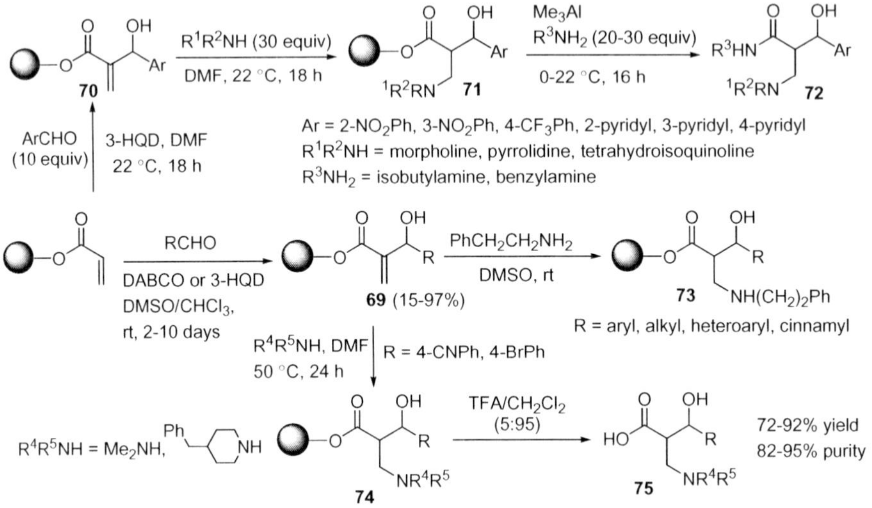

R¹R²NH (30 equiv)
DMF, 22 °C, 18 h
Me3Al
R³NH2 (20-30 equiv)
0-22 °C, 16 h
70
71
72
ArCHO (10 equiv)
3-HQD, DMF 22 °C, 18 h
Ar = 2-NO2Ph, 3-NO2Ph, 4-CF3Ph, 2-pyridyl, 3-pyridyl, 4-pyridyl
R¹R²NH = morpholine, pyrrolidine, tetrahydroisoquinoline
R³NH2 = isobutylamine, benzylamine
RCHO
DABCO or 3-HQD
DMSO/CHCl3,
rt, 2-10 days
69 (15-97%)
PhCH2CH2NH2
DMSO, rt
73
NH(CH2)2Ph
R = aryl, alkyl, heteroaryl, cinnamyl
R⁴R⁵NH, DMF
50 °C, 24 h
R = 4-CNPh, 4-BrPh
TFA/CH2Cl2
(5:95)
R⁴R⁵NH = Me2NH,
74
75
72-92% yield
82-95% purity

Scheme 2.30

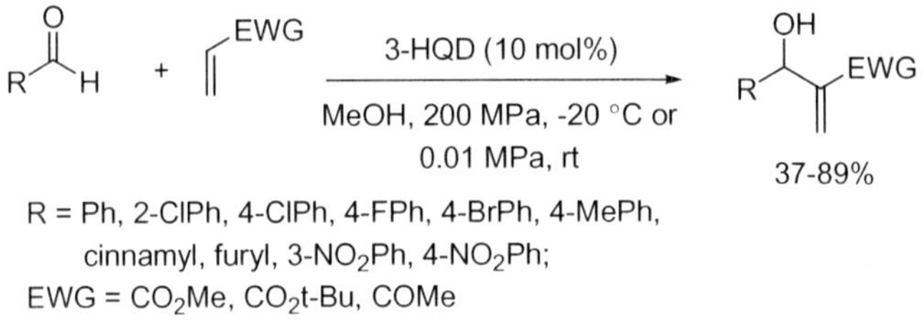

EWG
3-HQD (10 mol%)
MeOH, 200 MPa, -20 °C or
0.01 MPa, rt
37-89%
R = Ph, 2-ClPh, 4-ClPh, 4-FPh, 4-BrPh, 4-MePh,
cinnamyl, furyl, 3-NO2Ph, 4-NO2Ph;
EWG = CO2Me, CO2t-Bu, COMe

Scheme 2.31

Aggarwal and co-workers[36] have examined the correlation of the pK_a of various quinuclidine-based catalysts (Figure 2.1) with their reactivities in the MBH reaction. They found that quinuclidine (QD), which was previously reported as a poor catalyst, in protic solvents has the highest pK_a (11.3/H_2O) and is the most active catalyst for this reaction. They also observed that the combination of quinuclidine and methanol was the most efficient system for catalyzing the MBH reaction. Various activated olefins, including less reactive activated olefins such as vinyl sulfones, acrylamides and β-substituted α,β-unsaturated esters, have also been employed in this reaction, using quinuclidine as a catalyst (Scheme 2.32).[36]

In 1998, Rezgui and El Gaied were the first to carry out the DMAP-catalyzed MBH reaction of cyclohex-2-en-1-one derivatives **76** with formaldehyde in an aqueous medium to provide 2-(hydroxymethyl)cyclohex-2-en-1-one derivatives **77** (Scheme 2.33).[37] They noticed that this reaction could not proceed in the presence of DABCO. β-Substituted cyclohexenone derivatives did

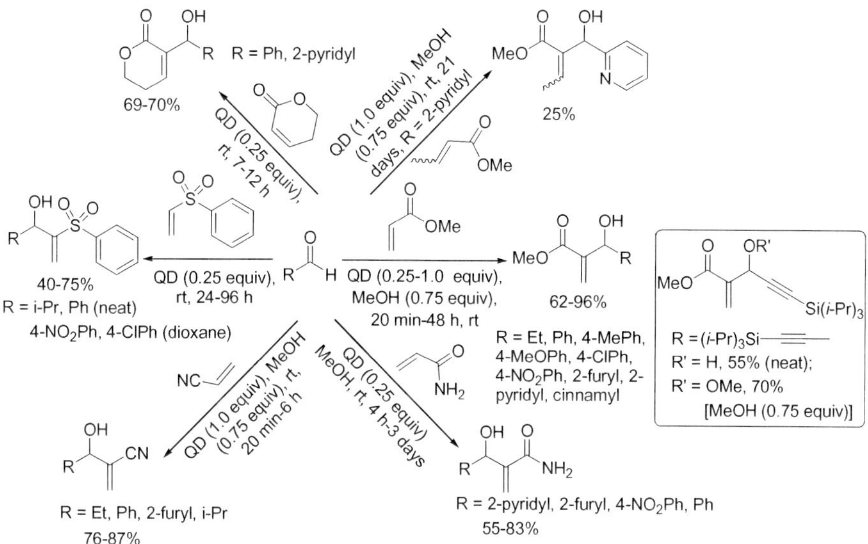

	quinuclidine (QD)	3-HQD	DABCO	3-AcOQD	3-ClQD	3-quinuclidinone
pK_a (in H_2O)	11.3	9.9	8.5	9.3	8.9	6.9
rate (%/min)	1.8	0.88	0.21	0.031	0.0082	0.0013
k_{rel}	9.0	4.3	1.0	0.15	0.04	0.006

Figure 2.1 pK_a of various quinuclidine-based catalysts.

Scheme 2.32

Scheme 2.33

not undergo a MBH reaction under these conditions either, which was attributed to β-C steric effects that prevented, in the first step of the MBH reaction, the Michael addition of DAMP to α,β-enones.

Subsequently, Kim and co-workers also used DMAP as a catalyst for the MBH reaction between cycloalkenones **78** and various aldehydes, in which no reaction was observed under the normal reaction conditions using DABCO as catalyst, to obtain the corresponding adducts **79** in moderate yields in aqueous THF (Scheme 2.34).[38]

The MBH reaction of *p*-methylquinols **80** with activated aromatic aldehydes has been investigated (Scheme 2.35).[39] Depending on the reaction conditions

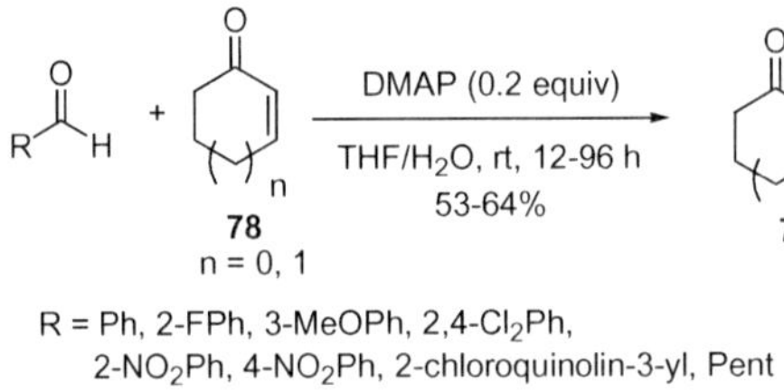

DMAP (0.2 equiv)
THF/H2O, rt, 12-96 h
53-64%
78
n = 0, 1
79
R = Ph, 2-FPh, 3-MeOPh, 2,4-Cl2Ph,
2-NO2Ph, 4-NO2Ph, 2-chloroquinolin-3-yl, Pent

Scheme 2.34

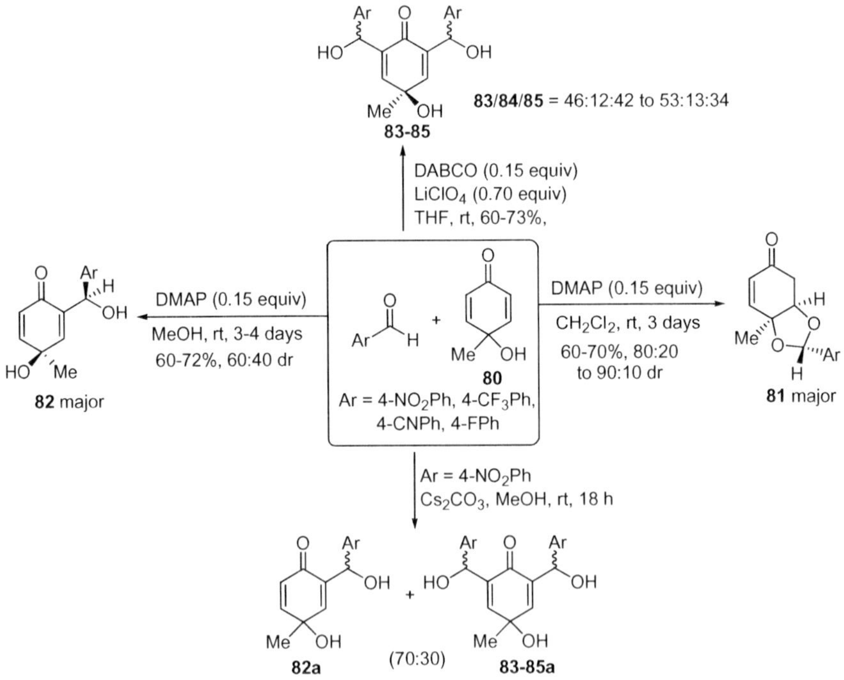

83/84/85 = 46:12:42 to 53:13:34
83-85
DABCO (0.15 equiv)
LiClO4 (0.70 equiv)
THF, rt, 60-73%,
DMAP (0.15 equiv)
MeOH, rt, 3-4 days
60-72%, 60:40 dr
82 major
DMAP (0.15 equiv)
CH2Cl2, rt, 3 days
60-70%, 80:20
to 90:10 dr
81 major
80
Ar = 4-NO2Ph, 4-CF3Ph,
4-CNPh, 4-FPh
Ar = 4-NO2Ph
Cs2CO3, MeOH, rt, 18 h
82a
(70:30)
83-85a

Scheme 2.35

(solvent and additives), a mono or double MBH adduct (**82–85**) and a fused 1,3-dioxolane (**81**) could be obtained in good chemical yields. The use of non-nucleophilic bases, such as Cs_2CO_3, to promote the reaction suggested that this reaction involves an autocatalytic mechanism.

1,8-Diazabicyclo[5.4.0]undec-7-ene (DBU), considered as a hindered and non-nucleophilic base, was shown to be a better catalyst for the MBH reaction, providing adducts at much faster rates than using DABCO or 3-HQD (Scheme 2.36).[40] The increased reactivity for this catalyst was attributed to stabilization of the intermediate β-ammonium enolate through conjugation, which increased its equilibrium concentration and resulted in significantly enhanced rates.

Subsequently, Kaye and Nocanda employed 2,2′-dithiobenzaldehyde **86** as an electrophile for MBH coupling with various activated olefins in the presence of DBU, to provide a convenient synthesis of benzothiopyran derivatives **87** (Scheme 2.37).[41]

Azizi and Saidi[42] found that *in situ* prepared iminium salts **88** [obtained *via* the treatment of aldehyde with (trimethylsilyl)dialkylamine] are very effective electrophiles in MBH reactions and react with methyl acrylate in the presence of a catalytic amount of DBU at ambient temperature to afford the corresponding adducts **89**, which can undergo conjugated addition with (trimethylsilyl) dialkylamines to give diamines **90** in good yields (Scheme 2.38).

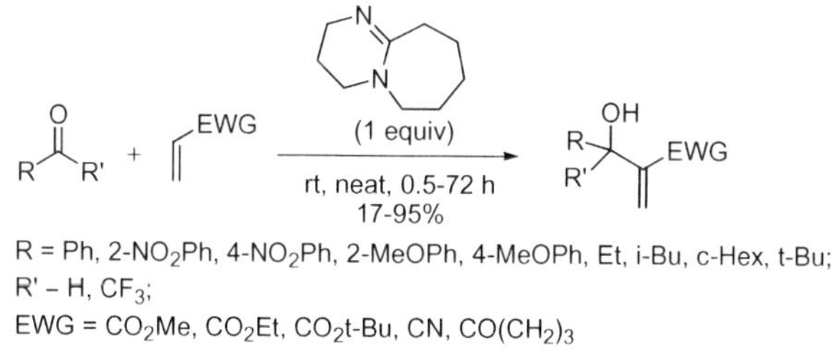
EWG
(1 equiv)
rt, neat, 0.5-72 h
17-95%
OH
EWG
R = Ph, 2-NO2Ph, 4-NO2Ph, 2-MeOPh, 4-MeOPh, Et, i-Bu, c-Hex, t-Bu;
R' - H, CF3;
EWG = CO2Me, CO2Et, CO2t-Bu, CN, CO(CH2)3

Scheme 2.36

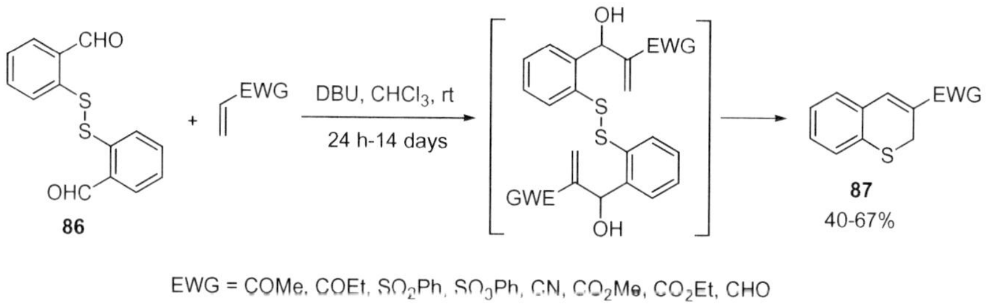
CHO
OHC
86
EWG
DBU, CHCl3, rt
24 h-14 days
OH
EWG
GWE
OH
EWG
87
40-67%
EWG = COMe, COEt, SO2Ph, SO3Ph, CN, CO2Me, CO2Et, CHO

Scheme 2.37

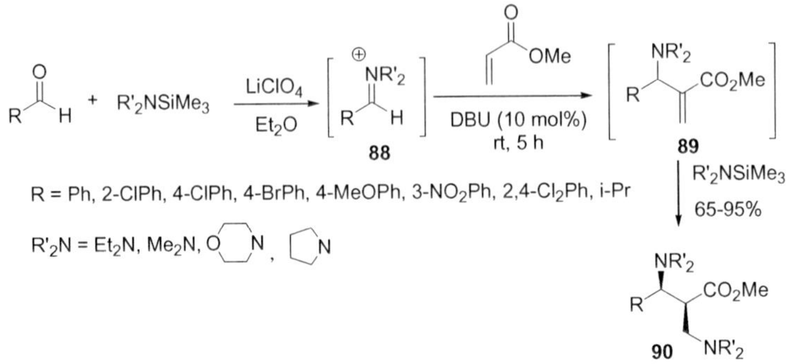

Scheme 2.38

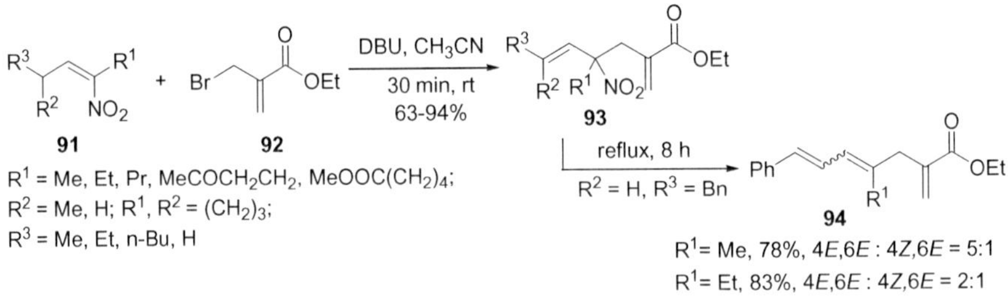

Scheme 2.39

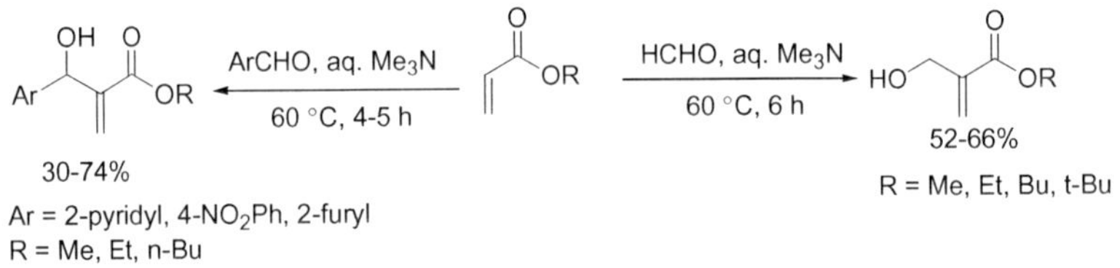

R^1 = Me, Et, Pr, MeCOCH$_2$CH$_2$, MeOOC(CH$_2$)$_4$;
R^2 = Me, H; R^1, R^2 = (CH$_2$)$_3$;
R^3 = Me, Et, n-Bu, H

R^1= Me, 78%, 4*E*,6*E* : 4*Z*,6*E* = 5:1
R^1= Et, 83%, 4*E*,6*E* : 4*Z*,6*E* = 2:1

Scheme 2.40

Ballini *et al* have reported a rapid diastereoselective MBH reaction by using nitroalkenes **91** as activated alkenes, ethyl 2-bromomethylacrylate **92** as electrophilic acceptor and DBU as base catalyst. Nitro dienes **93** were obtained in good yields and very short reaction times. Moreover, starting from appropriate nitroalkenes it is possible to realize a one-pot synthesis of trienic systems **94** (Scheme 2.39).[43]

Basavaiah *et al.* have demonstrated that aqueous trimethylamine, which is tertiary amine containing a minimum number of carbon atoms with the lowest possible molecular weight, mediated the MBH reaction of alkyl acrylates with paraformaldehyde and various reactive aromatic aldehydes (Scheme 2.40).[44]

However, the aqueous trimethylamine failed to mediate the MBH reaction of methyl acrylate with benzaldehyde. Subsequently, they found that methanolic trimethylamine is effective for coupling of benzaldehyde with methyl acrylate, indicating that methanolic trimethylamine is a better medium than aqueous trimethylamine for performing the MBH coupling of methyl acrylate with less reactive aldehydes. This methanolic trimethylamine was also used for the MBH reaction between representative aldehydes with activated alkenes, including methyl acrylates, acrylonitrile and acrolein, providing the corresponding adducts in good yields within reasonable reaction time (Scheme 2.41). Acenaphthenequinone **95**, a non-enolizable ketone, has also been employed as an electrophile for coupling with acrylonitrile in the presence of methanolic trimethylamine (Scheme 2.42).[45]

1-Benzopyran-4(4*H*)-one derivatives **97** have been successfully employed as novel activated alkenes in the methanolic trimethylamine catalyzed MBH reaction with heteroaromatic-aldehydes, nitrobenzaldehydes and isatin derivatives. The corresponding adducts **98**, derived from pyridine-2-carboxaldehyde, have been transformed into a novel indolizine-fused-chromone framework **100** (Scheme 2.43).[46]

Leadbeater *et al.*[47] were the first to report that tetramethylguanidine (TMG) is an efficient catalyst for the MBH reaction between various aldehydes and methyl acrylate, showing good activity with a range of aldehyde substrates and, unlike many other catalysts or catalyst mixtures, it can be used to good effect with simple aliphatic aldehydes (Scheme 2.44). Their attempts to use supported or derivatized TMG complexes as catalysts for the reaction were unsuccessful,

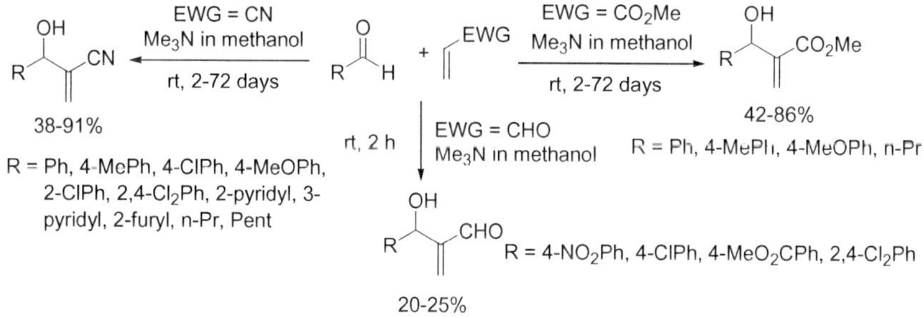

Scheme 2.41

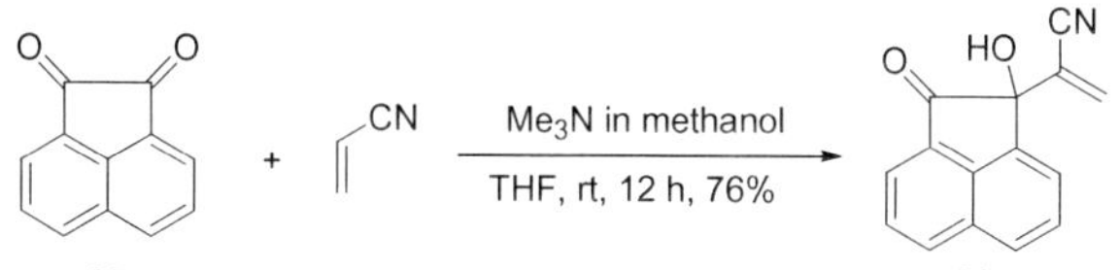

Scheme 2.42

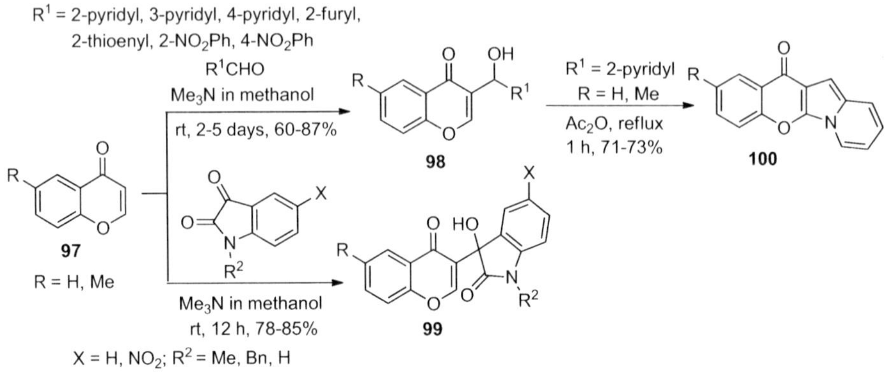

R^1 = 2-pyridyl, 3-pyridyl, 4-pyridyl, 2-furyl, 2-thioenyl, 2-NO$_2$Ph, 4-NO$_2$Ph

R^{1}CHO
Me$_3$N in methanol
rt, 2-5 days, 60-87%

R^1 = 2-pyridyl
R = H, Me
Ac$_2$O, reflux
1 h, 71-73%

Me$_3$N in methanol
rt, 12 h, 78-85%

97
R = H, Me

98

99

100

X = H, NO$_2$; R^2 = Me, Bn, H

Scheme 2.43

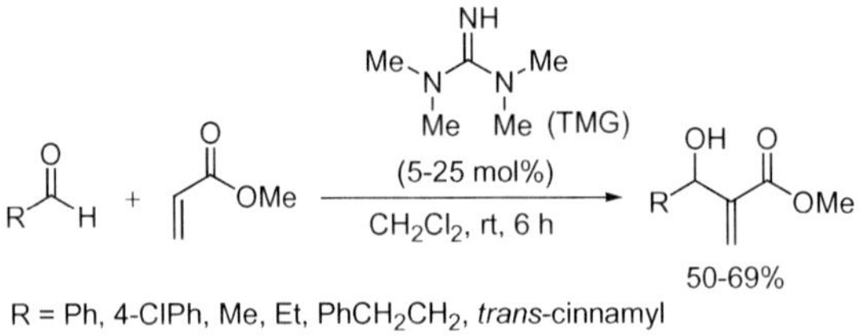

(TMG)
(5-25 mol%)
CH$_2$Cl$_2$, rt, 6 h

50-69%

R = Ph, 4-ClPh, Me, Et, PhCH$_2$CH$_2$, *trans*-cinnamyl

Scheme 2.44

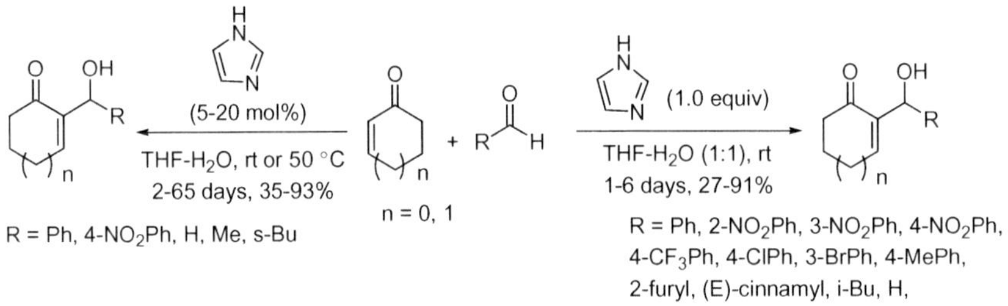

(5-20 mol%)
THF-H$_2$O, rt or 50 °C
2-65 days, 35-93%

n = 0, 1

(1.0 equiv)
THF-H$_2$O (1:1), rt
1-6 days, 27-91%

R = Ph, 4-NO$_2$Ph, H, Me, s-Bu

R = Ph, 2-NO$_2$Ph, 3-NO$_2$Ph, 4-NO$_2$Ph,
4-CF$_3$Ph, 4-ClPh, 3-BrPh, 4-MePh,
2-furyl, (E)-cinnamyl, i-Bu, H,

Scheme 2.45

suggesting that the presence of an amine hydrogen is key to the activity of TMG. They also examined the effect of temperature/solvent on the reaction rate.

Cheng *et al.*[48] have developed the first example of a stoichiometric weak Lewis base, imidazole, catalyzed MBH reaction between cyclic enones and aromatic aldehydes in aqueous media. At the same time, Gatri and El Gaied[49] also independently reported the MBH reaction of cycloalkenones with various aldehydes, including aliphatic and aromatic aldehydes, by using catalytic quantities imidazole as catalyst in aqueous media (Scheme 2.45).

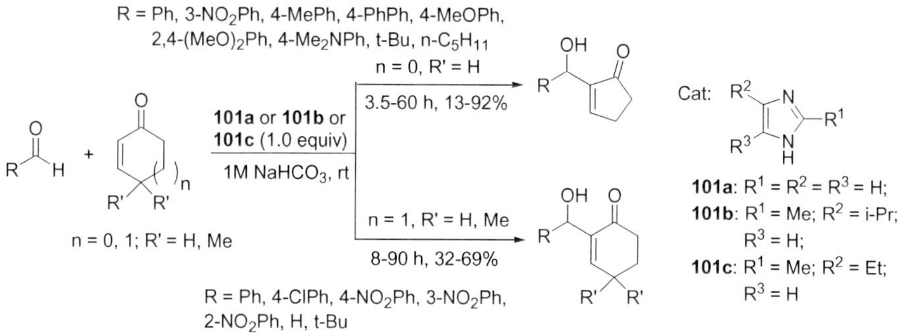

Scheme 2.46

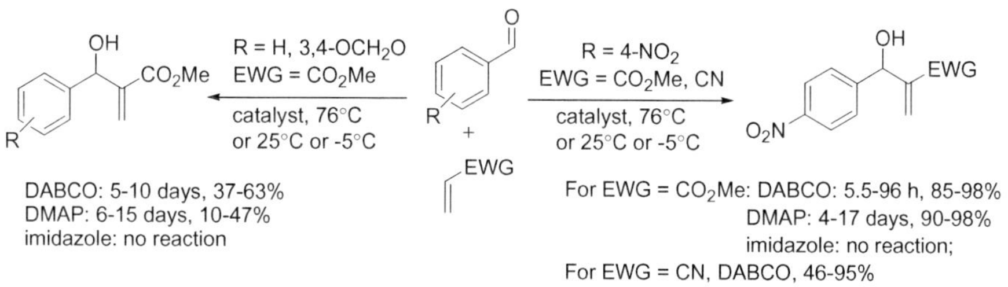

Scheme 2.47

With imidazoles **101** as catalysts, the MBH reaction of cyclic enones was greatly accelerated in basic water solution and bicarbonate solution has been shown to be the optimal reaction medium. The reaction runs much faster in sodium bicarbonate solution than in distilled water, and showed an obvious pH dependence. Notably, these conditions have been applied successfully to various aldehydes and cyclic enones, especially to unreactive and hindered aldehydes (Scheme 2.46). The apparent "enhanced basicity" of imidazoles accounted for the rate increase in alkaline solution.[50]

A comparative study has been made of DMAP, DABCO and imidazole as catalysts in the MBH reaction of methyl acrylate or acrylonitrile with aromatic aldehydes (Scheme 2.47). Using neat activated alkenes, where there is no hydrogen bonding or additives effects, DMAP and DABCO present similar catalytic activity at 76 °C in the reaction with *p*-nitrobenzaldehyde, and DMAP could be an option for DABCO. On the other hand, DABCO is better than DMAP when less reactive electrophiles are used. Imidazole, which exhibits catalytic activity in water media, does not show catalytic activity under solvent-free conditions. Rapid conversion using DMAP and DABCO at low temperature (at -5 °C) was observed, presumably due to an entropy controlled reaction.[51]

The Morita–Baylis–Hillman reactions of various conjugated nitroalkenes or nitrodienes **102** with activated non-enolizable carbonyl compounds such as glyoxylate, trifluoropyruvate, pyruvaldehyde, oxomalonate, ninhydrin and formaldehyde proceed smoothly in the presence of DMAP (40–100 mol.%) in acetonitrile or imidazole (100 mol.%) in CHCl$_3$ or THF, providing multi-functional adducts in good to excellent yields (Scheme 2.48). In most cases, the reactions catalyzed by DMAP in acetonitrile were superior to the imidazole-catalyzed reactions in terms of the rate of reaction and the isolated yields of the MBH adducts. The primarily catalytic role attributed to DMAP and imidazole in these reactions, *vis-à-vis* other MBH catalysts such as DABCO, is resonance stabilization of the initial zwitterionic intermediates. Whereas the (*E*) isomers are the major or exclusive products in the cases of glyoxylate, pyruvaldehyde and formaldehyde, the (*Z*) isomers predominate in the cases of tri-fluoropyruvate and ninhydrin. Interestingly, oxomalonate forms (*E*) isomers with aromatic nitroalkenes and (*Z*) isomers with aliphatic ones.[52]

Novel α-hydrazino-α,β-unsaturated nitroalkenes **111**, which exhibit dynamic phenomenon on the NMR time scale, have been synthesized in excellent yields *via* the imidazole or DMAP mediated aza-MBH reaction of nitroalkenes with azodicarboxylates (Scheme 2.49).[53]

Inspired by the intramolecular stabilization of the neighboring cationic character by non-bonding electron pairs on the nitrogen or phosphorus atom[54] and DBU catalyzed-MBH reaction, Kim *et al.* have developed *N,N,N',N'*-tetramethyl-1,3-propanediamine (TMPDA) as an efficient catalyst for the MBH reaction of cycloalkenones (Scheme 2.50). The increased reaction rate was attributed to the stabilizing effect of the zwitterionic intermediate *via* ion–dipole interaction.[55] Unfortunately, the application of TMPDA in the reaction of an acyclic alkene such as acrylonitrile (74%) and methyl vinyl ketone (55%) under the same reaction conditions gave the products in relatively low yields compared with previously reported methods.

Inexpensive and commercially available *N*-methylmorpholine (NMM),[56] urotropine[56,57] and *N*-methylpiperidine (NMP)[58] have been utilized effectively as new base catalysts for the MBH reaction at ambient temperature in aqueous dioxane (1 : 1) or DMSO to afford the corresponding adducts in good to excellent yields (Scheme 2.51). Significant rate enhancement has been observed.

As another kind of tertiary amine, hexamethylenetetramine (HMT) has proved to be an efficient catalyst in the MBH reaction between various aromatic aldehydes and methyl acrylate or acrylonitrile (Scheme 2.52). The corresponding adducts were obtained in good to high yield by use of 0.1 equiv or 1.0 equiv of HMT at room temperature or at 60 °C. The large amounts of HMT and high temperature can significantly shorten the reaction time. Some of the MBH adducts showed highly potent biological activity.[59]

Cheng *et al.* have found that the azoles, which were inactive in neutral aqueous media, can be activated in alkaline solution and catalyzed effectively the MBH reaction involving cyclic enones. Notably, the reaction conditions were suitable for a range of cyclic enones and aldehydes, and facilitated the coupling of unreactive aldehyde with cyclic enones. Both 1,2,3-triazole (**112**)

R = 4-MeOPh, 2,4-(MeO)2Ph, 2,5-(MeO)2Ph, 2-furyl, 3-furyl
106
106'
43-85% (106/106' = 71:29 to 100:0)
R = 4-MeOPh, 3,4-(MeO)2Ph, 3,4,5-(MeO)3Ph, 4-Me2NPh, 2-furyl
or
(n-Bu)i-Pr
107
45-57%
R = i-PrCH2, n-BuCH2
105
R1 = Me: DAMP (0.4 equiv) / CH3CN:
2-7 days, 26-35%;
Im (1.0 equiv) / THF:
2-7 h, 20-36%;
R1 = H, Ph: no reaction
104
104'
DAMP (0.4 equiv) / CH3CN: 5-10 min,
40-88% (104/104' = 0:100 to 100:0)
Im (1.0 equiv) / CHCl3: 10-15 min,
0-66% (104/104' = 0:100 to 100:0)
R = Cy, i-Pr; R1 = H, CF3, CO2Et; R2 = CO2Et, CO2Me, CO2Et
103
103'
(X = H, OMe, NO2; R2 = H, CF3; R3 = CO2Me, CO2Et, H)
DAMP (1.0 equiv) / CH3CN: 20-45 min, 63-88%
(103/103' = 0:100 to 100:0)
Im (1.0 equiv) / CHCl3: 5-16 h, 50-79%
(103/103' = 0:100 to 100:0)
DAMP (0.4 equiv)
CH3CN, rt, 5-20 min
DAMP (0.4-1.0 equiv)
CH3CN, rt, 30 min-10 h
108
63-80%
R = 4-MeOPh, 2,5-(MeO)2Ph, 4-Me2NPh, 3-furyl
102
R = cinnamyl
cat., solvent, rt
cat., solvent, rt
cat., solvent, rt
110
DAMP (0.4 equiv) / CH3CN: 5-30 min, 33-99%;
Im (1.0 equiv) / CHCl3: 2-30 h, 31-83%;
Im (1.0 equiv) / THF: 5-48 h, 25-80%
R = 4-MeOPh, 2,4-(MeO)2Ph, 2,5-(MeO)2Ph, 4-Me2NPh, 3,4-(MeO)2Ph, 3,4-(OCH2O)Ph, 3-MeO-4-HOPh, 4-ClPh, Ph, 2-propargyloxyPh, 2-allyloxyPh, 3,4,5-(OMe)3Ph, 2-furyl, 2-thienyl, 3-furyl, 3-thienyl, 3-indolyl
Ar = 4-MeOPh, 4-Me2NPh, 3,4-(MeO)2Ph, 4-ClPh, Ph, 2-allyloxyPh, 2-NO2Ph, 2-furyl, 2-thienyl, 3-furyl
109
109'
DAMP (0.5 equiv) / CH3CN: 15 min-2 h,
43-78% (109/109' = 55:45 to 100:0)
Im (1.0 equiv) / CHCl3: 5-24 h,
55-79% (109/109' = 63:37 to 100:0)

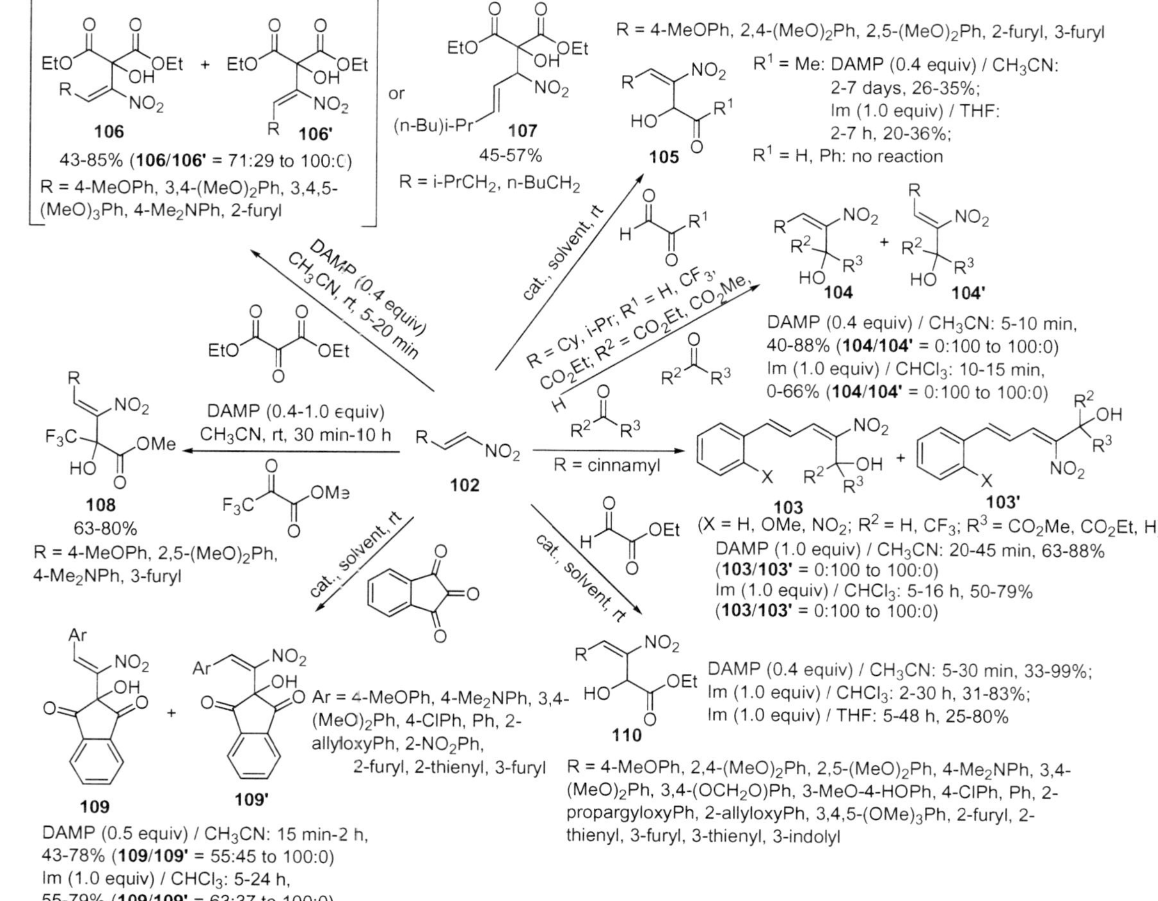

Scheme 2.48

R' = i-Pr, Et

R = 2-furyl, 2-thienyl, 3-furyl, 3-thienyl, 4-ClPh, Ph,
4-MeOPh, 3,4-(OCH$_2$O)Ph, 3,4-(MeO)$_2$Ph, 4-CF$_3$Ph,
4-NO$_2$Ph, 3-MeO-4-(HO)Ph, 4-(NMe$_2$)Ph,

imidazole: 0.5-24 h, 83-100%;
DMAP: 0.25-8 h, 43-87%

Scheme 2.49

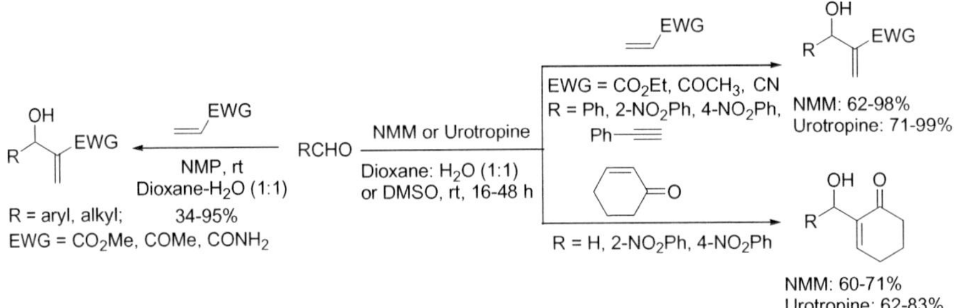

n = 1, 2
R = Ph, 4-MeOPh, 2-MeOPh, H, pent

Scheme 2.50

R = aryl, alkyl;   34-95%
EWG = CO$_2$Me, COMe, CONH$_2$

NMP, rt
Dioxane-H$_2$O (1:1)

NMM or Urotropine
Dioxane: H$_2$O (1:1)
or DMSO, rt, 16-48 h

EWG = CO$_2$Et, COCH$_3$, CN
R = Ph, 2-NO$_2$Ph, 4-NO$_2$Ph,
Ph—

NMM: 62-98%
Urotropine: 71-99%

R = H, 2-NO$_2$Ph, 4-NO$_2$Ph

NMM: 60-71%
Urotropine: 62-83%

Scheme 2.51

12-84%

0.1 equiv HMT
rt or 60 °C

ArCHO

1 equiv HMT
DMSO, 60 °C

20-90%

HMT

Ar = 2-NO$_2$Ph, Ph, 2-naphthyl, 4-OHPh, 4-MeOPh,
4-OH-3-MeOPh, 2,5-(MeO)$_2$Ph, 2-pyridinyl, 3-pyridinyl

Scheme 2.52

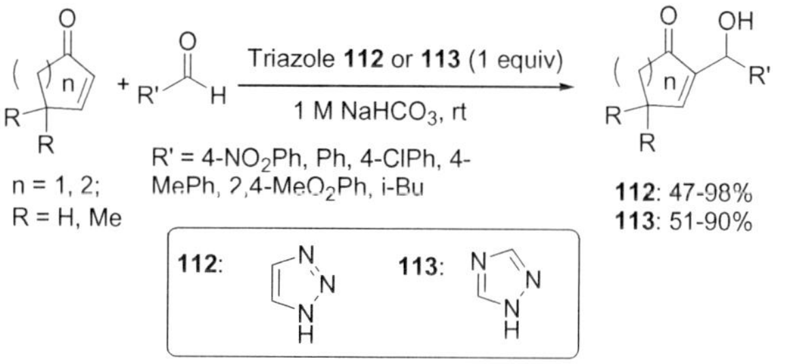

Scheme 2.53

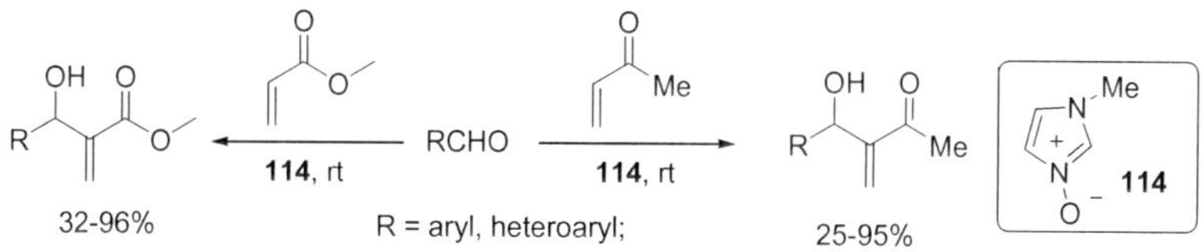

Scheme 2.54

and 1,2,4-triazole (**113**) promoted the MBH reaction smoothly; in most cases, the former showed better catalytic activity than the latter (Scheme 2.53).[60]

1-Methylimidazole 3-*N*-oxide (2 equiv) (**114**), derived from the oxidation of imidazole, was found to be a good promoter and can be used to replace strong Lewis bases for the MBH reaction of various activated aldehyde compounds in a non-solvent system, yielding α-(hydroxyalkyl)ketones or α-(hydroxyalkyl)acrylates in moderate to high yields (Scheme 2.54).[61]

N-Heterocyclic carbenes (NHCs) **115** have proved to be efficient catalysts for the aza-MBH reaction of cyclopent-2-en-1-one or cyclohex-2-en-1-one with various *N*-tosylarylimines to give the aza-MBH adduct in high yields (Scheme 2.55). Crossover experiments show the NHC can add to *N*-tosylarylimines in a reversible manner, which allows the addition of NHC to cyclic enones and thus catalyzes the aza-MBH reaction.[62]

Although the MBH reaction has been conducted in the presence of imidazolium-based ionic liquids to give the adducts in low yields due to the direct addition of deprotonated imidazolium salt to the aldehyde (see Chapter 1.7),[63] Cheng *et al.* have developed the imidazolium ionic liquid-bound quinuclidine (**116**) catalyzed-MBH reactions of aldehydes with various activated olefins, providing the corresponding adducts in moderate to high yields (Scheme 2.56). The optimal result was achieved when protic solvent methanol was used, which is similar to the accelerating effect in other Morita–Baylis–Hillman reactions using conventional catalysts.[64] To verify that the use of a hydroxyl containing ionic liquid alone may exert a similar accelerating effect, they further developed the hydroxyl ionic liquid (HIL)-bound quinuclidine **117** as a novel MBH catalyst. It showed better catalytic activity than other IL-immobilized catalysts that have no hydroxyl group attached to the IL scaffold, furnishing the

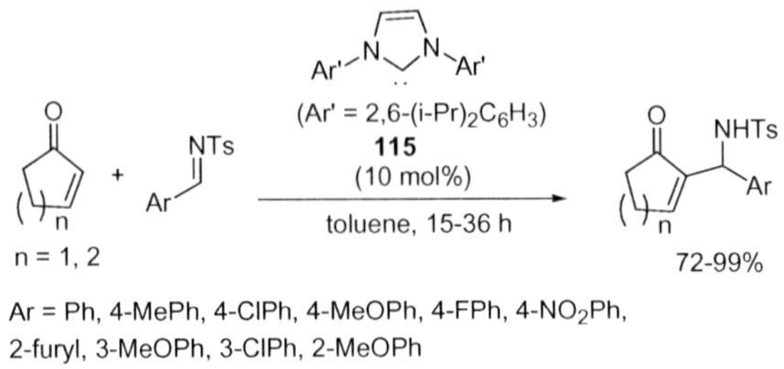

Scheme 2.55

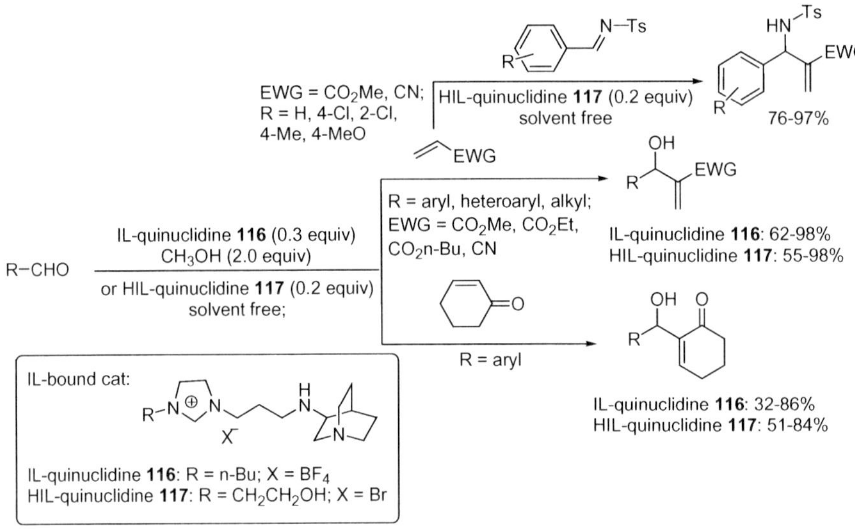

$(Ar' = 2,6-(i-Pr)_2C_6H_3)$
115
(10 mol%)
toluene, 15-36 h
n = 1, 2
72-99%

Ar = Ph, 4-MePh, 4-ClPh, 4-MeOPh, 4-FPh, 4-NO$_2$Ph,
2-furyl, 3-MeOPh, 3-ClPh, 2-MeOPh

Scheme 2.56

EWG = CO$_2$Me, CN;
R = H, 4-Cl, 2-Cl,
4-Me, 4-MeO

HIL-quinuclidine **117** (0.2 equiv)
solvent free
76-97%

R–CHO

IL-quinuclidine **116** (0.3 equiv)
CH$_3$OH (2.0 equiv)

or HIL-quinuclidine **117** (0.2 equiv)
solvent free;

R = aryl, heteroaryl, alkyl;
EWG = CO$_2$Me, CO$_2$Et,
CO$_2$n-Bu, CN

IL-quinuclidine **116**: 62-98%
HIL-quinuclidine **117**: 55-98%

R = aryl

IL-quinuclidine **116**: 32-86%
HIL-quinuclidine **117**: 51-84%

IL-bound cat:

IL-quinuclidine **116**: R = n-Bu; X = BF$_4$
HIL-quinuclidine **117**: R = CH$_2$CH$_2$OH; X = Br

corresponding adducts in good to excellent yields under solvent-free conditions.[65] The IL-supported quinuclidines **116** and **117** can be readily recovered and reused six times without significant loss of catalytic activity.[64,65]

Recently, a novel ionic catalyst, 1-butyl-4-aza-1-azoniabicyclo[2.2.2]octane chloride (**118**) based on 1,4-diazabicyclo[2.2.2]octane has been developed and applied in the MBH reaction, which occurred readily at room temperature to afford the corresponding adducts in good yields (Scheme 2.57). The ionic catalyst could be recycled for seven runs without diminution of its catalytic activity.[66]

2.2.2 Chiral Amine

2.2.2.1 Cinchona-derived Catalysts

As well as using a chiral auxiliary or enantiopure substrates, organic chemists have also directed their efforts at achieving an asymmetric version of the MBH

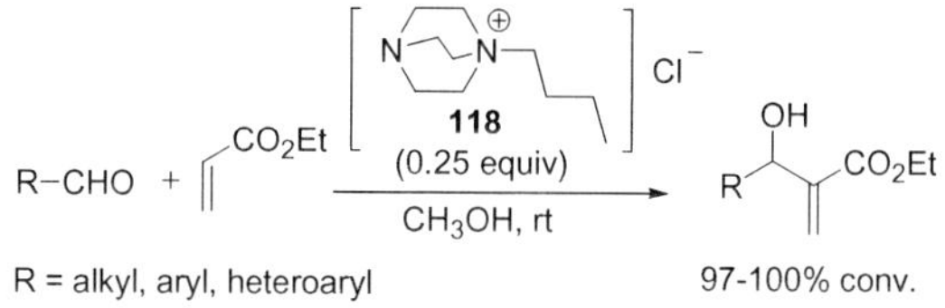

R–CHO + (CO$_2$Et) $\xrightarrow[\text{CH}_3\text{OH, rt}]{\substack{\textbf{118}\\(0.25\ \text{equiv})}}$ product

R = alkyl, aryl, heteroaryl

97-100% conv.

Scheme 2.57

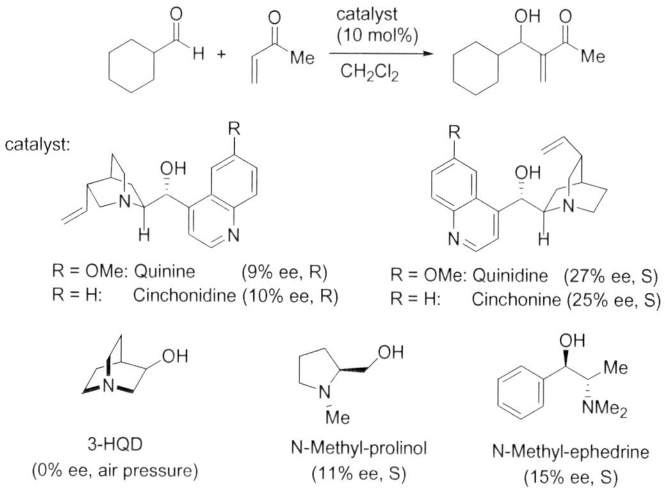

catalyst (10 mol%), CH$_2$Cl$_2$

catalyst:

R = OMe: Quinine (9% ee, R)
R = H: Cinchonidine (10% ee, R)

R = OMe: Quinidine (27% ee, S)
R = H: Cinchonine (25% ee, S)

3-HQD
(0% ee, air pressure)

N-Methyl-prolinol
(11% ee, S)

N-Methyl-ephedrine
(15% ee, S)

Scheme 2.58

reaction by using various chiral tertiary amine catalysts.[67] Owing to the importance of the proton donor capacity of the catalyst in the rate and selectivity of the MBH reaction, much attention has been paid to β-amino alcohol structures, such as the cinchona derivatives.[33,68,69] Markó *et al.*[68,70] first investigated the β-amino-alcohol catalyzed enantioselective MBH reaction of various aldehydes with MVK. Among the various β-hydroxy-amines screened, the cinchona alkaloids displayed the highest level of enantioselectivity, followed closely by ephedrine and proline derivatives for the MBH reaction of MVK and cyclohexyl carboxaldehyde (Scheme 2.58). As expected, quinine and cinchonidine gave lower but opposite enantioselectivities to quinidine and cinchonine. The crucial role played by the β-hydroxyl function was also noted, as the derivatization of quinidine into its *O*-acetyl analogue suppressed enantioselectivity. Notably, apart from the earlier discussed (*R*)-3-HDQ, which catalyzed the MBH reaction at atmospheric pressure (though with no enantioselectivity), other amino-alcohols all required high-pressure conditions. Moderate to good yields and up to 45% ee were obtained under the optimal conditions (Scheme 2.59).[33]

However, the most important progress in the development of cinchona derivative catalyzed asymmetric MBH reactions was the introduction of the

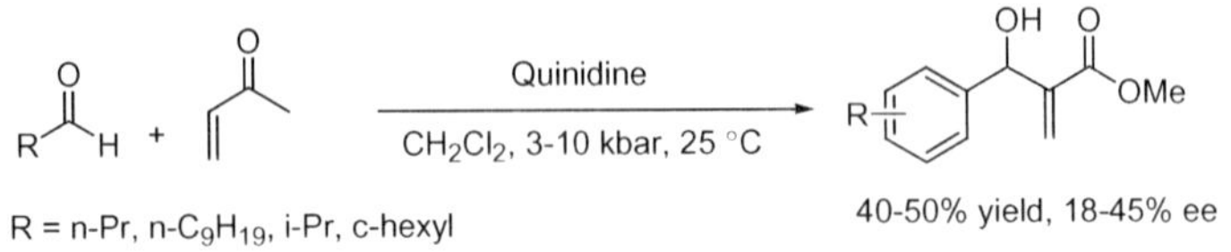

Scheme 2.59

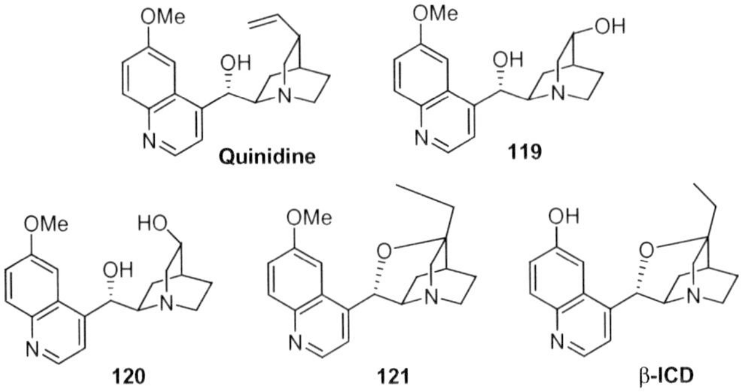

R = n-Pr, n-C$_9$H$_{19}$, i-Pr, c-hexyl

40-50% yield, 18-45% ee

Quinidine

119

120

121

β-ICD

Figure 2.2 Tertiary amines derived from cinchona alkaloids.

β-ICD (10 mol%)

DMF, -55 °C, 1-72 h

122

R = 4-NO$_2$Ph, Ph, trans-cinnamyl, Et,
i-Pr, i-Bu, c-Hex

(R)-**123**
31-58%, 91-99% ee

(S)-**124** [R = 4-NO$_2$Ph, (R)-]
0-25%, 4-85% ee

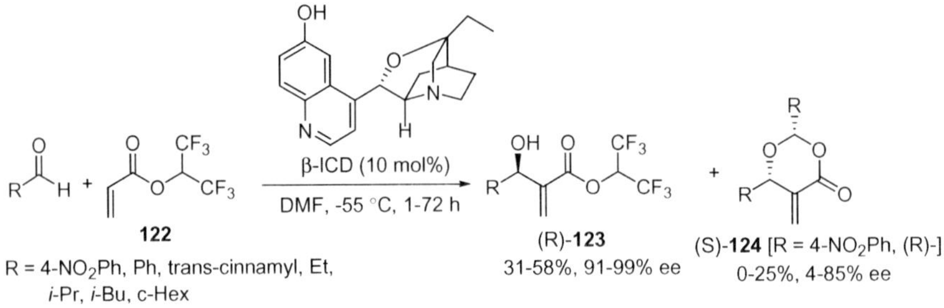

Scheme 2.60

quinidine-derived β-isocupreidine catalyst (β-ICD) by Hatakeyama's group. Hatakeyama and co-workers[71] investigated the effect of various tertiary amines (Figure 2.2) derived from cinchona alkaloids for enantioselective MBH reaction. β-ICD, derived from quinidine, was found to be the best catalyst for the MBH reaction between 1,1,1,3,3,3-hexafluoroisopropyl acrylate (**122**) and various aldehydes, providing the desired adducts **123** in good yields (31–58%) with excellent enantioselectivity (ee up to 99%) at − 55 °C in DMF (Scheme 2.60). However, limitations were observed with bulky aldehydes such as butyraldehyde, where dimerization of the acrylate occurred.

The fact that the enantioselectivity of the by-product dioxanone **124** was found to be considerably lower, or even of opposite configuration to the

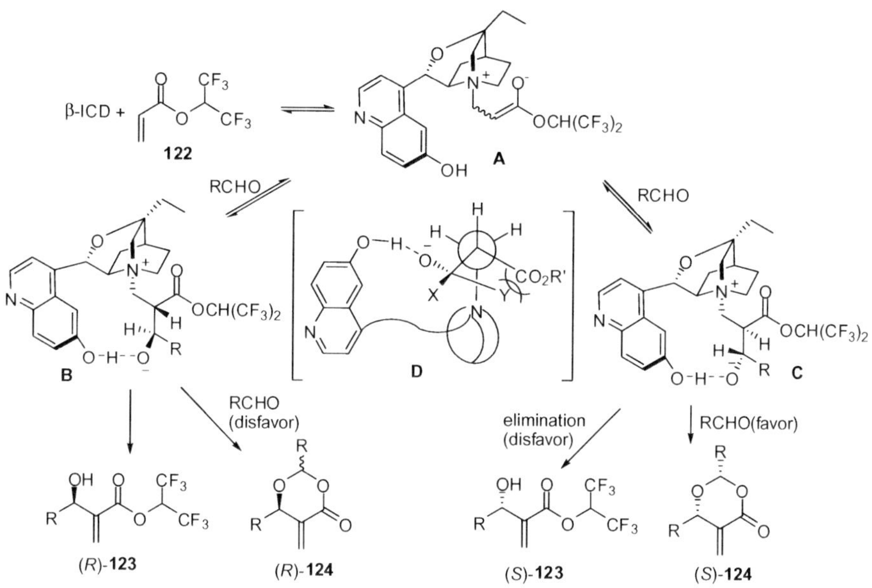

Scheme 2.61

"normal" adduct **123**, can be rationalized by the different mechanism of the stereo-determining proton abstraction steps of these products, as outlined in Scheme 2.61. Michael addition of β-ICD to acrylate **122** forms enolate **A**, which in turn undergoes an aldol reaction with an aldehyde to furnish two betaine intermediates, **B** and **C**, stabilized by intramolecular hydrogen bonding between the oxy anion and the phenolic OH. The conformations of **B** and **C** are nearly ideal for the subsequent *E*2 or *E*1cb reaction process for stereoelectronic reasons, as depicted in Newman projection **D**. However, intermediate **C** suffers severe steric interactions between Y and the ester and quinuclidine moieties [see **D** (X = H, Y = substituent)], and thus it undergoes reaction with a second aldehyde molecule rather than elimination to form dioxanone **124**. In contrast, intermediate **B** undergoes facile elimination to produce (*R*)-adduct **123** with regeneration of the catalyst because of less steric hindrance [see **D** (X = substituent, Y = H)]. The irregular ee values observed for dioxanones **124** can be explained on the basis of the reactivity of the starting aldehyde. Thus, as the reactivity of the aldehyde increases, the rate of formation of (*R*)-dioxanone **124** from intermediate **B** also increases in competition with elimination, which gives (*R*)-adduct **124**, leading to a decrease of the (*S*)-selectivity of **124**.

Subsequently, Hatakeyama *et al.*[72] successfully employed this methodology for the synthesis of the important biologically active molecules epopromycin B (**125**), 2-*epi*-epopromycin B (**126**) and (–)-mycestericin E (**127**), which involved a β-ICD-catalyzed asymmetric MBH reaction as the key step (Scheme 2.62).

In a continuation of his work, Hatakeyama *et al.* have synthesized a series of β-ICD-congeners (**128–136**, Figure 2.3) and examined their catalytic abilities in

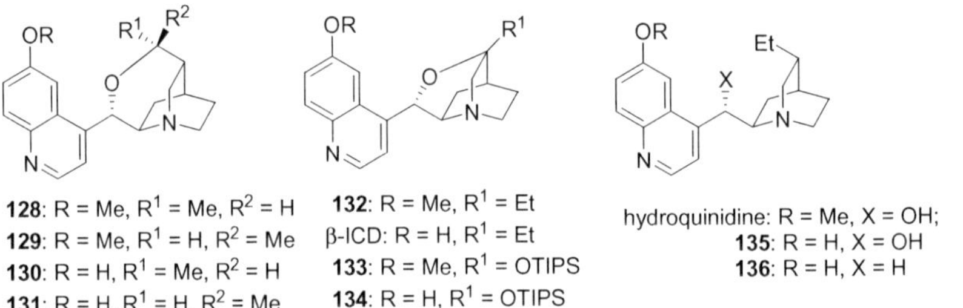

Scheme 2.62

Figure 2.3 β-ICD-congeners.

the MBH reaction of aldehydes with HFIPA (1,1,1,3,3,3-hexafluoroisopropyl acrylate) and its related fluorine-containing acrylates.[73] They proved that the cage-like tricyclic structure and the phenolic hydroxy of β-ICD as well as the branched structure of HFIPA are necessary to obtain a high level of asymmetric induction and rate acceleration (Schemes 2.63 and 2.64). This fact further supported the aforementioned mechanism in which intermediate **B** stabilized by hydrogen bonding would be responsible for the highly enantioselective production of (*R*)-enriched adducts (Scheme 2.61). In addition, they found that compound **134** was also able to serve as a chiral catalyst for asymmetric MBH reactions, suggesting that the C3 substituent on the quinuclidine ring exerts little effect on catalytic ability (Scheme 2.65). Notably, azeotropically dried β-ICD displayed remarkable catalytic ability. By this improved β-ICD–HFIPA method, aromatic aldehydes, except for very reactive *p*-nitrobenzaldehyde, can be converted into the corresponding MBH

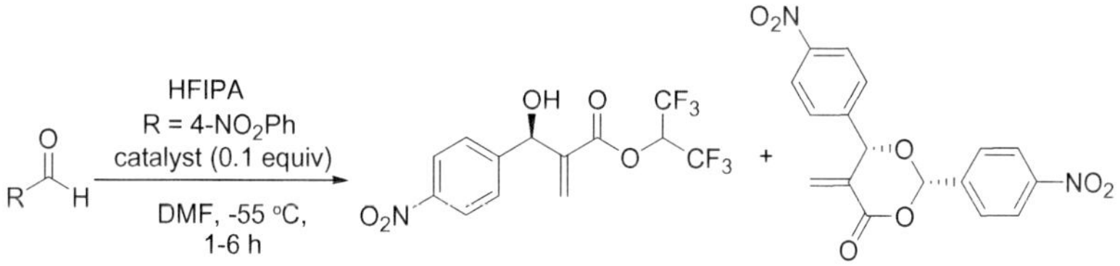

HFIPA
R = 4-NO$_2$Ph
catalyst (0.1 equiv)

DMF, -55 °C,
1-6 h

128: 6 h,	18%, nd.	29%, 45% ee
129: 6 h,	24%, 36% ee	32%, 32% ee
130: 1 h,	59%, 89% ee	19%, 38% ee
131: 1 h,	51%, 89% ee	18%, 26% ee
132: 1 h,	74%, 10% ee	7%, n.d.
133: 6 h,	2%, 3% ee	9%, 11% ee
134: 7 h,	**41%, 93% ee**	**20%, 51% ee**
135: 3 h,	0	26%, 4% ee
136: 3 h,	0	27%, 2% ee
β-ICD: 1 h	**58%, 91% ee**	**11%, 4% ee**

Scheme 2.63

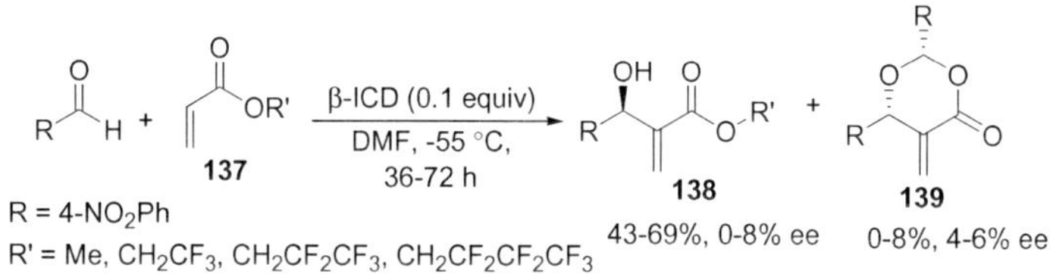

β-ICD (0.1 equiv)
DMF, -55 °C,
36-72 h

137

R = 4-NO$_2$Ph
R' = Me, CH$_2$CF$_3$, CH$_2$CF$_2$CF$_3$, CH$_2$CF$_2$CF$_2$CF$_3$

138
43-69%, 0-8% ee

139
0-8%, 4-6% ee

Scheme 2.64

134 (0.1 equiv)
DMF, -55 °C,
7-61 h

123
15-71%, 92-100% ee

124
0-21%, 17-61% ee

R = 4-NO$_2$Ph, Ph, trans-cinnamyl, PhCH$_2$CH$_2$, c-hex

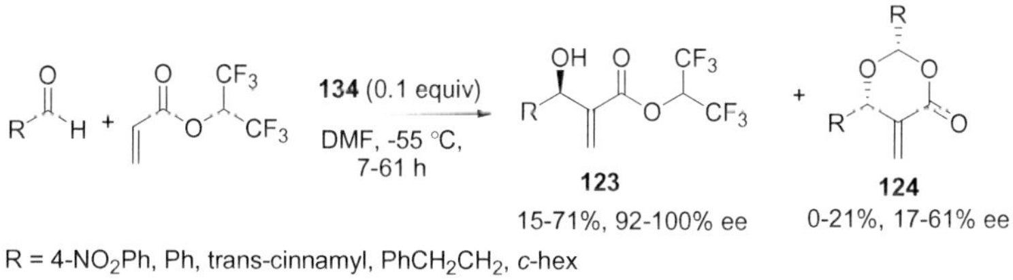

Scheme 2.65

adducts in >94% ee without concomitant formation of undesired dioxanones (Scheme 2.66).[73]

On the basis of this work, Hatakeyama *et al.* have investigated the β-ICD-catalyzed MBH reaction of chiral *N*-Boc-α-amino aldehydes **140** and HFIPA (Scheme 2.67). The reaction proceeded smoothly without racemization and exhibited the match–mismatch relationship between the substrate and the catalyst, which was similar to that of the synthesis of epopromycin B from *N*-Fmoc-L-leucinal with HFIPA.[72a] In the case of acyclic amino aldehydes,

undried: 1-72 h,   21-58%, 91-100% ee   11-29%, 4-76% ee
dried: 1.5-120 h,   23-82%, 94-99% ee   17-22%, rac to 65% ee

R = 4-NO$_2$Ph, Ph, trans-cinnamyl, 4-MeOPh, 1-naphthyl, 2-naphthyl, PhCH$_2$CH$_2$, *c*-Hex

Scheme 2.66

140

R = Me$_2$CHCH$_2$, Me$_2$CH, Me; R' = H;
R, R' = (CH$_2$)$_3$, CH$_2$OCMe$_2$

acyclic: L (matched): 63-77% (syn) [>98% ee]
          D (mismatched): 10-45% (syn:anti=0:100-5:95)
                              [--, nd to >98% ee]
cyclic: D (matched): 67-73% (anti) [>98% ee]
          L (mismatched): 10-11% (syn:anti=100:0-94:6)
                              [>98% ee, --]

Scheme 2.67

17-71%, 33-92% ee

R = Ph, 4-ClPh, 4-BrPh,
4-NO$_2$Ph, PhCH$_2$CH$_2$

β–ICD (0.1 equiv)
THF, 24-120 h
-20 °C-rt

β–ICD (0.1 equiv)
DMF or THF,
36-120 h, additive

no addititve: 24-78%, 7-49% ee
LiOTf or D-Proline: 31-88%,
                          26-49% ee

R = 4-ClPh, 4-BrPh, 4-NO$_2$Ph, PhCH$_2$CH$_2$

Scheme 2.68

L-substrates show excellent *syn* selectivity and high reactivity in contrast to D-substrates. In contrast, in the case of cyclic amino aldehydes, D-substrates rather than L-substrates show excellent *anti* selectivity and high reactivity.[74]

We have developed the β-ICD-mediated asymmetric MBH reaction of activated aromatic aldehydes with MVK.[75] Higher temperatures and longer reaction times were required due to the lower reactivity of MVK derivatives than HFIPA, resulting in allyl alcohols in poor to good enantioselectivities (Scheme 2.68). Because this catalytic asymmetric reaction is highly dependent on the substrates, including the aldehydes and the Michael acceptors, the use of

Brønsted or Lewis acid additives increases the reaction rate, yield and selectivity in some cases. The highest ee achieved is 49% with (R) configuration at the newly formed stereogenic centre. For α-naphthyl acrylate as substrate, the β-ICD catalyzed asymmetric MBH reaction can be achieved with 92% ee.

β-ICD is also an efficient and remarkably general catalyst in aza-MBH reactions, which has promoted the addition of various electron-deficient olefins such as acrylates, enones, enals and acrylonitrile to activated aromatic aldimines **142**.[76,77] We first developed the highly enantioselective aza-MBH reaction of aromatic aldimines with MVK/methyl acrylate/acrylonitrile by using β-ICD as catalyst, providing thus by far the highest ee values for the aza-MBH reaction involving MVK or methyl acrylate as Michael acceptor (Scheme 2.69).[77] Concerning aliphatic imines, however, satisfactory results were not obtained because of their extremely labile nature. Complicated unidentified products rather than normal MBH adducts were obtained under the above conditions.

At almost the same time, Hatakeyama *et al.* investigated the β-ICD catalyzed aza-MBH reaction of various aryl diphenylphosphinoyl imines **146** and 1,1,1,3,3,3-hexafluoroisopropyl acrylate (HFIPA), affording the (S)-product **147**, in up to 97% yields with high ees, in DMF at low temperature (Scheme 2.70).[76] Notably, the absolute stereochemistry of the aza-MBH adducts is opposite to that of adducts obtained from the analogous MBH reaction. With β-ICD as catalyst, imines give rise to (S)-enriched adducts, in contrast to aldehydes, which afford (R)-products. A plausible mechanism was proposed. Moreover, the utility

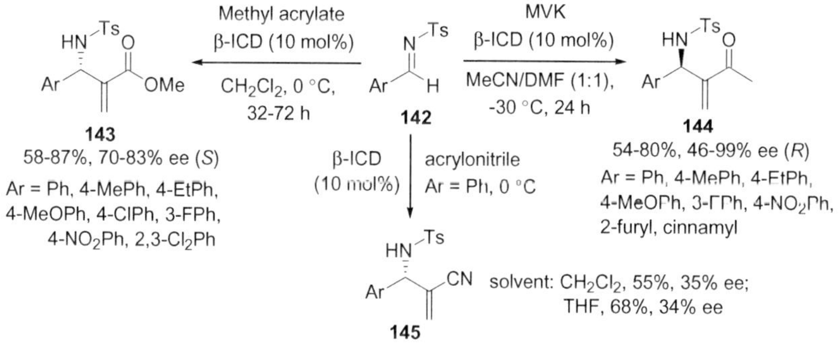

Scheme 2.69

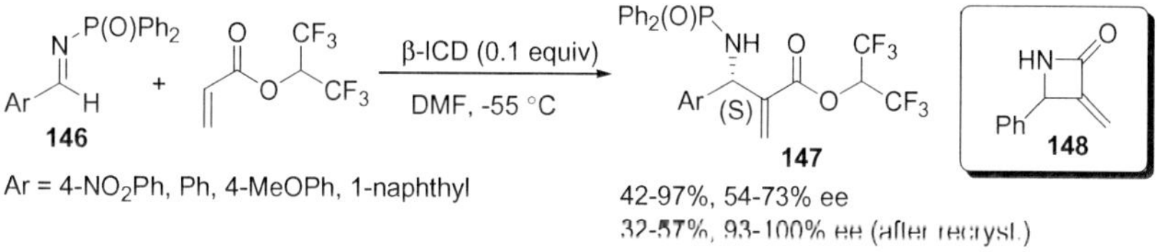

Scheme 2.70

of this methodology was further exemplified for the synthesis of β-lactam **148** by a sequence of transformations.

Because a different stereochemistry for the aza-MBH reaction involving different Michael acceptors was observed, in a continuation of our work, we reinvestigated systematically the reaction of *N*-sulfonated imines with various activated olefins, including ethyl vinyl ketone (EVK), acrolein, phenyl acrylate and α-naphthyl acrylate. An interesting inversion of absolute configuration between the adducts derived from MVK or EVK and those from acrolein, methyl acrylate, phenyl acrylate or α-naphthyl acrylate was observed, indicating that the substitution patterns of the olefin may alter or even invert this trend.[77b] Similar to the addition to HFIPA, the β-ICD-mediated addition of methyl, phenyl and naphthyl acrylates **149** to *N*-sulfonyl imines afforded adducts **150** with an (*S*) configuration, which is opposite to that observed with aldehydes (Scheme 2.71).[77]

The aza-MBH reaction of *N*-tosyl arylaldimines with acrolein or acrylonitrile catalyzed by β-ICD afforded invariably the (*S*)-enriched adducts **145** and **151**, respectively. Acrylonitrile is less reactive than acrolein, and its reaction required a higher temperature and afforded products (**145**) in lower chemical yields and ee (Scheme 2.72).[77]

Notably, the β-ICD mediated aza-MBH reaction of *N*-tosyl arylaldimines with ethyl or methyl vinyl ketones afforded products with (*R*)-absolute configuration in MeCN–DMF (1 : 1) mixtures at low temperature (–30 °C), which

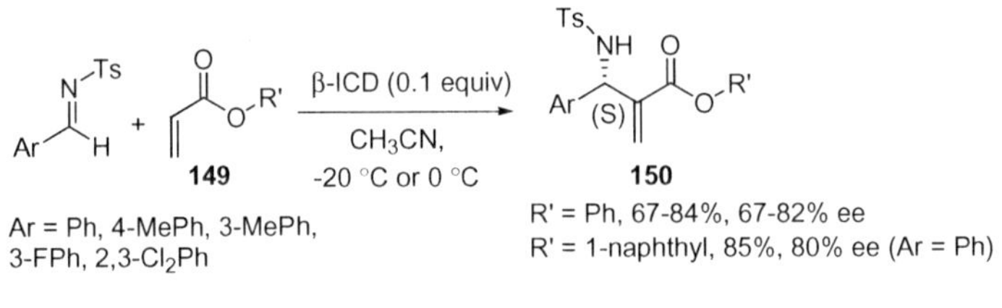

Scheme 2.71

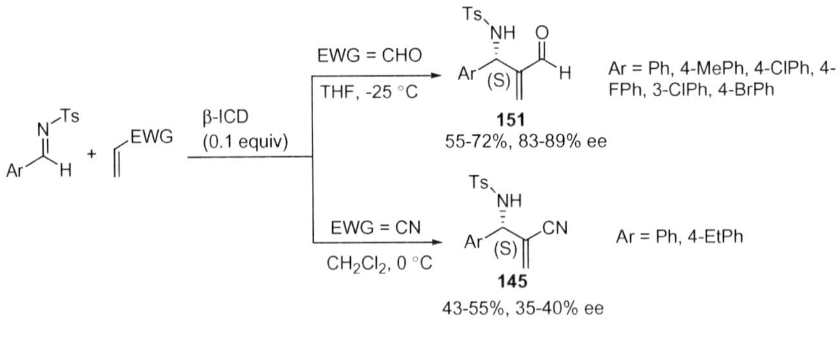

Scheme 2.72

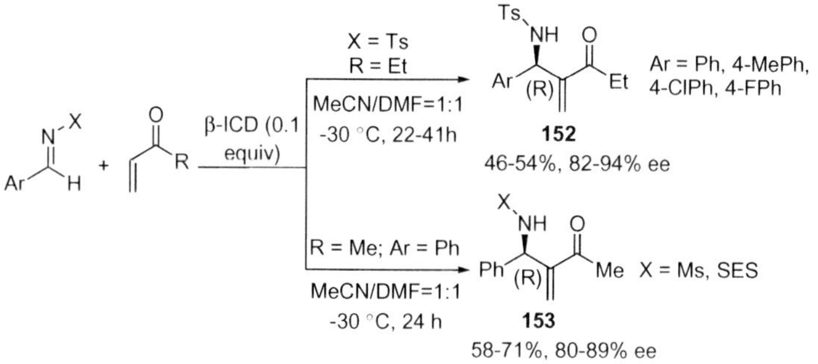

Scheme 2.73

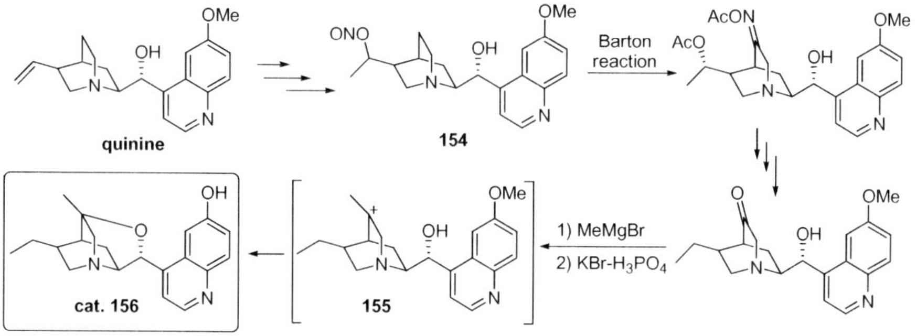

Scheme 2.74

is opposite to that observed in the related aza-MBH reaction with acrylonitrile, acrolein or acrylates.[77] *N*-Mesyl or *N*-SES-protected imine afforded similar results (Scheme 2.73).

As mentioned above, the β-ICD catalyzed MBH reaction has attracted considerable interest due to its fascinating high efficiency. However, one serious drawback of this method is that it cannot be applied to the synthesis of products with the opposite absolute configuration because the required enantiomer of β-ICD is not easily available. As one solution to this problem, Hatakeyama *et al.*[78] have investigated the synthesis of catalyst **156**, a pseudo-enantiomer of iso-cupreidine (β-ICD), from quinine by employing a Barton reaction of nitrosyl ester **154** and acid-catalyzed cyclization of carbinol **155** as key steps (Scheme 2.74). Using **156** as a chiral amine catalyst, the MBH reaction of various aldehydes with HFIPA afforded the corresponding (*S*)-enriched adducts in high optical purity (>91% ee) in contrast to the β-ICD-catalyzed reaction, which affords (*R*)-enriched adducts, indicating that catalyst **156** can serve as an enantiocomplementary catalyst of β-ICD in the asymmetric MBH reaction (Scheme 2.75).

To resolve poor substrates, such as acrylates and aliphatic imines in the aza-MBH reaction, Masson and Zhu *et al.* have developed a novel β-ICD-amide

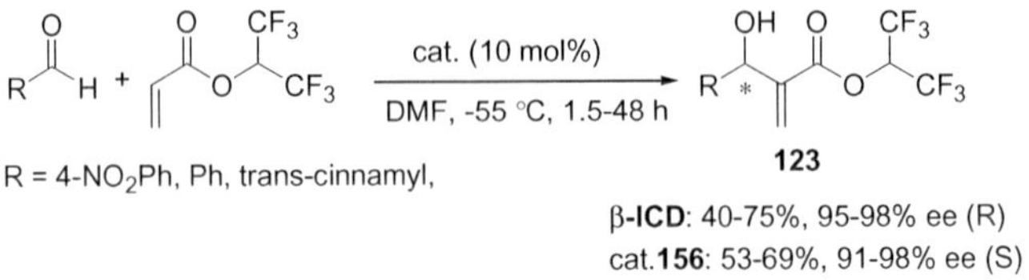

R = 4-NO$_2$Ph, Ph, trans-cinnamyl,

123

β-**ICD**: 40-75%, 95-98% ee (R)
cat.**156**: 53-69%, 91-98% ee (S)

Scheme 2.75

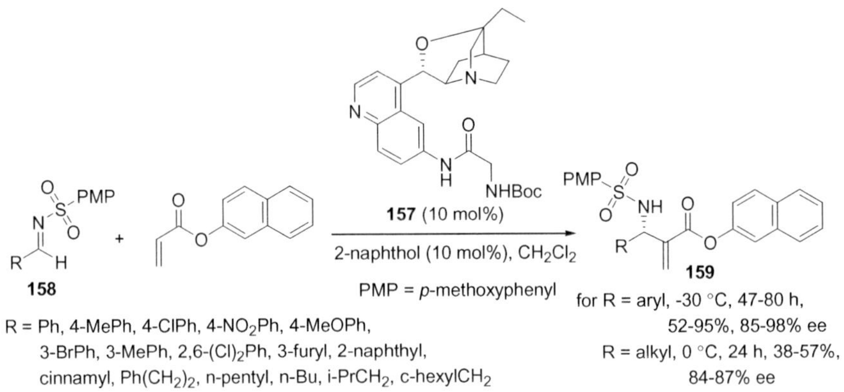

158

157 (10 mol%)

2-naphthol (10 mol%), CH$_2$Cl$_2$

PMP = *p*-methoxyphenyl

159

R = Ph, 4-MePh, 4-ClPh, 4-NO$_2$Ph, 4-MeOPh,
3-BrPh, 3-MePh, 2,6-(Cl)$_2$Ph, 3-furyl, 2-naphthyl,
cinnamyl, Ph(CH$_2$)$_2$, n-pentyl, n-Bu, i-PrCH$_2$, c-hexylCH$_2$

for R = aryl, -30 °C, 47-80 h,
52-95%, 85-98% ee
R = alkyl, 0 °C, 24 h, 38-57%,
84-87% ee

Scheme 2.76

TS-1

TS-2

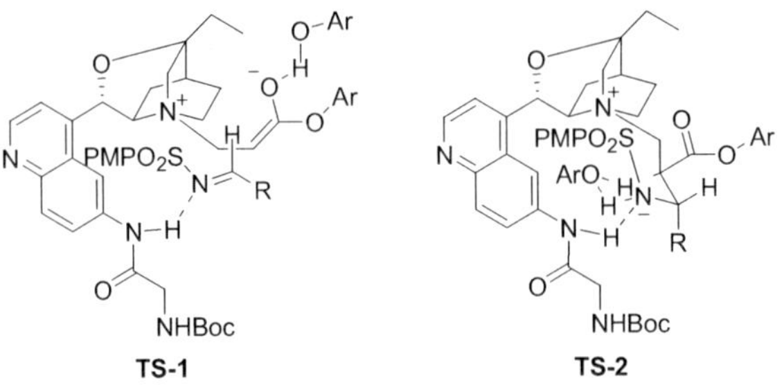

Figure 2.4 Transition states in an aza-MBH reaction leading to adducts **159** (Scheme 2.76).

bifunctional catalyst (**157**), which in combination with β-naphthol served as a highly effective dual catalyst for the asymmetric aza-MBH reactions.[79] High yields and enantioselectivity were uniformly observed in the case of aromatic imines. In addition, the aliphatic *N*-sulfinyl imines were successfully employed for the first time in the aza-MBH reaction, leading to the corresponding adducts **159** in over 84% ee (Scheme 2.76). It was assumed that the pairing of cooperative H-bonds is important and the nucleophilic addition of the (*Z*)-enolate onto the *re*-face of the (*E*)-imine *via* the less crowded transition state **TS-1** was held to account for the observed (*S*)-enantioselectivity in the adduct **159** (Figure 2.4).

On the basis of the above mechanistic assumption, the authors assumed that the β-ICD-amide–β-naphthol dual catalytic system should favor the (*S*)-aza-MBH product regardless of the nature of the Michael acceptors used and investigated the reaction between *N*-tosylimine **158** and alkyl vinyl ketone, which is known to provide the (*R*)-aza-MBH adduct. They developed a new β-ICD-amide (**160**) and found that an achiral protic additive was capable of inverting the β-ICD and β-ICD-amide (**160**) catalyzed enantioselective aza-MBH reaction between *N*-sulfonylimines and MVK/EVK, therefore providing another solution to the enantio-complementarity associated with this family of catalysts (Scheme 2.77).[80]

Although dimeric Sharpless ligands, as another kind of cinchona catalyst, showed impressive results in related organocatalytic transformations, they provided only limited success in asymmetric MBH reactions (Scheme 2.78).[81] These compounds can act as bifunctional catalysts in the presence of acid

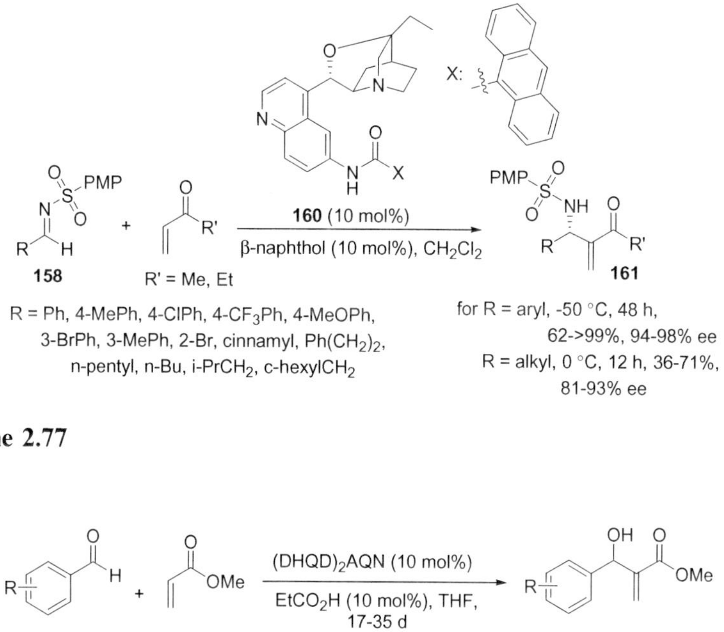

Scheme 2.77

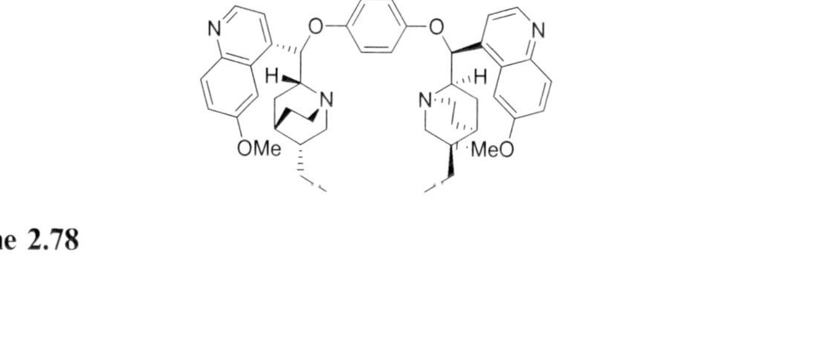

Scheme 2.78

additives; however, no satisfactory results have been obtained for the asymmetric MBH reaction. The (DHQD)$_2$AQN-catalyzed MBH reaction between methyl acrylate and electron-deficient aromatic aldehydes gave the adducts in low yields and lacked substrate generality, even though moderate ees up to 77% could be achieved. However, when (DHQD)$_2$PYR or (DHQD)$_2$PHAL was employed to mediate the same transformation only trace amounts of products were obtained. Notably, without acid, the reaction afforded the opposite enantiomer in a slow conversion.

2.2.2.2   *Unnatural Tertiary Amines*

To realize an efficient catalytic asymmetric MBH reaction, wherein a high level of asymmetric induction as well as desired rate acceleration is obtained, the appropriate combination of chiral amine catalyst and suitably activated alkene is required. Besides cinchona derivative catalysts, more chiral catalysts have been developed from achiral molecules such as DABCO, quinuclidine, indolizine or pyrrolizine-derived catalysts by introducing asymmetric functions. Hirama and co-workers have examined chiral C_2-symmetric 2,3-disubstituted 1,4-diazabicyclo[2.2.2]octanes such as 2,3-(dibenzoxymethyl)-DABCO (**162**) as a catalyst for the asymmetric MBH reaction between 4-nitrobenzaldehyde and MVK (Scheme 2.79).[82] Compared with achiral DABCO, chiral catalyst **162** showed diminished reactivity and the reaction required high pressure to ensure a reasonable reaction rate. The corresponding adduct was obtained in up to 47% ee.

Owing to aforementioned observation that an OH group suitably disposed on an amine catalyst exerts a marked effect on rate acceleration as well as asymmetric induction *via* stabilizing the oxy anion intermediate through hydrogen bonding,[33,68,70] a series of hydroxylated amines have been surveyed in the MBH reaction. Barrett *et al.*[83] have described an interesting enantiopure pyrrolizidine (**163**) derived from L-proline, which mediated the MBH reaction of alkyl vinyl ketones (MVK and EVK) with electron-deficient aromatic aldehydes to provide the desired adducts in good yield, albeit with modest enantioselectivity (21–72% ee) (Scheme 2.80). Notably, the addition of coordinating salts such as NaBF$_4$ or NaBPh$_4$ increased the selectivity of the MBH reaction. The salt effect was rationalized by the formation of two possible chelate intermediates, of which the intermediate **164** rather than the more

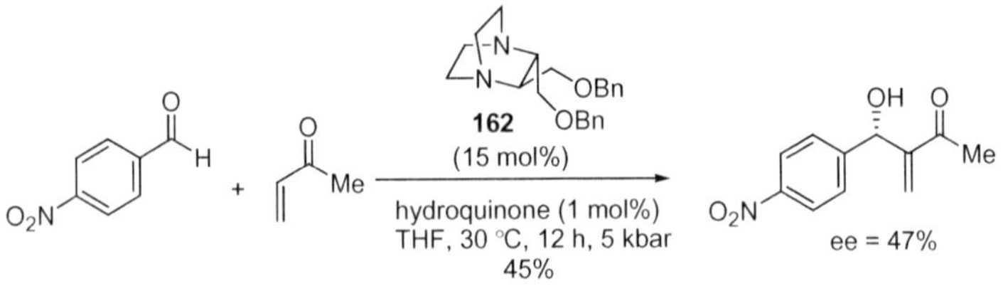

Scheme 2.79

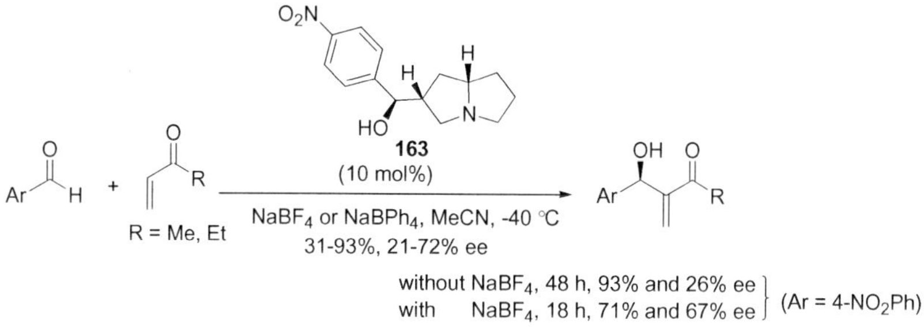

Scheme 2.80

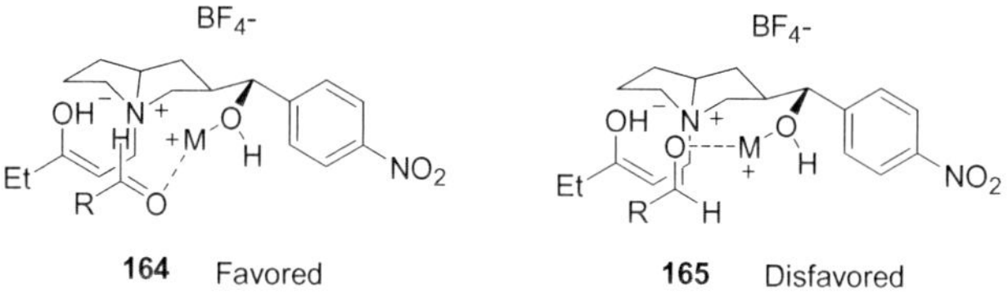

Figure 2.5 Two possible chelate intermediates arising through the salt effect on the reaction shown in Scheme 2.80.

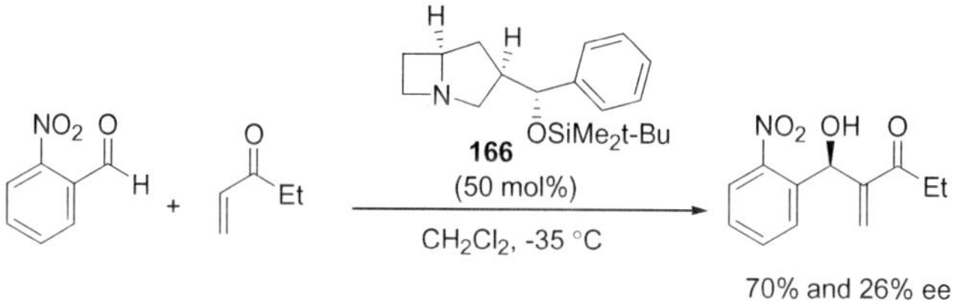

Scheme 2.81

sterically congested system **165** was preferred for steric reasons, leading to the (*R*)-product (Figure 2.5).

Subsequently, the same authors found that the structural analogue bicyclic azetidine derivative **166**, having a silylated alcohol on the lateral chain, showed increased reactivity in a similar transformation, albeit with low enantioselectivity (ee-value up to 26%), indicating that azabicyclo[3.2.0]heptane is more nucleophilic than azabicyclo[3.3.0]octane since the same MBH reaction with catalyst **163** was significantly slower (Scheme 2.81). This enhancement in catalytic activity can probably be attributed to increased pyramidalization of the nitrogen atom imposed by the four-membered ring.[84]

Similar observations were made in the *N*-methylprolinol (**167**) mediated addition of aromatic aldehydes to activated alkenes, which proceeded under atmospheric pressure in a 1,4-dioxane–water (1 : 1, v/v) solvent mixture at 0 °C,

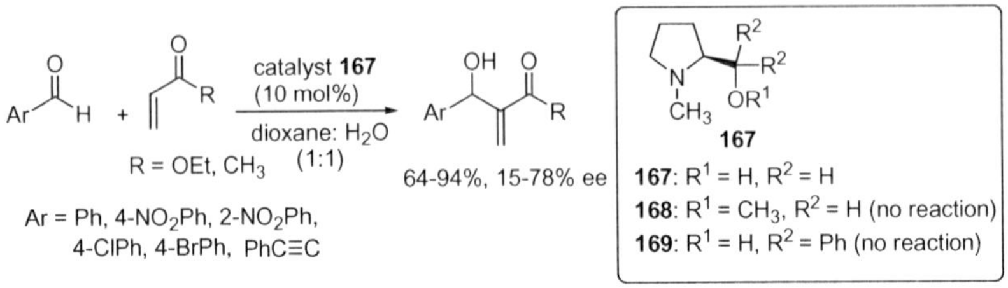

Scheme 2.82

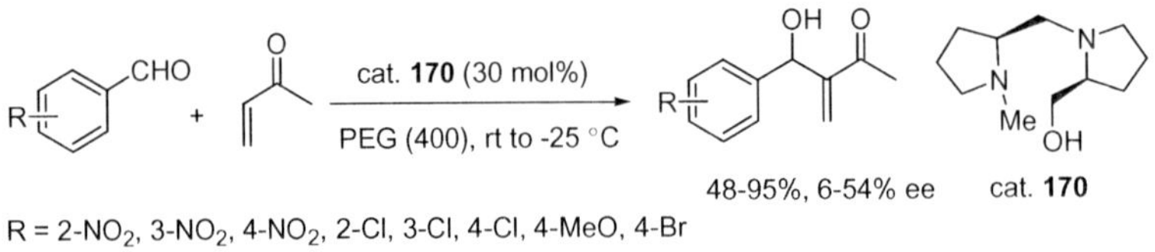

Scheme 2.83

and afforded product in good yields with moderate to good (*R*)-enantioselectivities (Scheme 2.82).[85] The catalytic activity was partially enhanced when the reaction was conducted at –10 °C in the presence of ethylene glycol, and was suppressed by methylating the free hydroxyl function (**168**) and by a bulky substituted group at the α-position (**169**). Correspondingly, the (*S*)-MBH adducts could be obtained by using *N*-methylprolinol derived from commercially available (*R*)-proline, thus providing scope for the facile generation of both enantiomers.

Another kind of chiral amino alcohol (**170**), derived from L-proline, has been synthesized and applied to the enantioselective MBH reaction of various aromatic aldehydes with MVK, affording the corresponding adducts in high yields with modest selectivities in poly(ethylene glycol) (PEG) 400 (Scheme 2.83). Notably, this catalytic system could be easily recovered and reused in three runs without significant changes in yields and selectivities.[86]

As described above, most amine catalysts for the highly asymmetric MBH reaction possess two functionalities: a basic amine moiety, such as a nucleophilic amine, and an acidic moiety such as a hydroxy or phenoxy group. However, Hayashi *et al.* were the first to demonstrate that chiral diamine **171** possessing two basic amine moieties without an acidic moiety can promote the MBH reaction of methyl vinyl ketone and various aldehydes enantioselectively, affording adducts in good yield with moderate to good enantioselectivity (up to 75% ee) (Scheme 2.84).[87] Although more experiments are needed to clarify the reaction mechanism, there is room for improvement in terms of enantioselectivity. The finding showed that not only amino alcohols but also diamines can promote the MBH reaction, which opened the way for the design of new asymmetric organic catalysts for this reaction.

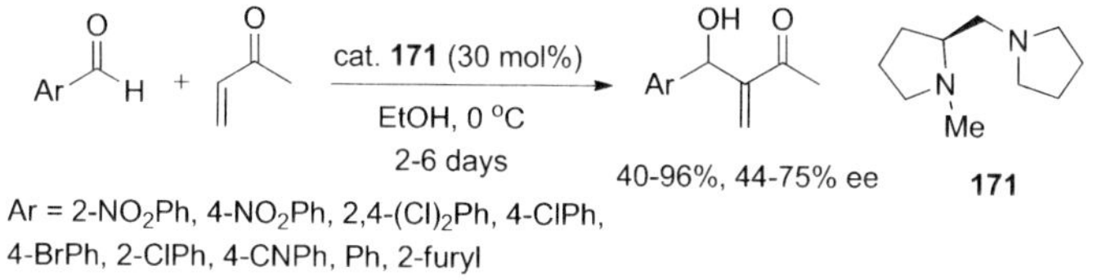

Scheme 2.84

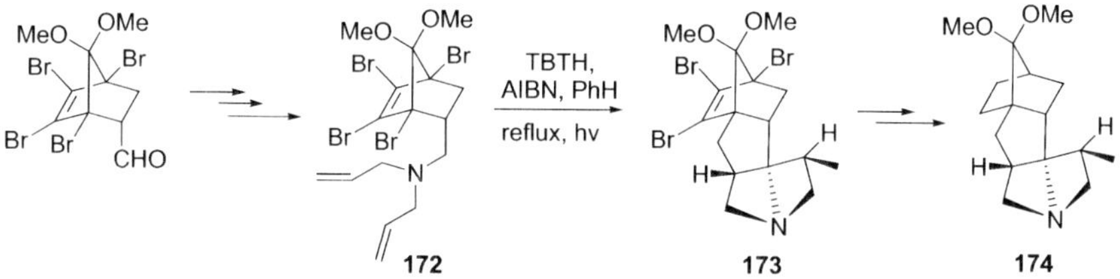

Scheme 2.85

Recently, a practical and expedient synthesis of racemic as well as optically pure antipodes of tetracyclic amines **174** was developed by Khan *et al*,[88] involving a stereoselective C7nC5x free-radical cascade protocol[89] from bis-allyl amine **172** starting material as key step (Scheme 2.85). Using 20 mol.% of the optically pure amine **174** along with *p*-nitrobenzaldehyde and methyl acrylate in MeOH under sonication conditions afforded the corresponding adduct in comparable yield,[11] albeit with the low asymmetric induction (8% ee).

Inspired by the use of chiral imidazolidinones as highly enantioselective catalysts for Diels–Alder, 1,3-dipolar cycloaddition and Friedel–Crafts reactions, Tan *et al*.[90] have synthesized a series of novel chiral imidazolines and examined their application in MBH reactions. Up to 54% ee and high yields were obtained by using stoichiometric amounts of imidazoline **175** for the MBH reactions of various aromatic aldehydes with unactivated acrylates. Furthermore, the imidazolines were also suitable promoters for reactions between aromatic aldehydes and alkyl vinyl ketone. Using 50 mol.% of imidazoline **176**, which bears a chiral methylnaphthyl group, afforded adducts in high yield with up to 78% ee (Scheme 2.86). These chiral imidazolines are readily prepared from commercially available amino alcohols and can be easily recovered for reuse without loss of product enantioselectivity.

During their studies on kinetic resolution (KR) of secondary alcohols, Connon *et al*. found that chiral pyridine catalyst **177** and its optimized analogue **178** promoted the synthetically useful KR of MBH adducts **179** derived from deactivated precursors (which were difficult to synthesize using catalytic asymmetric MBH reactions), allowing the convenient preparation of **179** in 62–90% ee and 82–97% ee, respectively (Scheme 2.87). This study also represents the first examples of effective non-enzymatic acylative KR of sec-sp^2-sp^2

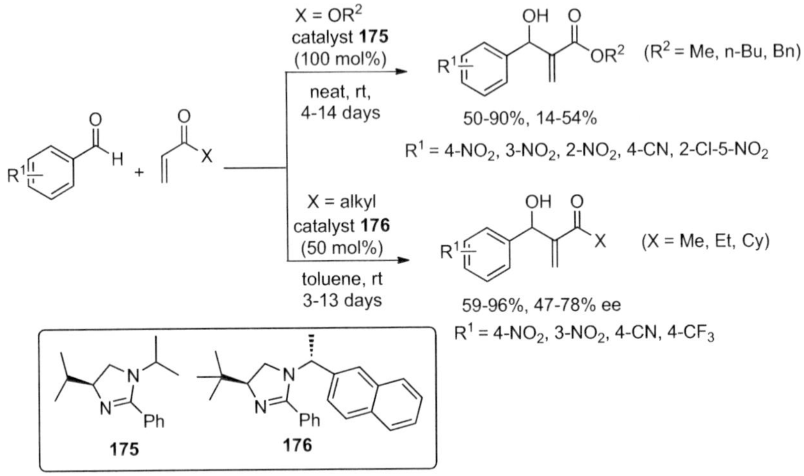

Scheme 2.86

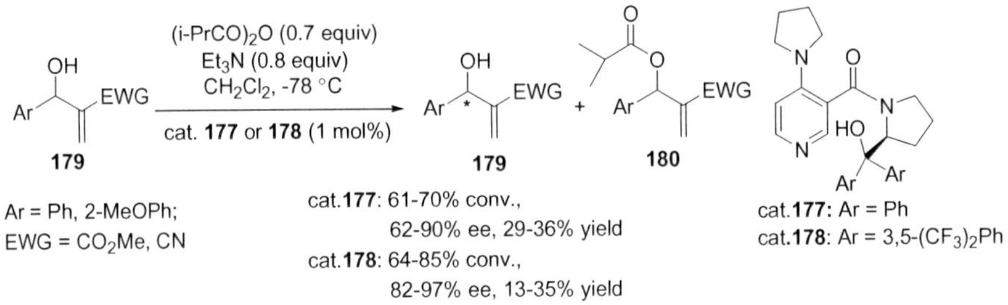

Scheme 2.87

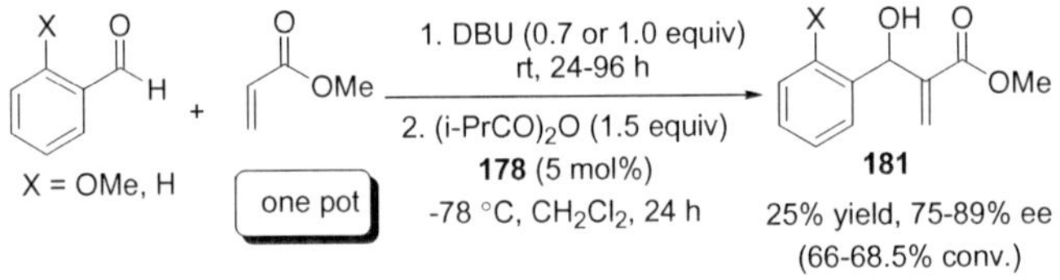

Scheme 2.88

carbinols. A novel one-pot synthesis–kinetic resolution process involving a DBU-catalyzed MBH reaction and subsequent **178**/DBU-mediated enantio-selective acylation has also been developed. Using this strategy, chiral MBH adduct **181** can be readily prepared in appreciable yield from the corresponding aldehyde and methyl acrylate precursors with high levels of enantiomeric excess in a convenient one-pot process (Scheme 2.88).[91]

2.2.2.3 Non-natural Tertiary Amine/Thiourea (Phenol) Catalysts

As an important kind of bifunctional catalyst, amine-thiourea (phenol) catalysts have received much attention in the asymmetric MBH reaction. On the basis of thiourea-based organocatalysts that have been widely used for effective activation of carbonyl groups, imines and nitro groups through efficient double hydrogen-bonding interactions, Wang *et al.* synthesized various amino-thiourea bifunctional catalysts (**182–185**) based on two well-studied chiral scaffolds, namely, *trans*-cyclohexane diamine and binaphthyl diamine, and examined their catalytic activity in the MBH reaction (Figure 2.6).[92] Among the catalysts screened, catalyst **183** has been demonstrated to promote the asymmetric MBH reactions of cyclohexenone with various aldehydes to afford highly functionalized, synthetically useful chiral allylic alcohols. Regardless of the length of linear aliphatic aldehydes, in every case, high enantioselectivities (80–83% ee) and good to high yields (71–84%) were achieved. More significantly, the more sterically demanding aldehydes gave the corresponding adducts in good yields (63–71%) with excellent enantioselectivities (90–94% ee). Aromatic aldehyde gave products in lower yield (55%) and lower ee (60%) (Scheme 2.89). However, cyclohexanediamine-derived amine thiourea **182**, which provided high enantioselectivities for the Michael addition[93] and aza-Henry reactions[94], showed poor activity in the MBH reaction.

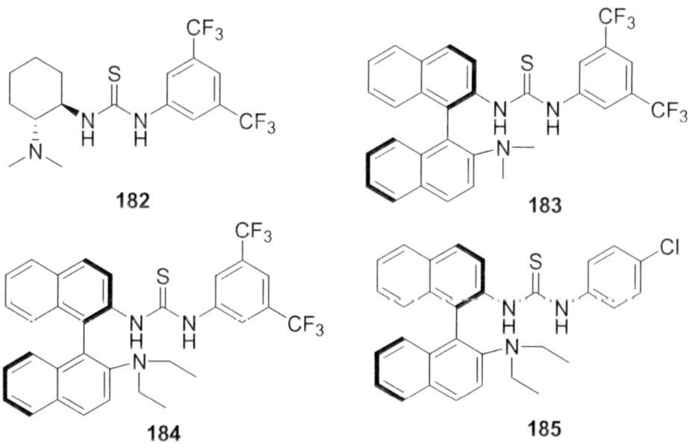

Figure 2.6 Some amino-thiourea bifunctional catalysts for the MBH reaction.

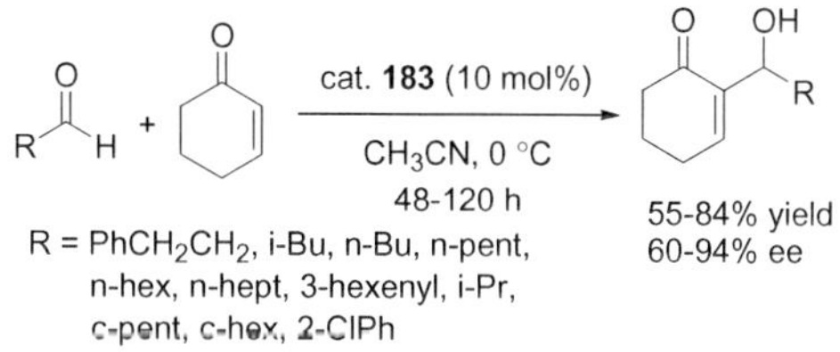

Scheme 2.89

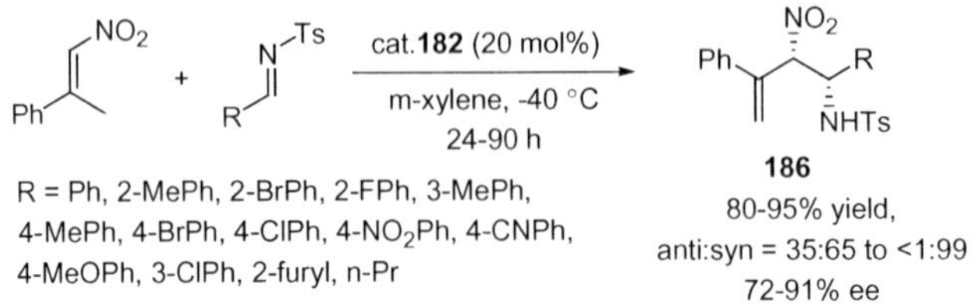

Scheme 2.90

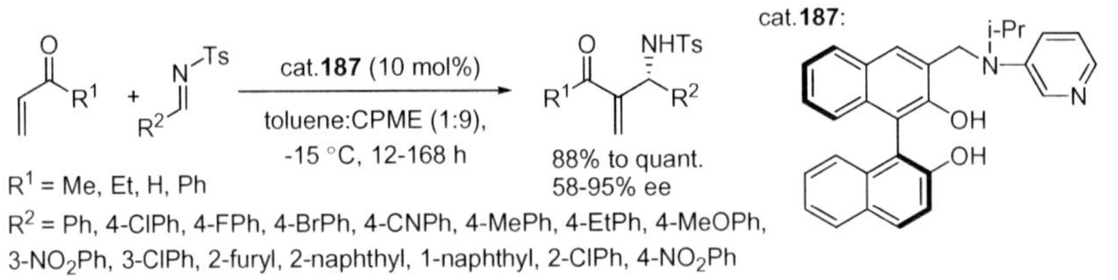

Scheme 2.91

Most recently, Xu *et al.* have demonstrated the first example of a diastereo- and enantioselective aza-MBH-type reaction by the asymmetric synthesis of β-nitro-γ-enamines *via* a (1*R*,2*R*)-diaminocyclohexane thiourea derivative (**182**) mediated tandem Michael addition and aza-Henry reaction in good yields (up to 95%) with high enantioselectivities (up to 91% ee) and diastereoselectivities (up to 1 : 99 dr) (Scheme 2.90); easily prepared *N*-tosylimines and nitroalkene are employed as the starting materials.[95]

As an outstanding representative of bifunctional amino-phenol catalysts, the BINOL-dimethylaminopyridine hybrid **187** was proved to promote efficiently the aza-MBH reaction with high enantioselectivity.[96] Under optimal reaction conditions, at –15 °C with a mixed solvent system consisting of toluene and cyclopentyl methyl ether (CPME) in a 1 : 9 ratio, the MBH adducts were obtained in high yield and good to excellent enantioselectivity (Scheme 2.91).[96,97] Notably, the reaction was sensitive to the structure of catalyst **187**, including the position of the Lewis base attached to BINOL, the substitution pattern of the amino group and the length of the spacer. The acid–base functionalities harmoniously cooperate to activate the substrate and fix the conformation of the organocatalyst, resulting in promotion of the reaction with high enantioselectivity.

2.3 Phosphine-catalyzed System

2.3.1 Achiral Phosphine

It is well known that the Morita–Baylis–Hillman reaction originated from the phosphine-catalyzed dimerization of activated olefins reported by Rauhut and

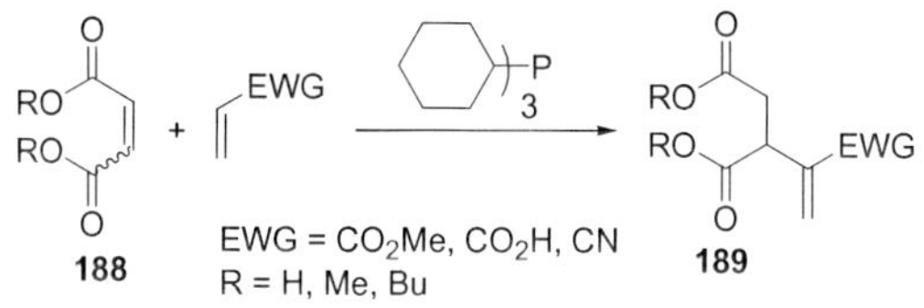

Scheme 2.92

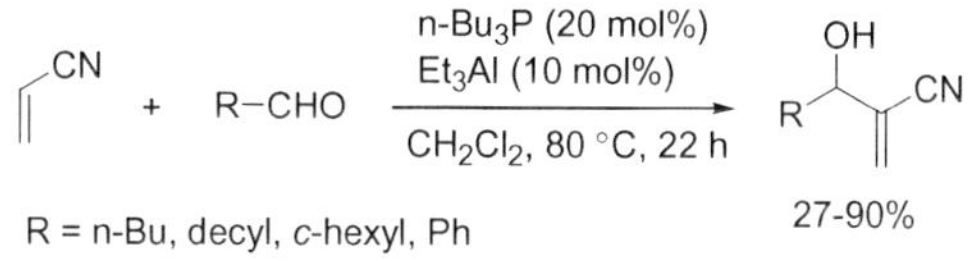

Scheme 2.93

Currier,[98] McClure[99] and Balzer and Anderson.[100] Since Morita and co-workers first reported the coupling reaction of activated alkenes with aldehydes or fumaric/maleic esters **188** in the presence of tricyclohexylphosphine to provide interesting multifunctional molecules **189** (Scheme 2.92),[101] many kinds of phosphines have been used as catalysts in various reactions involving the coupling of aldehydes and activated alkenes.[102,103]

Moreover, the use of trialkylphosphines can avoid the self-aldol reaction of aldehydes having an enolizable carbonyl, which is particularly important in the case of slowly reacting hindered aldehydes. Several trialkylphosphines have been tested in non-asymmetric reactions, and tributylphosphine was generally found to be the most effective catalyst. Kawanisi *et al.* have developed an efficient MBH reaction between aldehydes and acrylonitrile by employing PBu_3 in conjunction with triethylaluminium as catalyst, giving the corresponding adducts in good yields for aliphatic aldehydes, but with low yields for aromatic aldehydes (Scheme 2.93). The Lewis acid was believed to activate the aldehyde by coordination to the carbonyl oxygen, something that had not been investigated in the amine-catalyzed reaction.[103b]

During their studies on the design and synthesis of peptide mimetics, Bertenshaw and Kahn reported a three-component phosphine-catalyzed MBH reaction of activated olefins, aldehydes and amines, which provided a facile synthesis of 2-methylidene-3-aminopropanoates **190** based on the work of Morita (Scheme 2.94).[103a]

Leahy *et al.* have found that using tributylphosphine instead of amines can significantly accelerate the MBH reaction of propionaldehyde with methyl acrylate, and avoid the undesired formation of aldol products that are often generated in amine-catalyzed MBH reaction at room temperature, to afford the corresponding adduct in good yields (Scheme 2.95). Trimethylphosphine and 1,4-diphosphabicyclo[2.2.2]octane, the phosphorus analog of DABCO, do not catalyze the addition of propionaldehyde to methyl acrylate.[5]

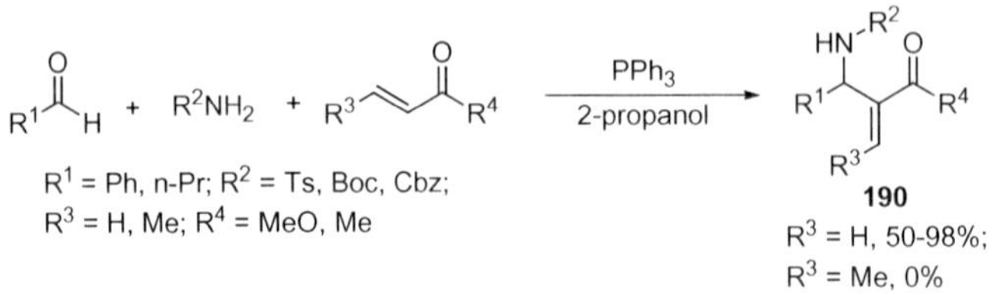

R^1 = Ph, n-Pr; R^2 = Ts, Boc, Cbz;
R^3 = H, Me; R^4 = MeO, Me

190
R^3 = H, 50-98%;
R^3 = Me, 0%

Scheme 2.94

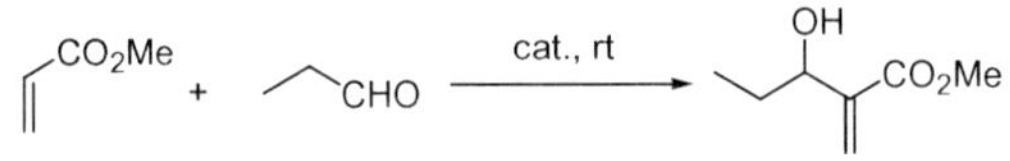

cat.	time	yield(%)
DABCO	10 d	84%
	2 d	<10%
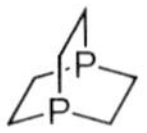	NR	
Me$_3$P	NR	
Bu$_3$P	2 d	80%

Scheme 2.95

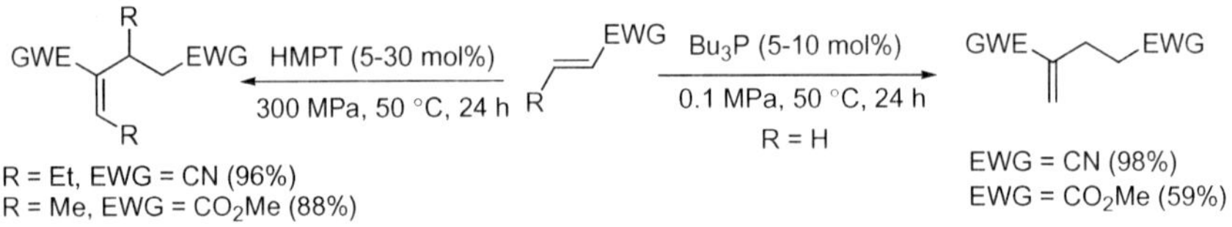

R = Et, EWG = CN (96%)
R = Me, EWG = CO$_2$Me (88%)

HMPT (5-30 mol%)
300 MPa, 50 °C, 24 h

Bu$_3$P (5-10 mol%)
0.1 MPa, 50 °C, 24 h
R = H

EWG = CN (98%)
EWG = CO$_2$Me (59%)

Scheme 2.96

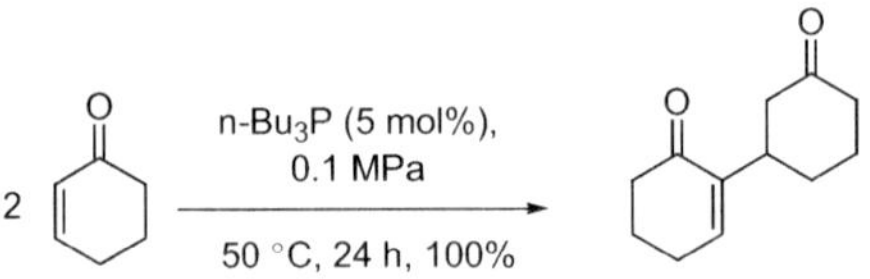

2

n-Bu$_3$P (5 mol%),
0.1 MPa

50 °C, 24 h, 100%

Scheme 2.97

Jenner[104] has reported phosphine-catalyzed dimerization of methyl acrylate and acrylonitrile under MBH conditions at ambient pressure (Scheme 2.96). He also noticed that the β-substituted derivatives generally require high pressures. However, cyclohex-2-en-1-one underwent dimerization smoothly under these conditions with 100% yield (Scheme 2.97).

Genski and Taylor[105] have employed successfully epoxy-activated alkene **191** for coupling with paraformaldehyde in the presence of Et_3Al/Bu_3P to provide the MBH adduct **192**, a highly functionalized molecule (containing epoxide and protected alcohol moieties in addition to the β-substituted enone), and transformed the adduct into the bioactive natural product (±)-*epi*-epoxydon (**193**) (Scheme 2.98).

Ito and Iguchi have developed an efficient tributylphosphine or dimethylphenylphosphine catalyzed MBH reaction of 2-cyclopenten-1-one and substituted 2-cyclohexen-1-ones with formalin or various aldehydes to give the corresponding adducts within a short period in excellent yields (Scheme 2.99).[106] The efficiency of the reaction (yield and time) was strongly dependent on the solvent and the best results were obtained with the $MeOH–CHCl_3$ solvent system.

The MBH reaction of aldehydes with acrylates were sluggish when using PPh_3 or PCy_3 as catalyst; however, the reaction can be mediated with the readily available, air-stable ferrocenyldialkylphosphine **194** to afford the corresponding adducts in high yields and short reaction times (Scheme 2.100).[107]

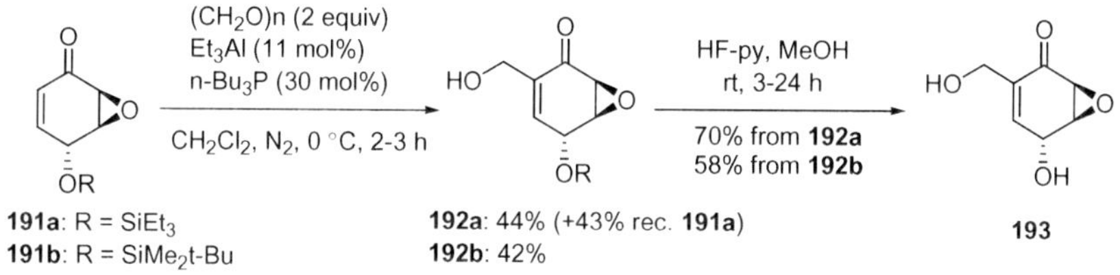

Scheme 2.98

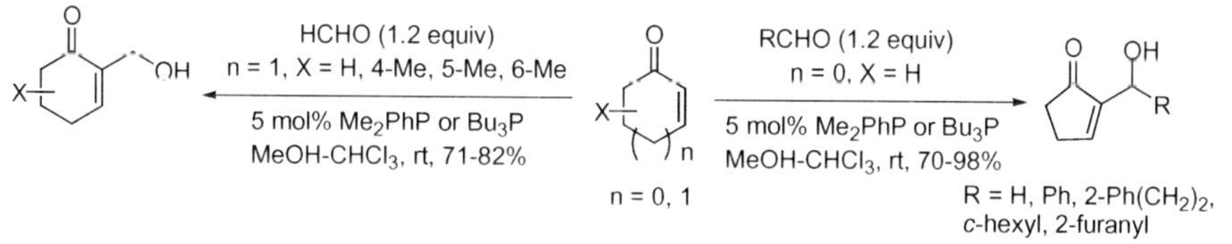

Scheme 2.99

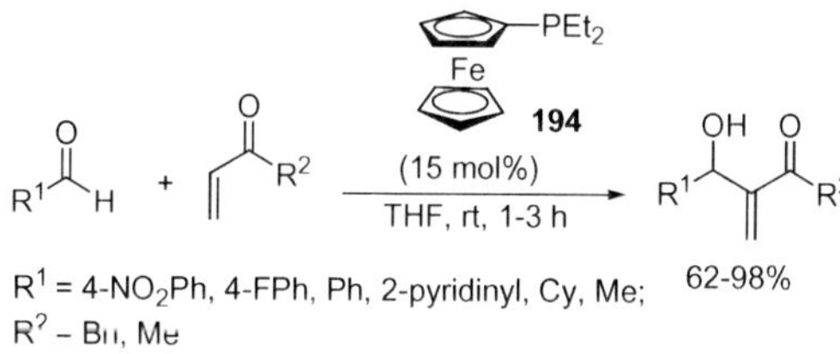

Scheme 2.100

During their study of $P(RNCH_2CH_2)_3N$-catalyzed 1,2-addition reactions of activated allylic synthons with aldehydes, Verkade *et al.* found that the reaction of allylic nitrile with aromatic aldehydes affords the MBH product (**195**) as the only product in the presence of $P(i\text{-}PrNCH_2CH_2)_3N$ **196** (Scheme 2.101). Unlike the nucleophilic pathway proposed for traditional MBH reactions, the reaction is proposed to proceed through the addition of an allylic anion to the aldehyde, affording an intermediate alkoxide anion that undergoes a facile 1,3-proton shift.[108]

In the past decade, our research group has investigated systematically the Lewis base catalyzed aza-MBH reaction. First, we examined the Lewis base effects in the aza-MBH reaction of cyclohex-2-en-1-one and cyclopent-2-en-1-one with imines in the presence of DMAP, Bu_3P and DBU in various solvents.[109] We found that the reaction of cyclopent-2-en-1-one or cyclohex-2-en-1-one with *N*-arylidene-4-methylbenzenesulfonamides in the presence of DMAP can be greatly accelerated to afford the normal MBH adducts **197** and **198**, respectively, in good or very high yields (Scheme 2.102). Moreover, using PBu_3 as a Lewis base in the reaction of *N*-benzylidene-4-methylbenzenesulfonamide with cyclopent-2-en-1-one, the normal MBH adducts **197** could be obtained in

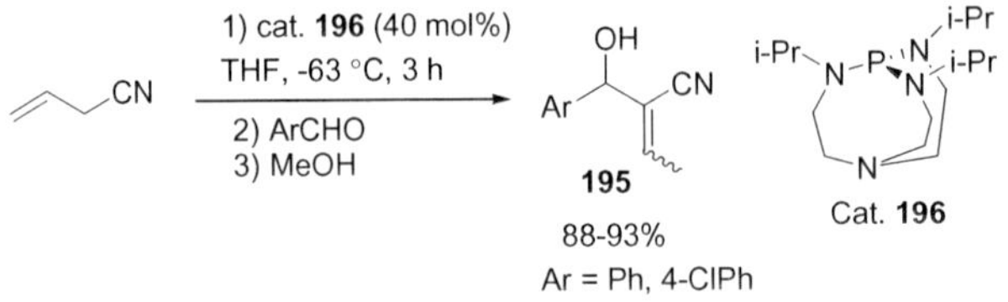

Scheme 2.101

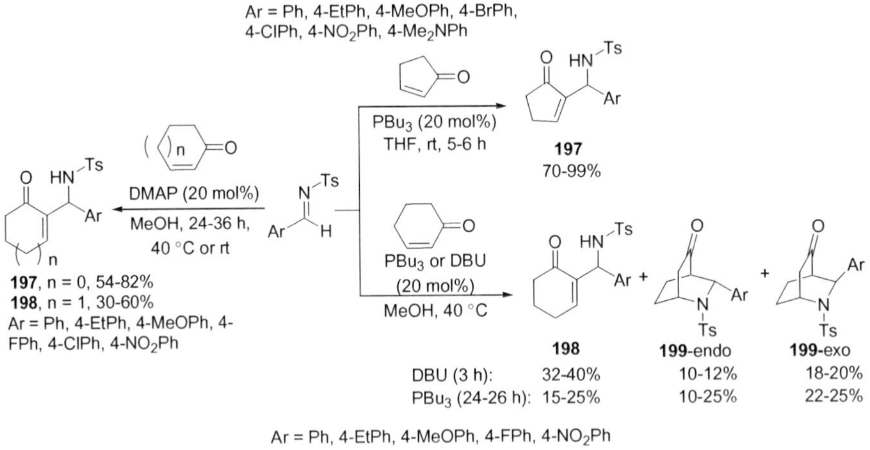

Scheme 2.102

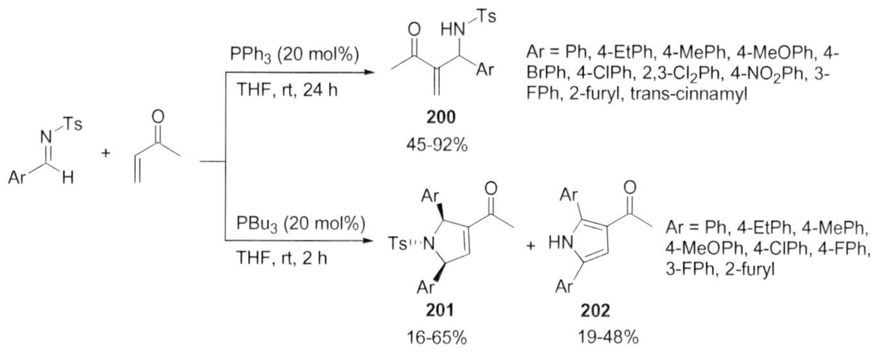

Scheme 2.103

very high yields within 6 h; however, using PBu$_3$ or DBU as a Lewis base in the corresponding reaction with cyclohex-2-en-1-one also produce, besides the normal MBH adduct **198**, abnormal MBH adduct 3-aryl-2-[(4-methylphenyl)sulfonyl]-2-azabicyclo[2.2.2]octan-5-one **199** (Scheme 2.102).

Subsequently, we further examined the Lewis base effects in the aza-MBH reaction between MVK and *N*-arylidene-4-methylbenzenesulfonamides using various amine and phosphine bases.[110] We found that DMAP, DABCO and dppe were effective for the aza-MBH reaction, and PPh$_3$ was the best Lewis base, giving the normal MBH adducts **200** in good to very high yields (Scheme 2.103). In addition, with Bu$_3$P as Lewis base, the abnormal MBH adducts **201** and **202** were formed under the same reaction conditions (Scheme 2.103).

When we extended the activated olefins to cyclohept-2-en-1-one and cyclooct-2-en-1-one we found that the reaction is very complicated, because the Lewis bases, solvents and the ring-size of the cyclic enone can all significantly affect the MBH reaction rate and even the reaction product. The reaction of cyclohept-2-en-1-one with *N*-arylidene-4-methylbenzenesulfonamides afforded the usual MBH adducts **203** along with abnormal adducts **204**, whereas the corresponding reaction of cyclooct-2-en-1-one provided different aldol products (**205–207**) depending upon the Lewis base employed (no MBH adduct was formed in any of the cases) (Scheme 2.104).[111] Moreover, the formation of aldol products **209** along with the usual MBH adducts **208** has been observed in the reaction between cyclopent-2-en-1-one and aromatic aldehydes in the presence of Bu$_3$P (Scheme 2.105).

In the MBH type reaction of chiral non-racemic *N*-sulfinimines **210** with cyclopent-2-en-1-one we found that, in the presence of a catalytic amount of dimethylphenylphosphine (PhPMe$_2$), a diastereoselective reaction could be achieved in toluene at room temperature to give the normal MBH adducts **211** in good yields with high diastereoselectivities (Scheme 2.106).[112]

Subsequently, we employed *N*-arylidenediphenylphosphinamide **212** as an electrophile and systemically investigated the aza-MBH reaction with various

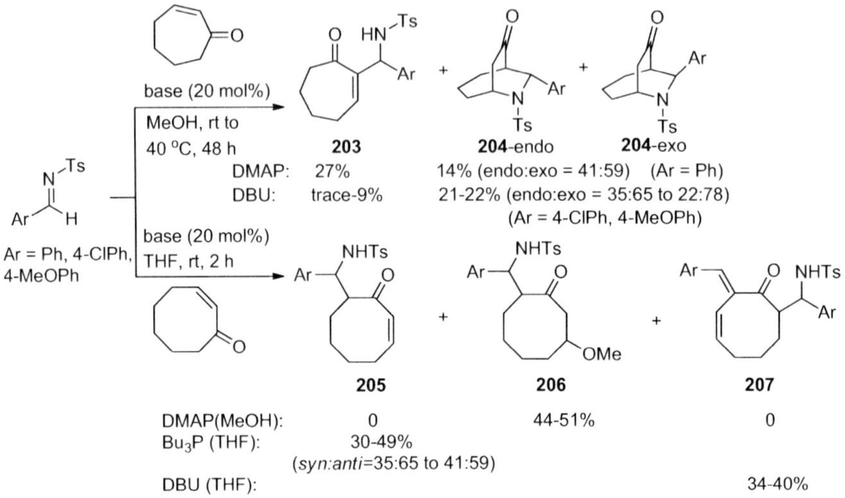

Scheme 2.104

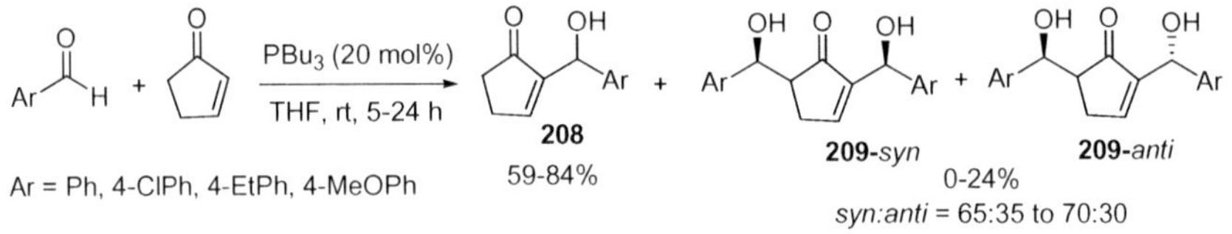

Scheme 2.105

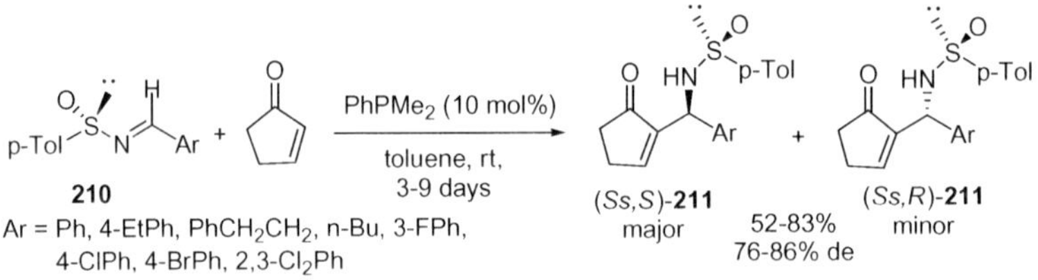

Scheme 2.106

activated alkenes, such as methyl vinyl ketone (MVK), methyl acrylate, acrylonitrile, ethyl vinyl ketone (EVK), phenyl vinyl ketone (PVK), phenyl acrylate,
2-cyclopenten-1-one and 2-cyclohexen-1-one, in the presence of Lewis base.[113]
Similar Lewis base and solvent effects were observed. Using PPh$_3$ as Lewis base
in the reaction of **212** with MVK and EVK in DMF, THF or MeCN, the
normal MBH adducts **213** were formed in very high yields. In the MBH
reaction of **212** with methyl acrylate, the Lewis base Ph$_2$PMe must be used to
obtain a high yield of the MBH adduct **214**. In contrast, for the reaction of **212**

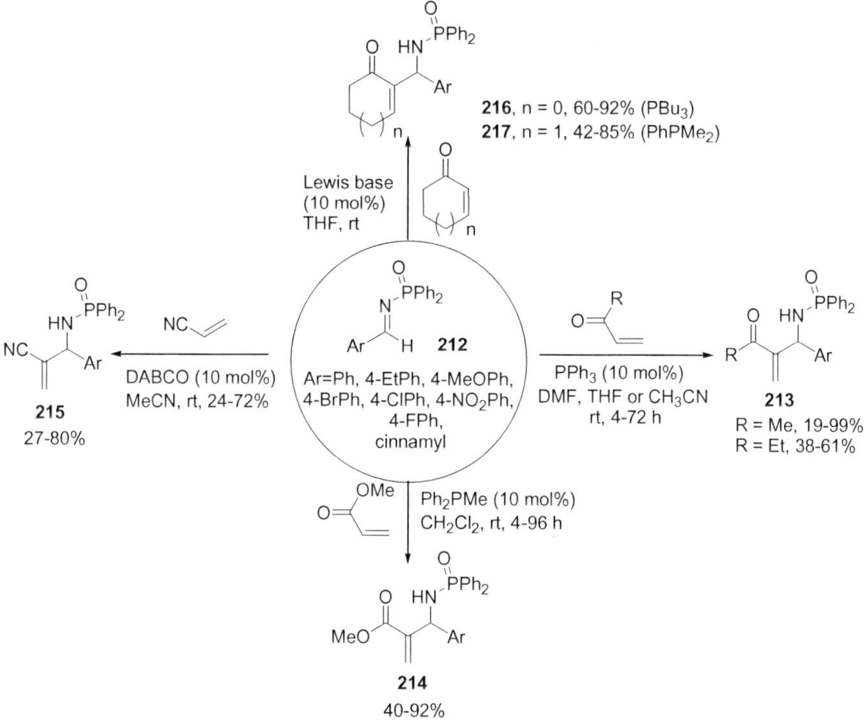

Scheme 2.107

with acrylonitrile, DABCO is the best Lewis base, giving the corresponding MBH adducts **215** in high yields. For the cyclic enones, such as 2-cyclopenten-1-one, the best results were obtained by using PBu$_3$ as catalyst, and in the case of 2-cyclohexen-1-one the strong Lewis base PhPMe$_2$ gave the best results (Scheme 2.107).[113]

In addition, we also found that upon using stronger Lewis bases such as PPhMe$_2$ or PR$_3$ (R = Bu, Me) as promoters in the reaction of *N*-arylidenediphenylphosphinamides **212** with methyl acrylate, phenyl acrylate or PVK, the double aza-MBH adducts **218** and **219** and the abnormal aza-MBH adducts **220** or **221**, derived from further reactions of the double adducts, can be obtained either as the major products or as the sole products depending on the reaction conditions (Scheme 2.108). Initial investigations of a catalytic asymmetric version of this reaction in the presence of chiral amine and phosphine Lewis bases have been disclosed (Scheme 2.109).[113a]

Later, we developed an interesting aza-MBH reaction of various ethyl (arylimino)acetates **222** with MVK under mild conditions. We found that abnormal aza-MBH adducts **223** could be formed in the presence of DABCO (30 mol.%) and the corresponding adducts **224** could be obtained in the presence of PPh$_3$ (30 mol.%) in moderate to good yields in acetonitrile under

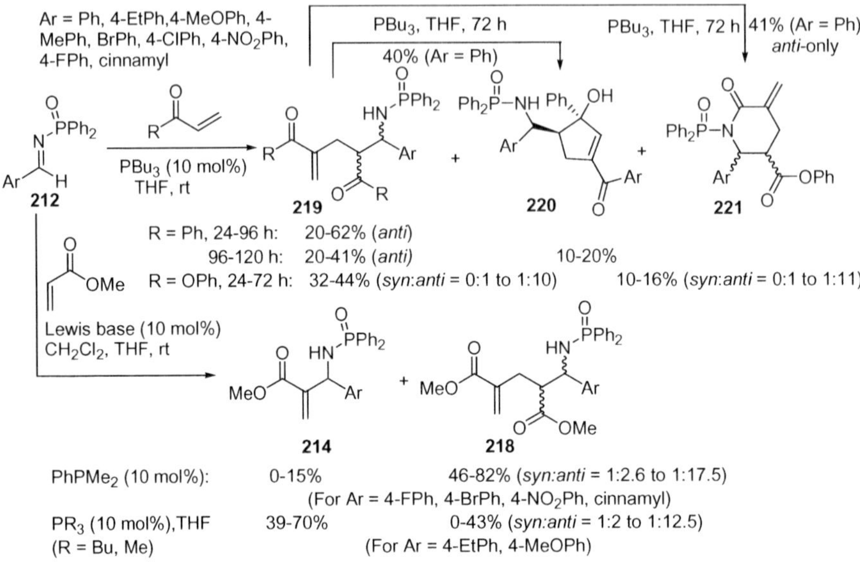

Scheme 2.108

Scheme 2.109

mild conditions. The adducts **224** can be transformed into adducts **223** in the presence of DABCO or Et$_3$N in 90% and 52% yields, respectively, under standard reaction conditions (Scheme 2.110).[114]

Most recently, a new bifunctional phosphine catalyst, (2′-hydroxybiphenyl-2-yl)-diphenylphosphane **225**, has been developed for the selective aza-MBH reaction and aza-MBH domino reaction of *N*-sulfonated imines with acrolein under mild conditions in moderate to excellent yields (Scheme 2.111).[115] The possible catalytic cycle of the domino reaction and aza-MBH reaction has been proposed. The aza-MBH/Michael/aldol/dehydrate domino reaction also provides an efficient method to synthesize tetrahydropyridine

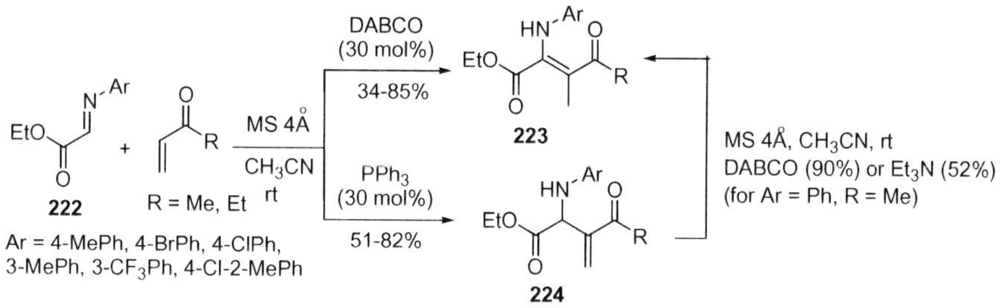

Scheme 2.110

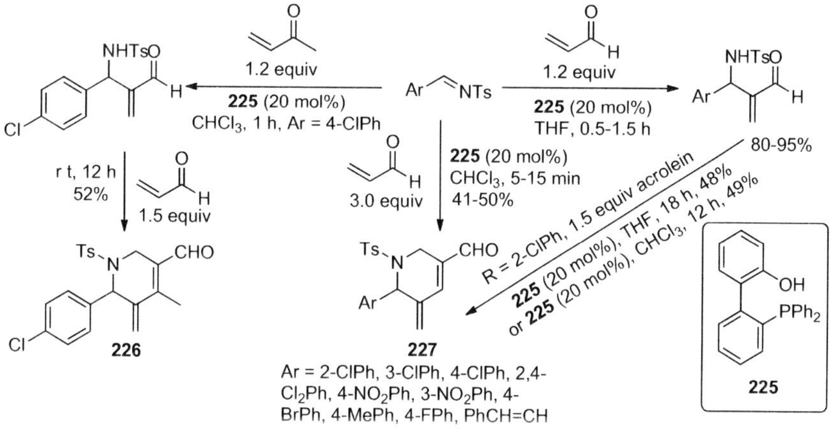

Scheme 2.111

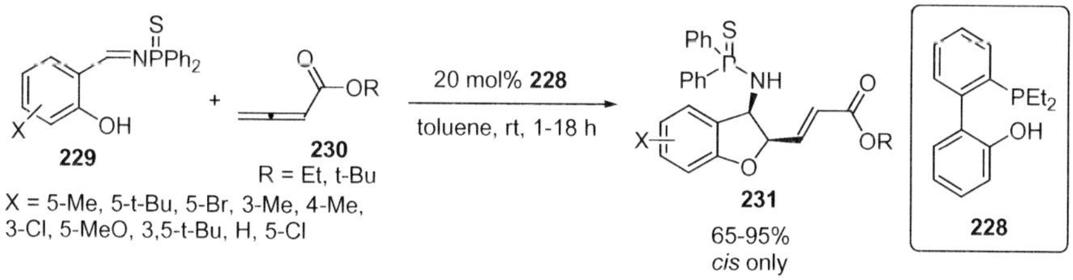

Scheme 2.112

derivatives **226** and **227**, which can be used as building blocks in organic synthesis (Scheme 2.111).[115]

Subsequently, catalyst **228**, the analogue of catalyst **225**, was found to be an effective catalyst for highly stereoselective synthesis of *cis*-2,3-dihydrobenzofurans **231** *via* an aza-MBH/umpolung addition domino reaction of salicyl *N*-thiophosphinyl imines **229** with electron-deficient allenes **230** (Scheme 2.112). It was assumed that dual activation of both nucleophile and

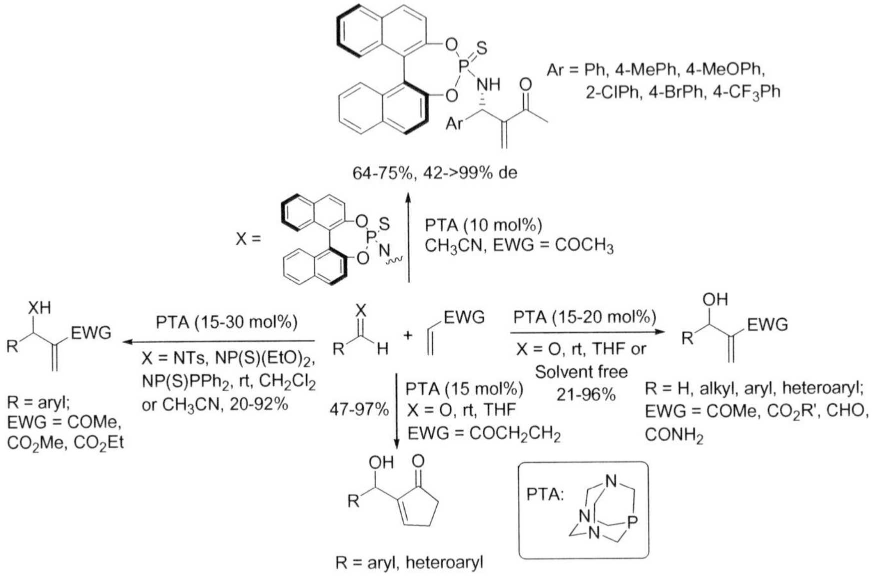

Ar = Ph, 4-MePh, 4-MeOPh,
2-ClPh, 4-BrPh, 4-CF$_3$Ph

64-75%, 42->99% de

PTA (10 mol%)
CH$_3$CN, EWG = COCH$_3$

X =

XH

PTA (15-30 mol%)

X = NTs, NP(S)(EtO)$_2$,
NP(S)PPh$_2$, rt, CH$_2$Cl$_2$
or CH$_3$CN, 20-92%

R = aryl;
EWG = COMe,
CO$_2$Me, CO$_2$Et

47-97%

PTA (15 mol%)
X = O, rt, THF
EWG = COCH$_2$CH$_2$

PTA (15-20 mol%)

X = O, rt, THF or
Solvent free
21-96%

R = H, alkyl, aryl, heteroaryl;
EWG = COMe, CO$_2$R', CHO,
CONH$_2$

R = aryl, heteroaryl

PTA:

Scheme 2.113

electrophile by the bifunctional catalyst accounts for the observed high reactivity and stereoselectivity.[116]

The air-stable and readily available 1,3,5,-triaza-7-phosphaadamantane (PTA) was found to be a convenient and efficient nucleophilic trialkylphosphine organocatalyst for the (aza)-MBH reaction. Under the mediation of 15–30 mol.% of PTA, various electrophiles such as aldehydes[117] and imines[117b,118] readily undergo the (aza)-MBH reactions with various activated olefins, giving the corresponding adducts in fair to excellent yields. By systematic comparison with other structurally similar N,P-catalysts, it is concluded that the superiority of PTA in the above nucleophilic catalysis is attributable to its comparable nucleophilicity with that of trialkylphosphines.[117b] Moreover, the chiral-imine-induced diastereoselective aza-MBH reaction has also been realized with high diastereoselectivity (Scheme 2.113).[118]

2.3.2   Chiral Phosphine

In contrast to chiral amines, phosphine-based chiral catalysts were less developed for asymmetric MBH transformations. In the first asymmetric cycloisomerization reaction, the cyclopentenol derivative **233** was prepared from **232** in the presence of (–)-CAMP (Scheme 2.114).[103c] The low asymmetric control (14% ee) was attributed to the reversibility of the cyclization. Notably, this reaction is not suitable for the preparation of six-membered rings.

Various chiral phosphines (**CP1–CP6**) were further examined for intermolecular MBH reaction of pyrimidine 5-carboxaldehyde and methyl acrylate

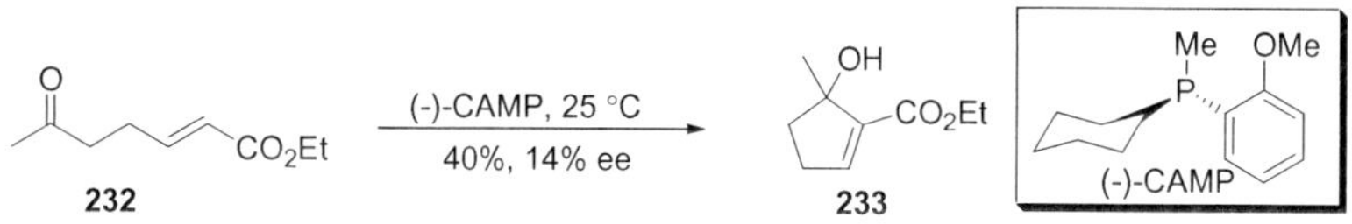

Scheme 2.114

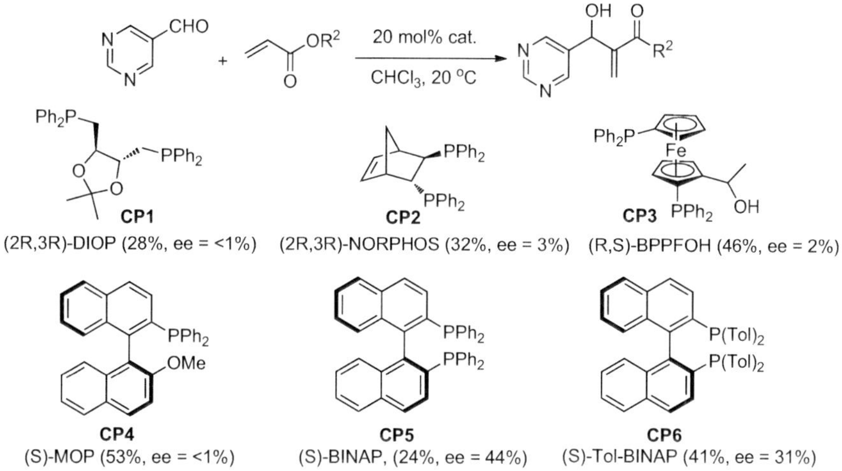

Scheme 2.115

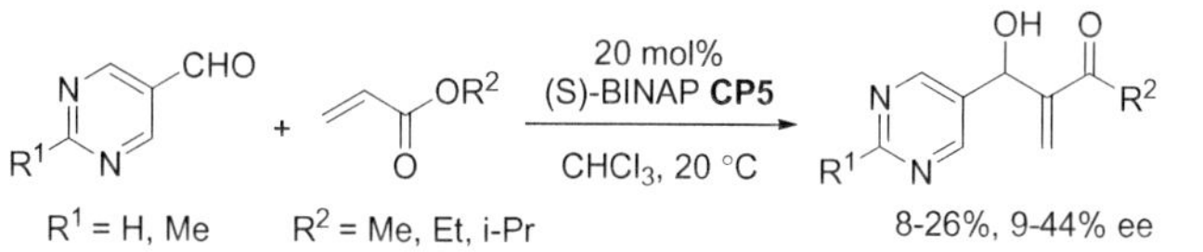

Scheme 2.116

(Scheme 2.115).[119] Among the phosphine catalysts screened, 2,2′-bis(diphenylphosphino)-1,1′-binaphthyl (BINAP) **CP5** was found to be the best catalyst and the corresponding products were obtained with up to 44% ee, which is comparable with those of reported enantioselective methods using chiral tertiary amines under high pressures (Scheme 2.115).[70,82b] For the MBH reaction of substituted pyrimidine 5-carboxaldehyde and other acrylates, the yield and ee were dependent on the bulk of the acrylate. The less bulky acrylate gave the higher yield and ee (Scheme 2.116).

Several D-mannitol-derived phosphines (**CP7–CP10**) catalyzed asymmetric MBH reactions have been investigated by Zhang and co-workers.[120] These phosphine catalysts did not offer any satisfactory enantioselectivities (2–19% ee); however, some rate acceleration was observed in the case of hydroxyl phospholane **CP9** (Scheme 2.117).

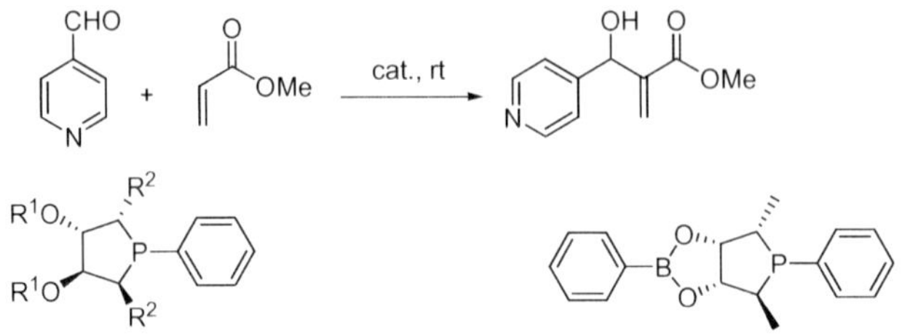

cat.: **CP7**: R^1 = Bn, R^2 = Me, [29%, ee = 19% (+)] **CP10**: 56%, ee = 18% (+)
 CP8: R^1 = Bn, R^2 = Bn, [18%, ee = 2% (-)]
 CP9: R^1 = H, R^2 = Me, [83%, ee = 17% (+)]

Scheme 2.117

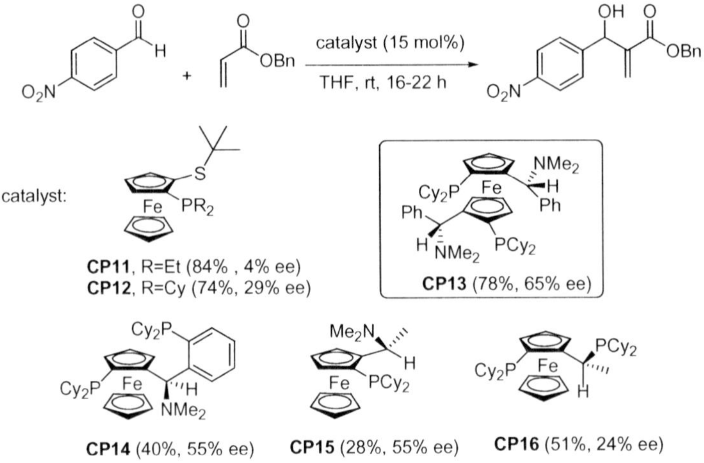

catalyst:

CP11, R=Et (84% , 4% ee)
CP12, R=Cy (74%, 29% ee)

CP13 (78%, 65% ee)

CP14 (40%, 55% ee)　　**CP15** (28%, 55% ee)　　**CP16** (51%, 24% ee)

Scheme 2.118

On the basis of their work on ferrocenyldialkylphosphines-catalyzed MBH reactions of aldehydes and acrylate (Section 2.3.1),[107] Carretero *et al.* have screened various planar or combined planar and central chirality ferrocenyldialkylphosphines **CP11–CP16** for the asymmetric MBH; the best enantioselectivities were obtained by using commercially available aminophosphinoferrocene **CP13** as chiral catalyst (up to 65% ee) (Scheme 2.118).[107]

In 2003, we first demonstrated that 1,1′-bi-2,2′-naphthol (BINOL)-derived chiral LBBA (Lewis base and Brønsted acid) bifunctional phosphine **CP17** (LB = PPh$_3$, BA = Ph-OH) could be used as an effective catalyst in asymmetric aza-MBH reaction of *N*-tosylimines with MVK and phenyl acrylate, affording the corresponding adducts in good yields with high ees (Scheme 2.119).[121] The addition of molecular sieves increased chemical yields because they removed the ambient moisture that caused the decomposition of *N*-sulfonated imines. The asymmetric induction of this catalyst is comparable to that of the quinidine

chiral LBBA bifunctional phosphine
OH
PPh2
CP17
(10 mol%)
R = Me, THF,
or MS 4Å
-30 °C or -20 °C
12-36 h
with MS: 82-96%, 79-92% ee
26-94%, 61-95%
HN–Ts
O
(S)
X
R = Et, THF,
MS 4Å
-30 °C or -20 °C
24 h to 7 days
HN–Ts
O
Et
(S)
X
17-78%
80-92% ee
X = H, 4-Me, 4-Et, 4-F, 4-Cl, 4-Br,
3-Cl, 3-F, 4-NO2, 3-NO2, 2-Cl,
2-NO2, (E)-PhCH=CH-
O
R
N–Ts
X
R = OAr (Ar = Ph, 1-
naphthyl, 2-naphthyl
40 °C, CH2Cl2, 12 h
HN–Ts
O
OAr
(S)
X
49-97% yields
52-77% ee
R = H, THF, rt
2-12 h
HN–Ts
O
H
(S)
X
74-99% yields
76-86% ee

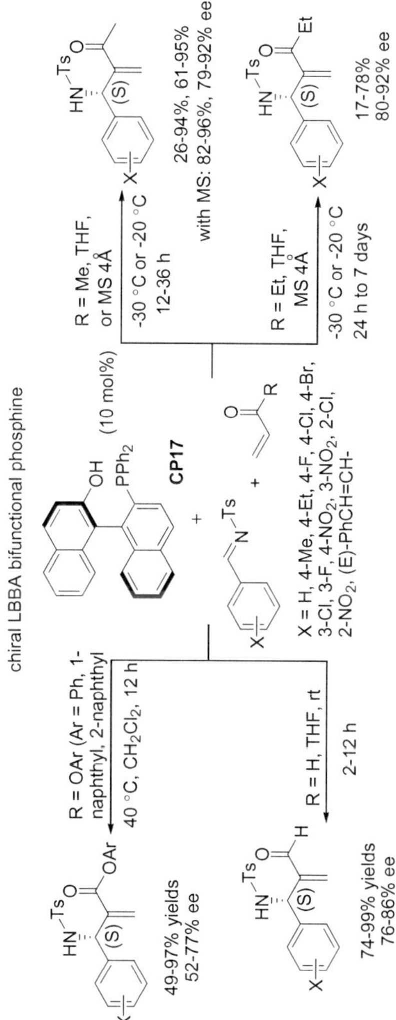

Scheme 2.119

derivatives.[77a] Notably, the presence of a phenolic hydroxyl group in catalyst **CP17** seems to be crucial for a good yield and high ee since replacing the phenolic hydroxyl group with a methoxy or an ethyl group gave the product in much lower yield and ee. Catalyst **CP17** was also tested for the addition of various activated aromatic aldimines with EVK, aryl acrylates and acrolein (Scheme 2.119).[121c] The *N*-Ms, SES and *p*-chlorobenzenesulfonyl activated aldimines afforded products in similar selectivity, while Ns derivatives decomposed under the reaction conditions. The replacement of MVK with EVK resulted in a diminished reactivity, with longer reaction times and lower chemical yields. Aza-MBH reactions of aryl acrylates and acrolein required higher reaction temperatures than those reactions with MVK or EVK. While phenyl or naphthyl acrylates afforded products in CH_2Cl_2 with ee-values up to 77%, the best solvent in the addition of acrolein to activated aldimines proved to be THF (Scheme 2.119).[121c]

We then conducted the first mechanistic studies of bifunctional chiral phosphine catalysis in the aza-MBH reaction through the ^1H and ^{31}P NMR spectroscopic tracing experiments, which revealed the bifunctional role of catalyst **CP17**.[121c] We have proposed a detailed mechanism to rationalize the stereochemistry of the adducts produced. The phosphorus center acts as a Lewis base (LB) to initiate the reaction, and the phenolic OH group acts as a Brønsted acid (BA) to stabilize the *in situ* formed key enolate intermediate **A** and the reaction intermediates **B** and **C** through hydrogen bonding; one of diastereomers (**B**) is the most favorable for the subsequent proton transfer and β-elimination, leading to the desirable enantioselective adduct (Scheme 2.120).

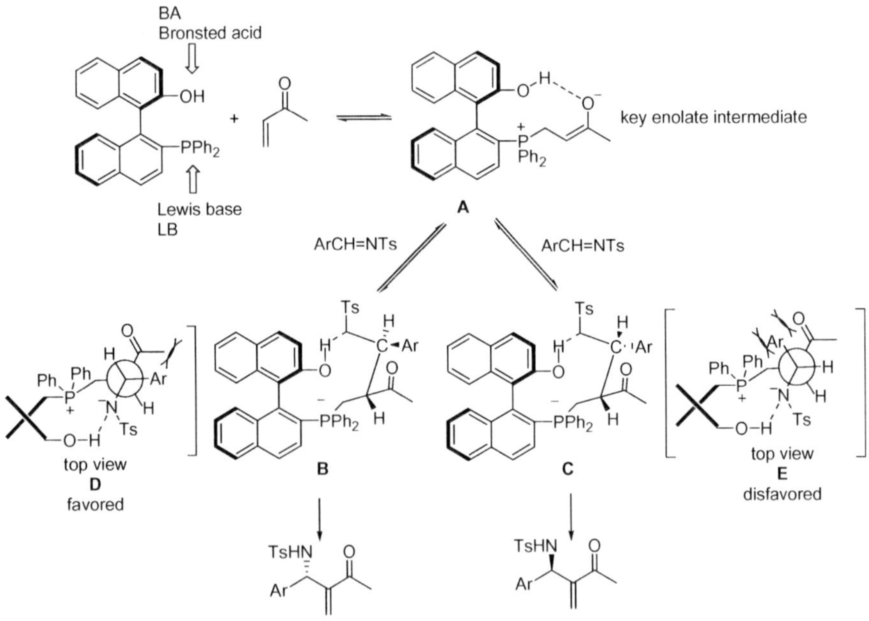

Scheme 2.120

Although triphenylphosphine either alone or in combination with protic additives could cause racemization of the aza-MBH product, to our delight the chiral catalyst **CP17** did not induce any racemization on a similar time scale.[122] The phenolic hydroxyl group in the catalyst might play an important role in preventing racemization of the product.

Catalyst **CP17** also shows good asymmetric induction for the aza-MBH reactions of ethyl (arylimino)acetates with MVK and EVK to give the corresponding adducts in moderate to good yields and good to high enantiomeric excesses under mild conditions (Scheme 2.121).[123]

However, catalyst **CP17** could not give good enantiomeric excess in the reaction of *N*-arylmethylidenediphenylphosphinamides with activated alkenes such as MVK, acrylonitrile or phenyl acrylate (Scheme 2.122).[113a] The reaction is sluggish in the case of acrylonitrile, and moderate yields and low ees were obtained with MVK and with phenyl acrylate.

Having identified **CP17** as an efficient catalyst for the aza-MBH reaction, we hypothesized that replacing the phenol group in catalyst **CP17** with other groups such as a (thio-)urea group might also give high catalytic activity and good asymmetric induction, because the acidic NH protons provide a good opportunity for hydrogen bond formation, which may stabilize certain intermediates.[124] Indeed, the chiral thiourea-phosphine **CP18** in combination with benzoic acid proved to be a very successful catalytic system for the aza-MBH reaction of *N*-tosylimines with MVK, PVK, EVK or acrolein (Scheme 2.123).[125] To the best of our knowledge, this was the first report on the synthesis and application of chiral phosphinothiourea catalysts in asymmetric catalysis.

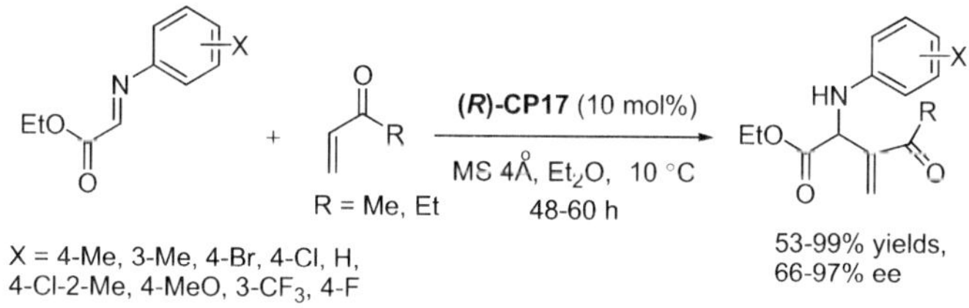
EtO
N
X
O
+
O
R
R = Me, Et
(R)-CP17 (10 mol%)
MS 4Å, Et2O, 10 °C
48-60 h
HN
X
R
EtO
O
O
53-99% yields,
66-97% ee
X = 4-Me, 3-Me, 4-Br, 4-Cl, H,
4-Cl-2-Me, 4-MeO, 3-CF3, 4-F

Scheme 2.121

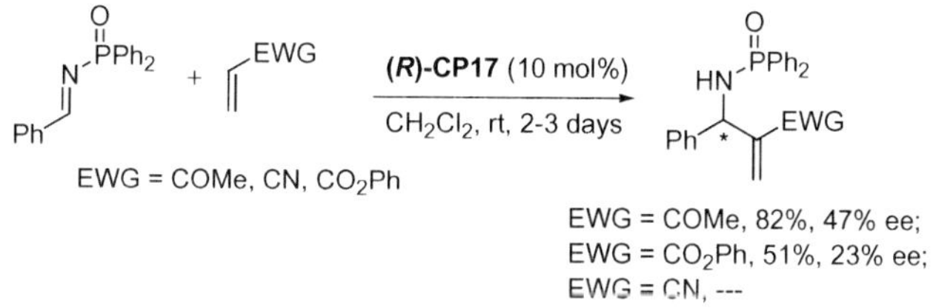
O
PPh2
N
Ph
+
EWG
EWG = COMe, CN, CO2Ph
(R)-CP17 (10 mol%)
CH2Cl2, rt, 2-3 days
O
HN
PPh2
Ph
*
EWG
EWG = COMe, 82%, 47% ee;
EWG = CO2Ph, 51%, 23% ee;
EWG = CN, ---

Scheme 2.122

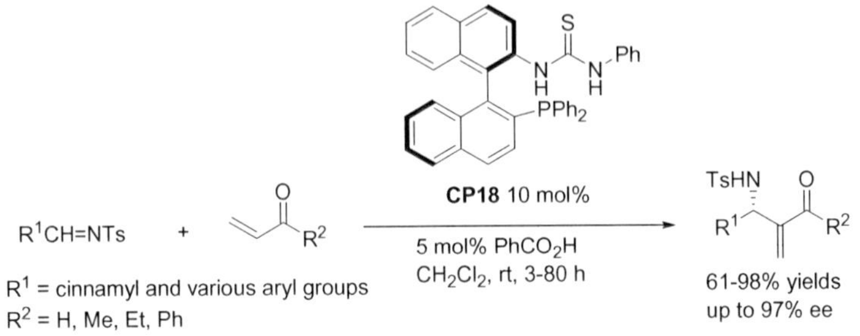

R^{1}CH=NTs + (acrylyl R^2)

R^1 = cinnamyl and various aryl groups
R^2 = H, Me, Et, Ph

CP18 10 mol%

5 mol% PhCO$_2$H
CH$_2$Cl$_2$, rt, 3-80 h

TsHN / R^1 / R^2

61-98% yields
up to 97% ee

Scheme 2.123

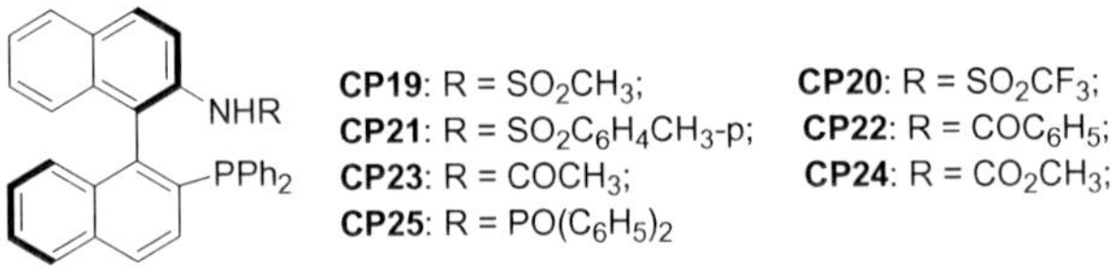

CP19: R = SO$_2$CH$_3$;　　**CP20**: R = SO$_2$CF$_3$;
CP21: R = SO$_2$C$_6$H$_4$CH$_3$-p;　**CP22**: R = COC$_6$H$_5$;
CP23: R = COCH$_3$;　　**CP24**: R = CO$_2$CH$_3$;
CP25: R = PO(C$_6$H$_5$)$_2$

Figure 2.7　Bifunctional chiral phosphine amides for use as catalysts.

Subsequently, to further improve the catalytic activity and enantioselectivity, we designed and synthesized a series of bifunctional chiral phosphine amides, **CP19–C25** (Figure 2.7), which may serve as efficient catalysts because they have a unique chiral environment and their acidic amide proton may act as an efficient hydrogen-bonding donor to interact with the substrate. Interestingly, the acetamide-phosphine **CP23** with a moderately acidic amide proton displayed the best asymmetric induction for the aza-MBH reaction of *N*-tosylimines with MVK or EVK, affording a high yield of up to 99% and excellent ee of up to 91% (Scheme 2.124).[126] However, the catalyst **CP20**, having a stronger acidic amide proton, could not give any product. Presumably, the strong acidic amide proton is transferred to zwitterionic enolate **A** in Scheme 2.120, leading to an inactive intermediate that cannot facilitate the catalytic cycle.

Subsequently, three sterically congested bifunctional chiral phosphane-amides (**CP26–CP28**, Figure 2.8) were further synthesized to evaluate the steric effect of asymmetric induction. This type of catalyst has similar chiral induction for the aza-MBH reaction of *N*-tosylimines with MVK at room temperature as catalyst **CP17** at − 30 °C. Catalysts **CP27** and **CP28** are indeed more effective than the less sterically hindered phosphane-monobenzamide **CP22**. In particular, **CP28** is more effective than **CP27** under optimal conditions, giving the corresponding adducts in higher ee values and similar chemical yields (Scheme 2.125).[127]

The nucleophilicity of the phosphorus center in the catalyst may affect catalytic activity. Thus, catalyst **CP29** was designed to test the nucleophilicity effect by exchanging the phenyl groups in catalyst **CP17** with methyl groups. The catalyst **CP29** was then examined in the aza-MBH reaction of

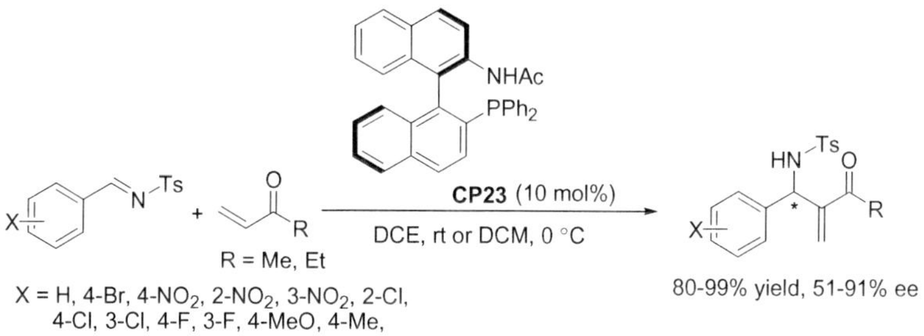

Scheme 2.124

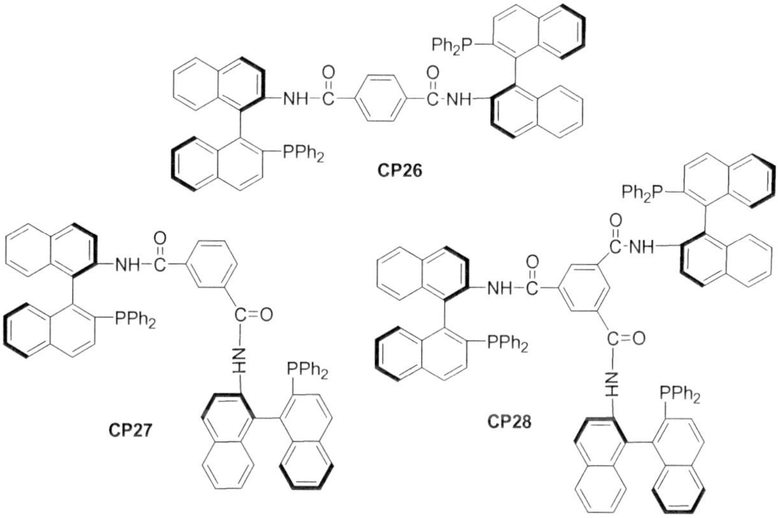

Figure 2.8 Sterically congested bifunctional chiral phosphane-amides for use as catalysts.

N-tosylimines with the less reactive olefins 2-cyclohexen-1-one and 2-cyclopenten-1-one (Scheme 2.126), which could not be catalyzed by **CP17**. The desired adducts were obtained in good yields and moderate enantiomeric excess.[128] This indicates that increasing the nucleophilicity of the reactive center improves the catalytic reactivity.

Having established the nucleophilicity effect, we further designed and developed a series of bifunctional chiral phosphine Lewis bases (**CP30–CP33**, Figure 2.9) bearing an alkyl group on the phosphorus atom to tune both the nucleophilicity of the phosphorus center and the steric hindrance.[129] We were pleased to find that catalysts **CP30–CP33** were very effective in the aza-MBH reaction of various *N*-tosylimines with MVK under mild and concise conditions to produce the corresponding adducts in good-to-excellent yields within relatively short reaction times. In particular, **CP32** was the most effective

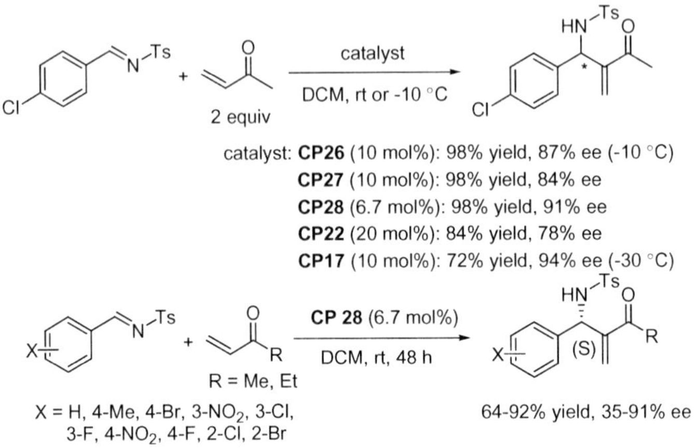

catalyst
DCM, rt or -10 °C
2 equiv
catalyst: CP26 (10 mol%): 98% yield, 87% ee (-10 °C)
CP27 (10 mol%): 98% yield, 84% ee
CP28 (6.7 mol%): 98% yield, 91% ee
CP22 (20 mol%): 84% yield, 78% ee
CP17 (10 mol%): 72% yield, 94% ee (-30 °C)
CP 28 (6.7 mol%)
DCM, rt, 48 h
R = Me, Et
X = H, 4-Me, 4-Br, 3-NO2, 3-Cl, 3-F, 4-NO2, 4-F, 2-Cl, 2-Br
64-92% yield, 35-91% ee

Scheme 2.125

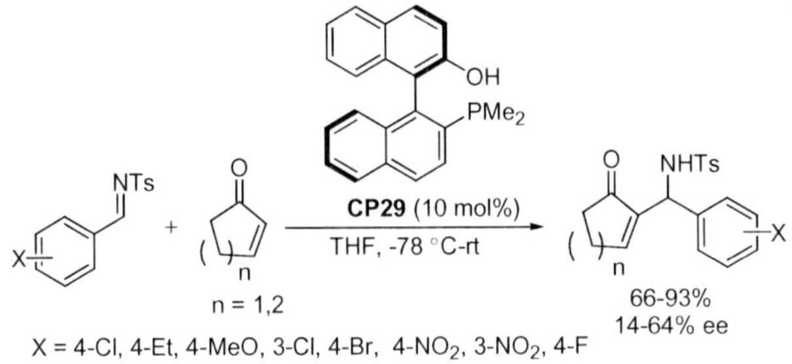

OH
PMe2
NTs
CP29 (10 mol%)
THF, -78 °C-rt
NHTs
n = 1,2
66-93%
14-64% ee
X = 4-Cl, 4-Et, 4-MeO, 3-Cl, 4-Br, 4-NO2, 3-NO2, 4-F

Scheme 2.126

OH
PPhR
CP30: R = Et
CP31: R = i-Pr
CP32: R = n-Bu
CP33: R = Cy

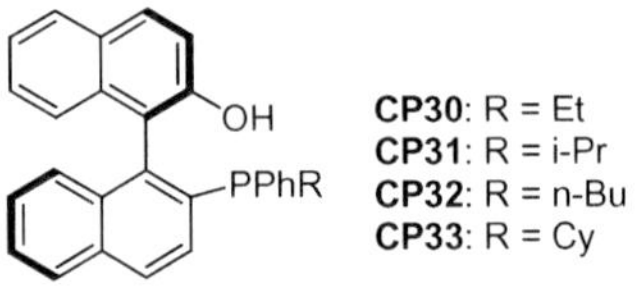

Figure 2.9 Bifunctional chiral phosphine Lewis bases for use as catalysts.

catalyst, giving the corresponding adducts in good-to-excellent yields and moderate-to-good enantioselectivities within only 1–5 h under mild conditions (Scheme 2.127).[129] To the best of our knowledge, this is the fastest asymmetric aza-MBH reaction reported thus far.

To investigate whether catalysts **CP30–CP33** were still effective for MBH reaction, we tested them in the reaction of aldehydes with activated alkenes. For comparison, the catalyst **CP17** and newly designed catalysts **CP34–CP38**

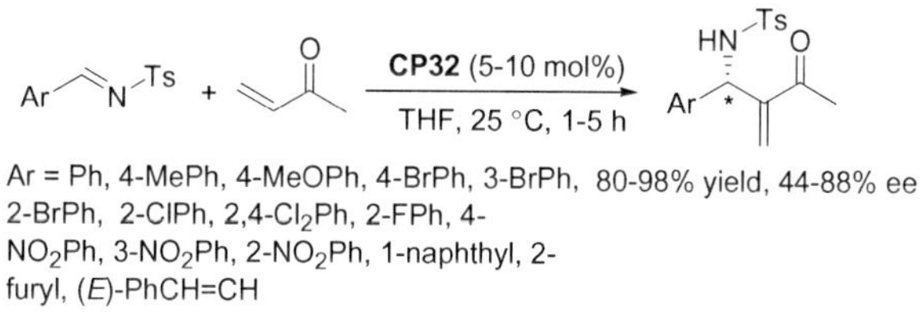

CP32 (5-10 mol%)
THF, 25 °C, 1-5 h
Ar = Ph, 4-MePh, 4-MeOPh, 4-BrPh, 3-BrPh, 2-BrPh, 2-ClPh, 2,4-Cl2Ph, 2-FPh, 4-NO2Ph, 3-NO2Ph, 2-NO2Ph, 1-naphthyl, 2-furyl, (E)-PhCH=CH
80-98% yield, 44-88% ee

Scheme 2.127

CP34: R = i-Bu
CP 35
CP36: R = H
CP37: R = Ph
CP38: R = Br

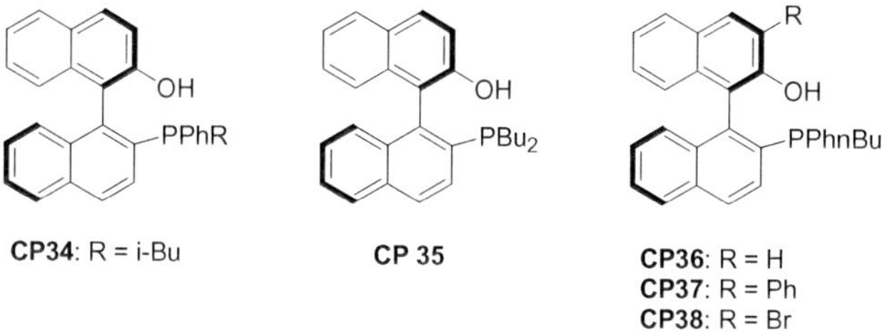

Figure 2.10 Newly designed catalysts for the MBH reaction.

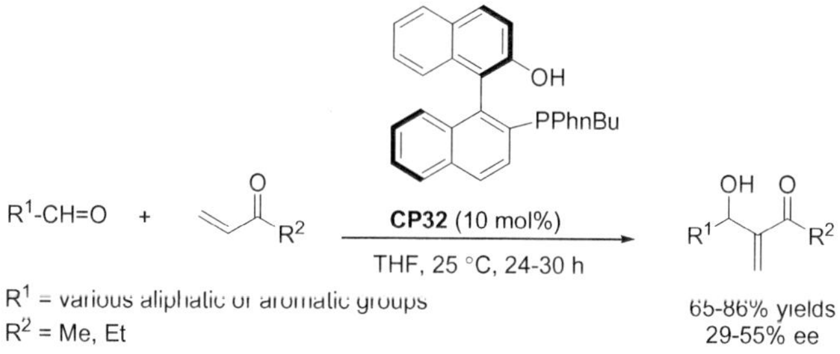

R1-CH=O +
CP32 (10 mol%)
THF, 25 °C, 24-30 h
R1 = various aliphatic or aromatic groups
R2 = Me, Et
65-86% yields
29-55% ee

Scheme 2.128

were also examined (Figure 2.10). Unfortunately, the effective catalyst **CP17** for aza-MBH reaction did not show catalytic activity for the reaction of 3-phenylpropanal and MVK. **CP32** was still the most effective catalyst with respect to a wide range of substrates, affording the corresponding products in good yields with moderate ees (Scheme 2.128).[130]

Inspired by the observation that the introduction of a long-chain alkyl group in various chiral ligands could improve the efficiency in homogeneous asymmetric catalysis,[131] we introduced the so-called "pony tails", long perfluoroalkane chains, on the naphthalene framework and, thereby, synthesized catalysts **CP39** and **CP40** (Figure 2.11). Indeed, catalyst **CP40** was more effective in the aza-MBH reaction of *N*-tosylimines with MVK than the previously reported original

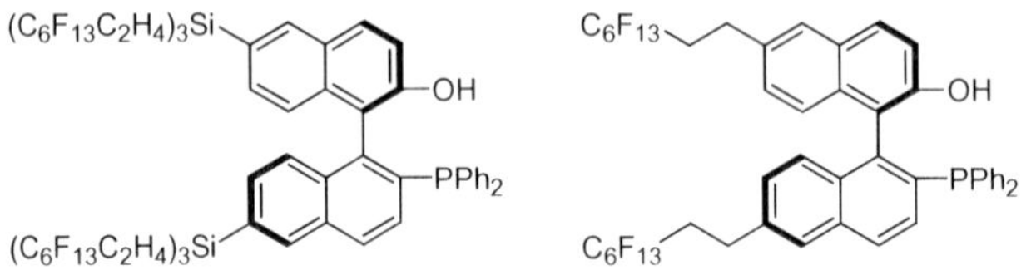
(C6F13C2H4)3Si
OH
PPh2
(C6F13C2H4)3Si
C6F13
OH
PPh2
C6F13

CP39 **CP40**

Figure 2.11 Catalysts with "pony tails".

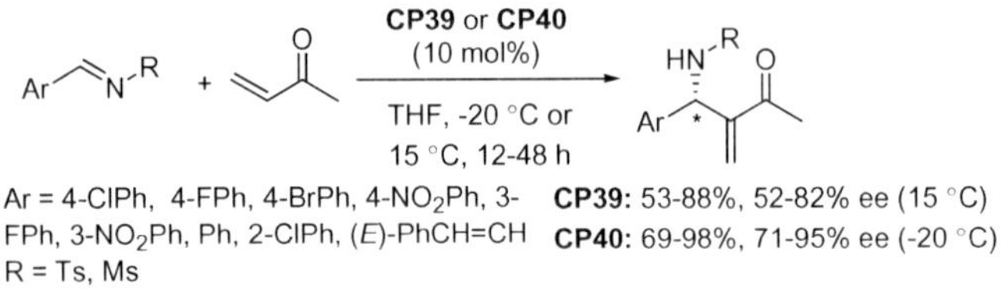
CP39 or CP40
(10 mol%)
THF, -20 °C or
15 °C, 12-48 h
Ar = 4-ClPh, 4-FPh, 4-BrPh, 4-NO2Ph, 3-FPh, 3-NO2Ph, Ph, 2-ClPh, (E)-PhCH=CH
R = Ts, Ms
CP39: 53-88%, 52-82% ee (15 °C)
CP40: 69-98%, 71-95% ee (-20 °C)

Scheme 2.129

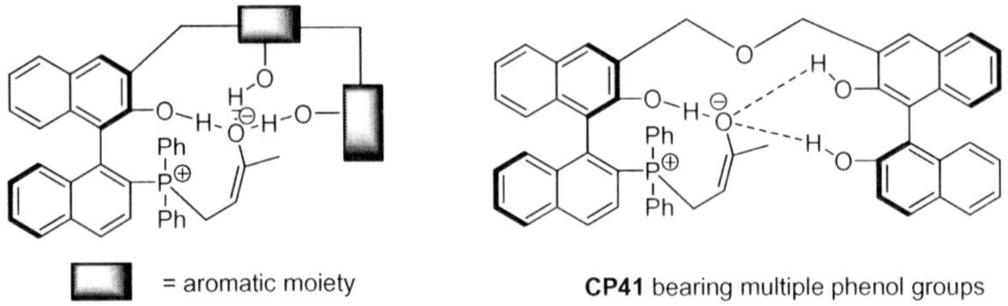
= aromatic moiety
CP41 bearing multiple phenol groups

Figure 2.12 Catalyst **CP41**, with multiple phenol groups so as to increase the number of hydrogen bond donors, gave improved catalytic activity and enantioselectivity.

chiral phosphine **CP17**.[121b,132] The performance of **CP39** was not so impressive, presumably due to the steric bulkiness (Scheme 2.129).

Another approach to improve the catalytic activity and enantioselectivity is to increase the number of hydrogen bond donors in the bifunctional chiral phosphines. Thus, **CP17** was modified to incorporate multiple phenol groups, and it was found that catalyst **CP41** (Figure 2.12) gave the best asymmetric induction. The corresponding adducts could be obtained in >90% ee and with good to excellent yields at −20 °C or room temperature (25 °C) in THF for most substrates, using MVK, EVK or acrolein as a Michael acceptor (Scheme 2.130).[133] The mechanism of this process was investigated by ^{31}P NMR spectroscopic analysis, which indicated that multiple phenol groups, namely, more hydrogen bond donors, can drive the equilibrium largely to the formation of the phosphonium enolate intermediate and stabilize the intermediate strongly

Ar-CH=NR^1 + (acrylate with R^2) → **CP41** (10 mol%), THF, -20-25 °C, 4-84 h → R^{1}HN product with R^2

Ar = Ph, 4-MePh, 4-EtPh, 3-FPh, 4-FPh, 4-ClPh, 2-ClPh, 3-ClPh, 4-BrPh, 2-NO$_2$Ph, 4-MePh, 3-NO$_2$Ph, 4-NO$_2$Ph, PhCH=CH

R^1 = Ts, Ms; R^2 = H, Me, Et

67-97% yields
72-96% ee

Scheme 2.130

R-CH=O + (methyl vinyl ketone) → **CP41** (10 mol%), THF, 10 °C, 3-7 d → product with OH

R = Ph, 4-ClPh, 2-ClPh, 2-NO$_2$Ph, 3-NO$_2$Ph, PhCH$_2$CH$_2$

16-71% yields
4-39% ee

Scheme 2.131

via intramolecular hydrogen bonding, which may be employed to rationalize the high enantioselectivities and yields achieved by **CP41**.

As well as being an effective catalyst in the aza-MBH reaction, catalyst **CP41** is also effective for the MBH reaction of aldehydes with MVK (Scheme 2.131).[134] If the two phenol groups on the second binaphthalene moiety were changed to methoxy groups, the resulting chiral phosphine could not catalyze the reaction under identical conditions. Therefore, it could be deduced that the multiple phenol groups played a significant role in accelerating the reaction rates, which is in line with our observations on the aza-MBH reaction.

On the basis of the same working hypothesis, Sasai *et al.* have functionalized the 3-position of BINOL with a series of aryl phosphines. Catalyst **CP42** was found to catalyze effectively the asymmetric aza-MBH reaction of *N*-tosylimines with vinyl ketones (Scheme 2.132).[135]

More recently, Ito *et al.* also reported biphenol-based bifunctional catalyst **CP43** for the aza-MBH reaction of *N*-tosylimines with MVK (Scheme 2.133).[136] High enantioselectivity (up to 96% ee) was achieved by **CP43** with a catalyst loading of 1 mol.%.

To recycle the catalyst, we have immobilized **CP17** on a series of dendrimers.[137] The dendrimer-immobilized catalyst **CP44** (Figure 2.13) afforded

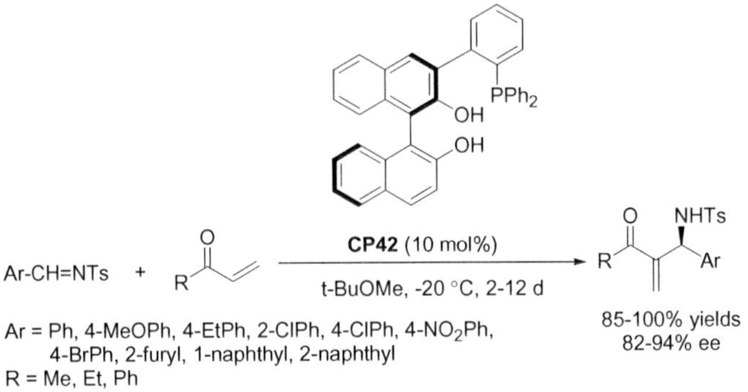

Scheme 2.132

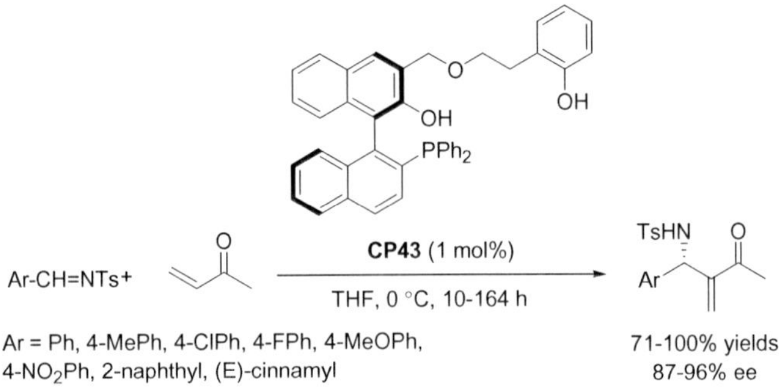

Ar-CH=NTs + (R-acryloyl) → product

CP42 (10 mol%)
t-BuOMe, -20 °C, 2-12 d

Ar = Ph, 4-MeOPh, 4-EtPh, 2-ClPh, 4-ClPh, 4-NO$_2$Ph,
4-BrPh, 2-furyl, 1-naphthyl, 2-naphthyl
R = Me, Et, Ph

85-100% yields
82-94% ee

Scheme 2.133

Ar-CH=NTs + (methyl vinyl ketone) → product

CP43 (1 mol%)
THF, 0 °C, 10-164 h

Ar = Ph, 4-MePh, 4-ClPh, 4-FPh, 4-MeOPh,
4-NO$_2$Ph, 2-naphthyl, (E)-cinnamyl

71-100% yields
87-96% ee

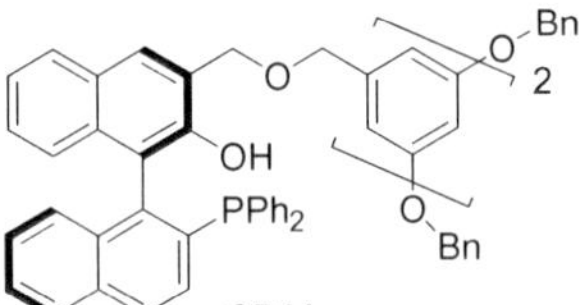

CP44

Figure 2.13 A dendrimer immobilized catalyst.

the better results than catalyst **CP17** for the aza-MBH reaction of *N*-sulfonyl imines with MVK, EVK or acrolein (Section 2.7).[138]

More recently, Liu and co-workers reported trifunctional organocatalyst-promoted counterion catalysis for aza-MBH reactions at ambient temperature.[139] Fast and enantioselective aza-MBH reactions between electron-deficient or electron-rich aromatic *N*-tosylimines and MVK were achieved at ambient temperature using asymmetric counterion-directed catalysis promoted by trifunctional organocatalysts **CP45** with a Brønsted base as the activity switch after protonation with benzoic acid (Scheme 2.134).

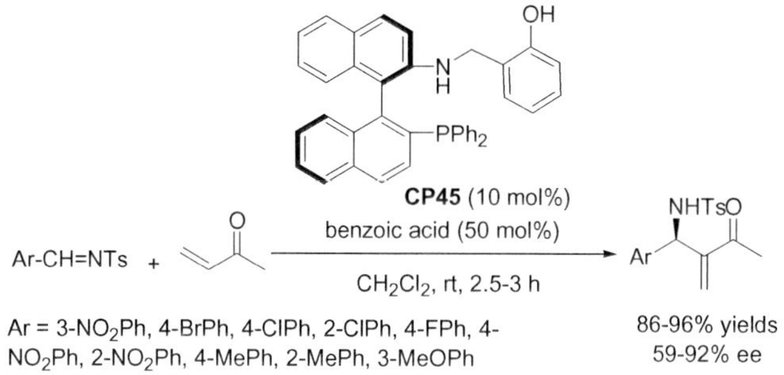

CP45 (10 mol%)
benzoic acid (50 mol%)

CH_2Cl_2, rt, 2.5-3 h

Ar-CH=NTs

Ar = 3-NO_2Ph, 4-BrPh, 4-ClPh, 2-ClPh, 4-FPh, 4-NO_2Ph, 2-NO_2Ph, 4-MePh, 2-MePh, 3-MeOPh

86-96% yields
59-92% ee

Scheme 2.134

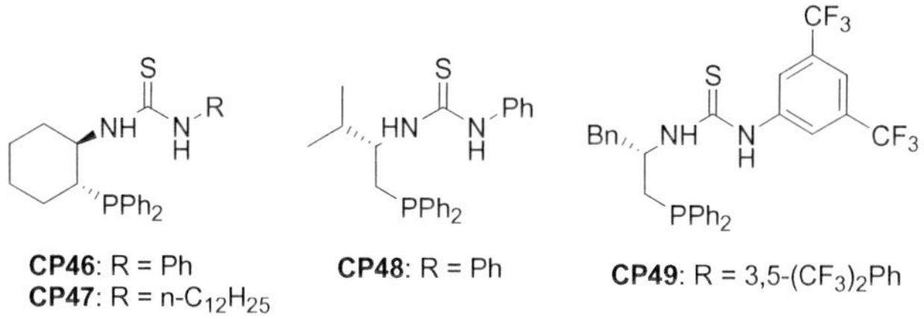

CP46: R = Ph
CP47: R = n-$C_{12}H_{25}$

CP48: R = Ph

CP49: R = 3,5-$(CF_3)_2$Ph

Figure 2.14   Some chiral phosphino(thio)ureas.

CP46 (10 mol%)
R = Me

$CHCl_3$, 13 °C,
15-180 min

15-91%, 87-94% ee

CP47 (8 mol%) or
CP48 (10 mol%)
THF, 25 °C

R = OR'

Ar = Ph, 4-CF_3Ph, 4-BrPh, 4-NO_2Ph, 3-NO_2Ph, 2-NO_2Ph, 4-ClPh, 2,4-Cl_2Ph, 2-furyl

(R' = Me, Et, n-Bu, t-Bu, Bn)

For **CP47**: 36-96%, up to 77% ee
For **CP48**: 11-96%, up to 83% ee

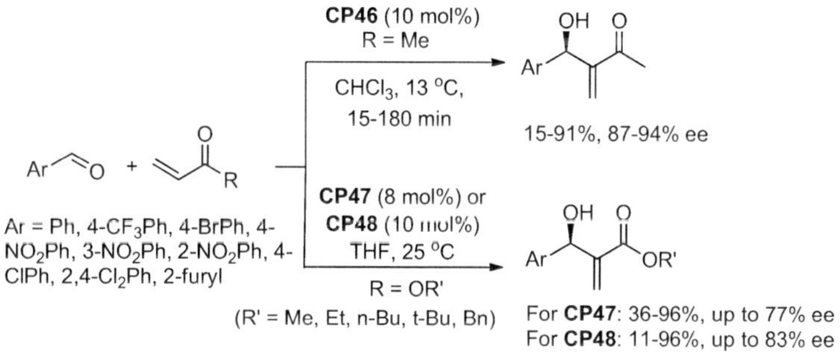

Scheme 2.135

Recently, Wu *et al.* have reported a series of chiral phosphinothioureas (**CP46**–**CP49**, Figure 2.14) derived from *trans*-2-amino-1-(diphenylphosphino) cyclohexane or L-amino acid.[140] They found that **CP46** was the best catalyst for the MBH reaction of various aromatic aldehydes with MVK, giving the products with excellent enantiomeric excesses under mild conditions in a relatively short reaction time (Scheme 2.135).[140a] However, phosphinothiourea **CP47**, containing a long aliphatic chain in the thiourea moiety, is more effective than **CP46** for the MBH reaction between acrylates and aldehydes. For various

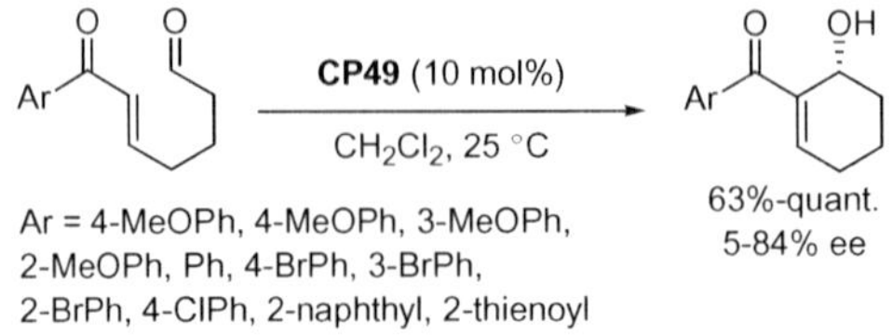

CP49 (10 mol%)

CH$_2$Cl$_2$, 25 °C

Ar = 4-MeOPh, 4-MeOPh, 3-MeOPh,
2-MeOPh, Ph, 4-BrPh, 3-BrPh,
2-BrPh, 4-ClPh, 2-naphthyl, 2-thienoyl

63%-quant.
5-84% ee

Scheme 2.136

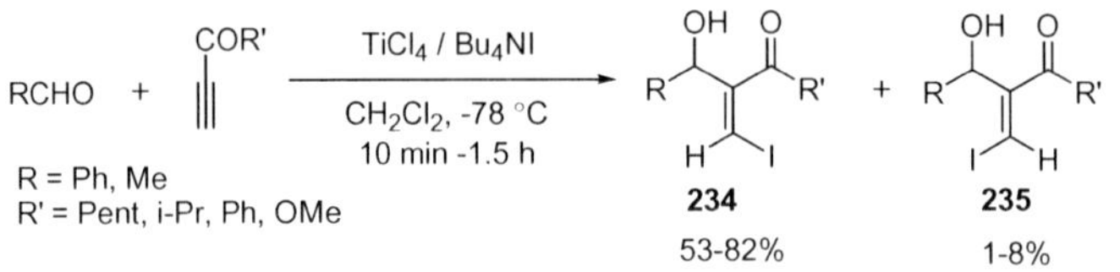

RCHO +

COR'

TiCl$_4$ / Bu$_4$NI

CH$_2$Cl$_2$, -78 °C
10 min -1.5 h

R = Ph, Me
R' = Pent, i-Pr, Ph, OMe

234
53-82%

235
1-8%

Scheme 2.137

commercially available acrylates, the corresponding adducts were obtained in good to excellent yields with up to 77% ee.[140b] The authors also found that the phosphinothio urea **CP48**, derived from L-valine, can also catalyze effectively the MBH reaction of acrylates and aldehydes, providing adducts in excellent yields and with a slightly higher enantioselectively (up to 83% ee).[140c]

Most recently, the same authors have reported the enantioselective intramolecular MBH reaction catalyzed by bifunctional phosphinothiourea **CP49**, which was derived from L-phenylalanine. The process afforded cyclic hydroxyl enones with up to 84% ee and good yields under mild conditions (Scheme 2.136).[140d]

2.4 TiCl$_4$-Lewis Acid Mediated Catalyzed System

The use of Lewis acids in MBH reactions at lower temperature in the presence of an additive furnishes either halo aldol[141] or halo methyl enone[142] as the major product while, at room temperature, the formation of the usual MBH adduct has also been reported.[143] However, recently, Iwamura *et al.*[144] have reported the formation of a hemiacetal through aqueous workup during a boron tribromide promoted MBH reaction of *p*-nitrobenzaldehyde with cyclohexenone. In an earlier observation, Li *et al.*[145] reported the formation of an unidentified side product (10%) in the TiCl$_4$-mediated MBH reaction of *p*-nitrobenzaldehyde with α,β-cycloalkenone.

Taniguchi *et al.*[146] were the first to examine the reaction of α,β-acetylenic ketones and aldehydes in the presence of various reagents such as TiCl$_4$/TMSI, TMSOTf/TMSI, TiCl$_4$/Bu$_4$NI, TMSI/Bu$_4$NF, Et$_2$AlI or TiI$_4$. The most effective catalyst was found to be TiCl$_4$/Bu$_4$NI, providing the compounds **234** and **235** (β-iodo MBH adducts) in good yields with a high (*Z*)-stereoselectivity (Scheme 2.137).

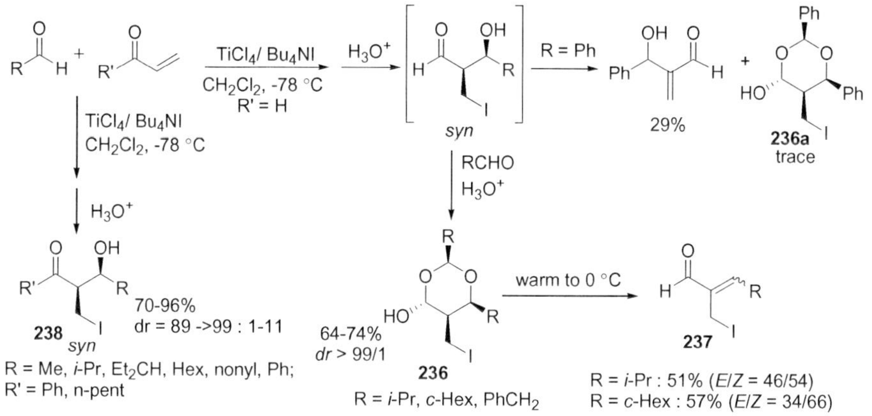

Scheme 2.138

Before Taniguchi's work, Oshima *et al.*[147] investigated the coupling reaction of acrolein and aliphatic aldehydes in the presence of TiCl$_4$/Bu$_4$NI, which provided cyclic hemiacetals **236** in high diastereoselectivity at $-78\,^\circ$C (Scheme 2.138). When the reaction mixture was warmed to $0\,^\circ$C, vinyl aldehydes **237** (dehydration products) was obtained. With benzaldehyde, a complex mixture containing the usual MBH adduct (29% isolated yield) along with trace amounts of cyclic hemiacetal **236a** was obtained. The authors also examined the reactions between alkyl vinyl ketones and aldehydes in the presence of TiCl$_4$/Bu$_4$NI, which provided iodo aldol products **238** in high *syn*-stereoselectivity (Scheme 2.138).

Li *et al.*[143b] have found that the MBH reaction of cycloalkenones and aldehydes proceeded smoothly in the presence of TiCl$_4$ with no need of a Lewis base, providing the desired adducts in moderate to good yield. In addition, this reaction can be controlled to give β-halogenated aldol products **240** as the major product by simply choosing an α,β-unsaturated *N*-acylbenzoxazoline (**239**) as activated olefin (Scheme 2.139). These processes involved the conjugate addition of TiCl$_4$ to α,β-unsaturated substrates followed by carbonyl coupling, and a plausible mechanism was proposed, as presented in Scheme 2.140.

Subsequently, the same authors synthesized allyl multifunctionalized (*Z*)-keto allyl halides **241** *via* the TiX$_4$ or TiX$_4$/Bu$_4$NI-mediated vicinal difunctionalization of α,β-unsaturated ketones (at room temperature)[142] and thioesters (reflux)[148] with aldehydes in good to excellent yields and complete (*Z/E*) stereoselectivity (Scheme 2.141). The formation of halo aldol adducts **242** and **242′** was also observed in the case of vinyl ketones when the reaction was conducted at $0\,^\circ$C (Scheme 2.141).[149]

In a continuation of this work, the authors investigated the reaction between α,β-acetylenic ketones and aromatic aldehydes in the presence of TiCl$_4$ or TiBr$_4$ to provide the desired β-halo MBH adducts **243** (Scheme 2.142) (similar to those obtained by Kataoka).[150] Similar results were obtained by employing a

R = i-Pr, Pr, Hept, 4-NO$_2$Ph, 4-CF$_3$Ph

239

240

Ar = 2-NO$_2$Ph, 3-NO$_2$Ph, 4-NO$_2$Ph, 4-CF$_3$Ph

Scheme 2.139

Scheme 2.140

TiX$_4$ (0.5 equiv)
CH$_2$Cl$_2$, rt, 24 h, 62-94%

R^1 = Ph, naphth-1-yl, 2-NO$_2$Ph, 4-NO$_2$Ph,
4-CF$_3$Ph, 4-OMePh, 2-nitrocinnamyl, nonyl
R^2 = Me, Et

TiX$_4$ (1 equiv), Bu$_4$NI (0.25 equiv), CH$_2$Cl$_2$
rt, 1 h, then reflux, 24 h, 43-72%
R^1 = Ph, 4-(OMe)Ph, 4-ClPh, 4-BrPh,
nonyl, cinnamyl, 2-nitrocinnamyl
R^2 = SEt

241   X = Cl, Br

TiX$_4$ (0.26 equiv),
Bu$_4$NI (0.26 equiv)
CH$_2$Cl$_2$, rt, 24 h, 54-85%

R^1 = Ph, 2-NO$_2$Ph, 4-NO$_2$Ph,
4-OMePh, nonyl
R^2 = Me, Et

TiCl$_4$ (1.2 equiv)
CH$_2$Cl$_2$, 0 °C
2 h, 61-87%

242   **242'**

(±)   (±)

R^1 = Ph, 4-NO$_2$Ph, 4-ClPh, 2-MePh
4-BrPh, 4-FPh, 2-naphthyl, PhCH$_2$
R^2 = Et

syn/anti = 1.0/1.0 to 8.4/1.0

Scheme 2.141

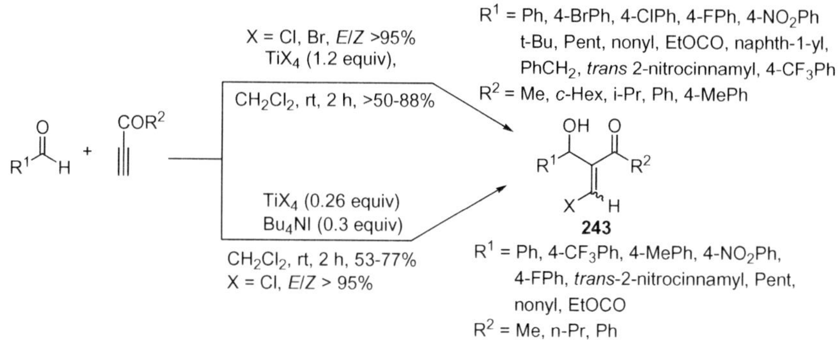

X = Cl, Br, *E/Z* >95%
TiX$_4$ (1.2 equiv),

CH$_2$Cl$_2$, rt, 2 h, >50-88%

TiX$_4$ (0.26 equiv)
Bu$_4$NI (0.3 equiv)

CH$_2$Cl$_2$, rt, 2 h, 53-77%
X = Cl, *E/Z* > 95%

R^1 = Ph, 4-BrPh, 4-ClPh, 4-FPh, 4-NO$_2$Ph
t-Bu, Pent, nonyl, EtOCO, naphth-1-yl,
PhCH$_2$, *trans* 2-nitrocinnamyl, 4-CF$_3$Ph
R^2 = Me, c-Hex, i-Pr, Ph, 4-MePh

243

R^1 = Ph, 4-CF$_3$Ph, 4-MePh, 4-NO$_2$Ph,
4-FPh, *trans*-2-nitrocinnamyl, Pent,
nonyl, EtOCO
R^2 = Me, n-Pr, Ph

Scheme 2.142

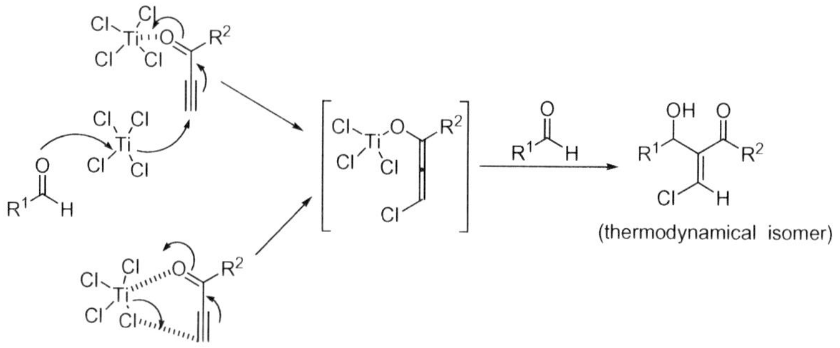

(thermodynamical isomer)

Scheme 2.143

mixture of Bu$_4$NI (0.3 equiv) and a substoichiometric amount of TiCl$_4$ (only 0.26 equiv). The usually expected iodinated products were not produced in this reaction. Scheme 2.143 shows the mechanism of the reaction.

The TiCl$_4$-mediated MBH reactions of alkyl vinyl ketones with various electrophiles such as α-keto esters, fluoromethyl ketones and aldehydes have been developed by Basavaiah *et al.* An interesting reaction trend was found in that α-keto esters and fluoro methyl ketones provided the usual MBH adducts **244** and **245**, whereas aldehydes provided the (*Z*)-allyl chlorides **246** exclusively (Scheme 2.144).[151] Subsequently, this strategy was also extended to aryl 1,2-diones **247** and **248**. Instead of the usual MBH adducts, an interesting class of functionalized fused furans **249** and **250** were obtained (Scheme 2.145).[152]

Our research group[153] has examined similar TiCl$_4$-mediated coupling of activated alkenes with various aldehydes in the presence of various Lewis bases such as amines (DBU, Et$_3$N and Et$_2$NH),[141a] quaternary ammonium salt (TBAB),[154] chalcogenides (Me$_2$S)[141b] and oxy-compounds (alcohols, ethers, and ketones)[155] to furnish either halo aldols **251** or allylic chlorides **252** as major product under different conditions. Notably, the usual MBH adduct **253**

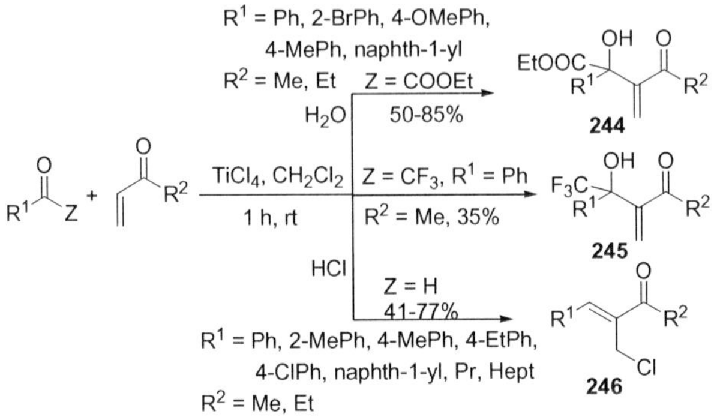

R^1 = Ph, 2-BrPh, 4-OMePh,
4-MePh, naphth-1-yl
R^2 = Me, Et   Z = COOEt
H$_2$O   50-85%

244

TiCl$_4$, CH$_2$Cl$_2$ Z = CF$_3$, R^1 = Ph
1 h, rt   R^2 = Me, 35%

245

HCl

Z = H
41-77%

246

R^1 = Ph, 2-MePh, 4-MePh, 4-EtPh,
4-ClPh, naphth-1-yl, Pr, Hept
R^2 = Me, Et

Scheme 2.144

247 + R, TiCl$_4$, rt, CH$_2$Cl$_2$, 3 h, 45-70% → **249**   R = Me, Et, Hex, t-Bu

248 + R, TiCl$_4$, rt, CH$_2$Cl$_2$, 3 h, 16-43% → **250**   R = Me, Et, t-Bu

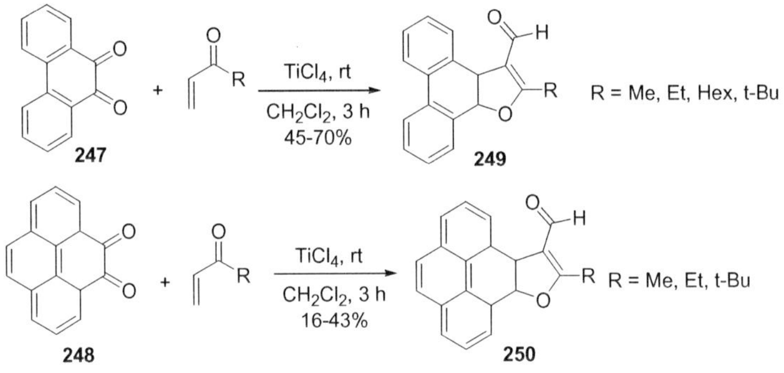

Scheme 2.145

can be obtained by treatment of halo aldol compounds **251** with base, such as DBU and triethylamine (Scheme 2.146). In the case of oxy-compounds, we also studied the applicability of enantiopure oxy-compounds **254–257** (Figure 2.15); however, the products were obtained in poor enantioselectivities.

Subsequently, we investigated the phosphine-mediated MBH reactions between MVK and aldehydes in the presence of Lewis acids (TiCl$_4$, ZrCl$_4$ or BCl$_3$), which provided the α-chloro aldol adducts **251** or allyl chlorides **252** depending on the reaction conditions.[141c] All tested Lewis acids (TiCl$_4$, ZrCl$_4$ and BCl$_3$) are effective for this reaction and the activities of Lewis acids for this reaction are TiCl$_4$ > BCl$_3$ > ZrCl$_4$. This reaction was initiated by chloride ion – which was produced by coordination of Lewis bases (PBu$_3$) to Lewis acids such as TiCl$_4$, BCl$_3$ and ZrCl$_4$ – attacking at the α,β-unsaturated ketone Michael acceptor. Poor enantioselectivity was observed when using chiral phosphine (*R*)-BINAP as catalyst (Scheme 2.147).

Moreover, we also examined the TiBr$_4$ or BBr$_3$ promoted MBH reaction between but-3-yn-2-one with aryl aldehydes.[156] The reaction temperature and amount of the Lewis acid employed can drastically affect the reaction products.

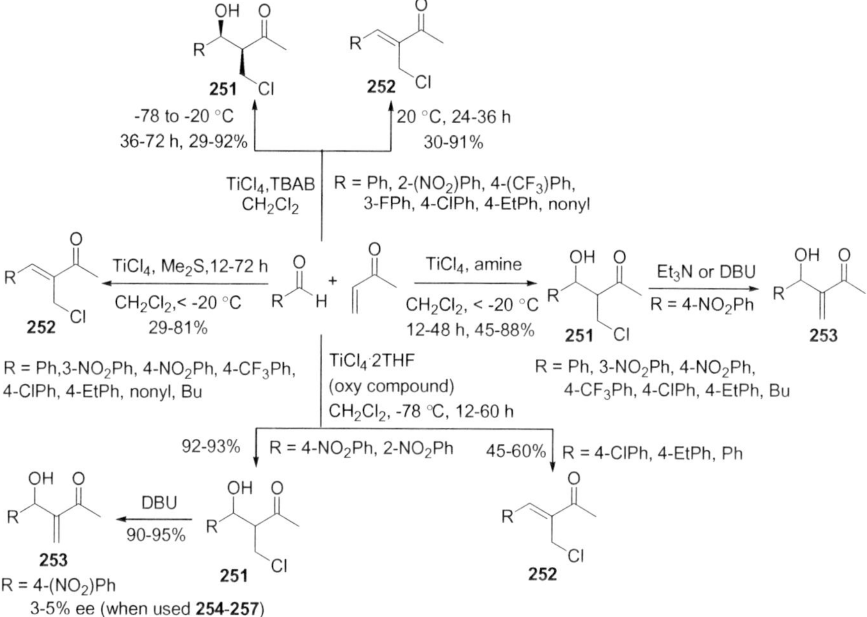

Scheme 2.146

Figure 2.15 Some enantiopure oxy-compounds.

Scheme 2.147

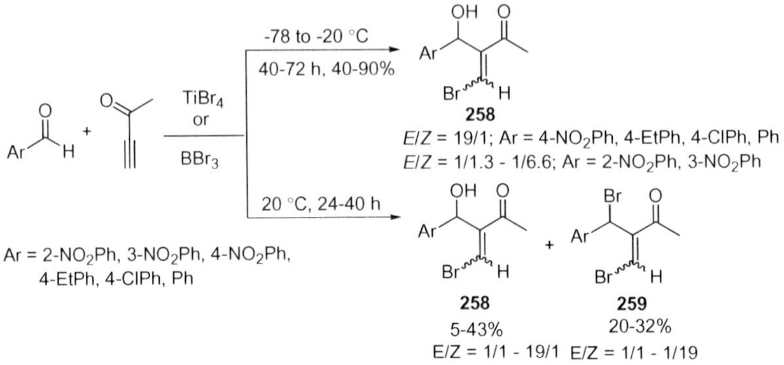

Scheme 2.148

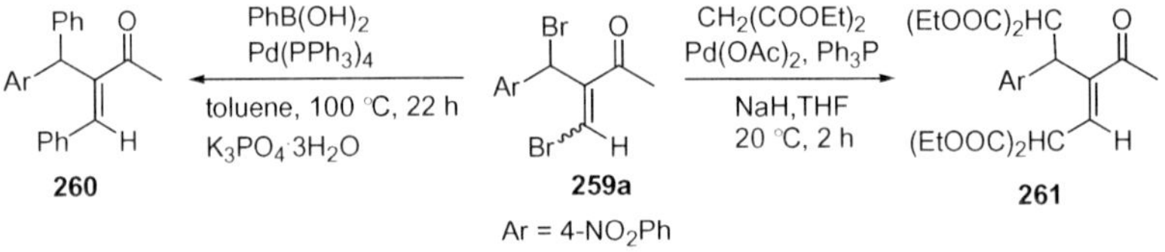

Scheme 2.149

At $-20\,^{\circ}$C, β-bromo MBH adducts **258** were formed with high (*E*)-selectivity [in the case of 2-(NO$_2$)Ph and 3-(NO$_2$)Ph, (*Z*)-selectivity is favored due to steric factors], whereas at room temperature both dibromides **259** and β-bromo MBH adducts **258** were obtained with a large excess of TiBr$_4$ (Scheme 2.148). In addition, the substituent on the phenyl ring can affect the (*E*) : (*Z*) ratio of products. The resulting dibromide **259a** [Ar = (4-NO$_2$)Ph] was subsequently subjected to Suzuki and Heck type coupling reactions to give **260** and **261**, respectively (Scheme 2.149).

Sato and co-workers have developed an elegant titanium alkoxide promoted MBH reaction between enantiopure acetylenic esters **262–264** (Figure 2.16) and aldehydes. A new chiral dimetallic species, the acetylenic ester titanium alkoxide complex with a camphor-derived auxiliary, enabled the preparation of β-trimethylsilylated MBH adducts **265** with high diastereoselectivity (Scheme 2.150).[157]

Balan and Adolfsson have reported that α-methylene-β-amino acid derivatives **267** are readily formed in a three-component one-pot MBH reaction of aryl aldehydes, sulfonamides and activated alkenes. The reaction is catalyzed efficiently by titanium isopropoxide [Ti-(OPr^i)$_4$] and 3-hydroxyquinuclidine (3-HQD) in the presence of molecular sieves, to afford the corresponding adducts in high yields with good to excellent chemoselectivities (Scheme 2.151).[158]

Batra and co-workers[159] have investigated the MBH reaction of isoxazole-5-carboxaldehydes **268** with α,β-unsaturated cyclic ketones in the presence of

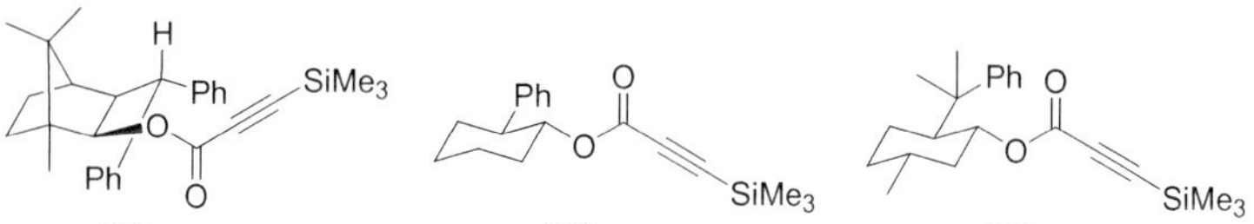

Figure 2.16 Enantiopure acetylenic esters employed in titanium alkoxide promoted MBH reactions.

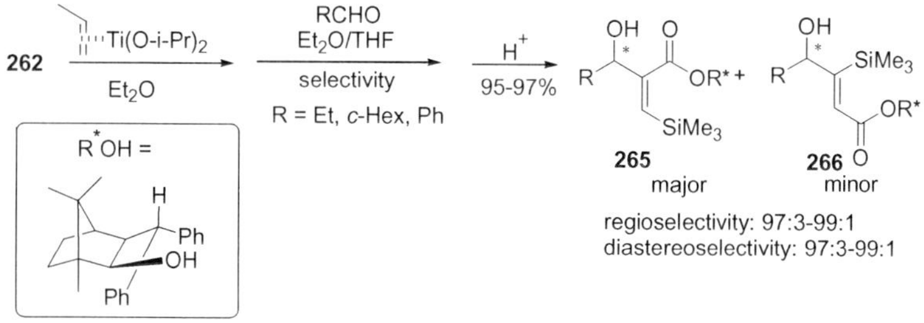

Scheme 2.150

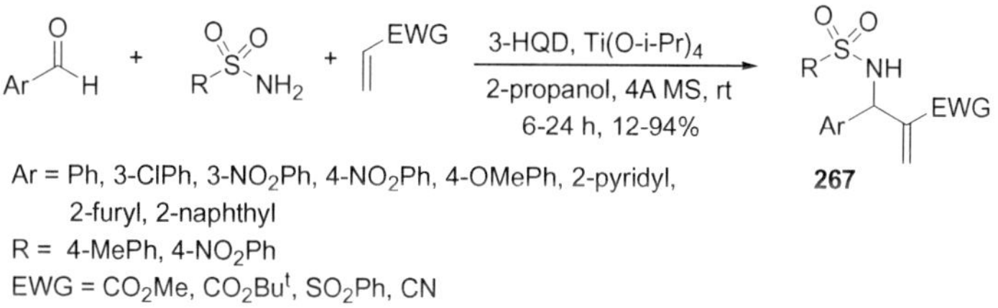

Scheme 2.151

TiCl$_4$. They observed that cyclopent-2-en-1-one provided the usual MBH adducts **269** along with allyl chlorides **270** in minor amounts, whereas cyclohex-2-en-1-one afforded the usual MBH adducts **271** along with hemiacetal **272** *via* aromatization of a cyclohexenone ring (Scheme 2.152). The authors described detailed synthetic and mechanistic aspects of the formation of hemiacetal.

Subsequently, Basavaiah *et al.* have investigated the steric factors in directing TiCl$_4$-mediated coupling between α-keto esters **273** and cyclohex-2-enone derivatives, which led to two different reaction pathways, affording MBH and aldol adducts. The reaction of α-keto esters with 5,5-dimethylcyclohex-2-enone provided the corresponding MBH adducts **274** exclusively, whereas a similar reaction of α-keto esters with cyclohex-2-enone furnished the corresponding aldol adducts **275** with high *syn*-diastereoselectivity as the major product, along with the MBH adducts **276** as the minor product. In addition, TiCl$_4$ mediated the reaction of α-keto esters with cyclopent-2-enone, leading also to the corresponding aldol adducts **277** with a *ca.* 88 : 12 *syn* : *anti* ratio (Scheme 2.153).[160]

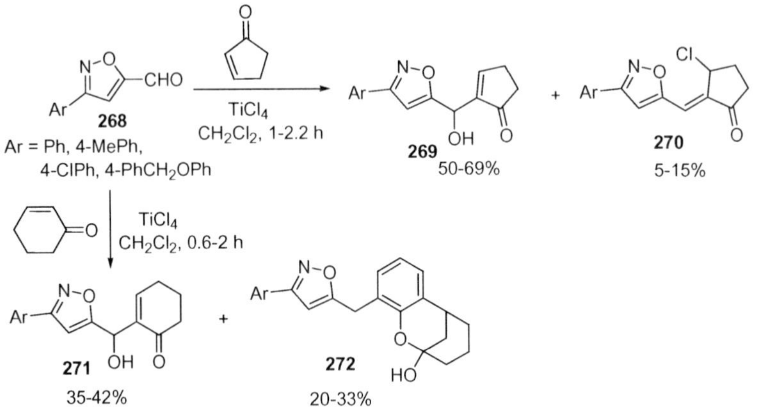

Scheme 2.152

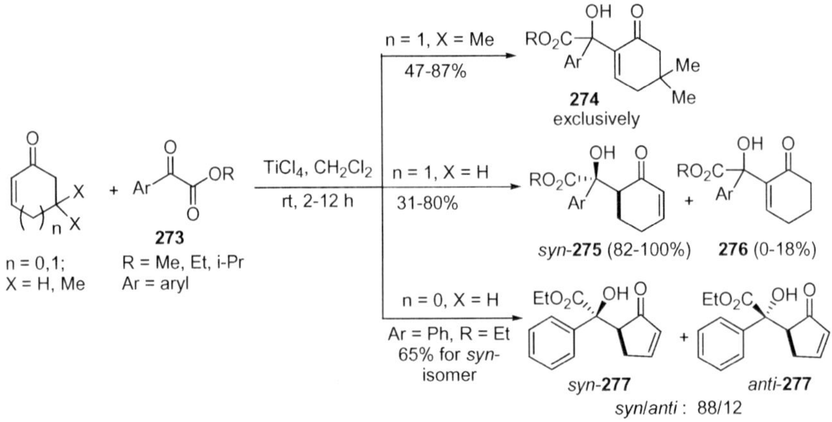

Scheme 2.153

The MBH reaction of C6 acyl protected enuloside **278** with various aromatic and aliphatic aldehydes in the presence of TiCl$_4$/TBAI yielded highly diastereo-enriched C3-branched deoxysugar derivative **279** and **280** in high yields, while reactions of unprotected enuloside and C(6) alkyl protected enulosides with aromatic aldehydes under the same conditions afforded the adducts in low yield with moderate selectivity (Scheme 2.154).[161]

Promoted by TiCl$_4$, α-oxoketene dithioacetals **281**, as a special kind of activated olefin bearing two α,β-dialkylthio substituents, reacts with aryl aldehydes to afford polyfunctionalized 1,4-pentadienes **283**, rather than normal MBH adduct **282** (Scheme 2.155).[162] In addition, a series of α-(1,3-dithiolan-2-ylidene)-β-amino carbonyl derivatives **284**, the aza-MBH adducts, were obtained in good to excellent yields from novel one-pot, three-component reaction of α-oxo

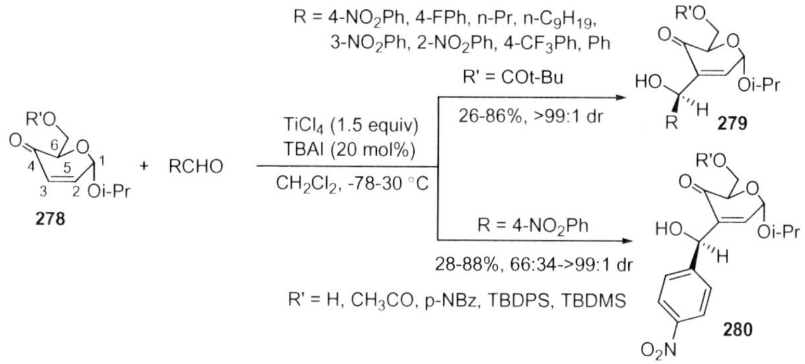

Scheme 2.154

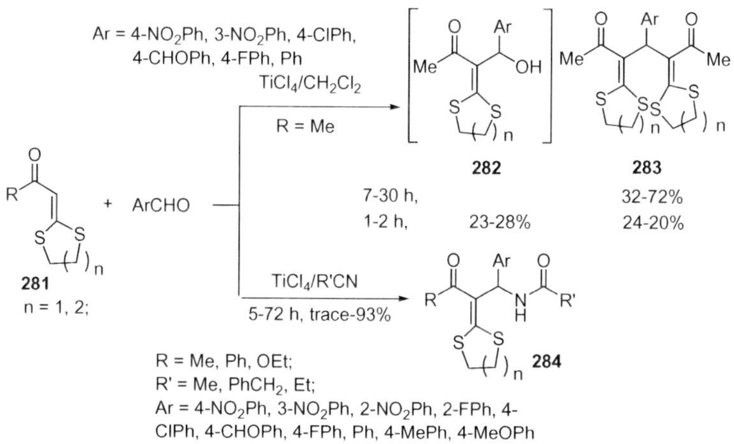

Scheme 2.155

cyclic ketene-S,S-acetals, aryl aldehydes and nitriles by sequential Morita–Baylis–Hillman and Ritter reactions (Scheme 2.155).[163]

Aminomethylbenzotriazoles **285**, as effective MBH electrophiles, react with ethyl acrylate in the presence of $TiCl_4$ at 20 °C to afford the corresponding benzotriazolated adducts **286** in 66–80% yield. The adducts **286** are readily transformed into the MBH olefins **287** by treatment with sodium hydride, demonstrating the convenience of *N*-(α-aminomethyl)benzotriazoles **285** as iminium ion equivalents with α-H atoms generated *in situ* (Scheme 2.156).[164]

2.5 Chalcogenide-Lewis Acid Mediated System[165]

Although the rates of MBH reactions can be improved significantly with the assistance of $TiCl_4$ with or without additive, the yields are generally only

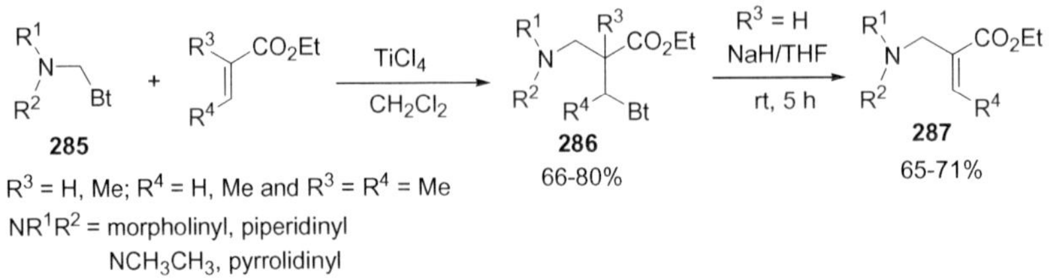

R1
N
R2
Bt
285
R3
CO2Et
R4
TiCl4
CH2Cl2
R1
N
R2
R3
CO2Et
R4
Bt
286
66-80%
R3 = H
NaH/THF
rt, 5 h
R1
N
R2
CO2Et
R4
287
65-71%
R3 = H, Me; R4 = H, Me and R3 = R4 = Me
NR1R2 = morpholinyl, piperidinyl
NCH3CH3, pyrrolidinyl

Scheme 2.156

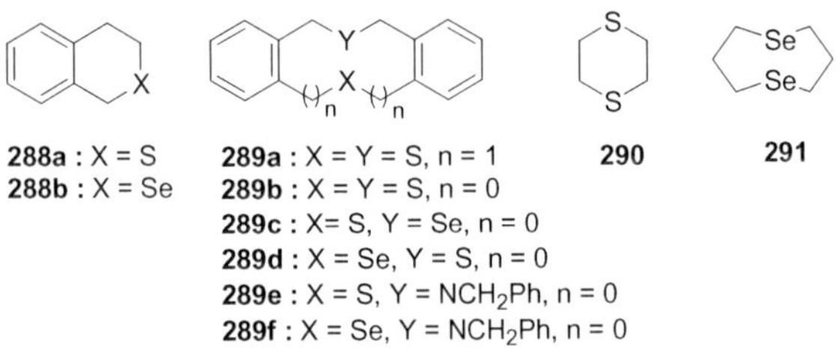

288a : X = S
288b : X = Se
289a : X = Y = S, n = 1
289b : X = Y = S, n = 0
289c : X= S, Y = Se, n = 0
289d : X = Se, Y = S, n = 0
289e : X = S, Y = NCH2Ph, n = 0
289f : X = Se, Y = NCH2Ph, n = 0
290
291

Figure 2.17 Sulfides and selinides used in the chalcogeno-MBH reaction of aldehydes.

moderate, and limitations are encountered in terms of the structures of the Michael acceptors and the aldehydes that undergo addition, with complex mixtures frequently produced. Inspired by the reaction mechanism of vinyl selenonium salts with nucleophiles to give the apparent substitution products with retention of configuration,[166] Kataoka *et al.* have developed an interesting tandem Michael aldol reaction of electron-deficient alkenes with electrophiles mediated by a chalcogenide and Lewis acid. The chalcogenide-Lewis acid mediated reactions produced the MBH adducts with or without work up with a base; therefore, this reaction can be used in a simliar way to the MBH reaction and can be called a "chalcogeno MBH reaction". We will now discuss this kind of reation in detail.

2.5.1 Chalcogenide–$TiCl_4$ Mediated System

Kataoka *et al.*[143a,167] were the first to investigate the chalcogeno-MBH reaction of aldehydes by the use of sulfides or selinides (Me_2S, PhSMe and **288**–**291**) (Figure 2.17) in the presence of Lewis acid. The reaction was applied to activated alkenes such as enones, including β-substituted derivatives, acrylonitrile, methyl acrylates and phenyl vinyl sulfone. Selinide **291** and Me_2S proved to be effective for this reaction; selinide **291** gave the best results due to the trans-annular interaction between the selenium atoms. They also examined the application of various Lewis acids, such as $BF_3 \cdot OEt_2$, $SnCl_4$, $AlCl_3$, $EtAlCl_2$, Et_2AlCl, $HfCl_4$ and $Hf(OTf)_4$, and found that $TiCl_4$ offered better results,

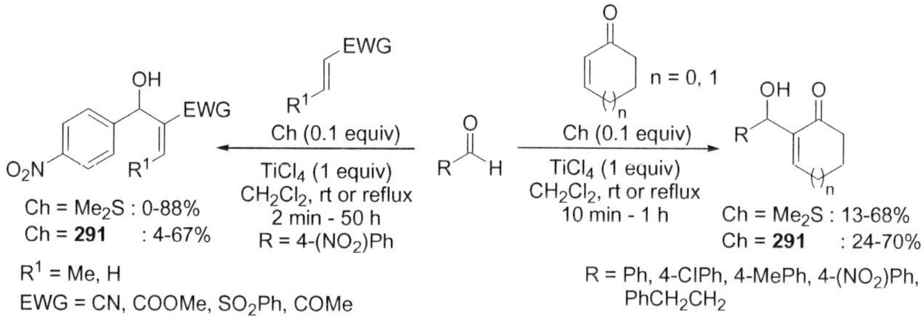

Scheme 2.157

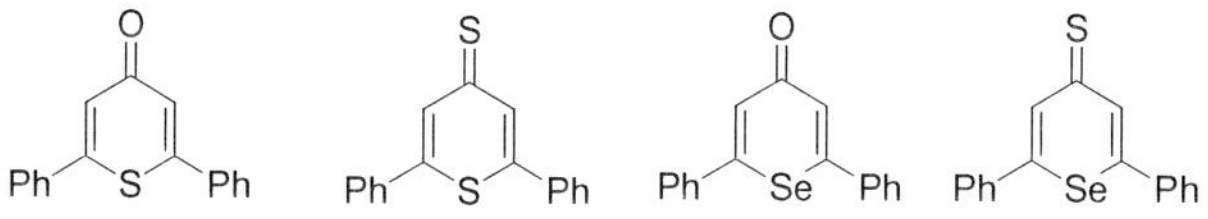

Figure 2.18 $4H$-Chalcogenoyran-4-ones and -4-thiones used as catalysts for the chalcogeno-MBH reaction.

providing the MBH adducts in moderate to good yields after purification of the raw products by preparative TLC (PTLC) on silica gel (Scheme 2.157).

Later, the same authors examined the applicability of 2,6-diphenyl-$4H$-chalcogenopyran-4-ones **292** and 2,6-diphenyl-$4H$-chalcogenopyran-4-thiones **293** (Figure 2.18) as a new series of catalysts for the MBH reaction. Both **292b** and **293a** are a more efficient kind of catalyst than Me_2S and offer better results in the presence of $TiCl_4$ (Scheme 2.158).[168]

Although the chalcogeno-MBH reaction using enones as activated alkenes proceeded smoothly under mild conditions, the reactions of methyl acrylate were somewhat more complicated. Thus, Kataoka *et al.* have extended this strategy to thioacrylates and envisaged that the use of acrylic thioesters would give better results than those obtained with methyl acrylate. The reaction of ethyl thioacrylate with aldehydes catalyzed by a catalytic amount of Me_2S in the presence of $TiCl_4$ proceeds smoothly and provides the chloro aldol products **294** as major products along with minor amounts of the usual MBH adducts **295**. Further treatment of the crude mixture with DBU or Et_2NH or $Ti(OPr^i)_4$ provided the desired MBH adducts **295** and the isopropyl esters of the MBH adducts **296**, respectively (Scheme 2.159). Notably, DABCO and Bu_3P were ineffective for the MBH reaction of a thioacrylate.[169]

On the basis of all these transformations, Kataoka *et al.* have reinvestigated the products and reaction mechanism of the chalcogeno-MBH reaction. They found that, when purifying the product from the reaction of *p*-nitrobenzaldehyde with methyl vinyl ketone by column chromatography on silica gel, chloromethyl aldol **251a** was obtained in 95% yield as a mixture of

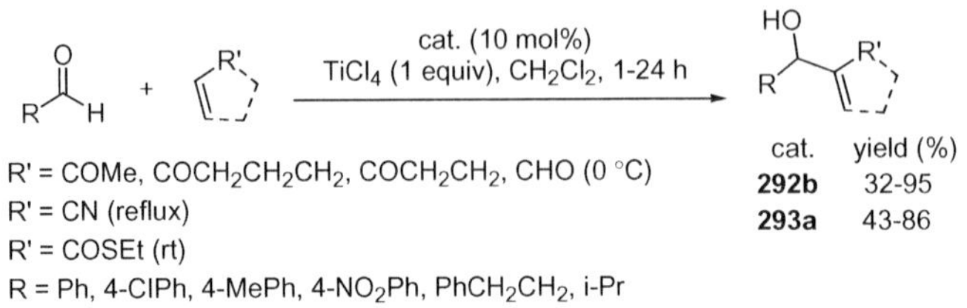

cat. (10 mol%)
TiCl4 (1 equiv), CH2Cl2, 1-24 h
R' = COMe, COCH2CH2CH2, COCH2CH2, CHO (0 °C)
R' = CN (reflux)
R' = COSEt (rt)
R = Ph, 4-ClPh, 4-MePh, 4-NO2Ph, PhCH2CH2, i-Pr
cat. yield (%)
292b 32-95
293a 43-86

Scheme 2.158

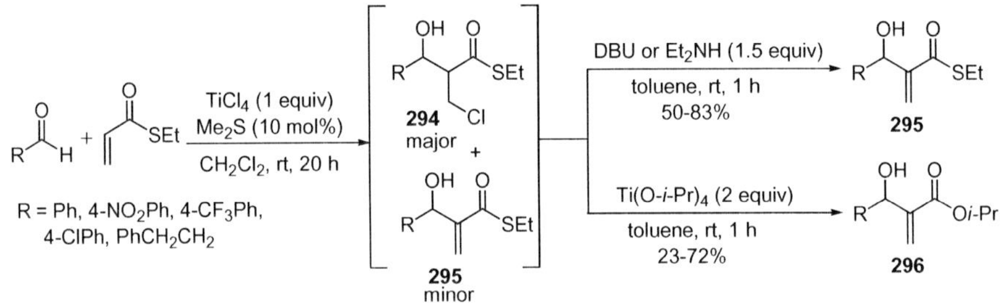

TiCl4 (1 equiv)
Me2S (10 mol%)
CH2Cl2, rt, 20 h
R = Ph, 4-NO2Ph, 4-CF3Ph, 4-ClPh, PhCH2CH2
294 major
295 minor
DBU or Et2NH (1.5 equiv)
toluene, rt, 1 h
50-83%
295
Ti(O-i-Pr)4 (2 equiv)
toluene, rt, 1 h
23-72%
296

Scheme 2.159

1. Ch, TiCl4, CH2Cl2
2. PTLC on silica gel
253a
Ch, TiCl4, CH2Cl2
DBU, toluene or
PTLC on silica gel
syn-251a
anti-251a
Ch: Me2S, 74% yield, syn:anti = 7:1
293a, 95% yield, syn:anti = 3:1

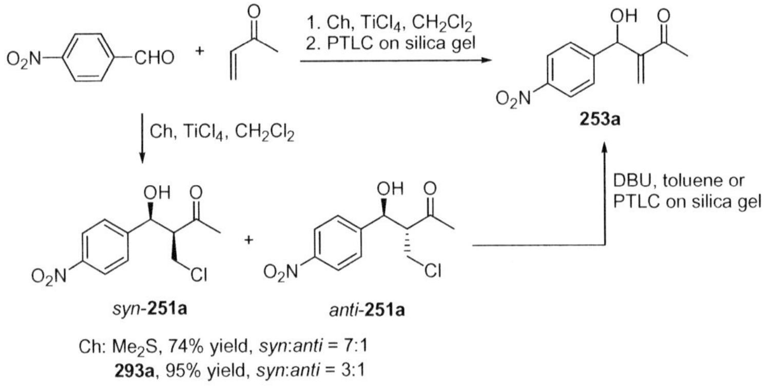

Scheme 2.160

two diastereoisomers in a ratio *syn* to *anti* of 3 : 1. Moreover, **251a** was submitted to PTLC on silica gel, and dehydrochlorination took place to afford MBH adduct **253a**. The chloride **251a** was also transformed into the MBH adduct **253a** upon treatment with DBU (Scheme 2.160).[170] Scheme 2.161 describes the mechanism proposed to explain the formation of different products under the different conditions.

During our studies on the chalcogeno-MBH reaction between aldehydes and methyl vinyl ketone (MVK), we found that this reaction can be drastically

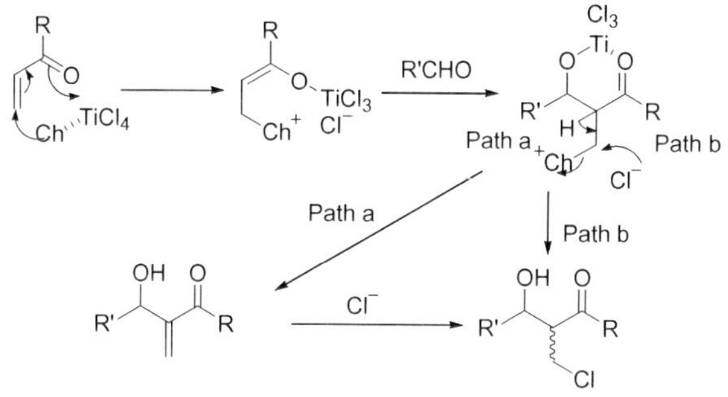

Scheme 2.161

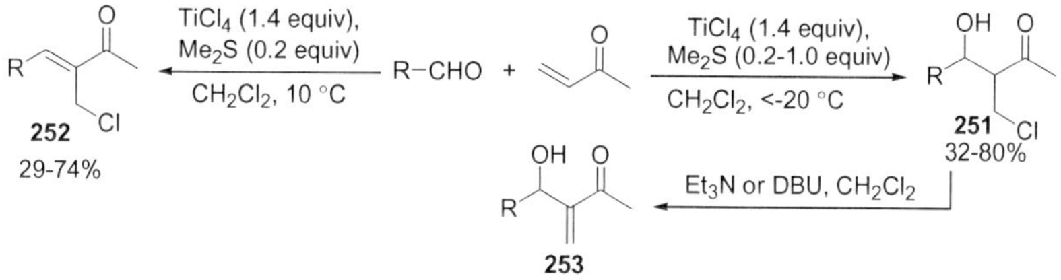

Scheme 2.162

affected by the reaction temperature and Lewis base.[141b] When the reaction was carried out at $<-20\,^{\circ}\mathrm{C}$ using methyl sulfide as a Lewis base in the presence of $TiCl_4$, the chlorinated compound **251** was obtained as the major product. However, if the reaction was carried out at room temperature $(10\,^{\circ}\mathrm{C})$ in the presence of $TiCl_4$, the elimination compound **252** [with (Z)-configuration] rather than **253** was formed as a major product (Scheme 2.162). Notably, this reaction was not affected by the Lewis base (Me_2S) as the same products were obtained in similar yields in the presence or absence of Lewis base (Me_2S). Concerning the reaction mechanism, we believe that at room temperature the chloride at the titanium metal can attack directly at the vinyl carbon atom, which combined with dehydration gives **252**.

Though the attempts to synthesize ethyl 2-hydroxy-2-phenyl-3-methylene-4 oxopentanoate *via* DABCO-catalyzed coupling of methyl vinyl ketone with ethyl phenylglyoxylate (α-keto ester) were unsuccessful, the chalcogeno-MBH reaction between α-keto esters and alkyl vinyl ketones proceeded smoothly under the catalytic influence of dimethyl sulfide in the presence of $TiCl_4$, affording 2-aryl-2-hydroxy-3-methylene-4-oxoalkanoates **297** in moderate to good yields (Scheme 2.163).[171]

Kataoka *et al.* have further extended the chalcogeno-MBH reaction to activated alkynes. Thus, the reactions of 3-butyn-2-one with aromatic aldehydes proceeded smoothly and gave (E)-β-halo-α-(α-hydroxybenzyl)enones **298** in high yields (Scheme 2.164).[172] In contrast, reactions of methyl propiolate

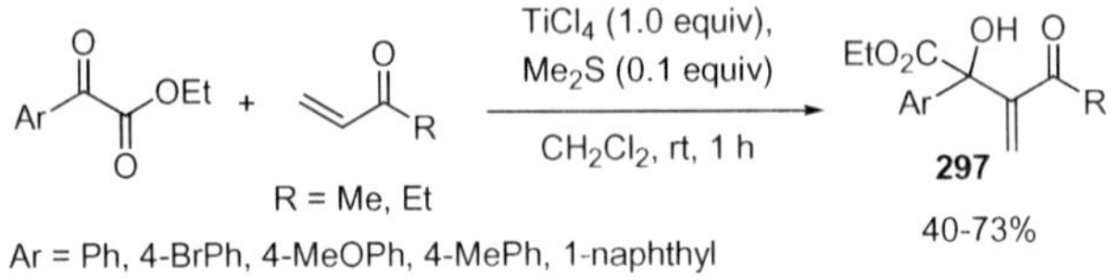

Scheme 2.163

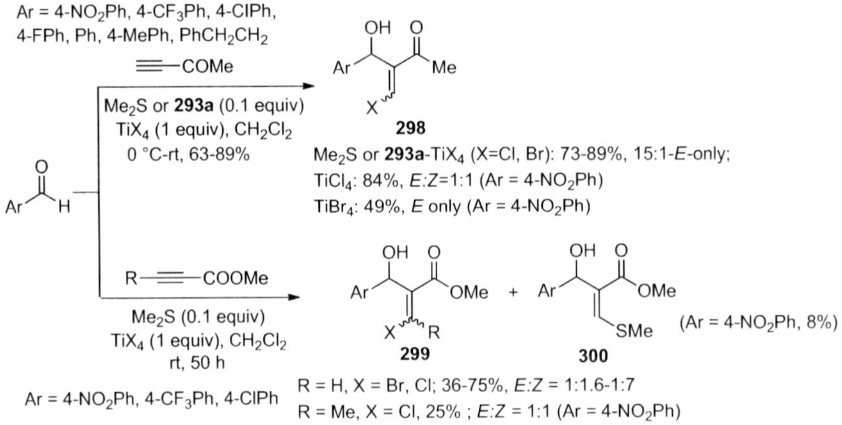

Scheme 2.164

provided 3-chloro-2-(α-hydroxybenzyl)acrylates **299** in 36–75% yields as a mixture of geometrical isomers ($E : Z = 1 : 1.6{-}1 : 7$). When a chalcogenide catalyst was not used, the isomer ratios were changed in the case of 3-butyn-2-one, and no product was formed in the reaction of methyl propiolate. In addition, a small amount of vinyl sulfide **300** was obtained in 8% yield from the reaction of methyl propiolate by using Me_2S–$TiBr_4$. All these findings indicate that a chalcogenide plays an important role in the reaction of electron-deficient alkynes (Scheme 2.164).[172b]

Moreover, Me_2S–$TiCl_4$ catalyzed the reaction of dimethyl acetylenedicarboxylate (DMAD) with *p*-nitrobenzaldehyde to afford dimethyl maleate derivative **301** (30%) and 5-oxo-2,5-dihydrofuran-3-carboxylate **302** (10%) (Scheme 2.165). When tetramethylthiourea was used instead of Me_2S, the yields of both **301** and **302** were significantly increased (to 40%).[172b]

Medvedeva *et al.* first investigated the substituted propynals **303** as aldehyde substrates involved MBH reactions. The initial attempt to conduct the MBH reaction in the presence of $TiCl_4$ was unsuccessful because of oligomerization of the reaction mixture. However, promotion with $TiCl_4$ and a catalytic amount of dimethyl sulfide in the chalcogeno-MBH reaction led to 2-[(Z,E)-chloromethylidene]-3-hydroxy-1-phenyl-5-organylpent-4-yn-1-ones **304** by tandem α-hydroxyethynylation/β-chlorination of 1-phenylprop-2-yn-1-one (Scheme 2.166).[173]

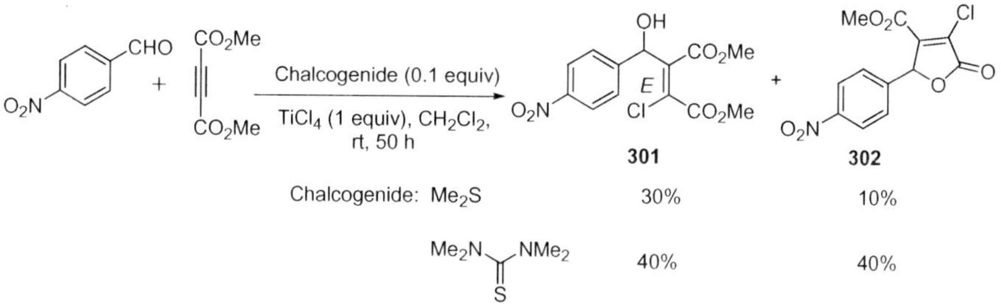

Scheme 2.165

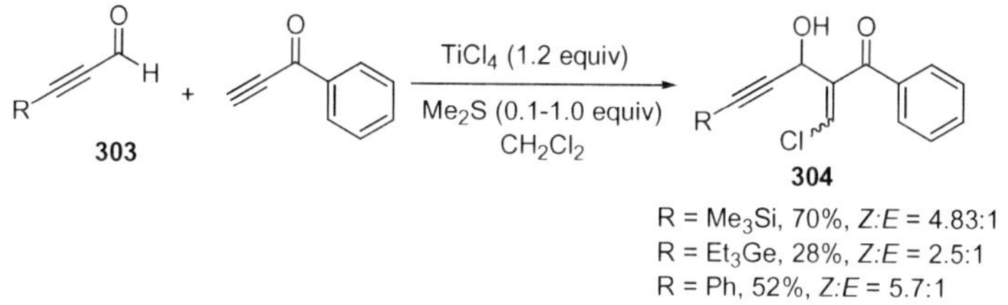

Scheme 2.166

Kataoka and co-workers[174] have also investigated the asymmetric version of the chalcogeno-MBH reaction by using various bifunctional catalysts containing chalcogenide and alcohol, ether or amine groups, **305–309**. Some of the hydroxyl chalcogenides gave the corresponding MBH adducts in very high yields but with low enantioselectivities (Scheme 2.167). The best enantioselectivity was obtained in the reaction between hydrocinnamaldehyde and MVK in the presence of 10-methylthioisoborneol **307b** (Scheme 2.168). Subsequently, they examined the application of several C_2-symmetric bidentate ligand–TiCl$_4$ complexes (including BINOL and bisoxazoline). However, the resulting adducts were obtained in low enantiomeric purities (maximum up to 7% ee).

Subsequently, Bauer and Tarasiuk[175] demonstrated the first example of a chalcogeno-MBH reaction using chiral glyoxylic acid derivatives, (–)-8-phenylmenthyl glyoxylate **310**, as an electrophile. The reaction with cyclic enones proceeds smoothly under the catalytic influence of dimethyl sulfide in the presence of TiCl$_4$, affording the desired adducts **311** in good yields with high diastereoselectivities (>95% de) (Scheme 2.169).

Later, Shaw and co-workers further employed sugar derived aldehydes **312** as electrophiles in the chalcogeno-MBH reaction (Scheme 2.170).[176] The resulting allyl chlorides **313** can be easily transformed into allylamines **314** *via* treatment with various amine (Et$_2$NH, pyrrolidine, piperidine and piperazine derivatives) (Scheme 2.170). They also evaluated these allyl chlorides and allylamines for their biological activity and found that (Z)-keto allyl chlorides possess antimycobacterial activity.[176b]

305a: X = S, n = 1 (34%, <1% ee)   **307a:** R^1 = Bn; R^2 = H (41%, 15% ee)
305b: X = S, n = 2 (49%, 3% ee)   **307b:** R^1 = Me; R^2 = H (95%, 2% ee)
305c: X = Se, n = 2 (44%, 3% ee)   **307c:** R^1= R^2 = Me (97%, 1% ee)

306 (26%, 1% ee)   **308** (44%, 8% ee)   **309** (37%, --)

Scheme 2.167

R = 4-CF$_3$Ph, 4-ClPh, 4-NO$_2$Ph, 3-pyridyl,   22-43%
4-pyridyl, PhCH$_2$CH$_2$ 8-74% ee

Scheme 2.168

310   n = 0, 1   **311**

76-78%, >95% *de*

Scheme 2.169

More recently, Verkade *et al.* have developed a highly active and generally effective proazaphosphatrane sulfide (**317**)/TiCl$_4$ catalyst system for the MBH reaction (Scheme 2.171). This protocol is applicable to the reaction of various aldehydes with activated alkenes such as cyclic or acyclic enones (including less reactive β-substituted derivatives), acrylonitrile and acrylates to give the corresponding adducts in good to high yields. It is believed that N→P intra-bridgehead interaction in **317** may play an important role in stabilizing a **317**/activated alkene/TiCl$_4$ intermediate in the reaction pathway.[177]

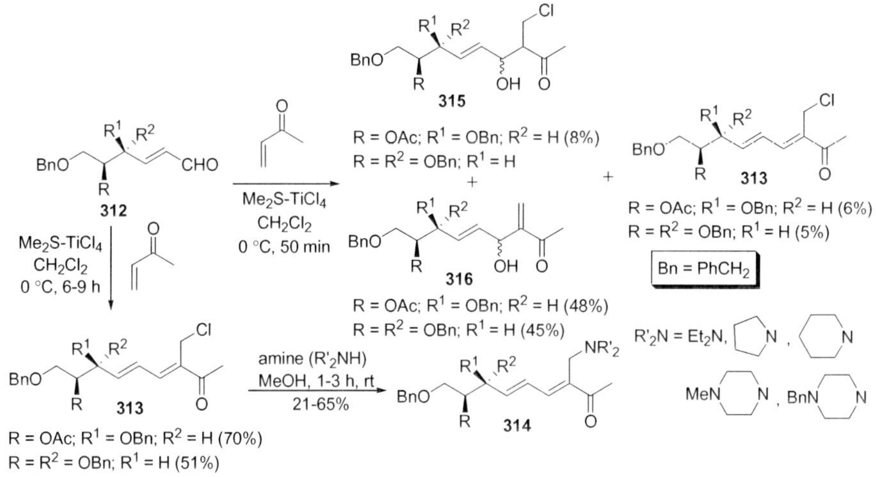

Scheme 2.170

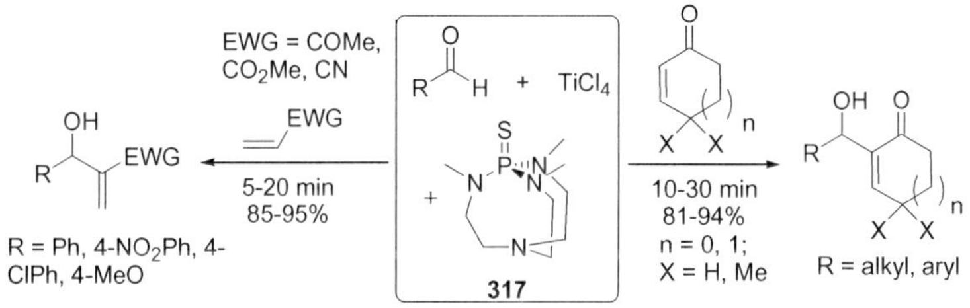

Scheme 2.171

2.5.2 Chalcogenide-mediated System in the Presence of Other Lewis Acids

As mentioned above, $BF_3 \cdot$ etherate was found to be ineffective for the chalcogenide-promoted MBH reaction[143a,167] However, Kataoka *et al.* envisaged that BBr_3 or BCl_3 would be reactive in this reaction because the bromide or chloride ion is more easily liberated *in situ* from the Lewis acid by the coordination of BBr_3 or BCl_3 with carbonyl oxygen than the fluoride ion. Thus, they developed the chalcogenide-MBH reaction between alkyl vinyl ketones and aldehydes in the presence of $BBr_3 \cdot SMe_2$ or $BCl_3 \cdot SMe_2$, which provided MBH adducts, aldol products and α-halomethyl enones when the reaction was quenched with $NaHCO_3$, and allyl halides or MBH adducts when the reaction was quenched with H_2O or trimethylamine (Scheme 2.172).[144] It was suggested that this reaction proceeds in a similar manner to the $TiCl_4$–Me_2S-mediated reactions, and the α-halomethylene aldols were formed initially, but the aldol–boron complexes generated *in situ* were more stable than the aldol–titanium

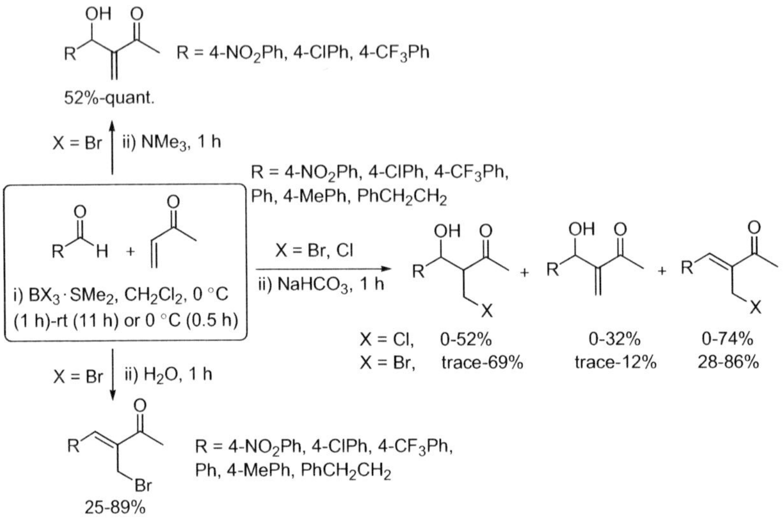

Scheme 2.172

complexes, which could not be completely decomposed by the treatment with water or a saturated aqueous sodium bicarbonate solution, thus affording α-halomethyl enones as major product.

They also observed that the reaction of 4-nitrobenzaldehyde with methyl acrylate or cyclohex-2-en-1-one or thioacrylate in the presence of $BBr_3 \cdot SMe_2$ gave exclusively MBH adducts **318** when quenched with $NaHCO_3$. A similar reaction of 4-nitrobenzaldehyde with ethyl thioacrylate in the presence of $BBr_3 \cdot SMe_2$ provided the corresponding allyl bromide **319** after aqueous workup. Interestingly, the reaction between 4-nitrobenzaldehyde and cyclohex-2-en-1-one in the presence of $BBr_3 \cdot SMe_2$ provided 4-nitrobenzylphenol **320** and acetal **321** (due to subsequent aromatization) after an aqueous workup (Scheme 2.173).[144]

Later, Goodman and co-workers found that the use of tetrahydrothiophene instead of Me_2S or 2,6-diphenyl-4*H*-thiopyran-4-one (**292a**) can efficiently promoted $BF_3 \cdot OEt_2$-mediated chalcogenide-MBH reaction of MVK with aldehydes, in which the sulfide directly participates *via* Michael addition to give α,β-unsaturated ketone. The reaction proceeds rapidly and is general with respect to several aldehydes, giving the MBH adducts in moderate yields after work up with triethylamine (Scheme 2.174). They also used enantiopure sulfide **322**, which provided the desired MBH adducts in up to 53% enantiomeric excess (Scheme 2.175).[178]

Indeed, the first example of other Lewis acid, except for $TiCl_4$ and BX_3, catalyzed chalcogenide-MBH reactions dates back to the 1980s, when Noyori *et al.* first demonstrated that the conjugated addition of a silyl selenide to an enone catalyzed by TMSOTf provided a silyl enol ether, which can be successfully engaged in subsequent Mukaiyama–retro-aldol reactions with various

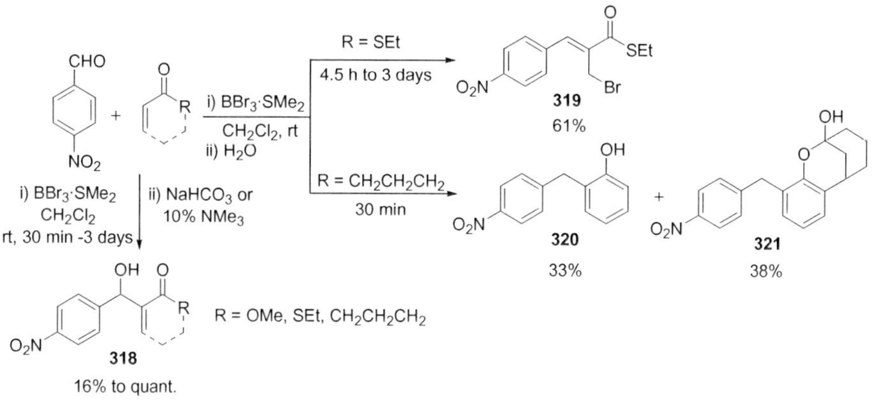

Scheme 2.173

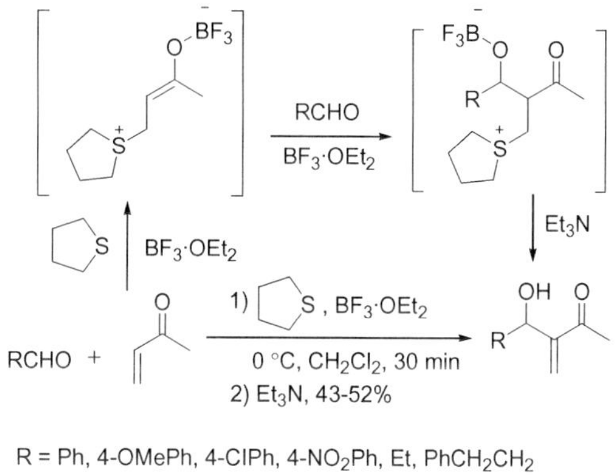

Scheme 2.174

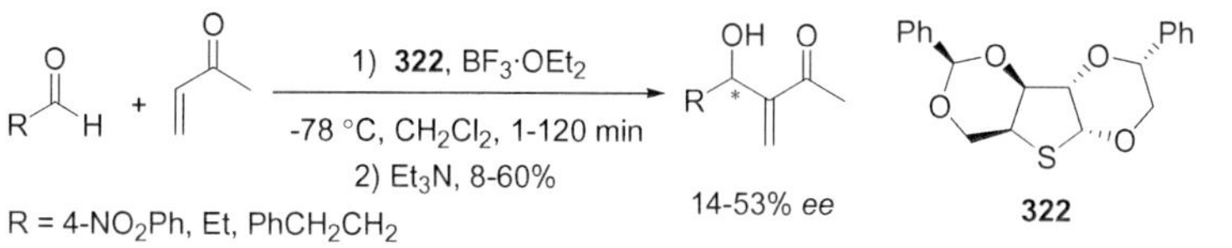

Scheme 2.175

acetals to form the corresponding MBH adducts.[179] Later, such a process was also proven by means of silyl amines (*via* **323c**), silyl sulfides (*via* **323d**) and pyridine (*via* **323e**). Scheme 2.176 gives an example.[180]

Basavaiah *et al.* have developed a similar domino Michael–Mukaiyama–retro-aldol process of cyclic enones with acetals, which conveniently were promoted by mixing a sulfide, TBDMSOTf, in the presence of the Hünig's base,

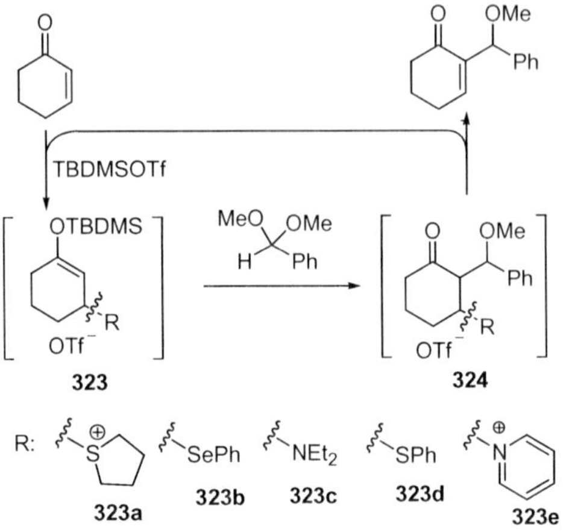

Scheme 2.176

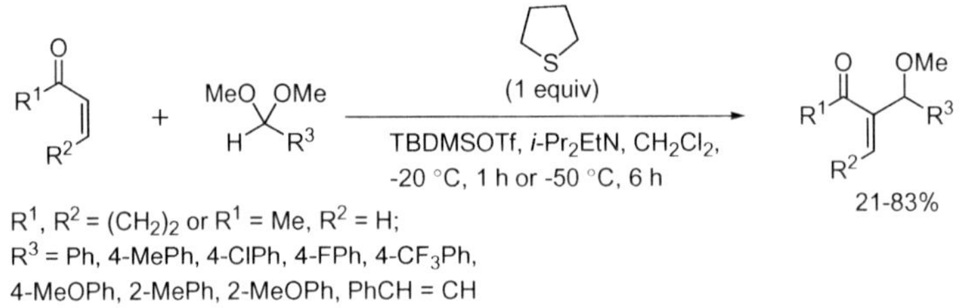

Scheme 2.177

to furnish the MBH adducts under very mild conditions within a reasonably short reaction time (Scheme 2.177). Notably, as an alternative chalcogenide-MBH reaction, this reaction requires an equal amount of both an enone and an acetal, which is different from usual MBH reaction, which requires an excess of either the aldehydes or the polymerizable enone.[181]

Recently, a new intramolecular hydroxyl sulfide (**325**) catalyzed MBH reaction, using alkyl halides or epoxides **327** as electrophiles and lactams or lactones **326** as substrates, was performed successfully under the basic conditions (Scheme 2.178). However, the procedure failed for aldehydes when a conventional acid (TiCl$_4$)-catalyzed procedure was used.[182]

2.5.3 Substrates Containing a Chalcogenide Group

Since the mechanism of chalcogenide-MBH reaction involved the Michael addition of chalcogenide to electron-deficient olefins and the intramolecular Michael addition reaction of a sulfide group to an enone moiety in an acidic medium is known,[183] Kataoka *et al.* envisaged that the tandem Michael aldol

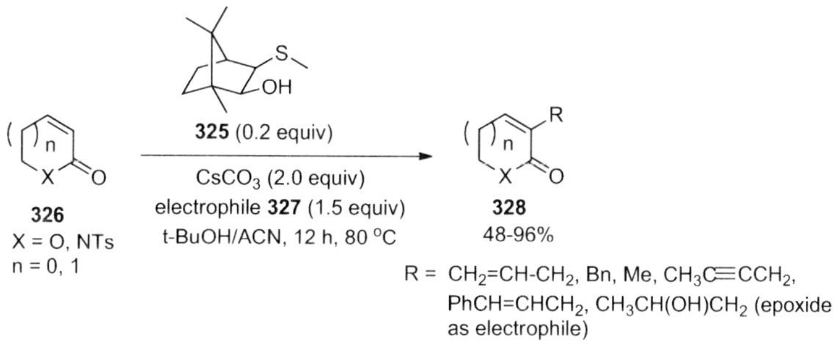

325 (0.2 equiv)

326
X = O, NTs
n = 0, 1

CsCO$_3$ (2.0 equiv)
electrophile **327** (1.5 equiv)
t-BuOH/ACN, 12 h, 80 °C

328
48-96%

R = CH$_2$=CH-CH$_2$, Bn, Me, CH$_3$C≡CCH$_2$,
PhCH=CHCH$_2$, CH$_3$CH(OH)CH$_2$ (epoxide
as electrophile)

Scheme 2.178

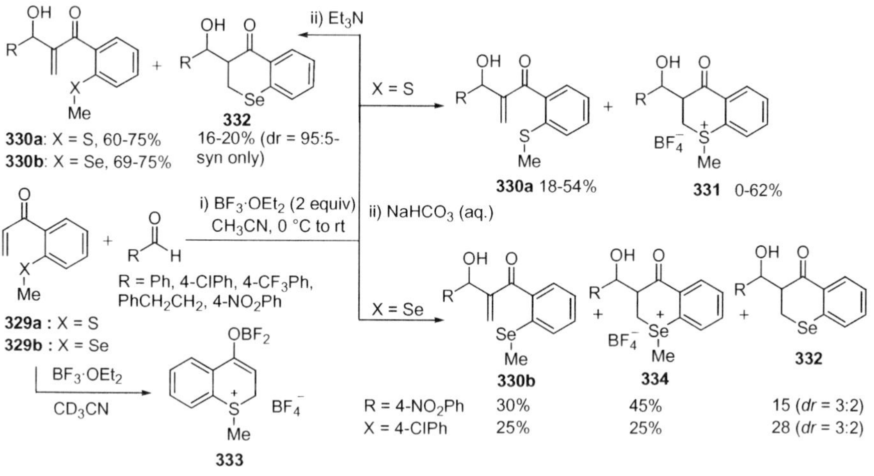

ii) Et$_3$N

330a: X = S, 60-75%
330b: X = Se, 69-75%

332
16-20% (dr = 95:5-
syn only)

X = S

330a 18-54%

331 0-62%

i) BF$_3$·OEt$_2$ (2 equiv)
CH$_3$CN, 0 °C to rt

ii) NaHCO$_3$ (aq.)

329a : X = S
329b : X = Se

R = Ph, 4-ClPh, 4-CF$_3$Ph,
PhCH$_2$CH$_2$, 4-NO$_2$Ph

X = Se

330b

334

332

BF$_3$·OEt$_2$

CD$_3$CN

333

R = 4-NO$_2$Ph   30%   45%   15 (dr = 3:2)
X = 4-ClPh   25%   25%   28 (dr = 3:2)

Scheme 2.179

reaction can proceed *via* a key step, namely, the intramolecular Michael addition of the chalcogenide to the enone moiety activated by a Lewis acid whose conjugate base had very low nucleophilicity. They developed a self-assisted (intramolecular) chalcogeno-MBH reaction of 2-(methylchalcogeno)phenyl vinyl ketone **329** with aryl aldehydes catalyzed by BF$_3$ · OEt$_2$. The quenching method influenced the products. Treatment of the reaction mixture with saturated NaHCO$_3$ gave the corresponding MBH adducts **330a** and onium salts **331**, whereas treatment with trimethylamine afforded MBH adducts **330a** in the case of sulfide **329a**, MBH adducts **330b** and the demethylated product **332** in the case of seleno congener **329b** (Scheme 2.179). Moreover, the sulfonio-enolate intermediate **333** was detected by ^{1}H NMR spectroscopic analysis of the reaction of **329a** and BF$_3$ · Et$_2$O in CD$_3$CN, indicating that the intermediate of the chalcogenide-MBH reaction was a boron enolate. Other Lewis acids such as TiF$_4$ (33% yield) or Yb(OTf)$_3$ (21% yield) also promoted the reaction.[184]

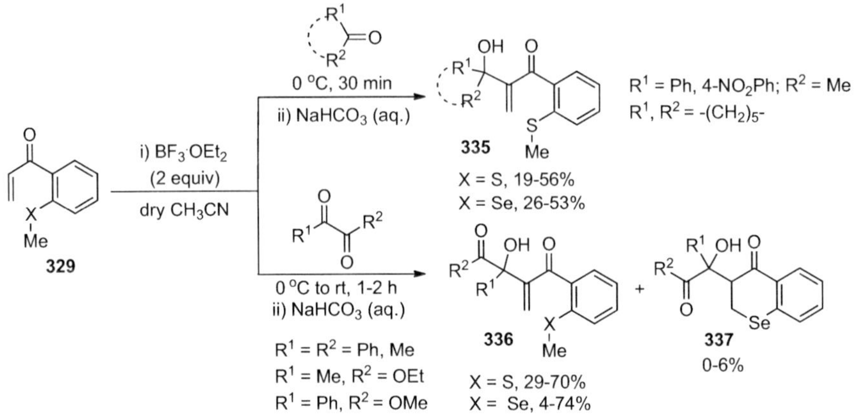

Scheme 2.180

Though the use of carbonyl compounds in MBH reactions is restricted it is, however, well known that boron enolates react with carbonyl compounds under mild reaction conditions. Ketones, α-diketones and α-keto esters, which hardly react in the traditional MBH reaction, similarly reacted with 2-(methylchalcogeno)phenyl vinyl ketone **329** in the presence of $BF_3 \cdot Et_2O$ to provide the corresponding MBH adducts **335** and **336** in moderate to good yields, together with a small amount of selenochromanones **337** as by-product in the case of seleno congener **329b** with α-keto esters (Scheme 2.180).[184b,c,185]

Instead of aldehyde, cyclic and acyclic acetals have also been used as electrophiles on treatment with a Lewis acid for the chalcogeno-MBH reaction. The corresponding MBH adducts **338** and **339** were obtained by the reaction of 2-(methylchalcogeno)phenyl vinyl ketone with benzaldehyde dimethyl acetals and trimethyl orthoformate in the presence of $BF_3 \cdot Et_2O$ in good yields. When the reaction mixture was worked up with a saturated $NaHCO_3$ solution instead of Et_3N, onium salts **340** were obtained together with MBH adducts **338** in the case of benzaldehyde dimethyl acetals (Scheme 2.181). Reactions with cyclic acetal gave MBH adducts **342** accompanied by dimeric products **343** (Scheme 2.181).[184c,186]

Given that an ynone reacts with $BF_3 \cdot Et_2O$ to form the boron allenolate as an intermediate, Kataoka *et al.* have developed an interesting intramolecular chalcogenide-MBH reaction involving the coupling of 1-(2-methylchalcogenophenyl)propynone **345** with representative aldehydes in the presence of $BF_3 \cdot OEt_2$ to afford 3-(hydroxyalkyl)chalcogenochromen-4-ones **346** (Scheme 2.182).[184c,187]

2.6 Co-catalyzed Systems

As stated in Chapter 1.7, the low reaction rates usually associated with the MBH reaction can be increased by pressure, the use of ultrasound, and by

329
OMe OMe
i) BF₃·OEt₂ (2 equiv)
CH₃CN, -40 °C
ii) Quench
338
X = S, 78% (Et₃N)
       43% (NaHCO₃)
X = Se, 79%
        36%
340 BF₄⁻
40% (Et₃N)
51% (NaHCO₃)
341
5 % (Et₃N)
    (NaHCO₃)
339
X = S, 89% (Et₃N) or 55% (sat. NaHCO₃)
X = Se, 65% (Et₃N) or 54% (sat. NaHCO₃)
CH(OMe)₃
i) BF₃·OEt₂ (3 equiv)
CH₃CN, 0 °C, 2 h
ii) Quench
342
X = S, 56%
       45%
X = Se, 39%
        45%
343
6% (Et₃N)
19% (NaHCO₃)
10%
344
19% (Et₃N)
7% (NaHCO₃)

Scheme 2.181

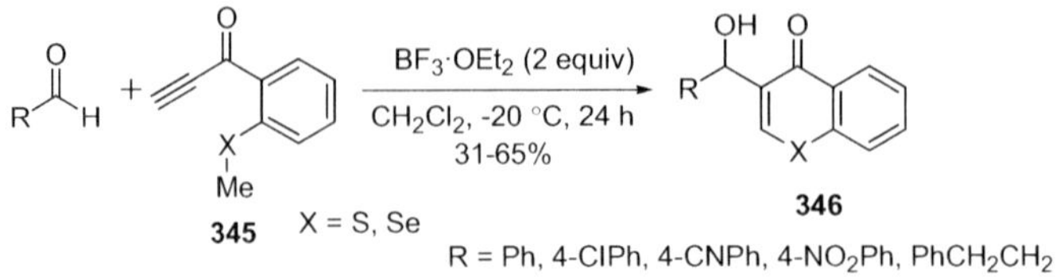

BF3·OEt2 (2 equiv)
CH2Cl2, -20 °C, 24 h
31-65%
X = S, Se
R = Ph, 4-ClPh, 4-CNPh, 4-NO2Ph, PhCH2CH2

Scheme 2.182

imidazole (30 mol%),
L-proline (30 mol%)
DMF, rt, 24-80 h
R = Ph, 4-EtPh, 4-BrPh, 4-ClPh, 2-NO2Ph, 3-NO2Ph, 4-NO2Ph, 2-pyridyl, 3-pyridyl, cinnamyl, n-Pr
30-90%, 5-10% ee

Scheme 2.183

microwave radiation, or by the addition of co-catalysts. Thus, various co-catalyst systems have been developed – in particular, mild Brønsted acid, such as proline, alcohol, phenol, (thio)urea, *etc.* and various Lewis acids.

2.6.1 Proline as Co-catalyst

Although chiral amine and phosphine catalysts continue to play a major role in asymmetric MBH reactions, the simultaneous use of a nucleophilic catalyst in combination with a conveniently chosen acid, such as proline, or proline-containing oligopeptides as co-catalyst has been emerging as an important alternative.[75,188] In these reactions, the co-catalyst accelerates the reaction and acts synergistically, allowing higher ee-values to be obtained than in the absence of the additive. Proline as a co-catalyst has an important effect on the observed stereochemistry of the product, leading to adducts with inverted absolute configuration in some cases.[75,189]

We, for the first time, investigated the proline-catalyzed MBH reaction between aldehydes and MVK in the presence of weak Lewis bases such as imidazole, Et_3N and DABCO.[188b] A rate acceleration of the reaction was clearly observed, with the usual MBH adducts provided in moderate to good yields (Scheme 2.183). However, enantioselectivities in these reactions were quite low (5–10%).[188b]

To improve the enantioselectivities, we then reinvestigated Hatakeyama's cinchona-derived catalyst, β-ICD, promoted MBH reaction of aldehydes with methyl vinyl ketone or (α)-naphthyl acrylate in the presence of co-catalyst. The enantioselectivities increased in some cases, and the absolute configuration of the MBH adducts can be inverted by the use of proline or lithium

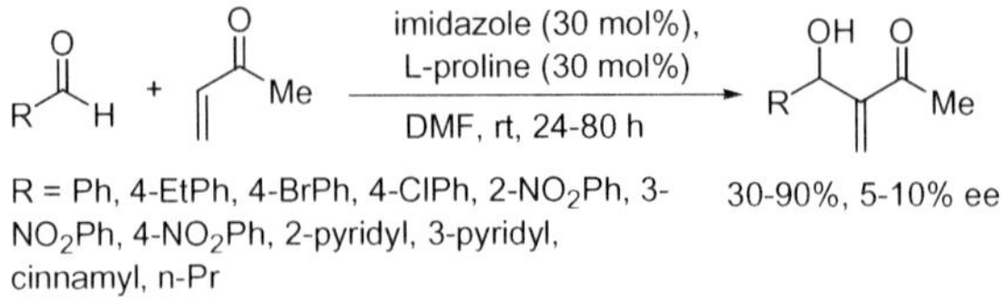

salt additives. When α-naphthyl acrylate and MVK were used as the substrates, the highest ees were achieved in 92% and 49%, respectively (Scheme 2.184).[75]

Miller and co-workers have developed an efficient co-catalytic system involving L-proline and peptides for the asymmetric ketone-based MBH reaction of MVK and aromatic aldehydes to provide the corresponding adducts with up to 81% ee.[189b,190] The reaction of MVK and benzaldehyde afforded the (*R*)-adduct in 78% ee, and the (*S*)-adduct in 31% ee, respectively, when pentapeptide **347** having a modified histidine side-chain was used as catalyst in the presence of L-proline and D-proline as co-catalysts, respectively, showing that the influence of the proline may be greater than that of the peptide in controlling the stereochemistry of the new center (Scheme 2.185). In control experiments, neither proline nor *N*-methylimidazole (NMI, 10 mol.%) is effective independently in terms of rate or enantioselectivity. Notably, when NMI and L-proline were employed together the reaction afforded a near-racemic product, indicating that the influence of the L-proline chirality without the peptide was minimal.

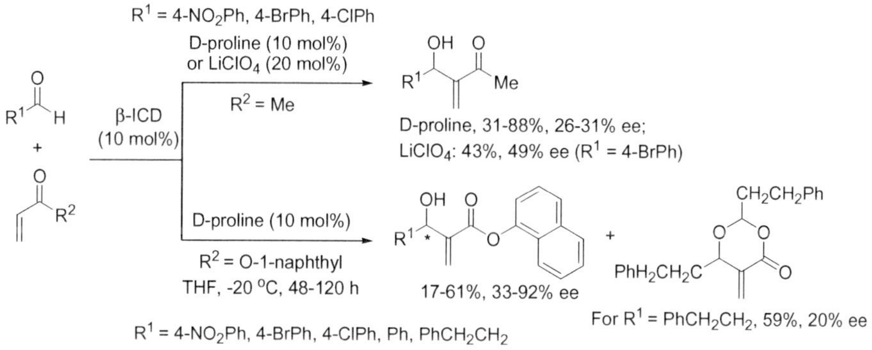

Scheme 2.184

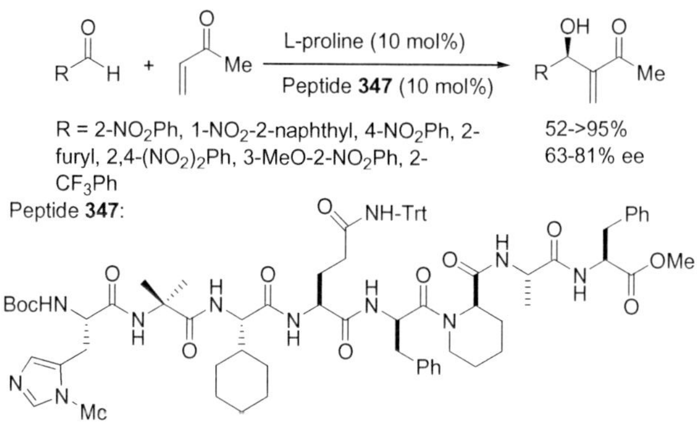

Scheme 2.185

Encouraged by these results, Miller *et al.* continued to investigate the enantioselective intramolecular MBH reaction. Similarly, both NMI and proline alone were not effective and gave a sluggish reaction. However, whereas the combination of NMI and proline provided good catalytic rates, but minimal ee, in the intermolecular reaction, this combination in the intramolecular case led to not only rate enhancement but also a product that exhibited an 80 : 20 enantiomeric ratio (60% ee). In contrast, whereas oligopeptides could be readily found to enhance the ee of the intermolecular process, a screen of approximately 160 peptide-based co-catalysts did not lead to an ee enhancement for the intramolecular reaction. Finally, by using a co-catalyst system involving *N*-methylimidazole and pipecolinic acid, they realized a catalytic asymmetric intramolecular MBH reaction in which an ee of 84% was achieved (Scheme 2.186).[189a]

Recently, Tomkinson *et al.* have reinvestigated the proline/imidazole catalyzed MBH reaction between methyl vinyl ketone and aldehydes, and found that the nature of the solvent is crucial for effective catalyst activity. Addition of small amounts of water brings about a more effective reaction, with a solvent mixture of DMF–H_2O (9 : 1) being optimal. Increasing or decreasing the amount of water present is detrimental to the observed yield. This solvent mixture is effective with various substrates, affording higher yields than in the absence of water (Scheme 2.187).[191]

Polystyrene-supported proline **348** has been used as co-catalyst (10 mol.%) with imidazole (10 mol.%) in the MBH reaction between methyl or ethyl vinyl ketone and aromatic aldehydes (Scheme 2.188). Recycling studies showed that the proline resin can be used up to five cycles in high isolated yields. This study was the first example of supported proline as a heterogeneous co-catalyst in the MBH reaction and broadened the scope of this catalytic material.[192]

Zhou's group has synthesized four types of chiral amines (**349–352**) starting from readily available chiral sources (Figure 2.19). These chiral amines in

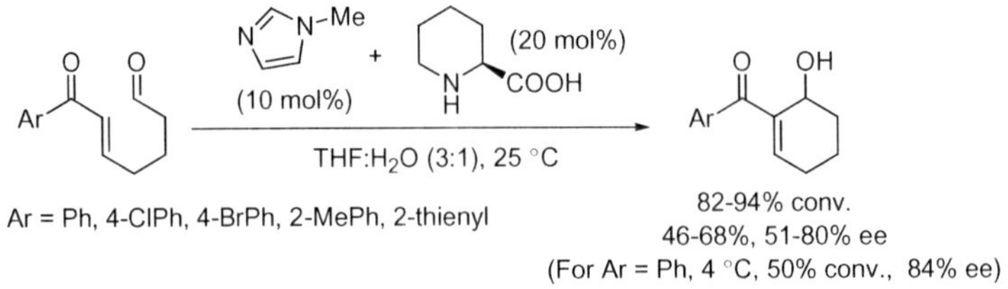

Scheme 2.186

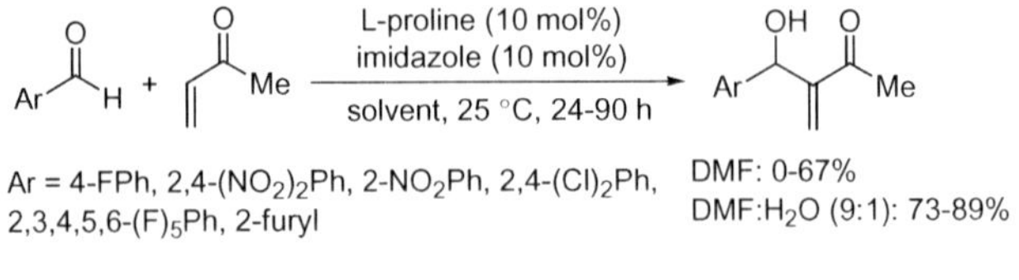

Scheme 2.187

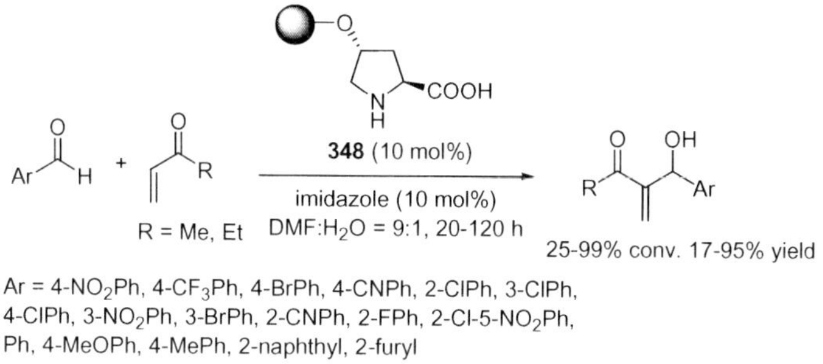

Scheme 2.188

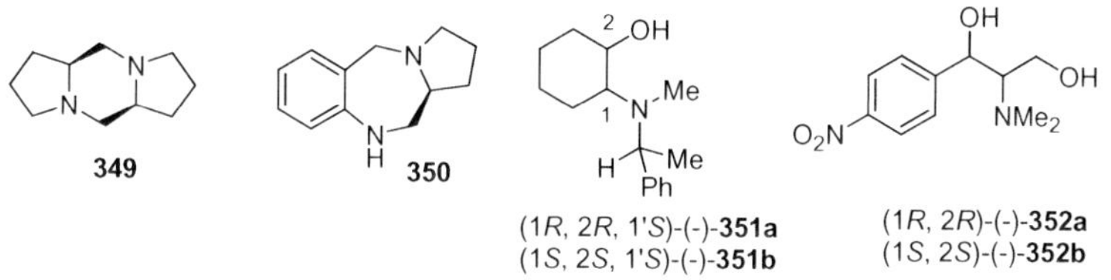

348 (10 mol%)

imidazole (10 mol%)
DMF:H$_2$O = 9:1, 20-120 h

R = Me, Et

25-99% conv. 17-95% yield

Ar = 4-NO$_2$Ph, 4-CF$_3$Ph, 4-BrPh, 4-CNPh, 2-ClPh, 3-ClPh,
4-ClPh, 3-NO$_2$Ph, 3-BrPh, 2-CNPh, 2-FPh, 2-Cl-5-NO$_2$Ph,
Ph, 4-MeOPh, 4-MePh, 2-naphthyl, 2-furyl

349

350

(1*R*, 2*R*, 1'*S*)-(-)-**351a**
(1*S*, 2*S*, 1'*S*)-(-)-**351b**

(1*R*, 2*R*)-(-)-**352a**
(1*S*, 2*S*)-(-)-**352b**

Figure 2.19 The four types of chiral amines synthesized by Zhou's group.

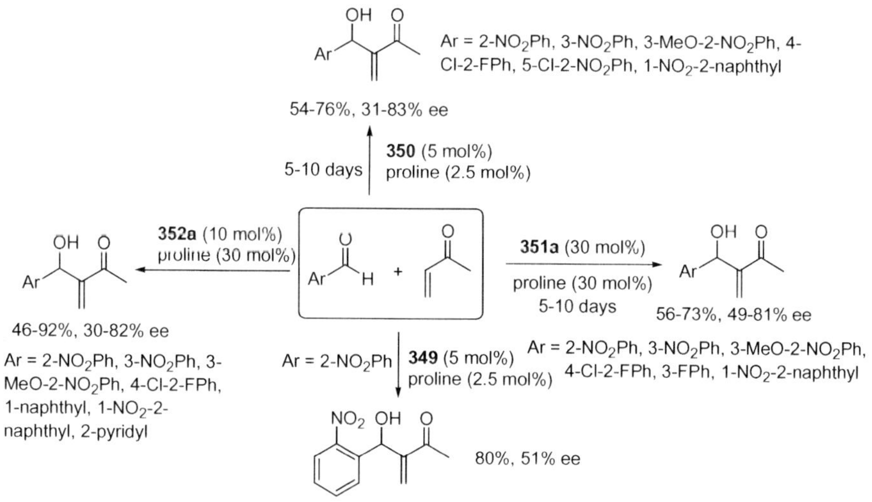

Ar = 2-NO$_2$Ph, 3-NO$_2$Ph, 3-MeO-2-NO$_2$Ph, 4-Cl-2-FPh, 5-Cl-2-NO$_2$Ph, 1-NO$_2$-2-naphthyl

54-76%, 31-83% ee

5-10 days **350** (5 mol%)
proline (2.5 mol%)

352a (10 mol%)
proline (30 mol%)

351a (30 mol%)
proline (30 mol%)
5-10 days

56-73%, 49-81% ee

46-92%, 30-82% ee

Ar = 2-NO$_2$Ph, 3-NO$_2$Ph, 3-MeO-2-NO$_2$Ph, 4-Cl-2-FPh,
1-naphthyl, 1-NO$_2$-2-naphthyl, 2-pyridyl

Ar = 2-NO$_2$Ph **349** (5 mol%)
proline (2.5 mol%)

Ar = 2-NO$_2$Ph, 3-NO$_2$Ph, 3-MeO-2-NO$_2$Ph, 4-Cl-2-FPh, 3-FPh, 1-NO$_2$-2-naphthyl

80%, 51% ee

Scheme 2.189

combination with L-proline have been found to be efficient co-catalysts for the asymmetric MBH reaction between MVK and aromatic aldehydes. The corresponding adducts were formed in reasonable chemical yields with good enantioselectivities (up to 83% ee) (Scheme 2.189). Moreover, parallel co-catalytic reactions with the two enantiomers of chiral amine **352** and L-proline revealed that the proline stereochemistry determines the configuration of the newly formed chiral center. In addition, the existence of the free hydroxy group in the amine enhanced the enantioselectivity of the reaction. Based on these findings, a plausible mechanism for this co-catalytic MBH reaction has been proposed.[193]

Most recently, Gruttadauria *et al.* investigated secondary amino acids such as proline, sarcosine, pipecolinic acid and homoproline in the presence of the sodium hydrogen carbonate catalyzed MBH reaction between methyl or ethyl vinyl ketone and aromatic aldehydes. Of these amino acids, proline was the most efficient catalyst and gave the corresponding adducts in good yields (Scheme 2.190). The obtained data shows that the proline may act as a bifunctional catalyst: the amino group attacks MVK to give the zwitterionic iminium species **A**, which undergoes intramolecular nucleophilic attack by the carboxylate group to give the bicyclic enamino-lactone species **B** as an intermediate. The hydrogen carbonate ion seems to provide hydrogen-bond assistance in the C–C bond formation step (Scheme 2.191). Quantum mechanical calculations support these mechanistic hypotheses.[194]

Recently, Tanaka and Barbas *et al.* have reported an imidazole and proline co-catalytic system for the reaction between enals and *N*-*p*-methoxyphenyl-(PMP)

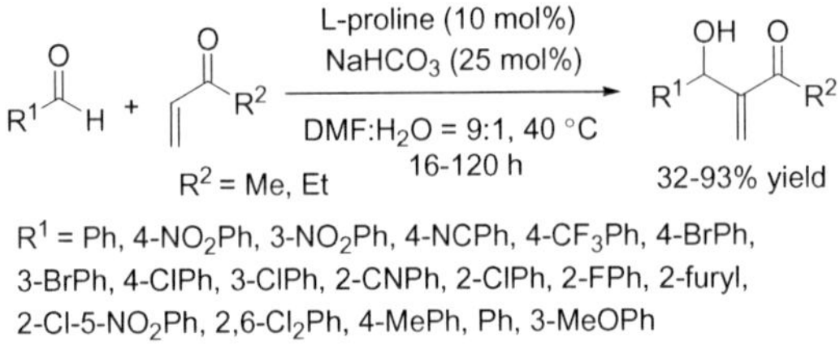

Scheme 2.190

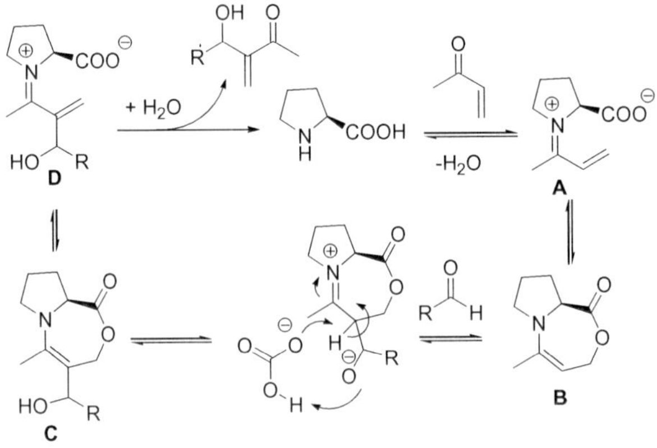

Scheme 2.191

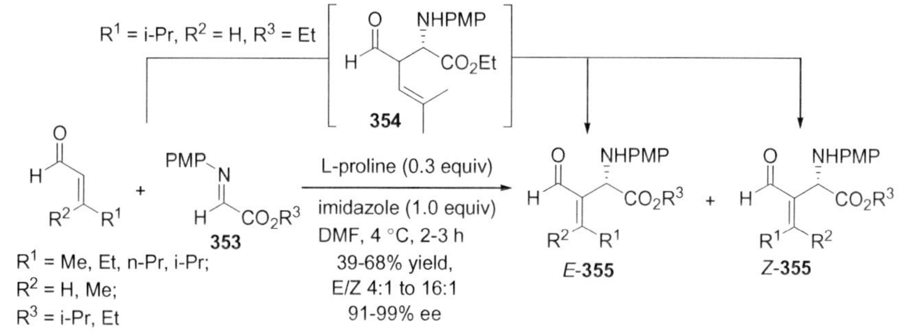

Scheme 2.192

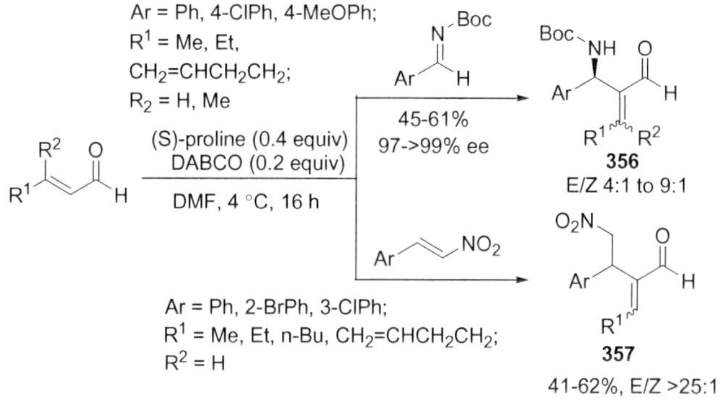

Scheme 2.193

protected α-imino glyoxylate **353** to afford highly enantiomerically enriched aza-MBH type products **355** (Scheme 2.192).[195] The reaction mechanism was proposed to involve a Mannich-type reaction followed by isomerization of the double bond. Subsequently, Córdova *et al.* developed a similar reaction between unmodified α,β-unsaturated aldehydes and *N*-Boc protected aryl imines by using the combined proline and DABCO as catalyst to give the corresponding compounds **356** in good yields with excellent chemo- and enantioselectivity (Scheme 2.193).[196] Encouraged by these results, they also demonstrated the first example of nitroolefins as electrophiles in MBH type reactions with unmodified enals to furnish the corresponding MBH Michael type adducts **357** with an α-alkylidene group in good yields and with excellent (*E*)-selectivity (Scheme 2.193).[197]

2.6.2 (Thio)ureas as Co-catalyst

Hydrogen-bonding interaction plays a crucial role in the molecular recognition and activation processes of various biologically important reactions that are mediated by enzymes and antibodies in living organisms. Recently, it has been shown that a hydrogen-bonding donor can be used as a general acid catalyst for various types of reactions in organic chemistry. The simultaneous use of urea or thiourea,[198] which possess a flat, readily modifiable and rotationally restricted structure with two mutually proximal N–H bonds available for hydrogen bond donation, as co-catalyst for the tertiary amine-promoted MBH reaction of methyl acrylate and various aldehydes was introduced by Connon group.[188a] To examine the catalysts' abilities to promote MBH processes, the relative rate constants of the reaction between methyl acrylate (10 equiv) and benzaldehyde were determined in the presence of DABCO (1 equiv) and catalysts. The results showed that the ureas **358** and **359** and thioureas **360** and **361** have similar abilities to powerful hydrogen bond donors (methanol or water) in accelerating this DABCO-promoted reaction (Scheme 2.194), indicating that both (thio)urea hydrogen atoms are involved in hydrogen bonding. Ureas **358** and **359** were superior to thioureas **360** and **361** in terms of stability and efficiency. Preliminary results implicate a mechanism involving binding to a zwitterionic intermediate/transition state *via* a Zimmerman–Traxler-type transition state **362** for addition of the resulting enolate anion to the aldehyde (Scheme 2.195).

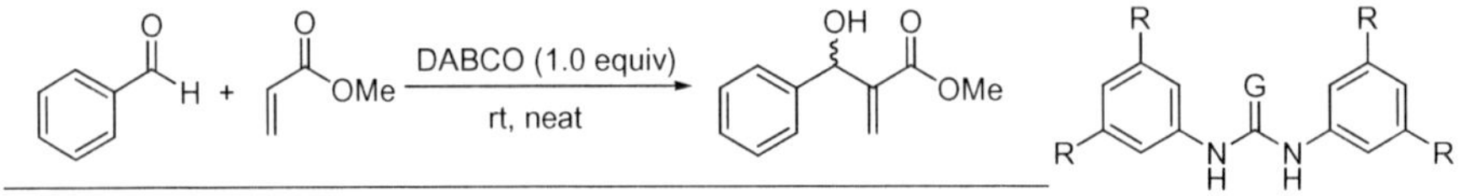

cat. (20mol%)	--	358	359	360	361	359*	MeOH*	H₂O*
k_{rel}:	1.0	5.4	6.7	5.7	3.7	9.4	2.5	1.6

*: 40 mol% catalyst was used

358: G = O, R = F
359: G = O, R = CF₃
360: G = S, R = F
361: G = S, R = CF₃

Scheme 2.194

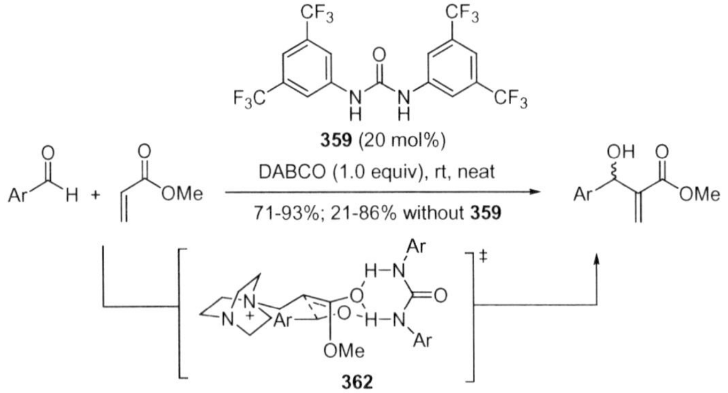

359 (20 mol%)

DABCO (1.0 equiv), rt, neat

71-93%; 21-86% without **359**

362

Scheme 2.195

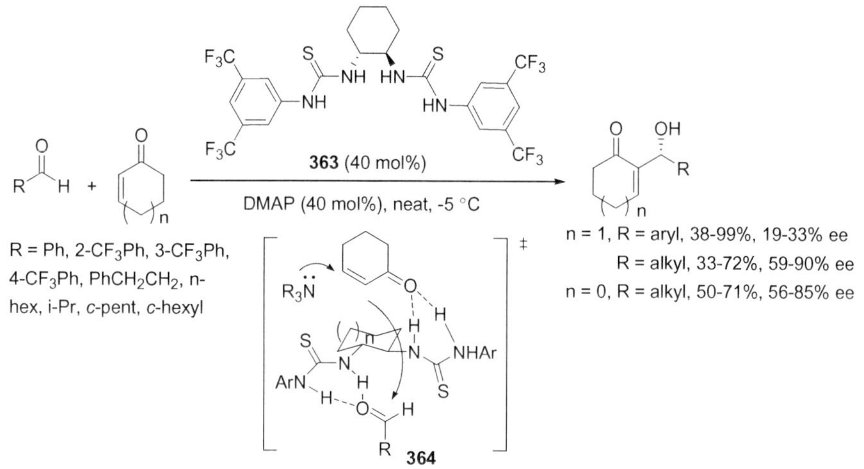

R = Ph, 2-CF$_3$Ph, 3-CF$_3$Ph, 4-CF$_3$Ph, PhCH$_2$CH$_2$, n-hex, i-Pr, c-pent, c-hexyl

363 (40 mol%)

DMAP (40 mol%), neat, -5 °C

n = 1, R = aryl, 38-99%, 19-33% ee
R = alkyl, 33-72%, 59-90% ee
n = 0, R = alkyl, 50-71%, 56-85% ee

364

Scheme 2.196

At almost the same time, Nagasawa's group developed the C$_2$-symmetric chiral bis-thiourea-catalyzed asymmetric MBH reaction of cyclic enones with aldehydes (Scheme 2.196).[199] They found that the bis-thiourea catalyst **363** promoted the reaction of cyclohexenone with benzaldehyde in the presence of an additive (0.4 equiv) such as DMAP or imidazole to afford the allylic alcohol (DMAP: − 5 °C, 88%, 33% ee; imidazole: room temperature, 40%, 57% ee). Higher enantioselectivities were obtained in the reaction with aliphatic aldehydes (up to 90% ee). A transition state **364**, in which both the aldehyde and the enone coordinate to the thiourea groups of **363** through hydrogen bonding interactions, was proposed to explain the stereochemistry of the product.

Later, Berkessel *et al.* developed a new and improved bis(thio)urea catalysts **365**, which was derived in one step from readily available isophorone-diamine (IPDA). Good yields and excellent enantiomeric excesses were obtained in the MBH reaction of cyclohexanecarbaldehyde with 2-cyclohexen-1-one by using bis-thiourea catalyst **365**, in combination with a novel base (N,N,N',N'-tetramethylisophorone-diamine, TMIPDA) in toluene (Scheme 2.197). However, with the exception of cyclohexanecarbaldehyde as substrate, DABCO provides superior yields and ee values than TMIPDA. Furthermore, it was shown for the first time that bis-(thio)ureas can activate Michael-acceptors besides 2-cyclohexen-1-one.[124c]

In addition, mono-thioureas **366**, which straightforwardly derived from commercially available enantiopure amino alcohols, have been found to promote the asymmetric MBH reaction of 2-cyclohexen-1-one with different aldehydes in the presence of triethylamine under solvent-free conditions. The corresponding allylic alcohols were obtained in good to high yields with up to 88% ee (Scheme 2.198).[200]

Although the MBH reactions of aliphatic aldehydes with 2-cyclohexen-1-one have achieved very high yields and ees by employing the organocatalysts

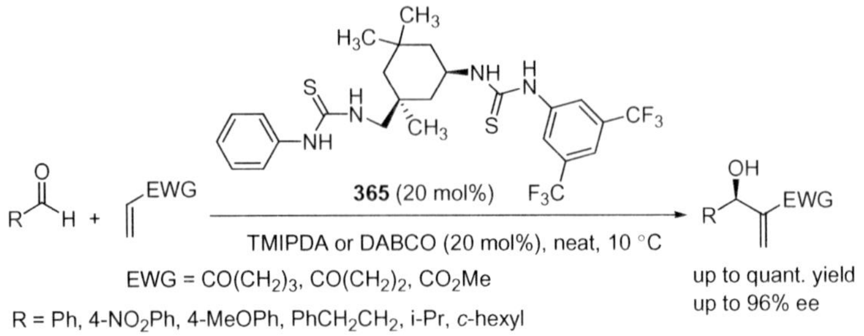

Scheme 2.197

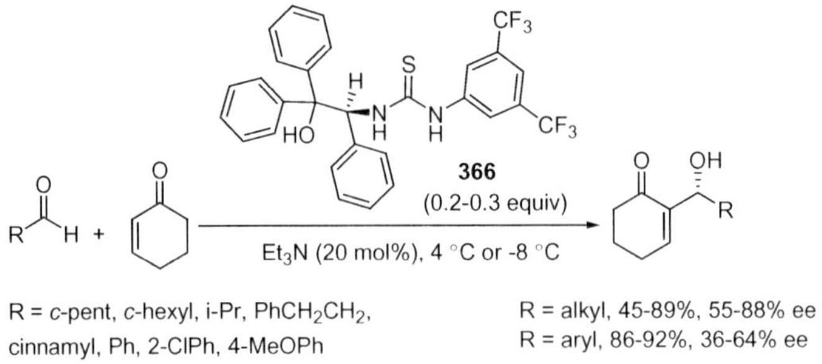

Scheme 2.198

mentioned above, using aromatic aldehydes usually afforded the corresponding products in moderate yields and moderate ees (the highest ee is 77%). Recently, we have reported that the improved bis(thio)urea organocatalyst **367**, derived from axially chiral (*R*)-5,5′,6,6′,7,7′,8,8′-octahydro-1,1′-binaphthyl-2,2′-diamine (H8-BINAM), is a fairly effective chiral organocatalyst for the enantioselective MBH reaction of aryl aldehydes with 2-cyclohexen-1-one, giving the corresponding adducts in up to 88% ee and in good to excellent yields (Scheme 2.199).[201]

Intrigued by the high activity of the elegant bis(thiourea) co-catalyst **363** reported by Nagasawa and co-workers,[199a] Clarke and Philp *et al.* have used electronic structure calculations to design the co-catalysts **368** and **369**, which can recognize a key transition state within the reaction manifold of the MBH reaction. These co-catalysts were capable of increasing the rate of the MBH reaction of 4-fluorobenzaldehyde and cyclohexanone through the hydrogen bond mediated recognition of both the nucleophile and the electrophile, and retain some activity even at very low (1 mol.%) loadings (Scheme 2.200).[202]

In screening thiourea catalysts for the asymmetric aza-MBH reaction, Jacobsen *et al.*[203] have developed a highly enantioselective catalytic aza-MBH reaction of various *N*-nosyl imines with methyl acrylate. High enantioselectivities

Ar = 4-NO$_2$Ph, 3-NO$_2$Ph, 4-ClPh, 2-ClPh, 3-ClPh, 2,3-Cl$_2$Ph, 2,4-Cl$_2$Ph, 4-BrPh, 2-BrPh, 3-BrPh, 4-FPh, 3-FPh, 3-CF$_3$Ph, 2-pyridinyl, Ph, 4-MePh, 4-EtPh

n = 1, 50-99%, 62-88% ee
n = 0, 43%, 60% ee (for Ar = 4-NO$_4$Ph)

thiourea **367**:

Scheme 2.199

bisthiourea:DMAP = 1:1.5

neat, -5 °C

catalyst	0 mol%	5 mol%	10 mol%	20 mol%
363	---	12		31
368	---	14	54	65
369	---	23		84

368

369

Scheme 2.200

(87–99% ee) were obtained by using chiral thiourea catalyst **370** and DABCO as the nucleophilic additive, albeit only in modest (25–49%) yields (Scheme 2.201). A mechanistic analysis based on the identity of isolated key intermediate, DABCO-acrylate-imine adduct **371**, as well as kinetic investigations and isotope studies was provided and the rationale for the observed limitations in yield was proposed.

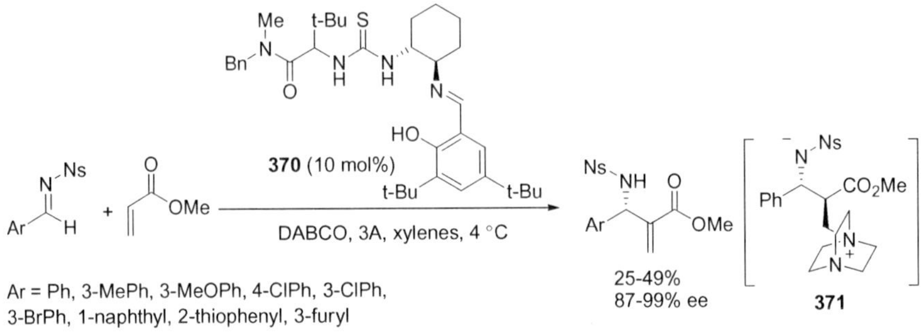

Scheme 2.201

Ar = Ph, 3-MePh, 3-MeOPh, 4-ClPh, 3-ClPh, 3-BrPh, 1-naphthyl, 2-thiophenyl, 3-furyl

25-49%
87-99% ee

2.6.3 Other Reagents as Co-catalysts

A remarkable acceleration of MBH reactions was observed in the presence of mild cooperative catalysts consisting of nucleophilic catalysts based on amine or phosphine and phenols or naphthols as Brønsted acids.[204] Ikegami *et al.* first demonstrated that the addition of 10 mol.% of racemic 1,10-bi-2-naphthol (BINOL) as a Brønsted acid co-catalyst can improve significantly the yield of tributylphosphine-catalyzed MBH reaction of 3-phenyl-1-propanal with cyclopentenone, from 23% to quantitative yield.[204a] Notably, none of methanol, benzoic acid or *p*-toluenesulfonic acid was efficient in this transformation. When the reaction was catalyzed by 2-hydroxy-2′-methoxy-1,1′-binaphthyl (**372**) or 2,2′-dimethoxy-1,1′-binaphthyl (**373**), the corresponding adduct was obtained in 80% and 24% yields, respectively, indicating that the synergetic action of both hydroxyls is beneficial to the reaction. DABCO, triphenylphosphine or dibutyl sulfide as Lewis bases were inefficient and afforded no products. Phenol and *p*-toluenesulfonamide, which has the same Brønsted acidity as phenol, were, in turn, effective (Scheme 2.202). The attempted asymmetric reaction with (*R*)-BINOL and tributylphosphine afforded low ee (< 10%).[204a] This process has been successfully applied to MBH reaction of various activated olefins with aldehydes, providing the corresponding adducts in moderate to excellent yields (Scheme 2.203).[204a] Furthermore, attracted by this procedure, Chapuis *et al.* have developed the facile synthesis of key the intermediate of *cis*-Hedione® and methyl jasmonate by cascade a MBH reaction and Claisen ortho ester rearrangement from cyclopent-2-en-1-one and the appropriate aldehydes.[205]

Fu *et al.*[206] modified Ikegami's procedure by converting air-sensitive trialkylphosphines into air-stable phosphonium salts, *via* protonation on phosphorus, which serve as an alternative catalyst to the corresponding phosphines (simple deprotonation under the reaction conditions by a Brønsted base liberates the trialkylphosphine) to perform the MBH reaction of cyclopent-2-en-1-one with 3-phenyl-1-propanal. The desired product was obtained with an isolated yield that is comparable to that furnished by Ikegami's procedure (Scheme 2.204).

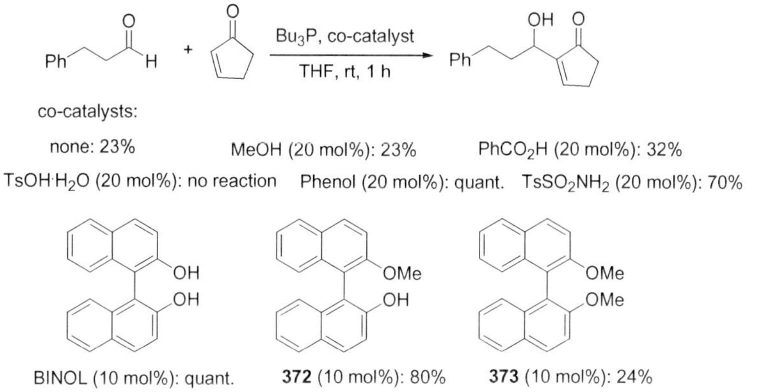
Bu3P, co-catalyst
THF, rt, 1 h
co-catalysts:
none: 23%
MeOH (20 mol%): 23%
PhCO2H (20 mol%): 32%
TsOH·H2O (20 mol%): no reaction
Phenol (20 mol%): quant.
TsSO2NH2 (20 mol%): 70%
BINOL (10 mol%): quant.
372 (10 mol%): 80%
373 (10 mol%): 24%

Scheme 2.202

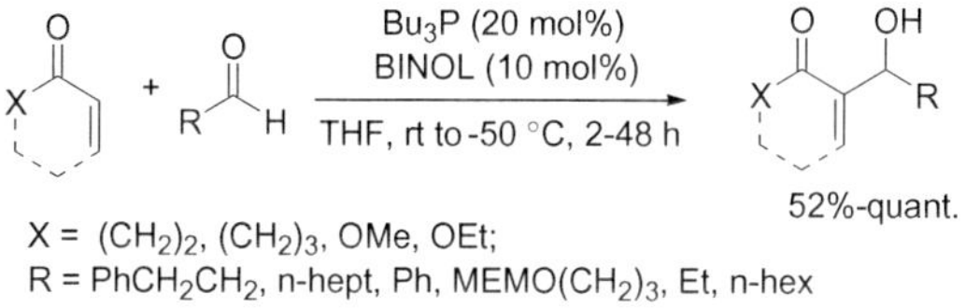
Bu3P (20 mol%)
BINOL (10 mol%)
THF, rt to -50 °C, 2-48 h
52%-quant.
X = (CH2)2, (CH2)3, OMe, OEt;
R = PhCH2CH2, n-hept, Ph, MEMO(CH2)3, Et, n-hex

Scheme 2.203

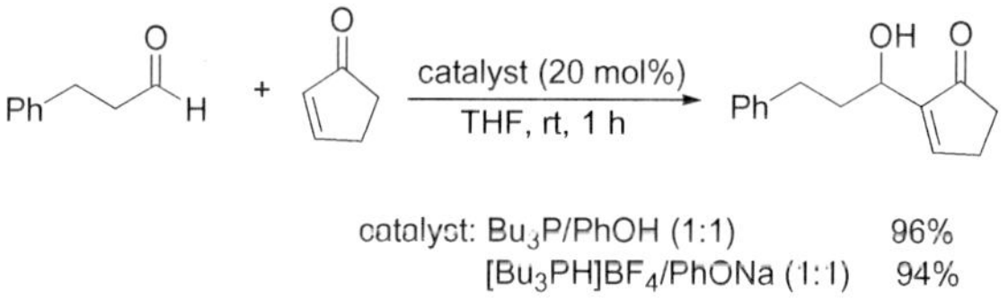
catalyst (20 mol%)
THF, rt, 1 h
catalyst: Bu3P/PhOH (1:1) 96%
[Bu3PH]BF4/PhONa (1:1) 94%

Scheme 2.204

During their studies on chiral Lewis acid-promoted MBH reactions, Schaus *et al.* have developed a highly enantioselective asymmetric MBH reaction involving the addition of cyclohexenone to aldehydes by using triethylphosphine as catalyst and binaphthol-derived chiral Brønsted acids as co-catalysts.[204b,c] Two structural features of the co-catalyst were important in achieving high enantioselectivity: saturation of the BINOL derivative and substitution at the 3,3′-positions (Scheme 2.205). When (*R*)-3,3′-diphenyl-H8-BINOL (**374a**) was used as the catalyst in the MBH reaction, the corresponding adduct was produced in 69% yield with 86% ee. However, when mesityl-catalyst **374b** was used, the product was formed in low yield and with low enantioselectivity. It was reasoned that the mesityl substituent restricted rotation about the biaryl bond of the co-catalysts:

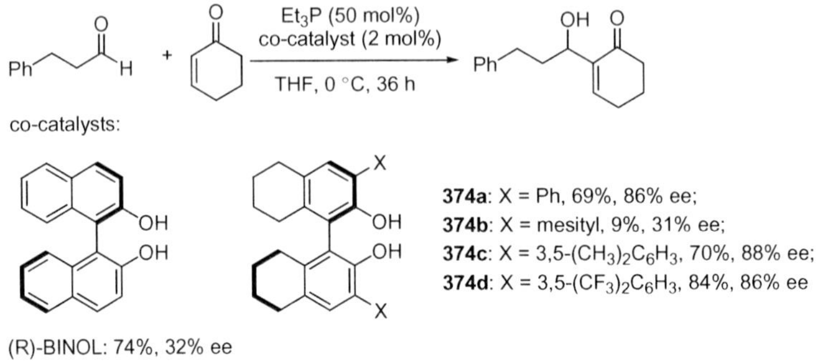

374a: X = Ph, 69%, 86% ee;
374b: X = mesityl, 9%, 31% ee;
374c: X = 3,5-$(CH_3)_2C_6H_3$, 70%, 88% ee;
374d: X = 3,5-$(CF_3)_2C_6H_3$, 84%, 86% ee

(R)-BINOL: 74%, 32% ee

Scheme 2.205

Et₃P (200 mol%)
374c (10 mol%)
THF, -10 °C, 48 h

70-82%, 92-96% ee
R = $CH_3CH_2CH=CHCH_2CH_2$, c-hexyl,
i-Pr, $CH(CH_2O)_2C(CH_3)_2$

THF, -10 °C, 48 h

Et₃P (200 mol%)
374c or 374d (10 mol%)

Et₃P (200 mol%)
374d (10 mol%)
THF, -10 °C, 48 h

74-88%, 82-90% ee
R = $PhCH_2CH_2$, $BnOCH_2CH_2$

R = Ph, PhCH=CH

39-40%, 67-81% ee

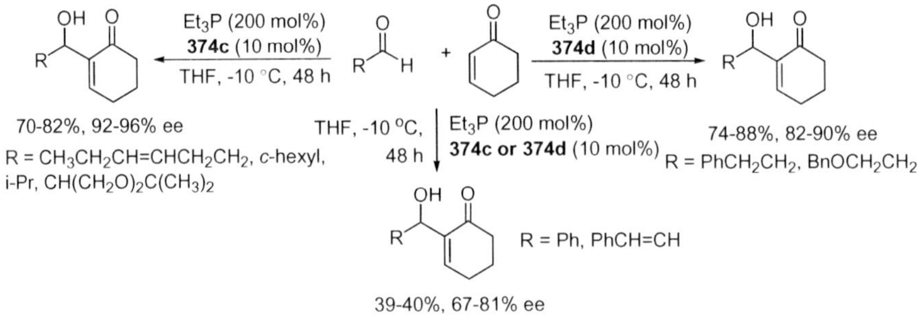

Scheme 2.206

3-substituent, a structural prerequisite for catalysis. The highest levels of enantioselectivity were achieved using (*R*)-3,3′-(3,5-dimethylphenyl)-H8-BINOL (**374c**) as the catalyst (88% ee), and employing 3,3′-[3,5-bis(trifluoromethyl)phenyl] catalyst **374d** resulted in the greatest levels of conversion (84% yield). Optimal results with the addition of aliphatic aldehydes were obtained with **374c**, while catalyst **374d** afforded the best results with more hindered aldehydes. Conjugated aldehydes such as benzaldehyde and cinnamaldehyde as acceptors resulted in low yields and enantioselectivity (Scheme 2.206).

Along with BINOL and binaphthol-derivatives, phenols are also commonly employed Brønsted acids and have been used extensively in MBH reactions as co-catalyst. Leadbeater *et al.* found that the use of phenol as a co-catalyst in the TMG-catalyzed MBH reaction led to significant rate acceleration (Scheme 2.207).[207] The catalytic system was efficient for aromatic aldehydes, but not for aliphatic aldehydes, reaching completion within 1 h at 48 °C. A rate acceleration over the solvent-free case was observed when using methanol and ethanol as solvents, but not to the same extent as with phenol.

We have found that in the presence of a catalytic amount of nitrophenol, the Lewis base triphenylphosphine can effectively promote the MBH reaction of aldehydes with methyl vinyl ketone to give the corresponding normal MBH

Scheme 2.207

Scheme 2.208

Scheme 2.209

adducts in good yields (Scheme 2.208). The mechanism has been investigated by ^{31}P NMR spectroscopy and the solvent and substituent effects have been examined.[208]

Pohmakotr *et al.* have developed a general and convenient method for the synthesis of 2-(hydroxyalkyl)-5-methylenecyclopentenones **377** *via* the MBH reaction of a masked 5-alkylidene-2-cyclopentenone (**375**) with aldehydes, catalyzed by tributylphosphine in the presence of phenol, to provide the corresponding MBH adducts **376**, followed by the flash vacuum pyrolysis (FVP) to give the resulting adducts **377** (Scheme 2.209).[209]

The aza-MBH reaction between acrylamide or *N*-arylacrylamide with *N*-tosylated imines has been accomplished by using DABCO as a catalyst and phenol as an additive in the absence of solvent (Scheme 2.210).[210]

As the remarkable rate acceleration under protic solvents such as water or methanol has been widely reported, Choo and Chong found that the MBH

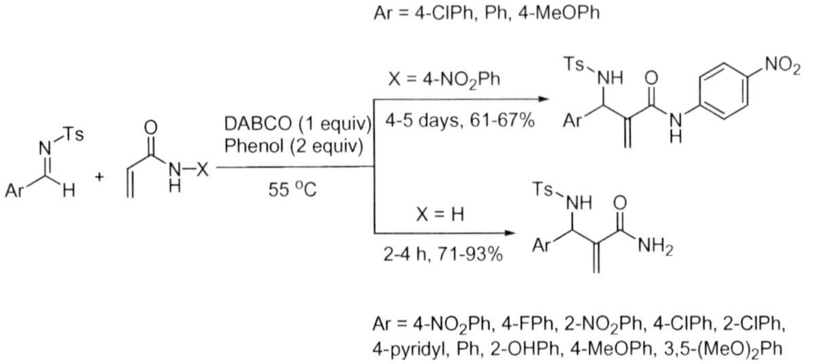

Scheme 2.210

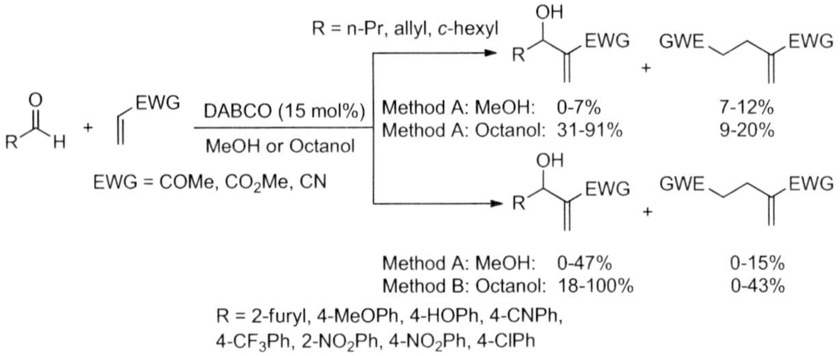

Scheme 2.211

reaction was greatly accelerated by the use of octanol as an additive. Under the octanol-accelerated MBH conditions, unactivated aldehydes such as aliphatic aldehydes and aromatic aldehydes with electron-withdrawing substituents were readily converted into the desired products in good to high yields. For example, aliphatic aldehydes that showed very slow transformation under the general MBH conditions (0–7% yield, Method A, Scheme 2.211) gave the desired products in moderate to high yield under the octanol-promoted reaction conditions (31–91% yield, Method B, Scheme 2.211). However, almost the same amount of the MVK dimer was obtained under both reaction conditions, which indicates that the production of the dimer is the result of the low reactivity of the aldehyde and does not depend on the reaction conditions.[211]

Recently, Matsubara *et al.* have developed the first example of a "catalytic" MBH reaction of methyl vinyl ketone with various aldehydes by using amphiphilic *N*-alkylimidazole derivative **378**, which bears an hydrophobic group, as organocatalyst in water without organic solvents.[212] This reaction is

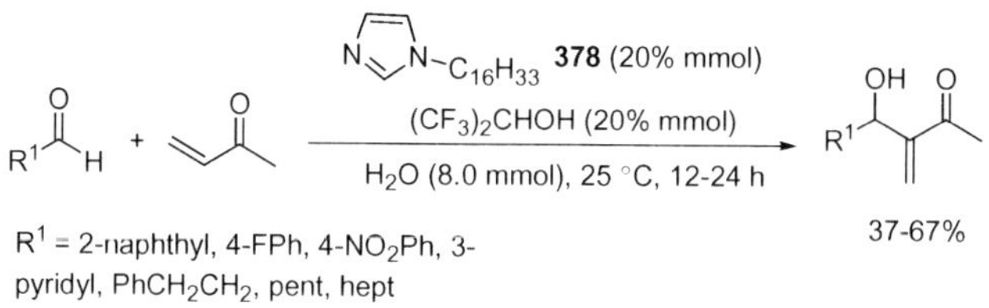

378 (20% mmol)
$(CF_3)_2CHOH$ (20% mmol)
H_2O (8.0 mmol), 25 °C, 12-24 h
37-67%
R^1 = 2-naphthyl, 4-FPh, 4-NO_2Ph, 3-pyridyl, $PhCH_2CH_2$, pent, hept

Scheme 2.212

imidazole (100 mol%)
anthranilic acid (AA) (10 mol%)
THF, rt
no additive: 19-57%
AA additive: 25-71%
R = 2-furyl, 2-thienyl, 3-furyl, 3-thienyl, 4-MeOPh, 2-NO_2Ph, 3,4-$(MeO)_2$Ph, 3-MeO-4-OHPh, 3,4-OCH_2O-Ph, 4-FPh, Ph

Scheme 2.213

accelerated by the addition of water and the yields can be improved by addition of a catalytic amount of Brønsted acid, 1,1,1,3,3,3-hexafluoropropan-2-ol, as co-catalyst (Scheme 2.212). It was assumed that the amphiphilic catalyst forms a hydrophobic field for an organic reaction near the boundary between the water and the organic compounds by self-assembly; such a reaction field constructed by an amphiphilic organocatalyst near the surface of water may be widely effective for the acceleration of organocatalytic reactions.

During their synthesis of α-hydroxymethylated conjugated nitroalkenes, Namboothiri and Panda found that the MBH reaction of various aromatic and heteroaromatic conjugated nitroalkenes with formaldehyde can proceed smoothly in the presence of stoichiometric amounts of imidazole, while no satisfactory results can be obtained by using general amine and phosphine-based catalysts, such as DABCO, DBU, DMAP, Et_3N, pyridine and $(Bu^n)_3P$. In addition, the yields were improved appreciably when anthranilic acid (AA) was used as the co-catalyst, affording the corresponding hydroxymethylated derivatives in moderate to good yields (Scheme 2.213).[213]

The Lewis acid catalyst is generally believed to activate electron-deficient alkenes and to facilitate the conjugate addition of the nucleophile catalyst. Consequently, various efforts have been made to accelerate this reaction sequence by using Lewis acid co-catalysts. Kobayashi *et al.* initially found that the MBH reaction of α,β-unsaturated carbonyl compounds with aldehydes was accelerated in the presence of a catalytic amount of 1,4-diazabicyclo[2.2.2] octane (DABCO) and lithium perchlorate in ether, affording the corresponding adducts in high yield (Scheme 2.214). A preliminary kinetic study revealed that the reaction using $LiClO_4$ in ether was 800-fold faster than that without $LiClO_4$.[214]

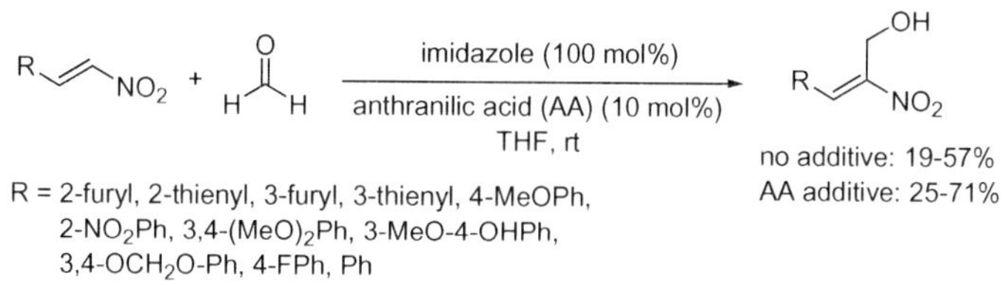

Scheme 2.214

Scheme 2.215

As stated before, the enantioselective MBH reaction of cyclopent-2-en-1-one with 3-phenyl-1-propanal resulted in low enantioselectivities ($<10\%$) when using (R)-BINOL and tributylphosphine. However, better results have been achieved using calcium chiral catalyst **379** along with Bu$_3$P, with the desired adduct provided in 56% enantiomeric purity (Scheme 2.215).[204a]

Later, Sasai *et al.* found that a combination of the heterobimetallic complex boron-lithium-mono(binaphthoxide) (BLB) (**380**) and (Bu^n)$_3$P is effective in promoting the enantioselective MBH reaction of cyclic enones with aldehydes *via* double activation to afford the adduct in good chemical yield with moderate to high enantioselectivity (Scheme 2.216).[215]

Namboothiri *et al.* have investigated the imidazole/LiCl-mediated MBH reaction by using activated alkenes, such as MVK and acrylate, as electrophiles to react with various aromatic and heteroaromatic nitroalkenes, providing novel α-substituted nitroalkenes in moderate yields. Normally, β-substituted activated alkenes do not react, or react sluggishly, in MBH type reactions. However, in the case of nitroalkenes, β-substituted ones were found to react satisfactorily.[216] Later, the same group also synthesized novel α-aminoalkylated conjugated nitroalkenes by imidazole/LiCl-mediated reaction of conjugated nitroalkenes with *N*-tosylimines with high yields (Scheme 2.217).[217]

Recently, Connell *et al.* have found that, by using a catalytic amount of 4-dimethylaminopyridine (DMAP) as a nucleophile in the presence of an equal amount of tetramethylethylenediamine (TMEDA) and MgI$_2$, MBH adducts can be obtained in good to excellent yields from various aromatic and aliphatic aldehydes and cyclic enones/enoates at room temperature after convenient reaction times (Scheme 2.218).[218]

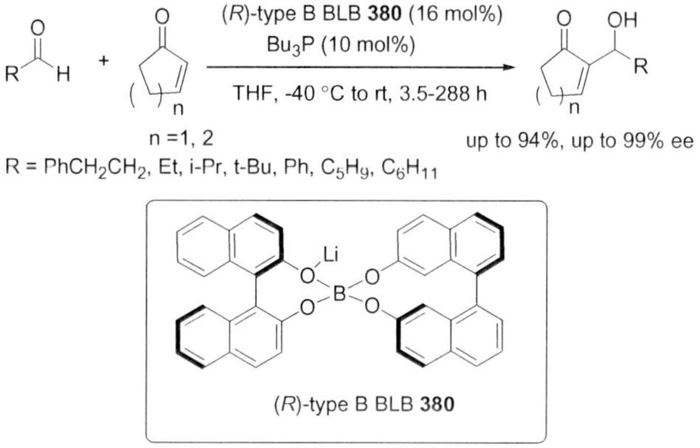
(R)-type B BLB 380 (16 mol%)
Bu3P (10 mol%)
THF, -40 °C to rt, 3.5-288 h
n =1, 2
up to 94%, up to 99% ee
R = PhCH2CH2, Et, i-Pr, t-Bu, Ph, C5H9, C6H11
(R)-type B BLB 380

Scheme 2.216

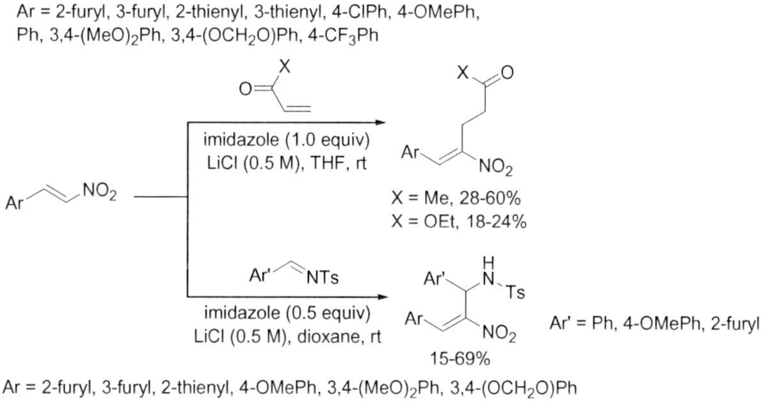
Ar = 2-furyl, 3-furyl, 2-thienyl, 3-thienyl, 4-ClPh, 4-OMePh,
Ph, 3,4-(MeO)2Ph, 3,4-(OCH2O)Ph, 4-CF3Ph
imidazole (1.0 equiv)
LiCl (0.5 M), THF, rt
X = Me, 28-60%
X = OEt, 18-24%
imidazole (0.5 equiv)
LiCl (0.5 M), dioxane, rt
15-69%
Ar' = Ph, 4-OMePh, 2-furyl
Ar = 2-furyl, 3-furyl, 2-thienyl, 4-OMePh, 3,4-(MeO)2Ph, 3,4-(OCH2O)Ph

Scheme 2.217

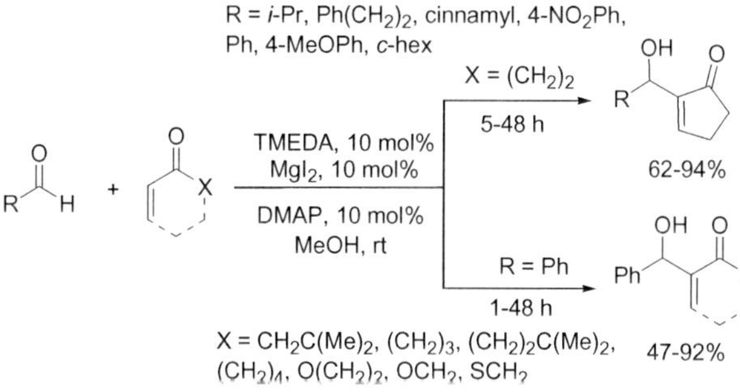
R = i-Pr, Ph(CH2)2, cinnamyl, 4-NO2Ph,
Ph, 4-MeOPh, c-hex
X = (CH2)2
5-48 h
62-94%
TMEDA, 10 mol%
MgI2, 10 mol%
DMAP, 10 mol%
MeOH, rt
R = Ph
1-48 h
47-92%
X = CH2C(Me)2, (CH2)3, (CH2)2C(Me)2,
(CH2)4, O(CH2)2, OCH2, SCH2

Scheme 2.218

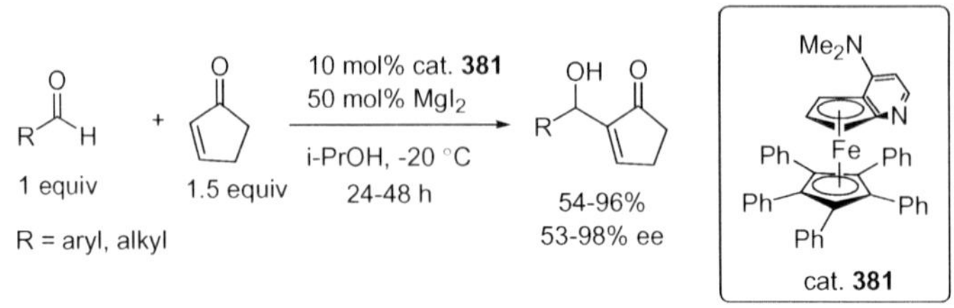

Scheme 2.219

Moreover, in conjunction with readily available MgI_2 as a co-catalyst, Fu's planar chiral DMAP derivative **381** has been applied successfully to an MBH reaction involving the addition of cyclopentenone to aromatic and aliphatic aldehydes, affording the corresponding adduct in good to excellent yields with moderate to excellent enantiomeric excesses (Scheme 2.219). This work first suggests that the scope of reactions catalyzed by Fu's planar chiral DMAP catalysts can be increased by employing a simple co-catalyst.[219]

Among various Lewis acids, lanthanide triflates $[Ln(OTf)_3]$ have been successfully utilized to promote the MBH reaction in the presence of a stoichiometric amount of Lewis base catalysts.[40,220] Aggarwal *et al.* first found that conventional Lewis acids, such as $BF_3 \cdot Et_2O$ and $TiCl_4$, led to a reduction of rates, and that $La(OTf)_3$ and $Sm(OTf)_3$ (5 mol.%) gave rate accelerations (k_{rel}) of approximately 4.7 and 4.9, respectively, in reactions between *tert*-butyl acrylate and benzaldehyde when using stoichiometric amounts of DABCO. At low loadings of DABCO (up to 10 mol.%), however, no reaction occurred due to association of DABCO with the metal. It was thought that the use of additional ligands to displace the DABCO from the metal would further increase the rate of MBH reaction. Of the ligands and metals tested, triethanolamine (50 mol.%) and $La(OTf)_3$ (5 mol.%) gave overall rate accelerations of between 23-fold and 40-fold, depending on the acrylates, and approximately five-fold for acrylonitrile by using stoichiometric DABCO as nucleophilic catalyst (Scheme 2.220); a simple acid wash procedure allowed recovery of the product from the reagents. While triethanolamine alone (80 mol.% was optimum) gave a rate acceleration of up to 22-fold, the acceleration is clearly greater when the lanthanum triflate is present (Scheme 2.220). The preferred system was demonstrated to be effective synthetically on various different acrylates and aldehydes that are more or less reactive than benzaldehyde, demonstrating the generality of this method. In addition, notably, reduced amounts of dimerized acrylates were obtained.[220a]

Subsequently, the same group found that $La(OTf)_3$ (5 mol.%) was also required to promote the DBU-catalyzed MBH reaction of pivaldehyde with methyl acrylate, as no adduct was obtained without it. Similarly, $La(OTf)_3$ enhanced the rate of the MBH reaction of pivaldehyde with cyclohex-2-en-1-one and gave a higher yield of corresponding adduct; in its absence a 44% yield of adduct was obtained after 4 days (Scheme 2.221). However, a less clean

	EWG = CO$_2$Me	EWG = CO$_2$Et	EWG = CO$_2$t-Bu	EWG = CN
catalyst system	k_{rel}	k_{rel}	k_{rel}	k_{rel}
DABCO	1	1	1	1
DABCO, La(OTf)$_3$, N(CH$_2$CH$_2$OH)$_3$	31.9	39.7	23	5.07
DABCO, N(CH$_2$CH$_2$OH)$_3$	19.2	22.0	18	2.78

Scheme 2.220

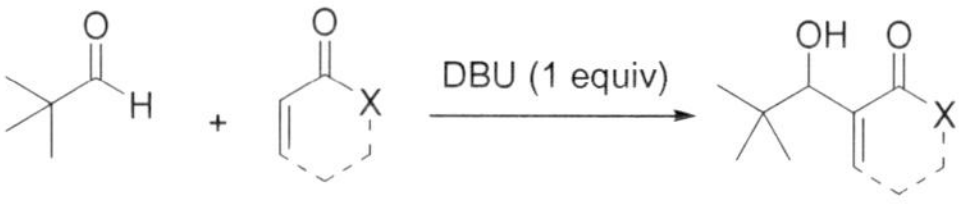

X = OMe, no La(OTf)$_3$: no reaction
La(OTf)$_3$ (0.05 equiv): 70 h, 20%;
X = (CH$_2$)$_3$, no La(OTf)$_3$: 4 days, 44%;
La(OTf)$_3$ (0.05 equiv): 21 h, 75%;

Scheme 2.221

reaction was observed and the additional acceleration was minimal in the case of benzaldehyde.[40]

Later, during their studies on rate acceleration of MBH reactions in polar solvents, Aggarwal *et al.*[221] found that the use of 5 equiv of formamide gave faster rates than reactions conducted in water. An additional acceleration was achieved in the presence of Yb(OTf)$_3$ (5 mol.%) and formamide. The MBH reaction of various Michael acceptors with benzaldehyde and a range of electrophiles with ethyl acrylate were investigated under the new conditions (Scheme 2.222).

On the basis of above work, Shang *et al.* have developed a novel and efficient catalytic system for the MBH reaction between aromatic aldehydes and activated alkenes by a combination of Sc(OTf)$_3$ and 3-hydroxyquinuclidine (3-HQD) as co-catalyst and DMF as solvent, affording the corresponding adduct in moderate to high yield in a short time (Scheme 2.223).[222]

In addition, Adolfsson *et al.*[220b] have used DABCO or 3-HQD as catalyst, together with La(OTf)$_3$ and molecular sieves, to accomplish a three-component reaction with activated alkenes, aldehydes and tosylamines, affording aza-MBH adducts in moderate to good yields (Chapter 1.5).

Chen *et al.* first demonstrated that the asymmetric induction can be obtained by the aid of a chiral Lewis acid for the MBH reaction. By using camphor-derived chiral ligand **382**, which can be prepared from the condensation of

X = CO$_2$Me, CO$_2$Et, CO$_2$t-Bu, CN,
COMe, CO(CH$_2$)$_3$, CO(CH$_2$)$_2$

R = Ph
10 min-24 h

10-96%

1 equiv 3-HQD
0.05 equiv Yb(OTf)$_3$
5 equiv H$_2$NCHO

X = CO$_2$Et
2 h-8 days

R = 2-ClPh, 2-MePh, 4-MePh,
2-NO$_2$Ph, i-Pr, c-hex, t-Bu

18-99%

Scheme 2.222

0.2 equiv 3-HQD
0.05 equiv Sc(OTf)$_3$
DMF, rt, 10 min-36 h

EWG = CO$_2$Me, CO$_2$Et, CN

R = Ph, 4-MeOPh, 4-ClPh, 4-NO$_2$Ph,
2,4-Cl$_2$Ph, 2-furyl

23-98%

Scheme 2.223

DABCO (30 mol%)
La(OTf)$_3$ (3 mol%)
ligand **382** (6 mol%)

Ligand:

R = Me, Et, i-Pr, 4-MeOPh, Ph,
4-NO$_2$Ph, c-C$_6$H$_{11}$, Ph(CH$_2$)$_2$CH$_2$

R' = Me, 55-89%, 6-66% ee;
R' = t-Bu 25%, 70% ee
R' = Ph, 97%, 75% ee
R' = Bn, 50-93%, 65-95% ee;
R' = α-naphthyl, 35-88%, 70-95% ee

382

Scheme 2.224

(+)-ketopinic acid with the corresponding diamines under acidic conditions, good to high enantioselectivities can be obtained using 3 mol.% La(OTf)$_3$ as catalyst. In addition, when α-naphthyl acrylate is used as a Michael acceptor, the reaction is completed within 20 min with high stereoselectivity and in reasonable chemical yields (Scheme 2.224).[223]

To accomplish facile catalyst separation from the reaction mixture and recycling of the catalyst, Yi *et al.* have developed a highly efficient ytterbium perfluorooctane sulfonate [Yb(OPf)$_3$] catalyzed MBH reaction in the presence

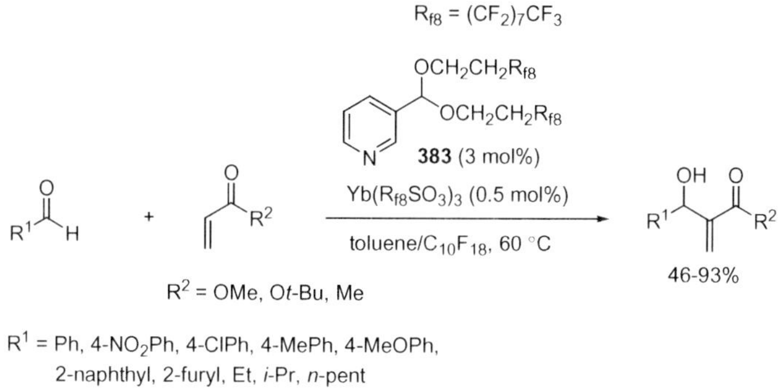

Scheme 2.225

of a catalytic amount of a novel perfluoroalkylated-pyridine (**383**) in a fluorous biphasic system (FBS) composed of toluene and perfluorodecalin. The new process can be carried out successfully without the use of a stoichiometric amount of Lewis base and the fluorous phase containing the active catalytic species is easily separated, and can be reused several times without significant loss of catalytic activity (Scheme 2.225).[224]

Kumar *et al.* have studied the salt effect on the MBH reaction, that is, salt solutions of water and "water-like" structured solvents, such as formamide and *N*-methylformamide, were shown to accelerate the MBH reaction in the presence of DABCO. Ethylene glycol, another structured solvent, and its salt solutions fail to make any impact on the reaction rates. The salts that are conventionally defined as salting-out or -in do not behave in a similar fashion when employed in the MBH reactions, an observation supported by solubility measurements. It seems that the cation, anion and the nature of the solvent and the reactants together ascertain whether a salt will enhance or retard the MBH reaction.[225]

Since imidazolium-based ionic liquids have been used as solvents for the MBH reaction with moderate success (Chapter 1.7), the phosphonium salt, having a similar high polarity, has also been selected as an attractive alternative co-catalyst for the DABCO-catalyzed MBH reaction, providing a much higher yield over the neat reaction or superior to that obtained in the imidazolium ionic liquid (Scheme 2.226). It was assumed that phosphonium salt could perform several roles in the reaction, which probably involve activating the alkene, activating the aldehyde and stabilizing the zwitterionic intermediate.[226]

Recently, the new acidic ionic liquid phenyl(butyl)ethyl-selenonium tetra-fluoroborate, [pbeSe]BF$_4$, was found to be a new acid co-catalyst and has been used successfully in the MBH reaction of several electron-deficient alkenes with aromatic and aliphatic aldehydes to give products in moderate to good yields and in relatively short reaction times under mild conditions (Scheme 2.227).[227]

Shea *et al.* have synthesized new achiral sulfamide **384**, phosphoric triamide **385** and thiophosphoric triamide **386** as hydrogen bond catalysts for the MBH reaction. Each of these three compounds showed activity and modest improvements (1.5-fold) for selected MBH reaction when compared to the

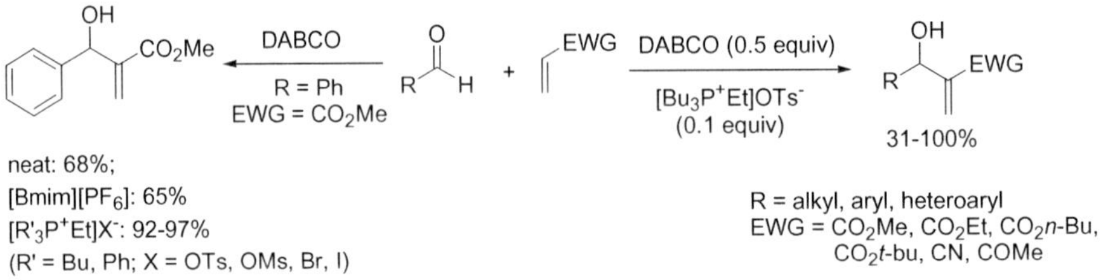

DABCO
R = Ph
EWG = CO2Me
DABCO (0.5 equiv)
[Bu3P+Et]OTs-
(0.1 equiv)
31-100%
neat: 68%;
[Bmim][PF6]: 65%
[R'3P+Et]X-: 92-97%
(R' = Bu, Ph; X = OTs, OMs, Br, I)
R = alkyl, aryl, heteroaryl
EWG = CO2Me, CO2Et, CO2n-Bu,
CO2t-bu, CN, COMe

Scheme 2.226

(5 mol%)
DABCO (100 mol%), CH3CN, rt
2-24 h
39-78%
R = Ph, 2-furyl, pent, hex
EWG = CO2Me, CN, COMe, CO(CH2)3

Scheme 2.227

co-catalyst (10 mol%)
DABCO, neat, rt, 48h
co-catalysts:
none
361
384
385
386
krel 0.5 1.0 1.5 1.1 1.3
% conv. 32 58 73 59 67
Ar = 3,5-(CF3)2C6H3

Scheme 2.228

corresponding thiourea catalyst **361** and, as such, suggest the possible development of new, more efficient MBH catalysts (Scheme 2.228).[228]

2.7 Polymer-supported Catalysts for the Morita–Baylis–Hillman Reaction

To improve the efficiency of the catalysts and to simplify the workup procedures, polymer-supported reagents and catalysts have also been adopted

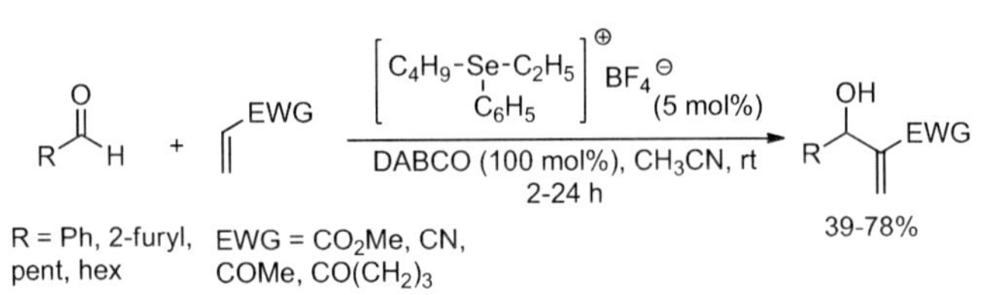

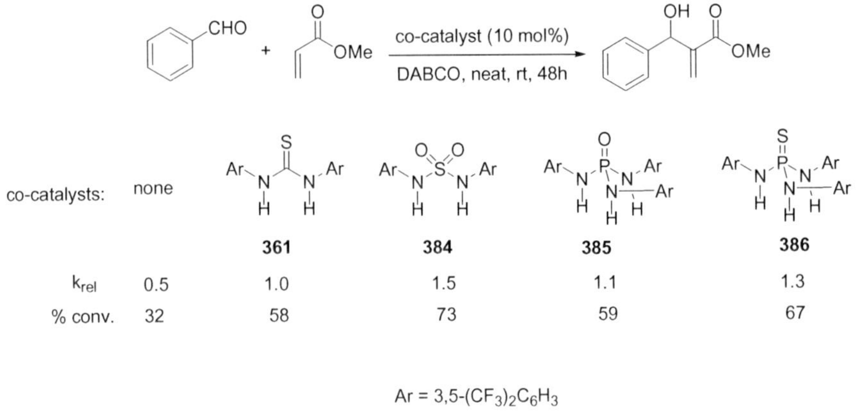

in MBH reactions. The polymer-supported catalysts PEG_{4600}-$(PPh_2)_2$ and polyDMAP were developed by us for reactions between N-tosylimines and methyl vinyl ketone or acrylates,[229] giving corresponding aza-MBH adducts in good yields. The polymer-supported catalysts could be recovered easily by filtration and the Lewis base PEG-$(PPh_2)_2$ can be reproduced by reduction with $LiAlH_4$ and $CeCl_3$ and then reused (Scheme 2.229).

At almost the same time, Corma and García *et al.* reported independently the stoichiometric polyDMAP-catalyzed MBH reaction of aromatic aldehydes and methyl vinyl ketone, which affords the corresponding adduct in moderate to good yields in DMF (Scheme 2.230).[230]

Subsequently, we and Toy *et al.* found that insoluble polystyrene-supported triphenylphosphine is an effective catalyst for the reactions between N-tosylimines and methyl vinyl ketone. The corresponding aza-MBH adducts were obtained in excellent yields and the catalysts could be reused.[231] Subsequently, the non-phosphane-bearing styrene aromatic rings have been functionalized with polar groups and a series of such catalysts were examined for their

R^1 = H, 4-Me, 4-MeO, 3-F, 4-F, 3-Cl, 4-Cl, 4-Br, 3-NO$_2$, 4-NO$_2$;
R^2 = Me, OMe, OPh

polyDMAP: dimethylaminopyridine, polymer-bound Lewis base:

PEG_{4600}-$(PPh_2)_2$: diphenylphosphine, polymer-bound Lewis base:

Scheme 2.229

R = n-Bu, Ph, 4-NO$_2$Ph, 2-FPh, 2-OH-3,5-(t-Bu)$_2$Ph, Bn

Scheme 2.230

10 mol% *JJ*-PPh$_3$

CH$_2$Cl$_2$, rt, 10-30 h,
R^2 = Me
63-92%

R^1 = H, 4-Me, 4-MeO, 3-NO$_2$, 4-F, 4-Cl, 4-Br

10 mol% *JJ*-OMe-PPh$_3$

THF, rt, R^2 = H, OPh
36-81%

R^1 = H, 4-Me, 4-Et, 4-MeO, 3-NO$_2$,
4-F, 4-Cl, 4-Br, 2,3-Cl$_2$, 4-NO$_2$

JJ-PPh$_3$: R^1 = R^2 = H;
JJ-OMe-PPh$_3$: R^1 = OMe, R^2 = H

Scheme 2.231

catalytic efficiency in a range of solvents.[232] It was observed that incorporation of 4-methoxystyrene into the polymer afforded the best support in terms of catalyst efficiency (Scheme 2.231).

A series of soluble, non-crosslinked bifunctional polymeric triphenylphosphine (**387** and **388**) and 4-dimethylaminopyridine reagents (**389** and **390**) have been prepared by our research group. Some of these polymeric reagents contained either alkyl alcohol or phenol groups on the polymer backbone. The use of these materials as organocatalysts in a range of intra- and intermolecular MBH reactions indicated that hydroxyl groups could participate in the reactions and accelerate product formation. In the cases examined, more acidic phenol groups were more effective than alkyl alcohol groups in catalyzing the reactions (Scheme 2.232).[233]

As an efficient catalyst for enantioselective aza-MBH reaction, bifunctional phosphine catalyst **CP17** has also been immobilized on a series of dendrimers.[137] The dendrimer-immobilized catalyst **CP44** was more effective than catalyst **CP17** for the aza-MBH reaction of *N*-sulfonyl imines with MVK, EVK or acrolein. The catalyst could be separated easily from the reaction mixture by simple filtration after the reaction and reused without obvious loss of activity (Scheme 2.233).[138]

However, our attempt to catalyze the MBH reaction by poly{styrene-*co*-3-(4-vinylbenzyloxy)-1-aza-bicyclo[2.2.3]octane} (**391**), prepared by the copolymerization of 3-(4-vinylbenzyloxy)-1-aza-bicyclo[2.2.2]octane and styrene, was not very successful. The use of this supported tertiary amine as a heterogeneous

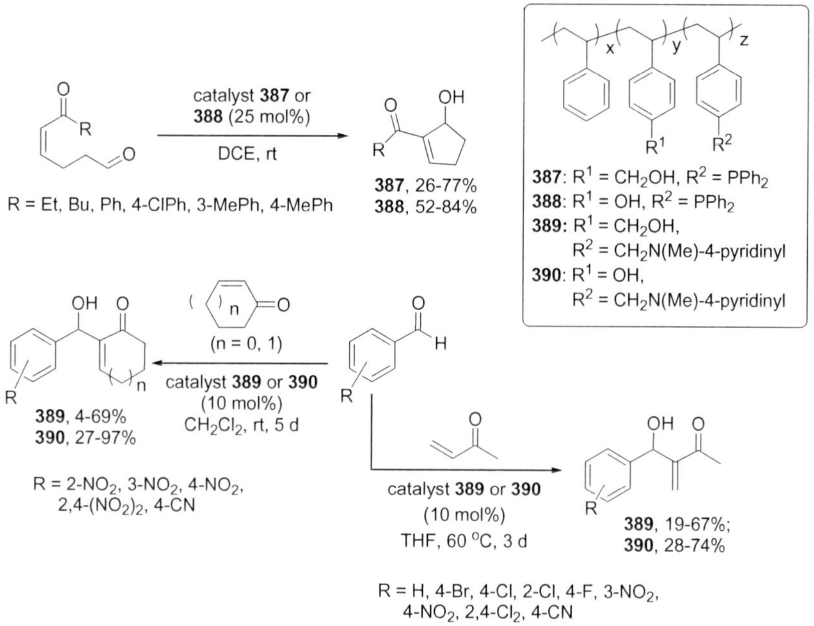

R = Et, Bu, Ph, 4-ClPh, 3-MePh, 4-MePh

catalyst **387** or **388** (25 mol%)

DCE, rt

387, 26-77%
388, 52-84%

387: R^1 = CH$_2$OH, R^2 = PPh$_2$
388: R^1 = OH, R^2 = PPh$_2$
389: R^1 = CH$_2$OH,
     R^2 = CH$_2$N(Me)-4-pyridinyl
390: R^1 = OH,
     R^2 = CH$_2$N(Me)-4-pyridinyl

(n = 0, 1)
catalyst **389** or **390**
(10 mol%)
CH$_2$Cl$_2$, rt, 5 d

389, 4-69%
390, 27-97%

R = 2-NO$_2$, 3-NO$_2$, 4-NO$_2$,
2,4-(NO$_2$)$_2$, 4-CN

catalyst **389** or **390**
(10 mol%)
THF, 60 °C, 3 d

389, 19-67%;
390, 28-74%

R = H, 4-Br, 4-Cl, 2-Cl, 4-F, 3-NO$_2$,
4-NO$_2$, 2,4-Cl$_2$, 4-CN

Scheme 2.232

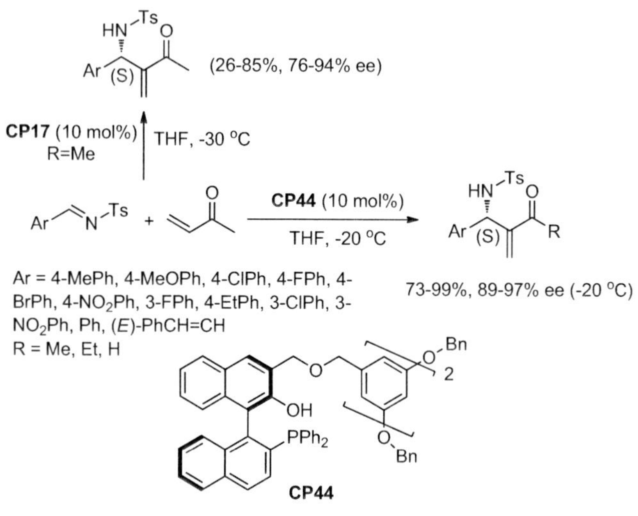

(26-85%, 76-94% ee)

CP17 (10 mol%) THF, -30 °C
R=Me

CP44 (10 mol%)
THF, -20 °C

73-99%, 89-97% ee (-20 °C)

Ar = 4-MePh, 4-MeOPh, 4-ClPh, 4-FPh, 4-BrPh, 4-NO$_2$Ph, 3-FPh, 4-EtPh, 3-ClPh, 3-NO$_2$Ph, Ph, (*E*)-PhCH=CH
R = Me, Et, H

CP44

Scheme 2.233

catalyst in both intra- and intermolecular (aza)-MBH reactions afforded the corresponding adducts in moderate yields (Scheme 2.234).[234]

Lin *et al.* have developed a new nucleophilic catalytic system for the MBH reaction that consists of dialkylaminopyridine-functionalized mesoporous silica nanospheres (DMAP-MSN). This material is an efficient heterogeneous catalyst for MBH reactions, exhibiting good reactivity, product selectivity and recyclability to give the corresponding adducts in moderate to excellent yields (Scheme 2.235).[235]

Cheng *et al.* found that a magnetic nanoparticle (MNP)-supported quinuclidine (**392**) can be evaluated as a recoverable MBH catalyst; it demonstrated comparable activity to that of DABCO and could be recycled simply with the assistance of an external magnet. The recycled catalyst could be reused seven times without significant loss of activity (Scheme 2.236).[236]

Zhao *et al.* have developed an efficient and recyclable hairy particle-supported 4-*N*,*N*-dialkylaminopyridine (DAAP) catalyst (**393**), which is efficient in catalyzing the MBH reaction of 4-nitrobenzaldehyde and methyl vinyl ketone

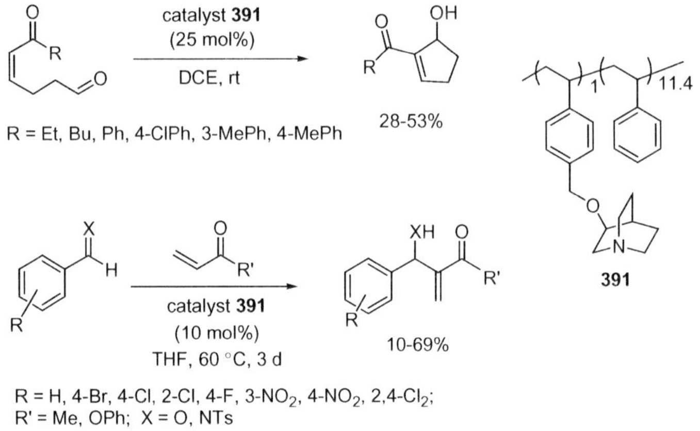

Scheme 2.234

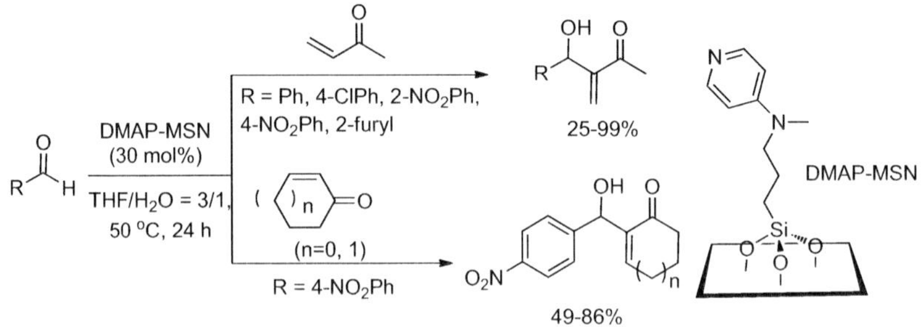

Scheme 2.235

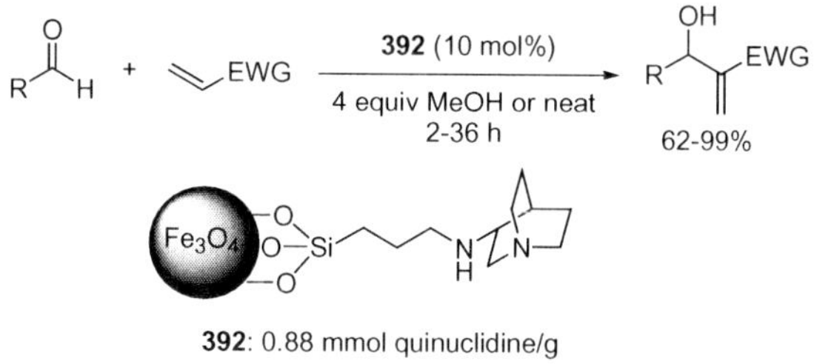

Scheme 2.236

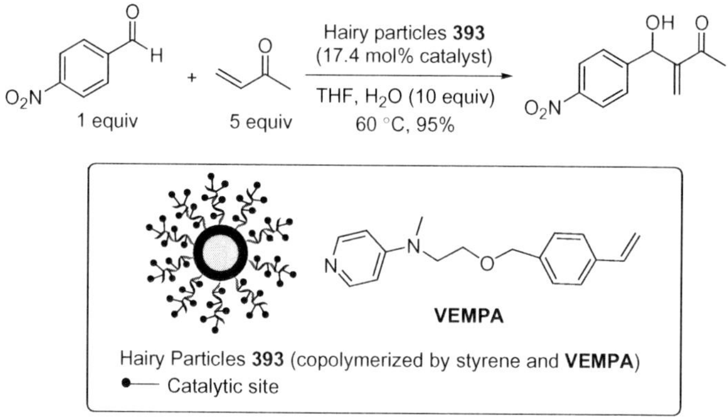

Scheme 2.237

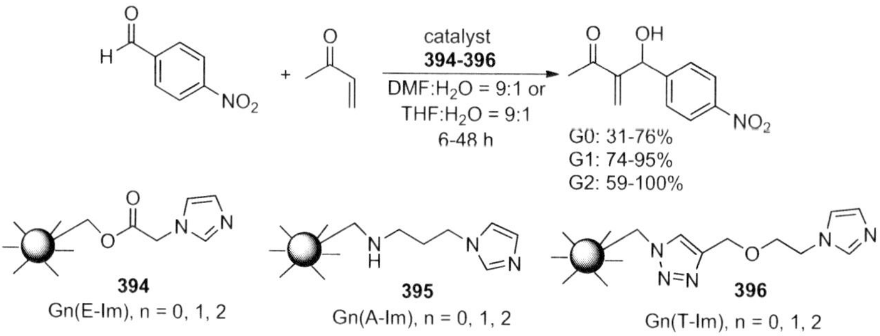

Scheme 2.238

at a low catalyst loading. The reaction was clean and no undesired side products were observed, which is in contrast to the same reaction catalyzed by the small molecule DMAP. The hairy particles were recycled six times with yields ≥ 90% (Scheme 2.237).[237]

Recently, Portnoy *et al.* have developed new modes of immobilization of *N*-alkylated imidazole-based catalysts to the regular and dendronized polymeric supports *via* dipolar cycloaddition, esterification or nucleophilic substitution;

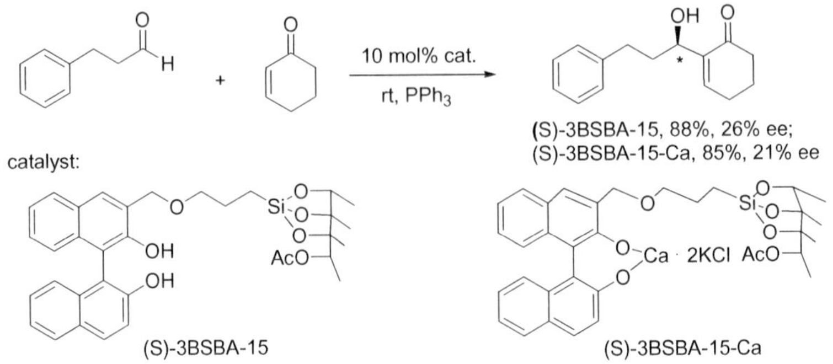

Scheme 2.239

these systems exhibited a remarkable positive dendritic effect for the MBH reaction between methyl vinyl ketone and aromatic aldehydes. The effect of water on the yield of the MBH reaction was also examined (Scheme 2.238).[238]

In addition, polymer-supported co-catalysts for MBH reactions have also been developed. Ji and Gao *et al.* have prepared a series of chiral BINOL functionalized mesoporous silicas by post grafting of organosilane derivatives of (*S*)-BINOL substituted at different positions of the binaphthyl backbone onto mesoporous silica SBA-15 and amorphous silica gel. The resulting chiral BINOL functionalized silicas were then used as asymmetric Brønsted acids to catalyze enantioselective MBH reaction of 3-phenylpropanal and cyclohexenone with tributylphosphine as Lewis base. The chiral BINOL functionalized silica 3BSBA-15 linked at the 3-position of binaphthyl backbone has shown higher ee and yield (26% ee, 88% yield) than the others, which are similar to the homogeneous catalyst (*S*)-BINOL (27% ee and 92% yield) as Brønsted acids catalyst. However, complex 3BSBA-15-Ca showed lower enantioselectivity (21% ee) than its homogeneous complex (*S*)-BINOL-Ca (32% ee) as ligand catalyst. 3BSBA-15 and 3BSBA-15-Ca can be reused with no significant decrease in enantioselectivity and yield (Scheme 2.239).[239]

References

1. A. B. Baylis and M. E. D. Hillman, *Chem. Abstr.*, 1972, **77**, 34174q (German Pat. 2155113, 1972).
2. S. E. Drewes and N. D. Emslie, *J. Chem. Soc., Perkin Trans. 1*, 1982, 2079.
3. H. M. R. Hoffmann and J. Rabe, *Angew. Chem., Int. Ed. Engl.*, 1983, **22**, 795.
4. (a) P. Perlmutter and C. C. Teo, *Tetrahedron Lett.*, 1984, **25**, 5951; (b) D. Basavaiah and V. V. L. Gowriswari, *Tetrahedron Lett.*, 1986, **27**, 2031; (c) H. Amri and J. Villieras, *Tetrahedron Lett.*, 1986, **27**, 4307; (d) J. S. Hill and N. S. Isaacs, *Tetrahedron Lett.*, 1986, **27**, 5007; (e) P. Auvray, P. Knochel and J. F. Normant, *Tetrahedron Lett.*, 1986, **27**, 5095.
5. S. Rafel and J. W. Leahy, *J. Org. Chem.*, 1997, **62**, 1521.
6. C. Yu, B. Liu and L. Hu, *J. Org. Chem.*, 2001, **66**, 5413.

7. C. Yu and L. Hu, *J. Org. Chem.*, 2002, **67**, 219.

8. X. Franck and B. Figadere, *Tetrahedron Lett.*, 2002, **43**, 1449.

9. D. Basavaiah and R. M. Reddy, *Indian J. Chem.*, 2001, **40B**, 985.

10. (a) W. P. Almeida and F. Coelho, *Tetrahedron Lett.*, 1998, **39**, 8609; (b) A. Masunari, E. Ishida, G. Trazzi, W. P. Almeida and F. Coelho, *Synth. Commun.*, 2001, **31**, 2127.

11. F. Coelho, W. P. Almeida, D. Veronese, C. R. Mateus, E. C. S. Lopes, R. C. Rossi, G. P. C. Silveira and C. H. Pavam, *Tetrahedron*, 2002, **58**, 7437.

12. W.-D. Lee, K.-S. Yang and K. Chen, *Chem. Commun.*, 2001, 1612.

13. T. Bosanac and C. S. Wilcox, *Chem. Commun.*, 2001, 1618.

14. A. Kamimura, Y. Gunjigake, H. Mitsudera and S. Yokoyama, *Tetrahedron Lett.*, 1998, **39**, 7323.

15. M. Shi and G.-L. Zhao, *Tetrahedron*, 2004, **60**, 2083.

16. J. Bacsa, P. T. Kaye and R. S. Robinson, *S. Afr. J. Chem.*, 1998, **51**, 47.

17. (a) P. T. Kaye and X. W. Nocanda, *J. Chem. Soc., Perkin Trans. 1*, 2000, 1331; (b) P. T. Kaye and X. W. Nocanda, *J. Chem. Soc., Perkin Trans.*, 2002, 1318.

18. S. K. Nayak, L. Thijs and B. Zwanenburg, *Tetrahedron Lett.*, 1999, **40**, 981.

19. K.-S. Yang and K. Chen, *Org. Lett.*, 2000, **2**, 729.

20. H. Sajiki, A. Yamada, K. Yasunaga, T. Tsunoda, M. F. A. Amer and K. Hirota, *Nucleic Acids Res. Supplement*, 2002, 13.

21. A. Patra, S. Batra, B. Kundu, B. S. Joshi, R. Roy and A. P. Bhaduri, *Synthesis*, 2001, 276.

22. S. Batra, S. K. Rastogi, B. Kundu, A. Patra and A. P. Bhaduri, *Tetrahedron Lett.*, 2000, **41**, 5971.

23. S. Batra, T. Srinivasan, S. K. Rastogi, B. Kundu, A. Patra, A. P. Bhaduri and M. Dixit, *Bioorg. Med. Chem. Lett.*, 2002, **12**, 1905.

24. (a) S. J. Garden and J. M. S. Skakle, *Tetrahedron Lett.*, 2002, **43**, 1969; (b) Y. M. Chung, Y. J. Im and J. N. Kim, *Bull. Korean Chem. Soc.*, 2002, **23**, 1651.

25. (a) P. V. Ramchandran, M. V. Ram Reddy and M. T. Rudd, *Chem. Commun.*, 2001, 757; (b) M. V. Ram Reddy, M. T. Rudd and P. V. Ramchandran, *J. Org. Chem.*, 2002, **67**, 5382.

26. D. Basavaiah, N. Kumaragurubaran and D. S. Sharada, *Tetrahedron Lett.*, 2001, **42**, 85.

27. D. Basavaiah, D. S. Sharada, N. Kumaragurubaran and R. Mallikarjuna Reddy, *J. Org. Chem.*, 2002, **67**, 7135.

28. C. H. Lee and K.-J. Lee, *Synthesis*, 2004, 1941.

29. S. W. Lee, C. H. Lee and K.-J. Lee, *Bull. Korean Chem. Soc.*, 2006, **27**, 769.

30. V. G. Nenajdenko, S. V. Druzhinin and E. S. Balenkova, *Mendeleev Commun.*, 2006, **16**, 273.

31. R. O. M. A. de Souza, V. L. P. Pereira, P. M. Esteves and M. L. A. A. Vasconcellos, *Tetrahedron Lett.*, 2008, **49**, 5902.

32. F. Ameer, S. E. Drewes, S. Freese and P. T. Kaye, *Synth. Commun.*, 1988, **18**, 495.

33. S. E. Drewew, S. D. Freese, N. D. Emslie and G. H. P. Roos, *Synth. Commun.*, 1988, **18**, 1565.

34. (a) K. Rolfing, M. Thiel and H. Künzer, *Synlett*, 1997, 325; (b) H. Richter and G. Jung, *Mol. Diversity*, 1998, **3**, 191; (c) S. K. Rastogi, P. Gupta, T. Srinivasan and B. Kundu, *Mol. Diversity*, 2000, **5**, 91.

35. Y. Hayashi, K. Okado, I. Ashimine and M. Shoji, *Tetrahedron Lett.*, 2002, **43**, 8683.

36. V. K. Aggarwal, I. Emme and S. Y. Fulford, *J. Org. Chem.*, 2003, **68**, 692.

37. F. Rezgui and M. M. El Gaied, *Tetrahedron Lett.*, 1998, **39**, 5965.

38. K. Y. Lee, J. H. Gong and J. N. Kim, *Bull. Korean Chem. Soc.*, 2002, **23**, 659.

39. M. C. Redondo, M. Ribagorda and M. C. Carreno, *Org. Lett.*, 2010, **12**, 568.

40. V. K. Aggarwal and A. Mereu, *Chem. Commun.*, 1999, 2311.

41. P. T. Kaye and X. W. Nocanda, *Synthesis*, 2001, 2389.

42. N. Azizi and M. R. Saidi, *Tetrahedron Lett.*, 2002, **43**, 4305.

43. R. Ballini, L. Barboni, G. Bosica, D. Fiorini, E. Mignini and A. Palmieri, *Tetrahedron*, 2004, **60**, 4995.

44. D. Basavaiah, M. Krishnamacharyulu and A. J. Rao, *Synth. Commun.*, 2000, **30**, 2061.

45. D. Basavaiah, A. Jaganmohan Rao and M. Krishnamacharyulu, *Arkivoc*, 2002, **vii**, 136.

46. D. Basavaiah and A. J. Rao, *Tetrahedron Lett.*, 2003, **44**, 4365.

47. N. E. Leadbeater and C. Van der Pol, *J. Chem. Soc., Perkin Trans. 1*, 2001, 2831.

48. S. Luo, B. Zhang, J. He, A. Janczuk, P. G. Wang and J.-P. Cheng, *Tetrahedron Lett.*, 2002, **43**, 7369.

49. R. Gatri and M. M. El Gaied, *Tetrahedron Lett.*, 2002, **43**, 7835.

50. S. Luo, P. G. Wang and J.-P. Cheng, *J. Org. Chem.*, 2004, **69**, 555.

51. R. O. M. A. de Souza and M. L. A. A. Vasconcellos, *Catal. Commun.*, 2004, **5**, 21.

52. (a) I. Deb, M. Dadwal, S. M. Mobin and I. N. N. Namboothiri, *Org. Lett.*, 2006, **8**, 1201; (b) I. Deb, P. Shanbhag, S. M. Mobin and I. N. N. Namboothiri, *Eur. J. Org. Chem.*, 2009, 4091.

53. M. Dadwal, S. M. Mobinb and I. N. N. Namboothiri, *Org. Biomol. Chem.*, 2006, **4**, 2525.

54. J. G. Verkade, *Acc. Chem. Res.*, 1993, **26**, 483.

55. K. Y. Lee, S. GowriSankar and J. N. Kim, *Tetrahedron Lett.*, 2004, **45**, 5485.

56. P. R. Krishna, E. R. Sekhar and V. Kannan, *Synthesis*, 2004, 857.

57. J. Cai, Z. Zhou, G. Zhao and C. Tang, *Org. Lett.*, 2002, **4**, 4723.

58. S.-H. Zhao and Z.-B. Chen, *Synth. Commun.*, 2005, **35**, 3045.

59. R. O. M. A. de Souza, Bruno A. Meireles, Lúcia C. S. Aguiar and M. L. A. A. Vasconcellos, *Synthesis*, 2004, 1595.

60. S. Luo, X. Mi, P. G. Wang and J.-P. Cheng, *Tetrahedron Lett.*, 2004, **45**, 5171.

61. Y.-S. Lin, C.-W. Liu and T. Y. R. Tsai, *Tetrahedron Lett.*, 2005, **46**, 1859.

62. L. He, T.-Y. Jian and S. Ye, *J. Org. Chem.*, 2007, **72**, 7466.

63. V. K. Aggarwal, I. Emme and A. Mereu, *Chem. Commun.*, 2002, **1**, 1612.
64. X. Mi, S. Luo and J.-P. Cheng, *J. Org. Chem.*, 2005, **70**, 2338.
65. X. Mi, S. Luo, H. Xu, L. Zhang and J.-P. Cheng, *Tetrahedron*, 2006, **62**, 2537.
66. Y. Cai, Y. Liu and G. Gao, *Monatsh. Chem.*, 2007, **138**, 1163.
67. (a) S. E. Drewes and G. H. P. Roos, *Tetrahedron*, 1988, **44**, 4653; (b) D. Basavaiah, P. Dharma Rao and S. R. Hyma, *Tetrahedron*, 1996, **52**, 8001; (c) E. Ciganek, in *Organic Reactions*, ed. L. A. Paquette, John Wiley & Sons, Inc., New York, 1997, **vol. 51**, p. 201; (d) D. Basavaiah, A. J. Rao and T. Satyanarayana, *Chem. Rev.*, 2003, **103**, 811; (e) P. Langer, *Angew. Chem., Int. Ed.*, 2000, **39**, 3049.
68. M. Bailey, I. E. Marko, W. D. Ollis and P. R. Rassmussen, *Tetrahedron Lett.*, 1990, **31**, 4509.
69. The use of AcOH as co-catalyst was described earlier: H. M. R. Hoffmann, *J. Org. Chem.*, 1988, **53**, 3701.
70. I. E. Markó, P. R. Giles and N. J. Hindley, *Tetrahedron*, 1997, **53**, 1015.
71. (a) Y. Iwabuchi, M. Nakatani, N. Yokoyama and S. Hatakeyama, *J. Am. Chem. Soc.*, 1999, **121**, 10219; (b) Y. Iwabuchi and S. Hatakeyama, *J. Synth. Org. Chem. Jpn; (Yuki Gosei Kagaku Kyokaishi, in Japanese)*, 2002, **60**, 1.
72. (a) Y. Iwabuchi, T. Sugihara, T. Esumi and S. Hatakeyama, *Tetrahedron Lett.*, 2001, **42**, 7867; (b) Y. Iwabuchi, M. Furukawa, T. Esumi and S. Hatakeyama, *Chem. Commun.*, 2001, 2030.
73. A. Nakano, S. Kawahara, S. Akamatsu, K. Morokuma, M. Nakatani, Y. Iwabuchi, K. Takahashi, J. Ishihara and S. Hatakeyama, *Tetrahedron*, 2006, **62**, 381.
74. A. Nakano, K. Takahashi, J. Ishihara and S. Hatakeyama, *Org. Lett.*, 2006, **8**, 5357.
75. M. Shi and J.-K. Jiang, *Tetrahedron: Asymmetry*, 2002, **13**, 1941.
76. S. Kawahara, A. Nakano, T. Esumi, Y. Iwabuchi and S. Hatakeyama, *Org. Lett.*, 2003, **5**, 3103.
77. (a) M. Shi and Y.-M. Xu, *Angew. Chem., Int. Ed.*, 2002, **41**, 4507 [the absolute configuration of 143 derived from methyl acrylate should be (S) rather than (R) as shown in this paper]; (b) M. Shi, Y.-M. Xu and Y.-L. Shi, *Chem. Eur. J.*, 2005, **11**, 1794.
78. A. Nakano, M. Ushiyama, Y. Iwabuchi and S. Hatakeyama, *Adv. Synth. Catal.*, 2005, **347**, 1790.
79. N. Abermil, G. Masson and J. Zhu, *J. Am. Chem. Soc.*, 2008, **130**, 12596.
80. N. Abermil, G. Masson and J. Zhu, *Org. Lett.*, 2009, **11**, 4648.
81. C. M. Mocquet and S. L. Warriner, *Synlett*, 2004, 356.
82. (a) T. Oishi and M. Hirama, *Tetrahedron Lett.*, 1992, **33**, 639; (b) T. Oishi, H. Oguri and M. Hirama, *Tetrahedron: Asymmetry*, 1995, **6**, 1241.
83. A. G. M. Barrett, A. S. Cook and A. Kamimura, *Chem. Commun.*, 1998, 2533.
84. A. G. M. Barrett, P. Dozzo, A. J. P. White and D. J. Williams, *Tetrahedron*, 2002, **58**, 7303.

85. P. R. Krishna, V. Kannan and P. V. N. Reddy, *Adv. Synth. Catal.*, 2004, **346**, 603.
86. Y.-P. Ding, Y. Ding, Z.-X. Shen, B. Li and Y.-W. Zhang, *Chin. J. Org. Chem.*, 2006, **26**, 1306.
87. Y. Hayashi, T. Tamura and M. Shoji, *Adv. Synth. Catal.*, 2004, **346**, 1106.
88. F. A. Khan and S. K. Upadhyay, *Tetrahedron Lett.*, 2008, **49**, 6111.
89. (a) A. J. McCarroll and J. C. Walton, *Angew. Chem., Int. Ed.*, 2001, **40**, 2224; (b) A. J. McCarroll and J. C. Walton, *J. Chem. Soc., Perkin Trans.*, 2001, 3215.
90. J. Xu, Y. Guan, S. Yang, Y. Ng, G. Peh and C.-H. Tan, *Chem. Asian J.*, 2006, **1**, 724.
91. C. O. Dálaigh and S. J. Connon, *J. Org. Chem.*, 2007, **72**, 7066.
92. J. Wang, H. Li, X. Yu, L. Zu and W. Wang, *Org. Lett.*, 2005, **7**, 4293.
93. (a) T. Okino, Y. Hoashi and Y. Takemoto, *J. Am. Chem. Soc.*, 2003, **125**, 12672; (b) T. Okino, Y. Hoashi, T. Furukawa, X. Xu and Y. Takemoto, *J. Am. Chem. Soc.*, 2005, **127**, 119; (c) Y. Hoashi, T. Okino and Y. Takemoto, *Angew. Chem. Int. Ed.*, 2005, **44**, 4032.
94. T. Okino, S. Nakamura, T. Furukawa and Y. Takemoto, *Org. Lett.*, 2004, **6**, 625.
95. X. Wang, Y.-F. Chen, L.-F. Niu and P.-F. Xu, *Org. Lett.*, 2009, **11**, 3310.
96. K. Matsui, S. Takizawa and H. Sasai, *J. Am. Chem. Soc.*, 2005, **127**, 3680.
97. K. Matsui, K. Tanaka, A. Horii, S. Takizawa and H. Sasai, *Tetrahedron: Asymmetry*, 2006, **17**, 578.
98. M. M. Rauhut and H. Currier, (American Cyanamide Co.) US Pat., 3074999, 1963; *Chem. Abstr.*, 1963, **58**, 11224a.
99. J. D. McClure, *US Pat.*, 3225083, 1965.
100. M. M. Balzer and J. D. Anderson, *J. Org. Chem.*, 1965, **30**, 1357.
101. K. Morita, Z. Suzuki and H. Hirose, *Bull. Chem. Soc. Jpn.*, 1968, **41**, 2815.
102. (a) Toyo Rayon Co. *French Pat.*, 1506132, 1967; *Chem. Abstr.*, 1969, **70**, 19613u; (b) Miyakoshi and S. Saito, *Nippon Kagaku Kaishii*, 1983, 1623; *Chem. Abstr.*, 1984, **110**, 156191g.
103. (a) S. Bertenshaw and M. Kahn, *Tetrahedron Lett.*, 1989, **30**, 2731; (b) T. Imagawa, K. Uemura, Z. Nagai and M. Kawanisi, *Synth. Commun.*, 1984, **14**, 1267; (c) F. Roth, P. Gygax and G. Frater, *Tetrahedron Lett.*, 1992, **33**, 1045.
104. G. Jenner, *Tetrahedron Lett.*, 2000, **41**, 3091.
105. T. Genski and R. J. K. Taylor, *Tetrahedron Lett.*, 2002, **43**, 3573.
106. H. Ito, Y. Takenaka, S. Fukunishi and K. Iguchi, *Synthesis*, 2005, 3035.
107. S. I. Pereira, J. Adrio, A. M. S. Silva and J. C. Carretero, *J. Org. Chem.*, 2005, **70**, 10175.
108. P. B. Kisanga and J. G. Verkade, *J. Org. Chem.*, 2002, **67**, 426.
109. M. Shi and Y.-M. Xu, *Chem. Commun.*, 2001, 1876.
110. M. Shi and Y.-M. Xu, *Eur. J. Org. Chem.*, 2002, 696.
111. M. Shi, Y.-M. Xu, G.-L. Zhao and X.-F. Wu, *Eur. J. Org. Chem.*, 2002, 3666.

112. M. Shi and Y.-M. Xu, *Tetrahedron: Asymmetry*, 2002, **13**, 1195.

113. (a) M. Shi and G.-L. Zhao, *Adv. Synth. Catal.*, 2004, **346**, 1205; (b) M. and G.-L. Zhao *Tetrahedron Lett.*, 2002, **43**, 4499.

114. J. Gao, G.-N. Ma, Q.-J. Li and M. Shi, *Tetrahedron Lett.*, 2006, **47**, 7685.

115. X. Meng, Y. Huang and R. Chen, *Chem. Eur. J.*, 2008, **14**, 6852.

116. X. Meng, Y. Huang and R. Chen, *Org. Lett.*, 2009, **11**, 137.

117. (a) Z. He, X. Tang, Y. Chen and Z. He, *Adv. Synth. Catal.*, 2006, **348**, 413; (b) X. Tang, B. Zhang, Z. He, R. Gao and Z. He, *Adv. Synth. Catal.*, 2007, **349**, 2007.

118. (a) X. Xu, C. Wang, Z. Zhou, X. Tang, Z. He and C. Tang, *Eur. J. Org. Chem.*, 2007, 4487; (b) A. Lu, X. Xu, P. Gao, Z. Zhou, H. Song and C. Tang, *Tetrahedron: Asymmetry*, 2008, **19**, 1886.

119. T. Hayase, T. Shibata, K. Soai and Y. Wakatsuki, *Chem. Commun.*, 1998, 1271.

120. W. Li, Z. Zhang, D. Xiao and X. Zhang, *J. Org. Chem.*, 2000, **65**, 3489.

121. (a) M. Shi and L.-H. Chen, *Chem. Commun.*, 2003, 1310; (b) M. Shi and L.-H. Chen, *Pure Appl. Chem.*, 2005, **77**, 2105; (c) M. Shi, L.-H. Chen and C.-Q. Li, *J. Am. Chem. Soc.*, 2005, **127**, 3790.

122. P. Buskens, J. Klankermayer and W. Leitner, *J. Am. Chem. Soc.*, 2005, **127**, 16762.

123. M. Shi, G.-N. Ma and J. Gao, *J. Org. Chem.*, 2007, **72**, 9779.

124. For (thio)urea derivative catalyzed reactions, see: (a) M. S. Taylor and E. N. Jacobsen, *Angew. Chem., Int. Ed.*, 2006, **45**, 1520; (b) S. J. Connon, *Chem. Eur. J.*, 2006, **12**, 5418; and (c) A. Berkessel, K. Roland and J. M. Neudörfl, *Org. Lett.*, 2006, **8**, 4195.

125. Y.-L. Shi and M. Shi, *Adv. Synth. Catal.*, 2007, **349**, 2129.

126. M.-J. Qi, T. Ai, M. Shi and G. Li, *Tetrahedron*, 2008, **64**, 1181.

127. X.-Y. Guan, Y.-Q. Jiang and M. Shi, *Eur. J. Org. Chem.*, 2008, 2150.

128. M. Shi and C.-Q. Li, *Tetrahedron: Asymmetry*, 2005, **16**, 1385.

129. Z.-Y. Lei, G.-N. Ma and M. Shi, *Eur. J. Org. Chem.*, 2008, 3817.

130. Z.-Y. Lei, X.-G. Liu, M. Shi and M. Zhao, *Tetrahedron: Asymmetry*, 2008, **19**, 2058.

131. J.-W. Han and T. Hayashi, *Chem. Lett.*, 2001, 976.

132. M. Shi, L.-H. Chen and W.-D. Teng, *Adv. Synth. Catal.*, 2005, **347**, 1781.

133. Y.-H. Liu, L.-H. Chen and M. Shi, *Adv. Synth. Catal.*, 2006, **348**, 973.

134. M. Shi, Y.-H. Liu and L.-H. Chen, *Chirality*, 2007, **19**, 124.

135. K. Matsui, S. Takizawa and H. Sasai, *Synlett*, 2006, 761.

136. K. Ito, K. Nishida and T. Gotauda, *Tetrahedron Lett.*, 2007, **48**, 6147.

137. For dendrimeric phosphines in asymmetric catalysis, see: A.-M. Caminade, P. Servin, R. Laurent and J.-P. Majoral, *Chem. Soc. Rev.*, 2008, **37**, 56.

138. Y.-H. Liu and M. Shi, *Adv. Synth. Catal.*, 2008, **350**, 122.

139. (a) J.-M. Garnier, C. Anstiss and F. Liu, *Adv. Synth. Catal.*, 2009, **351**, 331; (b) J.-M. Garnier and F. Liu, *Org. Biomol. Chem.*, 2009, **7**, 1272.

140. (a) K. Yuan, L. Zhang, H.-L. Song, Y. Hu and X.-Y. Wu, *Tetrahedron Lett.*, 2008, **49**, 6262; (b) K. Yuan, H.-L. Song, Y. Hu and X.-Y. Wu, *Tetrahedron*, 2009, **65**, 8185; (c) J.-J. Gong, K. Yuan and X.-Y. Wu,

Tetrahedron: Asymmetry, 2009, **20**, 2117; (d) J.-J. Gong, K. Yuan, H.-L. Song and X.-Y. Wu, *Tetrahedron*, 2010, **66**, 2439.

141. (a) M. Shi, J.-K. Jiang and Y.-S. Feng, *Org. Lett.*, 2000, **2**, 2397; (b) M. Shi and J.-K. Jiang, *Tetrahedron*, 2000, **56**, 4793; (c) M. Shi, J.-K. Jiang, S.-C. Cui and Y.-S. Feng, *J. Chem. Soc., Perkin Trans. 1*, 2001, 390; (d) Z. Han, S. Uehira, H. Shinokubo and K. Oshima, *J. Org. Chem.*, 2001, **66**, 7854.

142. G. Li, J. Gao, H. X. Wei and M. Enright, *Org. Lett.*, 2000, **2**, 617.

143. (a) T. Kataoka, T. Iwama, S.-i. Tsujiyama, T. Iwamura and S.-i. Watanabe, *Tetrahedron*, 1998, **54**, 11813; (b) G. Li, H. X. Wei, J. J. Gao and T. D. Caputo, *Tetrahedron Lett.*, 2000, **41**, 1.

144. T. Iwamura, M. Fujita, T. Kawakita, S. Kinoshita, S.-i. Watanabe and T. Kataoka, *Tetrahedron*, 2001, **57**, 8455.

145. L. R. Reddy and K. R. Rao, *Org. Prep. Proced. Int.*, 2000, **32**, 185.

146. (a) M. Taniguchi, T. Hino, L. R. Reddy and K. R. Rao, *Org. Prep. Proceed. Int.*, 2000, **32**, 185; (b) Y. Kishi, *Tetrahedron Lett.*, 1986, **27**, 4767.

147. S. Uehira, Z. Han, H. Shinokubo and K. Oshima, *Org. Lett.*, 1999, **1**, 1383.

148. S. H. Kim, H.-X. Wei, J. J. Gao and G. Li, *Molecules*, 2002, **7**, 89.

149. H.-X. Wei, S. Karur and G. Li, *Molecules*, 2000, **5**, 1408.

150. (a) H.-X. Wei, J. J. Gao and G. Li, *Tetrahedron Lett.*, 2001, **42**, 9119; (b) H.-X. Wei, S. H. Kim, T. D. Caputo, D. W. Purkiss and G. Li, *Tetrahedron*, 2000, **56**, 2397; (c) G. Li, H.-X. Wei, J. J. Gao and J. Johnson, *Synth. Commun.*, 2002, **32**, 1765.

151. D. Basavaiah, B. Sreenivasulu, R. Mallikarjuna Reddy and K. Muthukumaran, *Synth. Commun.*, 2001, **31**, 2987.

152. D. Basavaiah, B. Sreenivasulu and J. Srivardhana Rao, *Tetrahedron Lett.*, 2001, **42**, 1147.

153. (a) M. Shi, J.-K. Jiang and S.-C. Cui, *Molecules*, 2001, **6**, 852; (b) M. Shi and J.-K. Jiang, *J. Chem. Crystallogr.*, 1999, **29**, 1295.

154. M. Shi and Y.-S. Feng, *J. Org. Chem.*, 2001, **66**, 406.

155. M. Shi, J.-K. Jiang and S.-C. Cui, *Tetrahedron*, 2001, **57**, 7343.

156. M. Shi and C.-J. Wang, *Helv. Chim. Acta*, 2002, **85**, 841.

157. D. Suzuki, H. Urabe and F. Sato, *Angew. Chem., Int. Ed.*, 2000, **39**, 3290.

158. D. Balan and H. Adolfsson, *J. Org. Chem.*, 2002, **67**, 2329.

159. A. Patra, S. Batra, B. S. Joshi, R. Roy, B. Kundu and A. P. Bhaduri, *J. Org. Chem.*, 2002, **67**, 5783.

160. D. Basavaiah, B. Sreenivasulu and A. Jaganmohan Rao, *J. Org. Chem.*, 2003, **68**, 5983.

161. (a) R. Sagar, C. S. Pant, R. Pathak and A. K. Shaw, *Tetrahedron*, 2004, **60**, 11399; (b) M. Saquib, M. K. Gupta, R. Sagar, Y. S. Prabhakar, A. K. Shaw, R. Kumar, P. R. Maulik, A. N. Gaikwad, S. Sinha, A. K. Srivastava, V. Chaturvedi, R. Srivastava and B. S. Srivastava, *J. Med. Chem.*, 2007, **50**, 2942.

162. Y.-B. Yin, M. Wang, Q. Liu, J.-L. Hu, S.-G. Sun and J. Kang, *Tetrahedron Lett.*, 2005, **46**, 4399.

163. S. Sun, Q. Zhang, Q. Liu, J. Kang, Y. Yin, D. Li and D. Dong, *Tetrahedron Lett.*, 2005, **46**, 6271.

164. A. R. Katritzky, M. S. Kim and K. Widyan, *Arkivoc*, 2008 (iii), 91.

165. Review, see: T. Kataoka and H. Kinoshita, *Eur. J. Org. Chem.*, 2005, 45.

166. (a) T. Kataoka, S. Watanabe, K. Yamamoto, M. Yoshimatsu, G. Tanabe and O. Muraoka, *J. Org. Chem.*, 1998, **63**, 6382; (b) T. Kataoka, Y. Banno, S. Watanabe, T. Iwamura and H. Shimizu, *Tetrahedron Lett.*, 1997, **38**, 1809; (c) S. Watanabe, K. Yamamoto, Y. Itagaki and T. Kataoka, *J. Chem. Soc. Perkin Trans.*, 1999, 2053.

167. T. Kataoka, T. Iwama and S.-i. Tsujiyama, *Chem. Commun.*, 1998, 197.

168. T. Iwama, H. Kinoshita and T. Kataoka, *Tetrahedron Lett.*, 1999, **40**, 3741.

169. (a) T. Kataoka, T. Iwama, H. Kinoshita, S. Tsujiyama, Y. Tsurukami, T. Iwamura and S. Watanabe, *Synlett*, 1999, 197; (b) T. Kataoka, T. Iwama, H. Kinoshita, Y. Tsurukami, S. Tsujiyama, M. Fujita, E. Honda, T. Iwamura and S.-i. Watanabe, *J. Organomet. Chem.*, 2000, **611**, 455.

170. T. Kataoka, H. Kinoshita, T. Iwama, S.-i. Tsujiyama, T. Iwamura, S.-i. Watanabe, O. Muraoka and G. Tanabe, *Tetrahedron*, 2000, **56**, 4725.

171. D. Basavaiah, K. Muthukumaran and B. Sreenivasulu, *Synlett*, 1999, 1249.

172. (a) T. Kataoka, H. Kinoshita, S. Kinoshita, T. Iwamura and S.-i. Watanabe, *Angew. Chem., Int. Ed.*, 2000, **39**, 2358; (b) S. Kinoshita, H. Kinoshita, T. Iwamura, S. Watanabe and T. Kataoka, *Chem. Eur. J.*, 2003, **9**, 1496.

173. A. S. Medvedeva, M. M. Demina, P. S. Novopashin, G. I. Sarapulova and A. V. Afonin, *Mendeleev Commun.*, 2002, **12**, 110.

174. (a) T. Kataoka, T. Iwama, S.-i. Tsujiyama, K. Kanematsu, T. Iwamura and S.-i. Watanabe, *Chem. Lett.*, 1999, 257; (b) T. Iwama, S.-i. Tsujiyama, H. Kinoshita, K. Kanematsu, Y. Tsurukami, T. Iwamura, S.-i. Watanabe and T. Kataoka, *Chem. Pharm. Bull.*, 1999, **47**, 956.

175. T. Bauer and J. Tarasiuk, *Tetrahedron: Asymmetry*, 2001, **12**, 1741.

176. (a) R. Pathak, A. K. Shaw and A. P. Bhaduri, *Tetrahedron*, 2002, **58**, 3535; (b) R. Pathak, C. S. Pant, A. K. Shaw, A. P. Bhaduri, A. N. Gaikwad, S. Sinha, A. Srivastava, K. K. Srivastava, V. Chaturvedi, R. Srivastava and B. S. Srivastava, *Bioorg. Med. Chem.*, 2002, **10**, 3187.

177. J. You, J. Xu and J. G. Verkade, *Angew. Chem. Int. Ed.*, 2003, **42**, 5054.

178. L. M. Walsh, C. L. Winn and J. M. Goodman, *Tetrahedron Lett.*, 2002, **43**, 8219.

179. M. Suzuki, T. Kawagishi and R. Noyori, *Tetrahedron Lett.*, 1981, **22**, 1809.

180. (a) S. Kim, Y. G. Kim and J. H. Park, *Tetrahedron Lett.*, 1991, **32**, 2043 (pyridine); (b) M. Hojo, M. Nagayoshi, A. Fujii, T. Yanagi, N. Ishibashi, K. Miura and A. Hosomi, *Chem. Lett.*, 1994, 719 (silyl amine); (c) A. G. M. Barrett and A. J. Kamimura, *J. Chem. Soc., Chem. Commun.*, 1995, 1755 (an asymmetric version with a silyl sulfide or selenide); (d) F. Wang and R. Zibuck, *Synlett*, 1998, 245 (pyridine).

181. J. S. Rao, J.-F. Brière, P. Metznera and D. Basavaiah, *Tetrahedron Lett.*, 2006, **47**, 3553.

182. I. S. del Villar, A. Gradillas, G. Dominguez and J. Perez-Castells, *Org. Lett.*, 2010, **12**, 2418.

183. (a) V. G. Nenajdenko, M. V. Lebedev and E. S. Balenkova, *Synlett*, 1995, 1133; (b) V. G. Nenajdenko, M. V. Lebedev and E. S. Balenkova, *Tetrahedron Lett.*, 1995, **36**, 6317; (c) M. V. Lebedev, V. G. Nenajdenko, M. V. Lebedev and E. S. Balenkova, *Synthesis*, 2001, **1**, 2124.

184. (a) T. Kataoka, S. Kinoshita, H. Kinoshita, M. Fujita, T. Iwamura and S.-i. Watanabe, *Chem. Commun.*, 2001, 1958; (b) H. Kinoshita, S. Kinoshita, Y. Munechika, T. Iwamura, S.-i. Watanabe and T. Kataoka, *Eur. J. Org. Chem.*, 2003, 4852; (c) T. Kataoka and H. Kinoshita, *Phosphorus, Sulfur Silicon Relat. Elem.*, 2005, **180**, 989.

185. T. Kataoka, H. Kinoshita, S. Kinoshita and T. Iwamura, *J. Chem. Soc., Perkin Trans. 1*, 2002, 2043.

186. H. Kinoshita, T. Osamura, S. Kinoshita, T. Iwamura, S.-i. Watanabe, T. Kataoka, G. Tanabe and O. Muraoka, *J. Org. Chem.*, 2003, **68**, 7532.

187. T. Kataoka, H. Kinoshita, S. Kinoshita and T. Iwamura, *Tetrahedron Lett.*, 2002, **43**, 7039.

188. (a) D. J. Maher and S. J. Connon, *Tetrahedron Lett.*, 2004, **45**, 1301; (b) M. Shi, J.-K. Jiang and C.-Q. Li, *Tetrahedron Lett.*, 2002, **43**, 127.

189. (a) C. E. Aroyan, M. M. Vasbinder and S. J. Miller, *Org. Lett.*, 2005, **7**, 3849; (b) J. E. Imbriglio, M. M. Vasbinder and S. J. Miller, *Org. Lett.*, 2003, **5**, 3741.

190. M. M. Vasbinder, J. E. Imbriglio and S. J. Miller, *Tetrahedron*, 2006, **62**, 11450.

191. H. J. Davies, A. M. Ruda and N. C. O. Tomkinson, *Tetrahedron Lett.*, 2007, **48**, 1461.

192. F. Giacalone, M. Gruttadauria, A. M. Marculescu, F. D. Anna and R. Noto, *Catal. Commun.*, 2008, **9**, 1477.

193. (a) H. Y. Tang, G. F. Zhao, Z. H. Zhou, Q. L. Zhou and C. C. Tang, *Tetrahedron Lett.*, 2006, **47**, 5717; (b) H. Y. Tang, P. Gao, G. F. Zhao, Z. H. Zhou, L. N. He and C. C. Tang, *Catal. Commun.*, 2007, **8**, 1811; (c) H. Tang, G. Zhao, Z. Zhou, P. Gao, L. He and C. Tang, *Eur. J. Org. Chem.*, 2008, 126.

194. M. Gruttadauria, F. Giacalone, P. L. Meo, A. M. Marculescu, S. Riela and R. Noto, *Eur. J. Org. Chem.*, 2008, 1589.

195. N. Utsumi, H. Zhang, F. Tanaka and C. F. Barbas III, *Angew. Chem. Int. Ed.*, 2007, **46**, 1878.

196. J. Vesely, P. Dziedzic and A. Córdova, *Tetrahedron Lett.*, 2008, **48**, 6900.

197. J. Vesely, R. Rios and A. Córdova, *Tetrahedron Lett.*, 2008, **49**, 1137.

198. Y. Takemoto, *Org. Biomol. Chem.*, 2005, **3**, 4299.

199. (a) Y. Sohtome, A. Tanatani, Y. Hashimoto and K. Nagasawa, *Tetrahedron Lett.*, 2004, **45**, 5589; (b) Y. Sohtome, N. Takemura, R. Takagi, Y. Hashimoto and K. Nagasawa, *Tetrahedron*, 2008, **64**, 9423.

200. A. Lattanzi, *Synlett*, 2007, 2106.

201. M. Shi and X.-G. Liu, *Org. Lett.*, 2008, **10**, 1043.

202. C. E. S. Jones, S. M. Turega, M. L. Clarke and D. Philp, *Tetrahedron Lett.*, 2008, **49**, 4666.

203. I. T. Raheem and E. N. Jacobsen, *Adv. Synth. Catal.*, 2005, **347**, 1701.

204. (a) Y. M. A. Yamada and S. Ikegami, *Tetrahedron Lett.*, 2000, **41**, 2165; (b) N. T. McDougal and S. E. Schaus, *J. Am. Chem. Soc.*, 2003, **125**, 12094; (c) N. T. McDougal, W. L. Trevellini, S. A. Rodgen, L. T. Kliman and S. E. Schaus, *Adv. Synth. Catal.*, 2004, **346**, 1231; (d) P. M. Pihko, *Angew. Chem. Int. Ed.*, 2004, **43**, 2062.

205. C. Chapuis, G. H. Büchib and H. Wüest, *Helv. Chim. Acta*, 2005, **88**, 3069.

206. M. R. Netherton and G. C. Fu, *Org. Lett.*, 2001, **3**, 4295.

207. R. S. Grainger, N. E. Leadbeater and A. M. Pàmies, *Catal. Commun.*, 2002, **3**, 449.

208. M. Shi and Y.-H. Liu, *Org. Biomol. Chem.*, 2006, **4**, 1468.

209. M. Pohmakotr, S. Thamapipol, P. Tuchinda, S. Prabpai, P. Kongsaeree and V. Reutrakul, *J. Org. Chem.*, 2007, **72**, 5418.

210. Z.-G. Wu, G.-F. Zhou, J.-Y. Zhou and W. Guo, *Synth. Commun.*, 2006, **36**, 2491.

211. K.-S. Park, J. Kim, H. Choo and Y. Chong, *Synlett*, 2007, 395.

212. (a) K. Asano and S. Matsubara, *Synlett*, 2009, 35; (b) K. Asano and S. Matsubara, *Synthesis*, 2009, 3219.

213. R. Mohan, N. Rastogi, I. N. N. Namboothiri, S. M. Mobin and D. Panda, *Bioorg. Med. Chem.*, 2006, **14**, 8073.

214. M. Kawamura and S. Kobayashi, *Tetrahedron Lett.*, 1999, **40**, 1539.

215. K. Matsui, S. Takizawa and H. Sasai, *Tetrahedron Lett.*, 2005, **46**, 1943.

216. M. Dadwal, R. Mohan, D. Panda, S. M. Mobinc and I. N. N. Namboothiri, *Chem. Commun.*, 2006, 338.

217. N. Rastogi, R. Mohan, D. Panda, S. M. Mobin and I. N. N. Namboothiri, *Org. Biomol. Chem.*, 2006, **4**, 3211.

218. A. Bugarin and B. T. Connell, *J. Org. Chem.*, 2009, **74**, 4638.

219. A. Bugarin and B. T. Connell, *Chem. Commun.*, 2010, **46**, 2644.

220. (a) V. K. Aggarwal, A. Mereu, G. J. Tarver and R. McCague, *J. Org. Chem.*, 1998, **63**, 7183; (b) D. Balan and H. Adolfsson, *J. Org. Chem.*, 2001, **66**, 6498.

221. V. K. Aggarwal, D. K. Dean, A. Mereu and R. Williams, *J. Org. Chem.*, 2002, **67**, 510.

222. Y. Shang, D. Wang and J. Wu, *Synth. Commun.*, 2009, **39**, 1035.

223. K. S. Yang, W. D. Lee, J. F. Pan and K. Chen, *J. Org. Chem.*, 2003, **68**, 915.

224. W.-B. Yi, C. Cai and X. Wang, *J. Fluorine Chem.*, 2007, **128**, 919.

225. A. Kumar and S. S. Pawar, *Tetrahedron*, 2003, **59**, 5019.

226. C. L. Johnson, R. E. Donkor, W. Nawaz and N. Karodia, *Tetrahedron Lett.*, 2004, **45**, 7359.

227. E. J. Lenardão, J. de Oliveira Feijó, S. Thurow, G. Perin, R. G. Jacob and C. C. Silveira, *Tetrahedron Lett.*, 2009, **50**, 5215.

228. A. A. Rodriguez, H. Yoo, J. W. Ziller and K. J. Shea, *Tetrahedron Lett.*, 2009, **50**, 6830.
229. J.-W. Huang and M. Shi, *Adv. Synth. Catal.*, 2003, **345**, 953.
230. A. Corma, H. García and A. Leyva, *Chem. Commun.*, 2003, 2806.
231. L.-J. Zhao, H. S. He, M. Shi and P. H. Toy, *J. Comb. Chem.*, 2004, **6**, 680.
232. L.-J. Zhao, C. K.-W. Kwong, M. Shi and P. H. Toy, *Tetrahedron*, 2005, **61**, 12026.
233. C. K.-W. Kwong, R. Huang, M. Zhang, M. Shi and P. H. Toy, *Chem. Eur. J.*, 2007, **13**, 2369.
234. M.-J. Zhang, F.-F. Yu, H. H. Song, C. K.-W. Kwong, M. Shi and P. H. Toy, *J. East China Univ. Sci. Tech (Natural Science Edition)*, 2009, **35**, 80.
235. H.-T. Chen, S. Huh, J. W. Wiench, M. Pruski and V. S.-Y. Lin, *J. Am. Chem. Soc.*, 2005, **127**, 13305.
236. S. Luo, X. Zheng, H. Xu, X. Mi, L. Zhang and J.-P. Cheng, *Adv. Synth. Catal.*, 2007, **349**, 2431.
237. B. Zhao, X. Jiang, D. Li, X. Jiang, T. G. O'Lenick, B. Li and C. Y. Li, *J. Polym. Sci.: Part A: Polym. Chem.*, 2008, **46**, 3438.
238. K. Goren and M. Portnoy, *Chem. Commun.*, 2010, **46**, 1965.
239. X. Wang, P. Han, X. Qiu, X. Ji and L. Gao, *Catal Lett.*, 2008, **124**, 418.

Transformations of Functional Groups in Morita–Baylis–Hillman Adducts

MEI-XIN ZHAO, YIN WEI AND MIN SHI

3.1 Introduction

The Morita–Baylis–Hillman (MBH) reaction has received much attention from synthetic chemists as it provides versatile molecules containing a miminum of three functional groups, *i.e.* hydroxy (or amino), alkene and electron-withdrawing groups such as carbonyl or nitrile groups. Since these functional groups are in close proximity, they are expected to undergo various regio- and stereoselective transformations through fine tuning of these groups either individually or two together. During the past 30 years, the application of MBH adducts has been investigated extensively and several organic transformation methodologies have been developed.

3.2 Transformations of an Hydroxyl Group

3.2.1 Esterification and Etherification

Acetates of Morita–Baylis–Hillman (MBH) adduct **2** can be stereoselectively prepared by the reaction of acetyl chloride in the presence of a base[1] or, alternatively, acetic anhydride in conc. H_2SO_4.[2] However, acetates **2** derived from aromatic aldehydes (R = Ar) prefer to undergo isomerization to generate the thermodynamically more stable acetate **3**, either by intramolecular rearrangement

RSC Catalysis Series No. 8
The Chemistry of the Morita–Baylis–Hillman Reaction
By Min Shi, Fei-Jun Wang, Mei-Xin Zhao and Yin Wei

Published by the Royal Society of Chemistry, www.rsc.org

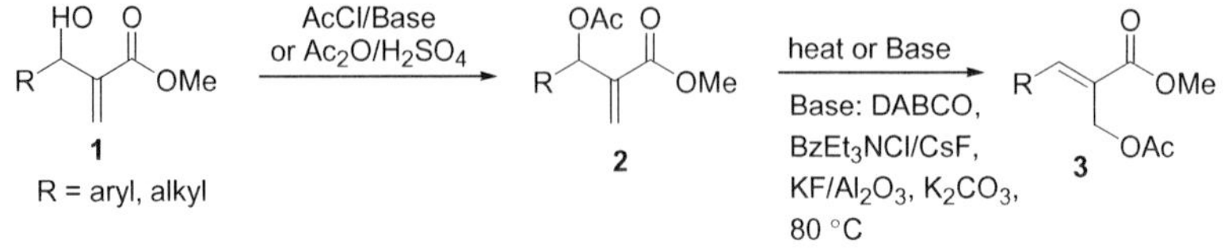

Scheme 3.1

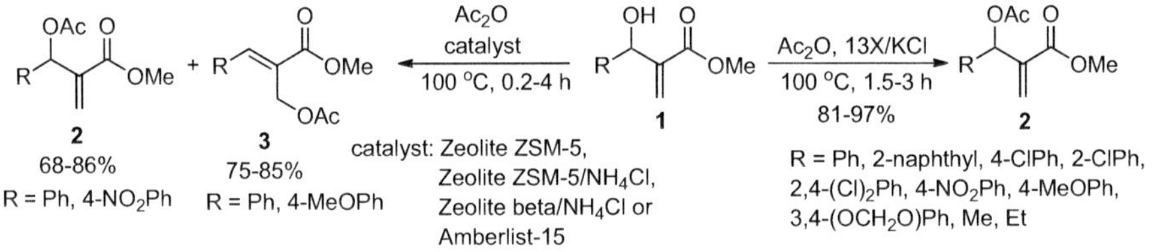

Scheme 3.2

or by nucleophilic catalysis by DABCO[2,3] or several other reagents such as benzyltriethylammonium chloride with cesium fluoride, potassium fluoride on alumina, or potassium carbonate at 80 °C (Scheme 3.1).[3] Consequently, the use of acetic anhydride in basic medium is frequently avoided due to the release of acetate ion as the by-product, which accelerates the allylic isomerization to **3**.[2,3]

Recently, Sá *et al.* have reported an efficient and environmentally benign regioselective synthesis of acetates of MBH adducts **2** in high yields employing a combination of acetic anhydride and an appropriate solid catalyst under solvent-free conditions.[4] Both the catalyst and the substrate influence the product distribution. The basic catalysis was responsible for the selective formation of unrearranged acetate **2**, whereas acid catalysis and substrates containing electron-donor groups were involved in the isomerization to **3** (Scheme 3.2).

Acetates **2** are reactive intermediates that can be used in S_N2' substitution reactions with C-, N-, O-, S- and P-nucleophiles, leading to multi-substituted alkenes as advanced synthetic scaffolds (Section 3.8).

Treatment of MBH addcut **1** with carboxylic acids under Mitsunobu conditions gave almost exclusively the S_N2' products **4**, rather than S_N2 product **5**.[5] Weak and bulky carboxylic acids and low temperatures favor S_N2' addition. Although the reaction conditions were effective for alkyl substituted derivatives, the addition of Et_3N to the Mitsunobu conditions was necessary to improve the S_N2' : S_N2 ratios for the vinyl and phenyl derivatives (Scheme 3.3).[5c] More recently, it was found that the nucleophilic substitution reaction mediated by triphenylphosphine linked to non-crosslinked polystyrene **6** led to a significantly more regioselective transformation. Tri-substituted alkenes **4** were obtained almost quantitatively *via* a highly regioselective S_N2' Mitsunobu reaction (Scheme 3.3).[6]

α-Hydroxyalkylacrylic acids **7**[7] and MBH adduct **8** containing an ester group in the side chain[8] can be cyclized to give the corresponding lactones **9** and **10** *via* intramolecular esterifications (Scheme 3.4).

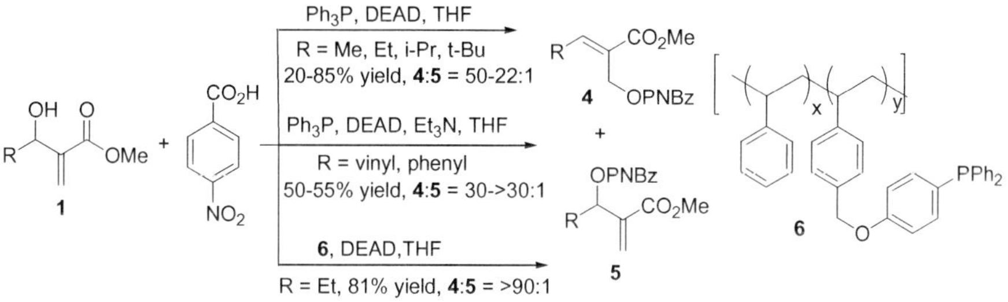

Scheme 3.3

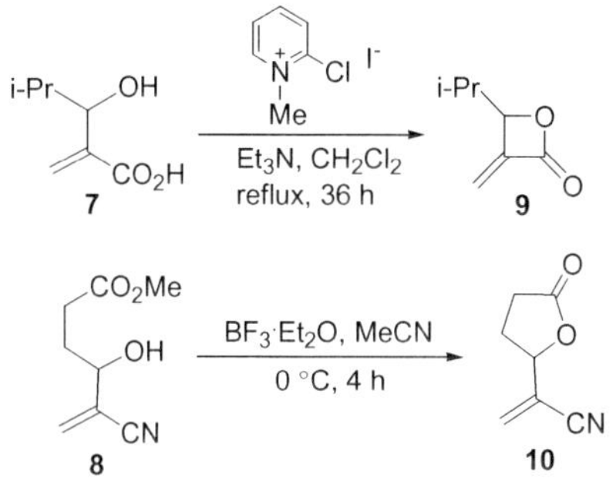

Scheme 3.4

The transformation of an MBH adduct into stereoselectively allyl phosphonates was initially reported by Morita *et al.*,[9] and then Janecki and Bodalski[10] reported that MBH adducts **11** and **12** can react with dialkyl chlorophosphites in the presence of triethylamine to give the corresponding phosphonates **13**, which underwent the Arbuzov rearrangement upon heating to produce 2-methoxycarbonylallylphosphonates **14** with high stereoselectivity ($Z:E = 95:5$) and 2-cyanoallylphosphonates **15** with moderate stereoselectivity ($E:Z = 60:40$–$75:25$) in moderate to high yields (Scheme 3.5).

The allyl phosphonates **14** and **15** are synthetically attractive precursors of substituted 1,3-butadienes. For instance, allyl phosphonate **17** derived from the MBH adducts **16** has been utilized efficiently in the synthesis of stereochemically pure substituted trienes and tetraenes **18** by Wittig–Hornor reaction (Scheme 3.6).[11]

Subsequently, Fields[12] reported a convenient synthesis of phosphonothrixin (**21**) (an important natural product) from the corresponding (*Z*)-allyl phosphonate [methyl (*Z*)-2-(diethoxyphosphorylmethyl)but-2-enoate] (**20**), derived from the MBH adduct (methyl 3-hydroxy-2-methylenebutanoate, **19**), in an overall yield of 24% following the reaction sequence shown in Scheme 3.7.

More recently, Kumar and Swamy reported the transformation of the MBH adduct **22** – which is derived from fresh ferrocenecarboxaldehyde[13] or aromatic

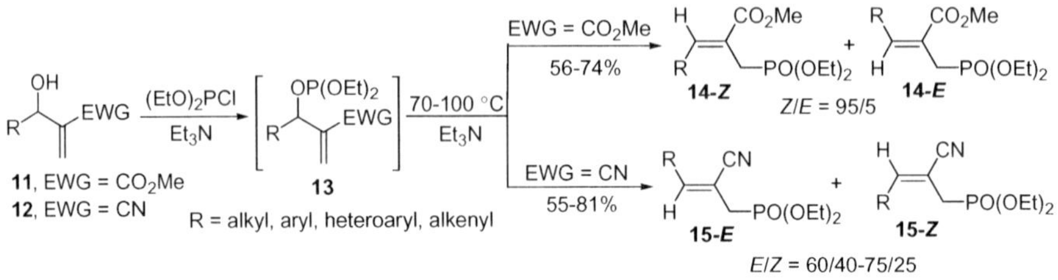

Scheme 3.5

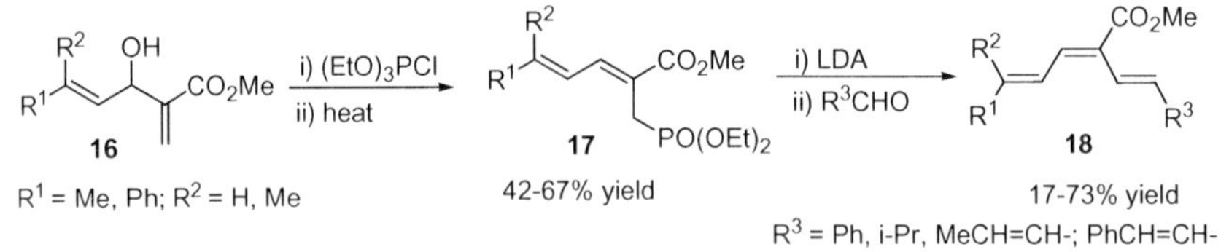

Scheme 3.6

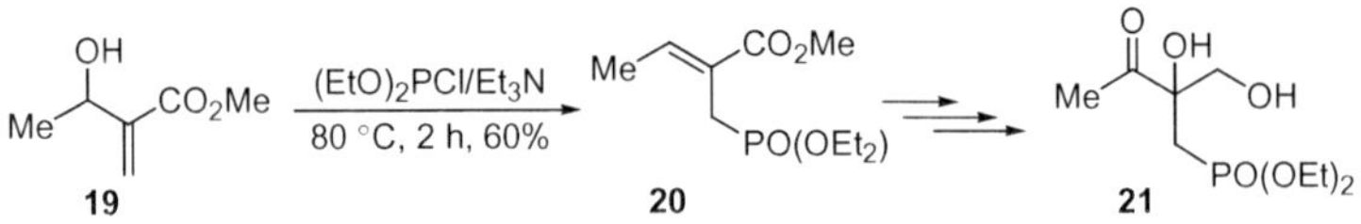

Scheme 3.7

aldehydes[14] and acrylonitrile or acrylates – into the corresponding allyl phosphonates **23**, using cyclic chlorophosphite, which underwent an Arbuzov rearrangement upon heating to afford allyl phosphonates **24**. The allyl phosphonate **24** was further converted into substituted butadiene derivatives **25–27** by the Horner–Wadsworth–Emmons reaction (Scheme 3.8).[13,14]

In addition, the hydroxyl group of a MBH adduct can be easily transformed into *N-p*-toluenesulfonyl carbamates **28** or *N*-acyl carbamates **29** by the reaction with the corresponding isocyanoates.[15] Treatment of **28** or **29** with DABCO in CH$_2$Cl$_2$ provided the corresponding 2-methylene-3-(*p*-toluenesulfonyl) amino esters **30** or 2-methylene-3-acylamino esters **31**, respectively, in good yields. However, treatment of **28** with DBU gave allylamine derivatives, that is, ethyl (2*E*)-3-aryl-2-(*p*-toluenesulfonylaminomethyl)propenoates **32** exclusively. Acetates of the MBH adducts were transformed into the corresponding secondary allylamines **33** *via* treatment with tosylamine in the presence of DABCO (Scheme 3.9).[15]

The racemic imidates **35** and carbamates **36**, which were obtained from an MBH adduct, can be transformed into enantioenriched amides **37** and amines **38** *via* the regio- and enantioselective [1,3]-sigmatropic *O*- to *N*-rearrangement directly or through a decarboxylation catalyzed by cinchona alkaloids.

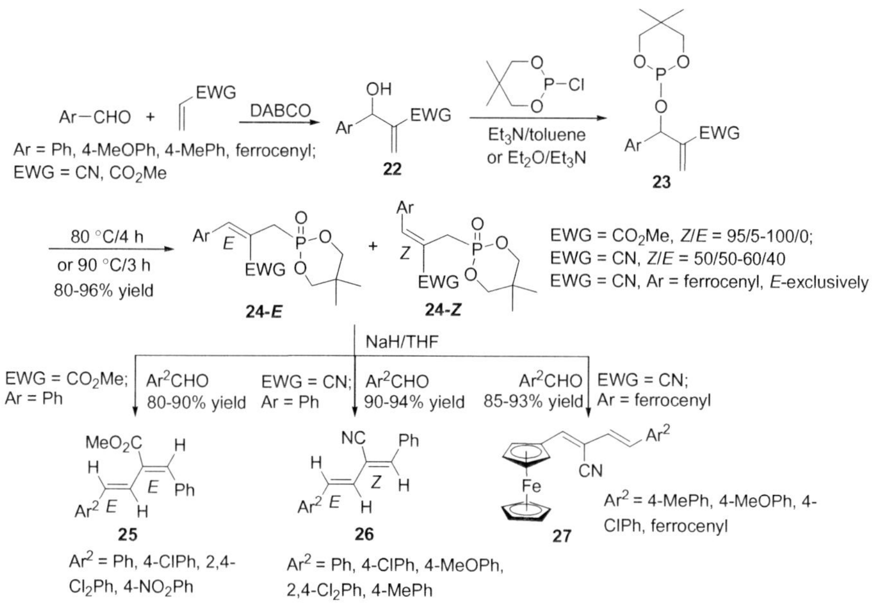

Scheme 3.8

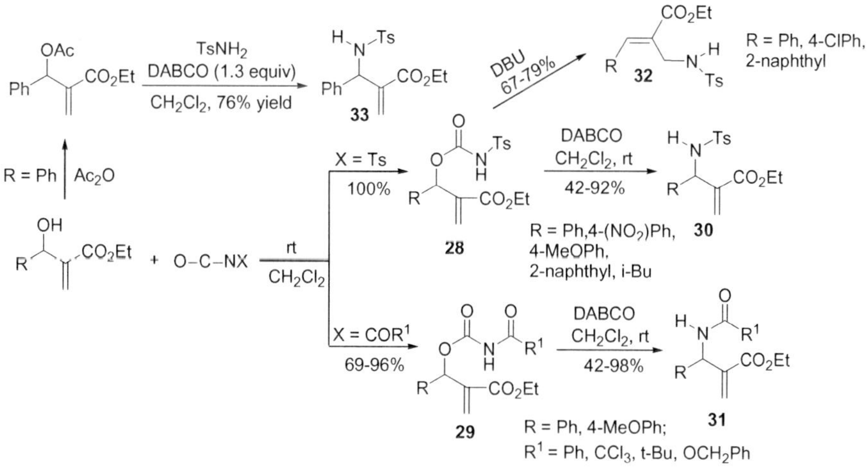

Scheme 3.9

The latter transformation can also be performed as a one-pot reaction from the MBH product **34** to the rearranged amine **38** with high enantioselectivity. These reactions provide an alternative method for obtaining enantioenriched aza-MBH adduct, that is, β-amino acid derivative, in good yield (Scheme 3.10).[16]

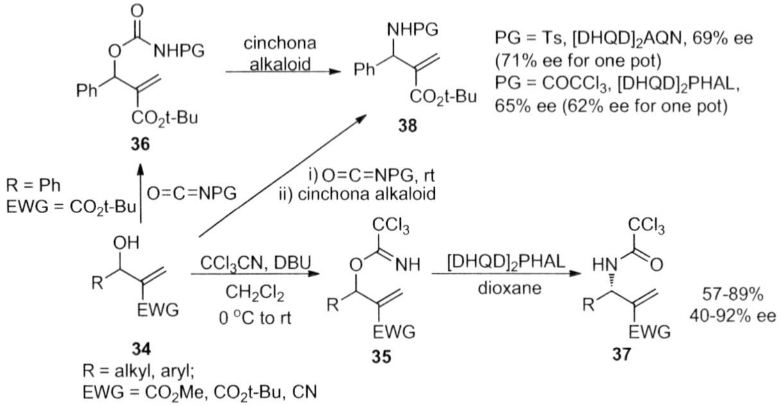

Scheme 3.10

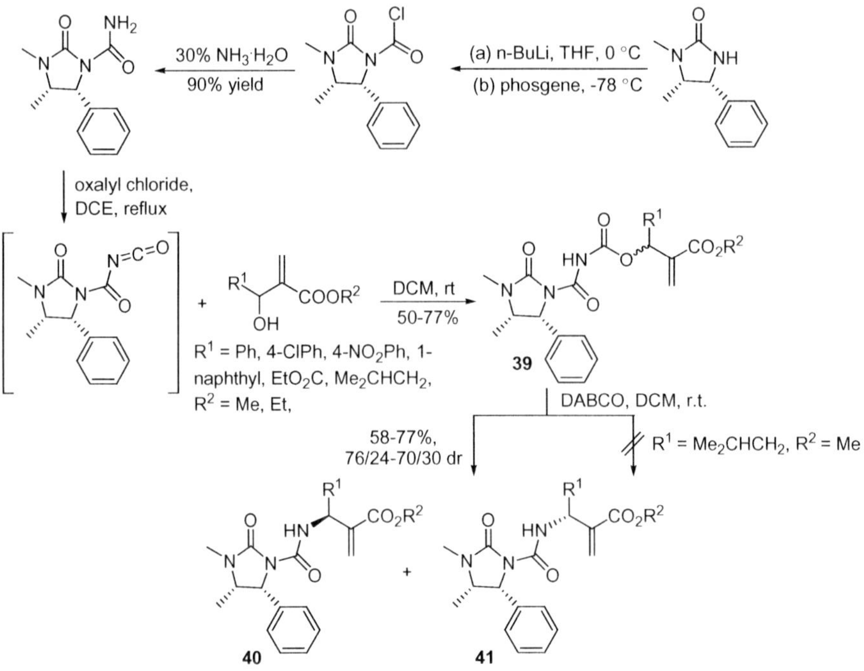

Scheme 3.11

Recently, Orena *et al.* have reported a similar reaction, which used chiral *N*-(4*S*,5*R*)-1-(3,4-dimethyl-2-oxo-5-phenylimidazolidine)carbonyl carbamates **39** as chiral auxiliaries for the stereoselective synthesis of chiral derivatives of the aza-MBH adducts **40** and **41** (Scheme 3.11).[17]

Sulfamate esters **44**, obtained by the reaction of the Burgess reagent **43** with MBH adducts **42**, can undergo pyrolysis with elimination of SO_3 to provide the

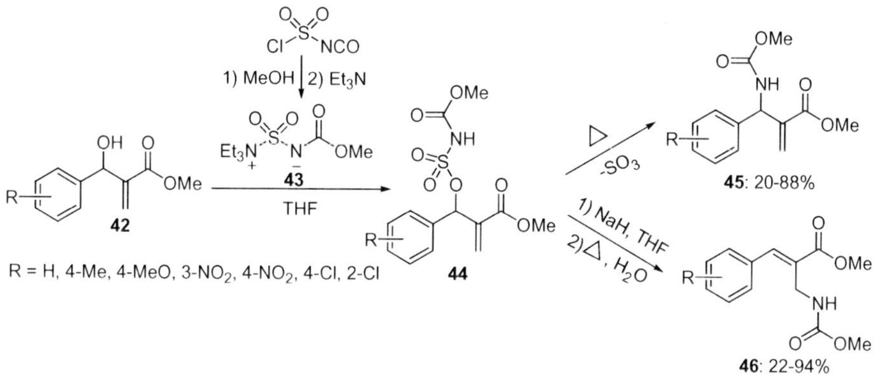

Scheme 3.12

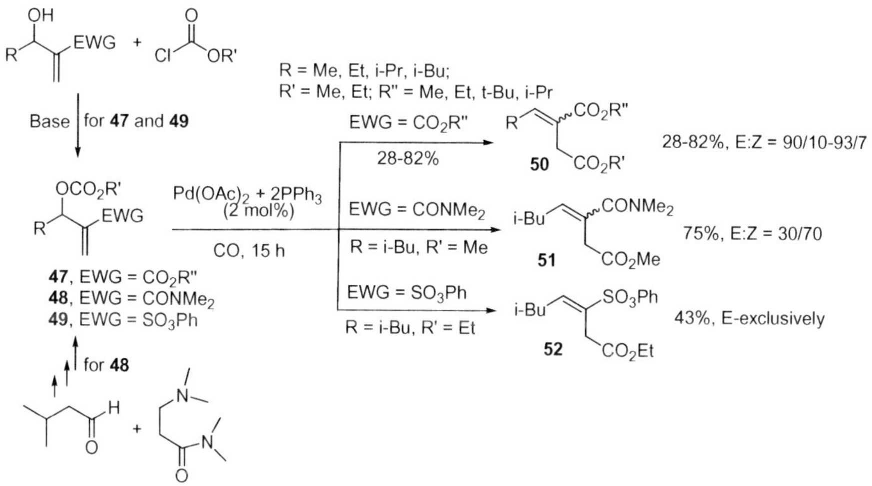

Scheme 3.13

corresponding directly substituted carbamates **45** in excellent yields *via* a S_N pathway, and the alcohol moiety was displaced to form the urethane. However, the carbamates **46** were obtained by treatment of sulfamate ester **44** with NaH, followed by heating at 80 °C and hydrolysis, which involves a substitution with an allylic rearrangement (Scheme 3.12).[18]

A series of carbonates (**47–49**) – derived from carbonate formation of the corresponding MBH adduct (for **47** and **49**) or indirect synthesis from 3-methylbutanal and 3-(dimethylamino)-*N,N*-dimethylpropanamide following a sequence of transformations that includes aldol reaction, carbonate formation and β-elimination of N-oxide (for **48**) – have been used in palladium(0)-catalyzed stereoselective carbonylation reactions to give alkylidenesuccinate and anaolgues in moderate to good yields (Scheme 3.13).[19] While the carbonates **47**

(EWG $= CO_2R''$) gave the corresponding (*E*)-alkylidene-succinates **50** predominantly, carbonate **48** (R $= i$-Bu, EWG $=$ CONMe$_2$) exhibited (*Z*)-selectivity (*E* : *Z* $=$ 3 : 7). In contrast, carbonate **49** (R $= i$-Bu, EWG $=$ SO$_3$Ph) gave stereochemically pure (*E*)-configured product **52** in moderate yields (Scheme 3.13).

Allyl alcohols, including MBH adducts, upon treatment with triethyl orthoacetate in the presence of propionic acid are known to undergo the Johnson–Claisen rearrangement.[20] However, an interesting observation in this respect is that MBH adducts **53** on treatment with triethyl orthoacetate in the presence of HClO$_4$–SiO$_2$ afford the corresponding allyl ethyl ethers **54**, while in the presence of NaHSO$_4$–SiO$_2$ or I$_2$–SiO$_2$ as a heterogeneous catalyst, they can undergo the Johnson–Claisen rearrangement to form ethyl alk-4-enoates **55** stereoselectively in high yield.[21] The allyl ethyl ethers **54**, which contain an ester moiety, have been produced with the (*E*)-configuration while those containing a nitrile moiety had the (*Z*)-configuration. In contrast, the ethyl alk-4-enoates **55**, with an ester group at C4, were obtained with a high (*E*)-selectivity when R is an aryl group and with high (*Z*)-selectivity when R is an alkyl group. However, alkenes **55** with a nitrile group at C4 were formed with (*Z*)-selectivity (Scheme 3.14). Therefore, two different types of trisubstituted alkenes are produced in a stereoselective manner using two different heterogeneous catalysts.

The hydroxy group of a MBH adduct can also be transformed into the corresponding silyl-ethers **56** and **57** by treatment with HMDS/I$_2$[22] or TBDMSCl/Li$_2$S,[23] respectively, under mild conditions in high yield. This reaction has been used to protect hydroxy groups and to synthesize multiple-point pharmacophores of natural and unnatural compounds (Scheme 3.15).

In addition, phenyl ethers **58**[24] and vinyl ethers **59**[20b,c,24,25] derived from MBH adducts have been subjected to the Claisen rearrangement to give the corresponding rearranged products **60** and **61** in good yields (Scheme 3.16).[25]

3.2.2 Halogenation

Stereoselective tranformations of MBH adducts into allyl halides have been well documented in the literature. 2-Fluoroalkylacrylates have been

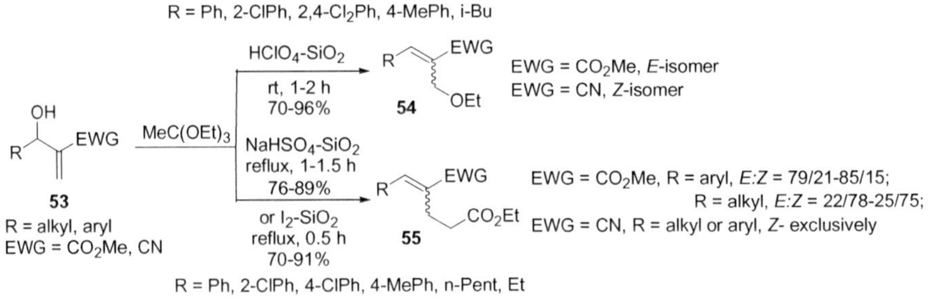

R = Ph, 2-ClPh, 2,4-Cl$_2$Ph, 4-MePh, i-Bu

HClO$_4$-SiO$_2$
rt, 1-2 h
70-96%

54

EWG = CO$_2$Me, *E*-isomer
EWG = CN, *Z*-isomer

MeC(OEt)$_3$

NaHSO$_4$-SiO$_2$
reflux, 1-1.5 h
76-89%

or I$_2$-SiO$_2$
reflux, 0.5 h
70-91%

55

EWG = CO$_2$Me, R = aryl, *E:Z* = 79/21-85/15;
R = alkyl, *E:Z* = 22/78-25/75;
EWG = CN, R = alkyl or aryl, *Z*- exclusively

53
R = alkyl, aryl
EWG = CO$_2$Me, CN

R = Ph, 2-ClPh, 4-ClPh, 4-MePh, n-Pent, Et

Scheme 3.14

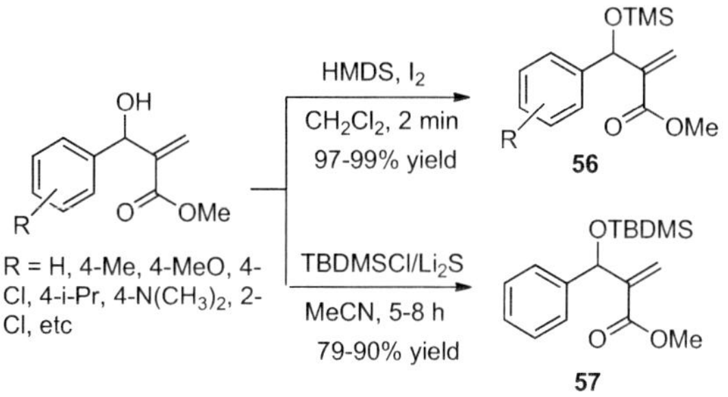

Scheme 3.15

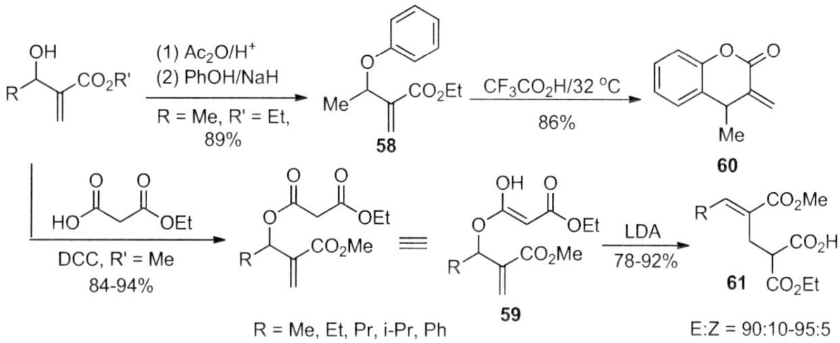

Scheme 3.16

obtained by the reaction of 2-hydroxyalkylacrylates, which are obtained from the MBH reaction of aliphalic aldehydes and acrylates, with diethyl[26] or dimethylaminosulfur trifluoride.[27] Ethyl 2-fluoromethylacrylate has also been prepared from the corresponding bromide with tetrabutyl-ammonium fluoride in hexamethylphosphoric triamide.[28] The phenyl substituted 2-hydroxyacrylates reacted with diethylaminosulfur trifluoride (DAST) to produce the corresponding allyl fluoride with *ca.* 5–10% competing allylic isomerization to the primary fluoride.[29] Some allylic fluorides (**62**) undergo 1,3-dipolar cycloaddition with a nitrone for the synthesis of enantiopure fluorine-containing isoxazolidines **63** and amino polyols **64** (Scheme 3.17).[26c]

Treatment MBH adducts with aluminium trichloride (from the acetate),[30] thionyl chloride,[28,31] phosgene[32] and *N*-chlorosuccinimide (NCS)/dimethyl sulfide,[33] respectively, furnishes chlorides **65** and **66**. With thionyl chloride in the presence of pyridine, the MBH adduct gives predominantly the unrearranged chloride **66**.[32] Equimolar mixtures of rearranged and unrearranged chlorides were obtained from ethyl 2-(1-hydroxyethyl)acrylate and ethyl 2-(1-hydroxybutyl)acrylate with hexachloroacetone/triphenylphosphine, while

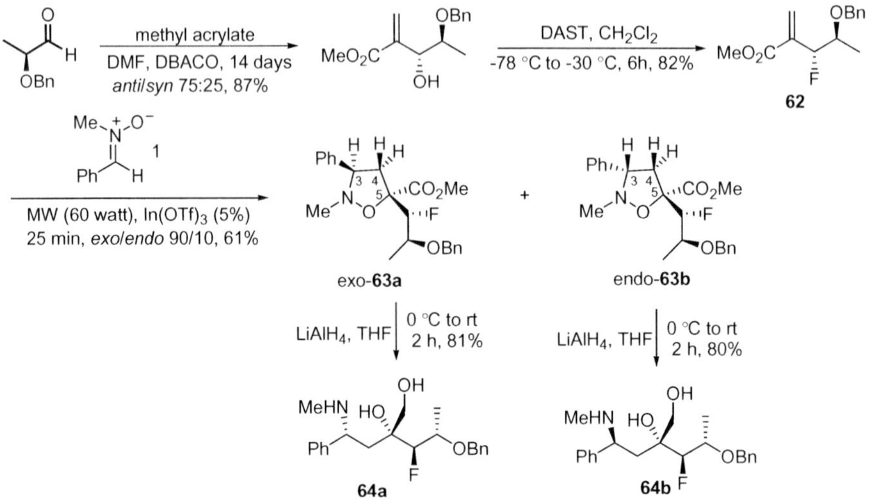

Scheme 3.17

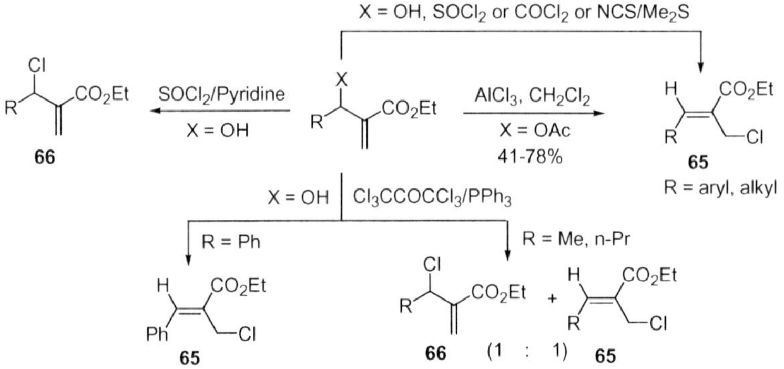

Scheme 3.18

only the rearranged chloride **65** was formed from ethyl 2-hydroxybenzyl-acrylate (Scheme 3.18).[34]

More recently, Chavan *et al.* have found that treatment of MBH adducts **67** with $Et_3N/MsCl$ only provided (2Z)-2-(chloromethyl)-3-arylprop-2-enoates **68** in moderate to high yields, and no mesylate derivatives **69** were observed (Scheme 3.19).[35]

Subsequently, Basavaiah *et al.* converted the MBH adducts **70** obtained from aldehydes and MVK into the (Z)-allyl chlorides or (Z)-allyl bromides **71** in moderate to good yields by using HX (X = Cl, Br) as halogen reagent (Scheme 3.20).[36]

MBH adducts have been chlorinated by Das *et al.* by treatment of MBH adducts **72** with a combination of readily available PPh_3 and Cl_3CCONH_2[37] or

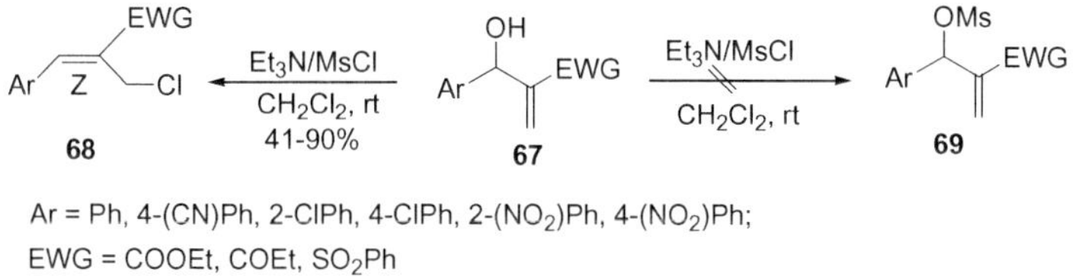

Ar = Ph, 4-(CN)Ph, 2-ClPh, 4-ClPh, 2-(NO_2)Ph, 4-(NO_2)Ph;

EWG = COOEt, COEt, SO_2Ph

Scheme 3.19

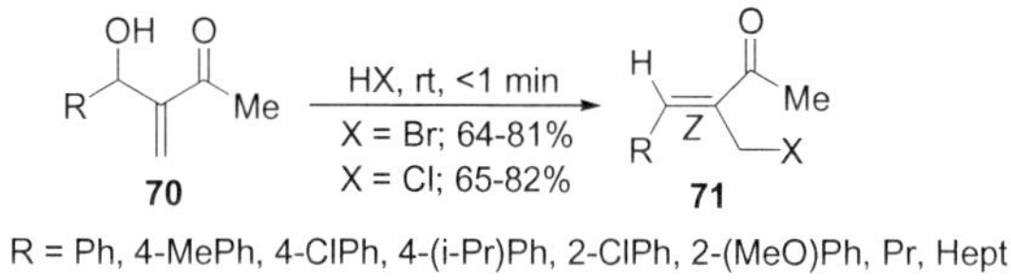

R = Ph, 4-MePh, 4-ClPh, 4-(i-Pr)Ph, 2-ClPh, 2-(MeO)Ph, Pr, Hept

Scheme 3.20

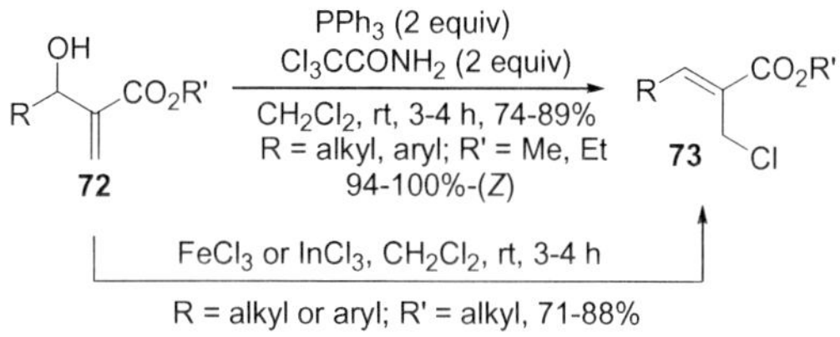

Scheme 3.21

by using $FeCl_3$ or $InCl_3$[38] as a reagent to treat MBH adducts **72** in CH_2Cl_2 at room temperature (Scheme 3.21). The mild and acid-free conditions, application of cheaper reagents, easy experimental procedure and high yields are notable advantages of these methods.

$FeCl_3$ and Yb(OTf)$_3$ have also been used for the conversion of acetates of MBH adducts (**74**) into trisubstituted allyl chlorides **75** and acetates **76**, respectively.[39] Notably, $FeCl_3$-mediated transformation resulted in the formation of allyl chlorides **75** exclusively as (*Z*)-isomers both on the ester as well as keto adducts, whereas Yb(OTf)$_3$ converted the adducts into allyl acetates **76** as *E* : *Z* mixtures [major (*E*)-isomer] (Scheme 3.22).

The allyl bromides could be formed by treatment of 2-hydroxyalkylacrylates with hydrogen bromide, with[40] or without[41] addition of concentrated sulfuric acid. This transformation can also be achieved by other reagents, such as *N*-bromosuccinimide/dimethyl sulfide,[33a,42] cupric bromide on silica,[43] phosphorus trimide[33b,44] and LiBr/H_2SO_4.[45] *N*-Bromosuccinimide/dimethyl sulfide has also been used to convert 2-hydroxyalkylvinyl phenyl sulfones into the allyl bromides.[46]

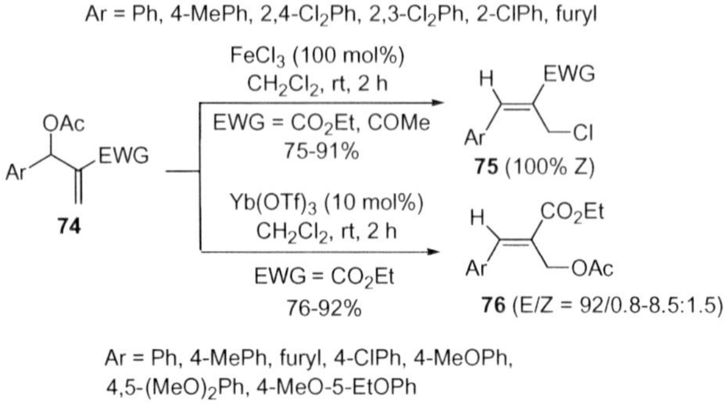

Ar = Ph, 4-MePh, 2,4-Cl$_2$Ph, 2,3-Cl$_2$Ph, 2-ClPh, furyl

74, OAc, Ar, EWG

FeCl$_3$ (100 mol%)
CH$_2$Cl$_2$, rt, 2 h

EWG = CO$_2$Et, COMe
75-91%

75 (100% Z)

Yb(OTf)$_3$ (10 mol%)
CH$_2$Cl$_2$, rt, 2 h

EWG = CO$_2$Et
76-92%

76 (E/Z = 92/0.8-8.5:1.5)

Ar = Ph, 4-MePh, furyl, 4-ClPh, 4-MeOPh,
4,5-(MeO)$_2$Ph, 4-MeO-5-EtOPh

Scheme 3.22

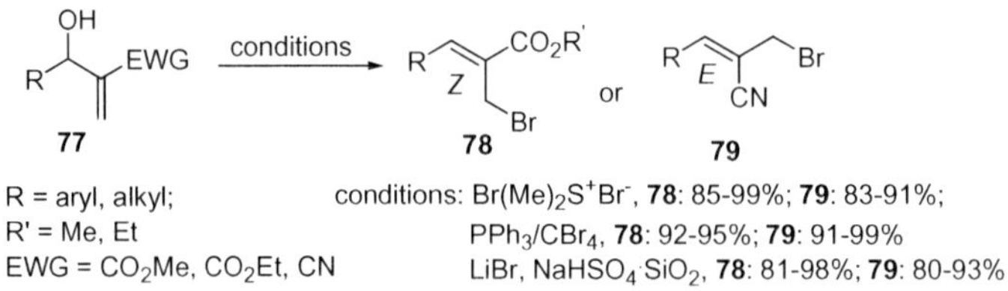

77, OH, R, EWG → **78**, R, CO$_2$R', Z, Br, or **79**, R, E, Br, CN

conditions

R = aryl, alkyl;
R' = Me, Et
EWG = CO$_2$Me, CO$_2$Et, CN

conditions: Br(Me)$_2$S$^+$Br$^-$, **78**: 85-99%; **79**: 83-91%;
PPh$_3$/CBr$_4$, **78**: 92-95%; **79**: 91-99%
LiBr, NaHSO$_4$·SiO$_2$, **78**: 81-98%; **79**: 80-93%

Scheme 3.23

Recently, some more effective methods, including bromodimethylsulfonium bromide (BDMS),[47] Appel agent (PPh$_3$/CBr$_4$)[48] and a combination of lithium bromide and using NaHSO$_4$·SiO$_2$ as a heterogeneous catalyst[49] have been utilized for high-yielding stereoselective synthesis of (Z)- and (E)-allyl bromides from different MBH adducts (**77**), containing COOMe, COOEt and CN functional groups, at room temperature. When the electron-withdrawing group (EWG) is an ester moiety such as COOMe or COOEt, the (Z)-isomer **78** is the major product; however, with a CN group, the (E)-isomer **79** predominates (Scheme 3.23). These methods have been applied successfully to the synthesis of the pheromone (E)-2,4-dimethyl-2-hexenoic acid[50] and naturally occurring bioactive fatty acid amides semiplenamides C and E[48] by a sequence reaction.

The synthesis of those 2-bromomethyl-2-aryl-acrylic acids (**84**) that are not accessible by the above routes due to failure of the MBH reaction can be accomplished by methoxide-induced addition of methyl acrylate to aromatic and heteroaromatic aldehydes (Scheme 3.24).[51] Thus, it may serve as an alternate method in cases where the MBH reaction is slow or fails altogether.

Rearranged iodides are formed from 2-hydroxyalkylacrylates with hydrogen iodide in phosphoric adid[34] or from the corresponding bromides with sodium iodide in acetone.[28,33b] Recently, Das and co-workers have developed the highly efficient stereoselective synthesis of (Z)- and (E)-allyl iodides by treatment of MBH adducts **85** with LiI/NaHSO$_4$·SiO$_2$,[49] PPh$_3$/I$_2$[52] or

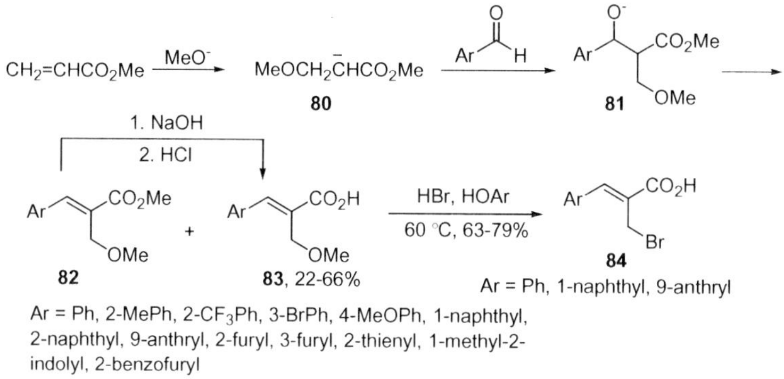

Scheme 3.24

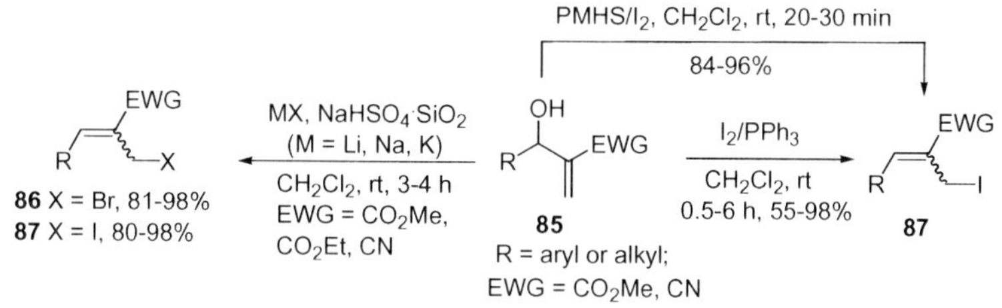

Scheme 3.25

polymethylhydrosiloxane (PMHS)/I_2 (Scheme 3.25).[53] MBH adducts possessing ester and nitrile moieties underwent the conversion readily and allyl iodides containing aryl as well as alkyl groups were obtained in high yields. Allyl iodides containing an ester moiety were formed with the (Z)-configuration while those possessing a nitrile moiety gave exclusively (E)-configured products.

The 2-haloalkyl derivatives of activated olefins are versatile intermediates[54] and have been widely used in organic synthesis by proceeding elimination,[55] reduction,[42d] substitution (as shown in Section 3.8) and organometallization with various metals,[56] *etc.* These reactions are discussed in the following sections.

3.2.3 Miscellaneous Transformations

The hydroxy group in a MBH adduct has been oxidized with Jones reagent, affording doubly activated olefins that are highly reactive.[57] However, pyridinium chlorochromate on silica as an oxidant is much less effective in the oxidation of MBH adducts.[57a] Methyl 2-hydroxyalkylacrylates have also been converted into the corresponding keto compounds by temporarily

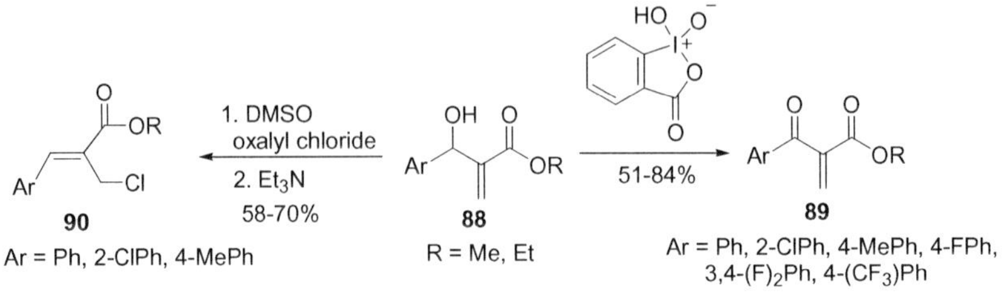

Scheme 3.26

protecting the double bond as the Diels–Alder adduct with anthracene.[58] Alkyl 2-hydroxyarylacrylates can be directly oxidized by Dess–Martin periodinane and the attempted Swern oxidation of the same adducts resulted in S_N2'-type substitution of the allylic hydroxyl group by chloride (Scheme 3.26).[59] Although the adduct of acetaldehyde with acrylonitrile could not be oxidized with several powerful oxidizing agents,[60] its Michael adduct with methanol can be oxidized with Jones reagent. Elimination of methanol is then accomplished by thermolysis with phosphorus pentoxide to give 2-acetylacrylonitrile, which polymerizes on exposure to air or moisture.[61]

The hydroxyl group has been eliminated from MBH products with acetic anhydride and pyridine under reflux[62] or by treatment of a derivative, such as the methanesulfonate, with a base. The obtained very reactive dienes often dimerize in a Diels–Alder fashion[63] but can be isolated[46c,64] or trapped with an external diene or dienophile[63a,65] or with an internal dienophile (Section 3.7).[66]

Kinetic resolutions of α-methylene-β-hydroxy esters (Morita–Baylis–Hillman products) and their acetate derivatives have been performed *via* enzymatic enantioselective transesterification with *Pseudomonas* AK,[67] lipase PS[68] and free or immobilized *Pseudomonas* sp. lipase (PSL) in poly(ethylene oxide) (PEO), silica gel and montmorillonite K10[69] using vinyl acetate as acylating agent. The corresponding enantiomerically pure (or enriched) MBH adducts were obtained for alkyl-substituted substrates; aryl-substituted hydroxy esters were inert under the experimental conditions. On the other hand, hydrolysis of racemic acetates of MBH adducts with lipase PLAP[70] and lipase PS or AK[68] can give the optically active alcohols in moderate to good enantioselectivities. Kinetic resolution has also been accomplished *via* enantioselective reduction catalyzed by horseradish peroxidase (HPR) to give the corresponding (R)-$(+)$-alcohols from alkyl-substituted racemic hydroperoxides with >99% ee (Scheme 3.27).[71]

3.3 Friedel–Crafts Reaction

MBH alcohols as electrophiles in a Friedel–Crafts reaction with benzene was initially reported by Foucaud *et al.*, leading to allylic rearrangement products **91** in the presence of K10 montmorillonite.[72] With phenols and $ZnBr_2$–SiO_2 as catalyst, coumarin derivatives **92** were formed in moderate yields (Scheme 3.28).[73]

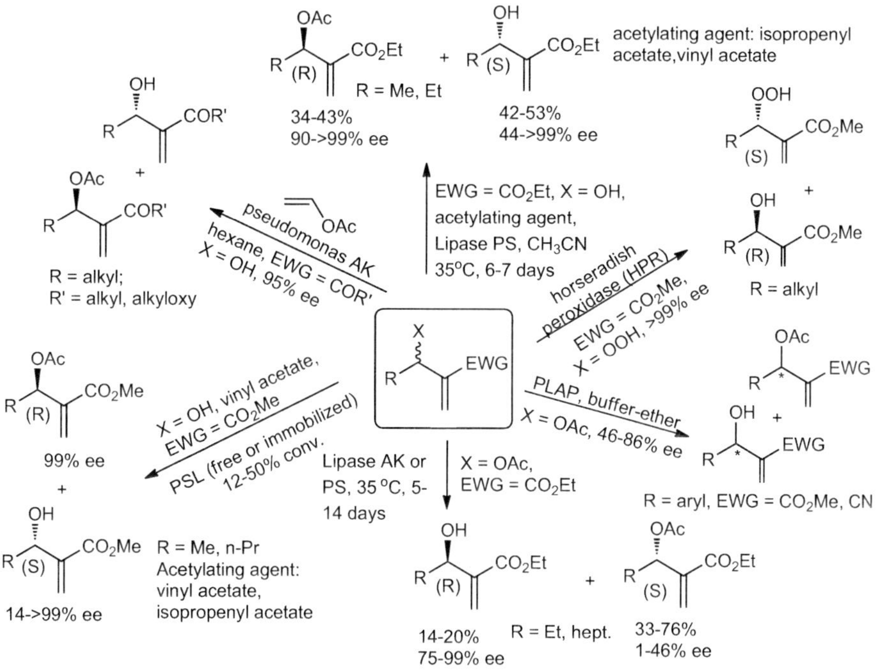

Scheme 3.27

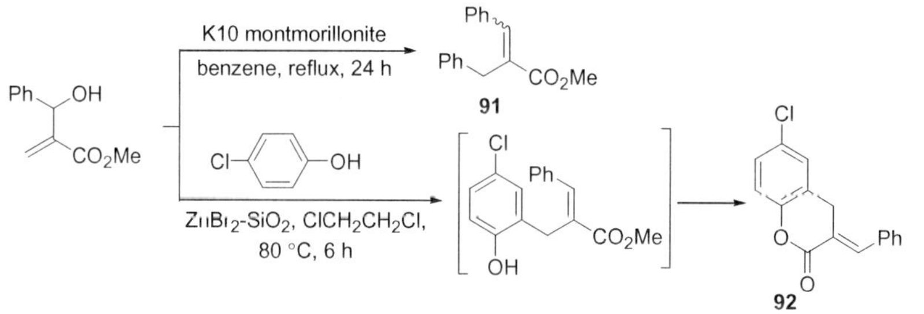

Scheme 3.28

Subsequently, MBH adducts were successfully utilized as novel stereodefined electrophiles in the Friedel–Crafts reaction with benzene in the presence of a Lewis acid[74] and sulfuric acid,[75] leading to the stereoselective synthesis of (*Z*)- and (*E*)-functionalized trisubstituted alkenes. Notably, MBH adducts obtained from acrylonitrile provide high (*Z*)-stereoselectivities, while adducts derived from methyl acrylate and aromatic aldehydes give high (*E*)-stereo-selectivities. When the MBH adducts drived from methyl acrylate and aliphatic aldchydes were involved in the Friedel–Crafts reaction, no significant stereo-selectivity was observed (Scheme 3.29).

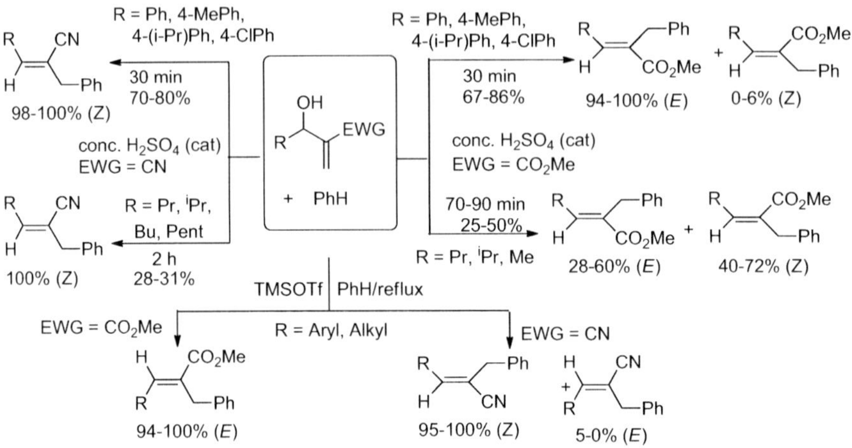

Scheme 3.29

Scheme 3.30

Acetates of MBH adducts have also been successfully utilized as novel stereodefined-electrophiles in the Friedel–Crafts reaction with benzene in the presence of AlCl$_3$,[30] leading to the stereoselective synthesis of (*Z*)- and (*E*)-functionalized trisubstituted alkenes. Attempts to perform an intramolecular Friedel–Crafts reaction in the absence of benzene, to obtain the desired indene derivatives, met with failure. However, this reaction provided a simple methodology for the synthesis of (*Z*)-allyl chlorides (Scheme 3.30)

Kim and co-workers have reported that the Friedel–Crafts reaction of aza-MBH adducts of *N*-tosylimine derivatives **93** with arenes catalyzed by sufuric acid provides a stereoselective methodology for the preparation of stereochemically defined trisubstituted olefins (Scheme 3.31).[76] They have also investigated the Friedel–Crafts reaction by using chlorobenzene and toluene as substrates; however, mixtures of *ortho* and *para* isomers were obtained.

The MBH adducts, derived from methyl acrylate and aldehydes, have been employed successfully to the general synthesis of 3-arylidene(alkylidene)chroman-4-ones **94**, which involved an intramolecular Friedel–Crafts reaction as

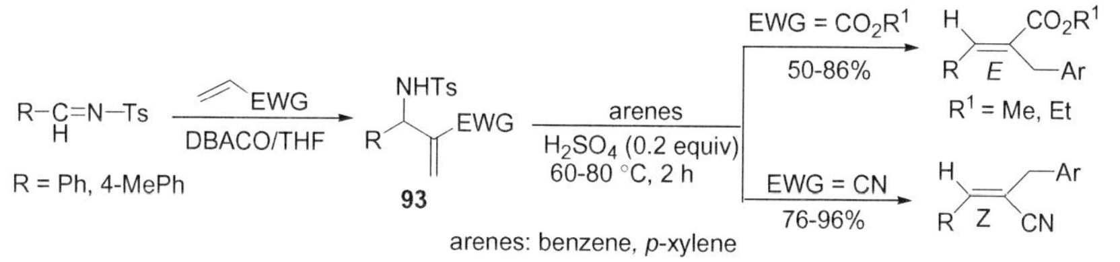

Scheme 3.31

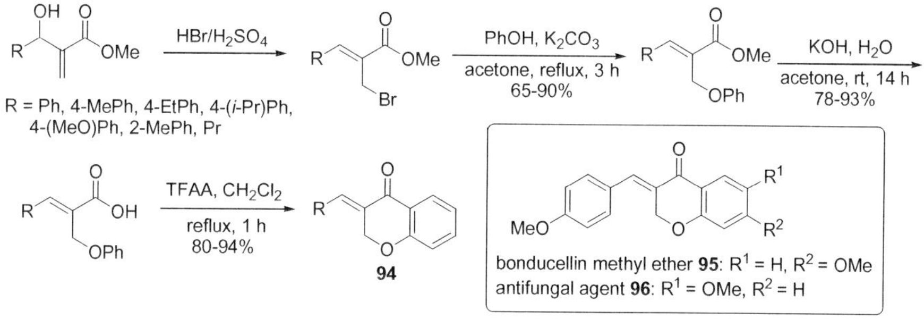

R = Ph, 4-MePh, 4-EtPh, 4-(*i*-Pr)Ph,
4-(MeO)Ph, 2-MePh, Pr

bonducellin methyl ether **95**: R^1 = H, R^2 = OMe
antifungal agent **96**: R^1 = OMe, R^2 = H

Scheme 3.32

Ar = Ph, 2-(CHO)Ph, 2-(CO₂Me)Ph, 2-(CO₂CH₂Ph)Ph, 2-CNPh,
3-(NHCOMe)Ph, 3-(MeO)Ph, 4-(MeO)Ph, 4-(CHO)Ph, 4-ClPh

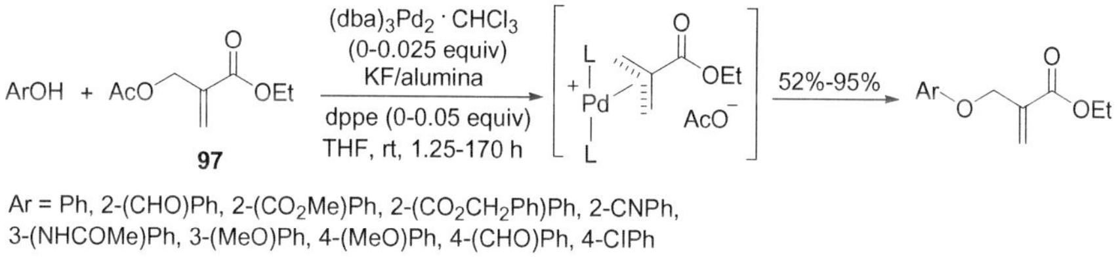

Scheme 3.33

the key step (Scheme 3.32).[77] This method was also applied to the synthesis of some natural products, such as bonducellin methyl ether (**95**) and antifugal agent **96** (Scheme 3.32).[77]

Muzart and co-workers[78] have studied the addition of various substituted phenols to the acetate of the MBH adduct **97** in the presene of a Pd(0) and/or KF/alumina as a catalyst. High yields and fast reaction were achieved when both reagents were used together (Scheme 3.33). They also observed the formation of a mixture of ethers in the case of the acetate of MBH adduct **98**, which was obtained from butyraldehyde and ethyl acrylate when Pd(0) and KF/alumina were employed, whereas in the absence of Pd(0) an S_N2' product was obtained predominantly (Scheme 3.34).

Attempts to obtain indene derivatives *via* the intramolecular Friedel–Crafts reaction of MBH adducts obtained from aromatic aldehydes and methyl

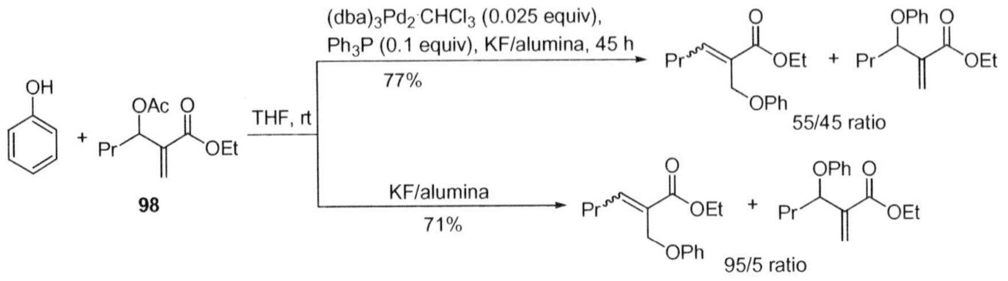

Scheme 3.34

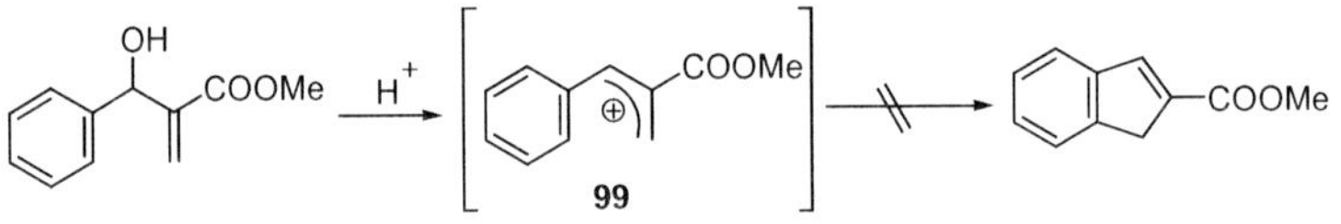

Scheme 3.35

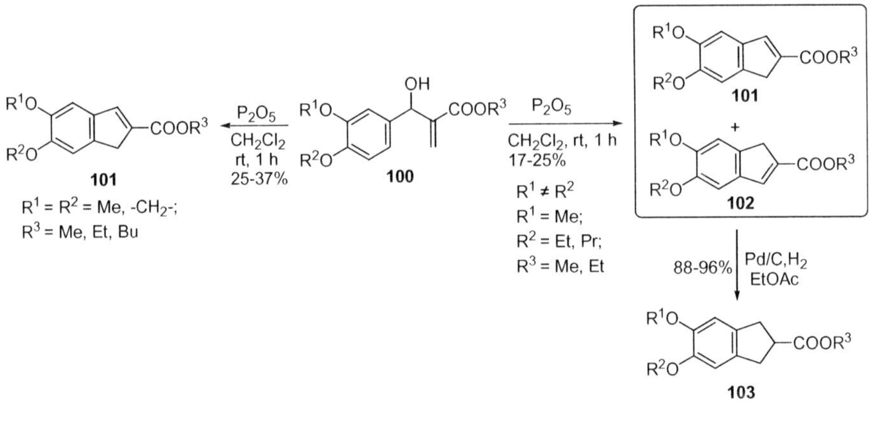

Scheme 3.36

acrylate with various reagents were unsuccessful, which may be due to lower stabilization of carbocation **99** because of the presence of an electron-withdrawing group (CO_2Me) (Scheme 3.35). However, the MBH adducts **100** containing electron-donating group(s) on aromatic ring underwent a facile intramolecular Friedel–Crafts reaction in the presence of P_2O_5, thus providing a convenient process for the synthesis of indene derivatives **101** and **102**. These indene derivatives were further hydrogenated to the corresponding indane derivatives **103** (Scheme 3.36).[79]

Subsequently, Basavaiah[80] and Kim[81] independantly developed a simple one-pot stereoselective transformation of alkyl 3-aryl-3-hydroxy-2-methylene-propanoates **104**, the MBH adducts obtained from acrylate and aromatic

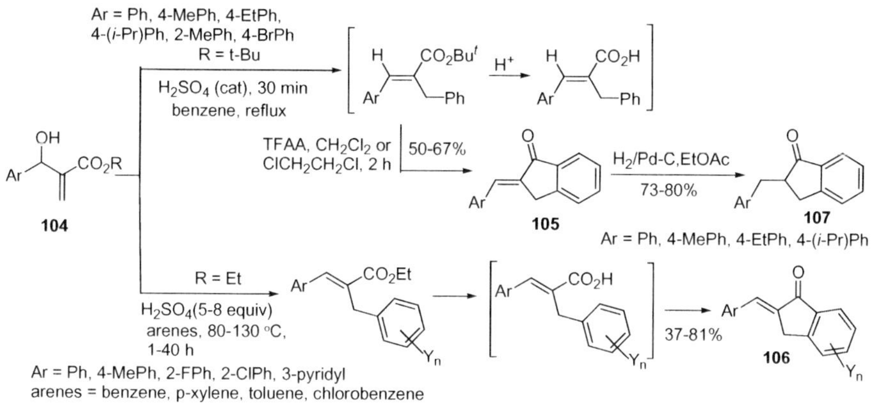

Scheme 3.37

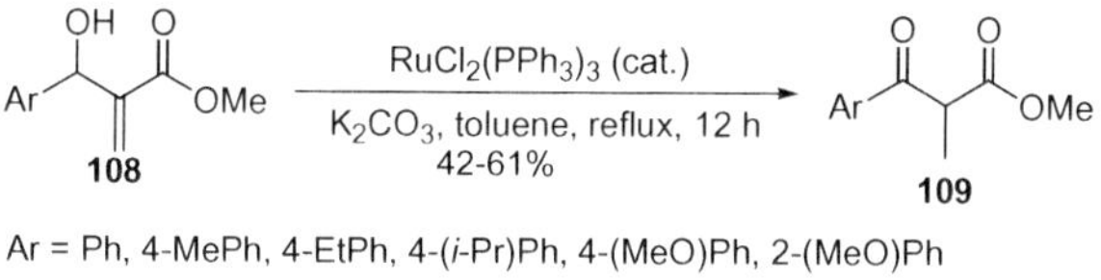

Scheme 3.38

aldehydes, into (*E*)-2-arylideneindan-1-ones **105** and **106** *via* successive inter- and intramolecular Friedel–Crafts reaction (Scheme 3.37). Some of these compounds were further transformed into the corresponding 2-arylmethyl-indan-1-ones **107** *via* catalytic hydrogenation in the presence of 5% Pd/C catalyst (Scheme 3.37).

3.4 Isomerization

The carbon–carbon double bond in the MBH adduct is easily isomerized under various conditions to give different products. Basavaiah and co-workers have studied the isomerization of methyl 3-aryl-3-hydroxy-2-methylenepropanoates **108**, the MBH adducts obtained from methyl acrylate and aromatic aldehydes, in the presence of $RuCl_2(PPh_3)_2$ and K_2CO_3 to form methyl 3-aryl-2-methyl-3-oxopropanoates **109** (Scheme 3.38). However, no desired product was obtained for the methyl 3-hydroxy-2-methylenehexanoate, the MBH adduct obtained from butyraldehyde and methyl acrylate, under the same conditions.[82]

In addition, they found the MBH adduct, α-methylene-β-hydroxyalk-anenitriles **110** (secondary allylic alcohols), can be conveniently isomerized into 3-aryl-2-(hydroxymethyl)prop-2-enenitriles **111** (primary allylic alcohols) *via* treatment with aqueous sulfuric acid (20%). These primary alcohols **111** can be further oxidized to the corresponding cinnamaldehydes **112** in the presence of PCC (Scheme 3.39).[83]

Kim *et al.* have reported a facile one-pot stereoselective synthesis of (*E*)-cinnamyl alcohols **113** *via* the treatment of MBH adducts, derived from aryl aldehydes and ethyl acrylate, with TFA. However, a similar reaction of MBH adducts derived from aryl aldehydes and acrylonitriles with TFA gave the (*E*)-allyl alcohols **111** in low yields (Scheme 3.40).[84] Since Basavaiah's method works well for nitrile-containing adducts [for (*E*)-selective nitriles] and the trifluoroacetic acid method works well with ester-containing adducts [for (*E*)-selective esters], these two methods are considered to be complementary for the preparation of stereochemically defined cinnamyl alcohols.

The MBH adducts have also been isomerized efficiently to the corresponding stereoselective (*E*)-cinnamyl alcohols **116** and **117** by treatment with Ac$_2$O in the presence of TMSOTf[85] or Amberlyst-15[86] followed by hydrolysis of the intermediate acetates **114** and **115** with K$_2$CO$_3$ in MeOH (Scheme 3.41).

In the presence of TMSOTf,[87] Bi(OTf)$_3$·4H$_2$O[88] or Pd(Ph$_3$P)$_4$,[89] the acetates of MBH adducts, such as methyl 3-acetoxy-3-aryl-2-methylenepropanoates and 3-acetoxy-3-aryl-2-methylenepropanitriles, have been smoothly converted into methyl (2*E*)-2-(acetoxymethyl)-3-arylprop-2-enoates and (2*E*)-2-(acetoxymethyl)-3-arylprop-2-enenitriles, respectively (Scheme 3.42). A remarkable reversal in stereochemical direction from ester to nitrile was observed and is consistent with earlier results.

Shanmugam *et al.* have demonstrated the usefulness of a montmorillonite K10 clay–microwave combination as an alternative, useful, speedy and efficient catalyst for the stereoselective isomerization of various acetates of MBH adducts to provide densely functionalized (*E*)-alkenes **118** in high yields (Scheme 3.43).[90] They also demonstrated the usefulness of the same catalyst system for a one-pot protection isomerization of various MBH adducts with trimethyl orthoformate and alcohols (Scheme 3.44).[91]

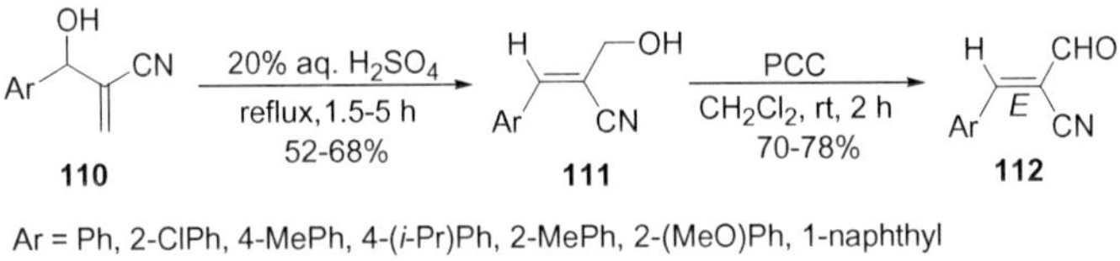

Scheme 3.39

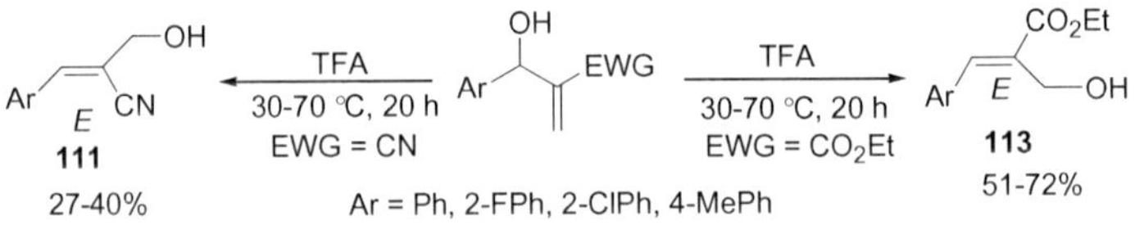

Scheme 3.40

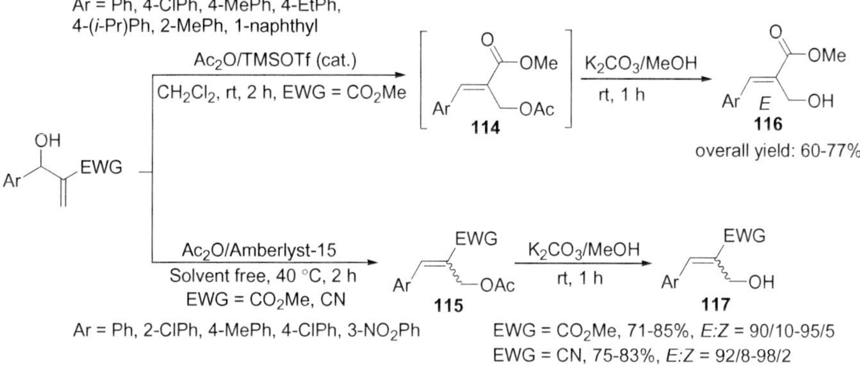

Scheme 3.41

Scheme 3.42

Scheme 3.43

Recently, Xu *et al.* have developed an efficient and stereoselective rearrangement catalyzed by only 1 mol.% gold(I) chloride/silver(I) trifluoromethanesulfonate of MBH acetates to afford 2-(acetoxymethyl)alk-2-enoates **119** under mild reaction conditions in good to high yields with 100% (*E*)-selectivity. For cyclohex-2-enone-derived MBH acetates, the reaction gave 2-alkylidenecyclohex-3-enones **120** in good yields (Scheme 3.45).[92]

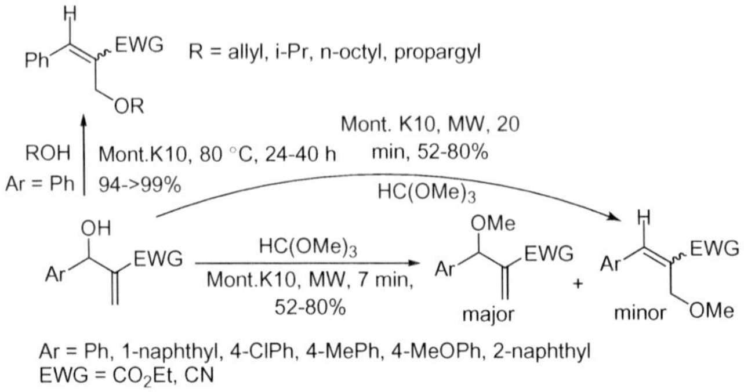

Scheme 3.44

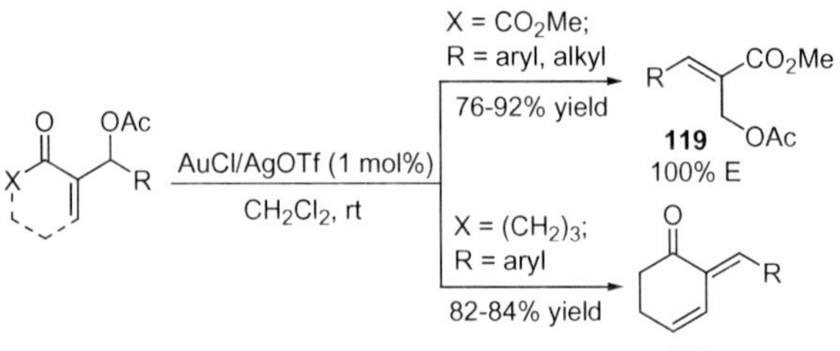

Ar = Ph, 1-naphthyl, 4-ClPh, 4-MePh, 4-MeOPh, 2-naphthyl
EWG = CO_2Et, CN

Scheme 3.45

3.5 Heck Reaction

MBH adducts have been utilized successfully as substrates for Heck coupling with various aryl bromides independently by Basavaiah,[93] Sundar and Bhat[94] and Kumareswaran and Vankar.[95] The acetates of the MBH adducts have also been used for a similar reaction to provide trisubstituted olefins **121** with (*E*)-stereoselectivities (Scheme 3.46).[95]

Later on, Kulakarni and Ganesan described a solid-phase synthesis of β-keto esters **122** *via* sequential MBH and Heck reactions. Subsequent hydrolysis of these keto esters **122** *via* treatment with TFA provided β-aryl ketone derivatives **123** in moderate yields (Scheme 3.47).[96]

Calo *et al.* have observed a very fast and efficient Heck reaction of aryl bromides with MBH adducts using Pd-catalyst **124** with benzothiazole carbene as a ligand in tetrabutylammonium bromide (TBAB) melt as a solvent to give the corresponding β-aryl ketones **123** in good yields (Scheme 3.48).[97]

Some β-aryl (aza)-MBH adducts, **125** and **126**, have been prepared *via* the Heck-type reaction of (aza)-MBH adducts and aryl iodide under the influence of $Pd(OAc)_2/TBAB/KOAc$ in refluxing CH_3CN in moderate yields as a cleanly separable (*E/Z*) mixture (Scheme 3.49).[98]

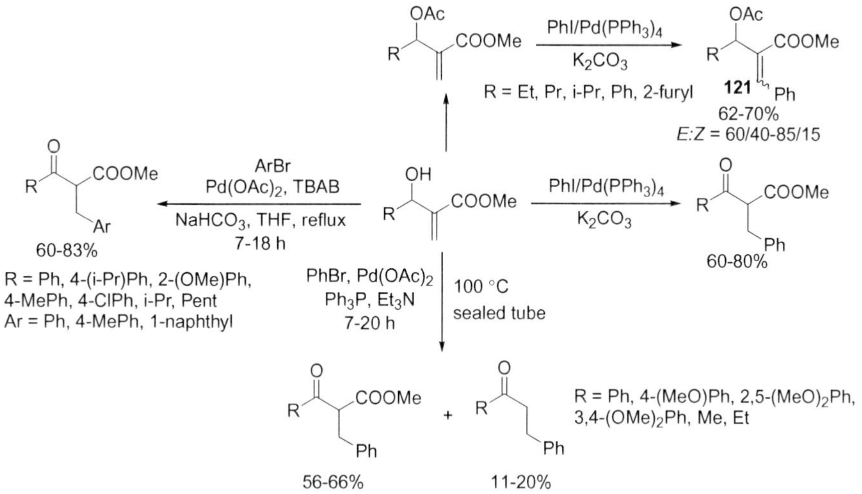

Scheme 3.46

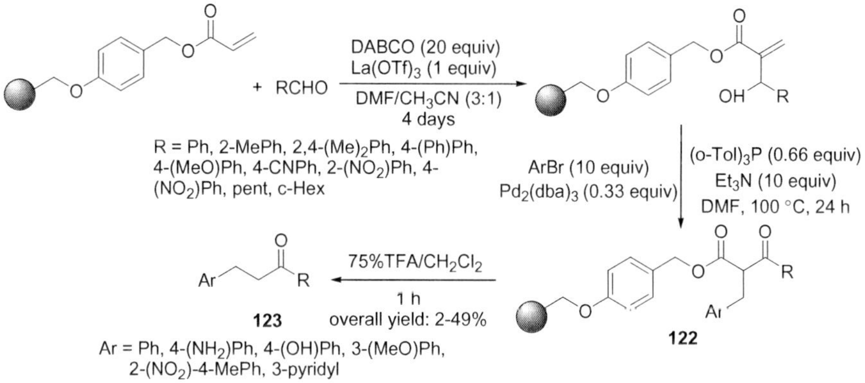

Scheme 3.47

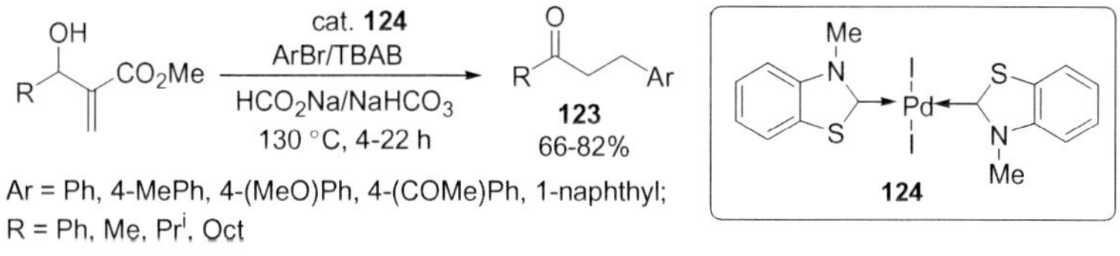

Scheme 3.48

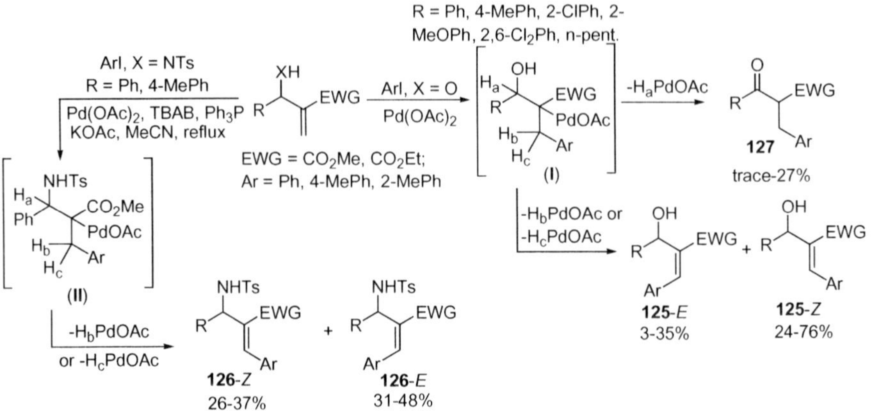

Scheme 3.49

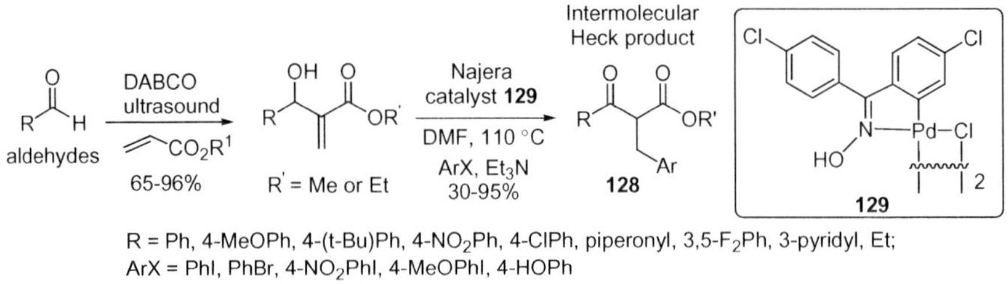

Scheme 3.50

Coelho *et al.* have developed an improved and highly efficient synthesis of several α-benzyl-β-ketoesters **128** by using MBH adducts as substrates for an intermolecular Heck reaction catalyzed by a Nájera oxime-derived palladacycle (**129**). These efficient catalytic conditions proved to be high selectivity and only provided the corresponding functionalized β-ketoesters **128** in high yields with no decarboxylation products (Scheme 3.50).[99]

Arenediazonium salts **130** have also been applied to a Heck reaction of MBH adducts to give α-(substituted-aryl)-β-keto esters **131** in the presence of palladium catalyst. The use of aerobic conditions, short reaction times and the tolerance towards many structurally diverse reactants offer several advantages over previous reports (Scheme 3.51).[100]

Lee *et al.* have developed a simple, two-step method for the synthesis of indanone derivatives **133** and **134** using an intramolecular Heck reaction of MBH adducts **132** of 2-iodobenzaldehyde in the presence of Pd(OAc)$_2$/(o-Tol)$_3$P/Et$_3$N (Scheme 3.52).[101] Interestingly, using MBH adduct derived from cyclohexenone as the substrate afforded 1-hydroxyfluorene (**135**) in 32% yield, presumably *via* a β-hydrogen elimination, dehydration and aromatization along with proton migration (Scheme 3.52).

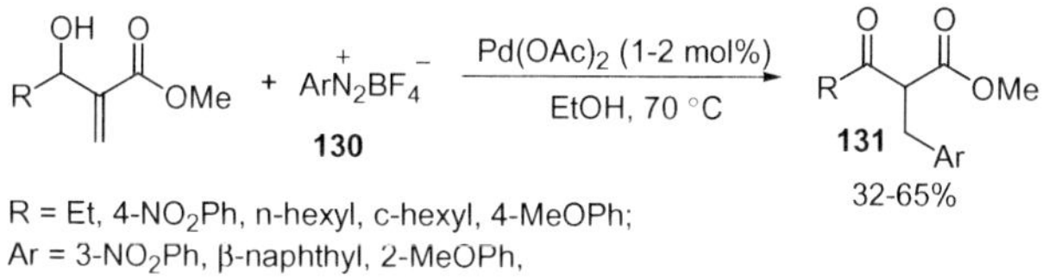

R = Et, 4-NO$_2$Ph, n-hexyl, c-hexyl, 4-MeOPh;
Ar = 3-NO$_2$Ph, β-naphthyl, 2-MeOPh,

Scheme 3.51

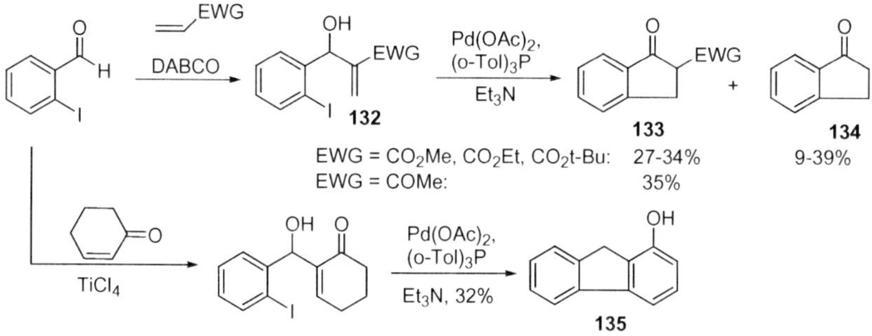

EWG = CO$_2$Me, CO$_2$Et, CO$_2$t-Bu: 27-34%
EWG = COMe: 35%

Scheme 3.52

X = I, Br; EWG = CO$_2$Me, COMe;
R^1 = H, F R^2 = Ph, 4-ClPh, 4-FPh, 2-furyl, 2-thienyl, Me

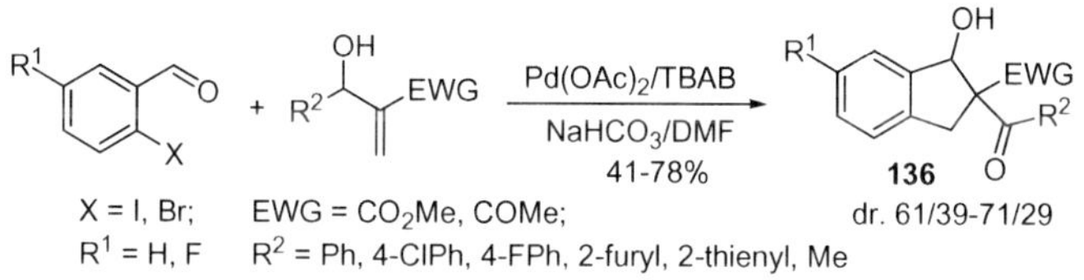

Scheme 3.53

Using ortho-halogenated aryl aldehydes and MBH adducts as substrates, 2-carbonyl-1-indanols **136** have been synthesized in moderate to good yields *via* a one-pot, palladium-catalyzed tandem Heck–aldol reaction. Various MBH adducts were examined to find the scope and limitations of this process (Scheme 3.53).[102]

3.6 Hydrogenation

MBH adducts have been employed successfully in various diastereo-/enantio-selective catalytic homogeneous hydrogenation processes by Brown and co-workers,[42e,103] Noyori and co-workers,[104] Sato *et al.*,[105] and Yamamoto and co-workers.[106] In all cases, *anti*-products were formed predominantly.

Brown and co-workers have studied the Rh complex **137** catalyzed hydrogenation of α-(hydroxyalkyl)-*N*-methoxyacrylamides, and Ru complex **138**

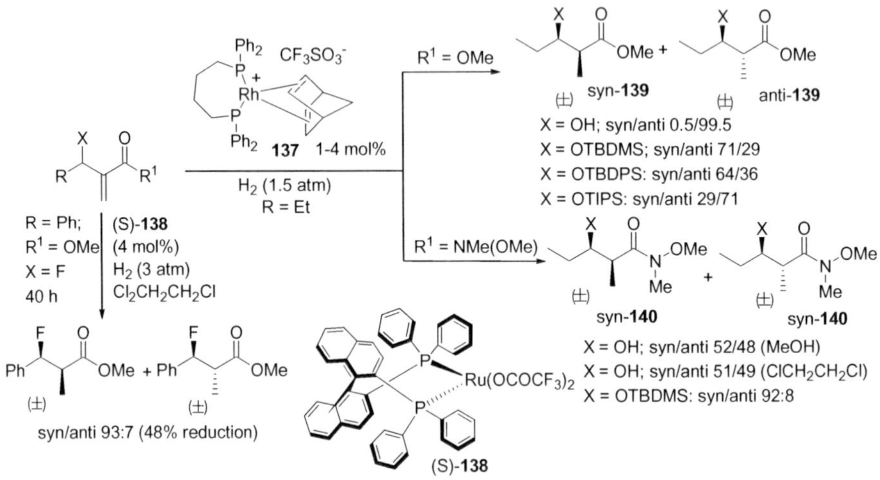

Scheme 3.54

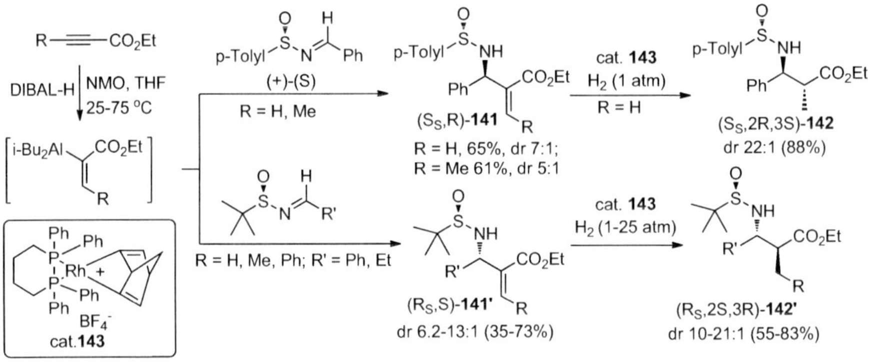

Scheme 3.55

catalyzed hydrogenation of α-(fluoroalkyl)-acrylates, which provided the corresponding *syn*-selective compounds (Scheme 3.54).[29] However, hydrogenation of the *N*-sulfinyl aza-MBH adducts **141**, obtained by the addition of vinylaluminium NMO reagents to *N*-(*p*-toluenesufinyl)- and *N*-(2-methyprop-2-yl sulfinyl)-derived sulfinimines from the least hindered direction in good diastereoselectivity (dr = 5 : 1 to 13 : 1), affords *anti*-α-substituted *N*-sulfinyl-β-amino esters **142** in good yield and high dr (10 : 1 to 22 : 1) with Rh(i) catalyst **143** (Scheme 3.55).[107]

Coelho and co-workers have also described a highly *syn*-diastereoselective heterogeneous catalytic hydrogenation of MBH adducts that depends on the protecting groups used on the hydroxyl group of these adducts (Schemes 3.56 and 3.57). For silylated MBH adducts, a high *syn* diastereoselectivity has been

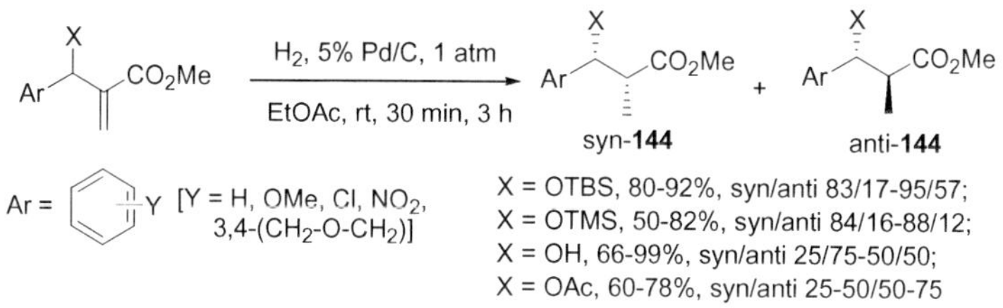

Scheme 3.56

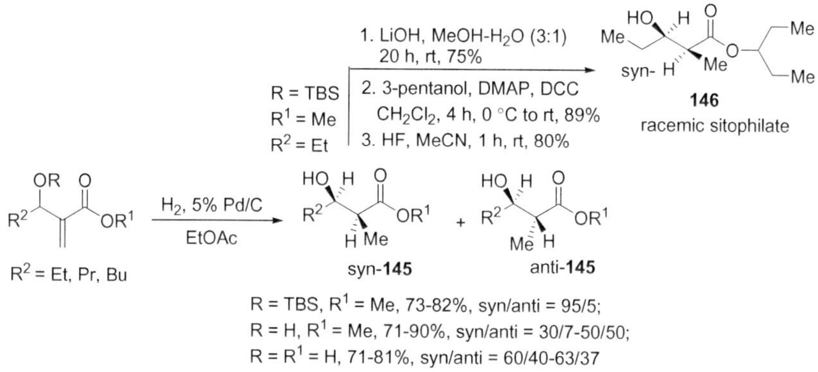

Scheme 3.57

observed. However, when the hydroxyl group of MBH adducts was unprotected or protected as acetate, a moderate *anti* selectivity was attained. Adducts protected as the methyl ether gave poor *syn* diastereoselectivity.[108] Because the preferential diastereoselectivity obtained for the silylated adducts is the opposite of that attained in homogenous catalytic hydrogenation conditions of this type of adduct, these results are complementary to those described for homogeneous catalytic hydrogenation reactions. In addition, the authors also successfully applied this methodology in the synthesis of racemic sitophilate (**146**) *via* the stereoselective heterogeneous catalytic hydrogenation reaction of methyl 3-(*tert*-butyldimethylsilyloxy)-2-methylene-pentanoate (Scheme 3.57).[109]

Bouzide has reported a highly *syn*-diastereoselective chelation-controlled heterogeneous hydrogenation of MBH adducts in the presence of palladium on carbon combined with $MgBr_2$ to afford the corresponding aldol derivatives in good yields (Scheme 3.58).[110]

Batra *et al.* have studied the catalytic hydrogenation of MBH adducts obtained from substituted 3-, 4- and 5-isoxazolecarboxaldehydes and their corresponding acetates in the presence of Raney-Ni and Pd–C. The hydrogenation of MBH adducts of substituted 5- and 3-isoxazolecarbaldehydes in the presence of Raney-Ni furnishes diastereoselectively *syn* enaminones **147** and **149**, respectively, over *anti*, and in the presence of boric acid as an additive

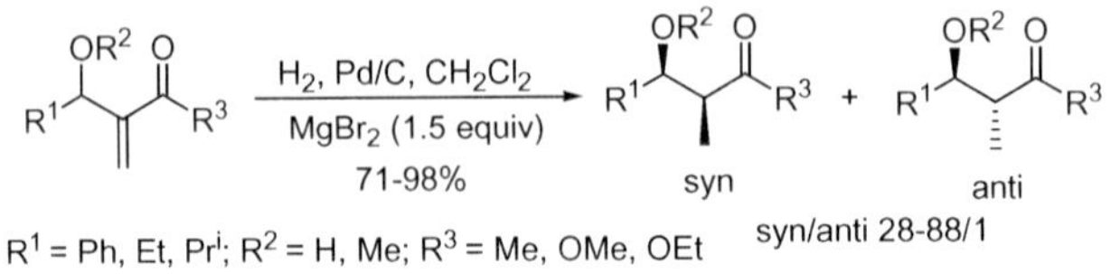
H2, Pd/C, CH2Cl2
MgBr2 (1.5 equiv)
71-98%
syn
anti
R1 = Ph, Et, Pri; R2 = H, Me; R3 = Me, OMe, OEt
syn/anti 28-88/1

Scheme 3.58

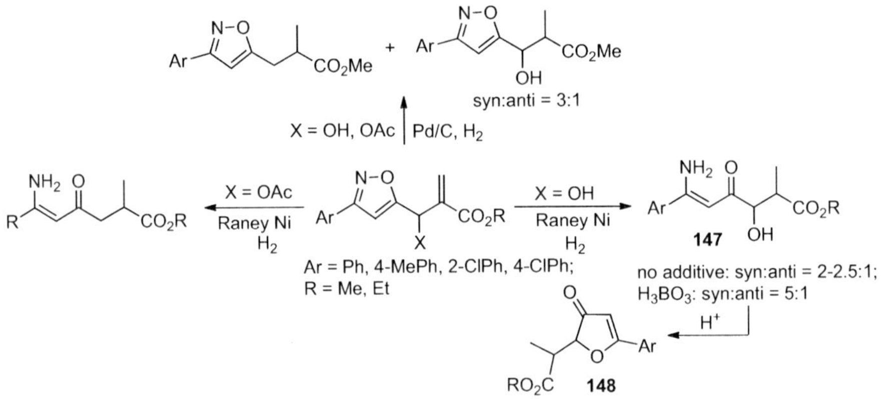
syn:anti = 3:1
X = OH, OAc
Pd/C, H2
X = OAc
Raney Ni
H2
X = OH
Raney Ni
H2
Ar = Ph, 4-MePh, 2-ClPh, 4-ClPh;
R = Me, Et
147
no additive: syn:anti = 2-2.5:1;
H3BO3: syn:anti = 5:1
H+
148

Scheme 3.59

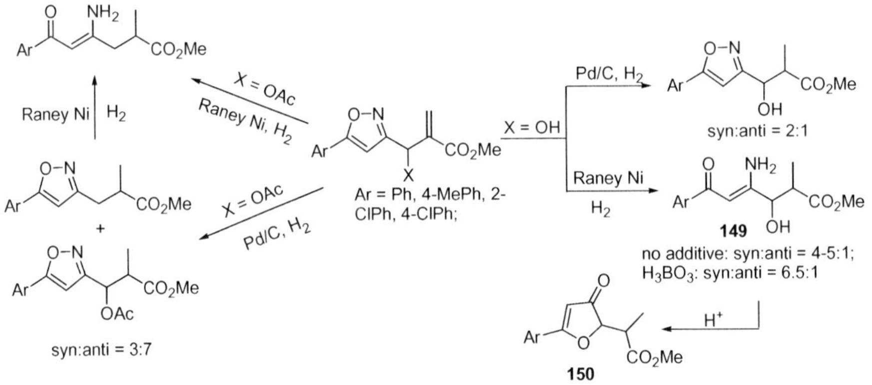
Raney Ni H2
X = OAc
Raney Ni, H2
X = OAc
Pd/C, H2
syn:anti = 3:7
Ar = Ph, 4-MePh, 2-ClPh, 4-ClPh;
X = OH
Pd/C, H2
Raney Ni
H2
syn:anti = 2:1
149
no additive: syn:anti = 4-5:1;
H3BO3: syn:anti = 6.5:1
H+
150

Scheme 3.60

further enhancement of diastereoselectivity in favor of *syn* isomer is observed. The Pd/C-promoted hydrogenation of these substrates is also diastereoselective in favor of the *syn* isomer but occurs without the hydrogenolysis of isoxazole-ring. It should be noted the hydrogenation of MBH acetate derived from 5-isoxazolecarboxaldehyde in the presence of Pd/C afforded *anti*-isomer as major product. The presence of boric acid as an additive in this

Scheme 3.61

Scheme 3.62

hydrogenation exhibits no pronounced effect on diastereoselectivity (Schemes 3.59 and 3.60). The Raney/Ni-mediated hydrogenation of MBH adducts of substituted 4-isoxazolecarbaldehydes yield pyridone derivatives **151**, while Pd/C-promoted hydrogenation of the same substrate is diastereoselective to afford the *anti* isomer of the resulting products **152** (Scheme 3.61). The enaminones **147** and **149** derived from MBH adducts of 3- and 5-isoxazolecarbaldehydes serve as versatile precursors for α'-hydroxy-1,3-diketones, which undergo acid-catalyzed ring-closure reaction to afford furanone derivatives **148** and **150**, respectively, in excellent yields (Schemes 3.59 and 3.60).[111]

Hu and Zheng *et al.* have reported the first asymmetric synthesis of chiral 2-substituted glutarates *via* a Rh-catalyzed enantioselective hydrogenation. After extensive ligand screening, the bidentate P-ligand BINAP and monodentate P-ligand FAPhos were found to show good enantioselectivities (94% ee and 92% ee, respectively) in the hydrogenation of dimethyl 2-methyleneglutarate. In contrast, the hydrogenation of 2-benzylideneglutarates **153** was more difficult; up to 81% ee was obtained by the use of a BoPhoz-type ligand **154** bearing a stereogenic P center in the phosphino moiety and a 4-CF$_3$ group in the phenyl ring of the aminophosphino moiety (Scheme 3.62). These

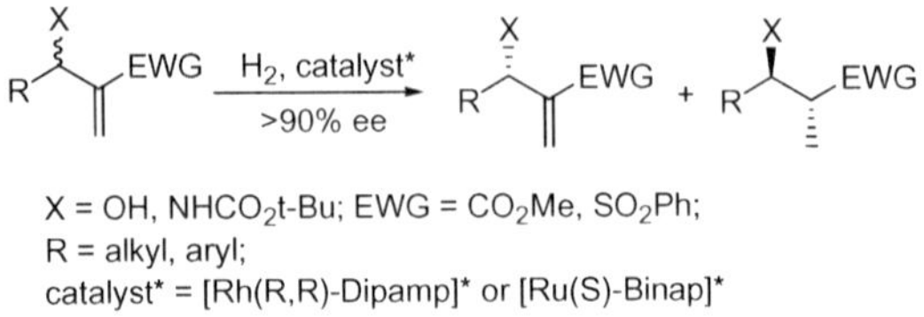

X = OH, NHCO$_2$t-Bu; EWG = CO$_2$Me, SO$_2$Ph;
R = alkyl, aryl;
catalyst* = [Rh(R,R)-Dipamp]* or [Ru(S)-Binap]*

Scheme 3.63

observations are similar to those for the hydrogenation of itaconate, in which the hydrogenation of β-substituted itaconic acid derivatives was found to be less efficient than that of the corresponding parent itaconic acid derivatives.[112]

The asymmetric hydrogenation methodology has also been used for the kinetic resolution of MBH alcohol to afford the enantiomerically pure (or enriched) MBH adducts (Scheme 3.63).[42e,103a–c,104,106]

3.7 Diels–Alder Reaction

In the mid-1980s, Hoffman and co-workers described a simple synthesis of racemic mikanecic acid *via in situ* Diels–Alder dimerization of the diene generated from *t*-butyl 2-bromomethylbut-2-enoate.[42b,113] The first example of a Diels–Alder reaction of a MBH adduct was the dimerization of 2-hydroxy-alkylenones[114] and the previously mentioned addition to anthracene.[58] The application of the MBH adducts as hetero dienes or precursors of dienes, and dienophiles for the Diels–Alder cycloaddition reactions was then initiated and expanded by the group of Hoffman. In a series of reports, Hoffman and co-workers have described the *in situ* Diels–Alder dimerization of various dienes (**157**), generated *via* stereoselective dehydration with MsCl-DABCO-DMAP of the corresponding MBH adducts **156**.[46c,63a,b] The elimination of water from MBH adducts always resulted in the exclusive formation of (*E*)-double bond. The dienes **157a** generated from adduct **156a** (EWG = SO$_2$Ph) were reasonably stable and allowed full characterizaion.[46c] In contrast, dienes **157b** (EWG = CO$_2$Me) and **157c** (EWG = COMe) dimerized spontaneously under dehydration conditions (Scheme 3.64). The dimerization was highly regioselective, *i.e. para*-selective. Stereoselectivity in the formation of dimers **158b** and **158c** from the dienes **157b** and **157c** was moderate while the dienes **157a** always gave *trans*-**158a** with regard to sulfonyl and alkyl groups.[46c] However, the alkenyl group in all these dimeric products is always *endo* oriented with respect to the roof-like cyclohexene ring (**159**).

The first enantioselective synthesis of mikanecic acid, (+)-**162**, a terpene dicarboxylic acid was achieved by Basavaiah *et al.*[63c] *via* a double stereo-differentiating asymmetric Diels–Alder reaction involving the same molecule as chiral diene and chiral dienophile generated *in situ*. Treatment of MBH adduct with MsCl/NEt$_3$ afforded the chiral dienes **161**, through a Diels–Alder reaction of the *in situ* generated chiral 1,3-butadiene-2-carboxylate **160**. Hydrolysis of the diesters then afforded the desired mikanecic acid **162** (25–74% ee) in good

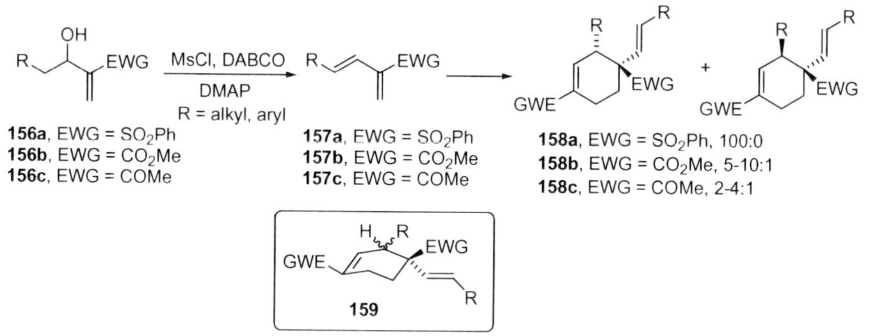

OH
R
EWG
MsCl, DABCO
DMAP
R = alkyl, aryl
156a, EWG = SO2Ph
156b, EWG = CO2Me
156c, EWG = COMe
157a, EWG = SO2Ph
157b, EWG = CO2Me
157c, EWG = COMe
GWE
+
158a, EWG = SO2Ph, 100:0
158b, EWG = CO2Me, 5-10:1
158c, EWG = COMe, 2-4:1
159

Scheme 3.64

OH
Me
CO2R*
MsCl
NEt3
160a-c
*RO2C
161a-c
i) KOH
ii) H+
iii) recryst.
HO2C
CO2H
162
92% ee
R* =
(a)
Pri2NO2S
(b)
(C6H11cyc)2NO2S
(c)
H

Scheme 3.65

O
CO2Me
+
DABCO
HO
163
MsCl, DABCO
DMAP
164
MeO2C
N
[4+2], rt
165
60% yield

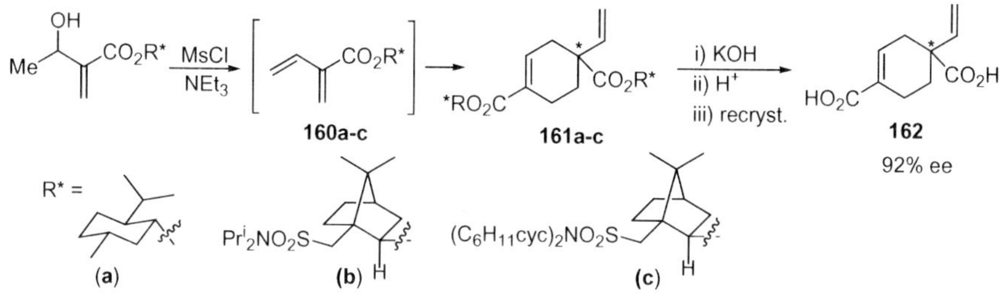

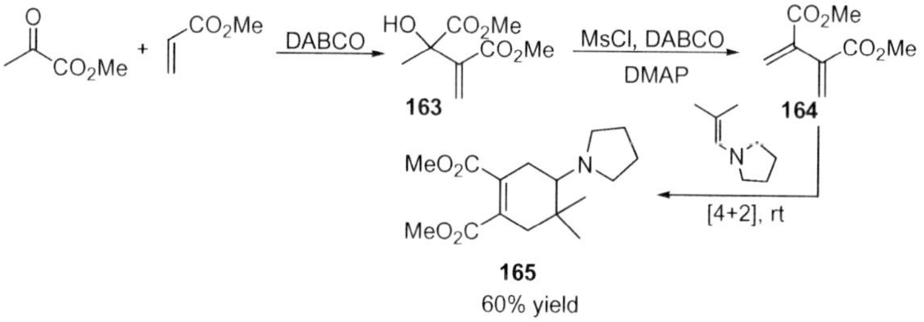

Scheme 3.66

yields. After recrystallization, the (+)-mikanecic acid was furnished in 92% ee (Scheme 3.65).

Compared with a previously cumbersome synthesis of 2,3-dimethoxycarbonyl-1,3-butadiene (**164**) from either 2,3-butanedione (four steps, 23% yield)[115] or acrylonitrile (eight steps),[116] Hoffman has devised a simple two-step preparation of **164** from the corresponding MBH adduct **163** *via* dehydration with MsCl-DABCO-DMAP (Scheme 3.66).[117] The diene **164** could undergo an

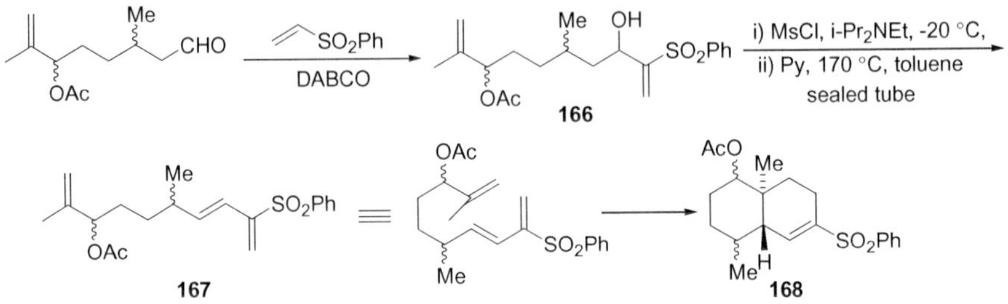

Scheme 3.67

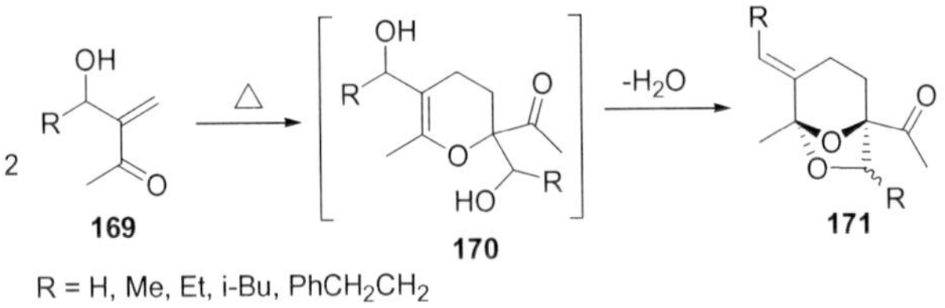

Scheme 3.68

inverse electron demand Diels–Alder cycloaddition reaction with pyrrolidi-noisobutene to give the adduct **165** in 60% yield.

Weichert and Hoffman[66] have synthesized the eudesmane precursor **168** *via* an inverse electron demand intramolecular [4 + 2] cycloaddition reaction of triene **167**, which in turn was generated *in situ* from the mesylate of the MBH adduct **166**. (Scheme 3.67).

When MBH adducts **169** are heated in a high-boiling aromatic hydrocarbon, they undergo an intermolecular dehydrative double cyclization to produce functionalized 6,8-dioxabicyclo[3.2.1]octanes **171** (Scheme 3.68), which are often present as the basic framework in several pheromones, *e.g.* frontalin, exo- and endo-brevicomins, α-multistriatin, *etc.*[114] However, the stereoselectivity in these processes was very poor.

α-Methylene-β-keto sulfones **173**[57b] and α-methylene-β-keto esters **174**,[57a] prepared from the corresponding MBH adducts **172** *via* a modified Jones oxidation procedure, are synthetically attractive intermediates and participate in various cycloaddition processes (Scheme 3.69). In addition, sulfone **173** has also been utilized efficiently in the synthesis of racemic frontalin (**175**) (Scheme 3.70).[57b]

Adam *et al.*[7b] have found that α-methylenepropiolactone **176**, which is obtained from the corresponding MBH adduct *via* hydrolysis followed by β-lactonization, can be used as a more reactive dienophile to react with various dienes *via* Diels–Alder reactions, affording the spiro β-lactones **177** and then the desired cyclic alkenes **178** upon pyrolysis (Scheme 3.71).

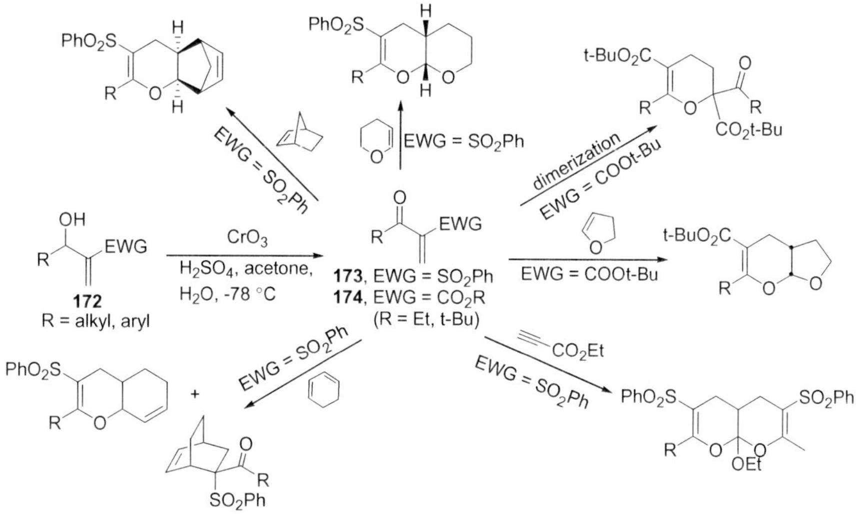

Scheme 3.69

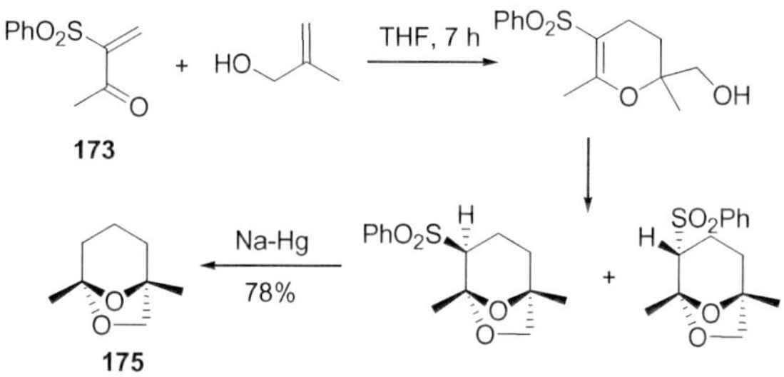

Scheme 3.70

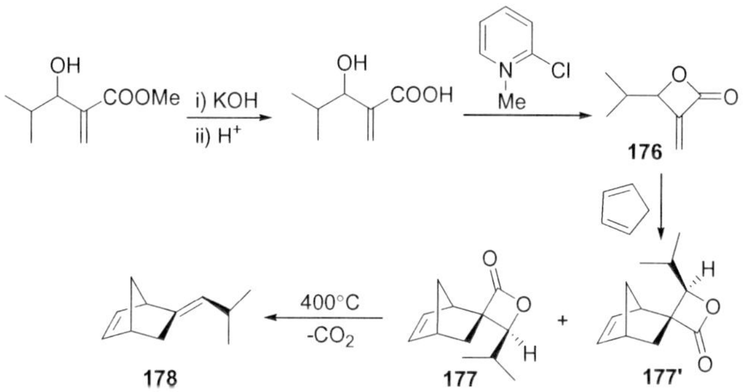

Scheme 3.71

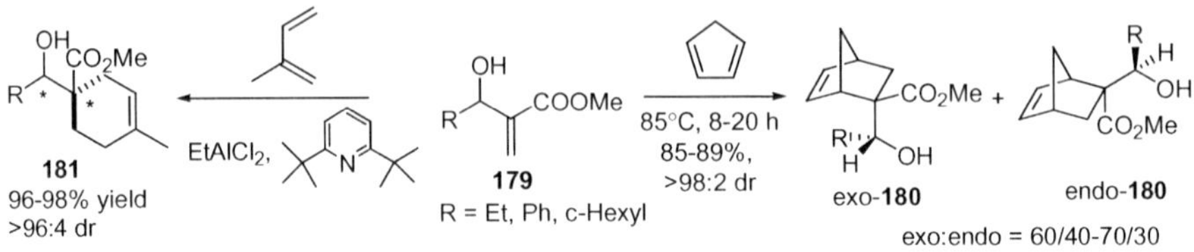

Scheme 3.72

Recently, Aggarwal *et al.* have found that MBH adducts **179** are excellent dienophiles in Diels–Alder reactions, providing essentially complete diastereocontrol with all dienes. Although *exo/endo* stereoisomers **180** were formed with cyclopentadiene and no regioisomers were obtained with isoprene, the emerging asymmetric MBH reaction coupled with these new Diels–Alder reactions rapidly builds up complex architectures in a stereocontrolled process from very simple and inexpensive starting materials, and this will no doubt find applications in synthesis (Scheme 3.72).[118]

3.8   Nucleophilic Addition

Allyl acetates, halides or sulfides derived from MBH adducts can undergo substitution reactions with various nucleophiles. These processes have been shown to proceed with high regio- and stereoselectivity. The various nucleophiles employed so far include C-, H-, N-, O-, S- and P-nucleophiles. These processes produce compounds with a trisubstituted olefininc moiety or a terminal olefin and may be regarded as an alternative pathway to the well-known Wittig-type reaction.

3.8.1   Carbon Nucleophiles

The reaction of allyl acetates, halides and sulfides derived from the MBH adducts with carbon nucleophiles has been well investigated. In fact, Drewes and co-workers first reported the utilization of the nucelophilic addition of a MBH adduct in the stereoselective synthesis of (2*E*)-integerrinecic acid, a natural product with a trisubstituted olefinic moiety.[42a] Subsequently, Drewes *et al.*[34,40d,119] have carried out stereo- and regioselective addition of carbon nucleophiles derived from ethyl acetoacetate, malonate and phenylacetylide derivatives to various MBH halides and MBH acetates.

Bauchat *et al.*[120] have carried out reactions of MBH acetates with carbanions generated from a 1,3-diketone, methyl cyanoacetate or nitroalkane by treatment with potassium carbonate or potassium fluoride on alumina to provide trisubstituted olefins **182** and **183** with (*E*)-selectivity. These products were subsequently transformed into useful γ-lactones **184** and δ-lactones **185**, respectively (Scheme 3.73).

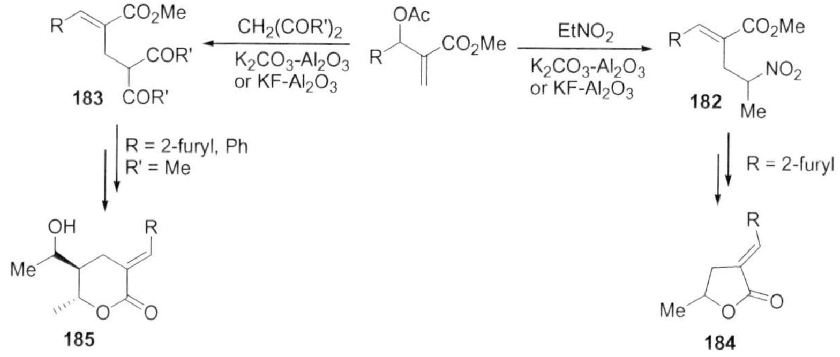

Scheme 3.73

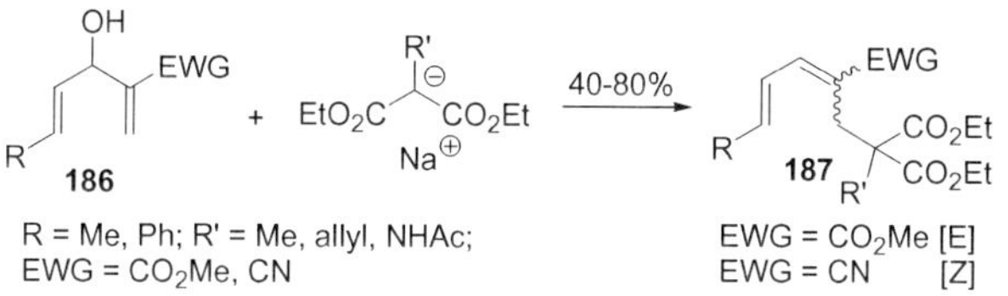

Scheme 3.74

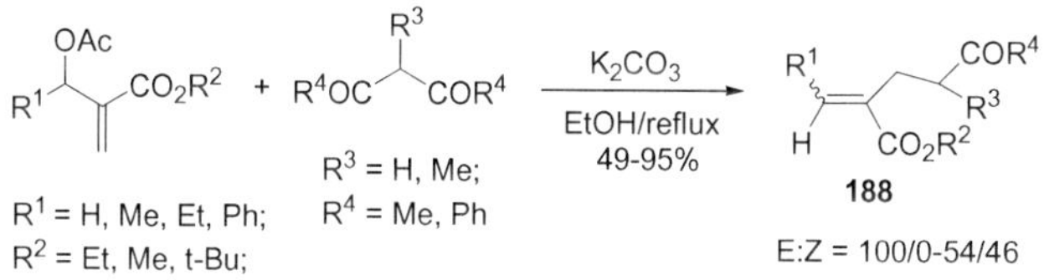

Scheme 3.75

Heerden and co-workers[121] have reported a convenient synthesis of multi-functional stereodefined dienes **187** using substituted diethyl malonate anions and MBH adducts **186**, which were derived from α,β-unsaturated aldehydes (Scheme 3.74).

Amri *et al.* have investigated the addition of MBH acetate by using 1,3-diketones as nucleophiles in the presence of K_2CO_3 to provide 1,5-ketoesters **188** in good yields (Scheme 3.75).[122] They then subsequently described a one-pot synthesis of (*E*)-4-alkylidene-2-cyclohexen-1-ones **189** *via* a cross coupling of acetates of the MBH adducts, derived from MVK or EVK, and aliphatic 1,3-diketones in the presence of K_2CO_3 in absolute ethanol at reflux temperature to give the desired product in moderate to good yields (Scheme 3.76).[123]

Rezgui and El Gaied[124] have reported an interesting synthesis of bicyclic dienones **190** in sequential and also in a one-pot process *via* the reaction of

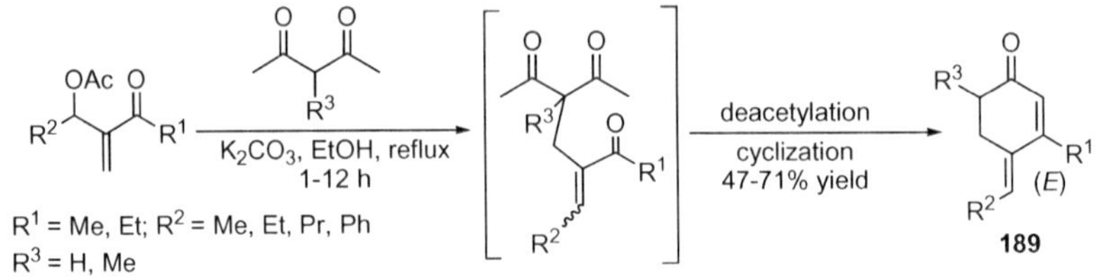

Scheme 3.76

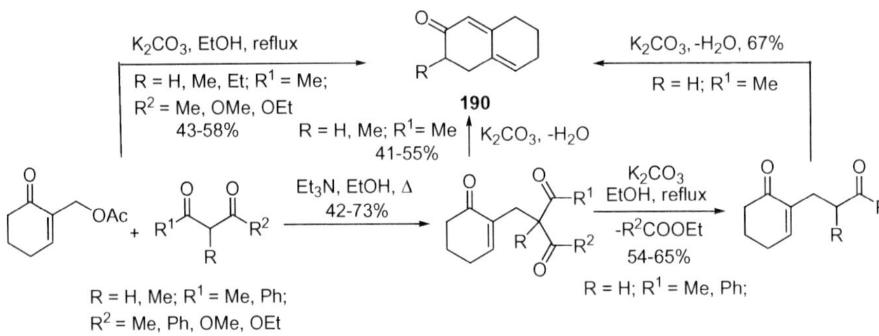

Scheme 3.77

2-(acetoxymethyl)cyclohex-2-enone with 1,3-dicarbonyl compounds in the presence of K_2CO_3 following the reaction sequence described in Scheme 3.77.

Kim *et al.* have transformed successfully MBH acetates into *o*-hydroxy-acetophenone derivatives[125] *via* treatment with 1,3-dicarbonyl compounds in the presence of K_2CO_3 (Scheme 3.78).

In the continuation of work, Kim *et al.* also investigated the reaction of MBH acetates with 1,3-dicarbonyl compound[126] to provide trisubstituted olefins **191** and **192** with (*E*)-selectivity from ester- and acetyl-containing substrates. The products can be further transformed into 4-arylidenecyclohexane-1,3-dione derivatives **193** (existing as an enol form in DMSO-d6 based on its ^{1}H NMR spectrum) by treatment with LiHDMS (Scheme 3.79).[126b] This methodology has also been applied to the synthesis of various 2,4,5-trisubstituted-1,4-pentadienes **196** following a reaction sequence that involves a Pd-catalyzed decarboxylation–elimination protocol as the key step under the conditions of a low loading of PPh_3 (Scheme 3.80).[127]

MBH diene adduct **197**, derived from methyl acrylate and crotonaldehyde, has been employed successfully in a palladium-catalyzed asymmetric allylic alkylation (AAA) reaction with Meldrum's acid to provide product **198** that can undergo further manipulations, as well as act as an intermediate in a one-pot tandem palladium AAA reaction. The versatility of these products for various synthetic manipulations was further illustrated by alkene metathesis to form the optically active cyclopentene **201** (Scheme 3.81).[128]

Amri and co-workers have prepared 2-methylenealkanoate **202** *via* the treatment of ethyl 2-acetoxymethylprop-2-enoate with Grignard reagents in the

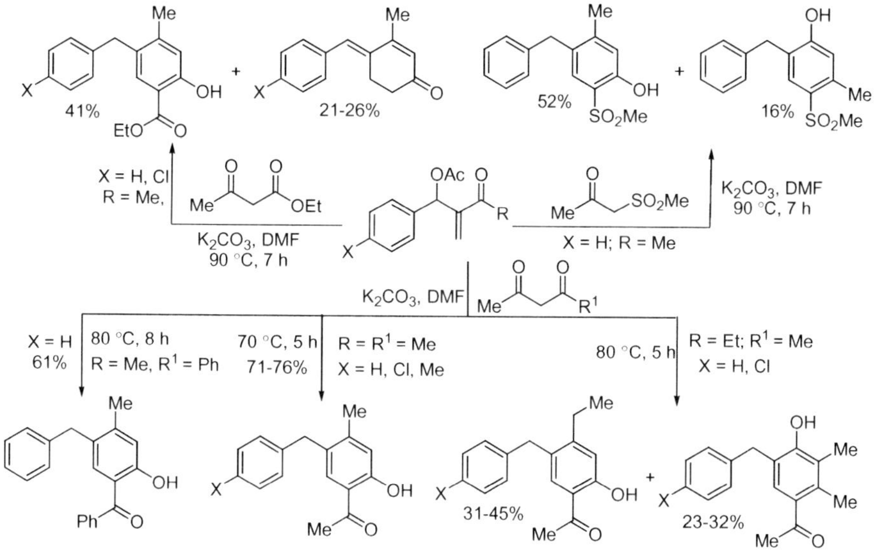

Scheme 3.78

presence of copper(I) salts.[129] They have also extended the same strategy to the synthesis of (2E)-alkenoates **203** by treating 2-methylene-3-acetoxy-alkanoates with di-n-butyllithium cuprate (Scheme 3.82).[130]

Basavaiah *et al.* have used Grignard reagents as nucleophiles and observed a remarkable reversal of stereoselectivity between esters and nitriles. Thus treatment of 3-acetoxy-2-methylenealkanoates **204** with Grignard reagents provided (2E)-alk-2-enoates **206**, while the similar reaction of 3-acetoxy-2-methylenealkanenitriles **205** produced (2Z)-alk-2-enenitriles **207** predominantly.[131] In addition, functionalized 1,4-dienes **208** have been obtained by coupling allylic MBH acetates and vinyl magnesium chloride at low temperature in the presence of a catalytic amount of LiCuBr$_2$ (3%) (Scheme 3.83).[132]

Some ene-ynamides (**210**) have been prepared *via* nucleophilic addition by using alkyne-Grignard reagents as C-nucleophiles starting from the MBH adducts. It was found that π-cation interactions could increase the acidity of the nearby protons of the triple bond of non-conjugated ene-ynamide **210** and could turn the isomerization into a more stable conjugated ene-ynamide form (**211**).[133] In addition, the initial resulting alkynyl moiety-containing MBH adducts **209** can undergo an intramolecular Friedel–Crafts alkenylation of triple bond-tethered methyl cinnamates to afford 9-phenyl-7H-benzocycloheptene derivatives **212** in good yields (Scheme 3.84).[134]

The addition of several trialkyl or triarylindium reagents to the acetates of MBH adducts proceeds readily under the catalysis of copper and palladium derivatives. The reactions of trialkylindiums are catalyzed efficiently by CuI whereas additions of triarylindiums produce better results with Pd(PPh$_3$)$_4$. The reactions with 3-acetoxy-2-methylenealkanoates provide (E)-alkenes **213**,

K$_2$CO$_3$, EtOH, 50 °C, 17 h, 74%
EWG = CO$_2$Et, R = CH$_3$
K$_2$CO$_3$, EtOH, rt, 1 h
heating
191
LiHMDS, 71%
OH
193
EWG = COCH$_3$, R = OEt
K$_2$CO$_3$, CH$_3$CN, rt, 1 h
DMAP, xylene reflux, 6 days
192
LiHMDS, 61%
Ph$_3$P=CHCO$_2$Et
THF, reflux, 15 h
aq. KCN, 60 °C, 10 h

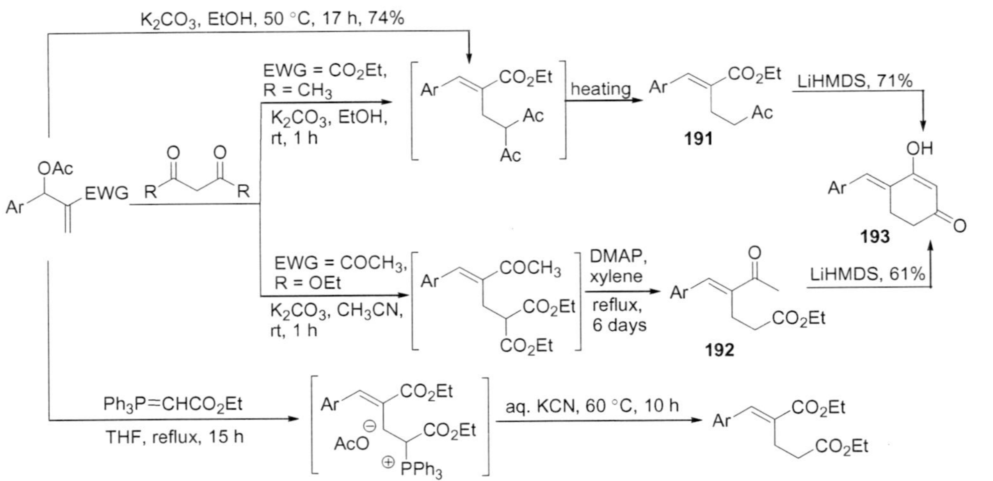

Scheme 3.79

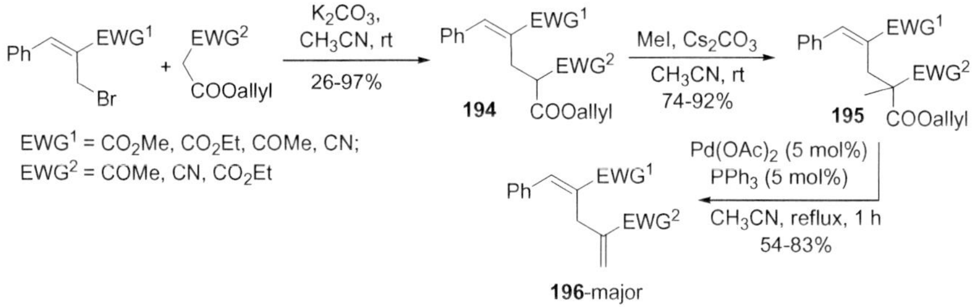

Scheme 3.80

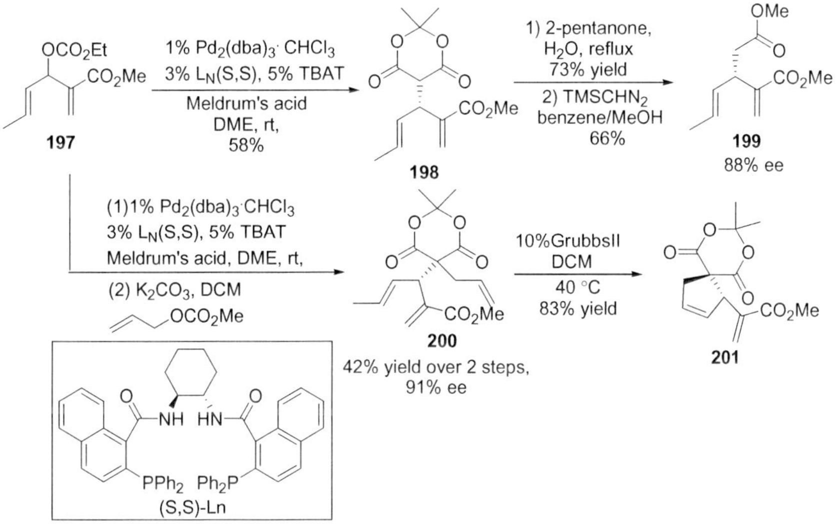

Scheme 3.81

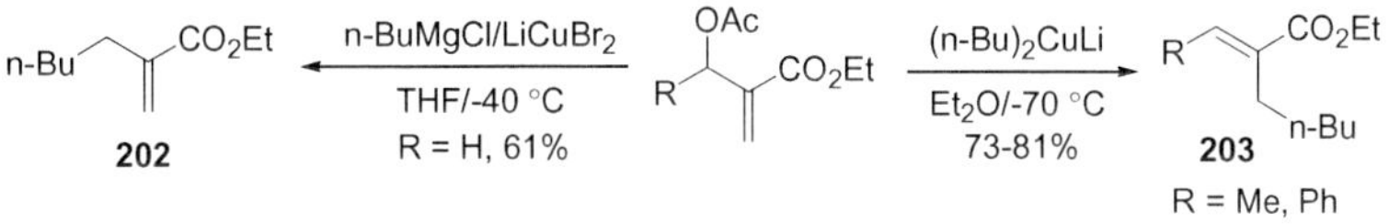

Scheme 3.82

whereas similar reactions with 3-acetoxy-2-methylenealkanenitriles lead to (Z)-alkenes **214** (Scheme 3.85).[135]

An efficient procedure for propenylation of MBH acetates in the presence of allyl bromide, zinc, copper iodide and silica gel has been reported by Srihari, leading to substituted 1, 5-dienes **215**, in good yields, that may find further use in synthetic chemistry.[136] Recently, it was found acetates of MBH adducts

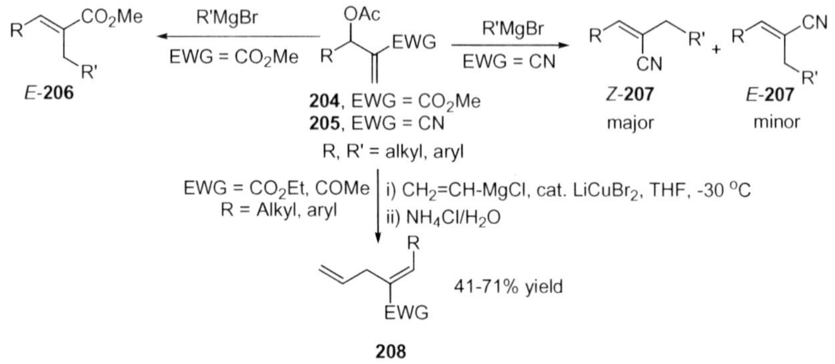

Scheme 3.83

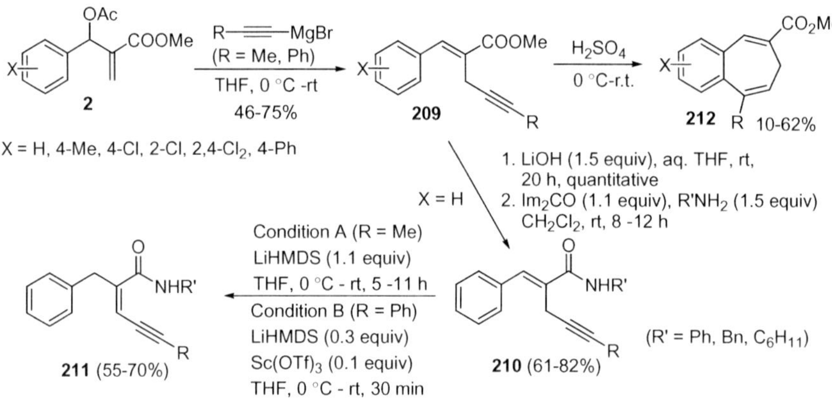

Scheme 3.84

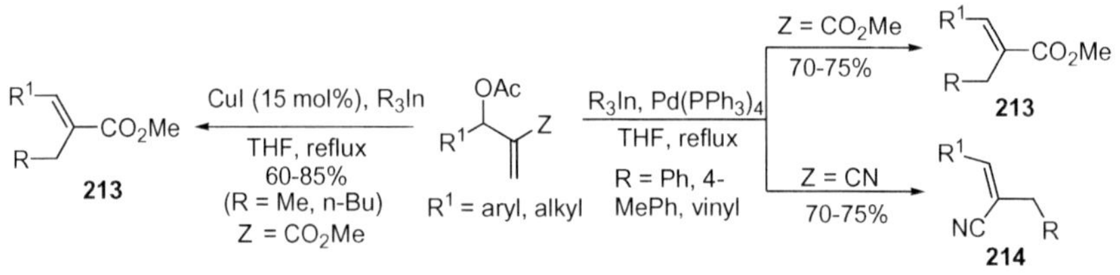

Scheme 3.85

derived from ethyl acrylate, methyl vinyl ketone and acrylonitrile can be coupled with allyltributylstannane using Pd(PPh$_3$)$_4$ or Pd(dba)$_2$ as catalyst at room temperature to afford the corresponding trisubstituted alka-1,5-dienes **215** in good to high yields (Scheme 3.86).[137]

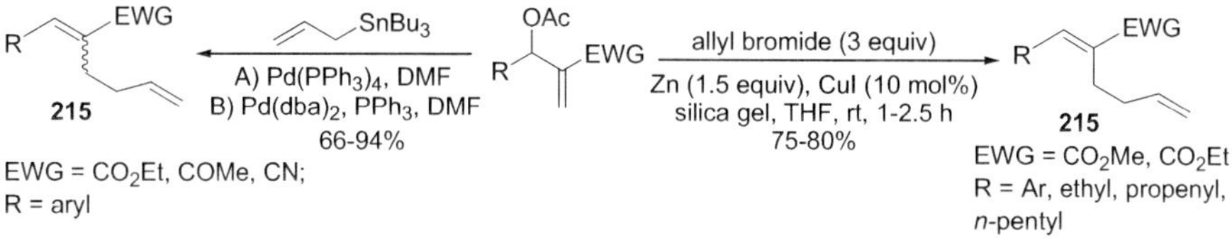

Scheme 3.86

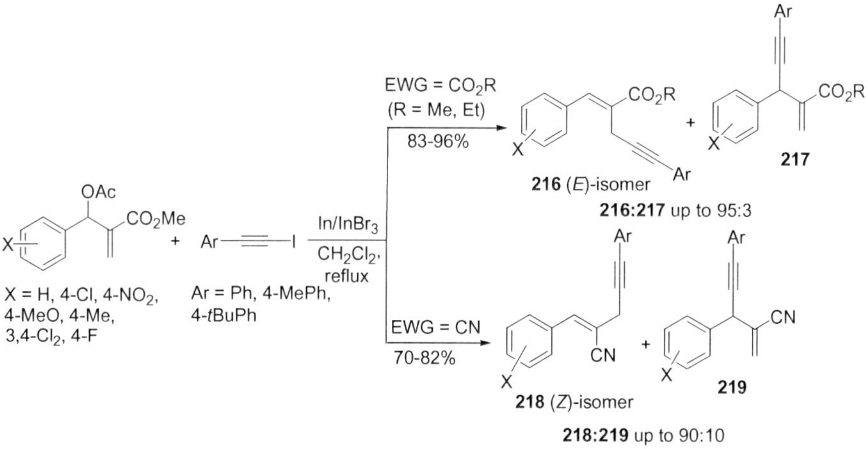

Scheme 3.87

MBH acetates undergo smooth alkynylation with aryl-substituted iodo-alkynes in the presence of indium metal to furnish 1,4-enynes in good to high yields with (*E*)-stereoselectivity for adducts derived from acrylates and with (*Z*)-stereoselectivity for adducts derived from acrylonitrile (Scheme 3.87).[138]

Das *et al.* have explored the reaction of MBH acetates with unactivated alkyl halides in the presence of Zn in saturated aqueous NH_4Cl solution to afford stereoselective trisubstituted olefins. The reactions of 3-hydroxy-2-methylene-alkanoates **220** gave (2*E*)-2-substituted-alk-2-enoates **222** exclusively, whereas the reactions of 3-hydroxy-2-methylene-alkanenitriles **221** afforded (2*Z*)-2-substituted-alk-2-enenitriles **223** as major products with high (*Z*)-selectivity (Scheme 3.88). The methodology was successfully applied to the synthesis of (2*E*)-2-butyloct-2-enal **224**, an alarm pheromone component of the African weaver ant, *Oecophylla longinoda*.[139]

Woodward *et al.* have reported a series of stereo- and enantioselective alkylations of non-symmetrical allylic electrophiles derived from MBH products. The allyl halides or allyl mesylate **225** derived from a MBH adduct are chemo- and regiospecifically transformed into β, β-disubstituted α-methylene-propionates **226** on treatment with either diorganozincs or organozinc halides in the presence of catalytic amounts of copper(I) salts (3–20 mol.%) in high

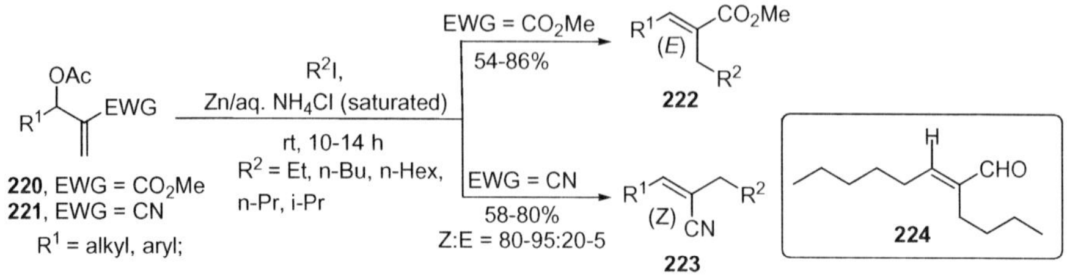

Scheme 3.88

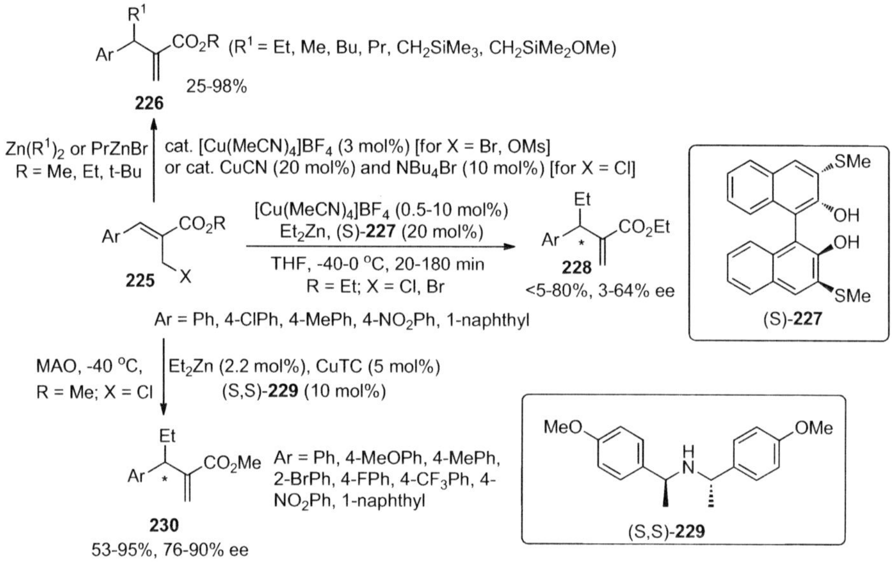

Scheme 3.89

yields.[140] In the presence of chiral thioether ligand **227**, copper-catalyzed asymmetric chemo- and regiospecific S_N2' addition of organozinc reagents to (Z)-allyl halides provided the corresponding products **228** in low to moderate ee.[141] After high-throughput ligand screening, the chiral secondary amines **229** were identified as alternative ligands for the copper-catalyzed highly enantio-selective alkylation of **MBH**-derived electron-deficient allylic chlorides with organozinc reagents in the presence of CuTC [copper(I) thiophene-2-carboxy-late] as the source of copper(I) and promotion by methylaluminoxane (MAO) of the zinc Schlenk equilibrium (Scheme 3.89).[142]

In a continuation of their work, they have also investigated the nickel-cat-alyzed enantioselective methylation of **MBH**-derived allylic chlorides and acetates **231** to afford α-chiral methylene carboxylate products **232** in the presence of the (Rp) planar chiral ferrophite ligands (Scheme 3.90).[143]

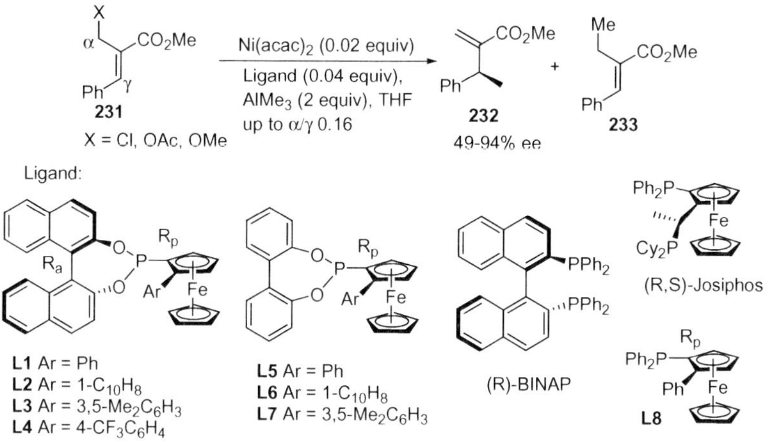

Scheme 3.90

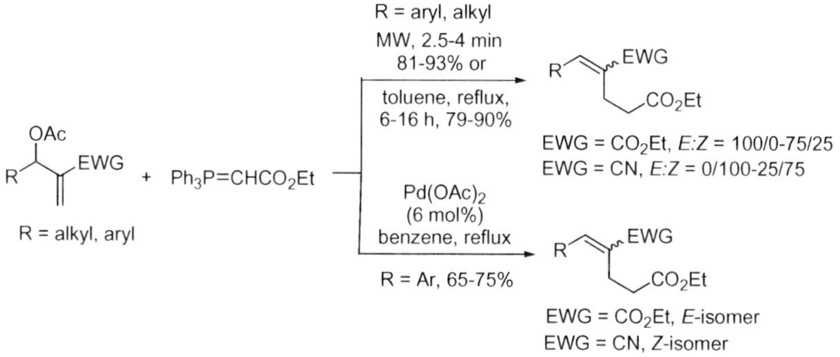

Scheme 3.91

In comparison to copper-catalyzed alkylation to afford only γ-alkylated products,[134,136] the Ni-ferrophite system delivered mixtures of γ-methylated **232** and α-methylated **233** products (with [**233**]/[**232**] down to 0.16). Although its rate is insufficient, moderate to good levels of stereoinduction were achieved (49–94% ee) at low catalyst loadings. The observed regio- and enantioselectivity were further rationalized by DFT and experimental studies that supported the idea of the selective formation of a limited number of energetically favored *anti* and *syn* π-allyl intermediates.[144]

Treatment of MBH acetates with stabilized ylide (ethoxycarbonylmethylene) triphenylphosphorane can provide trisubstituted olefins with (*E*)-selectivity for ester-containing substrate, whereas a mixture of (*Z*)- and (*E*)-isomers is obtained for nitrile-substituted acetate under thermal conditions or microwave irradiation (Scheme 3.91).[145] The reaction rates and yields are significantly improved by employing microwave irradiation. In addition, this reaction also can proceed smoothly in the presence of palladium acetate (Scheme 3.91).[146]

Amri and co-workers[147] have successfully transformed the acetates of the MBH adducts obtained from various activated alkenes into β-nitro-alkene derivatives *via* treatment with nitroalkanes. The resulting nitroalkane derivatives **234** and **235** were further transformed into the corresponding 1,4-diketones **236** and keto alkene derivatives **237** *via* a Nef reaction (Scheme 3.92).

Subsequently, Kim and co-workers[148] followed a similar strategy using different conditions to obtain 2-arylidene-4-nitroalkanoates **238** directly *via* an S_N2'-type allylic substitution (Path I, Scheme 3.93) and 2-methylene-4-nitroalkanoates **239** *via* treatment of *in situ* generated DABCO salt with nitroalkanes (Path II, Scheme 3.93). Finally, they transformed these adducts into the corresponding keto esters **240** and **241**, respectively, *via* a Nef reaction (Scheme 3.93).

Recently, Rezgui *et al.* have described a simple and direct allylic substitution of primary MBH alcohols **242** and **243** with various pronucleophiles under modified Taber's conditions (DMAP, toluene, reflux, 4Å molecular sieves), without the Pd catalysts/activating agents usually required for the

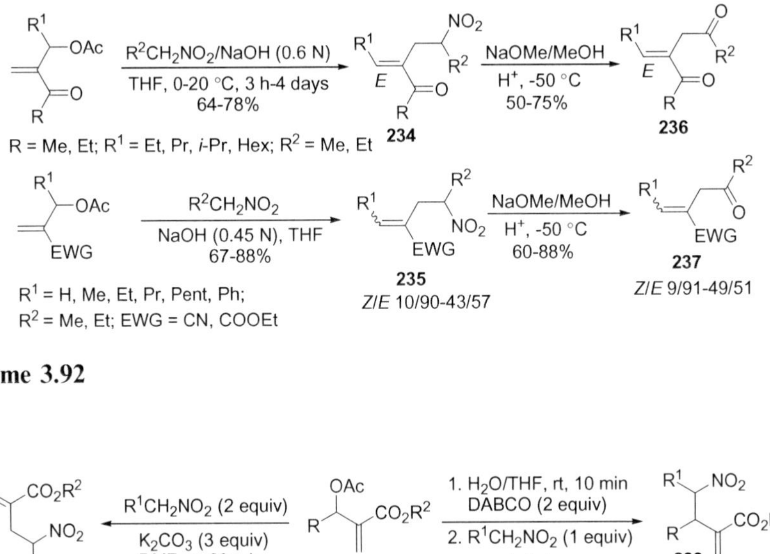

Scheme 3.92

Scheme 3.93

process, to afford *C*-allylation products **244–246** in moderate to good yields (Scheme 3.94).[149]

Fujimoto and co-workers[150] have reported a tandem Michael-intramolecular Corey–Chaykovsky reaction of the five-membered cyclic oxosulfonium ylide **247** with acetates of the MBH adducts in the presence of base, producing cycloheptene oxide derivatives **248** as a single stereoisomer. However, in the case of the six-membered oxosulfonium ylide **247′**, the cyclooctane oxide derivatives **249** were obtained as a mixture of stereoisomers in moderate yields (Scheme 3.95).

Kim and co-workers[151] have reported an interesting synthesis of tetra-substituted alkenes, that is, ethyl β-cyano-α-methylcinnamates and β-cyano-α-methylcinnamonitriles **250** with (*E*)-selectivities *via* a successive S_N2'-S_N2'-isomerization strategy as described in Scheme 3.96. However, they could obtain only ethyl 3-cyano-2-methyleneoctanoate (**251**) with a MBH adduct obtained from hexanal and ethyl acrylate (Scheme 3.97).

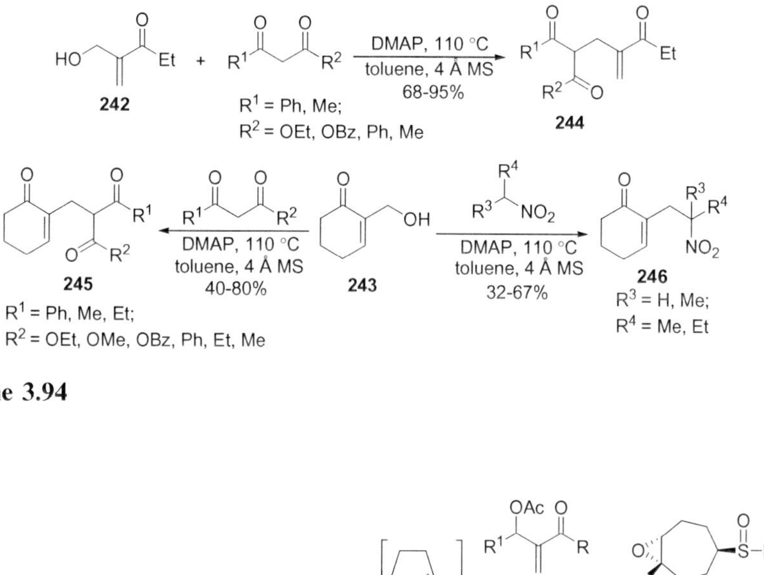

Scheme 3.94

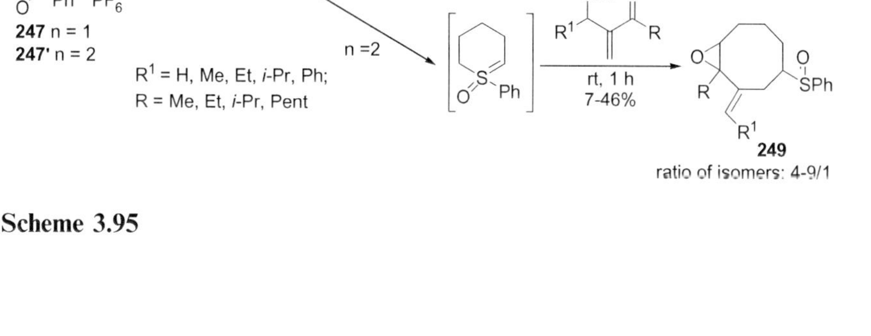

Scheme 3.95

Scheme 3.96

Scheme 3.97

Scheme 3.98

Methyl 2-cyanomethylcinnamates **252** and 2-(aryl-methylidene)succino-nitriles **253** have been obtained by treatment of readily available MBH acetates with KCN or NaCN *via* S_N2' nucleophilic substitution.[152] The resulting product **252** can be easily transformed into naphthalene **254** and benzylidene-succinimide derivatives **255** in good yields (Scheme 3.98).[152b]

Kabalka *et al.* have studied the cross-coupling reaction of potassium organo-trifluoroborates[153] and organosilanes[154] with acetates of MBH adducts in the presence of Pd catalysts to afford (*E*)-trisubstituted olefins as major products for 3-acetoxy-2-methylenealkanoates, whereas similar reactions with 3-acetoxy-2-methylenealkanenitriles lead to (*Z*)-alkenes (Scheme 3.99).

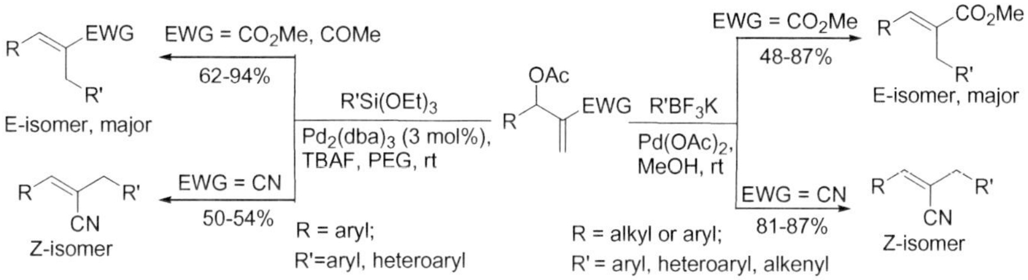

Scheme 3.99

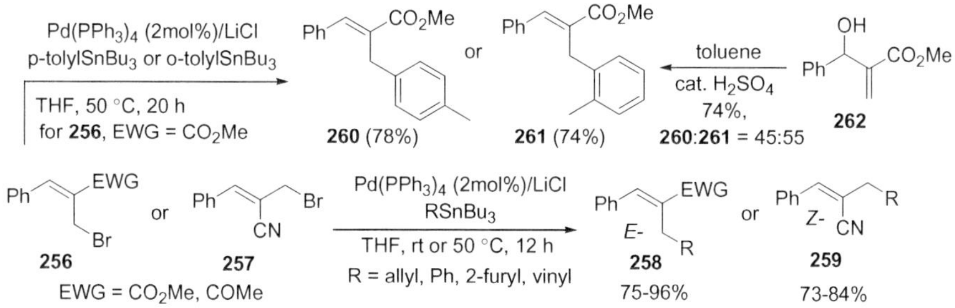

Scheme 3.100

Palladium-mediated cross-coupling reactions between the bromide of an MBH adduct (**256** or **257**) and organostannane compounds affords the corresponding allyl-, aryl- and vinyl-attached MBH adducts **258** or **259** in good yield (73–96%) by using Pd(Ph$_3$P)$_4$/LiCl (0.5 equiv) as catalytic system. As a palladium catalyst, Pd(Ph$_3$P)$_4$ was superior to Pd(OAc)$_2$ and the reaction time can be shortened by raising the reaction temperature [along with 5 mol.% of Pd(0) catalyst].[155] Notably, tolyl-substituted products **260** and **261** have been synthesized in good yields (74–78%) similarly by the reactions of **256** and *o*-tolyltributylstannane and *p*-tolyltributylstannane, while the traditional Friedel–Crafts reaction of the MBH adduct **262** and toluene under the influence of H$_2$SO$_4$ (0.2 equiv) afforded an *ortho/para* mixture in 74% yield (55 : 45 ratio by ^{1}H NMR) (Scheme 3.100).

Darses and co-workers have published a series of papers on the synthesis of stereo-defined trisubstituted alkenes by the coupling of readily available unreactive MBH adducts with either organoboronic acids or potassium trifluoro(organo)borates in the presence of a rhodium complex *via* a reaction pathway involving a 1,4-addition/β-hydroxy elimination mechanism.[156–158] Compared with the aforementioned Pd-catalyzed cross-coupling reaction, this reaction does not need the activation of the hydroxyl group with acetate or carbonate; therefore it is more desirable, particularly in terms of atom economy. For the MBH adducts derived from methyl acrylates, the initial reported catalyst, [{Rh-(cod)Cl}$_2$], was active at 50 °C for boronic acids[156] and at 70 °C for potassium trifluoro(organo)borates.[157] However, the reactions with other MBH adducts, particularly those derived from enones, gave lower yields, and the formation of by-products was observed. Recently, they achieved the reactions of either organoboronic acids or potassium trifluoro(organo)borates with a large variety of MBH adducts derived from α,β-unsaturated esters, ketones, amides and cyano derivatives under mild conditions in high yields by using [{Rh(cod)OH}$_2$] precursor as catalyst and without the need of additional phosphane ligands (Scheme 3.101).[158]

Recently, Kantam *et al.* have performed the cross-coupling reaction of MBH adducts with arylboronic acids by using rhodium-exchanged fluorapatite catalyst (RhFAP), which was prepared by treatment of fluorapatite (prepared by incorporating basic species F$^-$ in apatite *in situ* by coprecipitation) with an aqueous solution of RhCl$_3$. Various arylboronic acids and MBH adducts were converted into the corresponding trisubstituted olefins with high stereoselectivity, demonstrating the versatility of the reaction. The catalyst (RhFAP) was recovered quantitatively by simple filtration and reused with almost consistent activity (Scheme 3.102).[159]

MBH adducts smoothly undergo allylic nucleophilic substitution (S_N2') with allyltrimethylsilane in the presence of $BF_3 \cdot OEt_2$ under mild reaction conditions to afford 1,5-diene derivatives in good yields with high stereoselectivity (Scheme 3.103).[160]

Indole can also be used as a carbon nucleophile to react with MBH acetates to give trisubstituted olefins with (E)-selectivity for ester- and acetyl-containing substrates, whereas (Z)-isomers for nitrile-substituted acetate were obtained *via* S_N2'-type allylic substitution catalyzed by InBr$_3$[161] or I$_2$[162] (Scheme 3.104). The resulting products generated from 2-nitro-MBH adduct can proceed *via* a one-pot reductive cyclization to form indolylquinoline derivatives **263** in good yields (Scheme 3.104).[162]

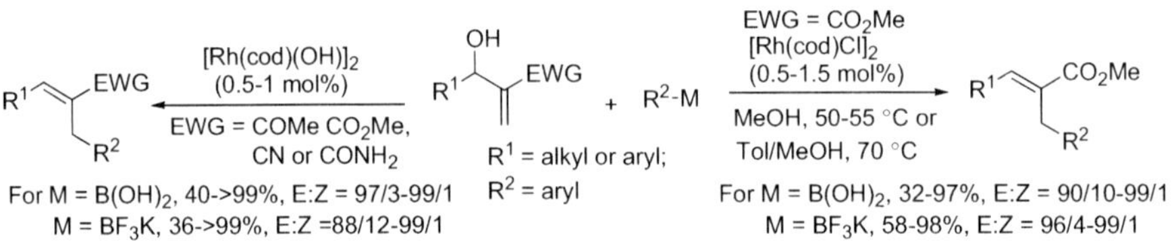

Scheme 3.101

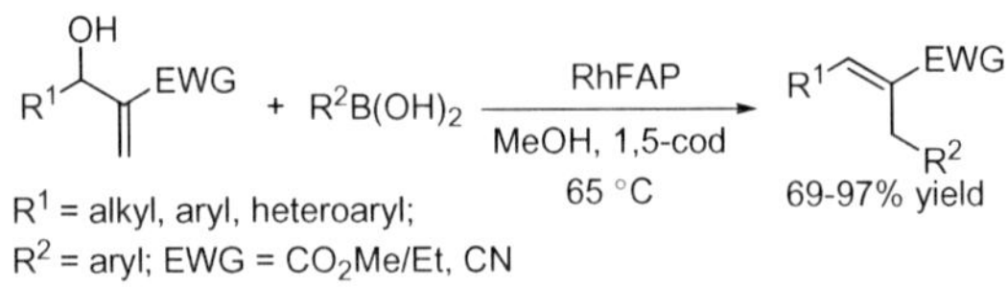

Scheme 3.102

EWG = CO₂Me or CO₂Et

76-84%

Z:E = 70-95:30-5

EWG = CN

78-81%

Z:E = 80-90:20-10

R = alkyl, vinyl, aryl, heteroaryl;
EWG = CO₂Me, CO₂Et, CN

BF₃·OEt₂

CH₂Cl₂, 0 °C

Scheme 3.103

EWG = CO₂Et

79-90%

E-isomer

R = alkyl, aryl, heteroaryl;
R' = H, 2-Me, 7-Et

EWG = CN

R = 4-MeOPh,
R'' = H

20 mol% InBr₃

ClCH₂CH₂Cl, reflux

30 mol% I₂
MeCN, rt

R = Ar; EWG = COMe,
73-89%

Z-isomer

Fe/AcOH
reflux

R = NO₂; R' = H

72-83%

263

X = H, NO₂; R' = H, 3-NO₂, 4-F
R'' = H, 2-Me, 2-Ph, 5-MeO, 5-Br, 7-Et, 2-CO₂Et

R'' = H, 2-Me, 5-MeO, 5-Br, 7-Et

Scheme 3.104

More recently, the allylic alkylation of MBH adducts catalyzed by a metal-free organic Lewis base has emerged as an attractive strategy to prepare multifunctional compounds. Upon exposure of MBH acetates to 2-trimethylsilyloxyfuran in the presence of substoichiometric quantities of triphenylphosphine, highly regio- and stereoselective substitution occurs to provide γ-butenolides **264** in high yields.[163] Moreover, as demonstrated by the substitution of MBH acetate, the absolute stereochemical course of these transformations can be controlled through the use of the (–)-8-phenylmenthol ester. Subsequently, our group accomplished the enantioselective asymmetric allylic substitution by using multifunctional chiral phosphine **266** as catalyst. Optically active γ-butenolides **264** were obtained in good to excellent yields and ees by using water as a co-additive (Scheme 3.105).[164]

Racemic MBH carbonates can be converted into densely functionalized compounds **267** featuring two adjacent stereocenters, one of them bearing four carbon substituents, by reaction with cyano esters in the presence of a catalytic amount of a modified cinchona alkaloid β-ICPD in high enantioselectivities and fair diastereoselectivities (Scheme 3.106).[165]

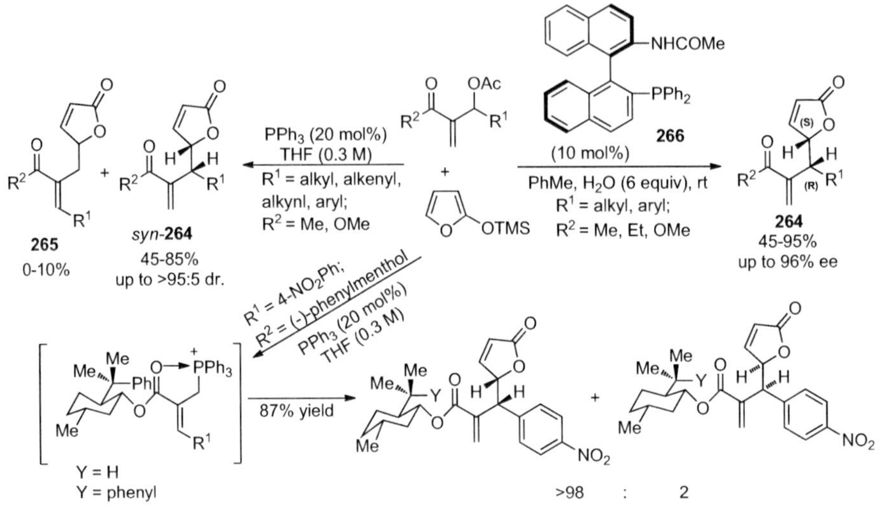

Scheme 3.105

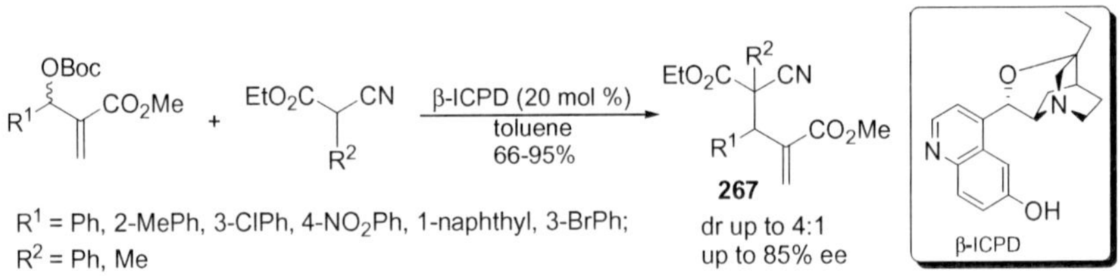

Scheme 3.106

Chen *et al.* have found that commercially available modified cinchona alkaloids [(DHQD)$_2$AQN] are outstanding catalysts for the asymmetric allylic alkylation (AAA) reaction of MBH carbonates with α,α-dicyanoolefins **268**[166] or 3-substituted oxindoles **269**.[167] Enantioenriched compounds **270** and **271**, respectively, with multifunctional frameworks bearing adjacent stereogenic centers were obtained in excellent stereoselectivities (Scheme 3.107).

Recently, the regio- and stereoselective allylic substitution of MBH acetates with various carbon-pronucleophiles catalyzed by 1,4-diazabicyclo[2.2.2]octane (DABCO) through a tandem S_N2'-S_N2' mechanism have been investigated by several groups. Yadav *et al.* investigated the reaction of MBH acetates with 2-phenyl-1,3-oxazol-5-one **272**, which affords N-protected α-amino acids **273** regioselectively in excellent yield (81–93%) and with high diastereoselectivity (>92%). This process involves a S_N2'-S_N2' reaction and water-driven ring-opening cascades in a one-pot procedure.[168] In addition, they also found that the product of MBH acetates with mercaptoacetyl transfer agent, 2-methyl-2-phenyl-1,3-oxathiolan-5-one (**274**), undergoes *in situ* selective hydrolysis with

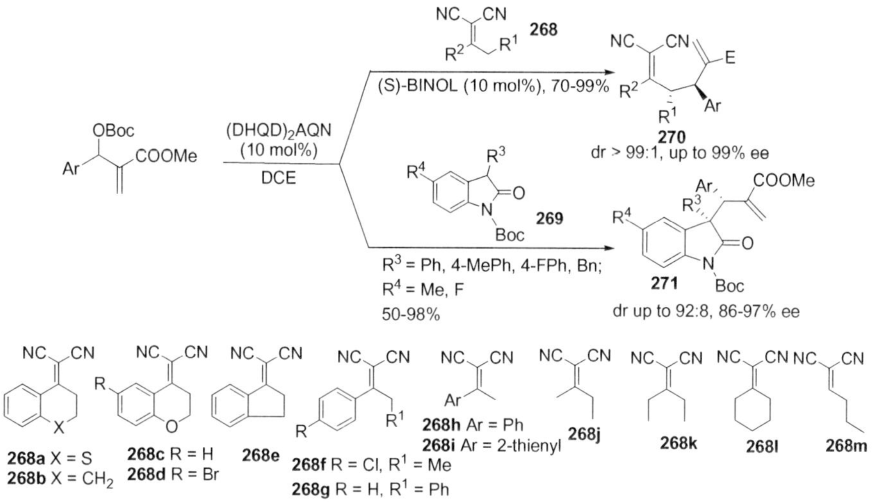

Scheme 3.107

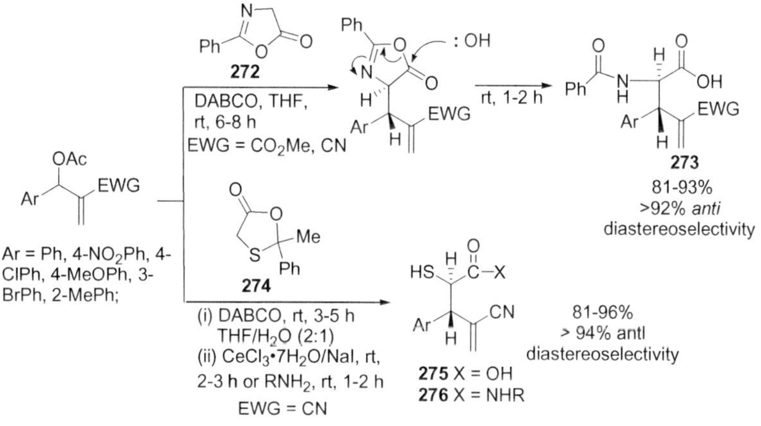

Scheme 3.108

water in the presence of $CeCl_3 \cdot 7H_2O/NaI$ and aminolysis of the 1,3-oxathiolan-5-one ring to afford functionally rich α-mercapto acids **275** and α-mercapto amides **276**, respectively, in excellent yield (81–96%) with high diastereoselectivity (>94%) (Scheme 3.108).[169]

An efficient DABCO-mediated regio- and diastereoselective substitution of MBH acetates with *p*-toluenesulfonylmethylisocyanide (TosMIC) has been developed by Krishna *et al. via* a tandem S_N2'-S_N2' mechanism. The products **277** were obtained in excellent yields (80–92%) with moderate diastereoselectivities (66–80%) in favor of the *syn*-isomer (Scheme 3.109).[170]

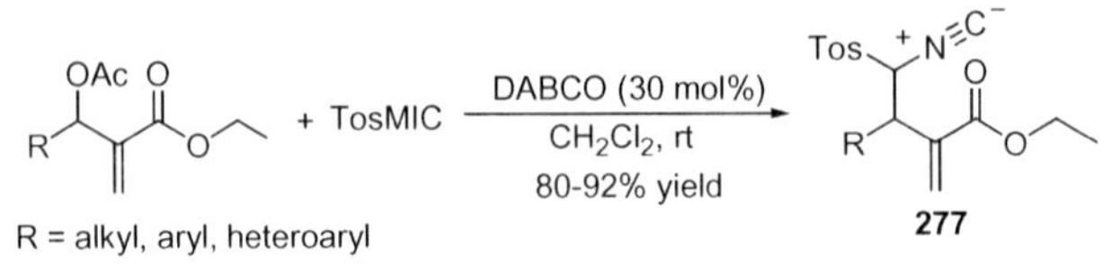

Scheme 3.109

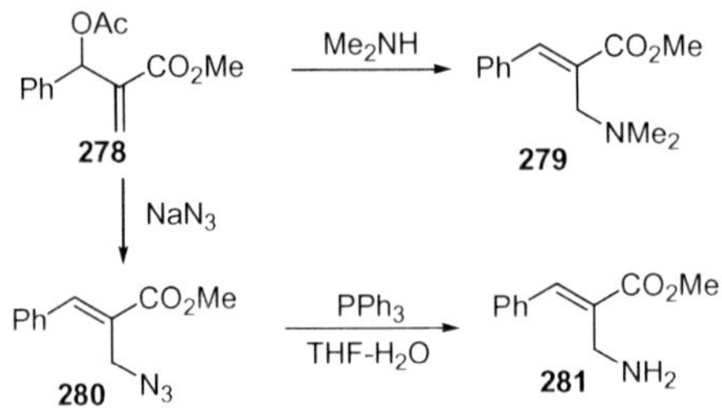

Scheme 3.110

3.8.2 Nitrogen Nucleophiles

MBH acetate **278** has been converted stereoselectively into the corresponding (*E*)-allyl amino compounds: treatment of acetate **278** with either primary or secondary amines produced (*E*)-allyl amine **279** predominantly, while treatment of **278** with sodium azide followed by reduction of the resulting azide **280** with PPh$_3$/THF/H$_2$O gave exclusively the primary (*E*)-allyl amine **281** (Scheme 3.110).[3]

Treatment of MBH acetates with excess ammonia acetate in anhydrous methanol affords (i) tertiary allyl amines **282** with an all-(*E*) configuration in the case of acrylates and (ii) secondary allyl amines **283** with all-(*Z*) configuration as major product in the case of acrylonitriles,[171] which is contrary to the case of primary allyl amines reported by Das *et al.*[172] However, the desired primary allylic amine **284** can be efficiently obtained from MBH acetates of acrylonitrile and methanol saturated with ammonia in a short period and on a multi-gram scale (Scheme 3.111).[173]

Drewes *et al.* have investigated the reaction of formyl imidazole with α-hydroxyaryl derivatives of methyl acrylate in the both the presence and absence of DABCO, with THF and methanol as solvent, respectively, to afford novel synthetic intermediates. By such control of the reaction conditions, either allylic substitution products **285** (THF, DABCO) or rearrangement (MeOH, no DABCO) products **286** could be obtained (Scheme 3.112).[174]

Kim and co-workers have elucidated an interesting reaction of DBN or DBU with the MBH acetates, which yielded lactam derivatives **287** in good yields (Scheme 3.113).[175]

DABCO salts of MBH adducts **288**, generated *in situ* from the corresponding acetate[176] or bromides,[177] can be used for the synthesis of region-retentive allyl

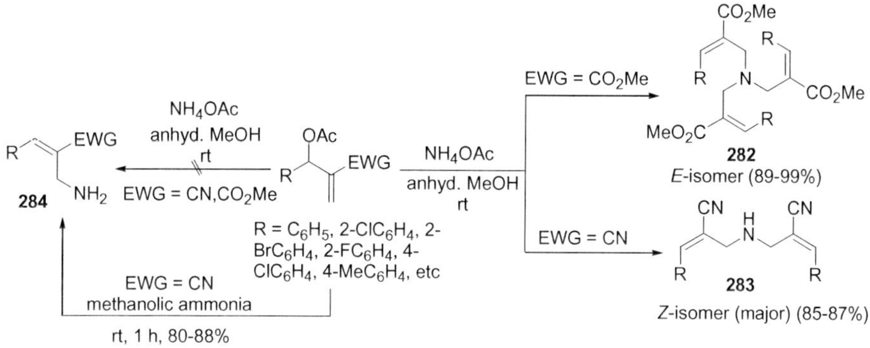

Scheme 3.111

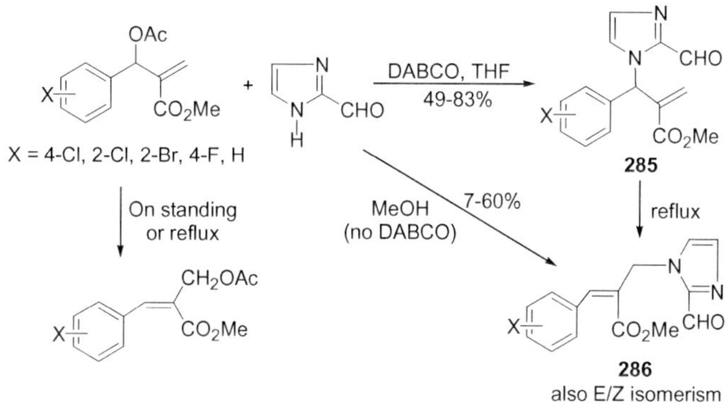

Scheme 3.112

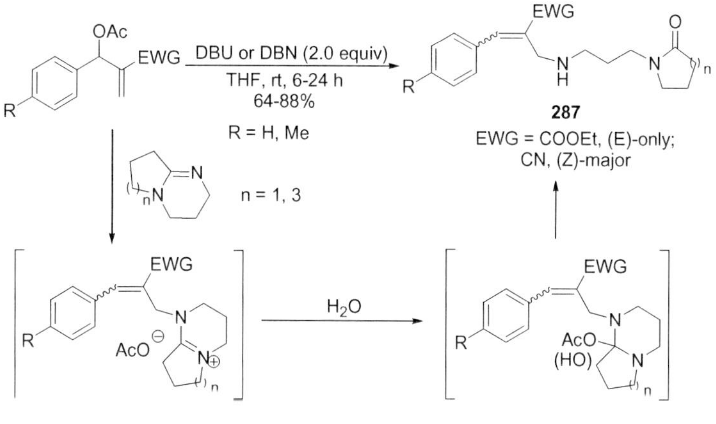

Scheme 3.113

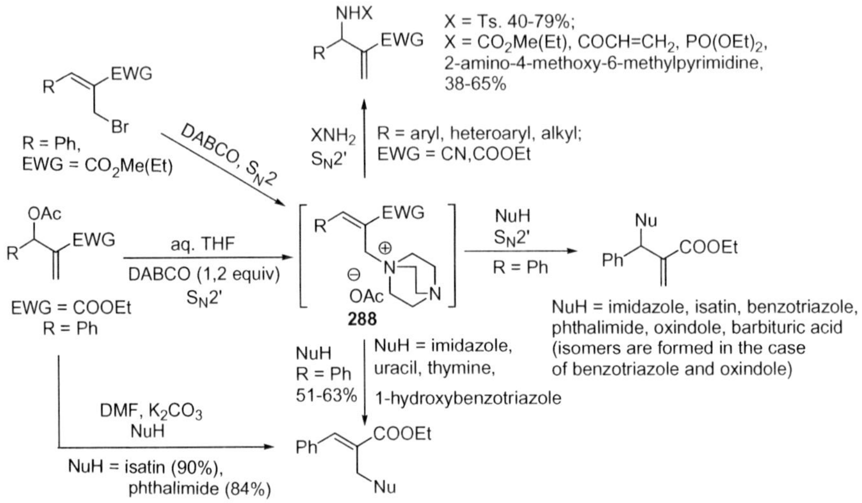

Scheme 3.114

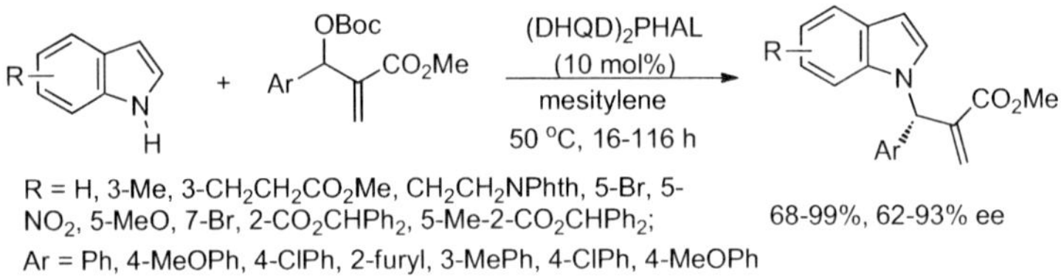

Scheme 3.115

substitution of MBH adducts bearing the N-substituent *via* successive S_N2'-S_N2' processes by treatment with various nucleophiles (Scheme 3.114). This approach is indicated by the "two-stage" regioretentive substitution of MBH acetates, wherein stoichiometric S_N2' displacement of the acetate by DABCO is followed by treatment of the resulting DABCO adduct with a heteroatom nucleophile that engages in a second S_N2' displacement of DABCO. Either allylic substitution products or rearrangement products could be obtained by controlling the reaction conditions.[176b]

Chen *et al.* developed the first asymmetric and chemoselective *N*-allylic alkylation of indoles with MBH carbonates. This method was realized by employing modified cinchona alkaloids (DHQD)$_2$PHAL as organocatalysts. Either electron-rich or electron-deficient indoles could be employed and moderate to excellent enantioselectivity was achieved (62–93% ee) (Scheme 3.115).[178]

Recently, Xia *et al.* have reexamined the reaction of MBH acetate with amines under both conventional stirring and ultrasound irradiation. They found the ultrasonic amination generally affords the target compounds in

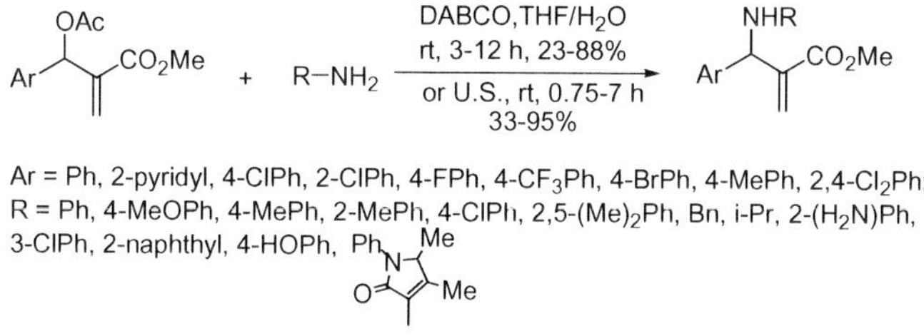
OAc
CO2Me
Ar
R–NH2
DABCO,THF/H2O
rt, 3-12 h, 23-88%
or U.S., rt, 0.75-7 h
33-95%
NHR
CO2Me
Ar
Ar = Ph, 2-pyridyl, 4-ClPh, 2-ClPh, 4-FPh, 4-CF3Ph, 4-BrPh, 4-MePh, 2,4-Cl2Ph;
R = Ph, 4-MeOPh, 4-MePh, 2-MePh, 4-ClPh, 2,5-(Me)2Ph, Bn, i-Pr, 2-(H2N)Ph,
3-ClPh, 2-naphthyl, 4-HOPh,
Ph
Me
Me

Scheme 3.116

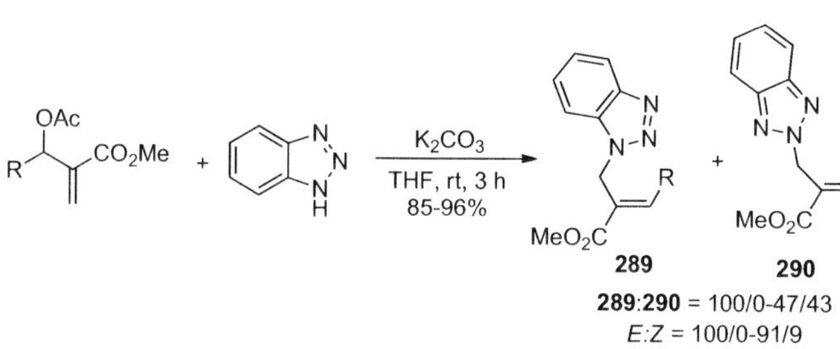
OAc
R
CO2Me
N
H
K2CO3
THF, rt, 3 h
85-96%
MeO2C
R
289
290
289:290 = 100/0-47/43
E:Z = 100/0-91/9

Scheme 3.117

improved yields with reduced reaction time and without the formation of any side products for the selected primary amines and acetates (Scheme 3.116).[179]

A convenient method has been developed towards the stereoselective synthesis of N-substituted benzotriazole derivatives under mild conditions. The good stereoselectivity, high yields and the simple procedure make this protocol attractive (Scheme 3.117).[180]

Iqbal and co-workers[181] have described an interesting solvent and substituent dependent regioselective (S_N2 or S_N2') Pd(0)-catalyzed synthesis of α-dehydro-β-amino esters *via* treatment of the acetates of MBH adducts with amines (Scheme 3.118). For proline-based cyclic secondary amine and serine methyl ester, the corresponding (*E*)-allylamines **291** (S_N2') were obtained exclusively in the presence of Pd(0) (catalyst) (Scheme 3.118).

Asymmetric allylic amination of allylic carbonates, prepared from racemic MBH adducts, proceeds in the presence of the Pd catalyst chiral diaminophosphine oxide (DIAPHOX) (**292**) and BSA to afford the corresponding chiral aza-MBH adduct derivatives **293** in excellent yield with up to 99% ee. The cyclic reaction products could be converted into various synthetically useful compounds, such as chiral cyclic β-amino acids (Scheme 3.119).[182]

Kim *et al.* have disclosed an efficient synthetic approach for *N*-tosyl aza-MBH adducts of cycloalkenones *via* the $FeCl_3$-mediated benzylic amination protocol from the easily available MBH adducts. The reactions of MBH adducts derived from 2-cyclohexen-1-one, 4,4-dimethyl-2-cyclohexen-1-one and 2-cyclopenten-1-one provided the corresponding aza-MBH adducts **294** in good yields (80–93%); the substituent of the aryl moiety had little effect on the

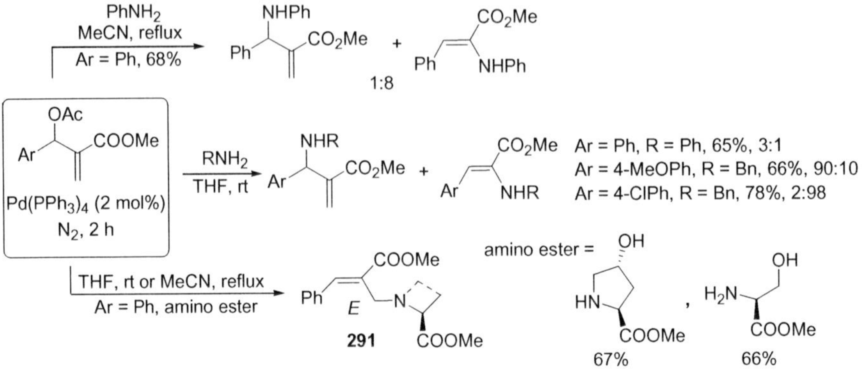

Scheme 3.118

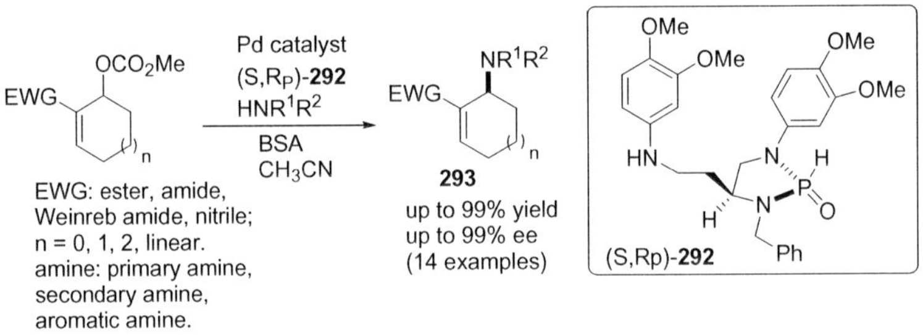

Scheme 3.119

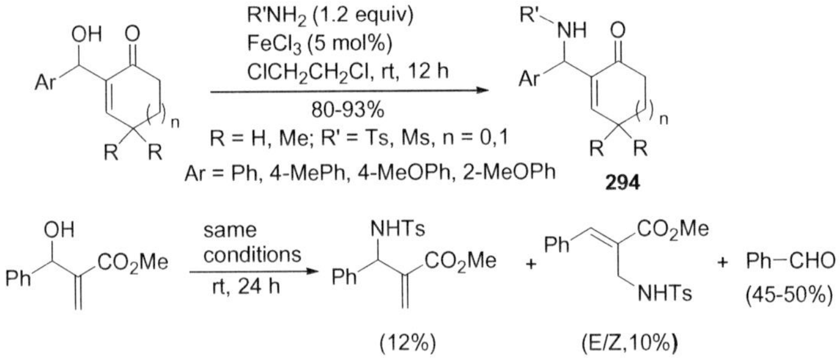

Scheme 3.120

reaction outcomes. However, the aliphatic chain-attached substrate and the acrylate-containing substrate cannot give the expected *N*-tosyl aza-MBH adducts (Scheme 3.120).[183]

α-Dehydro-β-amino esters **295** have been synthesized regioselectively from acetates of MBH adducts with amines in the presence of a catalytic amount of

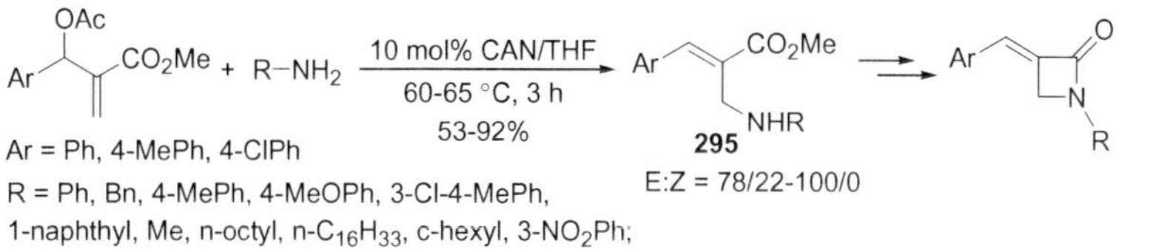

Ar = Ph, 4-MePh, 4-ClPh

R = Ph, Bn, 4-MePh, 4-MeOPh, 3-Cl-4-MePh,

1-naphthyl, Me, n-octyl, n-C$_{16}$H$_{33}$, c-hexyl, 3-NO$_2$Ph;

Scheme 3.121

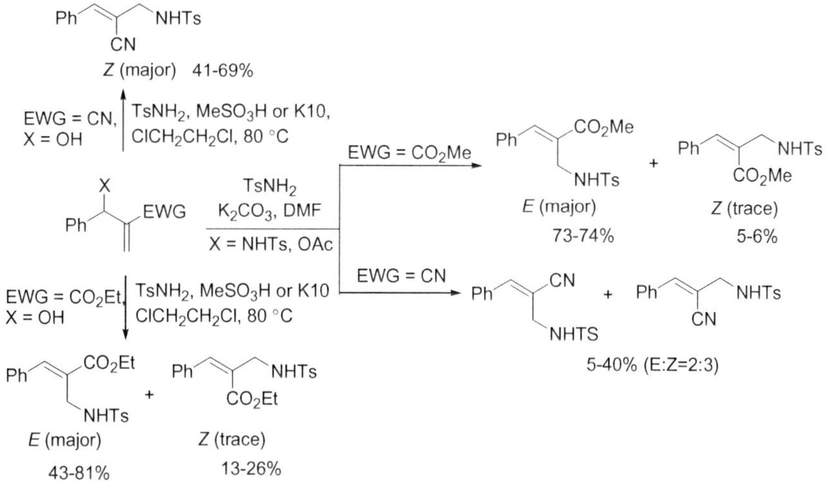

Scheme 3.122

ceric ammonium nitrate (CAN) in good yields. The polarity of the solvent does not affect the regioselectivity (Scheme 3.121).[184]

N-Tosyl *aza*-MBH adducts has been synthesized from the reaction of various MBH adducts and tosylamide. The (*E*)-isomers were obtained selectively (73%) together with only 6% of (*Z*)-isomer under basic conditions (DMF, K$_2$CO$_3$) for ester-containing substrate (Scheme 3.122). However, (*Z*)-isomers were obtained under acidic conditions in up to 26% for the MBH adducts having an ester moiety, and as major product for the MBH adducts having a nitrile moiety under the influence of MeSO$_3$H or montmorillonite K10.[185]

Ranu *et al.* have developed a mild and efficient method for the synthesis of α-dehydro-β-amino esters and nitriles by nucleophilic addition of amines to MBH acetates in water at room temperature without using any basic, acidic or metal catalyst. The reactions yielded only γ-addition products, providing, with high stereoselectively, (*E*)-isomers in the case of MBH adducts bearing a carboxylic ester moiety and (*Z*)-isomers for adducts containing CN functionality (Scheme 3.123).[186]

An interesting stereoselective transformation of MBH adducts into cinnamylamines *via* treatment with DMF–DMA has been described by Kim and co-workers (Scheme 3.124).[187]

X = H, 4-Cl, 3-Br, 4-MeO; EWG = CO$_2$Me, CN

R$_2$NH: i-Pr$_2$NH, BnNH$_2$, PhNH$_2$, 4-BrPhNH$_2$, 4-MeOPhNH$_2$,

Scheme 3.123

R = Ph, 2-FPh, 4-MePh, 4-(MeO)Ph

EWG = COOEt 10-24 h 39-41% X = NHTS

EWG = CN 1-2 h, 35-47%

DMF-DMA (2 equiv) DMF 70-80 °C

TsNHMe K$_2$CO$_3$, 12 h 45% R = Ph EWG = CN X = NHTs

R = Ph, 2-FPh, 4-MePh, 4-(MeO)Ph

R = Ph, 4-ClPh, 4-MePh

EWG = COOEt 69-73% X = OH

EWG = CN 63-68%

DMF-DMA (2 equiv) DMF, 80-90 °C, 2 h

R = Ph, 2-ClPh, 4-MePh

Scheme 3.124

MBH acetates can be stereoselectively transformed into (*E*)- or (*Z*)-allyl amides, in case of MBH adducts derived from acrylates or acrylonitriles, by reaction with nitriles in the presence of CH$_3$SO$_3$H[188] or Amberlyst-15[189] as a catalyst. In addition, 2-benzazepine derivatives **296** have been prepared from some MBH adducts containing an electron-rich aryl group *via* a convenient one-pot procedure involving tandem construction of C–N and C–C bonds (Scheme 3.125).[188]

Krische and co-workers have investigated the tertiary phosphine-catalyzed regiospecific allylic amination of MBH acetates through a tandem S_N2'-S_N2' mechanism by using phthalimide derivatives as nucleophiles.[190] When (*R*)-Cl-MeO-BIPHEP was used as a catalyst, the MBH adduct obtained from *p*-nitrobenzaldehyde and methyl acrylate reacted with phthalimide to give the allylic substituted product **297** in 80% yield with 56% ee (Scheme 3.126).[190a]

As continuation of this work, a planar chiral [2.2]paracyclophane mono-phosphine has been applied successfully as an organocatalyst in asymmetric allylic amination of MBH adducts. Up to 71% ee was achieved using (*R*)-**298** as a catalyst (Scheme 3.127).[191]

MBH acetates undergo smooth nucleophilic substitution with sodium azide in water under mild conditions to afford the corresponding ethyl 2-azido-methyl-3-phenylpropenoates **299** and 2-azidomethyl-3-phenylacrylonitriles **300** in excellent yields (Scheme 3.128).[152c]

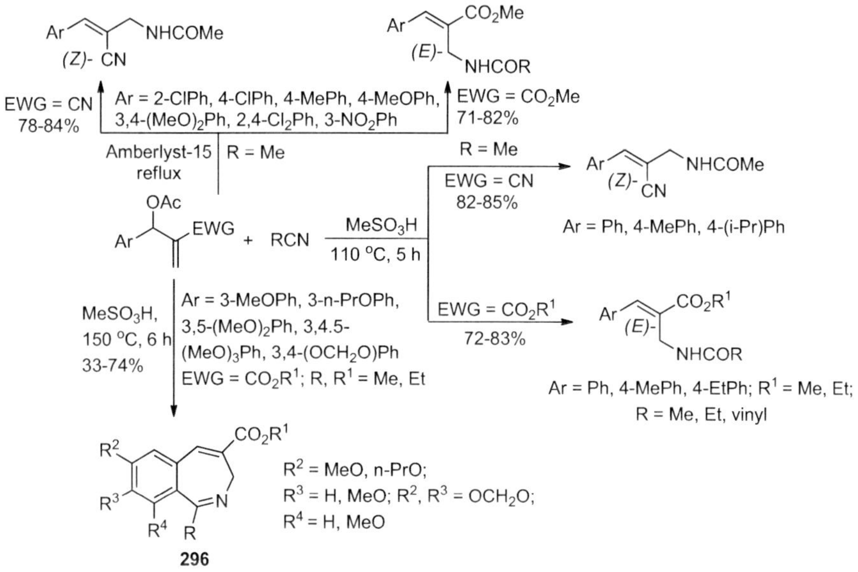

Scheme 3.125

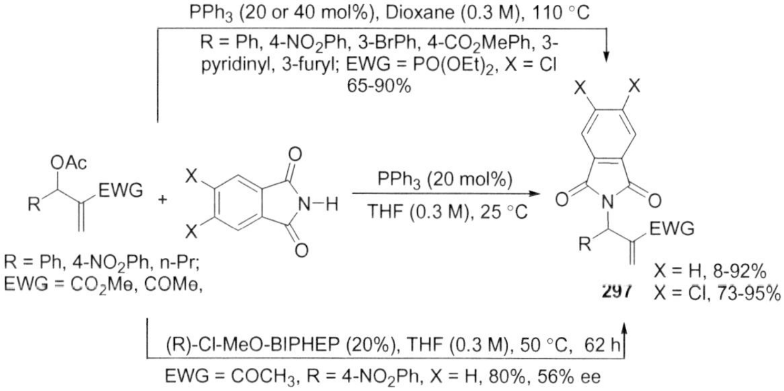

Scheme 3.126

Srihari and Jain *et al.* have developed a multicomponent addition reaction of MBH acetates with azide followed by a one-pot click reaction with nitriles to yield 1,5-disubstituted-tetrazoles. The (*E*)-isomer **301** was formed exclusively with MBH acetates containing an ester moiety, whereas nitrile-containing MBH acetates afforded mixture of diastereomers with the (*Z*) isomer **302** as the major product (Scheme 3.129).[192]

New important organic multifunctionalized cyclic six- and seven-membered allylic amines and azides have been obtained *via* regio- and diastereoselective reactions of cyclic MBH type adducts **303** with N-nucleophiles. We have found

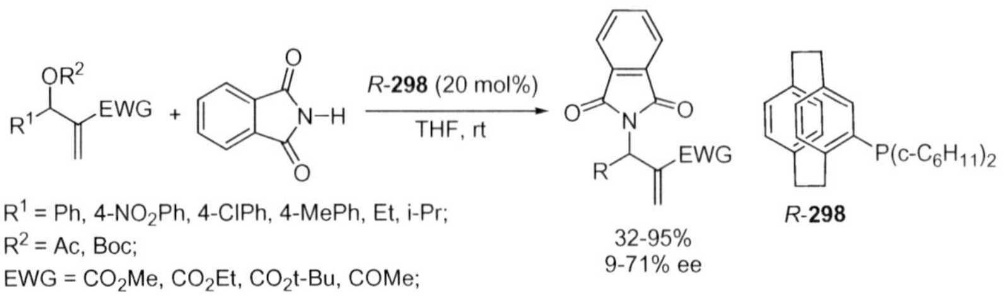

R^1 = Ph, 4-NO$_2$Ph, 4-ClPh, 4-MePh, Et, i-Pr;
R^2 = Ac, Boc;
EWG = CO$_2$Me, CO$_2$Et, CO$_2$t-Bu, COMe;

R-**298** (20 mol%)

THF, rt

32-95%
9-71% ee

R-**298**

P(c-C$_6$H$_{11}$)$_2$

Scheme 3.127

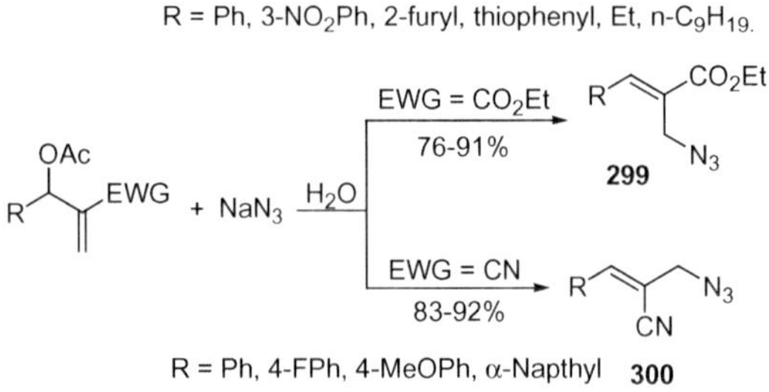

R = Ph, 3-NO$_2$Ph, 2-furyl, thiophenyl, Et, n-C$_9$H$_{19}$.

EWG = CO$_2$Et

76-91%

299

+ NaN$_3$

H$_2$O

EWG = CN

83-92%

R = Ph, 4-FPh, 4-MeOPh, α-Napthyl   **300**

Scheme 3.128

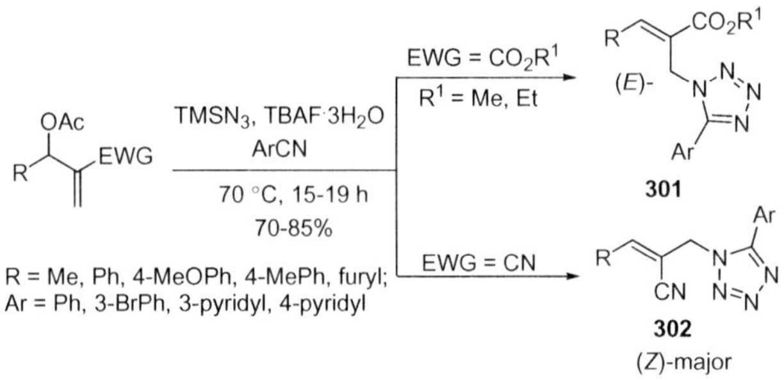

TMSN$_3$, TBAF·3H$_2$O

ArCN

70 °C, 15-19 h

70-85%

EWG = CO$_2$R^1

R^1 = Me, Et

(*E*)-

301

EWG = CN

302

(*Z*)-major

R = Me, Ph, 4-MeOPh, 4-MePh, furyl;
Ar = Ph, 3-BrPh, 3-pyridyl, 4-pyridyl

Scheme 3.129

that the reactions proceed by S_N2 or S_N2' processes exclusively, or by both processes simultaneously. The S_N2' process occurs with *anti* stereochemistry (Scheme 3.130).[193]

Lee *et al.* have developed a fast and easy access to methyl (*E*)-2-nitromethylcinnamates **304** from the corresponding MBH acetates with NaNO$_2$ in DMF. Several *o*-chloro-2-nitromethylcinnamates undergo an intramolecular aromatic substitution reaction, followed by rearrangement to yield coumarin derivatives **306** in moderate yields (Scheme 3.131).[194]

Nair *et al.* have investigated the reaction of MBH acetates with the Huisgen zwitterion to afford β-amino acid derivatives **307** and **308** in the presence of triphenylphosphine by invoking a S_N2' process (Scheme 3.132).[195]

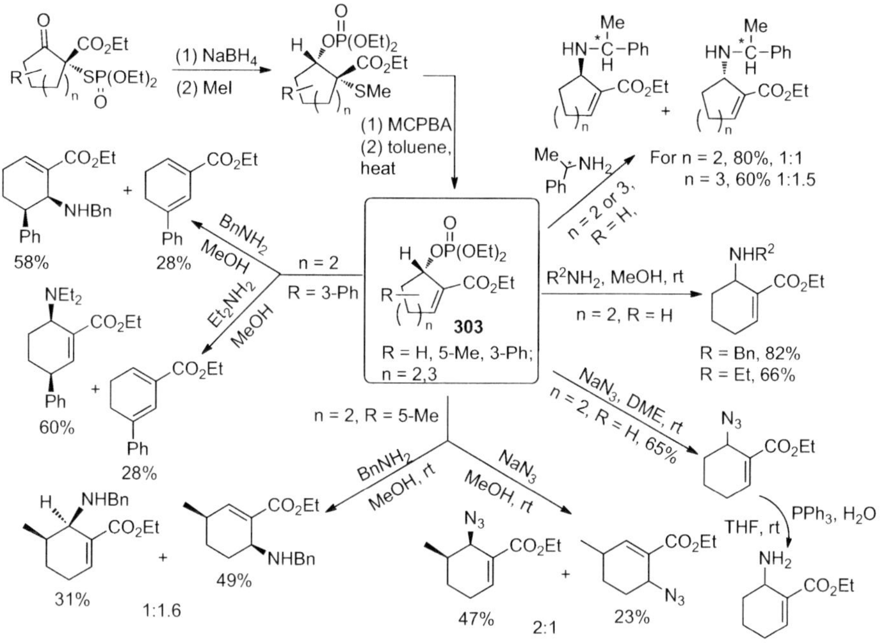

Scheme 3.130

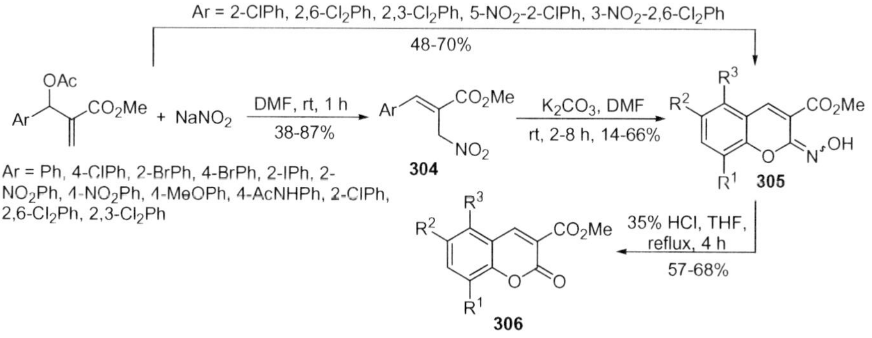

Scheme 3.131

Highly α-regioselective In(OTf)$_3$-catalyzed *N*-nucleophilic substitution of cyclic MBH adducts with aromatic amines has been developed. During the course of the reaction, allylic rearrangement from γ-product **310** to α-product **309** occurred. Utilizing this rearrangement reaction, α-regiocontrol of the *N*-nucleophilic substitution of cyclic MBH adducts could be realized. The synthetic method provides an efficient route to aromatic allylic amines bearing cyclic an unsaturated ketone unit (Scheme 3.133).[196]

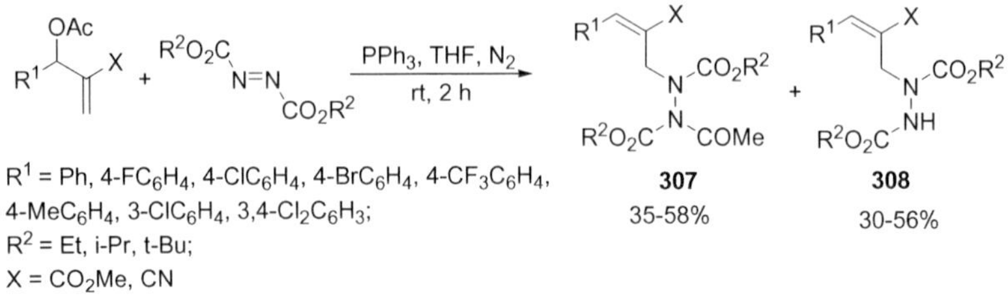

OAc
R1
X
R2O2C
N=N
CO2R2
PPh3, THF, N2
rt, 2 h
R1
X
N
CO2R2
R2O2C
N
COMe
307
35-58%
R1
X
N
CO2R2
R2O2C
NH
308
30-56%
R1 = Ph, 4-FC6H4, 4-ClC6H4, 4-BrC6H4, 4-CF3C6H4,
4-MeC6H4, 3-ClC6H4, 3,4-Cl2C6H3;
R2 = Et, i-Pr, t-Bu;
X = CO2Me, CN

Scheme 3.132

OH
O
R1
n
NH2
R2
In(OTf)3 (10 mol%)
THF/4A MS, reflux
62-80%, α/γ = 1:0-8:1
R2
NH
O
R1
n
α-309
O
R1
N
H
n
R2
γ-310
R1 = Ph, R2 = H,
n = 1
In(OTf)3 (10 mol%)
THF/4A MS, reflux
α/γ = 12:1
R1 = Ph, 4-FPh, 4-ClPh, 4-BrPh, 4-MePh, cyclohexyl;
R2 = H, 4-Me, 4-F, 3-MeO, 4-Br
n = 1, 2

Scheme 3.133

OAc
CO2Et
t-BuOO⊖
CO2Et
311
OOt-Bu
CO2Et
Br
t-BuOO⊖
OOt-Bu
CO2Et
312
benzene
heat
CO2Et
Me
O
Me
313

Scheme 3.134

3.8.3 Oxygen Nucleophiles

The reactions of *t*-butyl peroxylate anion with ethyl 2-methylene-3-acetoxy-butanoate and with ethyl 2-bromomethylbut-2-enoate have been studied by Mailard and co-workers, respectively (Scheme 3.134).[44b] They found that both reactions resulted in the γ-attack products **311** in low yields. However, the allyl *t*-butyl peroxides **312** were obtained in 45% yield when the reaction was carried out in the presence of poly(ethylene oxide) 400.[44b] Subsequently, **312** was converted into the corresponding glycidic ester **313** by heating in benzene.[197]

Recently, Chen *et al.* developed the first peroxo-asymmetric allylic alkylation of bulky hydroperoxyalkanes with MBH carbonates with catalysis by commercially available modified cinchona alkaloids. Peroxides **314** were generally

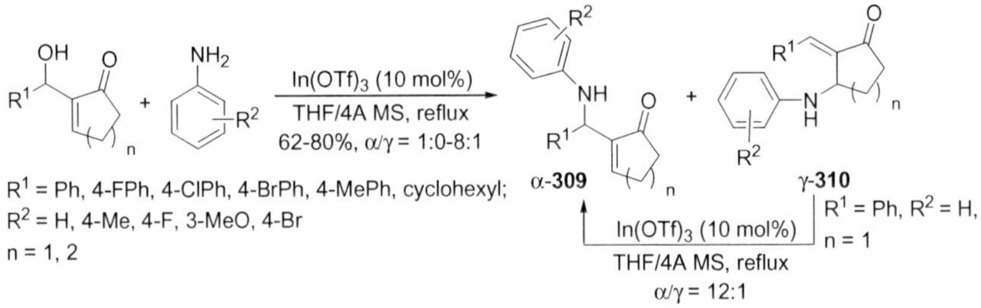

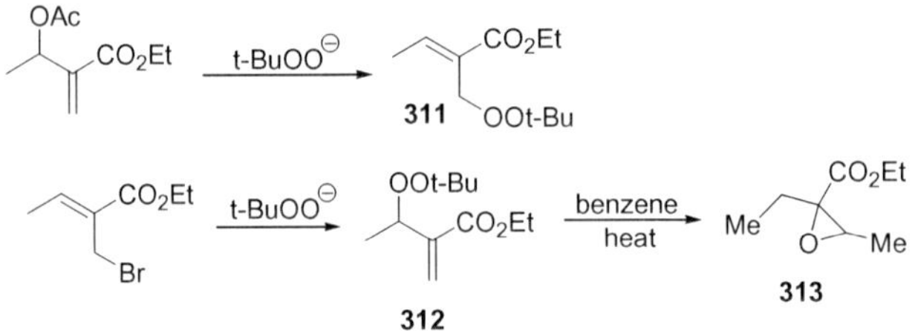

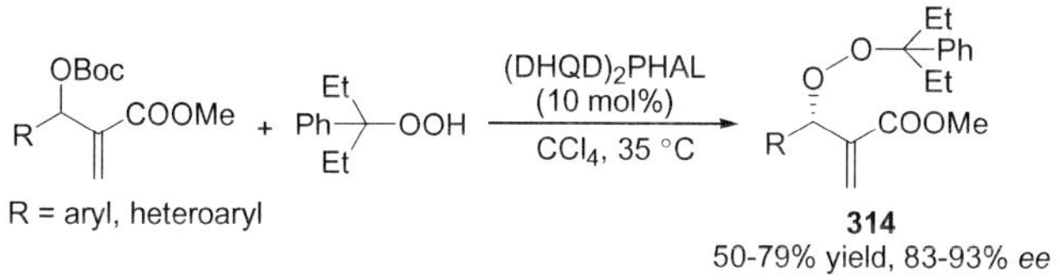

Scheme 3.135

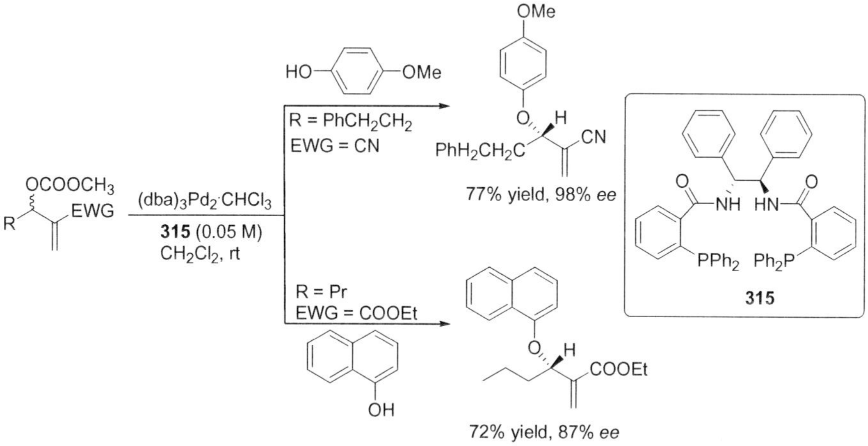

Scheme 3.136

obtained in fair to good yields with high enantioselectivities (83–93% ee), from which the corresponding α-methylene-β-hydroxy esters could be derived smoothly without affecting the optical purity (Scheme 3.135).[198]

Trost *et al.*[199] have reported elegant deracemization of MBH adducts *via* palladium-catalyzed dynamic kinetic asymmetric transformation (DYKAT) by using $(dba)_3Pd_2 \cdot CHCl_3$ and enantiopure ligands **315**. Aryl alcohols were found to be competent nucleophiles for this process. One representative example is described in Scheme 3.136. They also examined the influence of solvent, ligand and ligand concentration on regio- and enantioselectivity. Subsequently, they applied this methodology for the preparation of enantiomerically pure dihydrobenzofuran derivative **316**, which was further transformed into furaquinocin E (**317**) in good yields (Scheme 3.137).[200]

More recently, Trost *et al.* also examined the ability to use aliphatic alcohols as competent nucleophiles in the palladium-catalyzed DYKAT of MBH adducts. High yield and enantioselectivity were obtained for both the kinetic transformation and dynamic kinetic transformation to afford substituted pyran products. The utility of this method was further demonstrated in the context of a concise total synthesis of the gastrulation inhibitor (+)-hippospongic acid A (Scheme 3.138).[201]

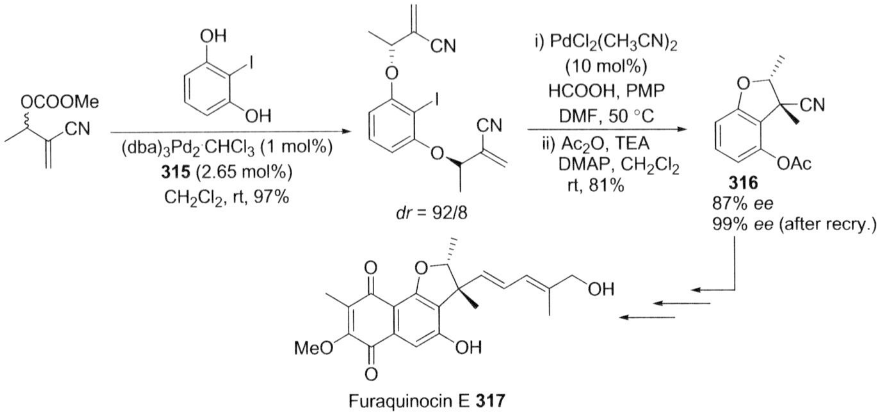
OCOOMe
CN
OH
I
OH
(dba)3Pd2·CHCl3 (1 mol%)
315 (2.65 mol%)
CH2Cl2, rt, 97%
CN
O
I
CN
O
dr = 92/8
i) PdCl2(CH3CN)2
(10 mol%)
HCOOH, PMP
DMF, 50 °C
ii) Ac2O, TEA
DMAP, CH2Cl2
rt, 81%
O
CN
OAc
316
87% ee
99% ee (after recry.)
O
O
OH
MeO
OH
O
Furaquinocin E 317

Scheme 3.137

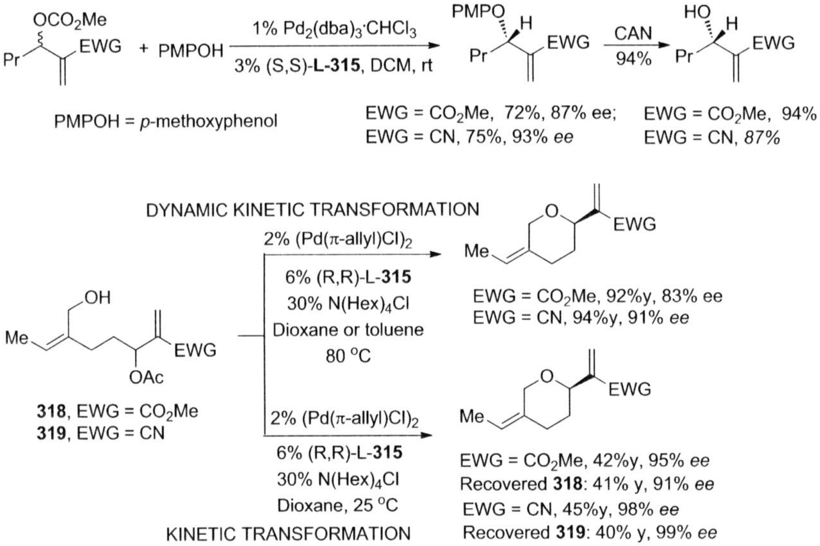
OCO2Me
EWG
Pr
+ PMPOH
1% Pd2(dba)3·CHCl3
3% (S,S)-L-315, DCM, rt
PMPO
H
EWG
Pr
CAN
94%
HO
H
EWG
Pr
PMPOH = p-methoxyphenol
EWG = CO2Me, 72%, 87% ee;
EWG = CN, 75%, 93% ee
EWG = CO2Me, 94%
EWG = CN, 87%
DYNAMIC KINETIC TRANSFORMATION
2% (Pd(π-allyl)Cl)2
6% (R,R)-L-315
30% N(Hex)4Cl
Dioxane or toluene
80 °C
O
EWG
Me
EWG = CO2Me, 92%y, 83% ee
EWG = CN, 94%y, 91% ee
OH
Me
EWG
OAc
318, EWG = CO2Me
319, EWG = CN
2% (Pd(π-allyl)Cl)2
6% (R,R)-L-315
30% N(Hex)4Cl
Dioxane, 25 °C
KINETIC TRANSFORMATION
O
EWG
Me
EWG = CO2Me, 42%y, 95% ee
Recovered 318: 41% y, 91% ee
EWG = CN, 45%y, 98% ee
Recovered 319: 40% y, 99% ee

Scheme 3.138

MBH diene adducts have also been employed successfully in a palladium-catalyzed asymmetric allylic alkylation reactions with various phenols in good regio- and enantioselectivity. These high enantioselectivities are even more substantial considering the ambiguity introduced by the additional double bond in the allylic system (Scheme 3.139).[128]

Enantiomerically enriched MBH alcohols **321** have been prepared in 25–42% yields with optical purities of 54–92% ee by using the combined concept of kinetic resolution during salt formation of racemic MBH acetates

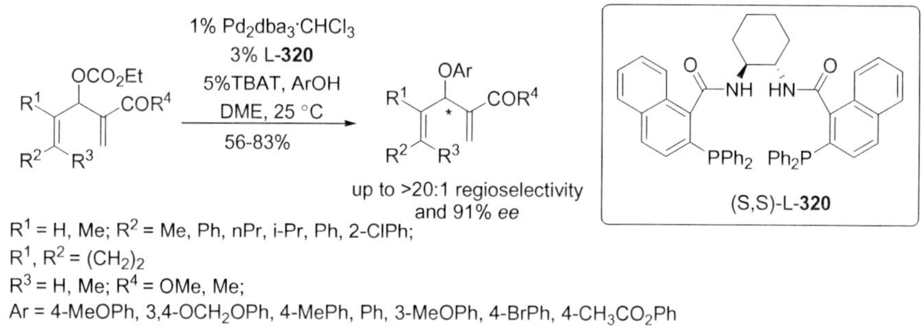

Scheme 3.139

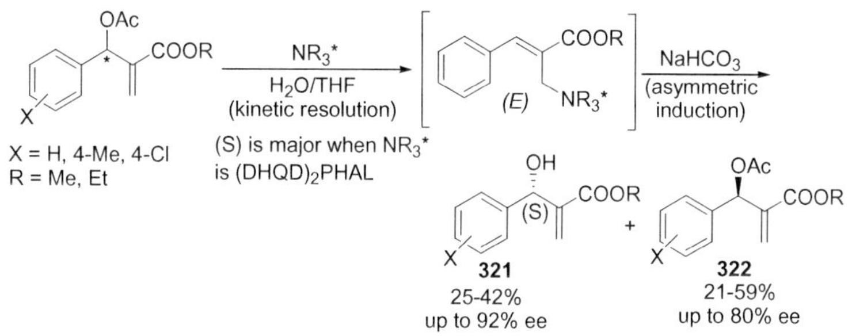

R^1 = H, Me; R^2 = Me, Ph, nPr, i-Pr, Ph, 2-ClPh;
R^1, R^2 = (CH$_2$)$_2$
R^3 = H, Me; R^4 = OMe, Me;
Ar = 4-MeOPh, 3,4-OCH$_2$OPh, 4-MePh, Ph, 3-MeOPh, 4-BrPh, 4-CH$_3$CO$_2$Ph

Scheme 3.140

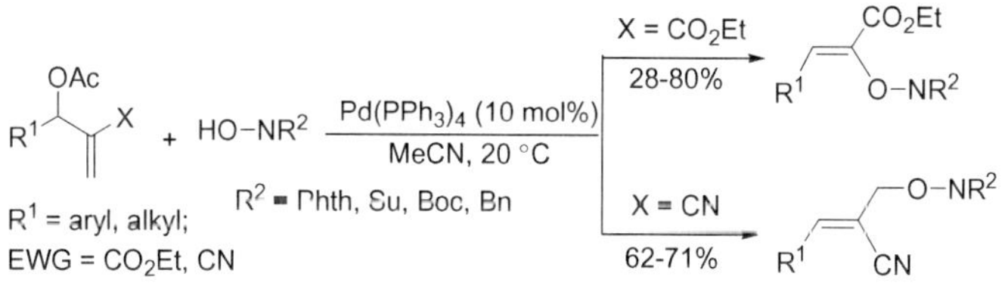

Scheme 3.141

with (DHQD)$_2$PHAL and the asymmetric induction during the S_N2' type reaction with sodium bicarbonate as a water surrogate (Scheme 3.140).[202]

Recently, Reddy *et al.* have described the regioselective Pd-catalyzed addition of hydroxylamine derivatives as the aminoxy equivalent of nucleophiles to MBH acetate adducts to afford the corresponding substituted allyloxy amine products with exclusive (*E*)-selectivity (Scheme 3.141).[203]

The FeCl$_3$-catalyzed nucleophilic substitution reaction of MBH adducts with alcohols has been reported by Chan[204] and by Jia *et al.*[205] Good to excellent yields (85–95%) and exclusive α-regioselectivities were obtained for cyclic MBH alcohols in the presence of FeCl$_3 \cdot$6H$_2$O as catalyst (Scheme 3.142).[204]

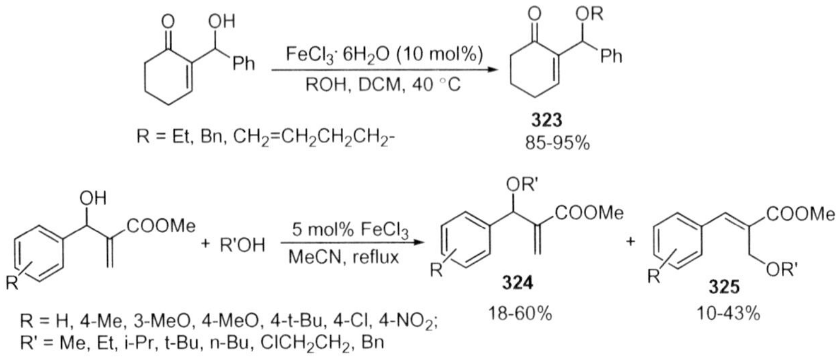

Scheme 3.142

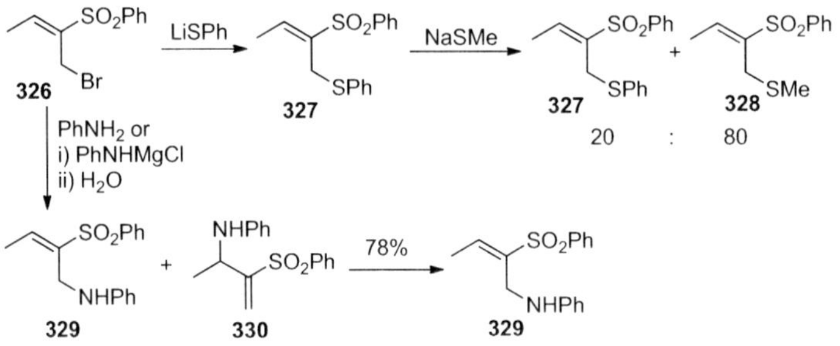

Scheme 3.143

However, for the MBH adducts derived from methyl acrylates and electron-rich aryl aldehydes, the corresponding products were obtained in good to excellent yields as mixtures of **324** and **325**, which can be easily separated by column chromatography (Scheme 3.142).[205]

3.8.4 Sulfur Nucleophiles

Knochel and co-workers[46a] have carried out a series of nucleophilic substitution reacion on (*E*)-allyl bromide using thiolate ions. The reaction between bromide **326** and lithium phenylthiolate yielded the (*E*)-sulfide **327** exclusively. However, treatment of (*E*)-sulfide **327** with the more nucleophilic reagent sodium methyl thiolate led to a 20 : 80 mixture of **327** and (*E*)-methyl sulfide **328**. They proposed that either a S_N2 mechanism is operative, or two S_N2' substitutions eventually resulted in the thermodynamically more stable sulfide (*E*)-methyl sulfide **328**. A similar reaction of **326** with aniline in excess or its chloromagnesium amide gave a mixture of S_N2 and S_N2' products **329** and **330** which slowly equilibrate to give S_N2 product **329** in 78% yield (Scheme 3.143).

Sodium sulfite and phenylsulfinate have been used as nucleophiles to produce the corresponding trisubstituted olefins **331** and **332**, respectively, when using bromo-MBH adducts as substrate.[40f,122,123,206] Compound **331** is an important synthon for orally active renin inhibitor A-72517 (Scheme 3.144).

Treatment of ethyl (*Z*)-2-benzylidene-3-bromopropionate with *t*-butylthiol provides ethyl (2*Z*)-2-benzylidene-3-(*t*-butylthio)propionate (**333**), which has been hydrolyzed and hydrogenated in the presence of chiral catalyst to produce (*S*)-2-benzyl-3-(t-butylthio)propionic acid (**334**) in good yields (Scheme 3.145).[207] This acid is a useful intermediate for inhibitors of renin and retrovirus protease.

Acetates of MBH adducts have also been established for the synthesis of sulfur-containing trisubstituted alkenes *via* a nucleophilic displacement protocol, in either S_N2 or S_N2' fashion with *S*-centered nucleophiles, such as thiolates,[208] arenesulfinate[209] and thiocyanate anions.[210] Significantly, these reactions are highly stereoselective for both acrylonitrile/acrylate ester-derived MBH acetates, but with a reversed stereochemical directive effect, that is, acrylonitrile-derived MBH acetates afford (*E*)-sulfur-containing trisubstituted alkenes **336** as the sole product, while acrylate ester-derived MBH acetates afford (*Z*)-olefins **335** exclusively under the same reaction conditions. Both (*Z*)- and (*E*)-allyl sulfides have been stereoselectively obtained from MBH acetates in a one-pot with treatment with benzenethiol in the presence of catalytic amounts of 15% aqueous NaOH and TBAI in DMSO at room temperature (Scheme 3.146).[208] The method has also been applied for the synthesis of (*Z*)-3-(4-methoxybenzylidene)thiochroman-4-one (**337**), a potent antifungal compound (Scheme 3.147).

Kabalka *et al.*[209] have reported the nucleophilic addition of sodium *p*-toluenesulfinate to an acetate of MBH adduct to afford trisubstituted olefins with (*Z*)-selectivity for acrylates and (*E*)-selectivity for acrylonitriles, and found that the reaction only proceeded in ionic liquids at high temperatures. Recently, Tian *et al.* have developed a catalyst-free alkylation reaction of sulfinic acids with *N*-(2-acyl) allylic sulfonamides *via* sp³ C–N bond cleavage to provide trisubstituted allyl sulfones with exclusive (*Z*)-selectivity (Scheme 3.148).[211]

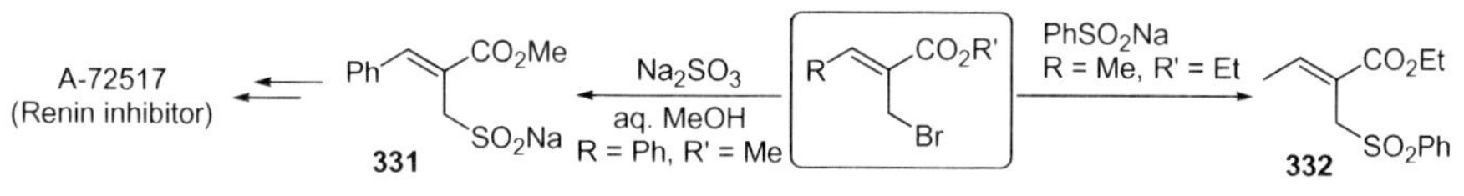

Scheme 3.144

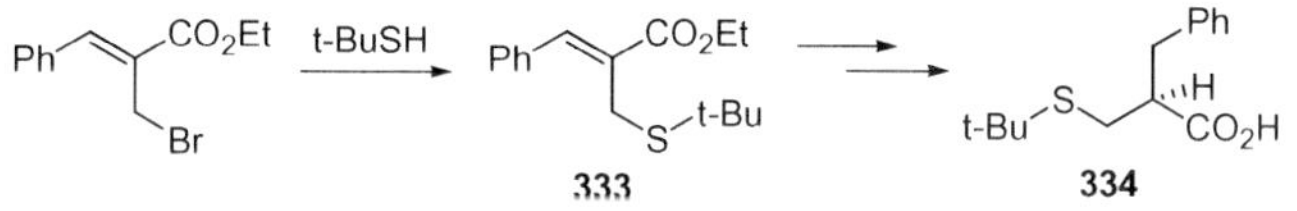

Scheme 3.145

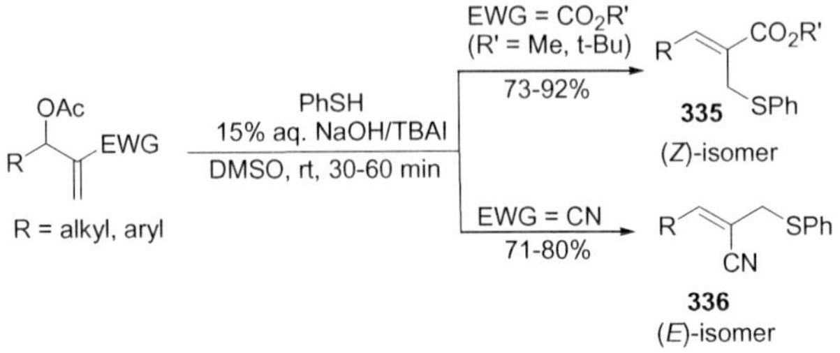

Scheme 3.146

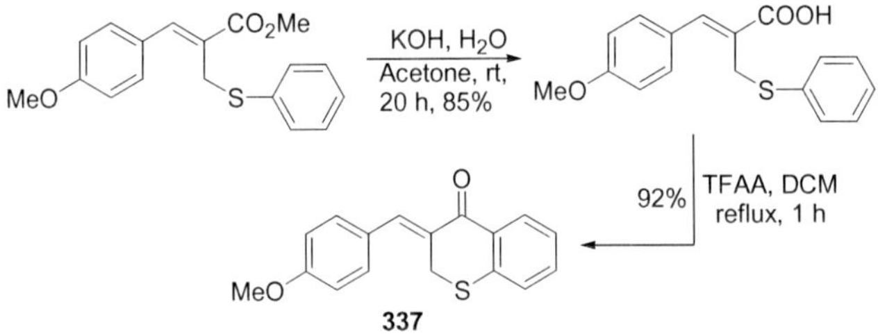

Scheme 3.147

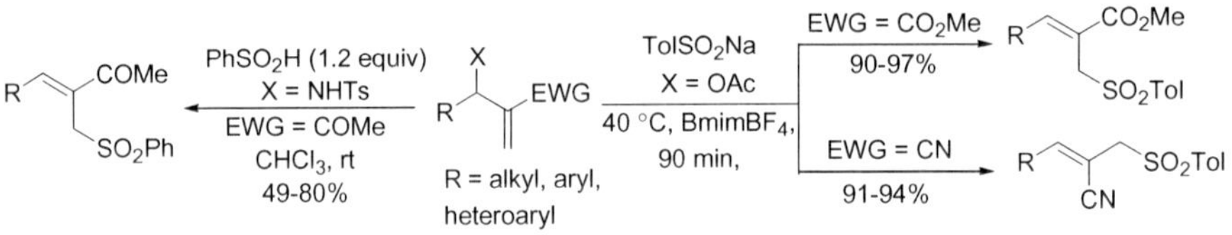

Scheme 3.148

It was found that intramolecular hydrogen bonding facilitates nucleophilic addition of sulfones to MBH adducts in a single step to acquire the substituted allyl sulfones[212] in poly(ethylene glycol) (PEG) as the solvent at high temperatures. However, Reddy *et al.*[213] have developed a general and practical synthesis of substituted allyl sulfones in both high yields and selectivity by the reaction of a MBH adduct with arylsulfonyl cyanide (Scheme 3.149).

A series of trisubstituted alkenes containing (Z)-allylthio moieties as the key structural units, that is, sodium (Z)-allyl thiosulfates **338**, symmetrical di[(Z)-allyl] sulfides **343** and di[(Z)-allyl] disulfides **339** and unsymmetrical diallyl sulfides **341** have been prepared in moderate to good yields *via* chemical transformations from the acetates of MBH adducts.[214] In addition, it was found that the Sm and a trace amount of I_2 could be used for the selective cleavage of the S–S or C–S bonds in sodium (Z)-allyl thiosulfates **338**, depending on the substituents (alkyl or aryl group), to give the corresponding

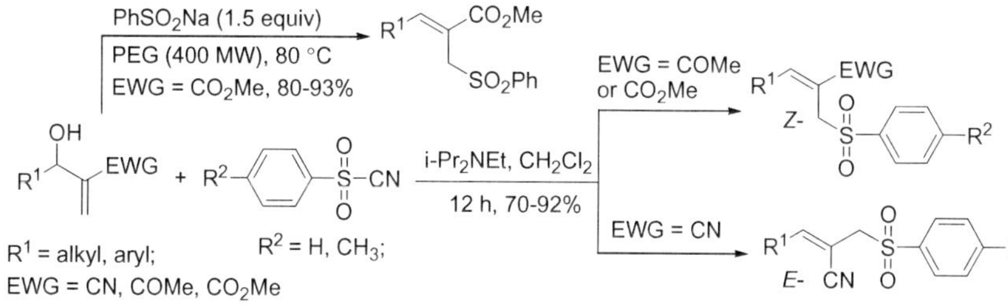

Scheme 3.149

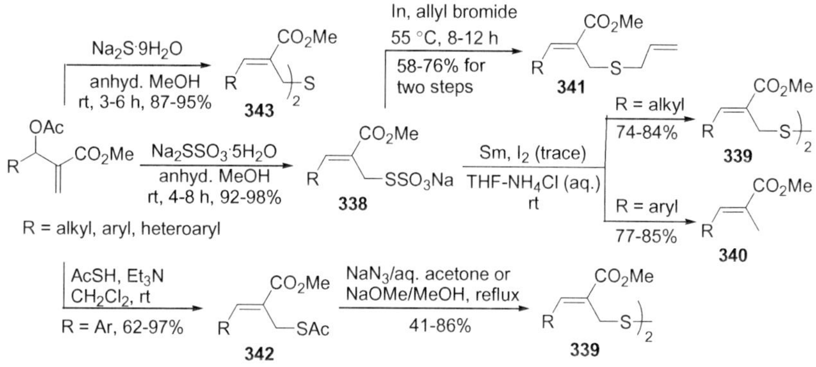

Scheme 3.150

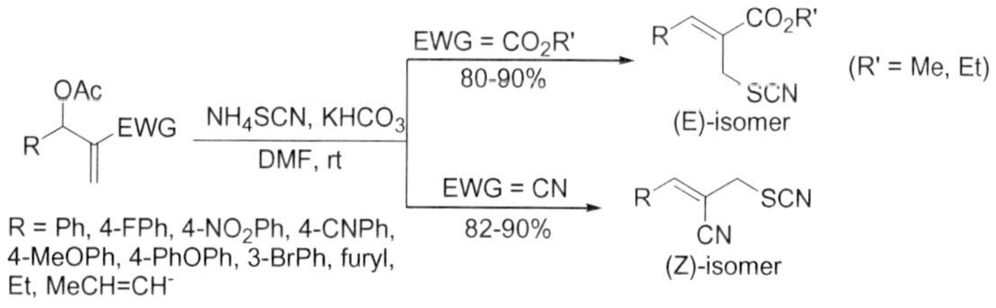

Scheme 3.151

di(Z-allyl) disulfides **339** or ($2E$)-methyl cinnamic esters **340**, respectively (Scheme 3.150).[214b,c]

Srihari *et al.* have examined the nucleophilic displacement of MBH acetates with ammonium thiocyanate using potassium hydrogen carbonate as base to afford (E)-3-substituted-2-(thiocyanatomethyl)acrylates and (Z)-3-substituted-2-(thiocyanatomethyl)acrylonitrile, respectively in high yields (Scheme 3.151).[210]

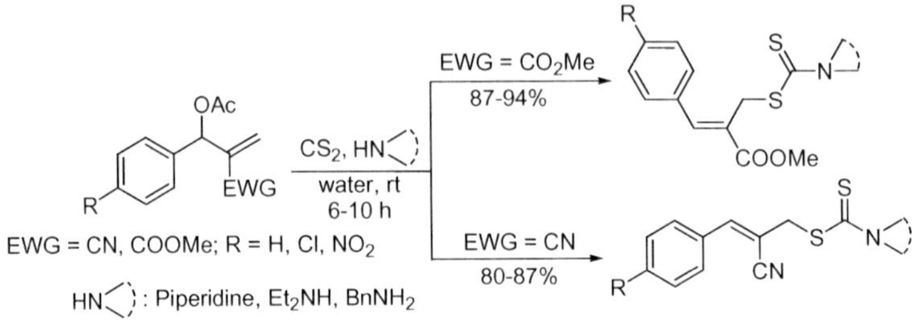

Scheme 3.152

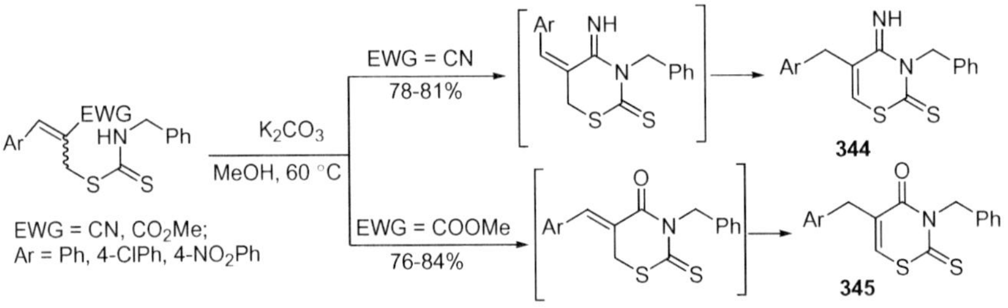

Scheme 3.153

Both (E)- and (Z)-allyl dithiocarbamates have been stereoselectively prepared in high yields from acetates of MBH adducts in catalyst-free one-pot three-component coupling reactions of carbon disulfide and amine in water under a mild and green procedure (Scheme 3.152). The reaction pathway involves the nucleophilic displacement (S_N2') of MBH acetates by dithiocarbamate anions. The utility of these allyl dithiocarbamates has been demonstrated in the synthesis of 3,5-dibenzyl-1,3-thiazines derivatives **344** and **345** (Scheme 3.153).[215]

Kaye and co-workers[216] have investigated the regio- and diastereoselectivity in the reaction of α-(1-hydroxyalkyl) acrylate derivatives with sodium methanethiolate. The hydroxy derivatives underwent conjugate addition with up to 66% de, whereas the acetoxy and bromo analogues underwent S_N2' and S_N2 reactions, respectively (Scheme 3.154).

3.8.5 Phosphorus Nucleophiles

Nucleophilic addition of triethyl phosphite to 3-acetoxy-2-methylenealkanenitriles and methyl 3-acetoxy-2-methylenealkanoates provides (2E)-2-(diethoxyphosphorylmethyl)alk-2-enenitriles and methyl (2Z)-2-(diethoxyphosphorylmethyl)alk-2-enoates, respectively, with good stereoselectivity (Scheme 3.155). Notably, nitriles provide allylphosphonates with high stereoselectivity ($E/Z = >90/<10$) particularly for aliphatic substituents (R = alkyl), thus complementing the work of Janecki and Bodalski.[10,217]

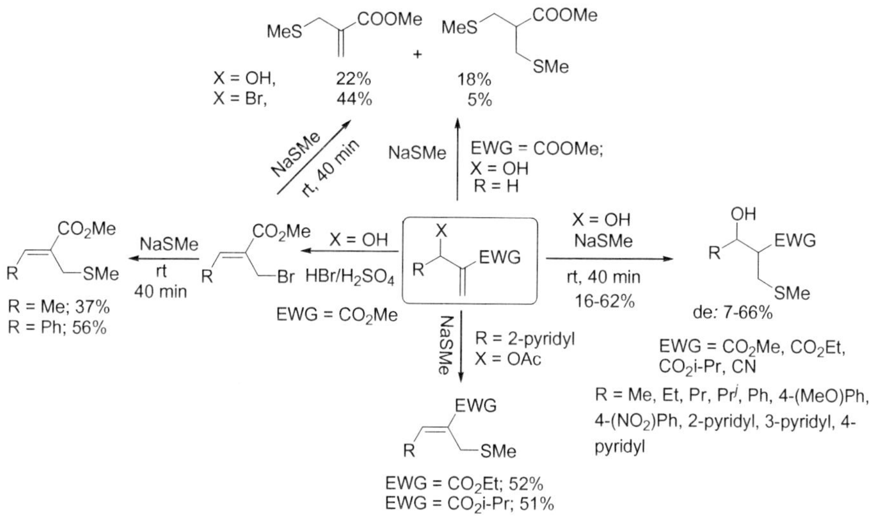

Scheme 3.154

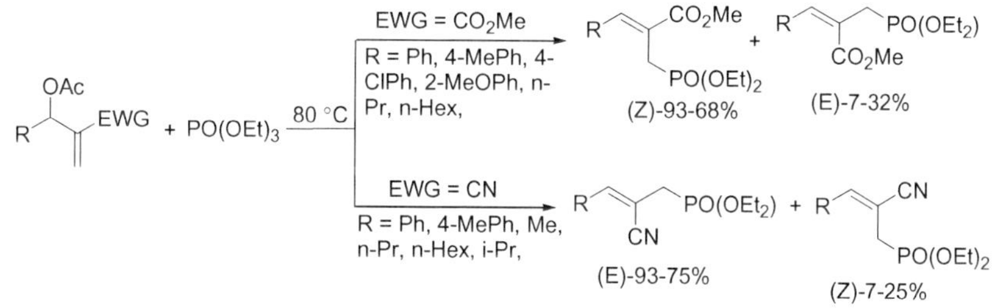

Scheme 3.155

Skowrońska and co-workers have performed the nucleophilic addition of a *P*-nucleophile such as trimethyl phosphite onto the cyclic MBH-type phosphonates **346** in boiling toluene to provide phosphonate **347** in 69% yield, whereas phosphonates **346** containing an additional alkyl or aryl substituent in the ring provided one regioisomer of the corresponding novel allylic phosphonate **348** in 70% yield, and allylic phosphonate **349** in 90% yield, exclusively, *via* an S_N2' process with *anti* stereochemistry (Scheme 3.156).[193]

Spilling *et al.* have elegantly transformed the acetates of MBH adducts obtained from acrylates into the corresponding allylic phosphonate **350** with high (*Z*)-selectivities. TFA cleavage of the *tert*-butyl ester and asymmetric hydrogenation of the unsaturated acid **351** yielded the phosphono-alkyl propanoic acid moiety **352**, which is commonly found in phosphonate- and phosphinate-based enzyme inhibitors (Scheme 3.157).[218]

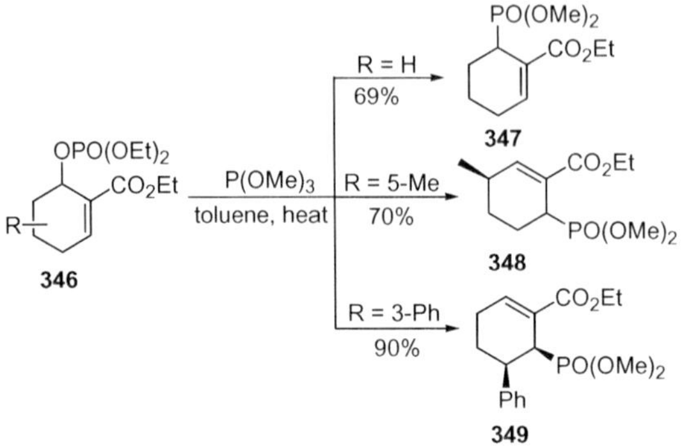

Scheme 3.156

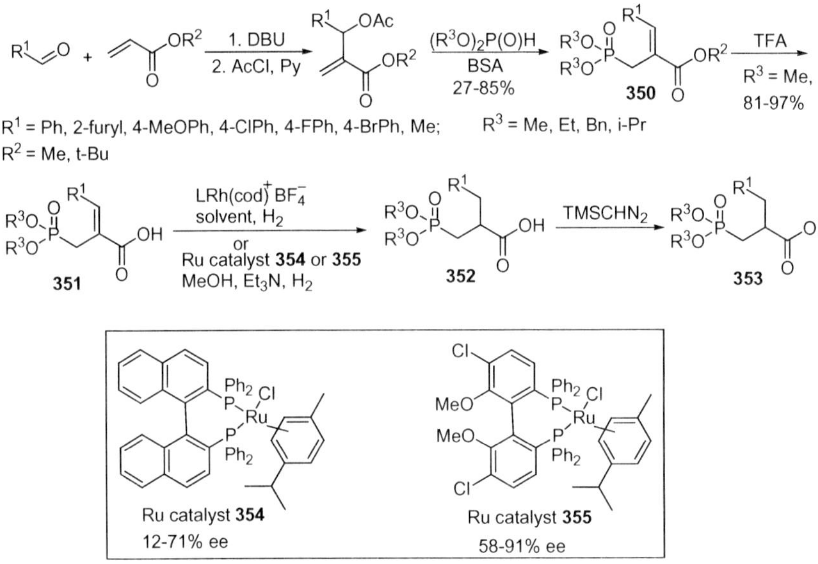

Scheme 3.157

Swamy has reported, in a ^{31}P NMR investigation, that the MBH adduct derived from methyl acrylate and benzaldehyde could react with [ClP-(μ-N-t-Bu)]$_2$ to lead to, mainly, the *cis*-alkoxy P(III) product **356** prior to rearrangement. However, upon heating the original reaction mixture at 150 °C for 20 min, only the *cis*-isomer **357** is formed quantitatively (Scheme 3.158).[219]

3.8.6   Other Nucleophiles

Hoffman and Rabe have successfully converted the α-methylene-β-acetoxy-alkanoates and 2-bromomethyl-2-alkenoates into (2E)-2-methylalkenoates and

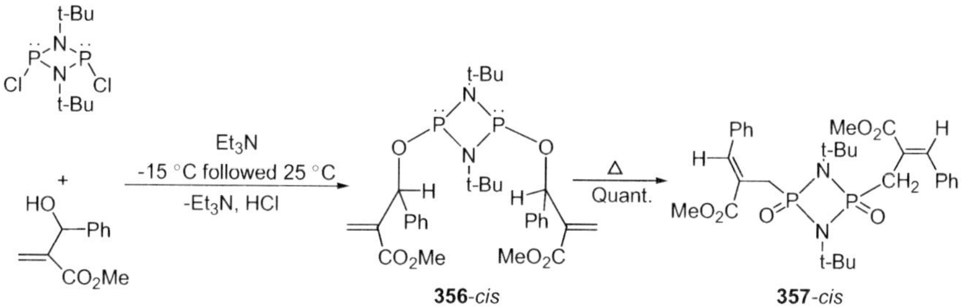

Scheme 3.158

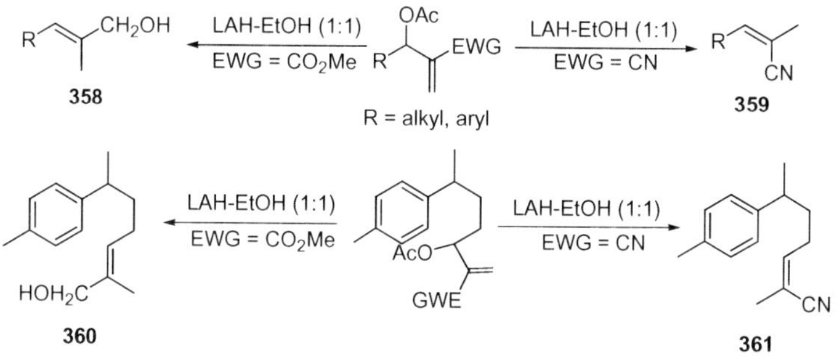

Scheme 3.159

2-methylenealkanoates, respectively, by treatment with LiEt$_3$BH.[42d,f] Later, Basavaiah *et al.* successfully reduced various MBH acetates with LiAlH$_4$.[220] The treatment of methyl α-methylene-β-acetoxyalkanoates with LAH–EtOH (1 : 1) gave exclusively (2E)-2-methylalk-2-en-1-ols **358**, while similar reaction of α-methylene-β-acetoxyalkanenitriles with LAH–EtOH led to remarkable reversal of stereochemistry of product, giving (2Z)-2-methylalk-2-enenitriles **359** (Scheme 3.159). The efficiency of this methodology was amply demonstrated by the synthesis of (E)-nuciferol (**360**) and a precursor (**361**) of (Z)-nuciferol (Scheme 3.159).[220]

As an alternative to LiAlH$_4$ reduction, acetates derived from various MBH adducts underwent reduction with HCOOH in the presence of Et$_3$N, Pd(OAc)$_2$ and dppe (or tri-isopropylphosphite) to yield the corresponding trisubstituted (Z)-olefins **362** in good yields with high regio- and stereoselectivity, without affecting the ester-group commonly present in the MBH products. The high (Z)-selectivity for aryl-containing MBH acetates was probably due to coordination between the aromatic substituent and the palladium intermediate, which stabilized the (Z)-conformation, whereas such coordination was not possible with MBH acetates containing aliphatic substituents (Scheme 3.160). Moreover, it was found the stereochemical outcome of reductions were *vis-à-vis* for LiAlH$_4$ reductions.[221]

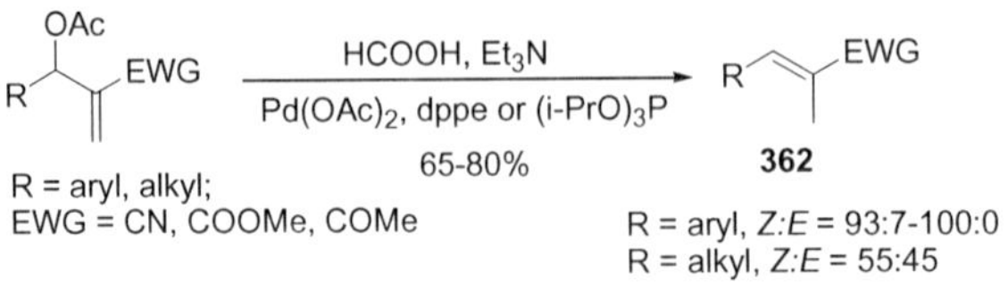

Scheme 3.160

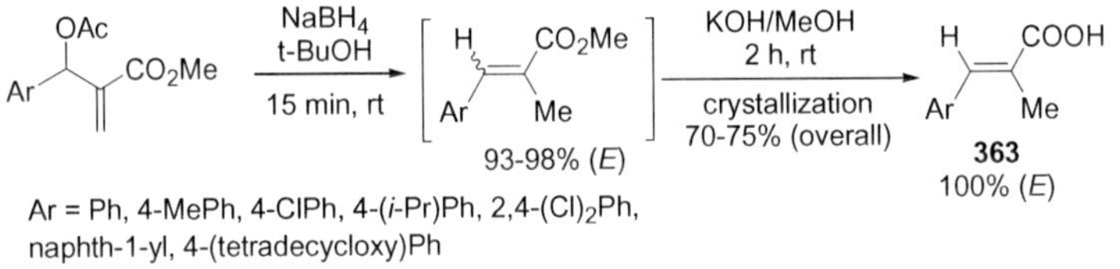

Scheme 3.161

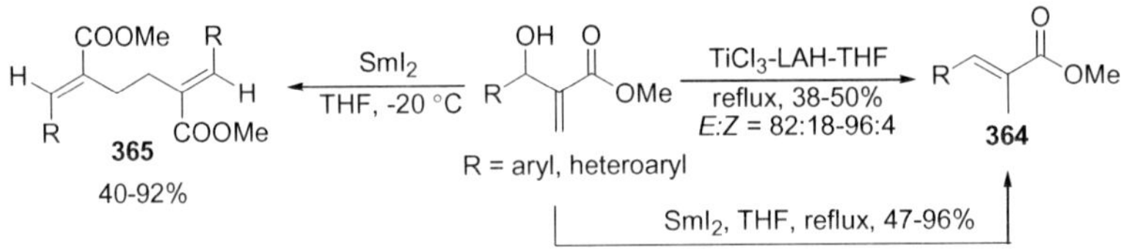

Ar = Ph, 4-MePh, 4-ClPh, 4-(*i*-Pr)Ph, 2,4-(Cl)$_2$Ph,
naphth-1-yl, 4-(tetradecycloxy)Ph

Scheme 3.162

Basavaiah *et al.* have described a simple and convenient synthesis of
(*E*)-α-methylcinnamic acids **363** *via* the nucleophilic addition of hydride ion
from NaBH$_4$ to the acetates of MBH adducts, followed by hydrolysis
and crystallization (Scheme 3.161). This methodology was utilized in the
synthesis of some bioactive molecules and precursor of some protease
inhibitors.[1b]

The MBH adduct shown in Scheme 3.162, prepared from aromatic aldehydes and methyl acrylate, can directly undergo smooth dehydroxylation with
concomitant olefin isomerization with low-valent titanium (LVT, prepared
from TiCl$_3$–LAH–THF) reagent to afford the trisubstituted alkenes **364** with
high (*E*)-selectivity. However, it is unsatisfactory in view of the low yield and
the purity of products.[222] More recently, Zhang *et al.* developed the samarium
diiodide-promoted hydroxyl elimination of MBH adducts to form trisubstituted alkenes **364** with total (*E*)-stereoselectivity in good to excellent yields.
This method also provided a new route to synthesizing a class of 1,5-hexadiene
derivatives **365** by temperature tuning (Scheme 3.162).[223]

MBH acetates were found to undergo a nucleophilic substitution reaction
when treated with magnesium bromide to produce the corresponding allyl
bromide, (*Z*)-**366** and (*E*)-**367**, with high stereoselectivity.[224] This reaction

could prove to be an attractive alternative for the synthesis of (*Z*)-allyl bromides **366** (Scheme 3.163).

Promoted by samarium metal in the presence of a catalytic amount of iodine, the MBH adducts underwent reductive elimination to form (*E*)-methylcinnamic ester derivatives **368**. However, allylic iodide derivatives, (2*Z*)-2-(iodomethyl)alk-2-enoates **369**, were obtained when the iodine was used in 1 : 1 ratio with metallic samarium. This method gave a new approach to the selective construction of stereo-defined trisubstituted alkenes with the simple Sm/I_2 system (Scheme 3.164).[225]

In the presence of samarium metal–trimethylsilyl chloride, (*Z*)-allyl selenides **370** were stereoselectively afforded by a one-pot reaction of diselenides with MBH adduct under mild conditions. Presumably, the diselenides are cleaved by the Sm–TMSCl system to form selenide anions, which then undergo S_N2 substitution of MBH adducts to produce the corresponding (*Z*)-allyl selenides **370** in high yields (Scheme 3.165).[226]

Kabalka *et al.* have investigated the Pd-catalyzed cross-coupling reaction of MBH acetates and bis(pinacolato)diboron to produce 3-substituted-2-alkoxycarbonyl allylboronates **371**,[227] which can be further transformed into stable allyl trifluoroborate salts **372** by addition of excess aqueous KHF_2.[227a] The allylboronate **371** and allyltrifluoroborate derivatives **372** react with aldehydes to afford functionalized homoallylic alcohols **373** and **374**, respectively, stereoselectively in the presence of Lewis acid (Scheme 3.166).[227a]

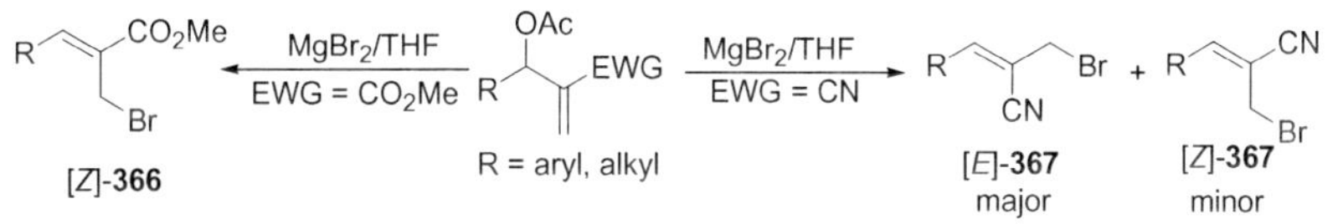

Scheme 3.163

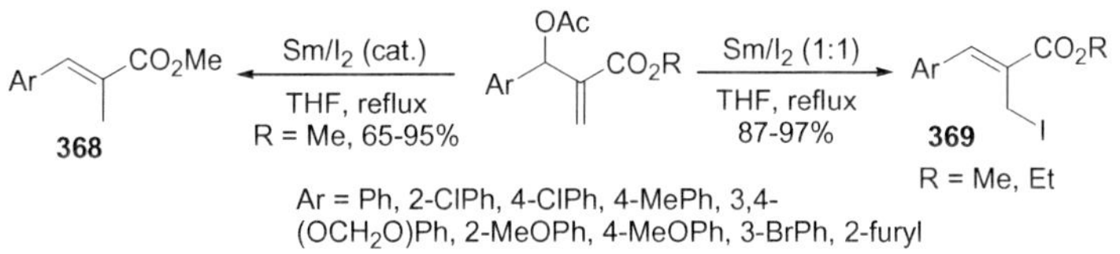

Scheme 3.164

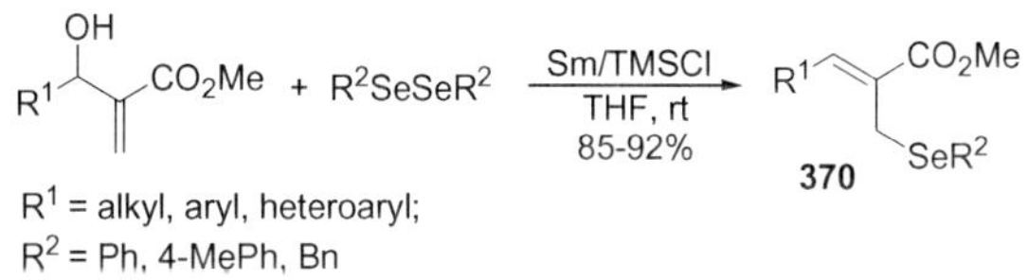

Scheme 3.165

Z = CO₂Me, COMe

Pd₂(dba)₃, 50 °C, toluene

371

BF₃·OEt or BF₃ on SiO₂

R¹CHO
Z = CO₂Me
56-84%

R = alkyl, aryl, heteroaryl;
R¹ = Aryl,

373

aq. KHF₂
56-72%

372

n-Bu₄I
(10 mol%)
89%

R = Ph, Z = CO₂Me
4-NO₂C₆H₄CHO

374

Scheme 3.166

376

Pd₂(dba)₃, toluene

R = alkyl, aryl, heteroaryl

Pd₂(dba)₃, toluene
X = Me, Ph

375

64-85%
Z:E = 92-100:8-0

64-87%
Z:E = 85-100:15-0

Scheme 3.167

In a continuation of their work, they also reported the cross-coupling reactions between MBH acetates and bimetallic reagents (Si-Si, Ge-Ge) catalyzed by a phosphine-free palladium complex. Substituted-2-carbonylallylsilanes **375** and allylgermanes **376** were isolated in high yields with regio- and stereoselectivity (Scheme 3.167).[228]

3.9 Radical Reactions

Free radical reactions of the MBH adducts have been well-studied by Giese and co-workers and by Kundig *et al.* Giese *et al.*[229] showed that the conjugate addition of radicals to either the silyl ethers **377** or 2-cyclohexyl-5-methylene-6-methyl-1,3-dioxane-4-ones **378** derived from the corresponding MBH adduct proceeds with high stereoselectivity (Scheme 3.168). Thus the silyl ethers **377** produced, after deprotection, the *erythro*-β-hydroxy esters **379** as major products along with the minor *threo*-isomers, whereas the 5-methylidene-1,3-dioxan-4-ones **378** yield only *threo*-β-hydroxy esters **381** (>99% de). Furthermore, Giese found that the stereoselectivity depended significantly upon the substituent at the radical center. A similar free radical addition on the corresponding silyl derivatives **382** (of the MBH adduct derived from acrylonitrile and phenyl vinyl sulfone) resulted in low selectivity (Scheme 3.169).[230]

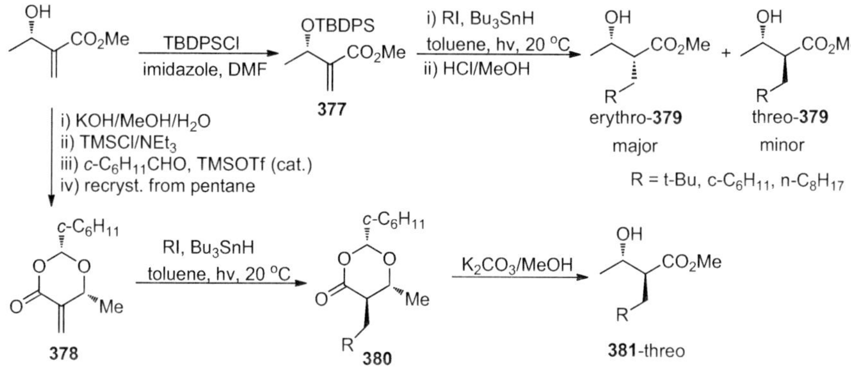

Scheme 3.168

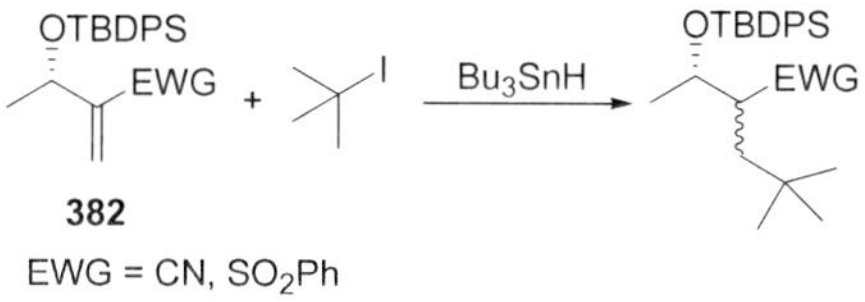

Scheme 3.169

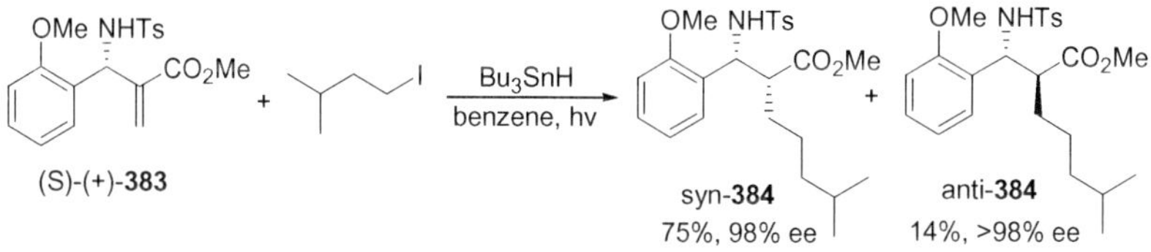

Scheme 3.170

Kundig and co-workers[231] have studied diastereoselective addition of alkyl radicals to enantiomerically pure aza-MBH adduct **383**, affording the *syn*-isomer **384** as major product (Scheme 3.170).

During their studies on Lewis acid controlled radical chemistry (chelation control), Guindon and Rancourt[232] have observed that the free radical reactions of methyl ethers of MBH alcohols **385** provide the corresponding products in high *syn*-diastereoselectivities in the presence of Lewis acid ($MgBr_2 \cdot OEt_2$), whereas a similar reaction provided *anti*-products predominately in the absence of Lewis acid. Scheme 3.171 shows a representative example.

Toru and co-workers[233,234] have elegantly demonstrated the role of intramolecular hydrogen bonding for diastereoselectivity as well as reactivity towards radical addition onto MBH adducts. Thus, the reaction of $(2S,S_S)$-α-(1-hydroxyethyl)vinyl sulfoxide **387** with alkyl radical and Bu_3SnH gave the

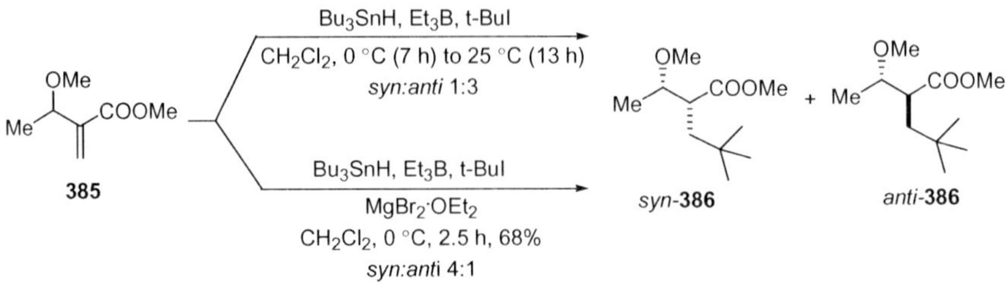

Scheme 3.171

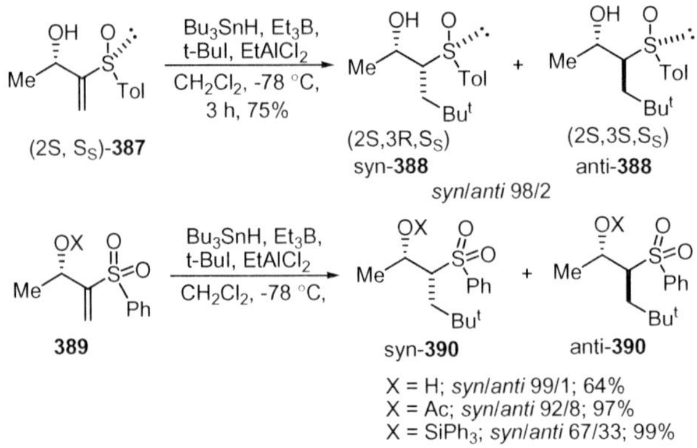

Scheme 3.172

addition–hydrogenation product **388** with high diastereoselectivities. However, no product was observed in the case of (2*R*,*S*s)-α-(1-hydroxyethyl)vinyl sulfoxide and *O*-protected (2*S*,*S*s)-α-(1-hydroxyethyl)vinyl sulfoxide.[233] Subsequently, they also used α-(1-hydroxyethyl) vinyl sulfone **389**[234] for free radical reactions with alkyl iodides, providing a *syn* adduct **390** predominately, and demonstrating the important role of intramolecular hydrogen bonding between the hydroxyl group and stereogenic sulfonyl oxygen for the stereoselectivity and reactivity (Scheme 3.172).

Mikami and co-workers have reported a new type of photochemical carbon skeletal rearrangement of α,β-unsaturated carbonyl compounds, which were prepared *via* MBH reaction, to 1,4-dicarbonyl compounds. The reaction was proposed to take place through the reaction pathway shown in Scheme 3.173. The reaction is initially irradiated to generate the radical **391**, which then forms the biradical **392** *via* 1,4-hydrogen abstraction by the assistance of an allylic hydroxy group in the photoexcited state of carbonyl compounds; in turn, this radical is coupled with **393**, which gives the important intermediate cyclopropanol **394**. The rearranged product **395** is eventually obtained by double tautomerization of enol and cyclopropanol portions in cyclopropanol **394**. This transformation can be regarded as the consequence of a high level of control

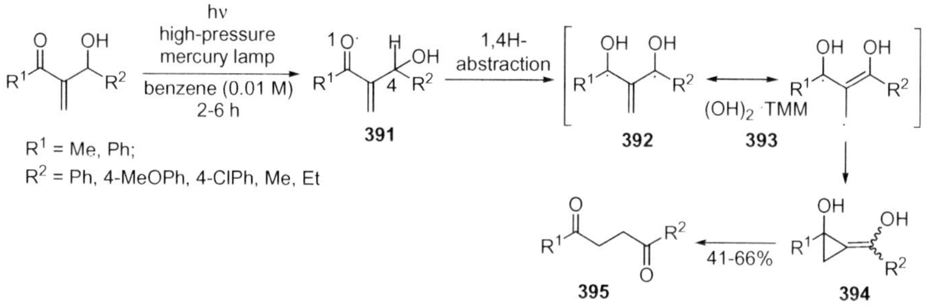

Scheme 3.173

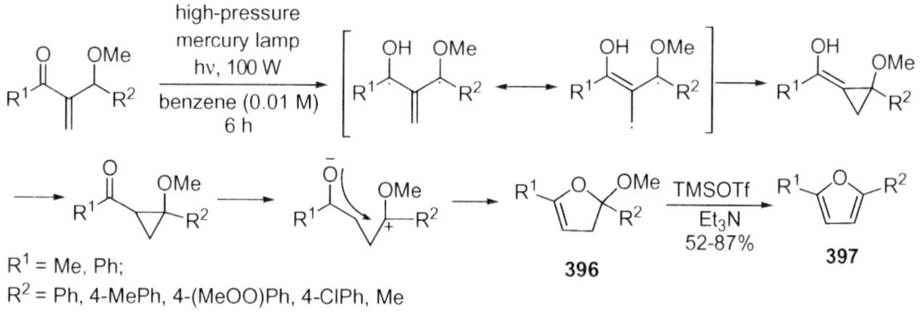

Scheme 3.174

over photochemical pathways by the introduction of an allylic alcohol "functionality" in the carbonyl substrates (Scheme 3.173).[235]

In the case of methyl ethers of MBH adducts, the reaction proceeds through β-C–H activation to provide the substituted dihydrofurans **396**, which have been subjected to *in situ* treatment with TMSOTf/Et$_3$N, affording substituted furan rings **397** in moderate to good yields (Scheme 3.174).[236]

A similar photoreaction of the MBH adduct derived from cyclohept-2-enone gives 1,4-diketones **398** and **399** without ring contraction, whereas the MBH adduct obtained from γ,γ-disubstituted cyclohex-2-enone gives 1,4-diketones **400** along with ring contraction under similar conditions (Scheme 3.175).[237]

Methyl ethers of vinylogous MBH adducts provide cyclopentene derivatives **401** under photochemical conditions, rather than cyclopropane (di-π-methane rearrangement) and dihydrofuran (β-hydrogen abstraction) derivatives (generally observed in non-vinylogous MBH products). Scheme 3.176 shows a representative example.[238]

Recently, the same group further investigated the enantiospecific carbon skeletal reorganization of β-substituted MBH adducts. Through triple binding by a C_2-symmetric chiral DPEN controller in a chiral γ-cyclodextrin (γ-CD) super cage in water, the chiral 1,4-dicarbonyl compounds **403** were obtained in

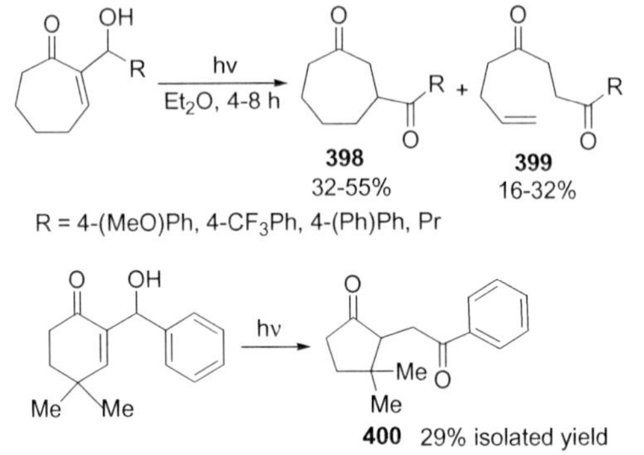

OH
R
hv
Et$_2$O, 4-8 h
R
R
398
32-55%
399
16-32%
R = 4-(MeO)Ph, 4-CF$_3$Ph, 4-(Ph)Ph, Pr
OH
hv
Me
Me
Me
Me
400 29% isolated yield

Scheme 3.175

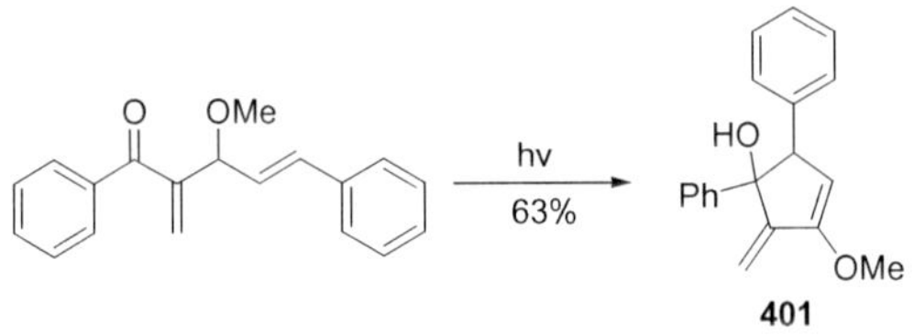

OMe
hv
63%
HO
Ph
OMe
401

Scheme 3.176

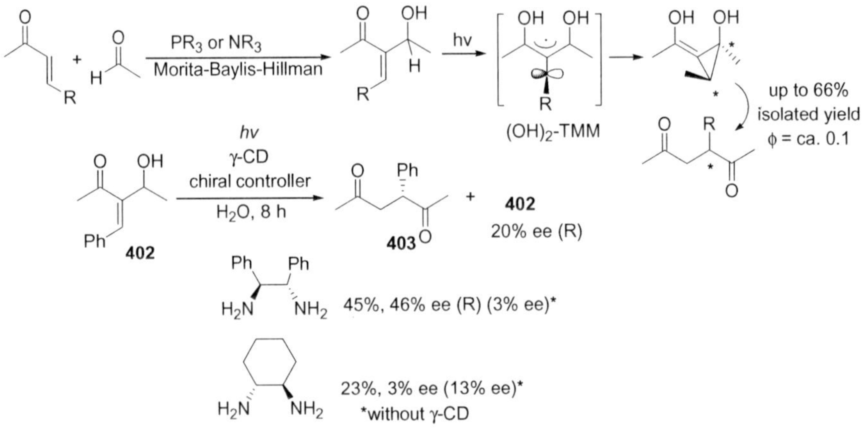

PR$_3$ or NR$_3$
Morita-Baylis-Hillman
hv
(OH)$_2$-TMM
up to 66% isolated yield
ϕ = ca. 0.1
hv
γ-CD
chiral controller
H$_2$O, 8 h
402
403
402
20% ee (R)
45%, 46% ee (R) (3% ee)*
23%, 3% ee (13% ee)*
*without γ-CD

Scheme 3.177

moderate yield with moderate enantioselectivity by photochemical synthesis (Scheme 3.177).[239]

Using titanocene(III) chloride (Cp_2TiCl) as the radical generator, the MBH adduct underwent smooth radical-induced condensation with activated bromo

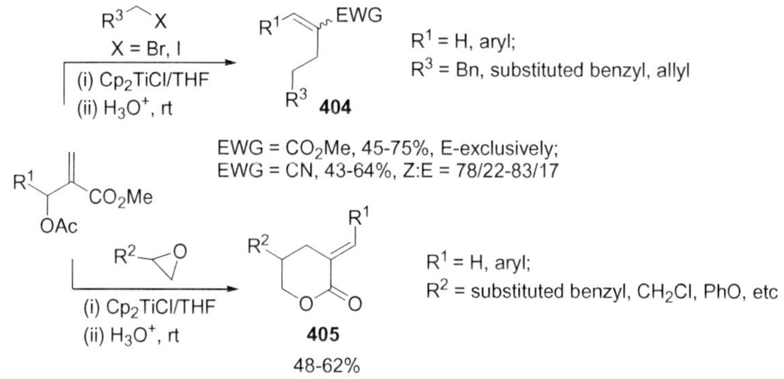

Scheme 3.178

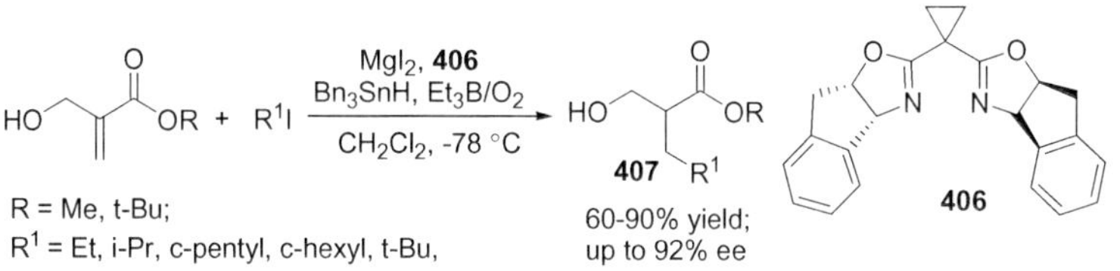

Scheme 3.179

compounds and epoxides. The reactions of activated bromo compounds with different electron-withdrawing group substituted MBH adducts afforded stereoselective trisubstituted alkenes **404**. Ester-containing adducts provided (*E*)-alkenes exclusively, whereas nitriles-containing adducts led to (*Z*)-alkenes as the major product. The reactions of epoxides with MBH adduct furnished α-methylene/arylidene-δ-lactones **405** in good yields *via* addition followed by *in situ* lactonization (Scheme 3.178).[240]

Using MgI_2–ligand **406** as the chiral Lewis acid, enantioselective radical alkylation of an MBH adduct furnished aldol products **407** in good yields and selectivities without protection of the hydroxyl group. An unexpected reversal in enantioselectivity is observed between methyl and *tert*-butyl esters, that is, the selectivity in the hydrogen atom transfer is dependent on the size of the ester substituent, with smaller substituents providing better enantioselectivity (Scheme 3.179).[241]

In addition, Chattopadhyay *et al.* have established a straightforward three-step synthetic route to dibenzo-fused nine-membered oxacycles **408** using sequential MBH reaction and radical cyclization on salicylaldehyde derivatives as described in Scheme 3.180. This methodology provides the possibility of obtaining synthetically challenging nine-membered dibenzo-heterocycles incorporating nitrogen or sulfur heteroatom.[242]

The allylic-allylic alkylation products **409** of α,α-dicyanoalkenes and MBH carbonates derived from methyl vinyl ketone (MVK) can undergo cyclization

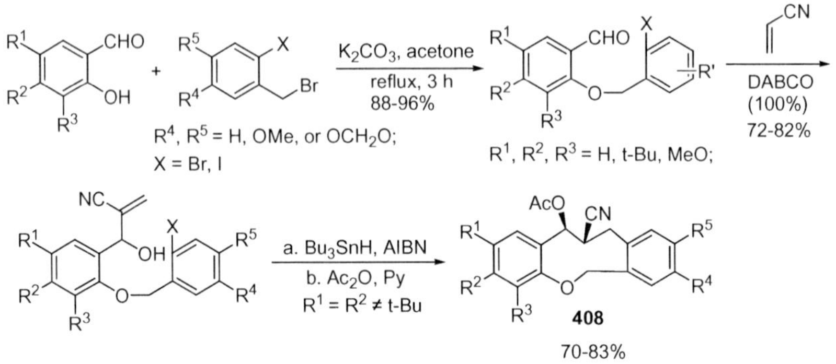

Scheme 3.180

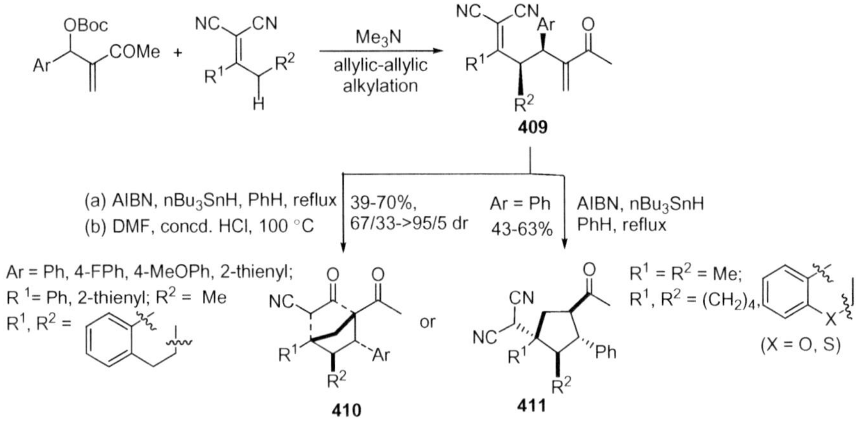

Scheme 3.181

reactions mediated by nBu$_3$SnH, affording cyclopentane derivatives **410** and **411** bearing multiple substituents with moderate to excellent diastereoselectivities (Scheme 3.181).[243]

3.10   Metathesis Reaction

Ring-closing metathesis (RCM) in MBH chemistry was initially reported by Paquette and Mendez-Andino for the synthesis of α-methylene-γ-lactones **412** fused to medium and large rings, starting from the MBH adduct, by the reaction sequence described in Scheme 3.182.[244]

Kim *et al.* have published a series of papers on the application of RCM for MBH derivatives to generate several heterocyclic scaffolds. Initially, they subjected the *O*- and *N*-allylic derivatives to RCM in the presence of Grubbs'

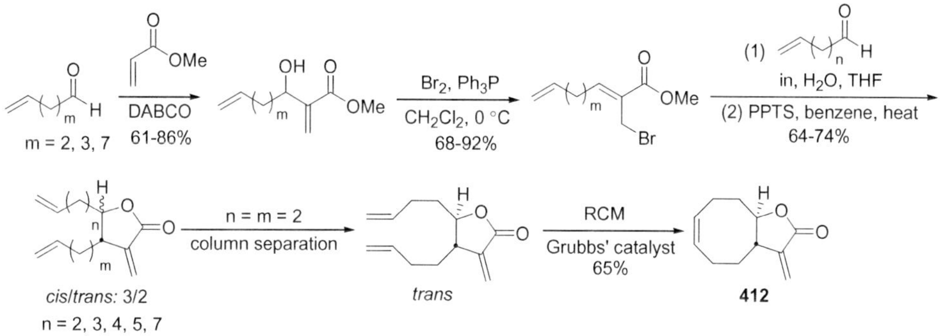

Scheme 3.182

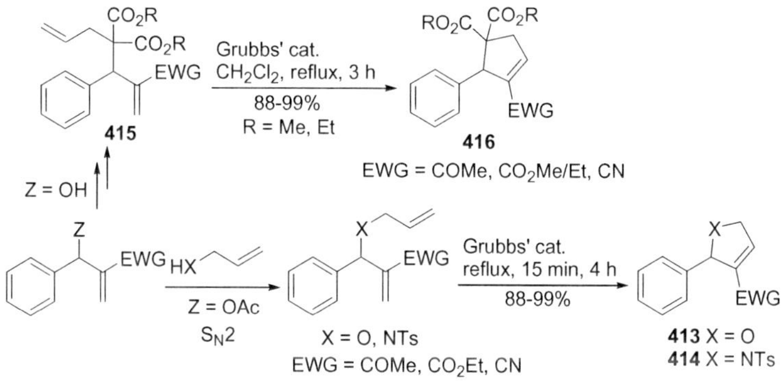

Scheme 3.183

second-generation catalyst to synthesize dihydrofurans **413** or dihydropyrroles **414** (Scheme 3.183).[245] Furthermore, they generated other allyl derivatives **415**, which served as useful synthons for the RCM reaction to yield cyclopentenes **416** in excellent yields (Scheme 3.183).[246] More recently, they have demonstrated the utility of RCM for the generation of furo[3,4-*c*]pyran **417** and pyrano[3,4-*c*]pyrrole rings **418** (Scheme 3.184).[247]

Balan and Adolfsson have reported the RCM of allyl amino derivatives **419**, originating from the aza-MBH products, in the presence of Grubbs' second-generation catalyst under microwave conditions to afford *N*-tosyldihydropyrrole **420** (Scheme 3.185).[248] Subsequently, Lamaty *et al.* demonstrated the synthesis of similar products (**421**) with an SES (2-trimethylsilyethyl sulfonyl) protecting group instead of a tosyl moiety. They showed that, in the presence of base, these dihydropyrroles (**421**) can be easily deprotected and aromatized to generate pyrroles **422** (Scheme 3.186).[249]

Later, in another report, they tested the reactivities of SES-protected pyrrolines and tosyl protected pyrrolines, respectively, for the preparation of

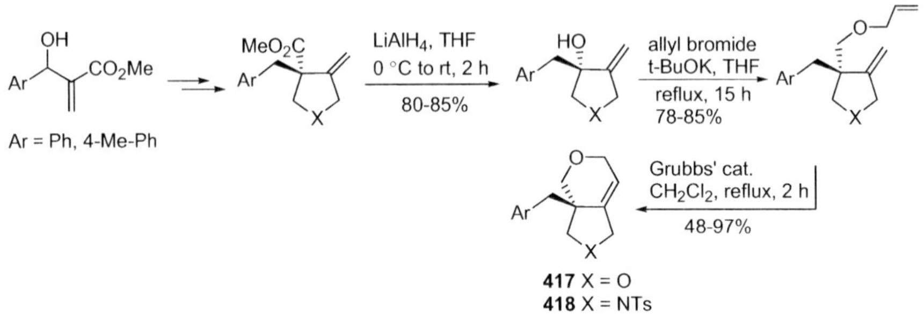

Ar = Ph, 4-Me-Ph

417 X = O
418 X = NTs

Scheme 3.184

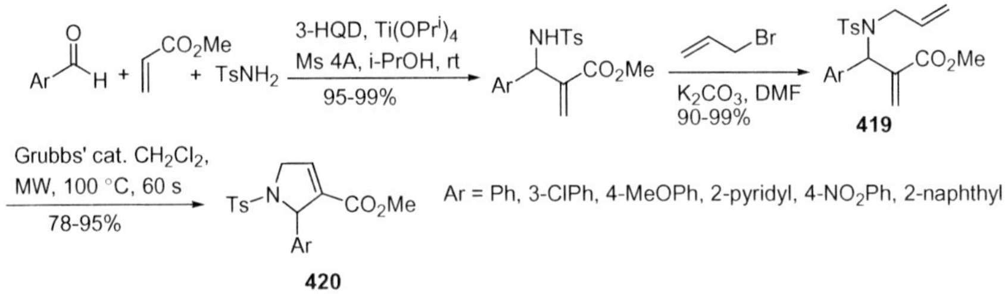

Ar = Ph, 3-ClPh, 4-MeOPh, 2-pyridyl, 4-NO$_2$Ph, 2-naphthyl

419

420

Scheme 3.185

421

422

Ar = Ph, 2-Br-Ph, 2-I-Ph, 3-F-Ph, 3,5-Me$_2$-Ph, 2-pyridyl

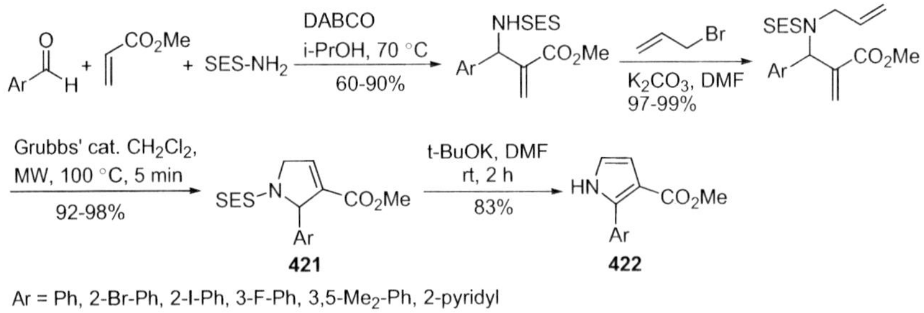

Scheme 3.186

nitrogen-containing five-membered rings obtained by the aza-MBH/alkylation/ RCM route. They observed that deprotection of tosyl-protected pyrrolines gave only pyrroles **422**, whereas deprotection of similar SES-protected derivatives furnished either pyrroles **422** or pyrrolines **423** depending on the deprotecting conditions. Indeed, they also hydrogenated the SES protected pyrrolines to yield pyrrolidines **424** with excellent diastereoselectivity (Scheme 3.187).[250]

Krafft *et al.* have described a new and efficient approach to functionalized hetero- and carbocyclic alkenols, which is a successful alternative route to the highly substrate dependent intramolecular MBH reaction. A quinuclidine-promoted MBH reaction of alkenyl aldehydes took place to generate the required

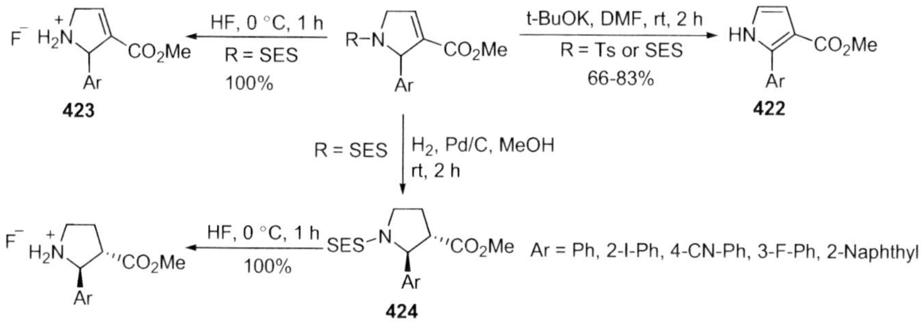

Scheme 3.187

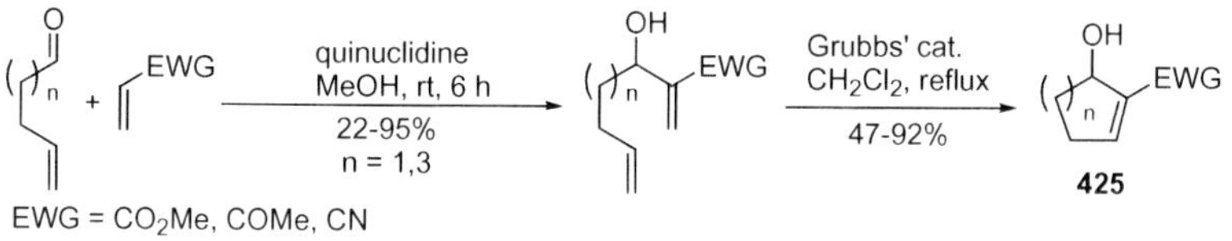

EWG = CO$_2$Me, COMe, CN

Scheme 3.188

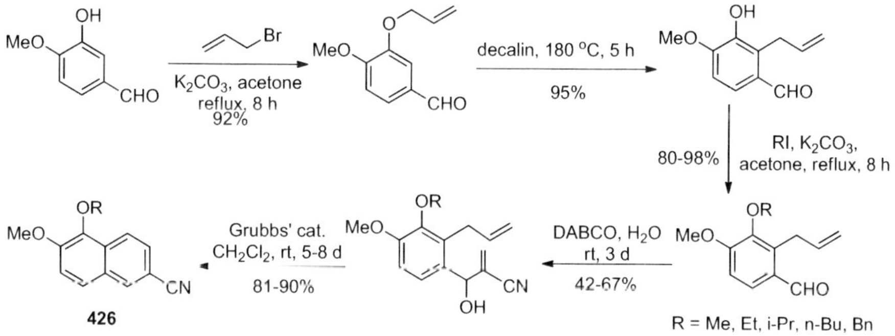

Scheme 3.189

synthon, which underwent an RCM reaction using Grubbs' second-generation catalyst, giving the cycloalkenols **425** in excellent yields (Scheme 3.188).[251] However, an attempt to generate eight- or nine-membered rings by this methodology failed, giving the dimerization product. More recently, Wang *et al.* have reported a novel, simple and eco-friendly method to provide cyanonaphthalenes **426** *via* a sequential Claisen rearrangement, MBH reaction and RCM in the presence of Grubbs' second-generation catalyst (Scheme 3.189).[252]

Sugar-based MBH adducts have been utilized efficiently to prepare a library of diverse α,β-unsaturated γ-lactones (**427** and **428**) by Krishna and Narsingam using the RCM reaction (Scheme 3.190).[253]

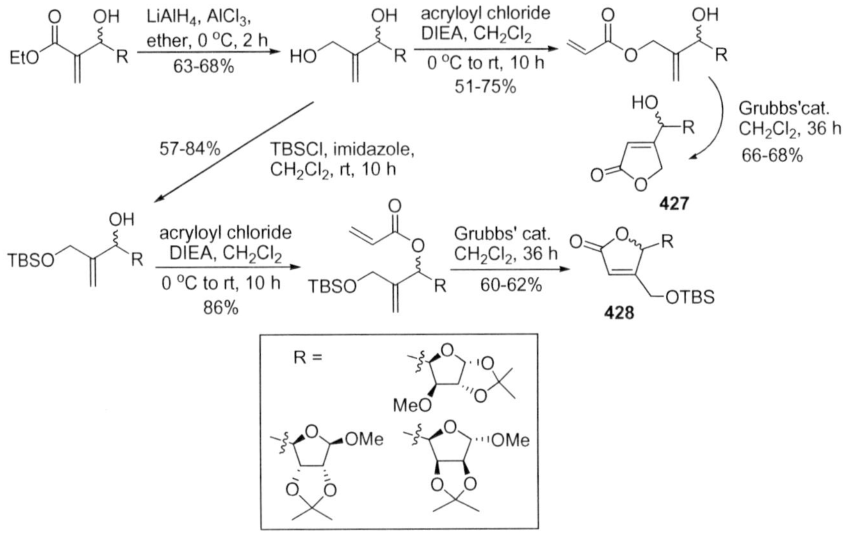

Scheme 3.190

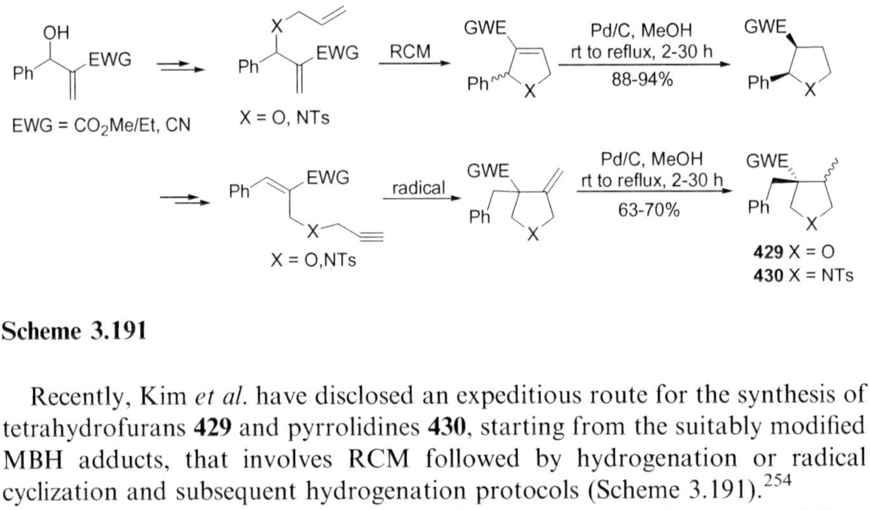

Scheme 3.191

Recently, Kim *et al.* have disclosed an expeditious route for the synthesis of tetrahydrofurans **429** and pyrrolidines **430**, starting from the suitably modified MBH adducts, that involves RCM followed by hydrogenation or radical cyclization and subsequent hydrogenation protocols (Scheme 3.191).[254]

Doddi and Vankar have described efficient syntheses of two pyrrolidine-based imino sugars, **431** and **432**, beginning from the MBH adduct of (*R*)-2, 3-*O*-isopropylideneglyceraldehydes. The key steps included regiospecific amination (*via* successive S_N2'-S_N2' displacement reactions), RCM and diastereospecific dihydroxylations (Scheme 3.192). These azasugars were reported to be moderate inhibitors of glycosidase.[255]

Cross-metathesis in MBH chemistry has rarely been reported. Lee and Kim have investigated the olefin metathesis from alkenyl MBH adducts using second-generation Grubbs catalyst.[256] Only the cross-metathesis (CM) products **434–436** were found, and no ring-closing metathesis (RCM) product **433** was

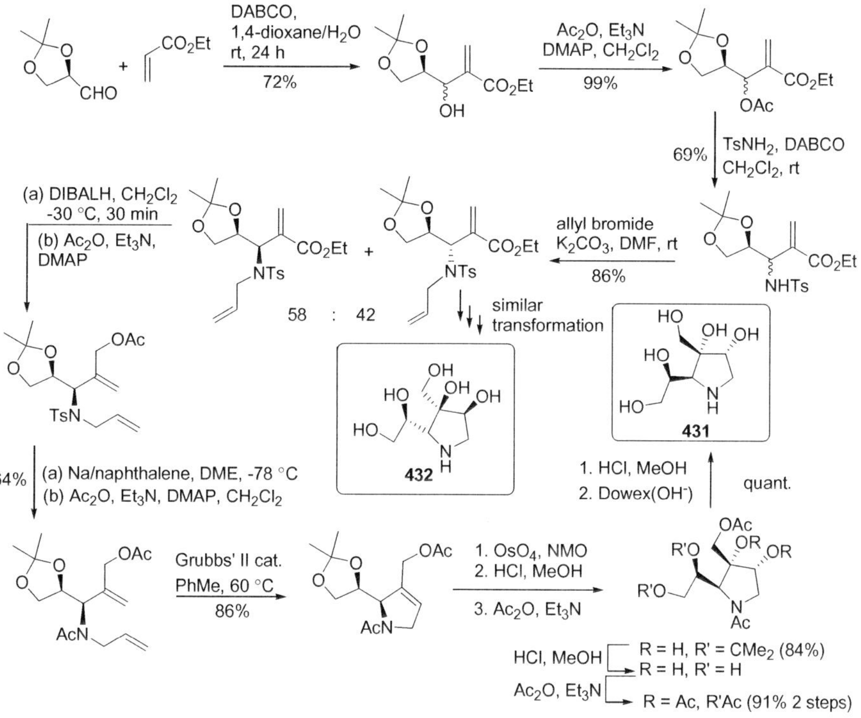

Scheme 3.192

detected, probably due to the high strain and steric effect in the metallacyclo-butane intermediates for RCM process using the second-generation Grubbs catalyst (Scheme 3.193). Computational studies provided consistent explanations for the experimental results.

The first examples of CM reactions of strained exocyclic enones have been reported by Howell *et al.*[257] CM reactions of α-methylene-β-lactone **437**, which was easily prepared from readily accessible MBH adducts, proceeded with high efficiency and diastereoselectivity. The unexpected predominant (*Z*)-stereo-chemistries of products **438** were observed, which was in contrast to the acquired high (*E*)-selectivity in the usual CM reactions of α,β-unsaturated carbonyl compounds, including 1,1-disubstituted enones, with simple alkenes. The obtained reaction outcomes were remarkable because of the strain inherent in the systems and the ring-opened MBH precursors did not undergo CM under similar conditions (Scheme 3.194).

3.11 Other Transformations

3.11.1 Epoxidation and Aziridination Reactions

The MBH adducts **441** and **442** were found to undergo *syn*-selective epoxidation under Sharpless epoxidation conditions, producing **444** and **445**, respectively.[258]

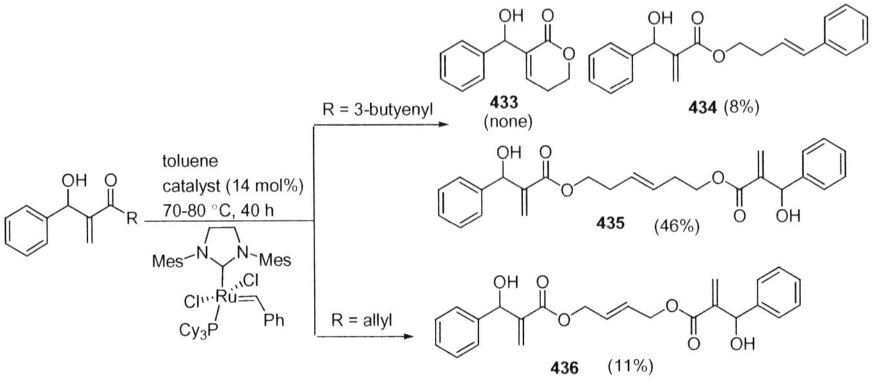

Scheme 3.193

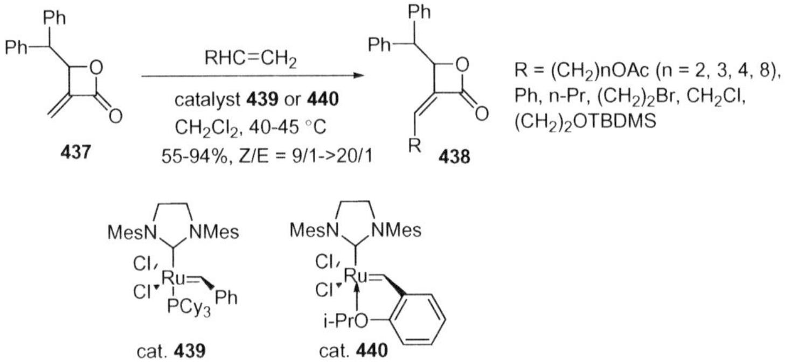

Scheme 3.194

However, there was no reaction with nitriles **443** (Scheme 3.195). The epoxidation was found to be less stereoselective (20–34%) under basic conditions (H_2O_2–NaOH, TBHP–NaOH). Such methodologies have been applied subsequently to the synthesis of the racemic upper-chain of clerocidin and terpenticin and to the asymmetric synthesis of (−)-mycestericin E, respectively.[258a,b,259]

Sodium hypochlorite converts the hydroxyl esters **446** and nitriles **447** into the corresponding keto epoxides **449** and **450** and the acetoxy ester **448** into acetoxy epoxide **451**, for which the diastereoselectivity was found to be 68% (Ar = Ph) and 48% (Ar = 4-ClC_6H_4) (Scheme 3.196).[260] The keto-epoxides derived from ester-containing MBH adducts **449** can be further transformed into 2-aroyloxypropenoic acid methyl esters **452** in the presence of triphenylphosphine.[261]

Lithium *t*-butyl peroxide has been used for stereoselective epoxidation of sulfones and their silyl derivative. While the free alcohols gave exclusively the *syn*-epoxides, the silyl ethers produced *anti*-epoxides as major products (Scheme 3.197).[262]

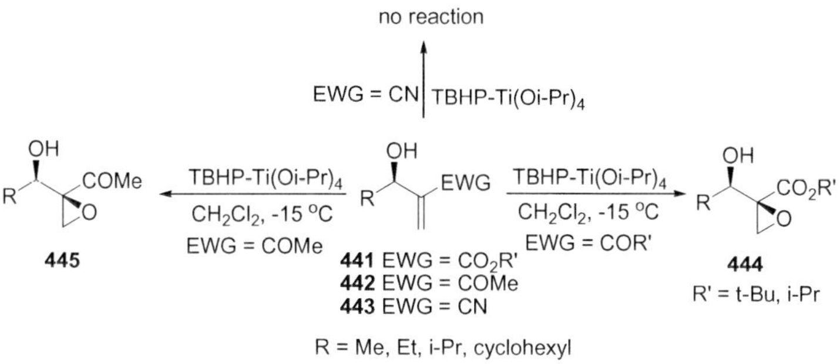

Scheme 3.195

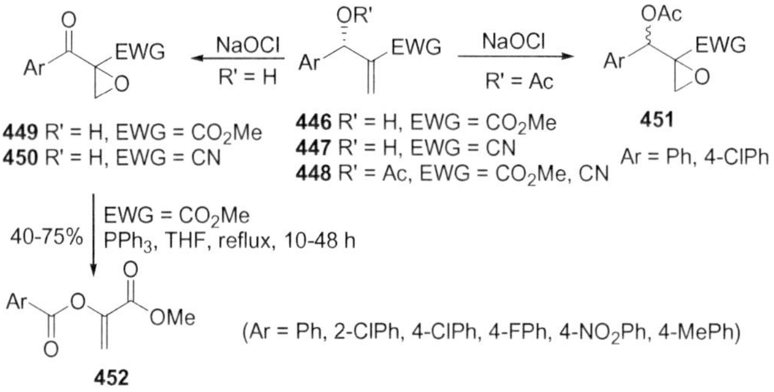

Scheme 3.196

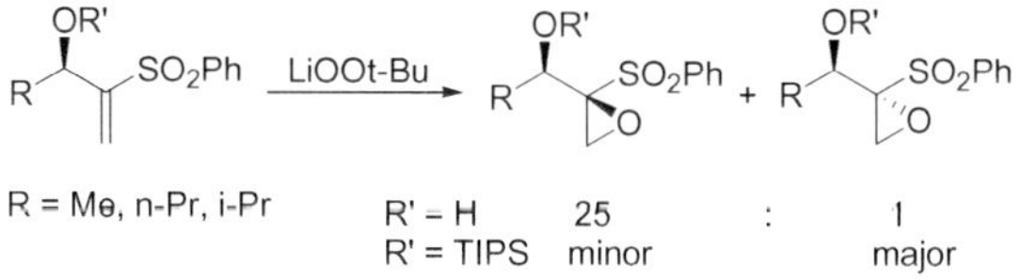

Scheme 3.197

Stereoselective aziridination of 3-hydroxy-2-methylene-4-methylpentanoate and its acetate using 3-acetoxyaminoquinazolin-4(3*H*)-one is described in Scheme 3.198.[263] However, the yields of the resulting aziridines are very poor.

MBH adducts have been transformed into acyloxiranes **453** by using iodosobenzene activated by a catalytic amount of KBr in water at room temperature. Iodosobenzene has been utilized here for two-fold oxidation of a secondary alcohol of a MBH adduct followed by subsequent epoxidation of the generated enone in a one-pot synthesis of an acyloxirane (Scheme 3.199).[264]

Complementarily to the *syn* diasteroselectivity obtained in the direct epoxidation of MBH adducts, mono-TBS protected allylic diols **454** derived from

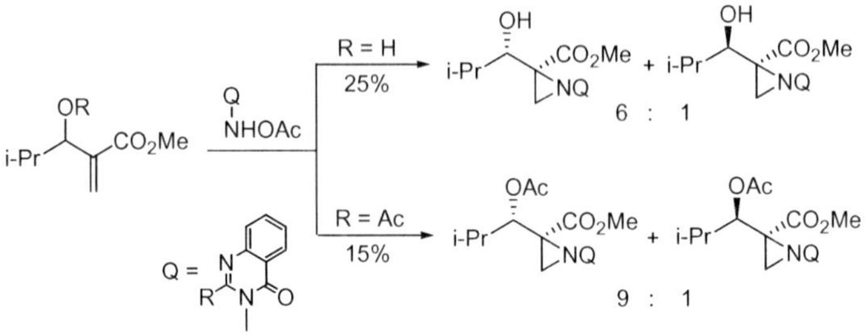

Scheme 3.198

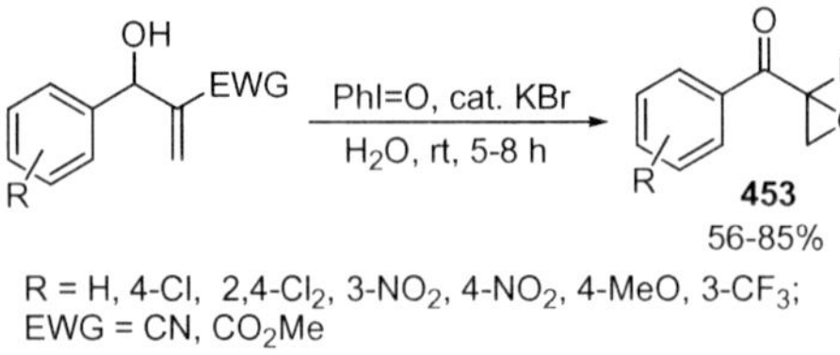

Scheme 3.199

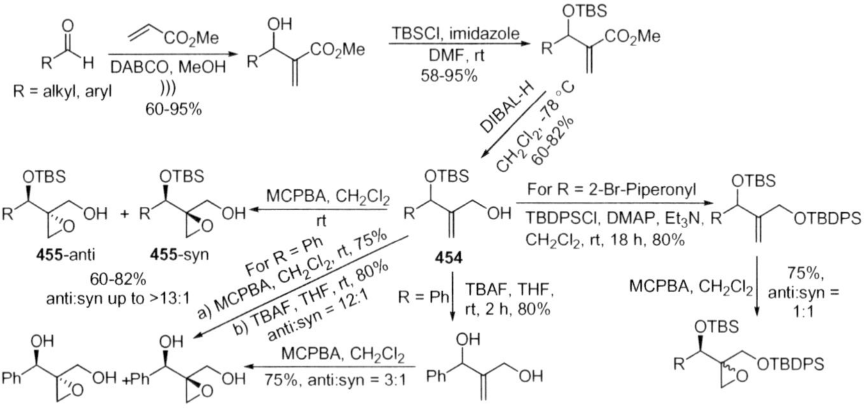

R = H, 4-Cl, 2,4-Cl$_2$, 3-NO$_2$, 4-NO$_2$, 4-MeO, 3-CF$_3$;
EWG = CN, CO$_2$Me

Scheme 3.200

MBH adducts undergo a highly *anti* diastereoselective epoxidation in the presence of MCPBA to give the corresponding adducts **455** in good yields (Scheme 3.200). It seems that the formation of an intramolecular hydrogen bond is responsible for the high *anti* diastereoselection.[265]

Yadav first reported a one-pot oxidative *anti*-Markovnikov bromohydroxylation and bromoalkoxylation of MBH adducts, accomplished at room temperature using LiBr as the bromine source and 2-iodoxybenzoic acid (IBX) as the oxidant. The process involves oxidation of MBH adducts with IBX to give β-ketomethylene compounds *in situ*, which undergo highly regioselective vicinal functionalization with $LiBr/H_2O$ or $LiBr/ROH$ to afford α-bromo-β-hydroxy **456** or α-bromo-β-alkoxy compounds **457**, respectively, in excellent yields. The α-bromo-β-hydroxy compounds **456** are readily transformed into epoxides **458** in aqueous NaOH solution (Scheme 3.201).[266]

A ligand-independent, osmium-catalyzed aminohydroxylation of MBH adduct has been developed by Sharpless *et al.* (Scheme 3.202).[267] The yield, rate and selectivity of this reaction are not affected by the addition of cinchona alkaloid ligands. The diastereoselectivity for the aminohydroxylation is influenced by the aldehyde-derived substituent, while the acrylate-derived substituent has a minimal effect. Various derivatives and close analogs of the MBH product-core failed to aminohydroxylate, emphasizing the unique reactivity of this class of olefins.

3.11.2 Michael Addition Reactions

Unmodified MBH adducts have been employed as Michael acceptors to react with oxygen, sulfur, nitrogen and carbon-centered nucleophiles, affording functionalized aldol products. Triphenylphosphine-catalyzed Michael addition of oximes onto MBH adducts provides easy access to a novel class of oxime functionalized aldol products (**459**). This demonstrates the first use of an oxygen-centered nucleophile in a Michael addition to MBH adducts, without touching any other functional group. Deprotection of oxime in **459**, by using molecular hydrogen (1 atm) and 10% Pd/C (cat.), afforded functionalized 1,3-diols **460** as potentially

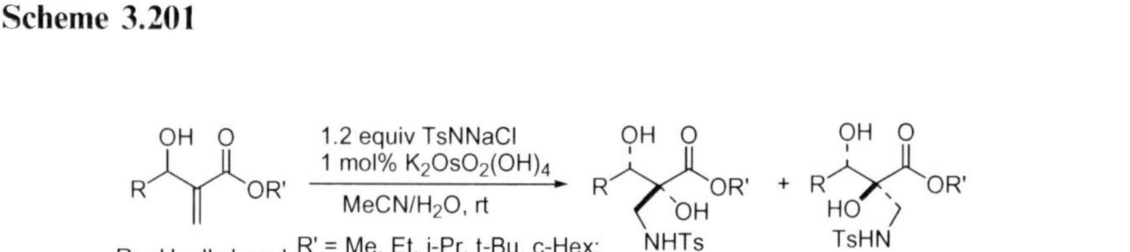

Scheme 3.201

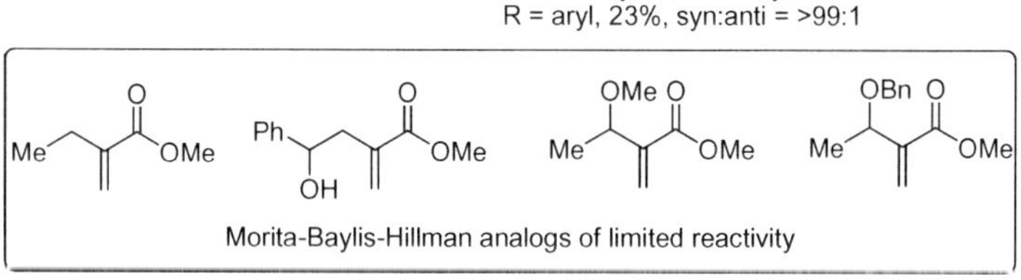

Scheme 3.202

useful synthons with optional choice of backbone.[268] Notably, when an oxime is allowed to react with a MBH adduct in the presence of the Lewis acid $CuBr_2$, isoxazoline **461** was obtained as two diastereomers in 32% overall yield. This result was otherwise expected as the 1,3-dipolar cycloaddition of MBH adduct with nitrile oxide, generated *in situ* from the respective oximes using NaClO as described before (Scheme 3.203).[269]

Kamimura and co-workers[270] have developed a diastereoselective methodology (up to 99%) for the synthesis of *syn*-β-hydroxy-α-thiomethyl carbonyl compounds *via* nucleophilic addition of ethanethiol (EtSH) to the TBS ether of MBH adducts in the presence of catalytic amounts of lithium thiolate (EtSLi). These adducts were further successfully transformed into β-lactams **462**. However, when a similar reaction was extended to the TBS ether of a MBH adduct obtained from acrylonitrile and acetaldehyde, the stereoselectivity was lost (Scheme 3.204).

Iminium salt **463**, prepared *in situ* from the aldehyde and an excess of the (trimethylsilyl)dialkylamine (2.5 equiv.) in a concentrated ethereal solution of

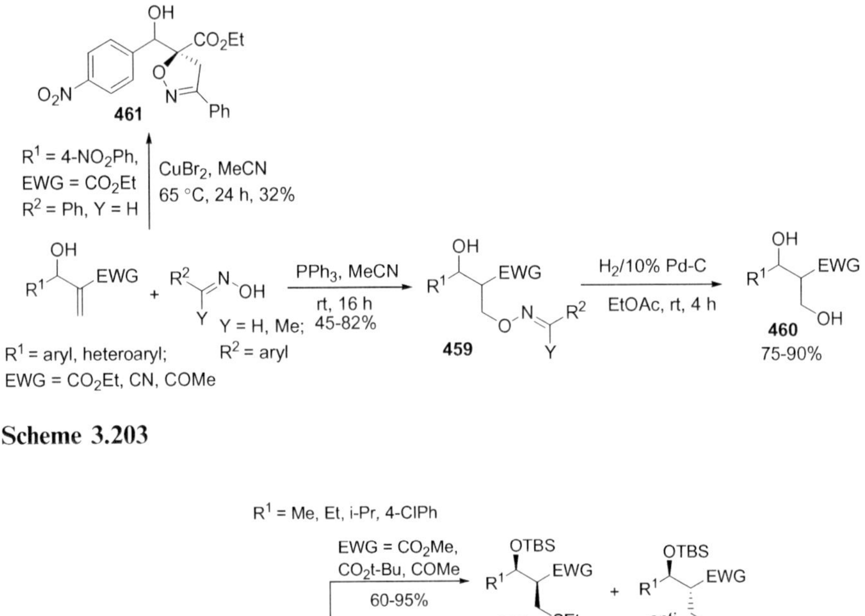

Scheme 3.203

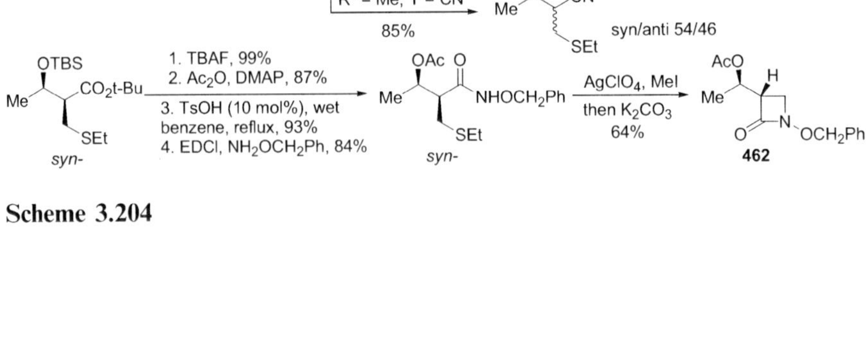

Scheme 3.204

lithium perchlorate, can react with methyl acrylate to form MBH adducts **464** in the presence of a catalytic amount of a tertiary amine such as 1,5-diazabicyclo[5.4.0]undec-5-ene (DBU). Conjugate addition of the (trimethylsilyl)dialkylamine to the MBH adduct **464** *in situ* then afforded diamine **465** in good yield (Scheme 3.205).[271]

3-Substituted-phenyl-5-isoxazolecarboxaldehydes have been identified as activated aldehydes for the generation of isoxazole-based combinatorial libraries in the solid (Scheme 3.206)[272] and solution phase (Scheme 3.207).[273] The highly functionalized isoxazole-based libraries have been synthesized in parallel format using the MBH reaction and Michael addition. With an objective of lead generation, the libraries have been evaluated for their biological activities *in vivo*.

Wang *et al.* have developed a new one-pot procedure for the sequential Morita–Baylis–Hillman and Michael additions to construct two carbon–carbon bonds with aromatic aldehyde, α,β-unsubstituted acrylate and a methide nucleophile catalyzed by DBU. This one-pot procedure will provide a useful tool for generating highly functionalized and diversified organic molecules (Scheme 3.208).[274]

Recently, oxidative conjugate addition of carbon, sulfur and phosphine-centered nucleophiles to MBH adducts has been reported. MBH adducts can

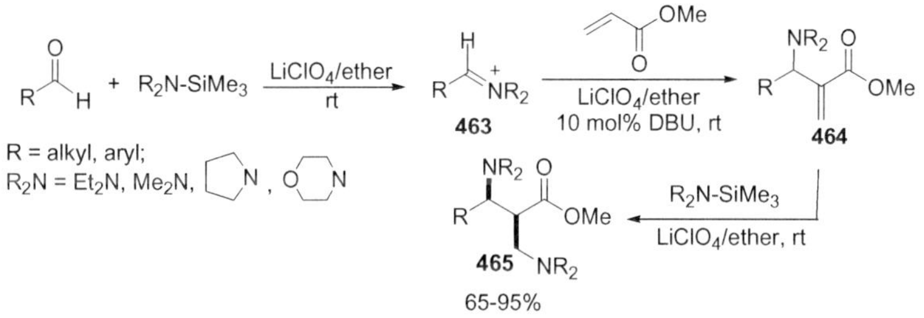

Scheme 3.205

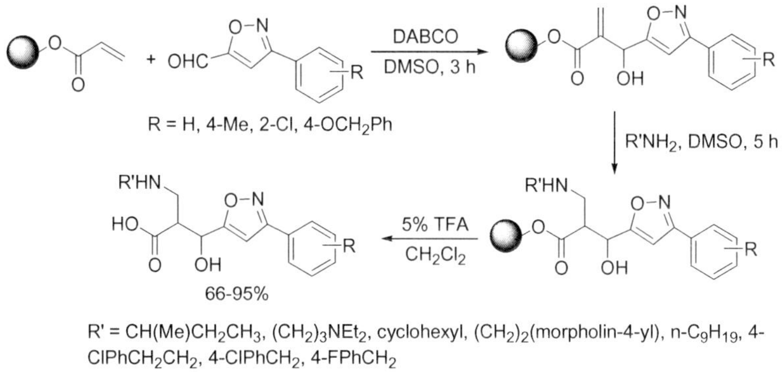

Scheme 3.206

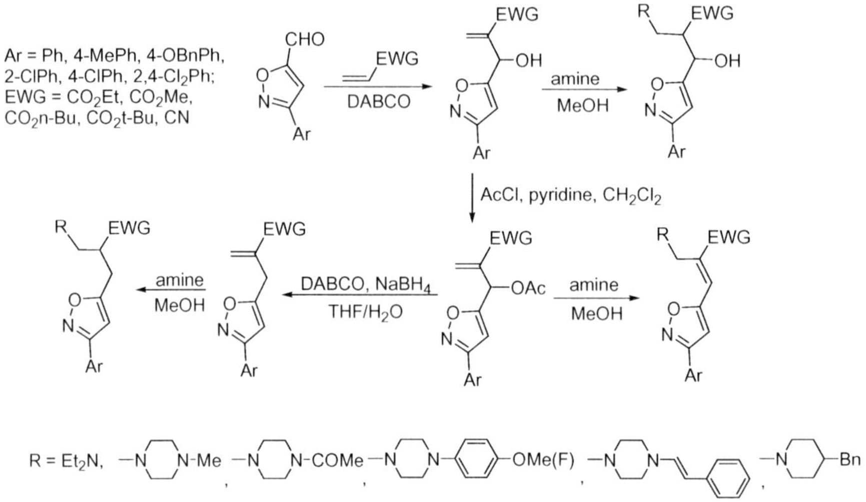

Scheme 3.207

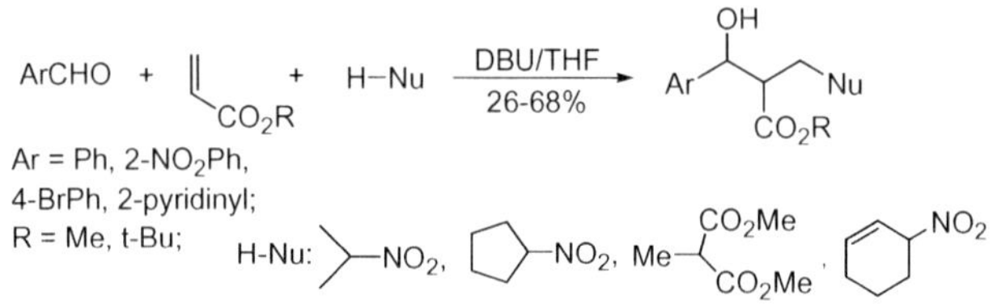

Ar = Ph, 4-MePh, 4-OBnPh, 2-ClPh, 4-ClPh, 2,4-Cl$_2$Ph;
EWG = CO$_2$Et, CO$_2$Me, CO$_2$n-Bu, CO$_2$t-Bu, CN

R = Et$_2$N, —N⟩⟨N-Me, —N⟩⟨N-COMe, —N⟩⟨N-⟨⟩-OMe(F), —N⟩⟨N-CH=CH-⟨⟩, —N⟩-Bn

Scheme 3.208

ArCHO + CH$_2$=C(CO$_2$R) + H–Nu $\xrightarrow[\text{26-68\%}]{\text{DBU/THF}}$ Ar-CH(OH)-CH$_2$-CH(Nu)(CO$_2$R)

Ar = Ph, 2-NO$_2$Ph, 4-BrPh, 2-pyridinyl;
R = Me, t-Bu;
H-Nu: (CH$_3$)$_2$C(H)–NO$_2$, cyclopentyl–NO$_2$, Me-C(H)(CO$_2$Me)(CO$_2$Me), cyclohexenyl–NO$_2$

undergo smooth, one-pot oxidative conjugate addition with indoles in the presence of 2-iodoxybenzoic acid (IBX) under neutral conditions to afford a new class of substituted indoles (**466**) in good yields.[275] As a continuation of this work, allyltrimethylsilane has also been used in the one-pot oxidative conjugate addition of MBH adducts using IBX/Sc(OTf)$_3$[276] or Dess–Martin periodinane (DMP)/BF$_3$·OEt$_2$[277] as catalytic system. These methods provide an efficient and attractive process for the preparation of homoallylated β-keto esters (Scheme 3.209).

Yadav *et al.* have developed the first example of ionic liquid-promoted one-pot oxidative conjugate hydrocyanation of MBH adducts with trimethylsilyl cyanide (TMSCN) (Scheme 3.210). This reaction involves an efficient regioselective addition of TMSCN to β-keto-α-methylenes and (*E*)-cinnamaldehydes, obtained from oxidation of MBH adducts with IBX/[bmim]Br or isomerization–oxidation with NaNO$_3$/[Hmim]HSO$_4$, respectively, to afford the corresponding thermodynamically more stable β-cyanated products **468** and **469**.[278]

A one-pot oxidative conjugate addition of sulfur-centered nucleophiles to MBH adducts has been developed by Yadav *et al.*[279] The reaction involves

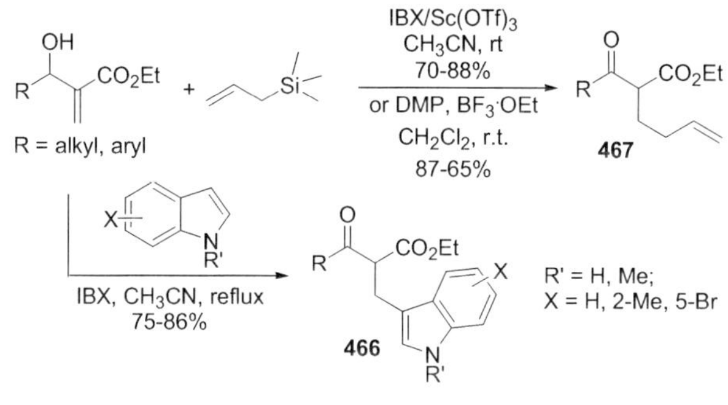

Scheme 3.209

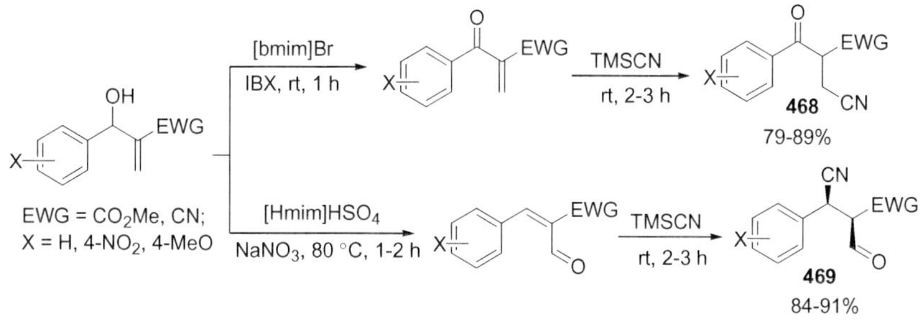

Scheme 3.210

Brønsted acidic ionic liquid [Hmim]HSO$_4$-mediated oxidation of MBH alcohols with NaNO$_3$ to give methyl (*E*)-α-formylcinnamates **470** or (*E*)-α-cyanocinnamaldehydes **471** followed by conjugate hydrothiocyanation/hydrosulfenylation with NH$_4$SCN/PhSH to afford the corresponding methyl β-thiocyanato (or β-phenylsulfenyl)-α-formylhydrocinnamates **472** (or **473**) or β-thiocyanato (or β-phenylsulfenyl)-α-cyanohydrocinnamaldehydes **474** (or **475**) diastereoselectively in good to high yields in a one-pot procedure (Scheme 3.211). Notably, ionic liquid [Hmim]HSO$_4$ could be easily recycled for further use without any loss of efficiency after isolation of the product.

Swamy has reported a simple transition metal-free hydro/hydrothiophosphonylation of MBH adducts, substituted allyl bromides, allenylphosphonates and alkynes, promoted by fluoride ion in an ionic liquid (Scheme 3.212).[280] It was the first time that clear-cut evidence was provided for fluoride activation of the phosphite *via* pentacoordinate phosphorus. The reaction of cyclic phosphites **476** with various MBH adducts **477** in the presence of (n-Bu)$_4$N$^+$F$^-$ (TBAF) in [bmim]$^+$[PF$_6$]$^-$ leads selectively to γ-hydroxyphosphonates **479**. In contrast, the corresponding reaction of **476a** with MBH acetates affords P-CH$_2$ allylphosphonates **481** with the elimination of acetic acid,[14] and the

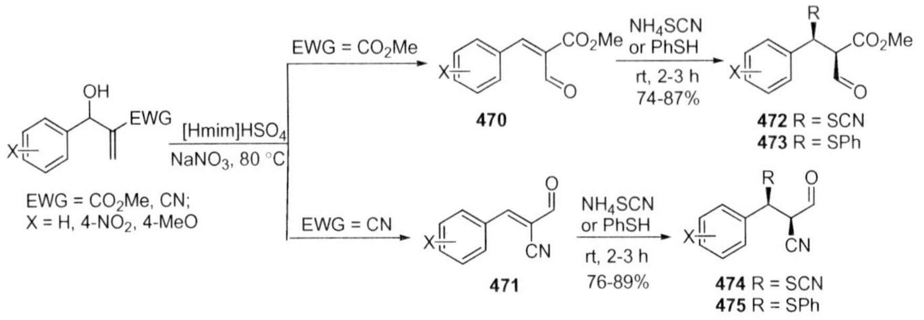

Scheme 3.211

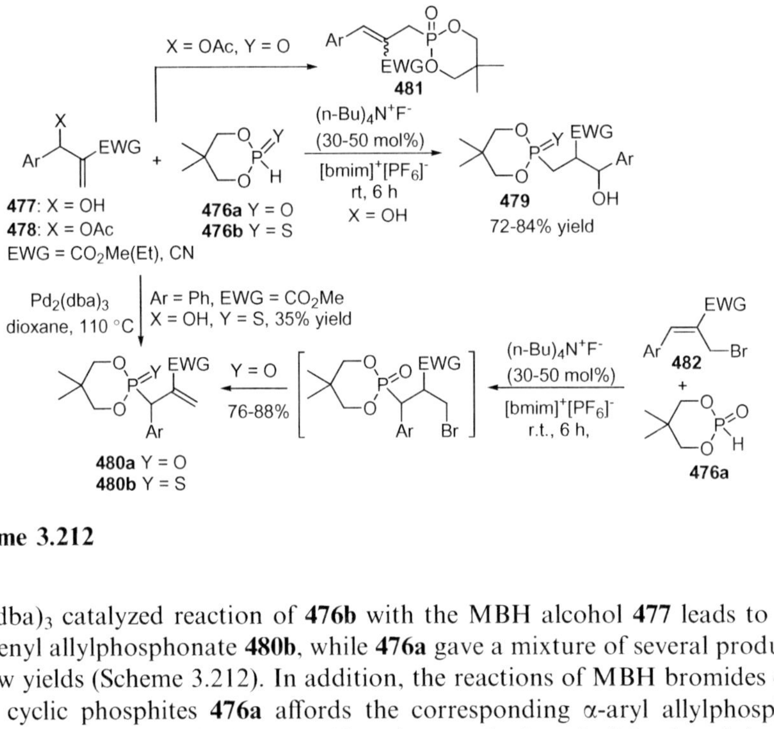

Scheme 3.212

Pd$_2$(dba)$_3$ catalyzed reaction of **476b** with the MBH alcohol **477** leads to the α-phenyl allylphosphonate **480b**, while **476a** gave a mixture of several products in low yields (Scheme 3.212). In addition, the reactions of MBH bromides **482** with cyclic phosphites **476a** affords the corresponding α-aryl allylphosphonate **480a** regioselectively. Notably, the synthesis of **480a** by Arbuzov rearrangement of phosphites (OCH$_2$CMe$_2$CH$_2$O)POCH$_2$C(EWG)]CHAr[14] is very difficult.

Recently, Darses *et al.* described for the first time a rhodium-catalyzed 1,4-addition of organoboranes to hindered MBH adducts, trisubstituted alkenes, to afford highly functionalized alkenes **483**, *via* addition of organoboranes and β-hydroxy elimination (Scheme 3.213). Moreover, preliminary results showed that enantio-enriched products **484** were easily accessible by using a mono-substituted chiral diene **485** as ligand, while chiral phosphane ligands were completely ineffective.[281]

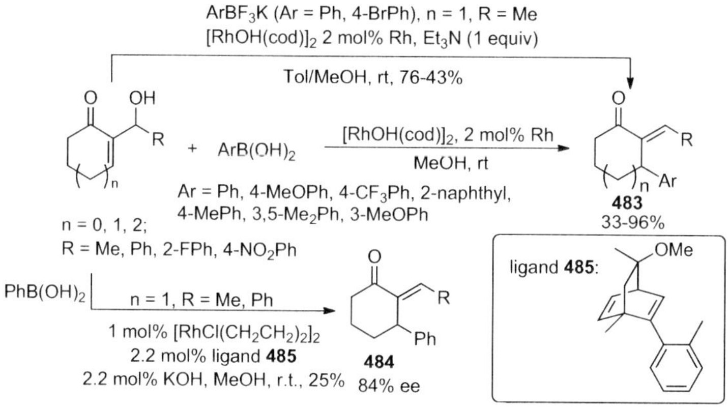

Scheme 3.213

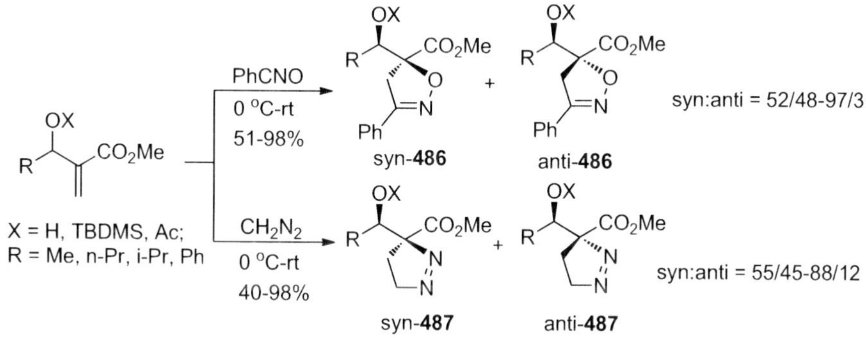

Scheme 3.214

3.11.3　Miscellaneous

MBH adducts and their derivatives derived from methyl acrylate and aldehydes undergo stereoselective cycloadditions with diazomethane and benzonitrile oxide to give the corresponding cycloadducts in good yields (Scheme 3.214).[1a] The stereochemical outcome can be explained by the so-called "inside alkoxy effect" theory.[282] However, in the case of diazomethane cycloadditions, electrostatic factors play a reduced role compared to the corresponding nitrile oxide reactions, while steric effects are of major importance in governing the stereoselectivity. This different behavior of the two 1,3-dipoles has been rationalized by analysis of the atomic charges, as calculated at the RHF/3-21G level of theory, for the transition structure of these reactions.[1a]

In contrast to the reported modes of reactions of dienes and allene with tropone, phosphine-catalyzed reaction of modified allylic compounds, including acetates, bromides, chlorides or *tert*-butyl carbonates derived from the MBH reaction with tropone, afforded [3 + 6] annulation products **488** in

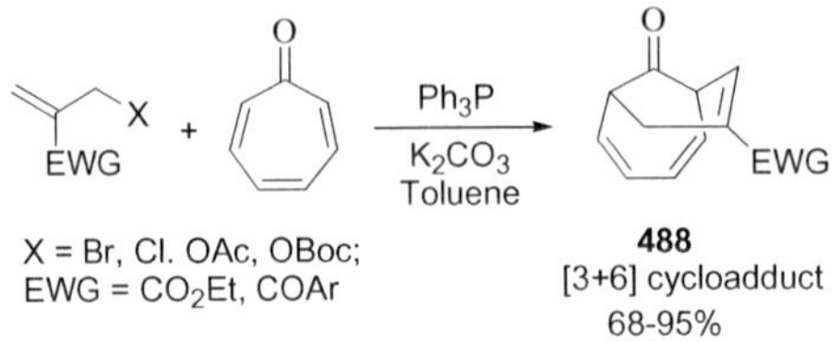

Scheme 3.215

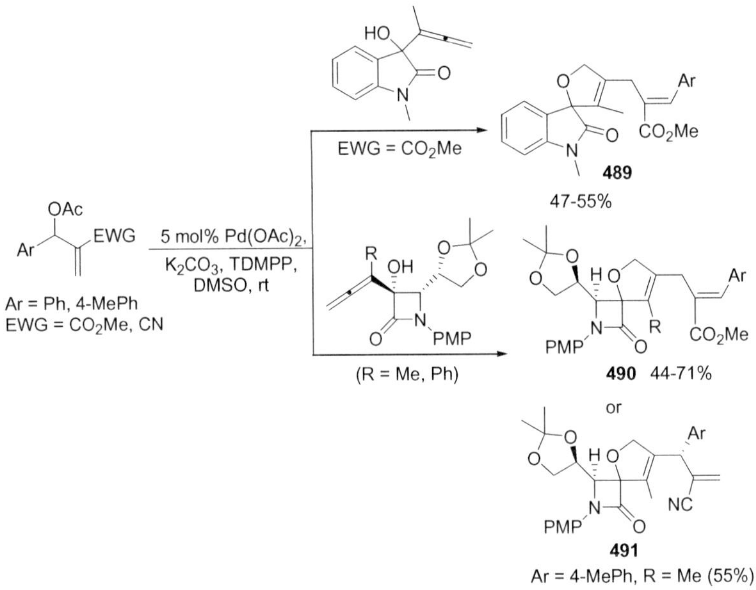

Scheme 3.216

excellent yields (Scheme 3.215). This method provides a simple and convenient method for constructing bridged nine-membered carbocycles.[283]

Alcaide and Almendros have developed a novel palladium-catalyzed domino heterocyclization/cross-coupling reaction of various α-allenols and MBH acetates, furnishing [(2,5-dihydrofuran-3-yl)methyl]acrylate derivatives **489** and **490** and the acrylonitrile **491** in moderate to good yields (Scheme 3.216).[284]

Ozonolysis of MBH adducts originating from aromatic aldehydes provides α-ketoesters **492** with different substitution patterns on the aromatic ring. Diastereoselective reduction of the α-ketoesters **492** affords the corresponding α,β-dihydroxy-esters **493** with excellent *anti* diastereoselectivity. This method provides an alternative approach for the synthesis of either α-ketoesters or α,β-dihydroxy esters (Scheme 3.217).[285]

The N-protected allylic amines **495** and α-methylene-β-amino acids **497** have been obtained from MBH acetate *via* tandem S_N2' substitution–Overman

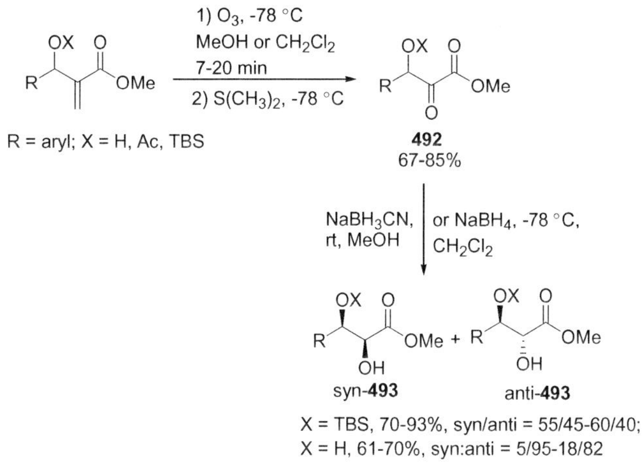

Scheme 3.217

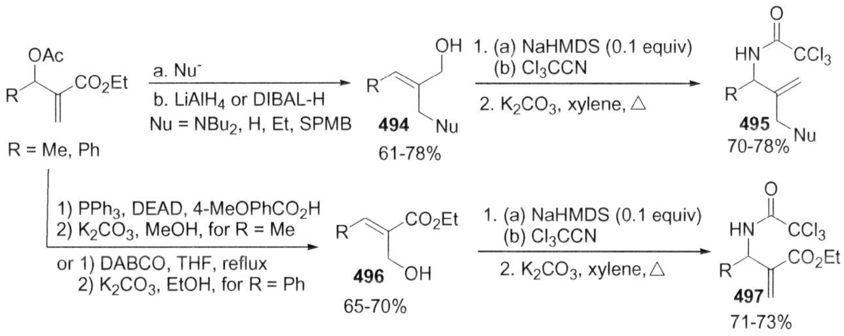

Scheme 3.218

rearrangement. This reaction sequence involves the S_N2' substitution of MBH acetate derived from methyl acrylate with various nucleophiles, followed by an *in situ* reduction of the esters to furnish primary allylic alcohols **494**, which can be converted into trichloroacetimidates and subjected to [3,3]-sigmatropic rearrangement to afford the allylic amides **495**. Utilizing the hydroxy group as the nucleophile furnished allylic hydroxy esters **496**, which have been converted into protected α-methylene-β-amino acids **497** *via* Overman rearrangement (Scheme 3.218).[286]

Ene-carbamates **499** can be prepared *via* a Curtius rearrangement of a hydrolyzed MBH adduct **498** followed by the reaction of intermediate vinyl isocyanates with an alcohol (methanol or *t*-butanol).[287] In addition, different acyloin (α-hydroxyketones) **500** have been obtained in good overall yields based on a Curtius rearrangement in the presence of water.[288] These methodologies have been demonstrated in the synthesis of some bioactive compounds (**501**and **502**) (Scheme 3.219).

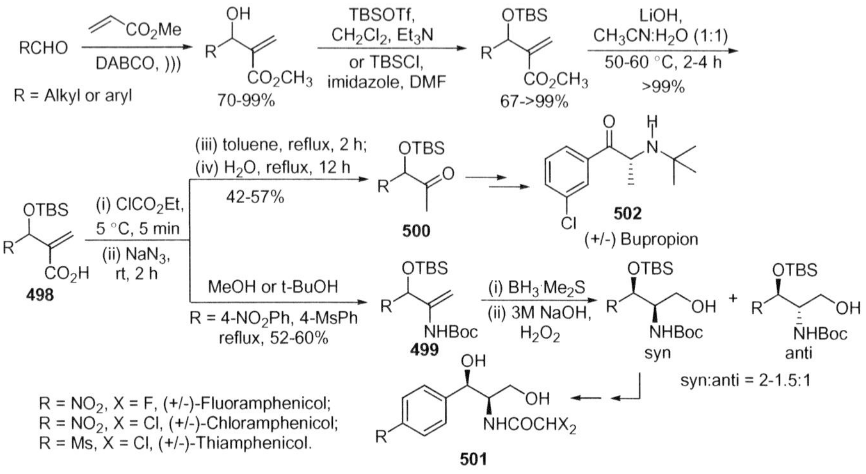

Scheme 3.219

The MBH adduct derived from acrylate reacts with $NaNO_2$–ceric ammonium nitrate (CAN) to form the corresponding β-nitro alcohols **503**, leading to β-nitro acrylic esters **504** in good to excellent yields with high (*Z*)-stereoselectivity *via* dehydration of their mesylates. The reaction of the corresponding MBH acetates with $NaNO_2$–CAN, however, gave an almost 1 : 1 mixture of nitro acrylic esters **505** and nitro alcohol **503′** in modest yields. Compound **505** was obtained as a mixture of (*E*)- and (*Z*)-isomers in which either the (*Z*)-isomer was the major product or the (*Z*)-isomer was the only product. Furthermore, mesylate-containing β-nitro acrylic esters **504** react with NaN_3 to form 2-cyano-3-substituted acrylic esters **506** in excellent yields (Scheme 3.220).[289]

A novel activation of the NC–H bond of MBH adducts of *N*-methylisatin (**507**) with various alcohols using CAN as a single-electron oxidizing agent has been developed. By treatment with primary alcohol, adducts bearing an ester group at the activated alkene **507a** afforded functionalized ethers **508**, while those with a nitrile **507b** afforded ethers **508** and nitrated aromatic products **509** in good yield. Notably, the reaction of secondary and tertiary alcohols (isopropanol and *t*-butanol) with MBH adduct **507a** did not yield any NC–H activated product – instead only the nitrated product **509** was obtained in good yields (Scheme 3.221).[290]

The MBH acetates of 2-cyclohexen-1-one undergo an unusual transformation into 2-arylmethylphenols **510** in the presence of K_2CO_3. Contrary to the Friedel–Crafts benzylation from phenols or benzyl phenyl ethers and Fries rearrangement of phenyl phenylacrylates, which suffers from the formation of mixtures of *ortho-/para*-orientation, this method provides an efficient synthesis of *ortho*-alkylphenols in high regioselectivity (Scheme 3.222).[291]

Howell *et al.* have described a general method for the synthesis of 3-alkylidene-2-methyleneoxetanes **513** by the methylenation of 3-alkylidene-2-oxetanones **512**, which in turn have been readily prepared from MBH type adducts (Scheme 3.223).

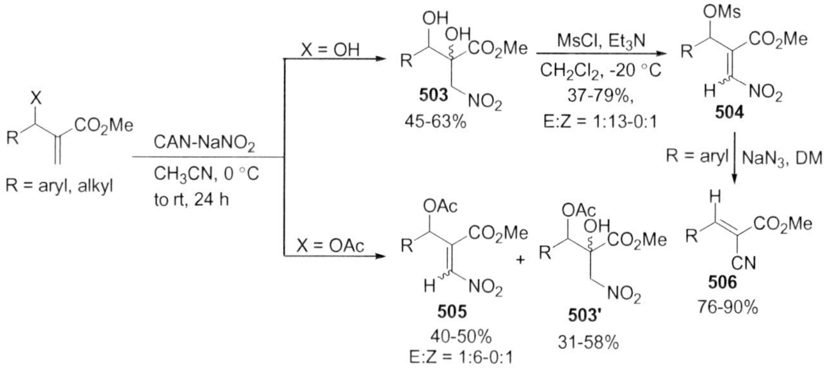

Scheme 3.220

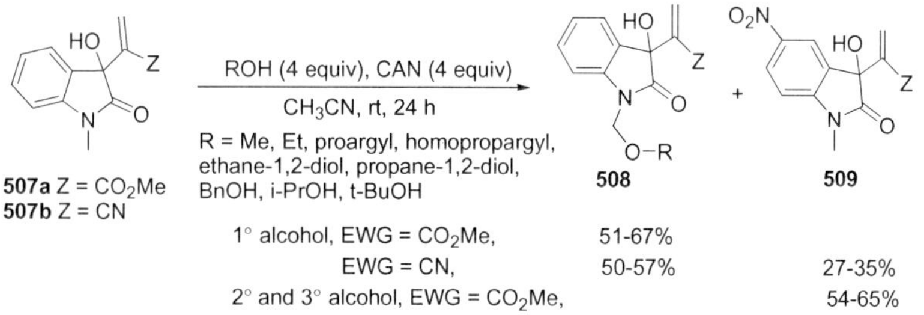

Scheme 3.221

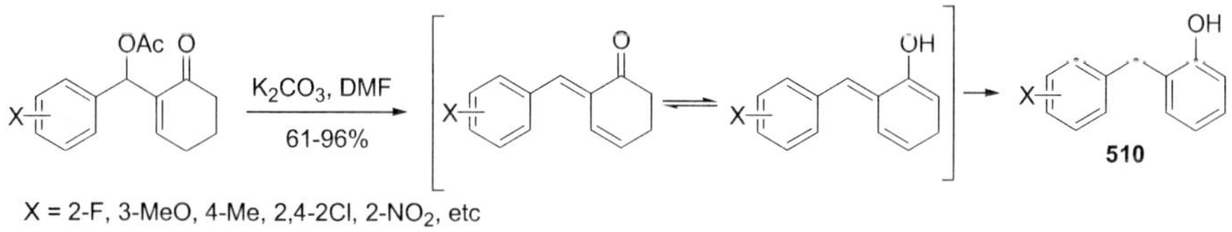

Scheme 3.222

Preliminary investigations of the reactivities of these unusual, strained heterocyclic compounds (**513**) have demonstrated their potential as synthetic scaffolds. The conversion of electron-rich 3-aryl-2-methylene-3-hydroxy acids **511** into electron-rich aryl allenes **514** at room-temperature is an unanticipated transformation that may have synthetic utility.[292]

Allyl halides **515**, obtained from MBH adducts, have been used as another carbon electrophile to react with various activated alkenes, providing a simple

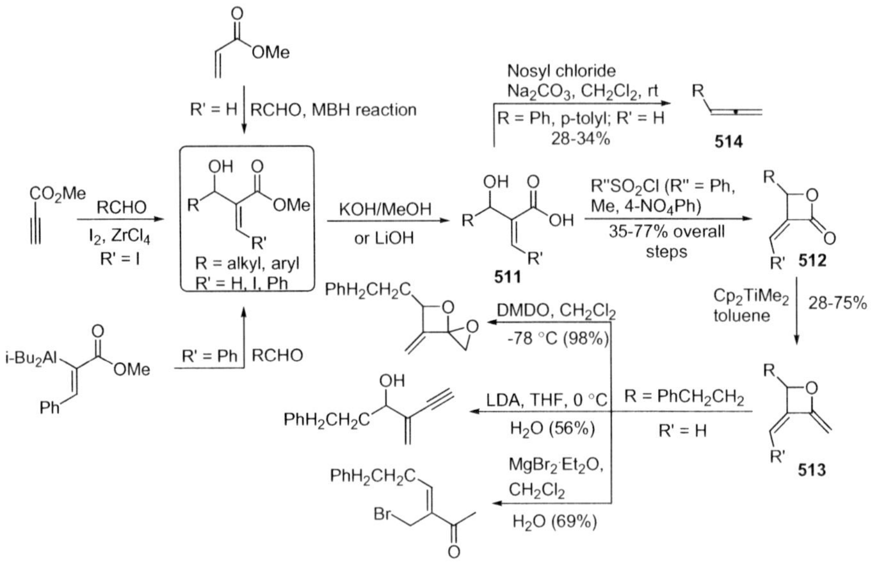

Scheme 3.223

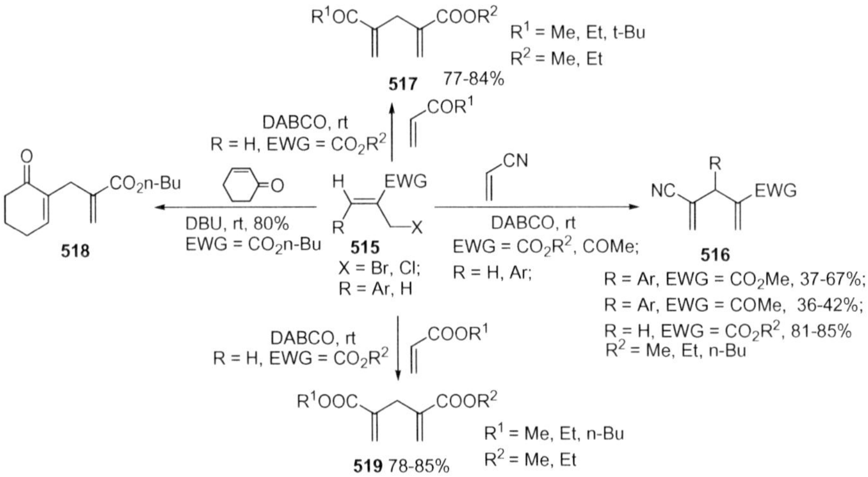

Scheme 3.224

synthesis of 2,4-functionalized 1,4-pentadienes **516–519** in the presence of DABCO or DBU (Scheme 3.224).[293]

1,4-bis-^{13}C-labeled (*E,E*)-2-ethoxycarbonyl-1,4-diphenylbutadiene (**520**) has been synthesized by MBH reaction of benzaldehyde with ethyl acrylate and subsequent rearrangement, followed by Wittig olefination with benzaldehyde, whereas on the basis of a double Horner–Wadsworth–Emmons (HWE)

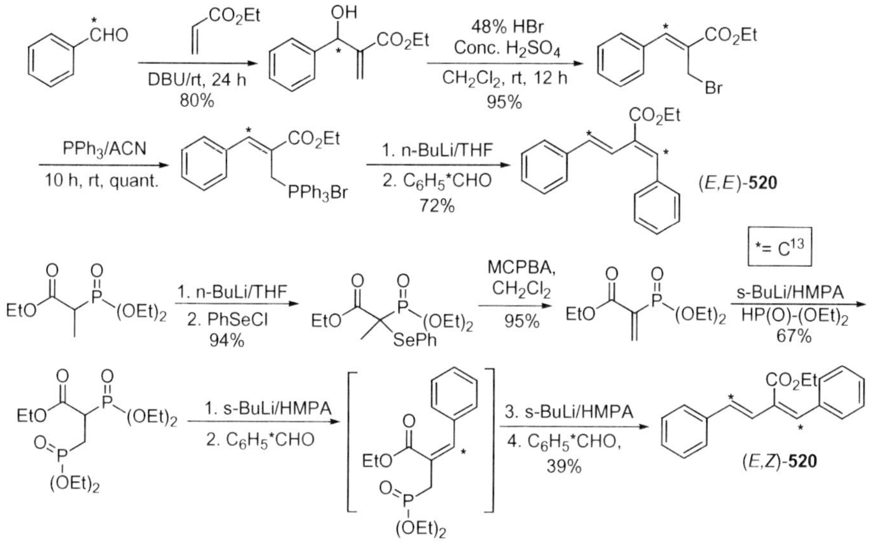

Scheme 3.225

strategy, only the (*Z,E*)-isomer was given in good yields. These routes provide stereoselective access to labeled compounds in good yields (Scheme 3.225).[294]

3.12 Conclusions

MBH adducts have been recognized as versatile intermediates for various stereo- and enantioselective transformations. However, the proximity and chemeospecificity of the functional groups in MBH adducts have not been fully explored and understood and the design of novel strategies for proper tuning of these functional goups is still required, so that novel reacton pathways and methodologies can be discovered for further applications in organic synthesis.

References

1. (a) R. Annunziata, M. Benaglia, M. Cinquini, F. Cozzi and L. Raimondi, *J. Org. Chem.*, 1995, **60**, 4697; (b) D. Basavaiah, M. Krishnamacharyulu, R. S. Hyma, P. K. S. Sarma and N. Kumaragurubaran, *J. Org. Chem.*, 1999, **64**, 1197; (c) V. Singh, G. P. Yadav, P. R. Maulik and S. Batra, *Tetrahedron*, 2006, **62**, 8731.
2. P. H. Mason and N. D. Emslie, *Tetrahedron*, 1994, **50**, 12001.
3. A. Foucaud and F. El Guemmout, *Bull. Soc. Chim. Fr.*, 1989, 403.
4. M. M. Sá, L. Meier, L. Fernandes and S. B. C. Pergher, *Catal. Commun.*, 2007, **8**, 1625.

5. (a) A. B. Charette and B. Côté, *Tetrahedron Lett.*, 1993, **34**, 6833; (b) A. B. Charette and B. Côté, *J. Am. Chem. Soc.*, 1995, **117**, 12721; (c) A. B. Charette, B. Côté, S. Monroc and S. Prescott, *J. Org. Chem.*, 1995, **60**, 6888.
6. A. B. Charette, M. K. Janes and A. A. Boezio, *J. Org. Chem.*, 2001, **66**, 2178.
7. (a) W. Adam, R. Albert, N. d. Grau, L. Hasemann, B. Nestler, E.-M. Peters, K. Peters, F. Prechtl and H. G. von Schnering, *J. Org. Chem.*, 1991, **56**, 5778; (b) W. Adam, V. O. N. Salgado, E.-M. Peters, K. Peters and H. G. von Schnering, *Chem. Ber.*, 1993, **126**, 1481.
8. P. Perlmutter and T. D. McCarthy, *Aust. J. Chem.*, 1993, **46**, 253.
9. K. Morita and Y. H. Miyaji, *Japan. Pat.*, 70 09526 (1970); *Chem. Abstr.*, 1970, **72**, 132968t.
10. T. Janecki and R. Bodalski, *Synthesis*, 1990, 799.
11. T. Janecki, *Synth. Commun.*, 1993, **23**, 641.
12. S. C. Fields, *Tetrahedron Lett.*, 1998, **39**, 6621.
13. K. S. Kumar and K. C. K. Swamy, *J. Organomet. Chem.*, 2001, **637–639**, 616.
14. C. Muthiah, K. S. Kumar, J. J. Vittal and K. C. K. Swamy, *Synlett*, 2002, 1787.
15. M. Ciclosi, C. Fava, R. Galeazzi, M. Orena and J. Sepulveda-Arques, *Tetrahedron Lett.*, 2002, **43**, 2199.
16. S. Kobbelgaard, S. Brandes and K. A. Jørgensen, *Chem. Eur. J.*, 2008, **14**, 1464.
17. G. Martelli, E. Marcucci, M. Orena and S. Rinaldi, *Molecules*, 2009, **14**, 2824.
18. M. Mamaghani and A. Badrian, *Tetrahedron Lett.*, 2004, **45**, 1547.
19. (a) S.-Z. Wang, K. Yamamoto, H. Yamada and T. Takahashi, *Tetrahedron*, 1992, **48**, 2333; (b) K. Yamamoto, R. Deguchi and J. Tsuji, *Bull. Chem. Soc. Jpn.*, 1985, **58**, 3397.
20. (a) W. S. Johnson, L. Werthemann, W. R. Bartlett, T. J. Brockson, T. Li, D. J. Faulkner and M. R. Petersen, *J. Am. Chem. Soc.*, 1970, **92**, 741; (b) D. Basavaiah and S. Pandiaraju, *Tetrahedron Lett.*, 1995, **36**, 757; (c) D. Basavaiah, S. Pandiaraju and M. Krishnamacharyulu, *Synlett*, 1996, **1**, 747.
21. (a) B. Das, A. Majhi and J. Banerjee, *Tetrahedron Lett.*, 2006, **47**, 7619; (b) B. Das, A. Majhi, K. R. Reddy and K. Venkateswarlu, *J. Mol. Catal. A: Chem.*, 2007, **263**, 273.
22. M. Mamaghani and A. Badrian, *Phosphorus, Sulfur Silicon Relat. Elem.*, 2004, **179**, 1181.
23. M. Mamaghani and A. Badrian, *Phosphorus, Sulfur Silicon Relat. Elem.*, 2004, **179**, 2429.
24. S. E. Drewes, N. D. Emslie, N. Korodia and G. Loizou, *Synth. Commun.*, 1990, **20**, 1437.
25. O. Mhasni and F. Rezgui, *Tetrahedron Lett.*, 2010, **51**, 586.
26. (a) L. A. Marshall, K. E. Steiner and G. A. Schiehser, *US Pat.*, 4,788,304 (1988); *Chem. Abstr.*, 1989, **110**, 135069q; (b) L. A. Marshall, K. E.

Steiner and G. A. Schiehser, *US Pat.*, 4,933,365 (1990); *Chem. Abstr.*, 1990, **113**, 126604v; (c) L. Bernardi, B. F. Bonini, M. Comes-Franchini, M. Fochi, M. Folegatti, S. Grilli, A. Mazzanti and A. Ricci, *Tetrahedron: Asymmetry*, 2004, **15**, 245.

27. A. W. Hall, D. Lacey, J. S. Hill and D. G. McDonnell, *Supramol. Sci.*, 1994, **1**, 21.

28. J. Villieras and M. Rambaud, *Synthesis*, 1982, 924.

29. E. Farrington, M. C. Franchinia and J. M. Brown, *Chem. Commun.*, 1998, 277.

30. D. Basavaiah, S. Pandiaraju and K. Padmaja, *Synlett*, 1996, 393.

31. K. Morita and H. Miyaji, *Japan. Pat.*, 70 17164 (1970); *Chem. Abstr.*, 1970, **73**, 66728g.

32. H. G. McFadden, R. L. N. Harris and C. L. D. Jenkins, *Aust. J. Chem.*, 1989, **42**, 301.

33. (a) K. Nishitani, M. Isozaki and K. Yamakawa, *Chem. Pharm. Bull.*, 1990, **38**, 28; (b) D. Colombani and P. Chaumont, *Macromol. Chem. Phys.*, 1995, **196**, 3643.

34. F. Ameer, S. E. Drewes, M. S. Houston-McMillan and P. T. Kaye, *J. Chem. Soc., Perkin Trans. 1*, 1985, 1143.

35. S. P. Chavan, K. S. Ethiraj and S. K. Kamat, *Tetrahedron Lett.*, 1997, **38**, 7415.

36. D. Basavaiah, R. Suguna Hyma, K. Padmaja and M. Krishnamachar-yulu, *Tetrahedron*, 1999, **55**, 6971.

37. B. Das, N. Chowdhury, K. Damodar and B. Ravikanth, *Helv. Chim. Acta*, 2007, **90**, 2037.

38. B. Das, J. Banerjee, N. Ravindranath and B. Venkataiah, *Tetrahedron Lett.*, 2004, **45**, 2425.

39. P. R. Krishna, V. Kannan and G. V. M. Sharma, *Synth. Commun.*, 2004, **34**, 55.

40. (a) F. J. Alvarez and K. M. O'Connor, *PCT Int. Appl.* WO 93 07886 (1993); *Chem. Abstr.*, 1993, **119**, 160821x; (b) S. E. Drewes, G. Loizou and G. H. P. Roos, *Synth. Commun.*, 1987, **17**, 291; (c) R. Buchholz and H. M. R. Hoffman, *Helv. Chim. Acta*, 1991, **74**, 1213; (d) F. Ameer, S. E. Drewes, J. S. Field and P. T. Kaye, *S. Afr. J. Chem.*, 1985, **38**, 35; (e) F. Ameer, S. E. Drewes, N. D. Emslie, P. T. S. Kaye and R. L. Mann, *J. Chem. Soc., Perkin Trans. 1*, 1983, 2293; (f) H. Mazdiyasni, D. B. Konopacki, D. A. Dickman and T. M. Zydowsky, *Tetrahedron Lett.*, 1993, **34**, 435; (g) C. Hubschwerien, R. Charnas, P. Angehrn, A. Furlenmeier, T. Graser and M. Montavon, *J. Antibiot.*, 1992, **45**, 1358; (h) S. E. Drewes and R. F. A. Hoole, *Synth. Commun.*, 1985, **15**, 1067; (i) H. M. R. Hoffmann and J. Rabe, *Angew. Chem., Int. Ed. Engl.*, 1985, **24**, 94.

41. C. M. Marson, J. H. Pink and C. Smith, *Tetrahedron Lett.*, 1995, **36**, 8107.

42. (a) S. E. Drewes and N. D. Emslie, *J. Chem. Soc., Perkin Trans. 1*, 1982, 2079; (b) H. M. R. Hoffman and J. Rabe, *Angew. Chem. Int. Ed. Engl.*, 1983, **22**, 795; (c) W. R. Roush and B. B. Brown, *J. Org. Chem.*, 1993, **58**,

2151; (d) H. M. R. Hoffman and J. Rabe, *J. Org. Chem.*, 1985, **50**, 3849; (e) J. M. Brown, A. P. James and L. M. Prior, *Tetrahedron Lett.*, 1987, **28**, 2179; (f) J. Rabe and H. M. R. Hoffman, *Angew. Chem. Int. Ed. Engl.*, 1983, **22**, 796.

43. A. Cruiec, A. Foucaud and C. Moinet, *New. J. Chem.*, 1991, **15**, 943.

44. (a) J. Villieras and M. Rambaud, *Org. Synth.*, 1988, **66**, 220; (b) C. Navarro, M. Degueil-Castaing, D. Colombani and B. Maillard, *Synth. Commun.*, 1993, **23**, 1025; (c) K. C. Reddy, H.-S. Byun and R. Bittman, *Tetrahedron Lett.*, 1994, **35**, 2679.

45. M. Ferreira, L. Fernandes and M. M. Sa, *J. Braz. Chem. Soc.*, 2009, **20**, 564.

46. (a) P. Auvray, P. Knochel and J. F. Normant, *Tetrahedron*, 1988, **44**, 6095; (b) P. Auvray, P. Knochel and J. F. Normant, *Tetrahedron Lett.*, 1986, **27**, 5095; (c) H. M. R. Hoffman, A. Weichert, A. M. Z. Slawin and D. J. Williams, *Tetrahedron*, 1990, **46**, 5591.

47. (a) B. Das, K. Venkateswarlu, M. Krishnaiah, H. Holla and A. Majhi, *Helv. Chim. Acta*, 2006, **89**, 1417; (b) L. H. Choudhury, T. Parvin and A. T. Khan, *Tetrahedron*, 2009, **65**, 9513.

48. B. Das, K. Damodar, N. Bhunia and B. Shashikanth, *Tetrahedron Lett.*, 2009, **50**, 2072.

49. B. Das, J. Banerjee and N. Ravindranath, *Tetrahedron*, 2004, **60**, 8357.

50. L. Fernandes, A. J. Bortoluzzi and M. M. Sá, *Tetrahedron*, 2004, **60**, 9983.

51. E. Ciganek, *J. Org. Chem.*, 1995, **60**, 4635.

52. B. Das, A. Majhi, J. Banerjee, N. Chowdhury and K. Venkateswarlu, *Tetrahedron Lett.*, 2005, **46**, 7913.

53. B. Das, H. Holla, Y. Srinivas, N. Chowdhury and B. P. Bandgar, *Tetrahedron Lett.*, 2007, **48**, 3201.

54. (a) H.-S. Byun, K. C. Reddy and R. Bittman, *Tetrahedron Lett.*, 1994, **35**, 1371; (b) R. Bittman, H.-S. Byun, K. C. Reddy, P. Samadder and G. Arthur, *J. Med. Chem.*, 1997, **40**, 1391.

55. H. M. R. Hoffman and J. Rabe, *Helv. Chim. Acta*, 1984, **67**, 413.

56. Conversion of the allylic bromides into organometallic species and reactions of the latter with aldeydes or ketones have been described involving the following metals: (a) zinc: (i) P. Auvray, P. Knochel and J. F. Normant, *Tetrahedron Lett.*, 1986, **27**, 5091; (ii) N. El Alami, C. Belaud and J. Villierás, *J. Organomet. Chem.*, 1987, **319**, 303; (iii) F. Lambert, B. Kirschlegar and J. Villierás, *J. Organomet. Chem.*, 1991, **406**, 71; (iv) F. Lambert, B. Kirschlegar and J. Villierás, *J. Organomet. Chem.*, 1991, **405**, 273. (b) Chromium: (i) ref. 40h; (ii) Y. Okuda, S. Nakatsukasa, K. Oshima and H. Nozaki, *Chem. Lett.*, 1985, 481. (c) Tin: (i) ref. 40h; (ii) L. Kovács, *Recl. Trav. Chim. Pays-Bas*, 1993, **112**, 471; (iii) P. Talaga, M. Schaeffer and C. Benzezra, J.-L. Stampf, *Synthesis*, 1990, 530. (d) Indium: C. J. Li and T. H. Chan, *Tetrahedron Lett.*, 1991, **32**, 7017. (e) Bismuth: L. Kovács, *Recl. Trav. Chim. Pays-Bas*, 1993, **112**, 471. (f) Palladium/tin (from the alcohols): Y. Masuyama, Y. Nimura and Y. Kurusu, *Tetrahedron Lett.*, 1991, **32**, 225.

57. (a) H. M. R. Hoffman, A. Gassner and U. Eggert, *Chem. Ber.*, 1991, **124**, 2475; (b) A. Weichert and H. M. R. Hoffman, *J. Org. Chem.*, 1991, **56**, 4098.

58. (a) K. Morita, Y. Kinoshita and R. Nakanishi, *Japan. Pat.*, 68 16364 (1968); *Chem. Abstr.*, 1969, **70**, 67642z; (b) K. Morita, Y. Kinoshita and R. Nakanishi, *Japan. Pat.*, 69 20,085 (1969); *Chem. Abstr.*, 1969, **71**, 123630d.

59. N. J. Lawrence, J. P. Crump, A. T. McGown and J. A. Hadfield, *Tetrahedron Lett.*, 2001, **42**, 3939.

60. J. S. Hill and N. S. Isaacs, *J. Chem. Res.* 1988, (S) 220, (M) 2641.

61. (a) Y. Kinoshita and K. Morita, *Japan. Pat.*, 71 05,134(1971); *Chem. Abstr.*, 1971, **75**, 36959g; (b) K. Morita and Y. Kinoshita, *Japan. Pat.*, 70 36,485(1971); *Chem. Abstr.*, 1971, 74, 63949q.

62. K. Morita, Z. Suzuki and H. Hirose, *Bull. Chem. Soc. Jpn.*, 1968, **41**, 2815.

63. (a) W. Poly, D. Schomburg and H. M. R. Hoffman, *J. Org. Chem.*, 1988, **53**, 3701; (b) H. M. R. Hoffman, U. Eggert and W. Poly, *Angew. Chem. Int. Ed. Engl.*, 1987, **26**, 1015; (c) D. Basavaiah, S. Pandiaraju and P. K. S. Sarma, *Tetrahedron Lett.*, 1994, **35**, 4227.

64. C. Grundke and H. M. R. Hoffman, *Chem. Ber.*, 1987, **120**, 1461.

65. M. Ohno, T. Azuma, K. Satoshi, Y. Shirakawa and S. Eguchi, *Tetrahedron*, 1996, **52**, 4983.

66. A. Weichert and H. M. R. Hoffman, *J. Chem. Soc., Perkin Trans. 1*, 1990, 2154.

67. K. Burgess and L. D. Jennings, *J. Org. Chem.*, 1990, **55**, 1138.

68. H. Hayashi, K. Yanagihara and S. Tsuboi, *Tetrahedron: Asymmetry*, 1998, **9**, 3825.

69. M. G. Nascimento, S. P. Zanotto, S. P. Melegari, L. Fernandes and M. Mandolesi Sá, *Tetrahedron: Asymmetry*, 2003, **14**, 3111.

70. (a) D. Basavaiah and P. Dharma Rao, *Synth. Commun.*, 1994, **24**, 917; (b) D. Bsasavaiah, *Arkivoc*, 2001, **viii**, 70.

71. W. Adam, U. Hoch, C. R. Saha-Moller and P. Schreier, *Angew. Chem., Int. Ed. Engl.*, 1993, **32**, 1737.

72. D. Saib and A. Foucaud, *J. Chem. Res. (S)*, 1987, 372.

73. A. Foucaud and N. Brine, *Synth. Commun.*, 1994, **24**, 2851.

74. S. Ravichandran, *Synth. Commun.*, 2001, **31**, 2345.

75. D. Basavaiah, M. Krishnamacharyulu, R. Suguna Hyma and S. Pandiaraju, *Tetrahedron Lett.*, 1997, **38**, 2141.

76. H. J. Lee, M. R. Seong and J. N. Kim, *Tetrahedron Lett.*, 1998, **39**, 6223.

77. D. Basavaiah, M. Bakthadoss and S. Pandiaraju, *Chem. Commun.*, 1998, 1639.

78. O. Roy, A. Riahi, F. Henin and J. Muzart, *Tetrahedron*, 2000, **56**, 8133.

79. D. Basavaiah, M. Bakthadoss and G. Jayapal Reddy, *Synthesis*, 2001, 919.

80. D. Basavaiah and R. Mallikarjuna Reddy, *Tetrahedron Lett.*, 2001, **42**, 3025.

81. H. J. Lee, T. H. Kim and J. N. Kim, *Bull. Korean Chem. Soc.*, 2001, **22**, 1063.

82. D. Basavaiah and K. Muthukumaran, *Synth. Commun.*, 1999, **29**, 713.

83. D. Basavaiah, N. Kumaragurubaran and K. Padmaja, *Synlett*, 1999, 1630.

84. H. S. Kim, T. Y. Kim, K. Y. Lee, Y. M. Chung, H. J. Lee and J. N. Kim, *Tetrahedron Lett.*, 2000, **41**, 2613.

85. D. Basavaiah, K. Padmaja and T. Satyanarayana, *Synthesis*, 2000, 1662.

86. B. Das, J. Banerjee, A. Majhi, N. Chowdhury, K. Venkateswarlu and H. Holla, *Indian J. Chem.*, 2006, **45B**, 1729.

87. D. Basavaiah, K. Muthukumaran and B. Sreenivasulu, *Synthesis*, 2000, 545.

88. T. Ollevier and T. M. Mwene-Mbeja, *Tetrahedron*, 2008, **64**, 5150.

89. J. Park, R. Heo, J.-Y. Kim, B. W. Yoo and C. M. Yoon, *Bull. Korean Chem. Soc.*, 2009, **30**, 1195.

90. P. Shanmugam and P. R. Singh, *Synlett*, 2001, 1314.

91. P. Shanmugam and P. Rajasingh, *Tetrahedron*, 2004, **60**, 9283.

92. Y. Liu, D. Mao, J. Qian, S. Lou, Z. Xu and Y. Zhang, *Synthesis*, 2009, 1170.

93. D. Basavaiah and K. Muthukumaran, *Tetrahedron*, 1998, **54**, 4943.

94. N. Sundar and S. V. Bhat, *Synth. Commun.*, 1998, **28**, 2311.

95. R. Kumareswaran and Y. D. Vankar, *Synth. Commun.*, 1998, **28**, 2291.

96. B. A. Kulkarni and A. Ganesan, *J. Comb. Chem.*, 1999, **1**, 373.

97. V. Calo, A. Nacci, L. Lopez and A. Napola, *Tetrahedron Lett.*, 2001, **42**, 4701.

98. (a) J. M. Kim, K. H. Kim, T. H. Kim and J. N. Kim, *Tetrahedron Lett.*, 2008, **49**, 3248; (b) J. M. Kim, S. H. Kim, S. H. Kim and J. N. Kim, *Bull. Korean Chem. Soc.*, 2008, **29**, 1583.

99. B. R. V. Ferreira, R. V. Pirovani, L. G. Souza-Filho and F. Coelho, *Tetrahedron*, 2009, **65**, 7712.

100. R. Perez, D. Veronese, F. Coelho and O. A. C. Antunes, *Tetrahedron Lett.*, 2006, **47**, 1325.

101. J. B. Park, S. H. Ko, W. P. Hong and K.-J. Lee, *Bull. Korean Chem. Soc.*, 2004, **25**, 927.

102. S. Tu, L.-H. Xu and C.-R. Yu, *Synth. Commun.*, 2008, **38**, 2662.

103. (a) J. M. Brown and I. Cutting, *J. Chem. Soc., Chem. Commun.*, 1985, 578; (b) J. M. Brown, *Angew. Chem., Int. Ed. Engl.*, 1987, **26**, 190; (c) J. M. Brown, *Chem. Soc. Rev.*, 1993, 25; (d) J. M. Brown, I. Cutting and A. P. James, *Bull. Soc. Chim. Fr.*, 1988, 211.

104. M. Kitamura, I. Kasahara, K. Manabe, R. Noyori and H. Takaya, *J. Org. Chem.*, 1988, **53**, 708.

105. S. Sato, I. Matsuda and M. Shibata, *J. Organomet. Chem.*, 1989, **377**, 347.

106. (a) K. Yamamoto, M. Takagi and J. Tsuji, *Bull. Chem. Soc. Jpn.*, 1988, **61**, 319; (b) M. Takagi and K. Yamamoto, *Tetrahedron*, 1991, **47**, 8869.

107. F. A. Davis, H. Qiu, M. Song and N. V. Gaddiraju, *J. Org. Chem.*, 2009, **74**, 2798.

108. (a) C. R. Mateus, W. P. Almeida and F. Coelho, *Tetrahedron Lett.*, 2000, **41**, 2533; (b) F. Coelho, W. P. Almeida, C. R. Mateus, L. D. Furtado and J. C. F. Gouveia, *Arkivoc*, 2003 (x), 443.

109. C. R. Mateus, M. P. Feltrin, A. M. Costa, F. Coelho and W. P. Almeida, *Tetrahedron*, 2001, **57**, 6901.

110. A. Bouzide, *Org. Lett.*, 2002, **4**, 1347.

111. R. Saxena, V. Singh and S. Batra, *Tetrahedron*, 2004, **60**, 10311.

112. Z.-C. Duan, X.-P. Hu, J. Deng, S.-B. Yu, D.-Y. Wang and Z. Zheng, *Tetrahedron: Asymmetry*, 2009, **20**, 588.

113. H. M. R. Hoffman and J. Rabe, *Helv. Chim. Acta*, 1984, **67**, 413.

114. N. Daude, U. Eggert and H. M. R. Hoffman, *J. Chem. Soc., Chem. Commun.*, 1988, 206.

115. W. J. Bailey, R. L. Hudson and E. T. Yates, *J. Org. Chem.*, 1963, **28**, 828.

116. (a) D. Bellus and C. D. Weis, *Tetrahedron Lett.*, 1973, **14**, 999; (b) D. Bellus, K. V. Bredow, H. Sauter and C. D. Weis, *Helv. Chim. Acta*, 1973, **56**, 3004; (c) P. Dowd and K. Kang, *Synth. Commun.*, 1974, 151.

117. (a) W. S. Johnson, L. Werthemann, W. R. Bartlett, T. J. Brockson, T. Li, D. J. Faulkner and M. R. Petersen, *J. Am. Chem. Soc.*, 1970, **92**, 741; (b) D. Basavaiah and S. Pandiaraju, *Tetrahedron Lett.*, 1995, **36**, 757; (c) D. Basavaiah, S. Pandiaraju and M. Krishnamacharyulu, *Synlett*, 1996, 747.

118. V. K. Aggarwal, A. Patin and S. Tisserand, *Org. Lett.*, 2005, **7**, 2555.

119. (a) F. Ameer, S. E. Drewes, M. S. Houston-McMillman and P. T. Kaye, *S. Afr. J. Chem.*, 1986, **39**, 57; (b) F. Ameer, S. E. Drewes, P. T. Kaye, G. Loizou, D. G. Malissar and G. H. P. Roos, *S. Afr. J. Chem.*, 1987, **40**, 35; (c) S. E. Drewes and B. J. Slater-Kinghorn, *Synth. Commun.*, 1986, **16**, 603.

120. P. Bauchat, E. Le Rouille and A. Foucaud, *Bull. Chim. Soc. Fr.*, 1991, 267; *Chem. Abstr.*, 1991, **115**, 91980b.

121. F. R. Heerden, J. S. Huyser and C. W. Holzapfel, *Synth. Commun.*, 1994, **24**, 2863.

122. I. Beltaief and H. Amri, *Synth. Commun.*, 1994, **24**, 2003.

123. A. Chamakh and H. Amri, *Tetrahedron Lett.*, 1998, **39**, 375.

124. (a) F. Rezgui and M. M. El Gaied, *Tetrahedron*, 1997, **53**, 15711; (b) F. Rezgui and M. M. El Gaied, *J. Chem. Res. (S)*, 1999, 510.

125. J. N. Kim, Y. J. Im and J. M. Kim, *Tetrahedron Lett.*, 2002, **43**, 6597.

126. (a) Y. J. Im, J. M. Kim and J. N. Kim, *Bull. Korean Chem. Soc.*, 2002, **23**, 1361; (b) Y. J. Im, C. G. Lee, H. R. Kim and J. N. Kim, *Tetrahedron Lett.*, 2003, **44**, 2987.

127. K. H. Kim, E. S. Kim and J. N. Kim, *Tetrahedron*, 2009, **65**, 5322.

128. B. M. Trost and M. K. Brennan, *Org. Lett.*, 2007, **9**, 3961.

129. H. Amri, M. Rambaud and J. Villieras, *J. Organomet. Chem.*, 1990, **384**, 1.

130. H. Amri, M. Rambaud and J. Villieras, *Tetrahedron*, 1990, **46**, 3535.

131. D. Basavaiah, A. K. D. Bhavani and P. K. S. Sarma, *J. Chem. Soc., Chem. Commun.*, 1994, 1091.

132. S. Khamri, T. Turki and H. Amri, *Synth. Commun.*, 2008, **38**, 3277.

133. S. Gowrisankar, K. Y. Lee, S. C. Kim and J. N. Kim, *Bull. Korean Chem. Soc.*, 2005, **26**, 1443.

134. S. GowriSankar, K. Y. Lee, C. G. Lee and J. N. Kim, *Tetrahedron Lett.*, 2004, **45**, 6141.

135. B. C. Ranu, K. Chattopadhyay and R. Jana, *Tetrahedron Lett.*, 2007, **48**, 3847.

136. P. Srihari, A. P. Singh, A. K. Basak and J. S. Yadav, *Tetrahedron Lett.*, 2007, **48**, 5999.

137. J. Park, Kwon, B. Young, K. Yang, H. Rhee and C. M. Yoon, *Synthesis*, 2010, 661.

138. J. S. Yadav, B. V. S. Reddy, N. N. Yadav, A. P. Singh, M. Choudhary and A. C. Kunwar, *Tetrahedron Lett.*, 2008, **49**, 6090.

139. B. Das, J. Banerjee, G. Mahender and A. Majhi, *Org. Lett.*, 2004, **6**, 3349.

140. K. Biswas, C. Börner, J. Gimeno, P. J. Goldsmith, D. Ramazzotti, A. L. K. So and S. Woodward, *Tetrahedron*, 2005, **61**, 1433.

141. C. Börner, J. Gimeno, S. Gladiali, P. J. Goldsmith, D. Ramazzotti and S. Woodward, *Chem. Commun.*, 2000, 2433.

142. P. J. Goldsmith and S. Woodward, *Angew. Chem. Int. Ed.*, 2005, **44**, 2235.

143. A. Novak, R. Fryatt and S. Woodward, *C. R. Chim.*, 2007, **10**, 206.

144. A. Novak, M. J. Calhorda, P. J. Costa and S. Woodward, *Eur. J. Org. Chem.*, 2009, 898.

145. (a) Y. J. Im, J. E. Na and J. N. Kim, *Bull. Korean Chem. Soc.*, 2003, **24**, 511; (b) J. S. Yadav, B. V. Subba Reddy, A. K. Basak and A. Venkat Narsaiah, *J. Mol. Catal. A: Chem.*, 2007, **274**, 105.

146. S. K. Murthy, C. Rambabu, K. Vijeender, P. Bibhuti Bhusan and S. Chandrasekhar, *Synlett*, 2007, 494.

147. (a) A. Chamakh, M. M'hirsi, J. Villieras, J. Lebreton and H. Amri, *Synthesis*, 2000, 295; (b) S. Hbaieb and H. Amri, *J. Soc. Chim. Tunis.*, 2000, **4**, 671.

148. J. M. Kim, Y. J. Im, T. H. Kim and J. N. Kim, *Bull. Korean Chem. Soc.*, 2002, **23**, 657.

149. O. Mhasni and F. Rezgui, *Tetrahedron Lett.*, 2010, **51**, 586.

150. (a) H. Akiyama, T. Fujimoto, K. Ohshima, K. Hoshino, I. Yamamoto and R. Iriye, *Org. Lett.*, 1999, **1**, 427; (b) H. Akiyama, T. Fujimoto, K. Ohshima, K. Hoshino, Y. Saito, A. Okamoto, I. Yamamoto, A. Kakehi and R. Iriye, *Eur. J. Org. Chem.*, 2001, 2265.

151. Y. M. Chung, J. H. Gong, T. H. Kim and J. N. Kim, *Tetrahedron Lett.*, 2001, **42**, 9023.

152. (a) C. H. Lee, Y. S. Song, H. I. Cho, J. W. Yang and K.-J Lee, *J. Heterocyclic Chem.*, 2003, **40**, 1103; (b) W. P. Hong, H. N. Lim, H. W. Park and K.-J. Lee, *Bull. Korean Chem. Soc.*, 2005, **26**, 655; (c) J. S. Yadav, M. K. Gupta, S. K. Pandey, B. V. S. Reddya and A. V. S. Sarma, *Tetrahedron Lett.*, 2005, **46**, 2761.

153. G. W. Kabalka, B. Venkataiah and G. Dong, *Org. Lett.*, 2003, **5**, 3803.
154. G. W. Kabalka, G. Dong, B. Venkataiah and C. Chen, *J. Org. Chem.*, 2005, **70**, 9207.
155. S. Gowrisankar, S. H. Kim and J. N. Kim, *Bull. Korean Chem. Soc.*, 2009, **30**, 726.
156. L. Navarre, S. Darses and J.-P. Genet, *Chem. Commun.*, 2004, 1108.
157. L. Navarre, S. Darses and J.-P. Genet, *Adv. Synth. Catal.*, 2006, **348**, 317.
158. T. Gendrineau, N. Demoulin, L. Navarre, J.-P. Genet and S. Darses, *Chem. Eur. J.*, 2009, **15**, 4710.
159. M. L. Kantam, K. B. S. Kumar and B. Sreedhar, *J. Org. Chem.*, 2008, **73**, 320.
160. J. S. Yadav, B. V. Subba Reddy, S. S. Mandal, A. P. Singh and A. K. Basak, *Synthesis*, 2008, 1943.
161. J. S. Yadav, B. V. Subba Reddy, A. K. Basak, A. V. Narsaiah, A. Prabhakar and B. Jagadeesh, *Tetrahedron Lett.*, 2005, **46**, 639.
162. C. Ramesh, V. Kavala, B. R. Raju, C.-W. Kuo and C.-F. Yao, *Tetrahedron Lett.*, 2009, **50**, 4037.
163. C.-W. Cho and M. J. Krische, *Angew. Chem. Int. Ed.*, 2004, **43**, 6689.
164. Y.-Q. Jiang, Y.-L. Shi and M. Shi, *J. Am. Chem. Soc.*, 2008, **130**, 7202.
165. Dirk Jan V. C. van Steenis, T. Marcelli, M. Lutz, A. L. Spek, Jan H. van Maarseveen and H. Hiemstra, *Adv. Synth. Catal.*, 2007, **349**, 281.
166. H.-L. Cui, J. Peng, X. Feng, W. Du, K. Jiang and Y.-C. Chen, *Chem. Eur. J.*, 2009, **15**, 1574.
167. K. Jiang, J. Peng, H.-L. Cui and Y.-C. Chen, *Chem. Commun.*, 2009, 3955.
168. L. D. S. Yadav, V. K. Rai and S. Singh, *Synlett*, 2009, 1423.
169. L. D. S. Yadav and V. K. Rai, *Tetrahedron Lett.*, 2009, **50**, 2414.
170. P. R. Krishna and Y. L. Prapurna, *Synlett*, 2009, 2613.
171. V. Singh, R. Pathak, S. Kanojiya and S. Batra, *Synlett*, 2005, 2465.
172. B. Das, G. Mahender, N. Chowdhury and J. Banerjee, *Synlett*, 2005, 1000.
173. R. Pathak, V. Singh, S. N. Nag, S. Kanojiya and S. Batra, *Synthesis*, 2006, 813.
174. S. E. Drewes, M. M. Horn and N. Ramesar, *Synth. Commun.*, 2000, **30**, 1045.
175. Y. J. Im, J. H. Gong, H. J. Kim and J. N. Kim, *Bull. Korean Chem. Soc.*, 2001, **22**, 1053.
176. (a) J. N. Kim, H. J. Lee, K. Y. Lee and J. H. Gong, *Synlett*, 2002, 173; (b) J. H. Gong, H. R. Kim, E. K. Ryu and J. N. Kim, *Bull. Korean Chem. Soc.*, 2002, **23**, 789; (c) J. Li, X. Wang and Y. Zhang, *Tetrahedron Lett.*, 2005, **46**, 5233.
177. K. Y. Lee, T. H. Kim and J. N. Kim, *Bull. Korean Chem. Soc.*, 2004, **25**, 1966.
178. H.-L. Cui, X. Feng, J. Peng, J. Lei, K. Jiang and Y.-C. Chen, *Angew. Chem. Int. Ed.*, 2009, **48**, 5737.
179. S.-Q. Ge, Y.-Y. Hua and M. Xia, *Ultrason. Sonochem.*, 2009, **16**, 743.

180. J. Li, Y. K. Liu and Y. M. Zhang, *Chin. Chem. Lett.*, 2006, **17**, 877.

181. S. Rajesh, B. Banerji and J. Iqbal, *J. Org. Chem.*, 2002, **67**, 7852.

182. T. Nemoto, T. Fukuyama, E. Yamamoto, S. Tamura, T. Fukuda, T. Matsumoto, Y. Akimoto and Y. Hamada, *Org. Lett.*, 2007, **9**, 927.

183. K. Y. Lee, H. S. Lee and J. N. Kim, *Bull. Korean Chem. Soc.*, 2008, **29**, 1099.

184. M. Paira, S. K. Mandal and S. C. Roy, *Tetrahedron Lett.*, 2008, **49**, 2432.

185. H. S. Kim, H. S. Lee and J. N. Kim, *Bull. Korean Chem. Soc.*, 2009, **30**, 941.

186. S. Ghosh, R. Dey, K. Chattopadhyay and B. C. Ranu, *Tetrahedron Lett.*, 2009, **50**, 4892.

187. (a) H. J. Lee, H. S. Kim and J. N. Kim, *Tetrahedron Lett.*, 1999, **40**, 4363; (b) H. J. Lee, Y. M. Chung, K. Y. Lee and J. N. Kim, *Bull. Korean Chem. Soc.*, 2000, **21**, 843.

188. D. Basavaiah and T. Satyanarayana, *Chem. Commun.*, 2004, 32.

189. B. Das, A. Majhi, J. Banerjee and N. Chowdhury, *J. Mol. Catal. A: Chem.*, 2006, **260**, 32.

190. (a) C.-W. Cho, J.-R. Kong and M. J. Krische, *Org. Lett.*, 2004, **6**, 1337; (b) H. Park, C.-W. Cho and M. J. Krische, *J. Org. Chem.*, 2006, **71**, 7892.

191. T.-Z. Zhang, L.-X. Dai and X.-L. Hou, *Tetrahedron: Asymmetry*, 2007, **18**, 1990.

192. P. Srihari, P. Dutta, R. S. Rao, J. S. Yadav, S. Chandrasekhar, P. Thombare, J. Mohapatra, A. Chatterjee and M. R. Jain, *Tetrahedron*, 2009, **65**, 5569.

193. E. Krawczyk, K. Owsianik and A. Skowrońska, *Tetrahedron*, 2005, **61**, 1449.

194. W. P. Hong and K.-J. Lee, *Synthesis*, 2005, 33.

195. A. Jose, R. R. Paul, R. Mohan, S. C. Mathew, A. T. Biju, E. Suresh and V. Nair, *Synthesis*, 2009, 1829.

196. Y.-L. Liu, L. Liu, D. Wang and Y.-J. Chen, *Tetrahedron*, 2009, **65**, 3473.

197. (a) D. Colombani and B. Maillard, *J. Chem. Soc., Chem. Commun.*, 1994, 1259; (b) D. Colombani and B. Maillard, *J. Org. Chem.*, 1994, **59**, 4765.

198. X. Feng, Y.-Q. Yuan, H.-L. Cui, K. Jiang and Y.-C. Chen, *Org. Biomol. Chem.*, 2009, **7**, 3660.

199. B. M. Trost, H.-C. Tsui and F. D. Toste, *J. Am. Chem. Soc.*, 2000, **122**, 3534.

200. B. M. Trost, O. R. Thiel and H.-C. Tsui, *J. Am. Chem. Soc.*, 2002, **124**, 11616.

201. B. M. Trost, M. R. Machacek and H. C. Tsui, *J. Am. Chem. Soc.*, 2005, **127**, 7014.

202. J. N. Kim, H. J. Lee and J. H. Gong, *Tetrahedron Lett.*, 2002, **43**, 9141.

203. Ch R. Reddy, N. Kiranmai, G. S. K. Babu, G. D. Sarma, B. Jagadeesh and S. Chandrasekhar, *Tetrahedron Lett.*, 2007, **48**, 215.

204. X. Zhang, W. Rao, S. Philip and W. H. Chan, *Org. Biomol. Chem.*, 2009, **7**, 4186.

205. X. Jia, P. Zhao, X. Liu and J. Li, *Synth. Commun.*, 2008, **38**, 1617.
206. D. Colombani, C. Navarro, M. Degueil-Castaing and B. Maillard, *Synth. Commun.*, 1991, **21**, 1481.
207. M. Konno, Y. Yuasa, M. Harada, T. Miura and H, Kumobayashi, *Jpn. Kokai Tokyo Koho JP 06,172,300 (94,172,300); Chem. Abstr.*, 1995, **112**, 55724g.
208. B. Das, N. Chowdhury, K. Damodar and J. Banerjee, *Chem. Pharm. Bull.*, 2007, **55**, 1274.
209. G. W. Kabalka, B. Venkataiah and G. Dong, *Tetrahedron Lett.*, 2003, **44**, 4673.
210. P. Srihari, A. P. Singh, R. Jain and J. S. Yadav, *Synthesis*, 2006, 2772.
211. C.-R. Liu, M.-B. Li, D.-J. Cheng, C.-F. Yang and S.-K. Tian, *Org. Lett.*, 2009, **11**, 2543.
212. S. Chandrasekhar, B. Saritha, V. Jagadeshwar, Ch. Narsihmulu, D. Vijay, G. Dattatreya Sarma and B. Jagadeesh, *Tetrahedron Lett.*, 2006, **47**, 2981.
213. L. R. Reddy, B. Hu, M. Prashad and K. Prasad, *Angew. Chem. Int. Ed.*, 2009, **48**, 172.
214. (a) Y. Liu, X. Xu, H. Zheng, D. Xu, Z. Xu and Y. Zhang, *Synlett*, 2006, 571; (b) Y. Liu, H. Zheng, D. Xu, Zhenyuan Xu and Y. Zhang, *Synlett*, 2006, 2492; (c) Y. Liu, D. Xu, Z. Xu and Y. Zhang, *Heteroat. Chem.*, 2008, **19**, 188; (d) M. J. Cha, Y. S. Song and K.-J. Lee, *Bull. Korean Chem. Soc.*, 2006, **27**, 1900.
215. L. D. S Yadav, R. Patel and V. P. Srivastava, *Tetrahedron Lett.*, 2009, **50**, 1335.
216. P. O. Deane, J. J. Guthrie-Strachan, P. T. Kaye and R. E. Whittaker, *Synth. Commun.*, 1998, **28**, 2601.
217. D. Basavaiah and S. Pandiaraju, *Tetrahedron*, 1996, **52**, 2261.
218. P. A. Badkar, N. P. Rath and C. D. Spilling, *Org. Lett.*, 2007, **9**, 3619.
219. N. N. Bhuvan Kumar and K. C. Kumara Swamy, *Polyhedron*, 2007, **26**, 883.
220. D. Basavaiah and P. K. S. Sarma, *J. Chem. Soc., Chem. Commun.*, 1992, 955.
221. K. Pachamuthu and Y. D. Vankar, *Tetrahedron Lett.*, 1998, **39**, 5439.
222. U. Shadakshari and S. K. Nayak, *Tetrahedron*, 2001, **57**, 4599.
223. J. Li, W. Qian and Y. Zhang, *Tetrahedron*, 2004, **60**, 5793.
224. D. Bassavaiah, A. K. D. Bhavani, S. Pandiaraju and P. K. S. Sarma, *Synlett*, 1995, **3**, 243.
225. J. Li, H. Xu and Y. Zhang, *Tetrahedron Lett.*, 2005, **46**, 1931.
226. Y. Liu, D.-Q. Xu, Z.-Y. Xu and Y.-M. Zhang, *J. Zhejiang Univ. Sci. B*, 2006, **7**, 393.
227. (a) G. W. Kabalka, B. Venkataiah and G. Dong, *J. Org. Chem.*, 2004, **69**, 5807; (b) G. W. Kabalka, B. Venkataiah and G. Dong, *Tetrahedron Lett.*, 2005, **46**, 4209.
228. G. W. Kabalka, B. Venkataiah and G. Dong, *Organometallics*, 2005, **24**, 762.
229. (a) M. Bulliard, M. Zehnder and B. Giese, *Helv. Chem. Acta*, 1991, **74**, 1600; (b) M. Bulliard, H.-G. Zeitz and B. Giese, *Synlett*, 1991, 423.

230. B. Giese, M. Bulliard, J. Dickhaut, R. Halbach, C. Hassler, U. Hoffmann, B. Hinzen and M. Senn, *Synlett*, 1995, 116.

231. E. P. Kundig, L. He-xu and P. Romanens, *Tetrahedron Lett.*, 1995, **36**, 4047.

232. Y. Guindon and J. Rancourt, *J. Org. Chem.*, 1998, **63**, 6554.

233. N. Mase, S. Wake, Y. Watanabe and T. Toru, *Tetrahedron Lett.*, 1998, **39**, 5553.

234. N. Mase, Y. Watanabe and T. Toru, *Tetrahedron Lett.*, 1999, **40**, 2797.

235. S. Matsumoto, Y. Okubo and K. Mikami, *J. Am. Chem. Soc.*, 1998, **120**, 4015.

236. S. Matsumoto and K. Mikami, *Synlett*, 1998, 469.

237. K. Mikami and Y. Okubo, *Synlett*, 2000, 491.

238. K. Mikami and Y. Okubo, *Synlett*, 2000, 1135.

239. K. Mikami, S. Tanaka, T. Tonoi and S. Matsumoto, *Tetrahedron Lett.*, 2004, **45**, 6133.

240. S. K. Mandal, M. Paira and S. C. Roy, *J. Org. Chem.*, 2008, **73**, 3823.

241. M. P. Sibi and K. Patil, *Org. Lett.*, 2005, **7**, 1453.

242. S. P. Majhi, A. Neogi, S. Ghosh, A. K. Mukherjee, M. Helliwell and P. Chattopadhyay, *Synthesis*, 2008, 94.

243. J. Lei, H.-L. Cui, R. Li, L. Wu, Z.-Y. Ding and Y.-C. Chen, *Org. Biomol. Chem.*, 2010, **8**, 2840.

244. L. A. Paquette and J. Mendez-Andino, *Tetrahedron Lett.*, 1999, **40**, 4301.

245. J. M. Kim, K. Y. Lee, S. Lee and J. N. Kim, *Tetrahedron Lett.*, 2004, **45**, 2805.

246. K. Y. Lee, J. E. Na, J. Y. Lee and J. N. Kim, *Bull. Korean Chem. Soc.*, 2004, **25**, 1280.

247. S. Gowrisankar, K. Y. Lee and J. N. Kim, *Tetrahedron*, 2006, **62**, 4052.

248. D. Balan and H. Adolfsson, *Tetrahedron Lett.*, 2004, **45**, 3089.

249. V. Declerck, P. Ribiere, J. Martinez and F. Lamaty, *J. Org. Chem.*, 2004, **69**, 8372.

250. V. Declerck, H. Allouchi, J. Martinez and F. Lamaty, *J. Org. Chem.*, 2007, **72**, 1518.

251. M. E. Krafft, E.-H. Song and R. J. Davoile, *Tetrahedron Lett.*, 2005, **46**, 6359.

252. P.-Y. Chen, H.-M. Chen, L.-Y. Chen, J.-Y. Tzeng, J.-C. Tsai, P.-C. Chi, S.-R. Lic and E.-C. Wang, *Tetrahedron*, 2007, **63**, 2824.

253. P. R. Krishna and M. Narsingam, *J. Comb. Chem.*, 2007, **9**, 62.

254. S. Gowrisankar, H. S. Kim, H. S. Lee and J. N. Kim, *Bull. Korean Chem. Soc.*, 2007, **28**, 1844.

255. V. R. Doddi and Y. D. Vankar, *Eur. J. Org. Chem.*, 2007, 5583.

256. M. J. Lee, K. Y. Lee, J. Y. Lee and J. N. Kim, *Org. Lett.*, 2004, **6**, 3313.

257. R. Raju and A. R. Howell, *Org. Lett.*, 2006, **8**, 2139.

258. (a) M. Bailey, I. E. Marko, W. D. Ollis and P. R. Rasmussen, *Tetrahedron Lett.*, 1990, **31**, 4509; (b) M. Bailey, I. E. Marko and W. D. Ollis, *Tetrahedron Lett.*, 1991, **32**, 2687; (c) M. Bailey, I. Staton, P. R. Ashton, I. E. Marko and W. D. Ollis, *Tetrahedron: Asymmetry*, 1991, **2**, 495.

259. (a) Y. Iwabuchi, M. Furukawa, T. Esumi and S. Hatakeiyama, *Chem. Commun.*, 2001, 2030; (b) Y. Iwabuchi, M. Nakatami, M. Yokoyama and S. Hatakeiyama, *J. Am. Chem. Soc.*, 1999, **121**, 10219.

260. A. Foucaud and E. Rouille, *Synthesis*, 1990, 787.

261. E.-G. Han and K.-J. Lee, *Synth. Commun.*, 2009, **39**, 3399.

262. R. F. W. Jackson, S. P. Standen, W. Clegg and A. McCamley, *Tetrahedron Lett.*, 1992, **33**, 6197.

263. R. S. Atkinson, J. Fawcett, D. R. Russel and P. J. Williams, *J. Chem. Soc., Chem. Commun.*, 1994, 2031.

264. B. Das, H. Holla, K. Venkateswarlu and A. Majhi, *Tetrahedron Lett.*, 2005, **46**, 8895.

265. R. S. Porto, M. L. A. A. Vasconcellos, E. Ventura and F. Coelho, *Synthesis*, 2005, 2297.

266. L. D. S. Yadav and C. Awasthi, *Tetrahedron Lett.*, 2009, **50**, 715.

267. W. Pringle and K. B. Sharpless, *Tetrahedron Lett.*, 1999, **40**, 5151.

268. D. Bhuniya, S. Gujjary and S. Sengupta, *Synth. Commun.*, 2006, **36**, 151.

269. R. E. Sammelson, C. D. Gurusinghe, J. M. Kurth, M. M. Olmstead and M. J. Kurth, *J. Org. Chem.*, 2002, **67**, 876.

270. A. Kamimura, R. Morita, K. Matsuura, Y. Omata and M. Shirai, *Tetrahedron Lett.*, 2002, **43**, 6189.

271. N. Azizi and M. R. Saidi, *Tetrahedron Lett.*, 2002, **43**, 4305.

272. S. Batra, T. Srinivasan, S. K. Rastogi, B. Kundu, A. Patra, A. P. Bhaduri and M. Dixit, *Bioorg. Med. Chem. Lett.*, 2002, **12**, 1905.

273. S. Batra, A. K. Roy, A. Patra, A. P. Bhaduri, W. R. Surin, S. A. V. Raghavan, P. Sharma, K. Kapoor and M. Dikshit, *Bioorg. Med. Chem.*, 2004, **12**, 2059.

274. W. Wang and M. Yu, *Tetrahedron Lett.*, 2004, **45**, 7141.

275. J. S. Yadav, B. V. Subba Reddy, A. P. Singh and A. K. Basak, *Tetrahedron Lett.*, 2007, **48**, 4169.

276. J. S. Yadav, B. V. S. Reddy, A. P. Singh and A. K. Basak, *Tetrahedron Lett.*, 2007, **48**, 7546.

277. J. S. Yadav, B. V. S. Reddy, A. P. Singh and A. K. Basak, *Synthesis*, 2008, 469.

278. L. D. S. Yadav, C. Awasthi and A. Rai, *Tetrahedron Lett.*, 2008, **49**, 6360.

279. (a) L. D. S. Yadav, R. Patel and V. P. Srivastava, *Synthesis*, 2008, 1789; (b) L. D. S. Yadav, V. P. Srivastava and R. Patel, *Tetrahedron Lett.*, 2008, **49**, 3142.

280. E. Balaraman, V. Srinivas and K. C. K. Swamy, *Tetrahedron*, 2009, **65**, 7603.

281. T. Gendrineau, J.-P. Genet and S. Darses, *Org. Lett.*, 2010, **12**, 308.

282. K. N; Houk, H.-Y. Duh, Y.-D. Wu and S. R. Moses, *J. Am. Chem. Soc.*, 1986, **108**, 2754.

283. Y. Du, J. Feng and X. Lu, *Org. Lett.*, 2005, **7**, 1987.

284. B. Alcaide, P. Almendros, T. M. del Campo and M. T. Quirós, *Chem. Eur. J.*, 2009, **15**, 3344.

285. C. A. M. Abella, P. Rezende, M. F. L. de Souza and F. Coelho, *Tetrahedron Lett.*, 2008, **49**, 145.
286. P. V. Ramachandran, T. E. Burghardt and M. V. R. Reddy, *Tetrahedron Lett.*, 2005, **46**, 2121.
287. C. R. Mateus and F. Coelho, *J. Braz. Chem. Soc.*, 2005, **16**, 386.
288. G. W. Amarante, P. Rezende, M. Cavallaro and F. Coelho, *Tetrahedron Lett.*, 2008, **49**, 3744.
289. K. Jayakanthan, K. P. Madhusudanan and Y. D. Vankar, *Tetrahedron*, 2004, **60**, 397.
290. P. Shanmugam, V. Vaithiyanathan and B. Viswambharan, *Tetrahedron Lett.*, 2006, **47**, 6851.
291. K. Y. Lee, J. E. Na and J. N. Kim, *Bull. Korean Chem. Soc.*, 2003, **24**, 409.
292. I. Martínez, A. E. Andrews, J. D. Emch, A. J. Ndakala, J. Wang and A. R. Howell, *Org. Lett.*, 2003, **5**, 399.
293. (a) D. Basavaiah, N. Kumaragurubaran and D. S. Sharada, *Tetrahedron Lett.*, 2001, **42**, 85; (b) D. Basavaiah, D. S. Sharada, N. Kumaragurubaran and R. M. Reddy, *J. Org. Chem.*, 2002, **67**, 7135.
294. R. M. Crist, P. V. Reddy and B. Borhan, *Tetrahedron Lett.*, 2001, **42**, 619.

Morita–Baylis–Hillman Adducts or Derivatives for the Construction of Cyclic Frameworks

FEI-JUN WANG, YIN WEI AND MIN SHI

4.1 Introduction

Efficient construction of highly functionalized carbocycles and heterocycles with a defined configuration is of significant importance in the synthesis of many natural products, pharmaceutically active products, perfumes and dyes.[1] Much effort has been devoted to this area of research, and cyclo-formation reactions such as the (hetero) Diels–Alder reaction, transition-metal catalyzed ring-closing metathesis (RCM) and cycloisomerization have been well established as powerful ring-forming tools.

As mentioned in Chapter 3, the Morita–Baylis–Hillman (MBH) adducts containing three chemospecific groups, *viz.*, hydroxy (or amino), alkene and electron-withdrawing groups (EWG), can undergo numerous functional group transformations. These transformations can be tailored appropriately to generate an array of cyclic compounds directly from the MBH adducts. The pictorial depiction of different points of cyclization is delineated in Figure 4.1 with different shades showing various opportunities for the cyclization of MBH adducts.

This chapter excludes the direct transformations of MBH adduct into cyclic compounds. The synthesis of cyclic compounds by intramolecular MBH reactions has also been omitted.

RSC Catalysis Series No. 8
The Chemistry of the Morita–Baylis–Hillman Reaction
By Min Shi, Fei-Jun Wang, Mei-Xin Zhao and Yin Wei
© Min Shi, Fei-Jun Wang, Mei-Xin Zhao and Yin Wei 2011
Published by the Royal Society of Chemistry, www.rsc.org

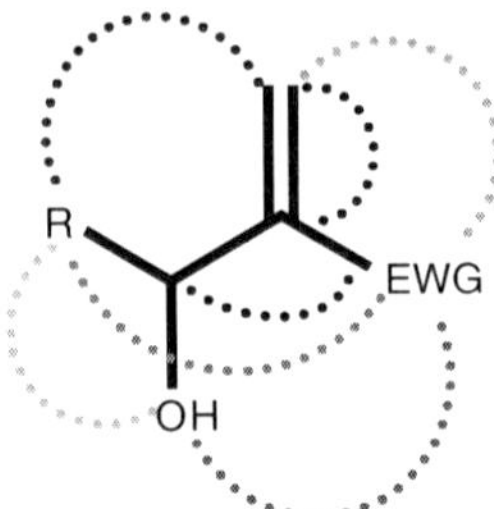

Figure 4.1 Pictorial depiction of different points of cyclization.

4.2 Synthesis of Cyclic Hydrocarbon Compounds

4.2.1 Cyclopropane Ring Systems

The construction of cyclopropane ring systems is of great interest for organic chemists due to the existence of such rings as a basic unit in a number of natural products and several active drug substances. Numerous accounts of the synthetic methods used to construct this unique structural unit have been reported.[2]

Recently, Lewis acidic hydrostannane was shown to be valuable for highly stereoselective homolytic hydrostannylation of allyl and homoallyl alcohols.[3] Using Et$_3$B–dry air as radical initiator, the hydrostannylation of Morita–Baylis–Hillman (MBH) adduct **1** took place smoothly, and subsequent butylation of the reaction mixture with BuLi gave γ-stannylated alcohol **2** in 64% yield with good *syn* diastereoselectivity (Scheme 4.1). Following treatment with pyridine and thionyl chloride, *syn*-**2** could be easily transformed into *trans*-1,2-disubstituted cyclopropane **3** in 85% yield.[4]

Isomerized bromo derivatives of MBH adducts of isatin have been utilized to synthesize 3-spirocycloalkylindolones[5] *via* reductive cyclization.[6] Bromo derivative **4** as a mixture of (*E*)- and (*Z*)-isomers – prepared from MBH adduct upon treatment with HBr embedded on silica gel under microwave irradiation – has been subjected to reduction with NaBH$_4$ at room temperature for 0.5 h (Scheme 4.2). Excellent yields of 3-spirocycloalkylindolones **5** were obtained, but with poor diastereoselectivities.

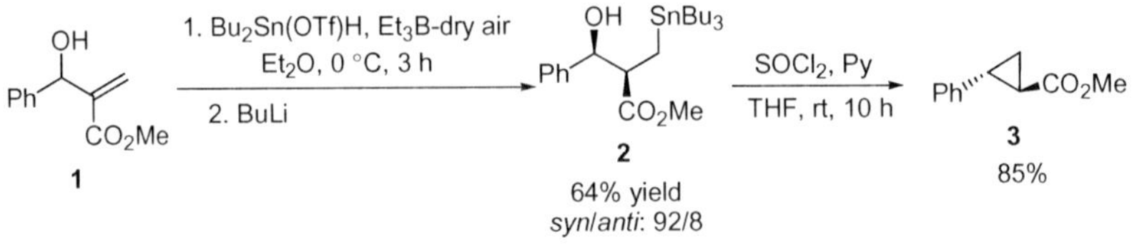

Scheme 4.1

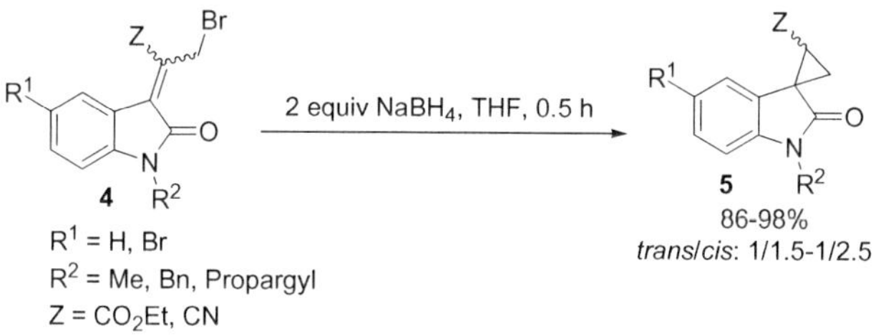

Br
Z
R1
O
N
4 R2
R1 = H, Br
R2 = Me, Bn, Propargyl
Z = CO2Et, CN
2 equiv NaBH4, THF, 0.5 h
Z
R1
O
N
5 R2
86-98%
trans/cis: 1/1.5-1/2.5

Scheme 4.2

4.2.2 Five-membered-ring Carbocyclic Compounds

MBH adducts or derivatives containing two olefin functional groups have been used to examine the catalytic activities of newly developed ruthenium catalysts in ring-closing metathesis (RCM).[7] It was found that the dimerization was much more favored than cyclization in most cases using compound **7** as the catalyst, while catalyst **6** showed the opposite results with the formation of cyclopentene derivatives **9** in moderate yields (Scheme 4.3). Moreover, TBS-protected MBH adduct **8** gave higher yields of product **9** than that of unprotected MBH adduct **8**. In addition, catalysts **6** and **7** showed similar catalytic activities in cyclization of MBH adduct **8** to obtain cyclohexene derivatives **9**. MBH derivative **12** was also subjected to RCM using the second-generation Grubbs catalyst in dichloromethane (Scheme 4.4).[8] The product **13** was obtained in 83% yield by alkene metathesis of the allyl group with the 1,2-disubstituted alkene, rather than with the olefin in the acrylate moiety. The bis-alkylated Meldrum's acid derivative **12** was obtained in 42% yield over two steps with 91% ee *via* two alkylation reactions from MBH acetate **11**.

Ylide reactions have become powerful tools for constructing cyclic compounds,[9] through many types of useful transformations (such as Wittig

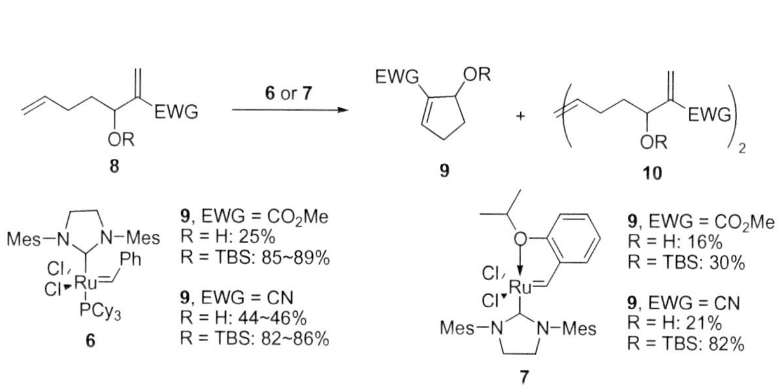

EWG
OR
8
6 or 7
EWG
OR
9
+
EWG
OR
2
10
Mes–N
N–Mes
Ph
Cl
Ru
PCy3
6
9, EWG = CO2Me
R = H: 25%
R = TBS: 85~89%
9, EWG = CN
R = H: 44~46%
R = TBS: 82~86%
Cl
Ru
Mes–N
N–Mes
7
9, EWG = CO2Me
R = H: 16%
R = TBS: 30%
9, EWG = CN
R = H: 21%
R = TBS: 82%

Scheme 4.3

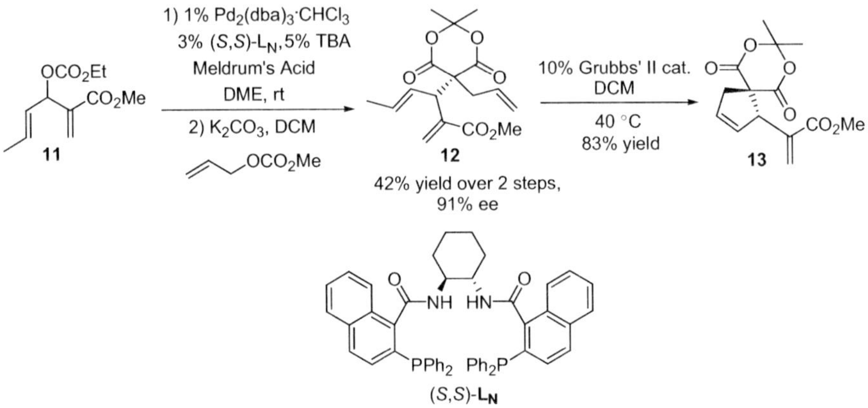

Scheme 4.4

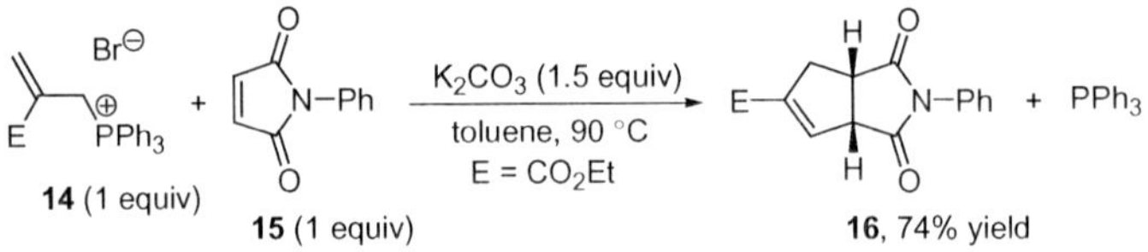

Scheme 4.5

reaction,[10] nucleophilic substitution,[11] *etc.*). The phosphonium bromide salt **14**, containing electron-withdrawing group E, reacts with *N*-phenylsuccinimide to afford the [3 + 2] annulation product **16** and triphenylphosphine in the presence of K_2CO_3 at 90 °C (Scheme 4.5).[12] Similar to the reports of phosphane-catalyzed isomerizations, α- and γ-additions and [3 + 2] cycloadditions of electron-deficient allenes or alkynes,[13] this [3 + 2] annulation can also achieved by a phosphane-catalyzed ylide reaction.[14]

A series of allyl bromides **17** as C3 component derived from MBH adducts containing electron-withdrawing group E have been examined in phosphane-catalyzed ylide reactions, and moderate to good yields of the [3 + 2] annulation products were obtained (Scheme 4.6). Both symmetrical and unsymmetrical terminal olefins as C2 components reacted smoothly to give annulation products with high selectivities. Moreover, both acetates **18** and *t*-butyl carbonates **19** could give the corresponding annulation products in moderate to good yields. Further investigation revealed that the key intermediate (ylide **20**) of this catalytic annulation reaction could be generated; the possible mechanism is also depicted in Scheme 4.6.

After screening phosphine catalysts, it was found that the more nucleophilic $EtPh_2P$ was the best catalyst for [3 + 2] annulation of allylic compounds, furnishing 2-substituted 1,1-dicyanoalkenes **22** in up to 95% yield (Scheme 4.7).[15]

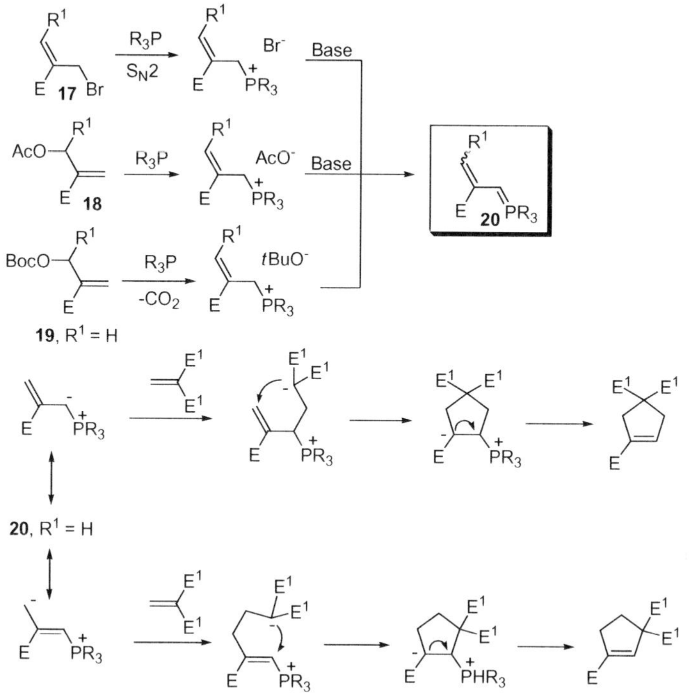

Scheme 4.6

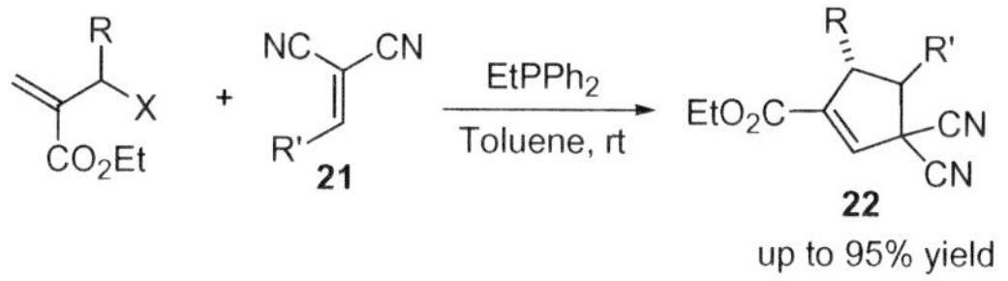

Scheme 4.7

MBH adduct **23**, containing a 1,6-diene unit, has been treated with a stoichiometric amount of HV(CO)$_4$(dppe) to give **24** in 70% yield after 3 h at room temperature.[16] This radical cyclization could be also initiated by CpCr(CO)$_3$H catalyst. Jones oxidation deleted the alcohol stereocenter and gave a separable mixture of two diastereomers, **25a** and **25b**, in 57% yield with a 3 : 2 ratio (Scheme 4.8). This synthetic strategy has also been applied in the construction of six-membered rings and the decalin framework.

t-Butyl 3-aryl-3-hydroxy-2-methylenepropanoates **26** have been treated with a catalytic amount of conc. H$_2$SO$_4$ (40 mol.%) in benzene under reflux for 30 min, followed by removal of the solvent and addition of a solution of trifluoroacetic anhydride (TFAA) in CH$_2$Cl$_2$ to give (*E*)-2-arylideneindan-1-ones **27** in moderate yields *via* an intermolecular and an intramolecular Friedel–Crafts reaction

Scheme 4.8

R = H, 4-Me, 4-Et, 4-*i*Pr,
2-Me, 4-Br

Scheme 4.9

Scheme 4.10

(Scheme 4.9).[17] Following the hydrogenation of compounds **27** in the presence of 5% Pd/C catalyst (40 psi) the corresponding products **28** could be obtained with good yields.

When MBH derivatives **29**[18] bearing the electron-donating groups such as Me and OMe at the aniline moiety were treated with polyphosphoric acid (PPA)[19] at 80–90 °C, 4-amino-2-benzylideneindan-1-ones **30** were obtained in 57–61% yields as major products, along with 1-amino-9a,10-dihydro-(4b*H*)indeno[1,2-*a*]inden-9-ones **31** in 13–18% yields (Scheme 4.10).[20] However, in the case of MBH derivatives **29** bearing groups such as H and Cl at the aniline moiety, 3-benzylidene-3,4-dihydro-1*H*-quinolin-2-ones **32** were obtained in good yields. It is interesting to note that such a subtle difference in the electron density at the aniline moiety caused such a strikingly different result.

By intramolecular Friedel–Crafts cyclization of **33** with 95% sulfuric acid in CCl$_4$ at room temperature for 0.5–4 h, 2-(9*H*-fluoren-9-yl)acrylic acid derivatives **34** have been obtained in 39–92% yields (Scheme 4.11).[21]

Intramolecular Heck coupling is a classical method for constructing cyclic products.[22] Taking an olefin functional group in MBH adducts into account, MBH adduct **35**, derived from *ortho*-halobenzaldehydes with a halogen atom such as I or Br at phenyl group, was designed to conduct intramolecular Heck coupling. Aza-MBH adducts **35** and **36** were subjected to a Pd(Ph$_3$)$_4$/triethyl-amine catalytic system in THF at 150 °C for 20 min, affording coupling products **37** and **38** in moderate yields (Scheme 4.12).[23] The acetate of MBH adduct **39**, with an iodine atom on the *ortho* position of the phenyl group, has also been subjected to the Pd-catalyzed intramolecular cross-coupling reaction. 1*H*-indene-2-carboxylic acid ethyl ester **40** was obtained in 40% yield in the presence of 10 mol.% [Pd(PPh$_3$)$_4$], 2.0 equivalents of In, 0.5 equivalents of InCl$_3$, 3.0 equivalents of LiCl and 2.0 equivalents of nBuNMe$_2$ in DMF (0.25 M) at 100 °C for 2 h, (Scheme 4.13).[24]

Larhed and others have demonstrated that the yield of Heck reaction product between two small molecules can be dramatically improved using a rapid microwave-assisted intermolecular reaction.[25] For example, the aryl bromide

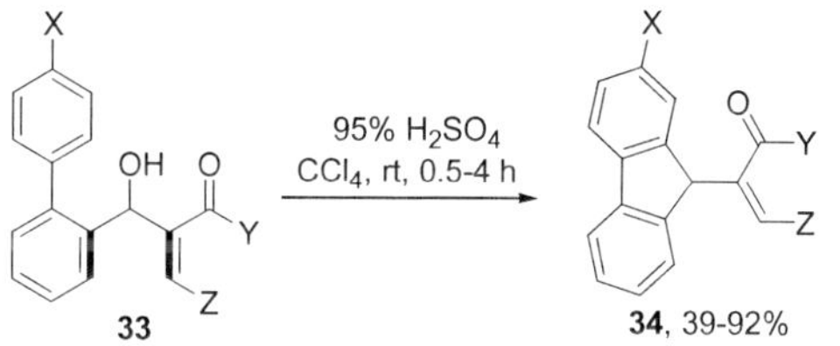

Scheme 4.11

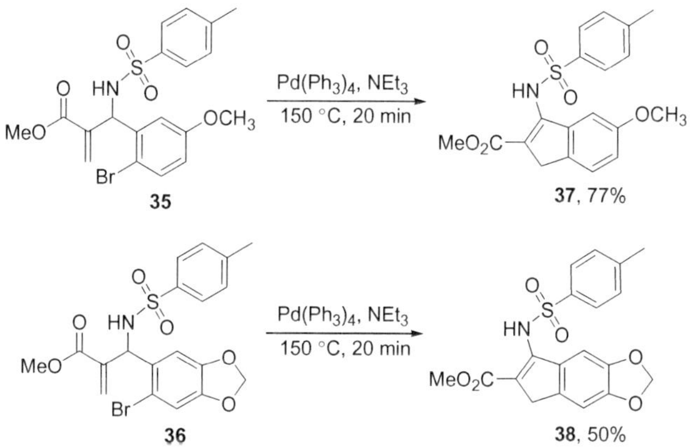

Scheme 4.12

Scheme 4.13

Scheme 4.14

Scheme 4.15

addition product (91% ee, **41**) can be converted into a disubstituted indanone in 10 min with 0.5 mol.% Pd(II) in DMF under microwave heating. Subsequent exposure of this diketone to methyl iodide in the presence of K_2CO_3 delivers indanone **42** with 70% ee and $>20:1$ dr (Scheme 4.14).[26] Another example clearly illustrates the role of microwave in Pd-catalyzed Heck reaction. When compound **43** was subjected to microwave irradiation (300 W) for 0.5 h at 100 °C instead of oil-bath heating at 120 °C for 21 h, the yield of product **44** increased significantly from 47% to 86% (Scheme 4.15).[27]

A palladium-catalyzed tandem Heck–aldol reaction in one-pot has been described to synthesize 2-carbonyl-1-indanol derivatives **47** from the reaction of *ortho*-halogenated aryl aldehyde **45** (X = I, Br) with MBH adduct **46**.[28] As shown in Scheme 4.16, 1-indanols **47** were produced as a mixture of diastereoisomers in a ratio of approximately 2 : 1 in moderate to good yields. Interestingly, MBH adduct derived from acrylonitrile could not afford the corresponding 1-indanol under identical conditions.

4.2.3 Six-membered-ring Carbocyclic Compounds

Since the Diels–Alder reactions of a MBH adduct were first disclosed by Hoffmann[29] and Basavaiah,[30] many papers based on using a MBH adduct as a

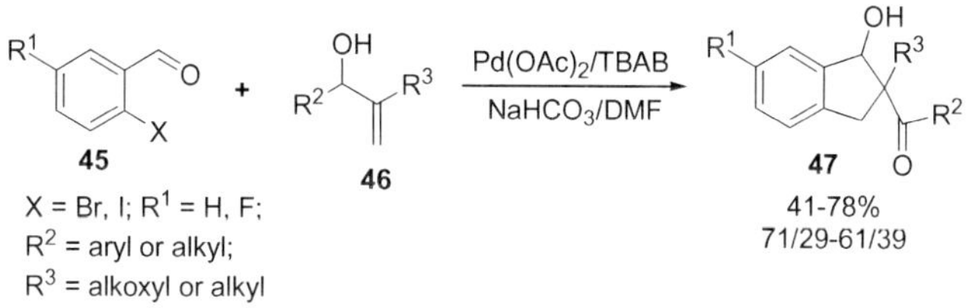

45

X = Br, I; R^1 = H, F;
R^2 = aryl or alkyl;
R^3 = alkoxyl or alkyl

46

$Pd(OAc)_2/TBAB$
$NaHCO_3/DMF$

47
41-78%
71/29-61/39

Scheme 4.16

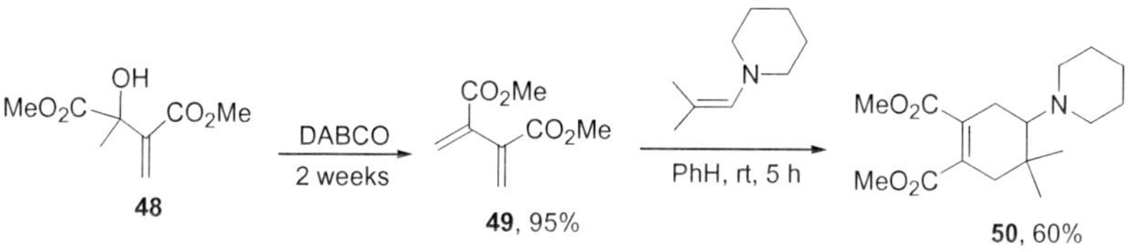

48

DABCO
2 weeks

49, 95%

PhH, rt, 5 h

50, 60%

Scheme 4.17

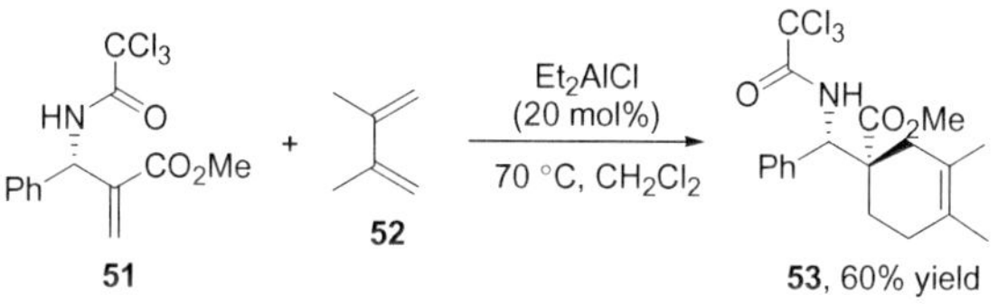

51

52

Et_2AlCl
(20 mol%)
70 °C, CH_2Cl_2

53, 60% yield

Scheme 4.18

dienophile have been extensively reported. For example, Ramachandran and co-workers have reported the Diels–Alder reaction of a MBH derivative in which compound **49** as an activated diene could be converted into multi-substituted cyclohexene derivative **50** in 60% yield (Scheme 4.17).[31] Moreover, chiral intermediate **51**, derived from [1,3]-sigmatropic rearrangement of the corresponding aza-MBH adduct, has been used as a dienophile to react with electron-rich diene **52**, affording the addition product **53** in 60% yield with high diastereoselectivity (>95 : 5) according to ^{1}H NMR spectroscopy (Scheme 4.18).[32]

Amri and co-workers have described a one-pot synthesis of 4-alkylidene-2-cyclohexen-1-ones **56** in 47–71% yields *via* a tandem three-step S_N2' substitution–deacetylation–cyclization series of reactions in the presence of anhydrous K_2CO_3 in absolute ethanol under reflux, using MBH acetates **54** as starting materials (Scheme 4.19).[33] Subsequently, decarboxylation products **57**, prepared from the reaction of MBH acetates and β-diketone or malonate esters, were reported to be treated with LHMDS (THF, 0 °C to room temp.) by Kim and co-workers,[34] affording 4-arylidenecyclohexane-1,3-dione derivatives **58** in 47–83% yields (Scheme 4.20). Similarly, when aza-MBH adducts **59a–c** were

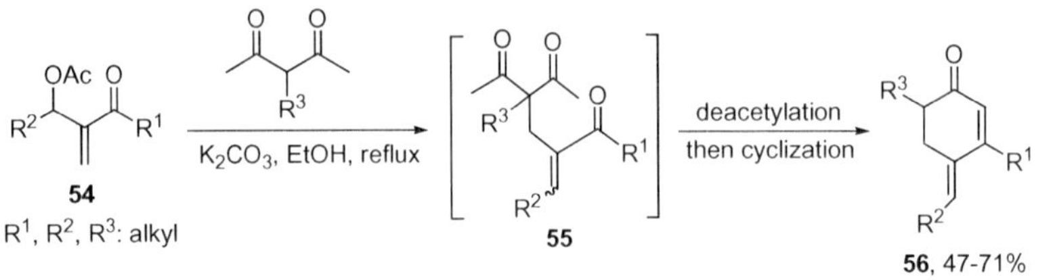

Scheme 4.19

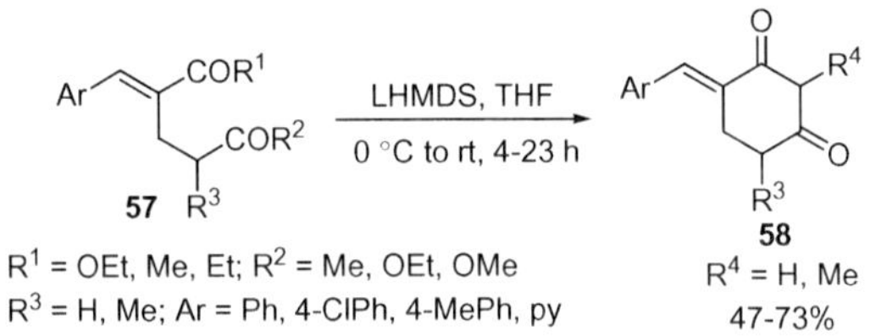

Scheme 4.20

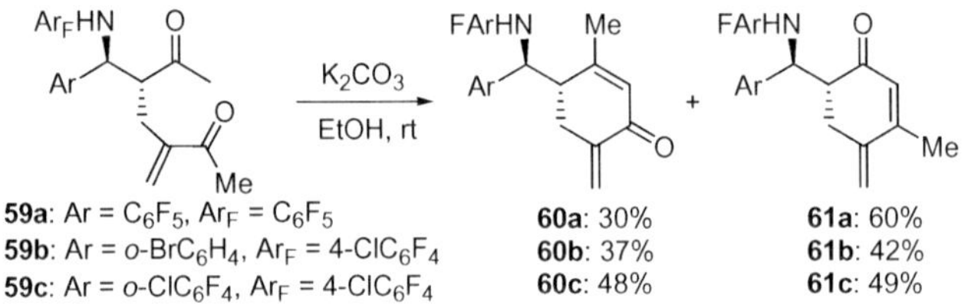

Scheme 4.21

treated with K_2CO_3 in absolute ethanol, the corresponding two isomers **60** and **61** were isolated in good yields (Scheme 4.21).[35]

The reactions of malonate derivatives **62** and Michael acceptor **63** have delivered cyclohexene derivatives **67** in moderate to good yields in the presence of DBU (Scheme 4.22).[36] This domino process involves the sequential Michael addition of **62** to the appropriate Michael acceptor **63** to give **64**, intramolecular aldol-type cyclization to **65**, dehydration to **66**, and DBU-promoted dealkoxycarbonylation to **67**. This method provides an efficient construction of highly functionalized cyclohexene derivatives starting from easily available MBH adducts.

Lu's group has developed a novel phosphine-catalyzed [3 + 3] annulation. In their first study, the reaction of the *t*-butyl allylic carbonate **68** and 2-(1-phenylethylidene)malononitrile **69** with PPh_3 (10 mol.%) in toluene under reflux, unfortunately, offered a non-cyclized product **70** in 95% yield (Scheme 4.23).[37] By screening the solvent effect, it was found that the reactions in polar solvents could give higher yields of [3 + 3] cycloadducts **71** than those

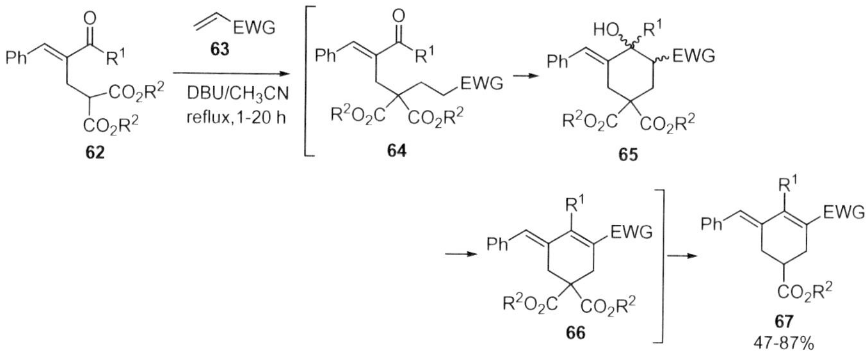

Scheme 4.22

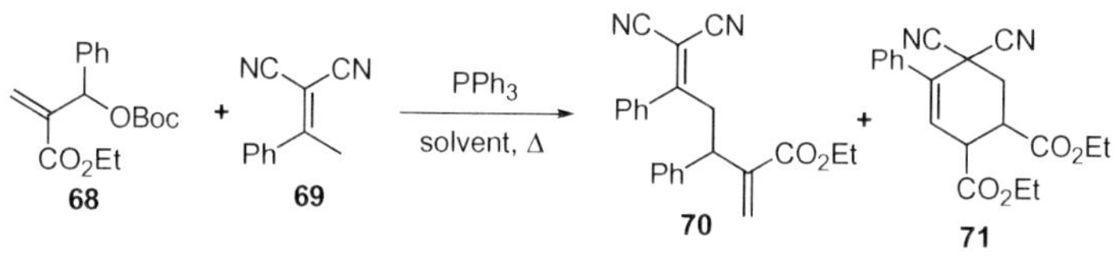

Scheme 4.23

in non-polar solvents. Under optimized conditions, a series of allylic *t*-butyl carbonates have been used to react with substituted alkylidenemalononitriles in refluxing *i*-PrOH to give functionalized cyclohexene derivatives **71** in 37–100% yields with moderate to good diastereoselectivity.

The S_N2' type compounds **72**, prepared from the reaction of MBH acetates and Grignard reagents, have been treated with 3 equiv. of H_2SO_4 at 60–70 °C, giving 3,4-dihydronaphthalenes **74** in moderate yields *via* an acid-catalyzed Friedel–Crafts-type reaction and subsequent acid hydrolysis of the ester moiety (Scheme 4.24).[38] However, when the reactions were conducted at reduced temperature (0–10 °C), 5,5-dimethyllactone derivatives **73** were obtained in 70–76% yields *via* acid-catalyzed lactonization. Such lactones could also be transformed into 3,4-dihydronaphthalenes **74** by treatment with H_2SO_4 in benzene at elevated temperature (60–70 °C).

With the development of the enantioselective allylic–allylic alkylation of α,α-dicyanoalkenes and MBH carbonates by dual organocatalysis of commercially available modified cinchona alkaloids and (*S*)-BINOL, Chen and co-workers have delivered an elegant construction of cyclohexene derivatives.[39] The intramolecular Michael reaction of allylic–allylic alkylation product **75a** could be cyclized to give the desired cyclohexene **76** in the presence of DBU (Scheme 4.25). In the presence of nucleophile $BnNH_2$, allylic compound **75b** furnished an unexpected cyclic product **77** rather than the formal double Michael adduct. Interestingly, the reaction of α,α-dicyanoalkene **79** and MBH carbonate **80** under optimized catalytic conditions directly afforded cyclohexene derivatives **81a–c** in

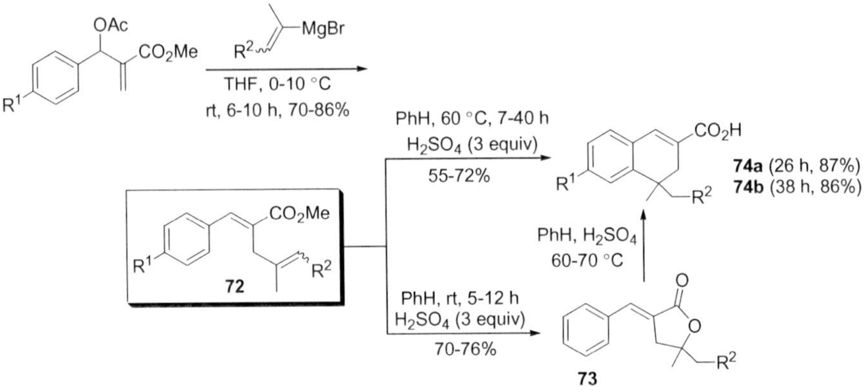

Scheme 4.24

moderate to good yields and excellent enantioselectivities with modest dr ratios *via* the expected allylic–allylic alkylation and subsequent domino intramolecular Michael reaction. Moreover, **81b** could be further converted into the diastereomerically pure diene **82** with elimination of HCN in the presence of KO*t*Bu.

β-Acetoxy-substituted enones **83** and **84** have been subjected to an intramolecular reductive cyclization using lithium in ammonia, which presents an effective way to synthesize substituted hydroindanones ($n = 1$) and decalones ($n = 2$) in moderate yields (Scheme 4.26).[40] In the process of cyclization, it was found that the ring size of substrates **83** and **84** has a key role in determining the stereoselectivity. This protocol represents a feasible method to construct the middle core of the clerodane family of natural products.[41]

Later, β-acetoxy-substituted cyclic enones **87** with various ring sizes were also examined for intramolecular Stetter cyclization. Under the optimized Stetter conditions (1.0 equiv. of thiazolium salt **88** and 1.2 equiv. of Et₃N in EtOH under reflux), bicyclic enedione **89** could be synthesized effectively (Scheme 4.27), but the yields mainly depended on the size of the ring in the substrate **87**.[42] Increasing the substrate ring size led to improved yields of **89**. Cycloheptenone and cyclooctenone derivatives **87c,d** afforded the desired enediones **89c,d** in 80% yields. To reduce the conjugated C–C double bond, McMurry's procedure[43] using TiCl₃ as a reductant was employed to give diketones **90** in excellent yields.

The electrohydrocyclization of MBH adducts **91** is another method used to construct the decalone skeleton. Compound **91** has been subjected to a constant current (cce) of 100 mA for 5 h using a tin anode and a platinum cathode in an undivided cell under an inert atmosphere of argon with 0.1 M tetraethylammonium chloride as the supporting electrolyte in aqueous acetonitrile (Scheme 4.28).[44] The reaction afforded cyclization product **93** in 69% yield. However, for substrate **92**, bearing an ether group on the tether, cyclization products **94** with retention of the alkoxy substituents were afforded. The same phenomenon was also found in the electrohydrocyclization of bis-enone

CO2Me
Ph
76, 95%
63:37 dr, 96%/94% ee
CN
NC
DBU (30 mol%)
DCM, rt, 15 h
CO2Me
Ph
NC CN
H
Ph
75a
BnNH2
MeOH, rt, 18 h
Ph
CO2Me
NC
Ph
CO2Me
75b, 94% ee
HN OMe
CO2Me
Ph
NC
Ph
77, 59%
96% ee
HCl/THF
rt, 3 h
MeO2C CO2Me
Ph
NC
Ph
78, 74%
94% ee
OBoc CO2Me
Ar
NC CN
+
79
80a, Ar = Ph
80b, Ar = p-Cl-Ph
80c, Ar = p-MeO-Ph
cat. (10 mol%)
(S)-BINOL (10 mol%)
DCE, 40 °C
18-24 h
CO2Me
Ar
CN
NC
81a, 69%, 69:31 dr., 94%/97% ee
81b, 85%, 72:28 dr., 92%/90% ee
81c, 64%, 71:29 dr., 96%/93% ee
KOtBu, rt, 0.5 h
for 81b
CO2Me
p-Cl-Ph
NC
82, 44%, 92% ee
cat: (DHQD)2AQN = hydroquinidine (anthraquinone-1,4-diyl) diether

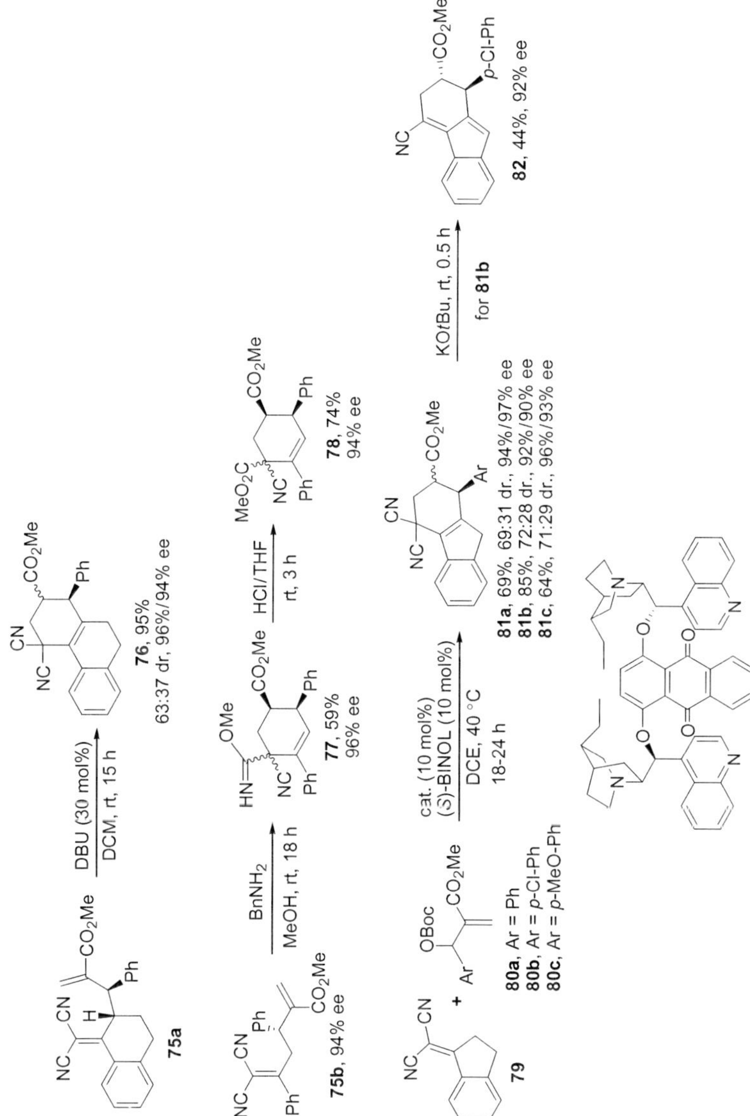

Scheme 4.25

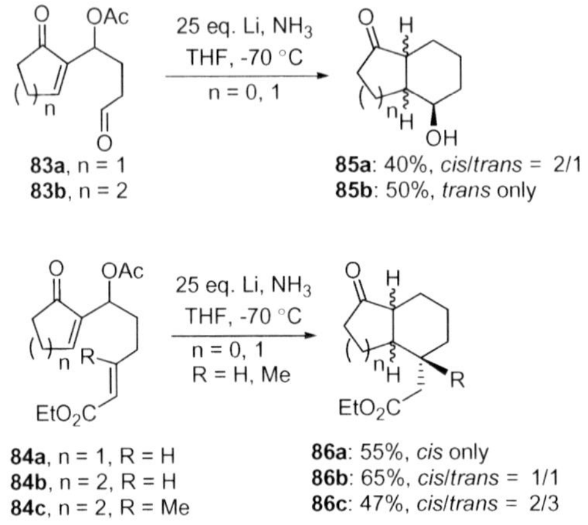

83a, n = 1
83b, n = 2

25 eq. Li, NH$_3$
THF, -70 °C
n = 0, 1

85a: 40%, *cis/trans* = 2/1
85b: 50%, *trans* only

84a, n = 1, R = H
84b, n = 2, R = H
84c, n = 2, R = Me

25 eq. Li, NH$_3$
THF, -70 °C
n = 0, 1
R = H, Me

86a: 55%, *cis* only
86b: 65%, *cis/trans* = 1/1
86c: 47%, *cis/trans* = 2/3

Scheme 4.26

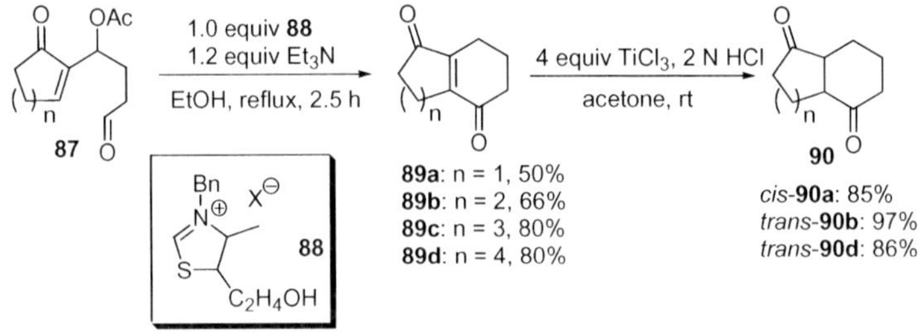

87

1.0 equiv **88**
1.2 equiv Et$_3$N

EtOH, reflux, 2.5 h

88

89a: n = 1, 50%
89b: n = 2, 66%
89c: n = 3, 80%
89d: n = 4, 80%

4 equiv TiCl$_3$, 2 N HCl

acetone, rt

90

cis-**90a**: 85%
trans-**90b**: 97%
trans-**90d**: 86%

Scheme 4.27

systems **95** and **96**, which respectively furnished tricyclic products **97–100** in moderate yields.

4.2.4 Aromatic Compounds

MBH acetates **101** furnish naphthalenes **102** upon treatment with primary nitroalkanes (2 equiv) in the presence of K$_2$CO$_3$ (3 equiv) in DMF (Scheme 4.29).[45] Regardless of the electron-withdrawing group on the MBH acetates, ethoxycarbonyl- and acetyl-naphthalenes **102** were obtained in good yields. The disclosed reaction mechanism may involve tandem nucleophilic addition–elimination reaction (S_N2'), intramolecular S_NAr reaction and elimination of nitrous acid. When using the MBH acetate derived from acrylonitrile, only trace amounts of the corresponding naphthalene were isolated. A similar phenomenon has also been disclosed by Murthy and co-workers.[46] MBH acetates **103**, prepared from substituted 2-chloronicotinaldehydes, were used to react with nitroethane to obtain multi-substituted quinolines **104** in high yields

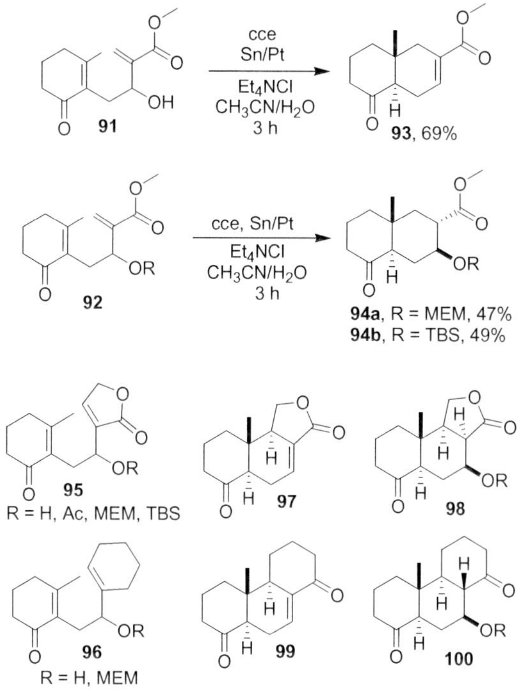

Scheme 4.28

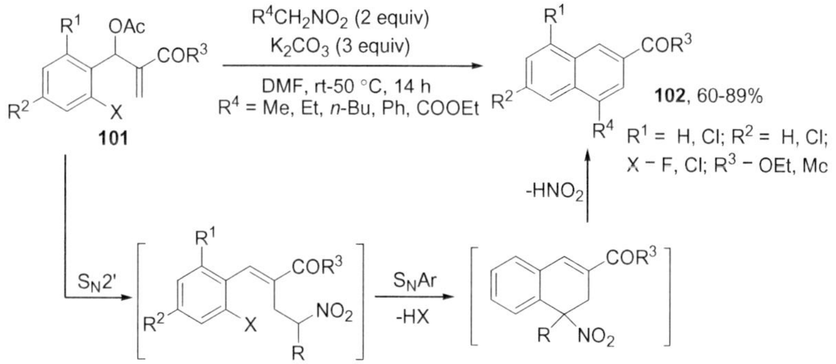

Scheme 4.29

in the presence of K_2CO_3 in DMF at 50–60 °C (Scheme 4.30). When the same substrates were reacted with ethyl cyanoacetate, higher temperatures (110–125 °C) and longer times (~ 10 h) were required to give the substituted 8-cyanoquinolines **105**, in only 50–60% yields.

Using $BF_3 \cdot OEt_2$ as a catalyst, MBH acetate **106** reacts with readily available α-EWG ketene-(S,S)-acetals **107** to afford effectively S_N2'-type product **108** in

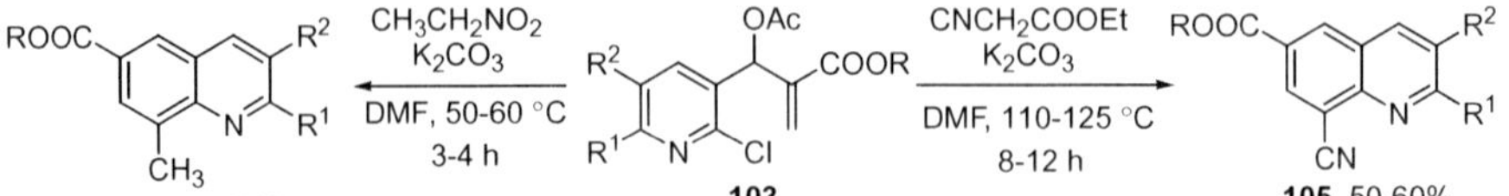

104, 80-92%   **103**   **105**, 50-60%

CH₃CH₂NO₂ / K₂CO₃ / DMF, 50-60 °C / 3-4 h

CNCH₂COOEt / K₂CO₃ / DMF, 110-125 °C / 8-12 h

Scheme 4.30

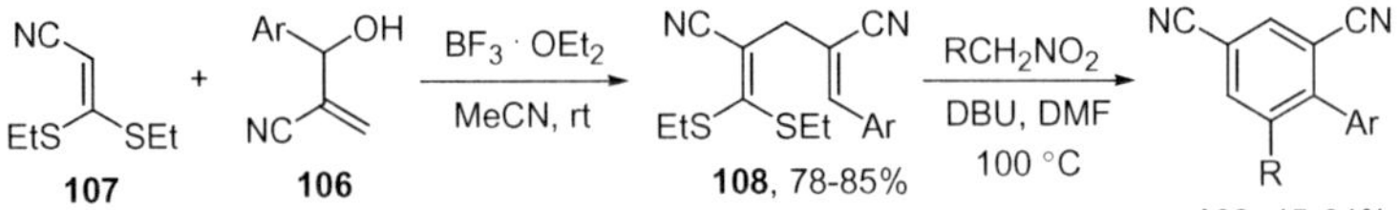

107   **106**   **108**, 78-85%   **109**, 45-64%

BF₃·OEt₂ / MeCN, rt

RCH₂NO₂ / DBU, DMF / 100 °C

Ar = 4-MeOPh, 4-EtOPh, 3,4-O₂CH₂Ph
R = Me, Et, CO₂Et

Scheme 4.31

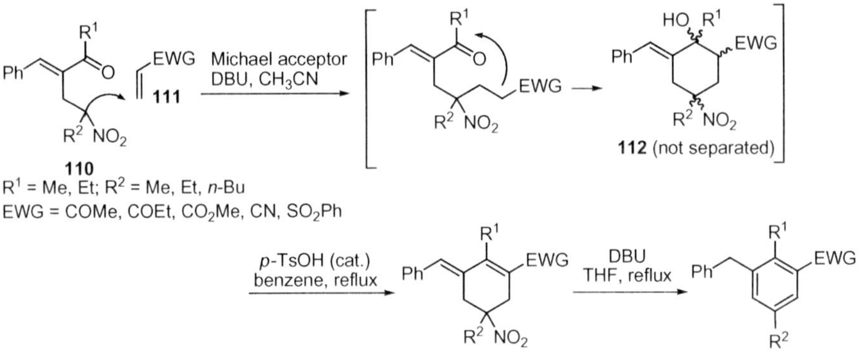

110
R¹ = Me, Et; R² = Me, Et, *n*-Bu
EWG = COMe, COEt, CO₂Me, CN, SO₂Ph

Michael acceptor **111** / DBU, CH₃CN

112 (not separated)

p-TsOH (cat.) / benzene, reflux

113, 41-83%

DBU / THF, reflux

114, 78-91%

Scheme 4.32

good yields. Based on the previous development of a [5C + 1C] annulation strategy,[47] the annulation reactions of 1,4-dienes **108** with nitroalkanes were examined. Unsymmetrical biaryls **109** were obtained in moderate yields from the one-pot annulation-aromatization (Scheme 4.31).[48] This process involves a sequence of intermolecular Michael, intramolecular Michael-S_NV reaction and subsequent aromatization.

The S_N2'-type compounds **110** as four-carbon units, prepared from MBH acetates and primary nitroalkanes, have been utilized to prepare poly-substituted benzenes *via* a new and regioselective [4 + 2] benzannulation protocol (Scheme 4.32).[49] Upon treatment with DBU, the reaction of compounds **110** and Michael acceptor **111** afforded six-membered ring intermediates **112**; after usual workup, this crude mixture was subjected to dehydration conditions (*p*-TsOH in benzene under reflux) and the dehydration product **113** was obtained in good overall yield. Final aromatization was conducted under refluxing conditions in the presence of DBU in THF, affording benzene derivatives **114** in 78–91% yields.

Compounds **115** with an ester group at R^1 have been subjected to aldol-type or Dieckmann cyclization by treatment with *t*-BuOK in THF to give six-membered ring intermediates **116** in 63–71% yields; the use of DBU as base did not give the desired product **116** (Scheme 4.33).[50] Aromatization could then be achieved to furnish polysubstituted phenol derivatives **117** in 44–76% yields in the presence of DBU. However, the reactant **115** containing a cyano group at R^1 did not undergo cyclization under identical conditions.

Based on the above methods for the construction of an aromatic ring from MBH adducts, a [3 + 3] annulation strategy has also been developed to synthesize polysubstituted phenol derivatives. Active methylene compounds **119** as C3 component were heated with MBH acetates **118** as the 1,3-dielectrophilic in DMF at 70–90 °C, furnishing *ortho*-hydroxyacetophenone derivatives **120** in moderate yields (Scheme 4.34).[51] When EWG was an ester group, the product (*E*)-4-alkylidene-2-cyclohexen-1-one (**121**) was isolated in appreciable amounts in the case of ethyl acetoacetate. This result suggests that under the same reaction conditions decarbethoxylation took place more readily than the corresponding deacetylation. The reaction of 1,3-dinucleophilic reagent **122** as another C3 component with MBH acetate **123**, at 50–60 °C in DMF for 5 h, afforded multisubstituted phenol derivatives **124** in 40–59% yields (Scheme 4.35).[52] In contrast,

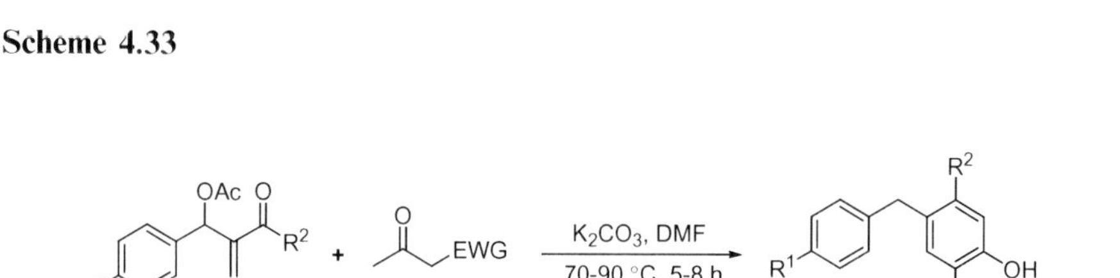

R^1 = CO$_2$Me, CO$_2$Et
R^2 = Me, Et, *n*-Bu
EWG = COMe, COEt, CO$_2$Me, CO$_2$Et

Scheme 4.33

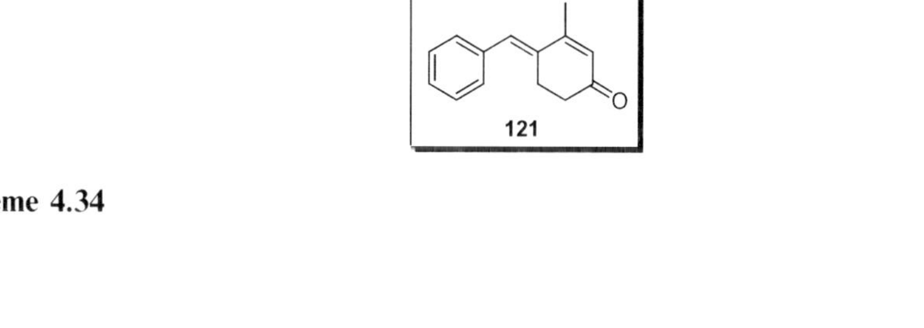

R^1 = H, Cl, Me
R^2 = Me, Et

EWG = COMe, CO$_2$Et, SO$_2$Me, COPh

Scheme 4.34

122 R^1 = OMe, OEt, Ph

123 R^2 = H, Cl, Me; R^3 = Me, Et

K$_2$CO$_3$, DMF 50-60 °C, 5 h

124 40-59%

125

126 R = Me, Et

1. *t*-BuOK (1.0 equiv) THF, rt, 5 h
2. *p*-TsOH (20 mol%) benzene, reflux, 20 h 41-55%

127

Scheme 4.35

128 R^1 = H, Cl, Me; R^2 = Me, Et

129

1) K$_2$CO$_3$ (3.0 equiv) DMF, rt, 2 h

2) *p*-TsOH (1.0 equiv) PhH, reflux, 7 h

130, 34-41%

K$_2$CO$_3$ (3.0 equiv) DMF, 50-60 °C 30 min

131, 56-64%

Scheme 4.36

the reaction of **126** with 1,3-diphenylacetone (**125**), under *t*-BuOK/THF conditions instead of the above conditions (K$_2$CO$_3$, DMF), proceed by cyclization, *via* an aldol reaction, to give, under typical dehydration conditions (*p*-TsOH, benzene, reflux), 2,6-diarylphenol derivatives **127** in 41–55% yields. Moreover, the use of 1,3-dinitroalkanes **129** as the 1,3-dinucleophilic components has also been developed, to synthesize poly-substituted nitrobenzenes **131** from the reaction with MBH acetates **128** (Scheme 4.36).[53]

With an efficient synthetic protocol to synthesize polysubstituted α-pyrones, including chromen-2-one, from MBH adduct in hand, various substituted α-pyrones **132** have been subjected to reaction with dimethyl acetylenedicarboxylate (DMAD) to obtain aromatic compounds **133** in excellent yields *via* Diels–Alder reaction (Scheme 4.37).[54] Naphthalenes and phenanthrenes could be further obtained from some of the aromatic compounds by DDQ oxidation.

Manganese(III)- and Ce(IV)-induced intramolecular homolytic malonylation of malonate derivatives with an aromatic ring at the δ-position has been utilized to construct tetrahydro- or dihydro-naphthalene frameworks.[55] S_N2'-type adducts **134** of MBH acetates were used to investigate the Mn(III)-mediated

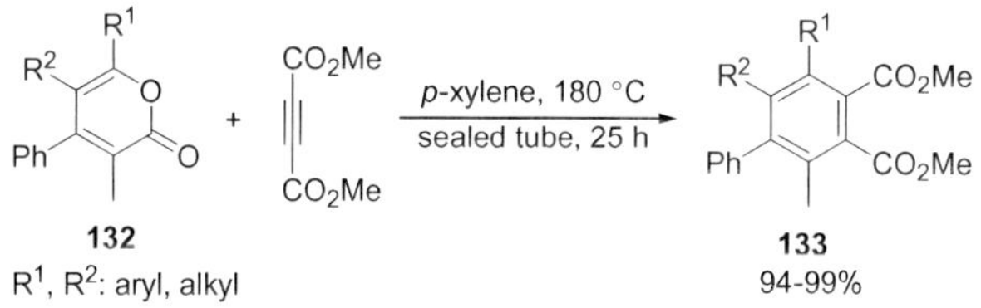

Scheme 4.37

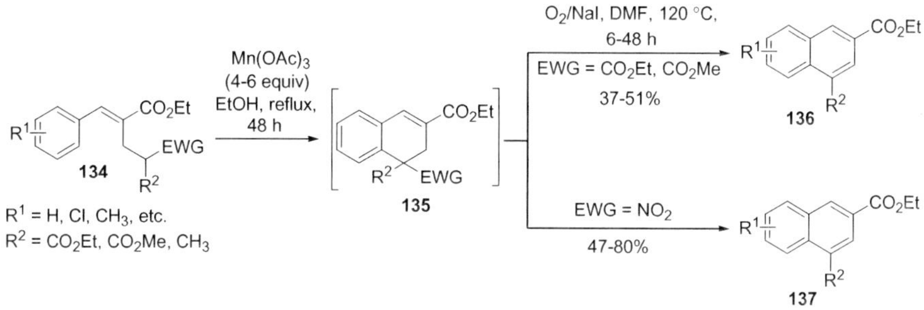

Scheme 4.38

oxidative free-radical cyclization reaction (Scheme 4.38).[56] Treatment of compounds **134** (EWG = ester group) with $Mn(OAc)_3$ (6 equiv) in refluxing ethanol afforded the dihydronaphthalene derivatives **135**; these crude products were then further subjected to a NaI/O_2 system in DMF to give naphthalenes **136** in low to moderate yields. However, extending the scope to other substrates, **134** (EWG = NO_2), furnished relatively high yields of naphthalenes **137** without the use of an NaI/O_2 system. Moreover, the electronic properties of the R^1 substituent in the aromatic ring also have a strong influence on the yield of naphthalene derivatives **136** and **137**. Later, another method for the synthesis of naphthalenes **136** was disclosed by Kim (Scheme 4.39).[57] In this case, treatment with trifluoroacetic acid (TFA) (0.5 mL) and 3.0 equiv. of H_2SO_4 at 40–50 °C afforded naphthalenes **139** from compounds **138** in good yields. Substrate **138c** has also been converted into naphthalene **139c**, in 39% yield.

With the development of the Friedel–Crafts type ring-opening reaction (FCRO) of *N*-tosylaziridines with arene compounds,[58] Kim and co-workers reported the intramolecular FCRO reaction of *N*-tosylaziridines.[59] Aziridine derivatives **142** could be prepared from the reaction of cinnamyl bromide **140** and *N*-tosylimine **141** in the presence of Me_2S and K_2CO_3 in CH_3CN at room temperature (Scheme 4.40). Such aziridine derivatives **142** were then subjected to benzene/H_2SO_4 (3 equiv)/refluxing conditions and within 2 h provided 1-arylnaphthalene derivatives **143** in 79–84% yields *via* the intramolecular FCRO reaction and concomitant elimination of *p*-toluenesulfonamide.

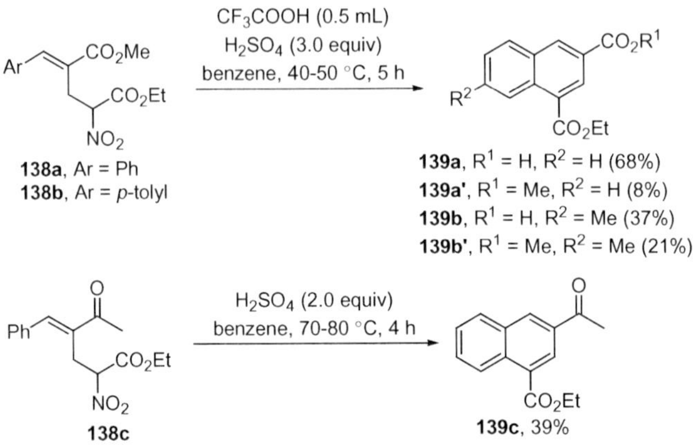

Scheme 4.39

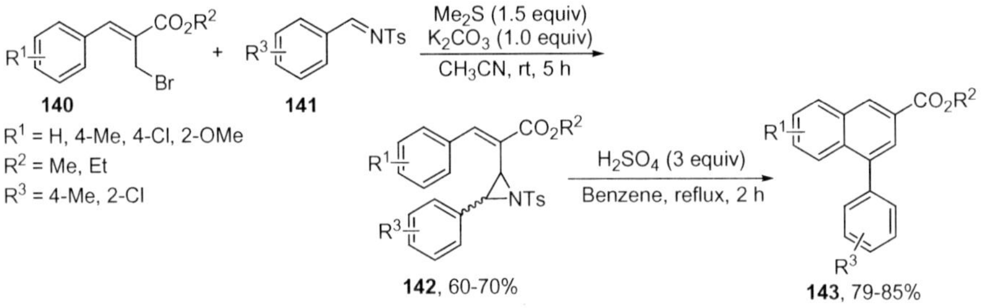

Scheme 4.40

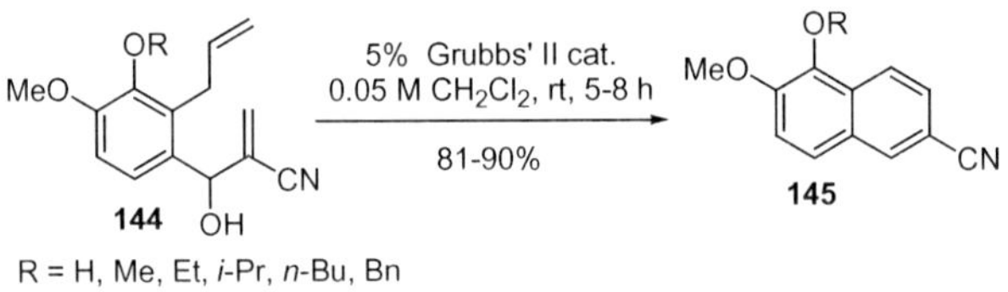

Scheme 4.41

RCM of MBH adducts **144** by Grubbs' catalyst (second generation) furnished cyclized products that underwent elimination of water to afford cyanonaphthalenes **145** in excellent yields (Scheme 4.41).[60]

4.2.5   Bridged Compounds

Activated olefin **146** as a dienophile, prepared from phenyl vinyl sulfone and acetaldehyde in the presence of DBACO, has been tested in intermolecular

Diels–Alder (D-A) reactions (Scheme 4.42).[61] The reactions of α-methylene-β-keto sulfone **146** with conjugated dienes **147** afforded the [4 + 2] carbon-bridged adducts **148** in moderate yields under mild conditions.

With the development of a divalent titanium reagent mediated cyclization of 2,7- and 2,8-enyn-1-ol derivatives,[62] enynes **149** have been subjected to the Ti-mediated cyclization to give interesting cyclic products.[63] Substrate **149a** was treated with 2.3 equiv. of titanium reagent [Ti(O-*i*-Pr)$_4$/2*i*-PrMgCl)] at $-40\,°$C to $-20\,°$C for 2 h, the mixture was then quenched by addition of H$_2$O, affording bicyclic compound **150** in a 95 : 5 diastereomeric ratio and 84% yield (Scheme 4.43). However, enynes **149** having a more sterically demanding ester group gave lower yields of **150**, along with higher yields of monocyclic compounds **151**. For example, substrate **149c** gave a 93% yield of **151c** along with trace amount of **150c**. Scheme 4.44 illustrates the synthetic method

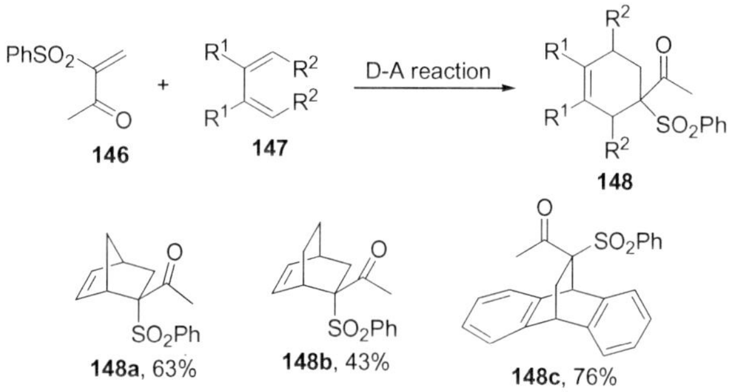

Scheme 4.42

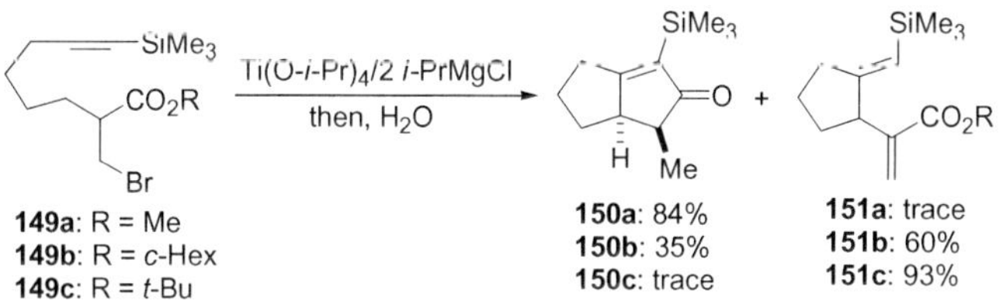

Scheme 4.43

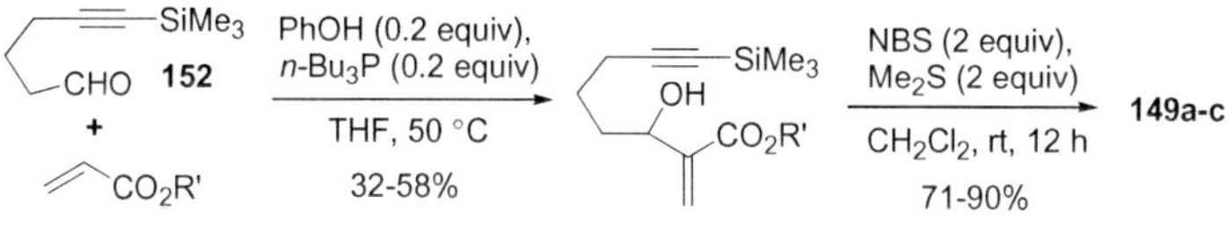

Scheme 4.44

employed to obtain substrates **149** for the cyclization reactions. Thus, compounds **149a–c** were synthesized by the MBH reaction of acrylic esters with 6-trimethylsilylhex-5-ynal **152** followed by bromination of the resulting alcohols; the bromination proceeded stereoselectively to yield the corresponding bromoesters as (Z)-isomers.

Under enyne cross-metathesis conditions, the intermolecular reaction of the α,ω-dienes **153**, derived from the MBH reaction, with different terminal alkynes **154** afforded triene intermediates that cyclized spontaneously under the reaction conditions to give substituted *cis*-hexahydro-1*H*-indenes **155** (Scheme 4.45), which can be further transformed into steroid analogues *via* TBS deprotection and oxidation. However, metathesis reactions starting with **156** only furnished trienes **157** [as (E/Z) mixtures] and no spontaneous intramolecular cycloaddition occurred. Even at elevated reaction temperatures, trienes **157** cyclized only slowly to give octahydronaphthalene diastereomers. With deprotection of the TBS and subsequent Dess–Martin oxidation, trienes **157** could be converted exclusively into *cis*-fused 7-substituted 6,7-dehydrodealone-1-one-10-carboxylic esters **158** in 50–60% yields. Moreover, c ross-metathesis of TBS-unprotected MBH adduct **159** with alkynes **154** along with treatment with Dess–Martin periodinane (DMP) in one pot could conveniently produce the corresponding bicyclic ketones **160** in moderate yields.[64]

Successive Michael reactions of cyclohexenones **161** with MBH derivatives **162** in the presence of K_2CO_3 and TBAB (n-$Bu_4N^+Br^-$) present a practical method for the construction of polyfunctionalized bicyclo[3.3.1]nonenones **163** (Scheme 4.46).[65] Such a construction can be carried out in both stepwise and one-pot reactions with similar regioselectivities. From *ab initio* calculations on the intramolecular Michael reactions, it was found that the regioselectivity was somewhat dependent on the substitution pattern of the reactants (R^2 or R^3).

Tropone is a molecule with interesting electronic properties, which can easily undergo nucleophilic attack at C(2) and C(7) according to frontier molecular orbital theory[66] to be converted into various cycloadducts.[67] Based on the discovery of phosphine-catalyzed [3 + 2] annulation, ethyl 2-bromomethyl-2-propenoate (**164**, X = Br, E = CO_2Et) and the C6 component (tropone **165**, 1.0 equiv) in toluene were added drop-wise with a syringe pump into a mixture of Ph_3P (10 mol.%) and K_2CO_3 (1.5 equiv) in refluxing toluene to give the [3 + 6] annulation product **166a**, which was isolated in 84% yield after chromatography (Scheme 4.47).[68] To evaluate the suitability of substrates, a series of the C3 components, including bromides, chlorides, acetates and *t*-butyl carbonates, were examined. Among them, the acetate derivative **164** is the most efficient C3 component, furnishing cycloadducts **166** in good to excellent yield.

Bromides **167** have been subjected to phosphine-catalyzed intramolecular cycloaddition in the presence of 20 mol.% of PPh_3 as catalyst and Cs_2CO_3 as base, giving the bicyclic compounds **168** under mild conditions.[69] In most cases, these bicyclo[3.3.0] compounds **168** were obtained with high

Hexahydro-1H-indenes 155
up to 90% yield
R = -CH2OAc, -(CH2)2CO2Me,
-CH(OAc)C3H7, Ph, 4-MeOPh
Triene intermediate
Ru-cat., CH2Cl2, rt or 45 °C,
24 h, then PhMe, 80 °C, 24 h
or Ru-cat., C6H6, 80 °C, 24 h
153a: n = 1; EWG = CO2Me; X = TBS
153b: n = 1; EWG = CN; X = TBS
154
158a: (60%; R = -CH2OAc)
158b: (54%; R = -(CH2)2CO2Me)
158c: (50%; R = -CH(OAc)C3H7)
158d: (53%; R = Ph)
1. TBAF, THF
2. DMP, CH2Cl2
rt
[4+2]
Trienes 157
60-90%
5 mol% Ru-cat.
CH2Cl2, 45 °C, 24 h
156a: n = 2; EWG = CO2Me; X = TBS
156b: n = 2; EWG = CN; X = TBS
154
Ru-cat.
160a: 50%
160b: 70%
1. Ru-cat., 45 °C
2. DMP, 25 °C
CH2Cl2
159
154

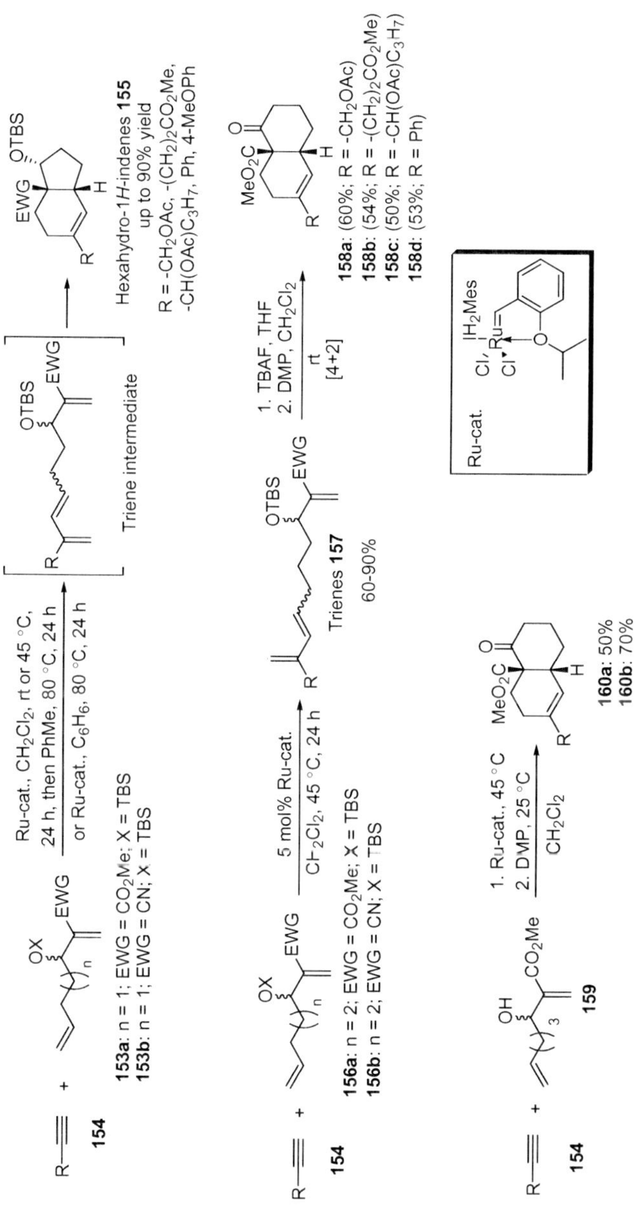

Scheme 4.45

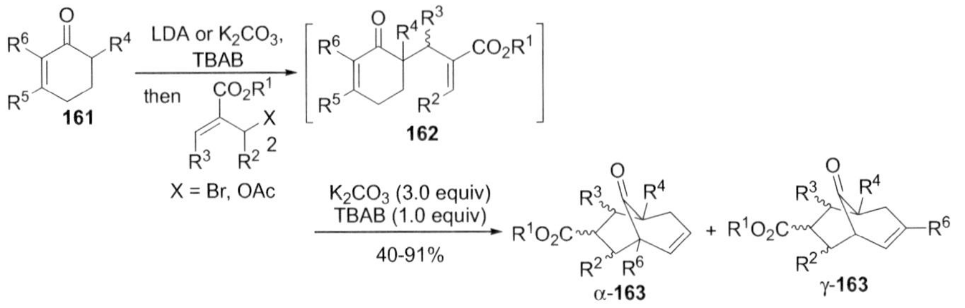

Scheme 4.46

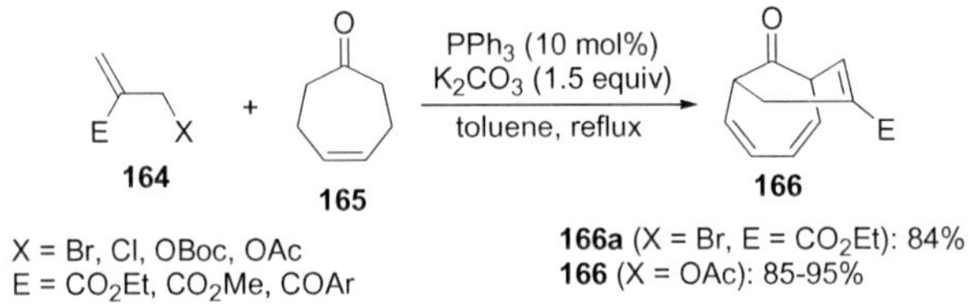

Scheme 4.47

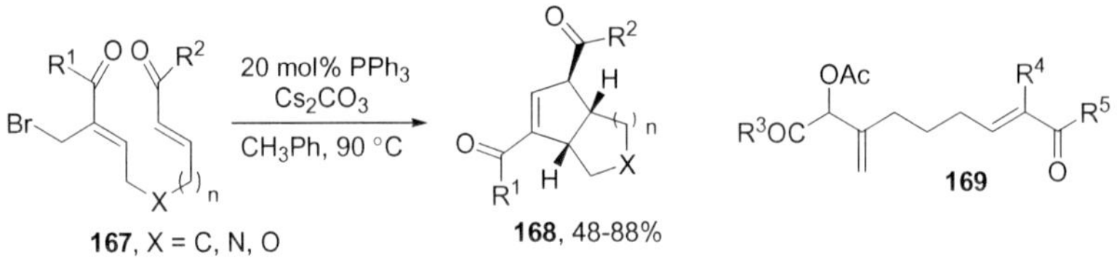

X = Br, Cl, OBoc, OAc
E = CO_2Et, CO_2Me, COAr

166a (X = Br, E = CO_2Et): 84%
166 (X = OAc): 85-95%

Scheme 4.48

167, X = C, N, O

168, 48-88%

169

diastereoselectivities in moderate to good yields; consequently, this represents an excellent intramolecular variant of the methodology for cycloaddition first developed by Krische and co-workers.[70] When X was a heteroatom, compounds **168** containing tetrahydropyrrole and tetrahydrofuran ring structures could also be synthesized. Moreover, the catalytic annulation reaction could occur smoothly not only from the bromides **167** but also from the corresponding acetates **169** (Scheme 4.48). In a further study, a strong base effect was found in the [3 + 2] cyclization of substrate **170a**.[71] Both benzobicyclo[4.3.0] compounds **171** and **172** could be synthesized selectively from the same starting material simply by the appropriate choice of base (Scheme 4.49). Moreover, using α-methyl α,β-unsaturated esters **170b** as substrates, benzobicyclo[4.3.0] compounds **173** bearing a quaternary carbon center were obtained in good yields. The authors also proposed a possible mechanism for the catalytic intramolecular ylide annulation (Scheme 4.50).

EtO$_2$C
COR^2
H
H
R^1
171
95-99% yield
171/172 > 90/10
EtO$_2$C
COR^2
H
H
R^1
172
82-96% yield
171/172 < 19/81
20 mol% PPh$_3$, Na$_2$CO$_3$
CH$_3$Ph, 80-90 °C
20 mol% PPh$_3$, Na$_2$CO$_3$
CH$_3$Ph, 80-90 °C
R = H
EtO$_2$C
Br
R^1
R
COR^2
170a: R = H
170b: R = Me
R^1 = 4-Cl, 4-Br, 4-OMe
R^2 = OMe, OEt, C$_6$H$_4$(p-Cl)
R = Me
20 mol% PPh$_3$, Na$_2$CO$_3$
CH$_3$Ph, 90 °C
EtO$_2$C
COR^2
H
H
R^1
173
78-87% yield
dr > 92/8

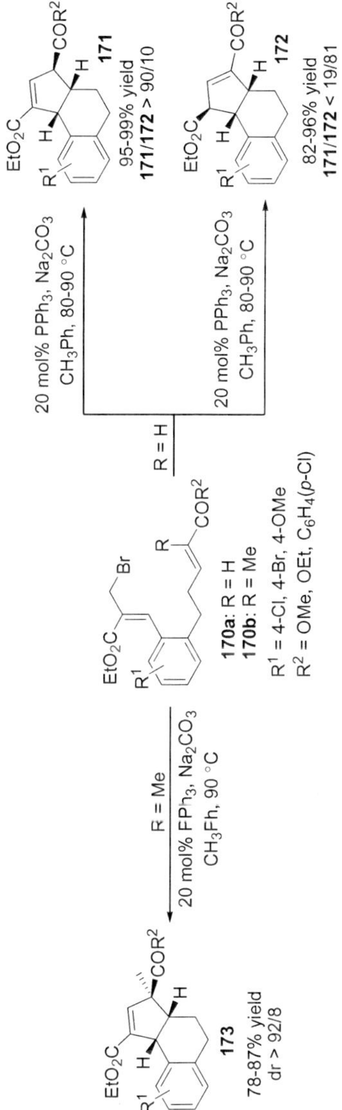

Scheme 4.49

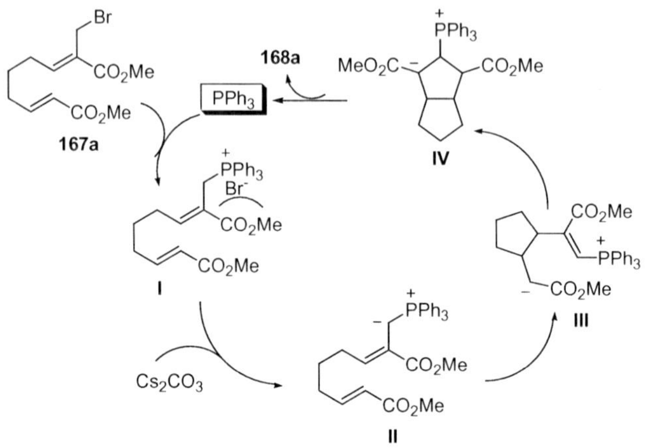

Scheme 4.50

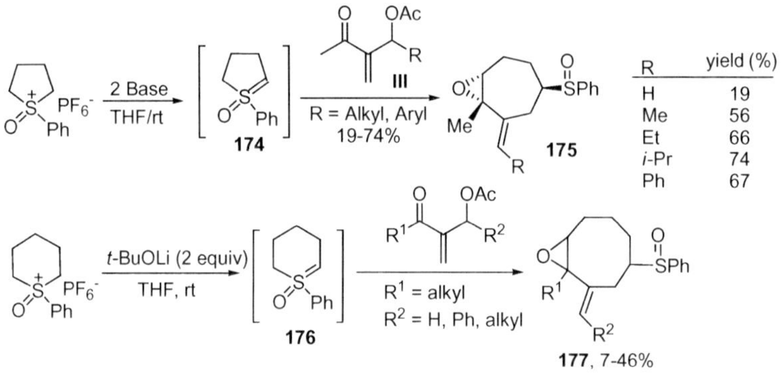

Scheme 4.51

4.2.6 Medium-sized-ring Carbocyclic Compounds

The Corey–Chaykovsky reaction of oxosulfonium ylides has provided a useful method for the synthesis of cyclopropane derivatives by Michael-type addition of the ylide to the activated carbon–carbon double bond.[72] The five-membered cyclic oxosulfonium ylide **174** was first developed to construct seven-membered rings (Scheme 4.51).[73] Cycloheptene epoxide derivatives **175**, which have a *trans* relationship between the phenyl sulfinyl group and the oxirane ring, were obtained with high stereoselectivity in 19–74% yield *via* a Michael-type addition of the ylide followed by elimination of the acetoxy group and an intramolecular Corey–Chaykovsky reaction. By applying this synthetic strategy, six-membered-ring oxosulfonium ylide **176** has been utilized to synthesize cyclooctene oxide derivatives **177** in relatively low yields.[74]

The starting material **178**, prepared from the reaction of MBH acetate and phenylethynylmagnesium bromide in THF at room temperature in the presence of CuI, has been treated with H_2SO_4 in CH_2Cl_2 to give, unexpectedly, 7*H*-benzocycloheptene derivative **180** in 41–62% yields (Scheme 4.52).[75] This method involves an intramolecular Friedel–Crafts alkenylation reaction of triple bond-tethered methyl cinnamates. However, the same reaction with methyl-group-substituted starting materials **179** gave low yields of 7*H*-benzocycloheptene derivatives **181** along with some by-products such as benzocycloheptene derivatives, naphthalene derivatives and two kinds of hydration products.

Compounds **182** bearing a 2-bromo group on the phenyl ring present at the 3-position of the isoxazole ring and a double bond on the carbon chain at C4 have been subjected to radical-promoted intramolecular cyclization using tributyltin hydride, affording novel isoxazolo-benzazulenes **183** in 40–48% yields along with dehalogenated products **184** (Scheme 4.53).[76]

The intramolecular conjugate displacement (ICD), in which the ring closure formally resembles both a conjugate addition and a S_N2' displacement, has been developed to achieve the construction of heterocycles.[77] An effective synthetic route to carbocycles has also been developed from the ICD of designed MBH adducts (Scheme 4.54).[78] When compound **185** was treated with DBU, ICD product **186** was obtained in moderate yield, but along with compound **187**, which can be further transformed into compound **186** in 75% yield in the presence of DBU for 12 h. Using an inorganic base such as Cs_2CO_3, **186** only was isolated in up to >99% yield from compound **185**. By extension of this strategy to designed substrates **188**, ICD products **189**

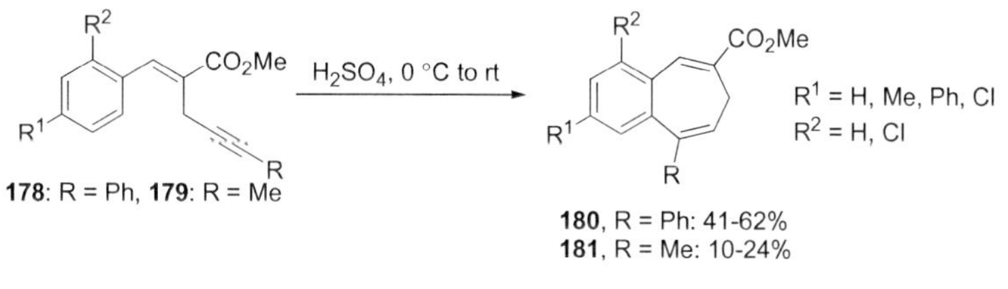

Scheme 4.52

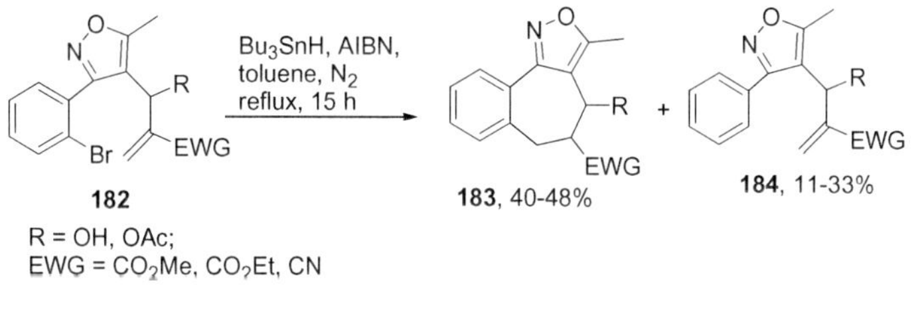

Scheme 4.53

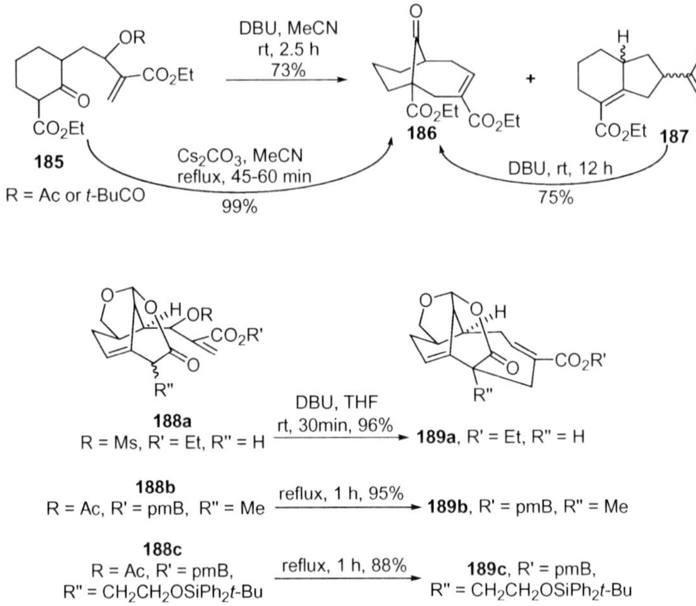

Scheme 4.54

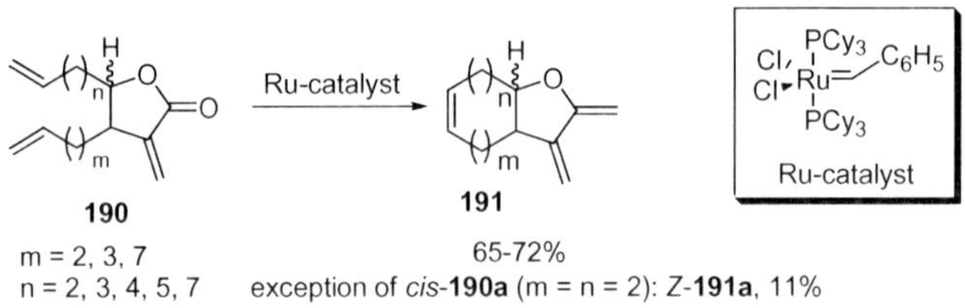

Scheme 4.55

containing an unusual and strained core framework related to the Ras farnesyl transferase inhibitors CP-225,917 and CP-263,114[79] have been synthesized in excellent yields. This method offers an effective route to the construction of such a framework, in a field that is still attractive to the organic synthetic community.

Intramolecular ring-closing metathesis has been used to construct cyclic products, especially for those compounds containing a medium- to macro-sized ring. The standard protocol utilized for all ring-closing metatheses involved 6 mM solutions of **190** in CH_2Cl_2. After the introduction of 30 mol.% of ruthenium catalyst under N_2, each reaction mixture was heated at 50 °C for 24 h (Scheme 4.55).[80] All of the substrates **190** with *cis* or *trans* geometry reacted smoothly and were efficiently converted into cyclized products. A mixture of (E)- and (Z)-isomers was generated in all successful cases in moderate yields, except for (*Z*)-cyclooctene **191a**, which was obtained ($m = n = 2$) in only 11% yield.

4.3 Synthesis of Oxygen-containing Heterocyclic Compounds

4.3.1 Oxiranes

Considerable attention has been paid to the preparation of epoxides and their ring-opening transformation. Intramolecular nucleophilic substitution reactions have furnished oxiranes from the corresponding MBH adducts. Scolastico and co-workers first described the transformation of compounds **192** into the vinyl epoxides **193**.[81] By optimizing the base used in the epoxide ring closure, phase-transfer conditions (*n*-Bu$_4$NOH, CH$_2$Cl$_2$) were found to give the respective epoxides in moderate yields from *trans*-**192** (Scheme 4.56). However, for the ring closure of *cis*-**192**, the ester substituent showed an important influence. *Cis*-**192a** with a *t*-butyl ester group gave a lower yield of *cis*-**193a** than that of *trans*-**192a**; and no epoxide was obtained for *cis*-**192b** with a methyl ester group.

Bromohydrins containing the MBH adduct moiety have been shown to afford the epoxide in moderate to good yields using K$_2$CO$_3$ or KF as a base.[82] Bromohydrins **194** prepared from IBX/LiBr-promoted oxidative bromohydroxylation of MBH adducts were transformed into the corresponding epoxides **195** in 84–94% yields by treatment with 10% aq NaOH at room temperature for 5–10 min (Scheme 4.57).[83]

Optically active epoxides are very attractive and versatile building blocks for organic synthesis and, therefore, the development of efficient routes to enantioenriched epoxides has been an active area of research for several decades.[84]

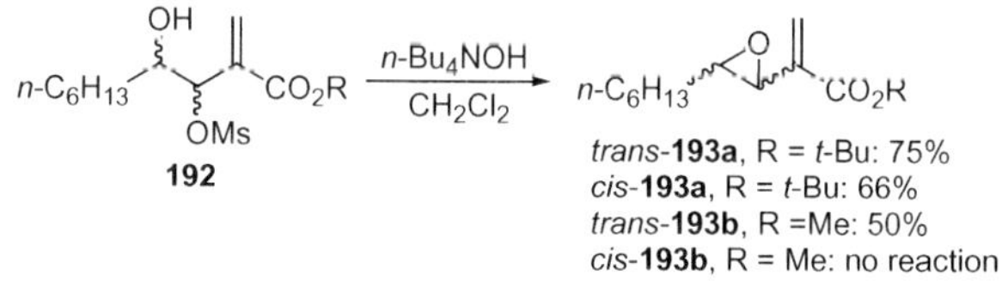

Scheme 4.56

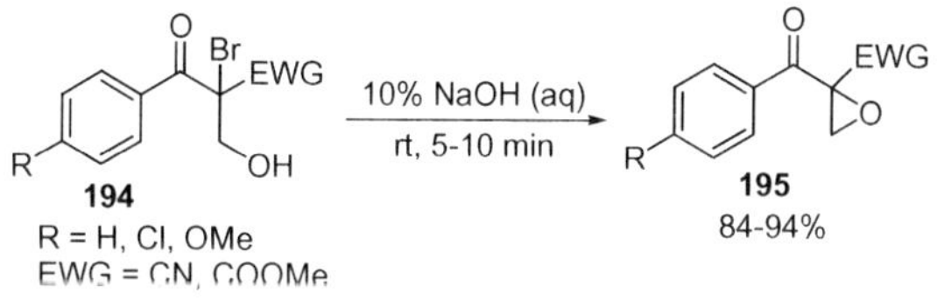

Scheme 4.57

With the development of enantiomerically pure vinyl epoxides bearing chiral sulfoxide group,[85] de la Pradilla has reported asymmetric nucleophilic epoxidation under the chiral induction of sulfoxide group.[86] The epoxidation of (S,S_S)-α'-hydroxy vinyl sulfoxides gave hydroxy sulfinyl oxiranes in high yields and up to 90% dr, but the steric hindrance of R^1 group has a strong influence on the diastereoselectivity of epoxide **197** (Scheme 4.58). For example, substrate **196c**, which has a bulky substituent ($R^1 = t$-Bu), gave a low dr (52%). However, for the epoxidation of (R,S_S)-**196**, diminished reactivities (0–59% yield of epoxide **197**) and very low diastereoselectivities were obtained.

As the nucleophilic epoxidation using t-BuOOM as oxidative reagents showed a substrate-dependent stereochemical outcome, de la Pradilla further investigated the metal-catalyzed electrophilic epoxidations of α'-hydroxy vinyl sulfoxides.[87] Upon treatment with the t-BuOOH/VO(acac)$_2$ oxidative system, *anti*-**198** or *syn*-**199** was obtained exclusively when the R^2 group of the substrates was alkyl or Ph group, respectively (Scheme 4.59).

Hatakeiyama *et al.*[88] have also reported a highly *syn*-diastereoselective epoxidation of MBH adducts using both Weitz–Scheffer and titanium-mediated oxidation procedures. As shown in Scheme 4.60, Ti(OPr^i)$_4$-mediated epoxidation of **200** was found to proceed with complete diastereoselectivity to furnish the *syn*-epoxy alcohol **201** in 73% yield.

Scheme 4.58

Scheme 4.59

Scheme 4.60

Using *m*-CPBA/CH$_2$Cl$_2$ as oxidative system, the epoxidation of silylated alcohols **202** gave the major *anti*-epoxide (*anti*-**203**) (Scheme 4.61).[89] The diastereoselectivities of epoxide **203** showed a strong substrate-dependent stereochemical outcome. The *anti* : *syn* ratio varied from 1.4 : 1 to 13 : 1 upon changing the R group. In addition, epoxidation of the unprotected alcohols furnished a lower dr than that of protected alcohol.

Myers and co-workers have further explored the epoxidation of MBH adducts under various oxidative systems.[90] As shown in Scheme 4.62, using the *t*-BuOOH/*t*-BuOK oxidative system, *anti*-**205** as major isomers were obtained; in contrast, Ti(OPr^i)$_4$-mediated epoxidation of **200** gave *syn*-**206** with a dr $\geq$ 20 : 1. The possible mechanism of this nucleophilic epoxidation process is discussed in their paper.

In addition, MBH adducts **207** have also been treated with iodosobenzene (PhI = O) in the presence of a catalytic amount of KBr in water at room temperature, furnishing the corresponding acyloxiranes **208** in moderate to good yields (Scheme 4.63).[91]

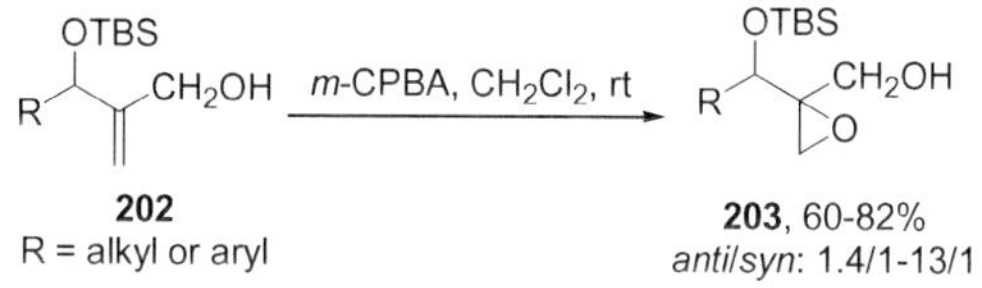

Scheme 4.61

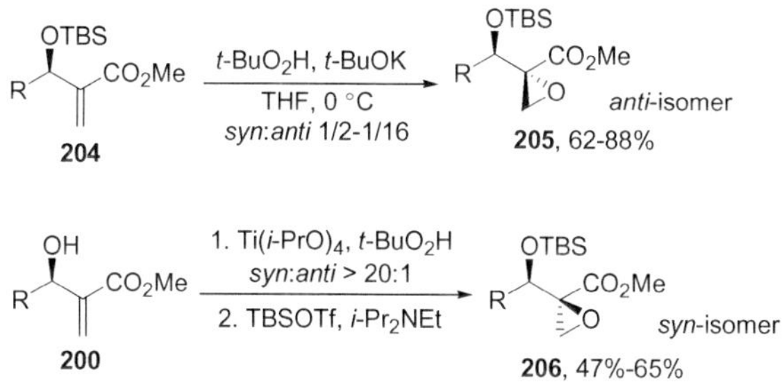

Scheme 4.62

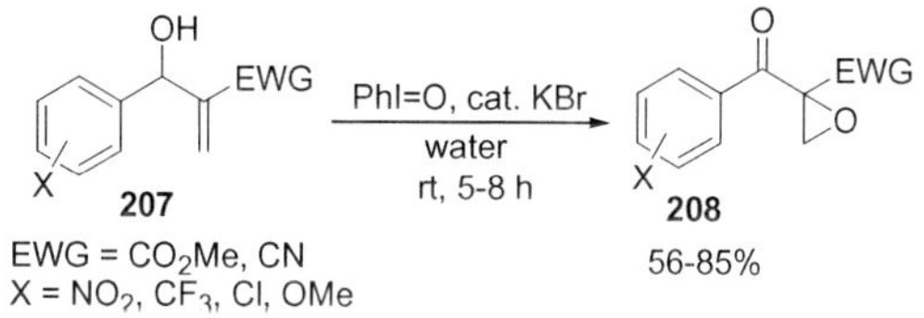

Scheme 4.63

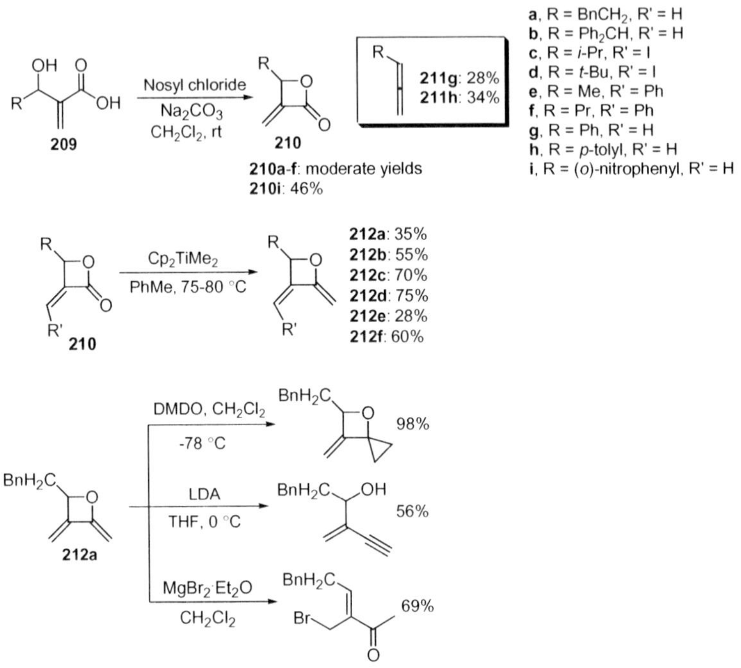

Scheme 4.64

4.3.2　β-Lactones

Acids **209**, prepared by hydrolysis of MBH adducts with KOH, have been subjected to lactonization under the promotion of various sulfonyl chlorides, affording α-alkylidene-β-lactones **210** in moderate yields when the R group was an alkyl substituted one.[92] While R was phenyl or *p*-methylphenyl group, aryl allene **211** was obtained as the sole product in low yield. However, for **209i**, bearing a nitro group on the phenyl section, lactone **210i** was obtained in 46% yield without the formation of aryl allene.

Moreover, the methylenation of lactones **210** with dimethyltitanocene was examined to synthesize 3-alkylidene-2-methyleneoxetanes **212**. The yields of **212** depended strongly on the steric hindrance at C4. Bulkier substituents at C4 led to higher yields, and up to 75% yield was acquired with lactones **210d**. Preliminary investigations of the reactivity of these unusual, strained heterocyclic compounds have been carried out. Using **212a** as a model substrate, several potential applications of such 3-alkylidene-2-methyleneoxetanes as synthetic scaffolds were demonstrated (Scheme 4.64).

4.3.3　δ-Butyrolactones

The potent biological action of many natural products isolated from plants and their ability to inactivate certain selected enzymes have been attributed to the presence of a α-methylene-δ-butyrolactone moiety, which is a part of several sesquiterpenes with interesting biological activity (cytotoxic, antifungal and antibacterial properties).[93] Therefore, numerous synthetic methods have been reported to prepare α-methylene-δ-butyrolactones,[94] which have served as valuable synthetic intermediates for the synthesis of such natural products and biologically important substances.

Acid-promoted lactonization of γ-hydroxy esters is an important way to construct lactones. For example, Scolastico and co-workers first reported the lactonization of MBH adducts **213** using 0.7 M HCl in AcOH–H$_2$O, obtaining α-methylene-β-hydroxy-δ-butyrolactones **215a,b** in moderate yields (Scheme 4.65).[95] Trifluoroacetic acid, CSA and sulfuric acid were also used in such lactonizations. In an alternative method, α-methylene-β-hydroxy-δ-butyrolactone **215c** was synthesized by the intramolecular MBH reaction in 62% yield (Scheme 4.66).[96] A series of α-alkylidene-δ-butyrolactones[97] and multi-substituted α-methylene-δ-butyrolactones[98] have been synthesized from MBH adducts (Figure 4.2).

To address the remarkable α-substituted effects, the Pd-catalyzed allylation of benzaldehyde with α,β- and β,γ-disubstituted allylic benzoates has been investigated. Using MBH adducts as substrates, α-methylene-δ-butyrolactones (*cis*-**217**) were obtained exclusively *via* Pd-catalyzed allylation and spontaneous

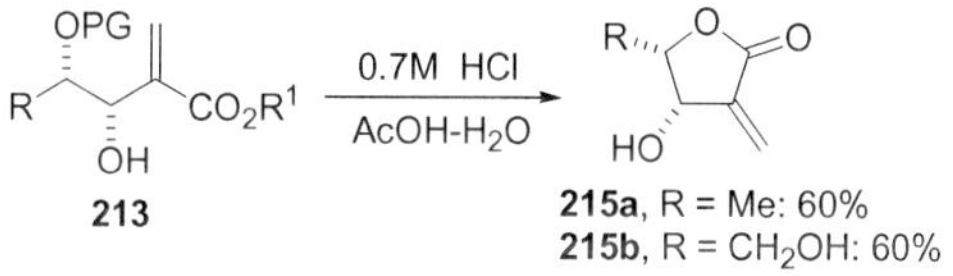

215a, R = Me: 60%
215b, R = CH$_2$OH: 60%

Scheme 4.65

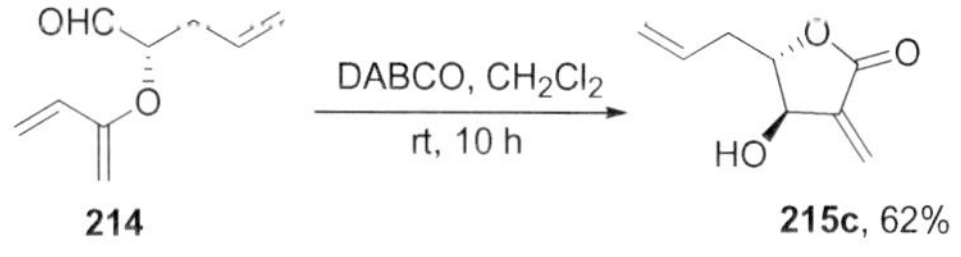

215c, 62%

Scheme 4.66

tulipalin B

R = (CH$_2$)$_9$-CH=CH$_2$
Listenolide A$_1$

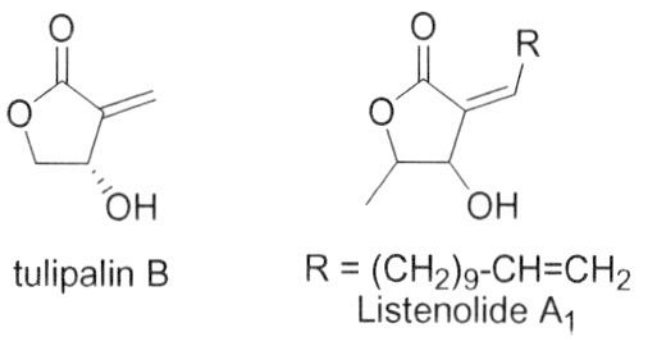

Figure 4.2   Structures of tulipalin B and Listenolide A$_1$.

cyclization, but the yields were unsatisfied.[99] Moreover, the steric hindrance of the ester group has a key influence on the yields of *cis*-**217**. For substrates **216a** and **216b**, higher yields of 47% and 57%, respectively, were obtained, while lower yields were obtained for substrates **216c** and **216d** (Scheme 4.67). Using SnCl$_2$ and PdCl$_2$-(PhCN)$_2$ as the catalytic system, mono- or disubstituted α-methylene-δ-butyrolactones **217** were easily synthesized from the reaction of MBH adducts with various aliphatic and aromatic aldehydes (Scheme 4.68).[100] In most cases, **217** with a high *cis* : *trans* ratio (85 : 15 to 100 : 0) was obtained in moderate to good yield; for 4-methoxybenzaldehyde, though, the *cis* : *trans* ratio was only 55 : 45 to 60 : 40.

Subsequently, MBH variants were also tested in the reaction with aldehyde to furnish β-substituted-α-methylene-γ-butyrolactones. For example, 2-carbomethoxyallylation of **218** with 2-carbomethoxyallyl bromide **219** in the presence of metallic tin and catalytic amounts of acetic and *p*-toluenesulfonic acids gave **220** in 70% yield (Scheme 4.69).[101] Compound **220** possesses extremely interesting properties with regard to its apoptosis-inducing ability in HL-60 cells.

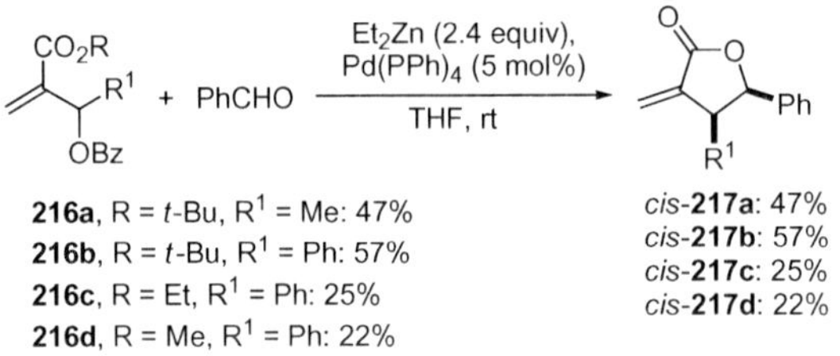

Scheme 4.67

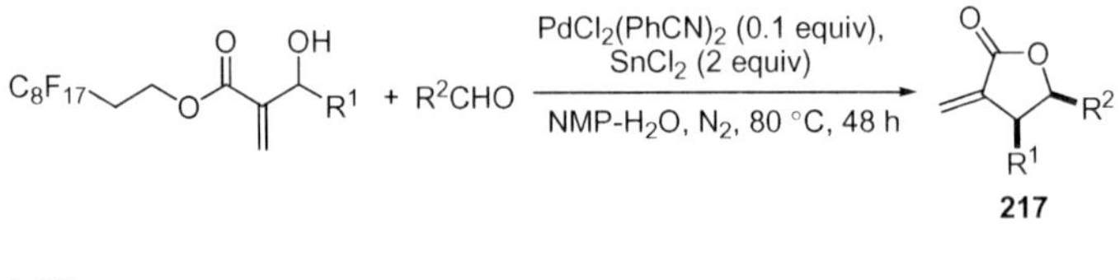

Scheme 4.68

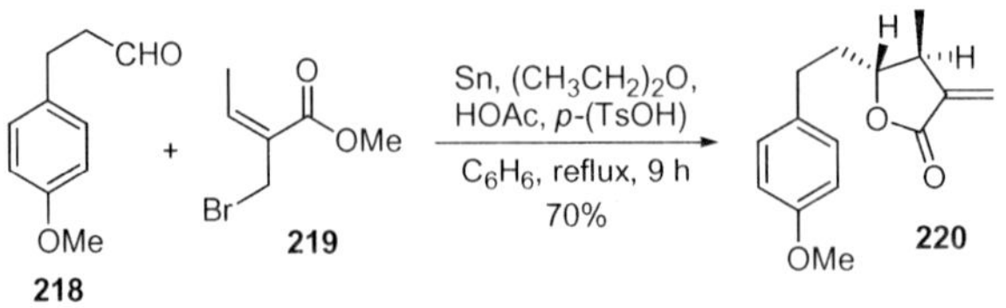

Scheme 4.69

Various allylboronate reagents **221** derived from a MBH adduct have been applied in the allylboration of aldehydes. Moderate to good yields of lactones **217a** with good *cis* : *trans* ratio were obtained (Scheme 4.70).[102] However, even when using chiral allylboronate reagents **222b** and **222c**, low enantioselectivities of up to 27% ee were obtained. This Brønsted acid-catalyzed allylboration was further applied in the total synthesis of eupomatilone-6 derived from the appropriate MBH adduct (Scheme 4.71);[103] eupomatilone-6 is a member of a structurally intriguing class of lignans isolated from the indigenous Australian shrub *Eupomatia bennettii*.[104] Improving the enantioselectivity of the lactone **217** remains a challenge.

When *syn*-homoallylic alcohols **224**, derived from MBH adducts, were subjected to intramolecular lactonization by treatment with CBr_4/PPh_3 or NBS at room temperature, unexpected *trans*-α-methylene-γ-lactones **225** were formed in 49–71% yields (Scheme 4.72).[105] In contrast, lactonization of **224** using *p*-toluenesulfonic acid produced the expected *cis*-methylene-γ-lactones in 94–99% yields in all cases.

Pure alcohol **226** (*anti* configuration) was easily converted into **227** in 52% yield by hydrolysis of the cyano group and *in situ* cyclization (Scheme 4.73).[106] Lactone **227** as a key intermediate[107] could be utilized for the construction of

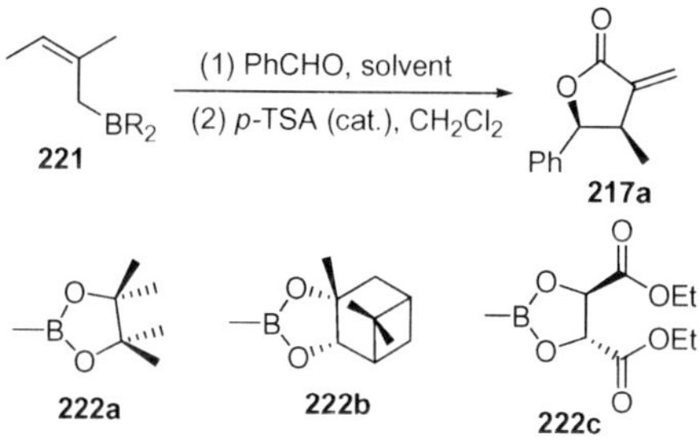

Scheme 4.70

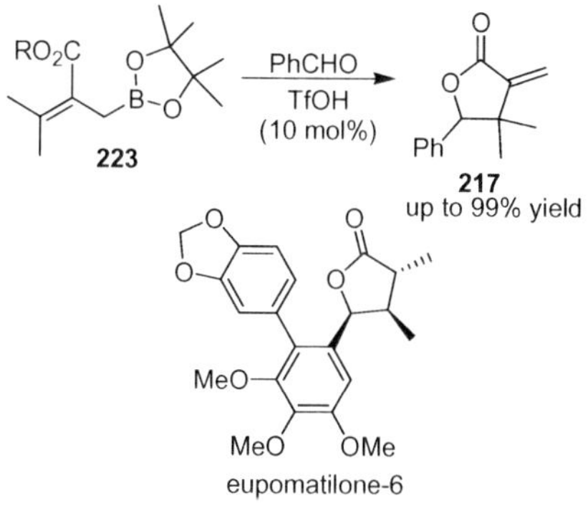

Scheme 4.71

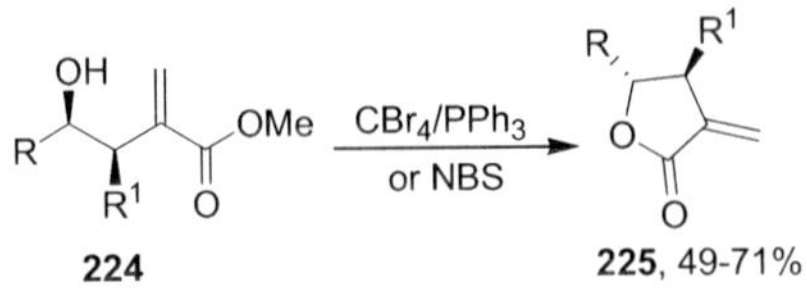

224

225, 49-71%

CBr$_4$/PPh$_3$ or NBS

Scheme 4.72

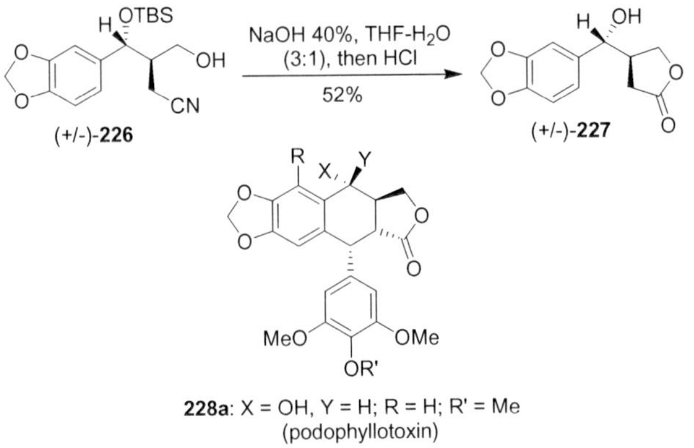

(+/-)-**226**

NaOH 40%, THF-H$_2$O (3:1), then HCl

52%

(+/-)-**227**

228a: X = OH, Y = H; R = H; R' = Me
(podophyllotoxin)

Scheme 4.73

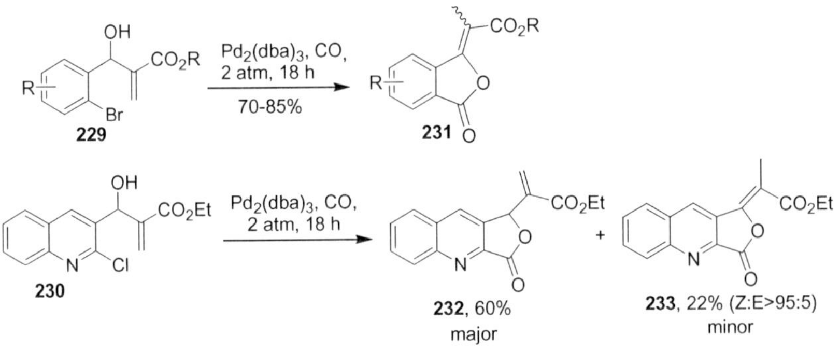

229

Pd$_2$(dba)$_3$, CO, 2 atm, 18 h

70-85%

231

230

Pd$_2$(dba)$_3$, CO, 2 atm, 18 h

232, 60%
major

233, 22% (Z:E>95:5)
minor

Scheme 4.74

podophyllotoxin (**228**) and related compounds, which display cytotoxic activities.[108]

Subsequently, a Pd-mediated cyclocarbonylation reaction[109] of MBH adducts was developed and then applied in the synthesis of quinoline–phthalide derivatives containing the lactone structure.[110] Adducts **229** are sequentially treated with Pd$_2$(dba)$_3$ and carbon monoxide to provide tetrasubstituted olefins **231** in 70–85% yield (Scheme 4.74). Phthalide **232** could be isolated as major product (60% yields) when adduct **230** was submitted to these conditions. Tetrasubstituted olefin **233** ($Z : E > 95 : 5$) was obtained as the minor product. Such compounds were subsequently investigated for their growth inhibitory properties against 60 different human tumor cell lines, and it was found that phthalide derivative **232** exhibits a potent antiproliferative effect on all cell lines.

The RCM protocol has been utilized to synthesize sugar-linked α,β-unsaturated δ-lactones.[111] With chiral induction of sugars **234**, derived from D-sugars, separable chiral MBH derivatives **235** and **236** were obtained mainly as the (*S*)-isomer. These derivatives were then were subjected to ring-closing metathesis using Grubbs' catalyst to afford chiral α,β-unsaturated δ-lactones **237** and **238** in moderate yields (Scheme 4.75). When **237a** and **238a** were subjected to hydrogenation in the presence of Pd-C at room temperature, butyrolactones **239a** and **240a** were obtained in quantitative yields with low chiral induction (6.7 : 3.3 dr) (Scheme 4.76). Even so, the method used to

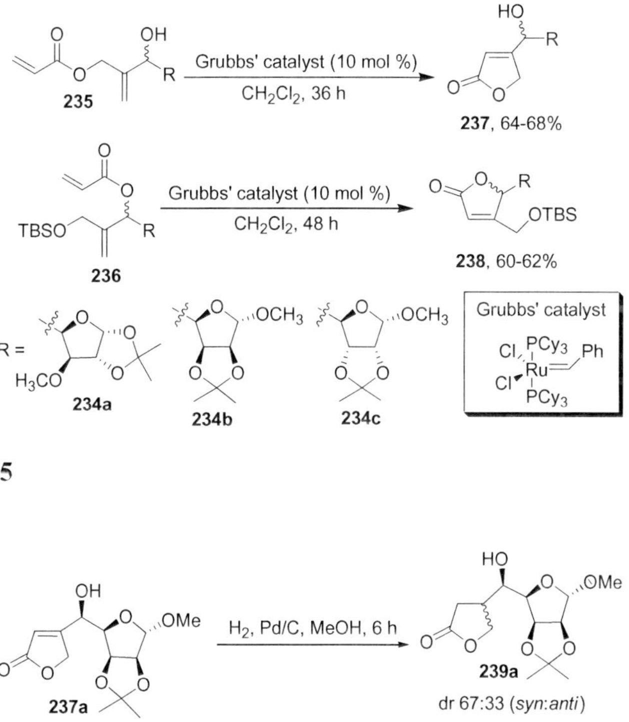

Scheme 4.75

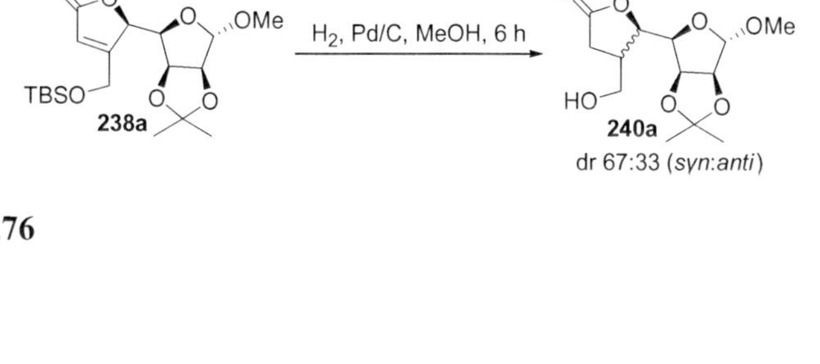

Scheme 4.76

construct these stereochemical and functional butyrolactones maybe applicable to the synthesis of many bioactive natural products and their natural-product-like relatives.

4.3.4 Furan-ring Derivatives

Grubbs and co-workers reported the first successful results of the inter-molecular cross-metathesis and intramolecular ring-closing metathesis with more reactive ruthenium catalyst for the α-functionalized olefins bearing ester, aldehyde, benzoyl or acetyl group.[112]

Another allyl group can be easily introduced into MBH adducts and, therefore, its intramolecular ring-closing metathesis is feasible. *O*-allyl derivatives **241**, derived from MBH adducts *via* the well-known consecutive S_N2'-S_N2' reaction, have been successfully subjected to the RCM reaction. The 2,5-dihydrofurans **242** were obtained in good yields using the commercially available second-generation Grubbs' catalyst (5 mol.%) upon reflux in dichloromethane for 15 min (Scheme 4.77).[113]

Subsequently, Donohoe and co-workers also reported the RCM reaction of MBH derivatives **243**, and the crude RCM products were further transformed, *via* elimination of methanol, into the fully aromatized system (a reaction promoted by adding acid). As shown in Scheme 4.78, multi-substituted furans **244** have been prepared in 59–81% yields in a one-pot process.[114]

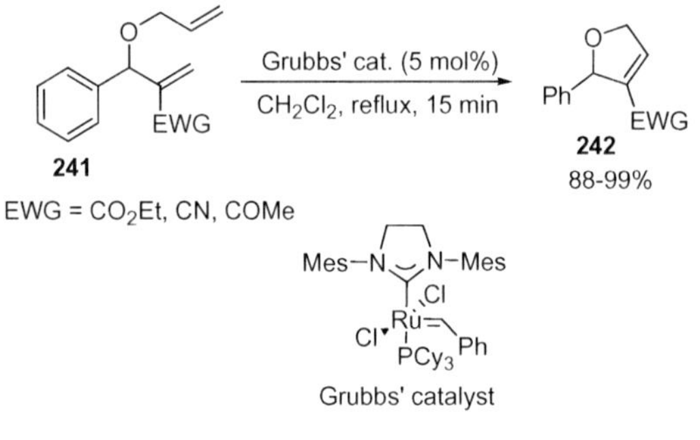

Scheme 4.77

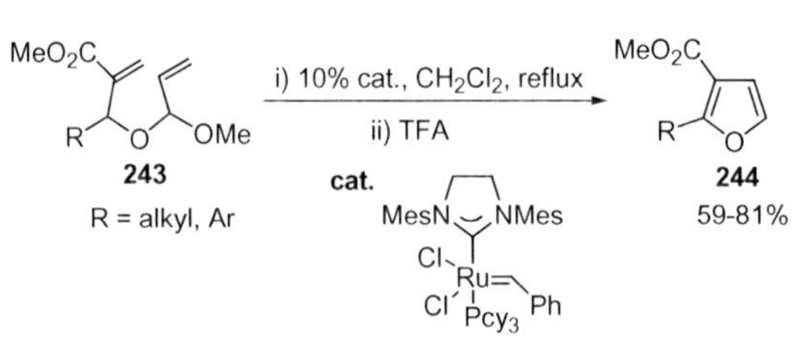

Scheme 4.78

A few reports have disclosed the construction of a tetrahydrofuran backbone through a radical reaction.[115] The propargyl derivatives of the MBH adduct **245** may be suitable substrates for the construction of lignan core structures by a radical cyclization protocol. Treatment of these derivatives with 1.5 equiv. of freshly distilled tri-*n*-butyltin hydride and a catalytic amount of azobisisobutyronitrile (AIBN) afforded crude vinylstannanes **246** through a 5-*exo*-trig cyclization process at 85 °C without any solvent under an inert atmosphere; these vinylstannanes **246** were then subjected to protiodestannylation with 1 M HCl in ether at room temperature for 4 h to give the cyclized products **247** in 90–97% yields (Scheme 4.79).[116] The reaction was general with respect to all substrates with electron-withdrawing or electron-donating substituents on aromatic moieties, affording the corresponding cyclized products **247** in excellent yields. Moreover, the treatment of the crude vinylstannane **246** with iodine in dichloromethane at 0 °C for 1 h afforded the iodo derivative **246** in quantitative yield. More recently, novel heteroaryl-substituted tetrahydrofurans **250** from furfuryl or thiophenyl derived MBH adducts **249** have been synthesized in excellent yields (Scheme 4.80).[117]

Kim and co-workers have also applied this radical cyclization to construct a furan skeleton from MBH derivatives. As shown in Scheme 4.81, the products **252** of radical cyclization were subjected to a sequence steps involving hydrolysis, halolactonization and spontaneous decarboxylation, and, thus, a facile synthetic method for the synthesis of 3,4-disubstituted 2,5-dihydrofurans **254**

Scheme 4.79

Scheme 4.80

Scheme 4.81

Scheme 4.82

was established.[118] Moreover, treatment of **254a** with DDQ conveniently generated furan **255** in 80% yield (Scheme 4.82). Later, through a series of steps including the reduction of ester, the introduction of an allyl group and the ring-closing metathesis (RCM) reaction, product **253a** was converted into bicyclic product **258** bearing a furo[3,4-*c*]pyran skeleton (Scheme 4.83).[119]

Racemic compound **262** can be cyclized into the hemiketal (±)-**263** as a 2 : 1 mixture of diastereoisomers by treatment with 3 M HCl; **262** was obtained from (±)-**261**, which in turn was easily prepared from compound **259** in three steps in 78% overall yield (Scheme 4.84).[120] Using chiral vinylamide (*R*)-**260** instead of **259** in this synthetic route afforded (*R*)-**263** in 47% yield from (*R*)-**261**. Moreover, reductive ozonolysis of **263** gave a solution of the hemiketal **264**.

Chiral adduct **265**, prepared by Pd-catalyzed dynamic kinetic asymmetric arylation of MBH adduct, has been subjected to a reductive Heck-type cyclization to give diastereomers of the dihydrobenzofuran derivative **266** (in an 8.3 : 1 ratio; major one depicted) in 72% yield without any racemization (Scheme 4.85).[121]

In more recent years, the nucleophilic and electrophilic epoxidations of α-hydroxy dienyl sulfoxides as versatile routes to highly functionalized sulfinyl and sulfonyl tetrahydrofurans have been studied in depth.[122] Fernández de la Pradilla's group has reported that α-hydroxy dienols bearing chiral sulfoxides (**267**), prepared from the reaction of chiral dienyl sulfoxide lithium compounds with freshly distilled aldehydes, were subjected to epoxidation using *m*-CPBA as oxidant, leading to monoepoxides **269** as main products in low to moderate yields and low stereochemical selectivities; subsequently, monoepoxides **269** were treated with catalytic CSA to afford 96–99% yields of predominantly 2,5-*cis* mixtures of sulfonyl dihydrofurans **270** (Scheme 4.86).[123] These highly functionalized sulfonyl tetrahydrofurans were also obtained in a

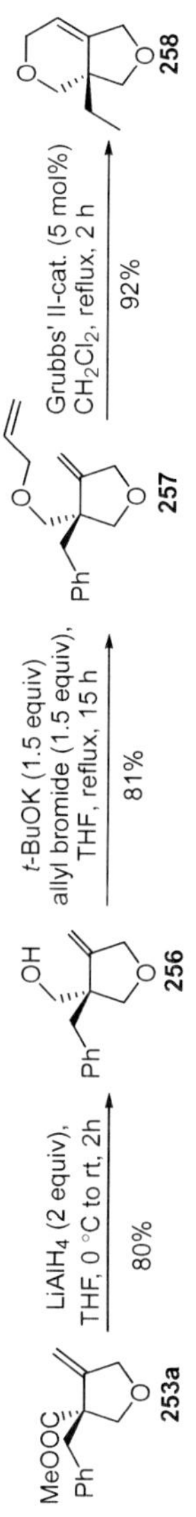
MeOOC
Ph
253a
LiAlH4 (2 equiv),
THF, 0 °C to rt, 2h
80%
OH
Ph
256
t-BuOK (1.5 equiv)
allyl bromide (1.5 equiv),
THF, reflux, 15 h
81%
Ph
257
Grubbs' II-cat. (5 mol%)
CH2Cl2, reflux, 2 h
92%
258

Scheme 4.83

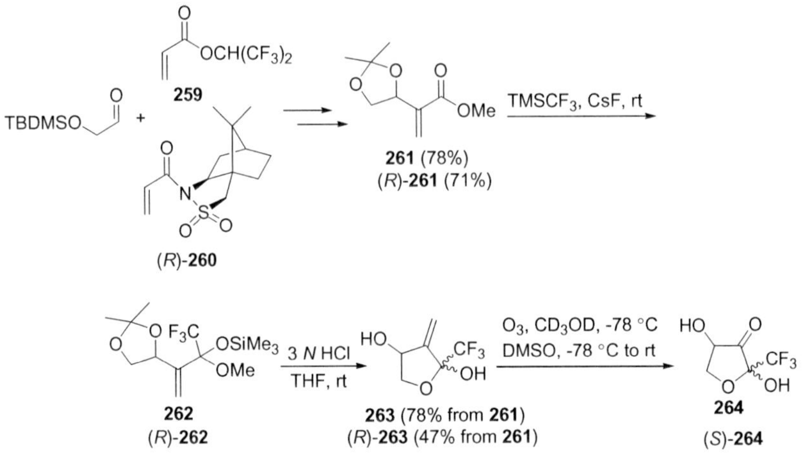

Scheme 4.84

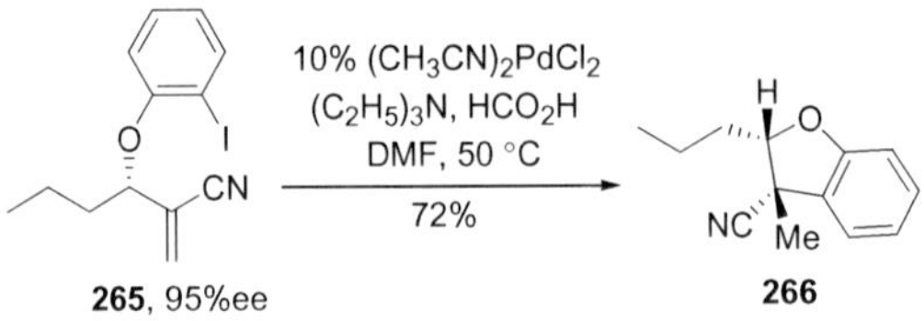

Scheme 4.85

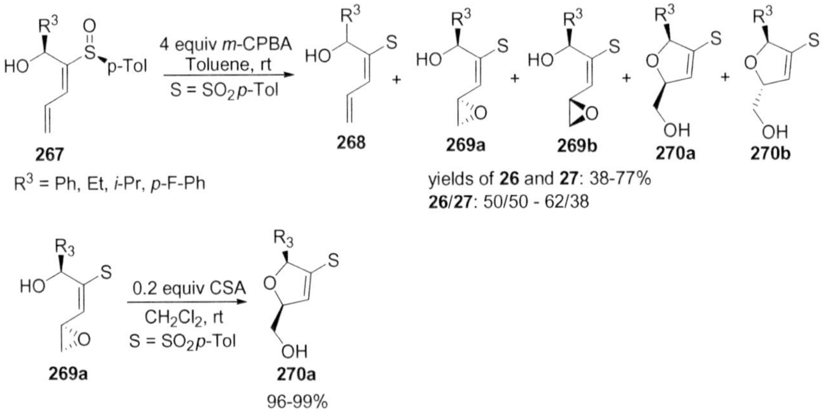

Scheme 4.86

one-pot sequence *via* nucleophilic epoxidation and subsequent acid-catalyzed cyclization. Considering the low selectivities associated with the use of *m*-CPBA, they developed a new route to the synthesis of 2,5-*trans* sulfonyl dihydrofurans **272** (Scheme 4.87).[124] Good selectivities (ratio of monoepoxides **271a** and **271b** up to 99 : 1) and moderate yields were achieved through the Katsuki–Jacobsen oxidation–epoxidation of acyclic α-silyloxy sulfinyl dienes. This methodology for providing 2,5-*trans*-substituted sulfonyl dihydrofurans in good selectivity has been applied successfully in the formal syntheses of natural products.

Recently, Alcaide and co-workers have disclosed the Pd-catalyzed domino cycloisomerization/cross-coupling of α-allenols and MBH acetates.[125] As shown in Scheme 4.88, several types of spiroheterocycles bearing a 2,5-dihydrofuran ring were synthesized in moderate to good yields.

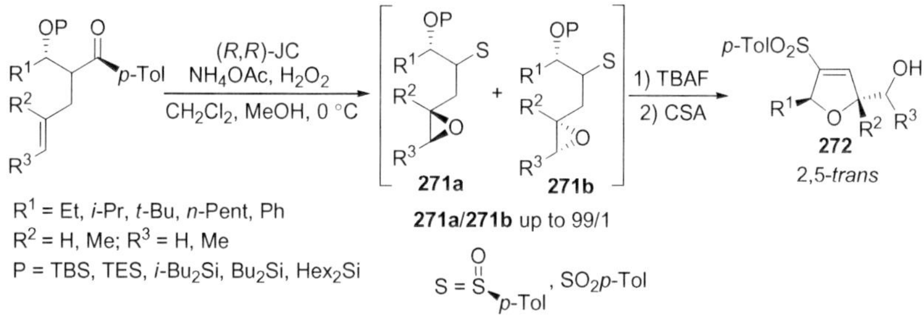

Scheme 4.87

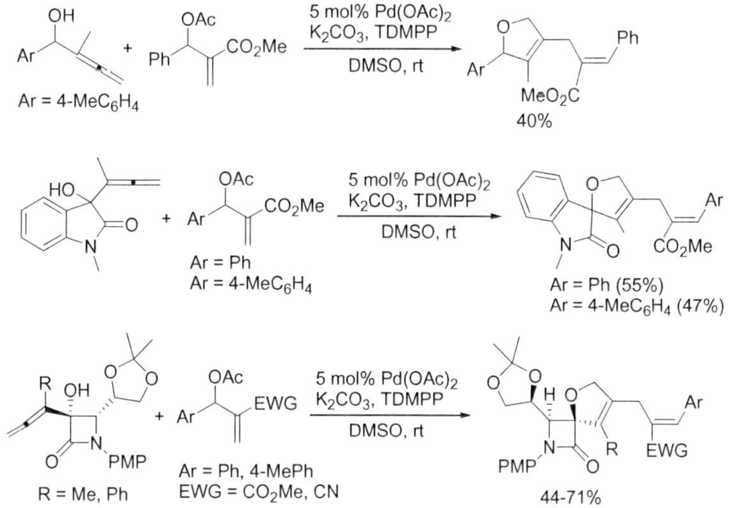

Scheme 4.88

4.3.5 Pyrans

The construction of a pyran skeleton from a MBH adduct *via* a hetero-Diels–Alder reaction was first described by Weichert and Hoffman.[126] 3-(Benzenesulfonyl)-3-buten-2-one (**273**) first served as a 1-oxa-1,3-butadiene unit, combining with a wide range of alkenes of graded nucleophilicity. As shown in Scheme 4.89, electron-rich 2π dienophiles, including sterically hindered 2-isopropylidene-1,3-dithiane, reacted very well with various alkenes, affording products **274** with a pyran skeleton in moderate to good yields. Utilizing this synthetic strategy, effective synthetic routes to some natural products such as the tricycle skeleton **275** of robustadial A and B,[127] frontalin **276**[128] and benzo-tricycle **277** were successfully established.

More recently, Maignan and Hayes[129] have reported the asymmetric hetero-Diels–Alder reaction of chiral (*S*)-3-*p*-tolylsulfinylbut-3-en-2-one (**278**) with simple enol thioethers **279**. Moderate yields of cycloaddition products **280** were obtained, but with low enantioselectivities (Scheme 4.90).

(2*E*)-2-Phenoxymethyl-3-phenylprop-2-enoic acids **283**, easily prepared from the reaction of allyl bromide **281** with phenol followed by hydrolysis, have been treated with TFAA in CH_2Cl_2 to provide (*E*)-3-benzylidene- or (*E*)-3-alkylidene-chroman-4-ones **284** in 80–94% yields (Scheme 4.91).[130] Whether R group in **283a–g** was alkyl or aryl, the TFAA-promoted Friedel–Crafts reaction showed good reaction activity. Interestingly, the (*E*)-3-benzylidenechroman-4-one moiety occupies a special place in the field of heterocycles as this skeleton is an integral part of many natural products, such as **285–287**,[131] and biologically active molecules such as **288**.[132] Therefore, the efficiency and generalization of

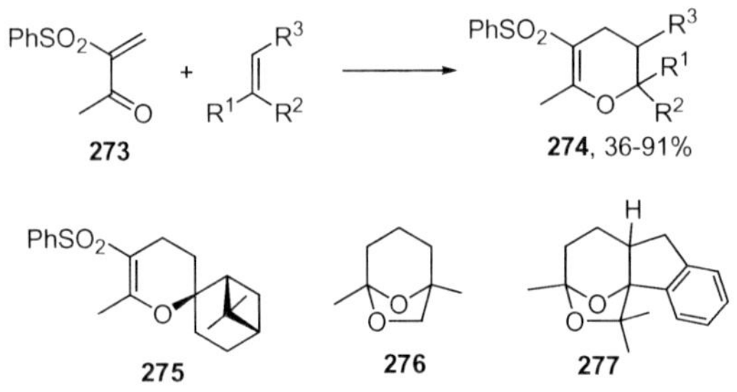

Scheme 4.89

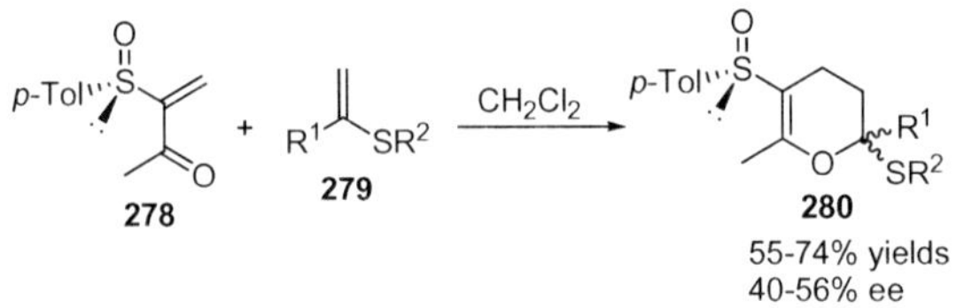

Scheme 4.90

281
OMe
Br
R

PhOH, K$_2$CO$_3$
acetone, reflux, 3 h

282
OMe
OPh
R
(65-90%)

KOH, H$_2$O
acetone, rt, 14 h

283
OH
OPh
R
(78-93%)

TFAA, CH$_2$Cl$_2$
reflux, 1 h

284
R
(80-94%)

R = Ph, p-MeC$_6$H$_4$, o-MeC$_6$H$_4$, p-EtC$_6$H$_4$, p-Pr-C$_6$H$_4$, p-MeOC$_6$H$_4$, Pr

Scheme 4.91

this synthetic methodology was further applied in the synthesis of such (*E*)-3-benzylidenechroman-4-ones. As shown in Scheme 4.92, the methyl ether of bonducellin (**284a**) and (*E*)-3-(4-methoxybenzylidene)-6-methoxychroman-4-one (**288**) as an antifungal agent have been synthesized.

The radical cyclization of enyne ethers **289** synthesized from the MK 10 clay-catalyzed reaction of MBH adducts and propargyl alcohol has afforded 6-*endo-trig* cyclization products **290** in 69–73% yields. Interestingly, enyne ethers **291** also underwent radical cyclization smoothly to furnish 2,4,5,5′-tetrasubstituted tetrahydropyrans **292** bearing different aryl substituents, in 69–73% yields, *via* 6-*exo-trig* cyclization (Scheme 4.93).[133] Along with the cyclization product, only less than 10% of reduced compounds were obtained. However, using compound **291′** as substrate gave a lower yield of 3,3′,4-trisubstituted cyclization product **292′** (45%), but together with a higher yield of the reduction product **293′** (40%).[134] Compound **292′** could be further bromolactonized to give product **294** in 59% yield under mild conditions using NBS, NaHCO$_3$ and LiOH in THF.

Later, Kim and co-workers reported allyltributylstannane (instead of *n*-Bu$_3$SnH) mediated radical cyclization (Scheme 4.94).[135] Two types of novel methyl 5-methylenetetrahydropyran-3-carboxylate derivatives, **297** and **298**, were synthesized stereoselectively starting from the MBH adducts. However, in the radical cyclization of substrate **296** bearing CN or CH$_2$OH as R group the expected compound **298** could not be obtained.

Dihydropyran derivatives can be synthesized facilely by a smooth oxidative Mukaiyama–Michael addition followed by a cyclization with silyl enol ethers in the presence of Dess–Martin periodinane (DMP) and pyridine under mild reaction conditions from MBH adducts in a one-pot process (Scheme 4.95).[136] Notably, these dihydropyrans were obtained exclusively as *cis*-isomers in good yields. Moreover, all the reactions worked very well, irrespective of whether MBH adducts were derived from aliphatic or aromatic aldehydes, and silyl enol ethers were derived from acetophenone, cyclohexanone or cyclopentanone.

Aryl alcohols are competent nucleophiles in the palladium-catalyzed dynamic kinetic asymmetric transformation (DYKAT) of racemic MBH derivatives. As an extension of this strategy, the palladium-catalyzed intramolecular DYKAT of MBH adducts was further explored.[137] As shown in Scheme 4.96, reactions were carried out in dioxane at 25 °C with chiral ligand (*R*,*R*)-L-**1**, affording **300** in up to 45% yields and 98% ee *via* a highly selective *kinetic* resolution; interestingly, when reactions were performed at 80 °C, up to 94% yield with 91% ee of **300** was obtained by the DYKAT process.

4.3.6 Pyran-2-ones

Considerable effort has been devoted to the development of effective approaches towards the synthesis of α-pyrones and related compounds, especially for α-methylene-δ-valerolactones.[138] The α-methylene-δ-valerolactone moiety is an important structural motif in a wide range of natural occurring biologically

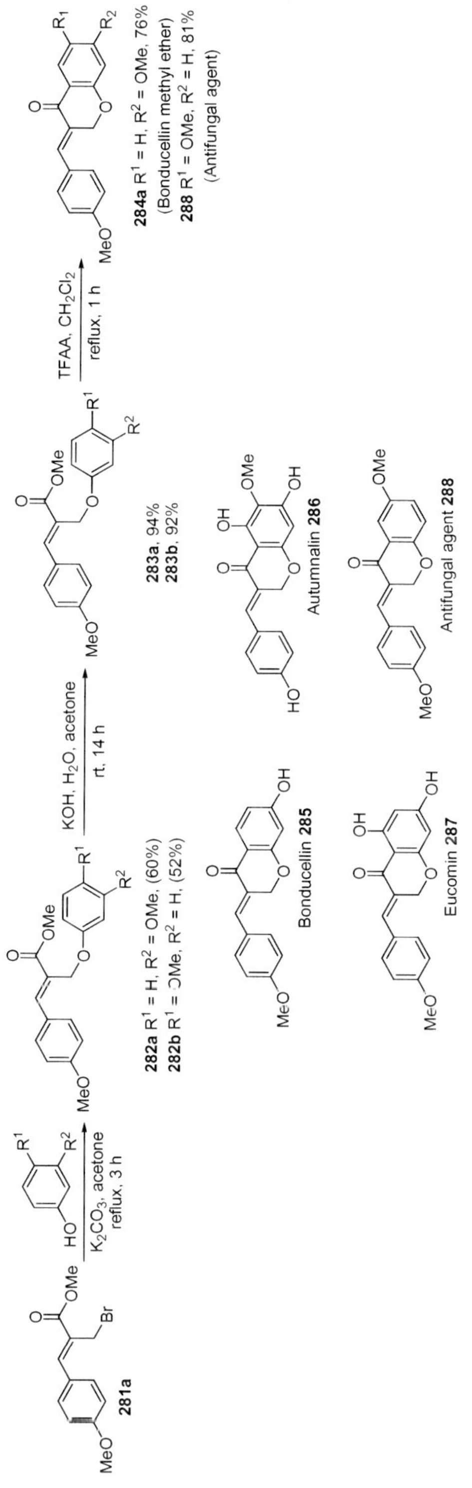

284a R^1 = H, R^2 = OMe, 76%
(Bonducellin methyl ether)
288 R^1 = OMe, R^2 = H, 81%
(Antifungal agent)

TFAA, CH$_2$Cl$_2$
reflux, 1 h

283a. 94%
283b. 92%

KOH, H$_2$O, acetone
rt, 14 h

282a R^1 = H, R^2 = OMe, (60%)
282b R^1 = OMe, R^2 = H, (52%)

K$_2$CO$_3$, acetone
reflux, 3 h

281a

Autumnalin 286

Antifungal agent 288

Bonducellin 285

Eucomin 287

Scheme 4.92

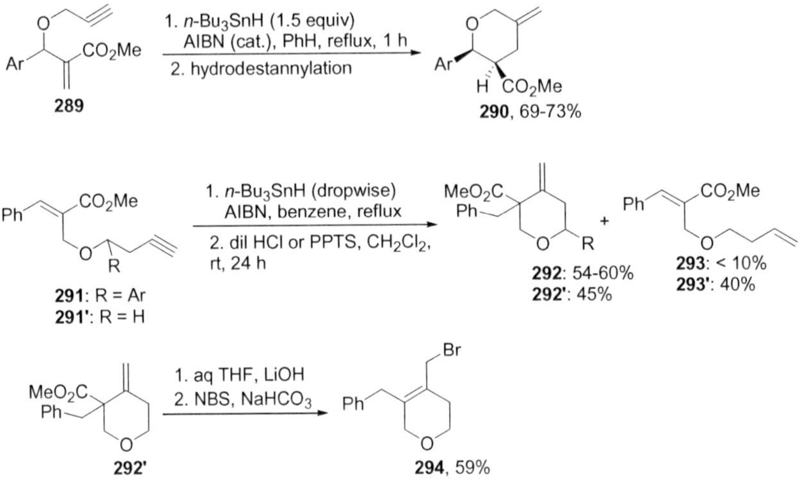

289 → **290**, 69-73%

1. *n*-Bu$_3$SnH (1.5 equiv) AIBN (cat.), PhH, reflux, 1 h
2. hydrodestannylation

291: R = Ar
291': R = H

1. *n*-Bu$_3$SnH (dropwise) AIBN, benzene, reflux
2. dil HCl or PPTS, CH$_2$Cl$_2$, rt, 24 h

292: 54-60%
292': 45%

293: < 10%
293': 40%

292' →
1. aq THF, LiOH
2. NBS, NaHCO$_3$
→ **294**, 59%

Scheme 4.93

295 →
n-Bu$_3$SnCH$_2$CH=CH$_2$ (4.0 equiv), AIBN (1.0 equiv), neat, 80 °C, 1 h
HCl, ether, 0 °C to rt, 1 h
→ **297**, 75-88%

296 →
n-Bu$_3$SnCH$_2$CH=CH$_2$ (4.0 equiv), AIBN (1.0 equiv), neat, 80 °C, 1 h
HCl, ether, 0 °C to rt, 1 h
→ **298**
R = CO$_2$Me: 81-82%
R = CN, CH$_2$OH: not isolated

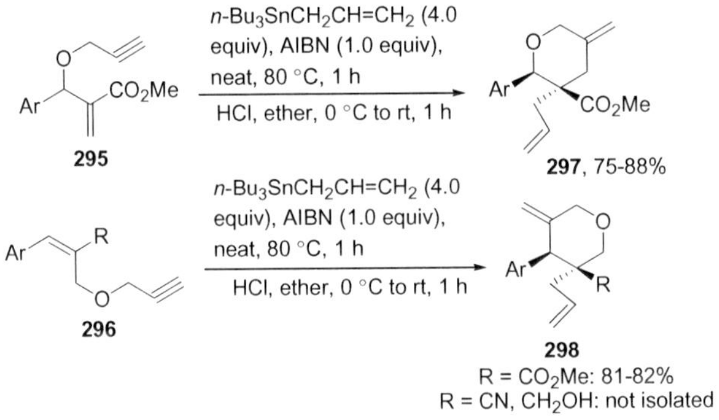

Scheme 4.94

active compounds, and is also a valuable synthetic intermediate towards natural products.[139]

MBH adducts and their derivatives have been recognized as a versatile intermediate to construct the α-pyrone skeleton. Ko and Cho reported the first example in the synthesis of gelastatin analogues **302** using MBH adducts and their derivatives as key intermediates.[140] MBH adducts (*E/Z*)-**301** were desilylated (Bu$_4$NF), which resulted in a concurrent lactonization to give (*E/Z*)-**302** (Scheme 4.97). Compounds (*E/Z*)-**302** as key intermediates have been used in the total synthesis of gelastatin analogues. Subsequently, the authors reported the mild acidic deprotection of the acetal group to yield the lactone **305**.[141] However, the reaction of MBH adduct **303** produced a mixture of two compounds, the methyl ester triol **304** and the lactone diol **305** in the presence of

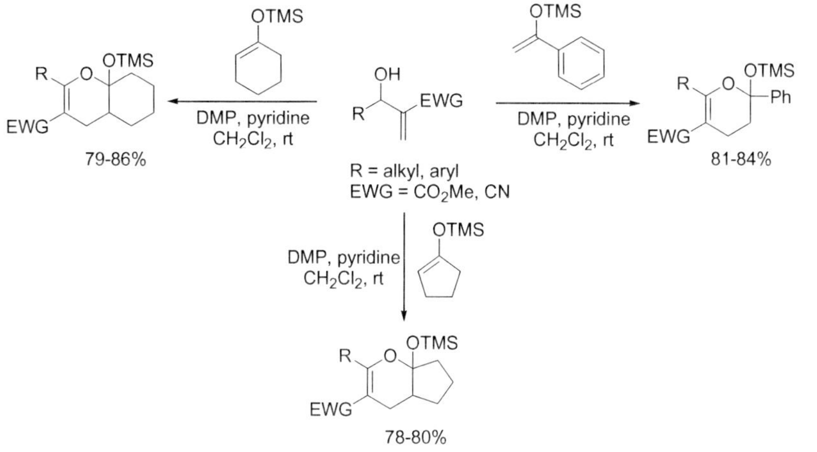

Scheme 4.95

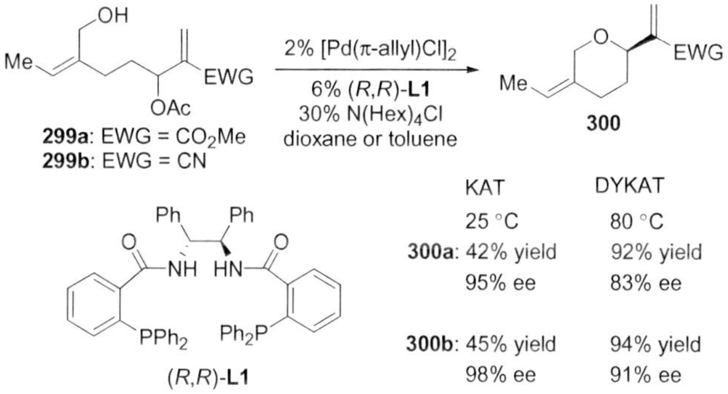

Scheme 4.96

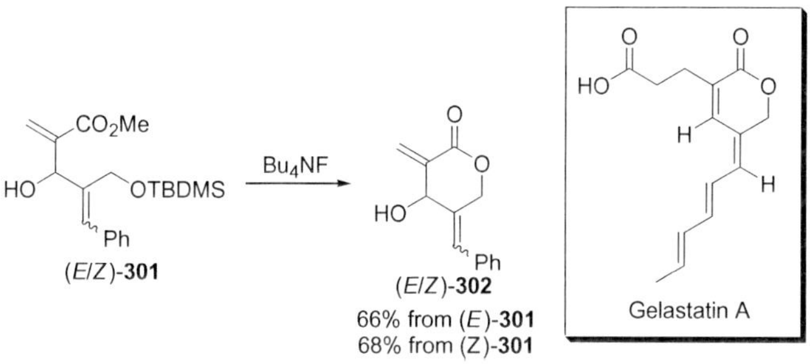

Scheme 4.97

0.005 M H₂SO₄. Fortunately, the methyl ester triol **304** could be transformed into lactone **305** with by treatment with TBAF, to give an overall yield of 72% of lactone **305** from the MBH adduct **303** (Scheme 4.98).

α-Methylene-δ-valerolactones have also been obtained in a one-pot procedure *via* the tandem deacetylation and saponification of the MBH diketo derivatives **306** upon treatment with 15% KOH (or NaOH) in methanol or 15% aqueous NaOH; subsequent reduction with NaBH₄ occurred in the presence of NaOH, followed by intramolecular cyclization in the presence of HCl (Scheme 4.99). High overall yields of lactone **309** were obtained yields from **306** (72–83%).[142] More recently, an efficient and practical synthesis giving excellent yields of substituted 3-methylene-pyran-2,6-diones **312** and 3-methylene-3,4-dihydropyran-2-ones **313** has been described starting from, respectively, the 1,5-pentane dicarboxylic acids **310** and monoketo acids **311**, under treatment of P₂O₅ in anhydrous toluene at room temperature (Scheme 4.100).[143]

A brilliant synthetic methodology leading to α-methylene/arylidene-δ-lactones in a one-pot manner has been delivered by Roy and co-workers, using titanocene(III)-mediated radical-induced addition of epoxides to MBH adducts.[144] As shown in Scheme 4.101, lactones **315** were obtained in 48–72% yields on treatment of epoxide **314** and MBH acetates with Cp₂TiCl in THF; the reaction proceeded by Michael addition followed by *in situ* lactonization. However, using 3-acetoxy-2-methylenenitrile instead of MBH acetates did not give the lactone **315**, but afforded a S_N2'-type product as an inseparable mixture of (*E*)- and (*Z*)-isomers in a ratio of 84 : 16 in 61% yield.

Scheme 4.98

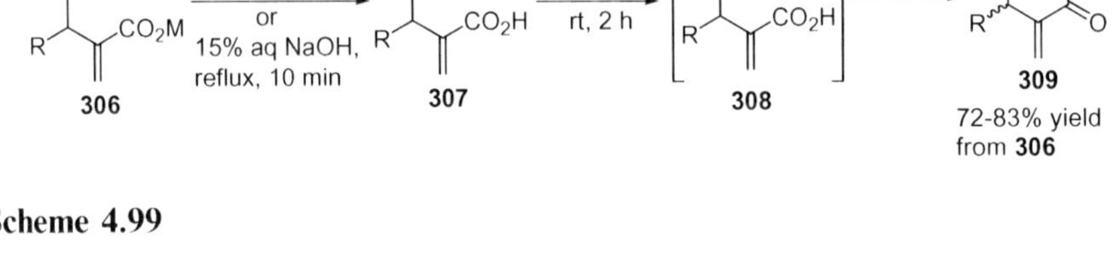

Scheme 4.99

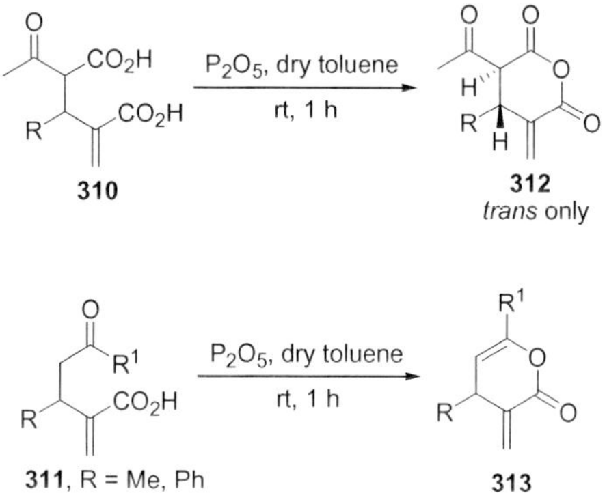

Scheme 4.100

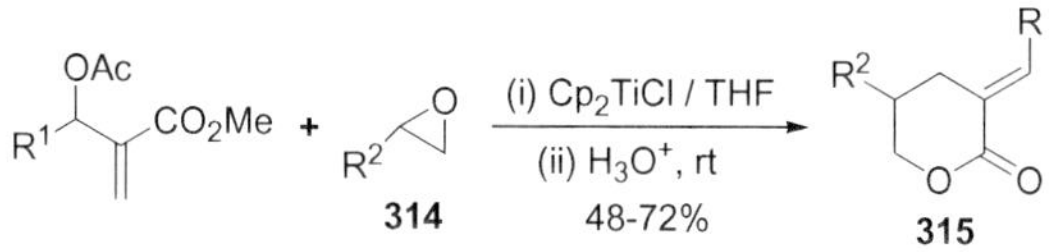

Scheme 4.101

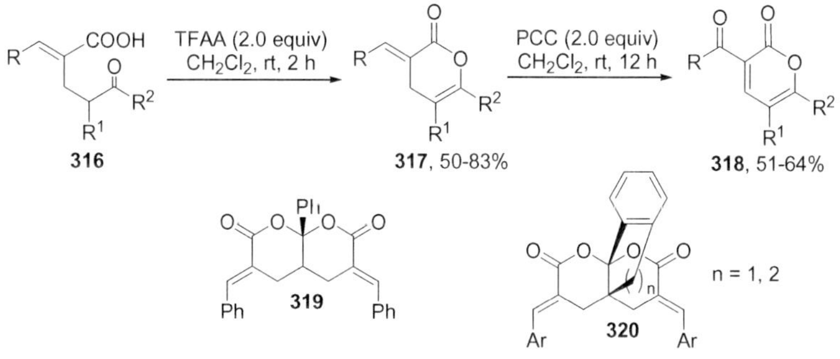

Scheme 4.102

The lactonization of MBH derivatives to give 3-arylidene-3,4-dihydropyran-2-one derivatives has been realized by treatment with TFAA in CH_2Cl_2 at room temperature.[145] As shown in Scheme 4.102, α-arylidene-δ-lactones **317** were obtained in 50–83% yields. Subsequent oxidation of **317** with PCC afforded the desired α-pyrones **318** in 51–64% yields. By application of this synthetic strategy, tricyclic compound **320**, previously reported by Basavaiah and Satyanarayana,[146] and bicyclic compound **319** could be prepared from easily available MBH adducts.

4.3.7 Coumarin Derivatives

Conjugate addition products **322** and **323** have been subjected to hydrogenolysis in the presence of 10% Pd/C catalyst (Scheme 4.103).[147] 3-Substituted coumarins **324** were obtained in 50–66% yields from substrates **322** through fission of the benzyl ether and spontaneous cyclization *via* acyl substitution. Using compounds **323** as substrates, piperidinyl derivatives **325** were obtained in 45–61% yields, and the de-aminated derivatives **326** were also acquired in 9–21% yields. Such a synthetic method may enable the synthesis of multi-substituted coumarins (2*H*-1-benzopyran-2-ones), which are widely distributed in nature and many of which exhibit pharmacological activities.[148]

2-Nitromethyl-*o*-chlorocinnamates **327a–c** have been employed as substrates in an attempt to produce the cyclic derivatives *via* an intramolecular S_NAr reaction.[149] As expected, 3-methoxycarbonylcoumarin 2-oxime (**329a**) was obtained in low yield (14%). Interestingly, dichlorocinnamates **327b** and **327c**, having another electron-attracting chlorine atom on the benzene ring, gave coumarin 2-oximes **329b** and **329c** in 66% and 50% yields, respectively. In contrast, the reaction of MBH acetates **328d** or **328e**, having a strong electron-withdrawing nitro group on the benzene ring, with $NaNO_2$ in DMF directly gave the corresponding coumarin 2-oximes **329d** in 70% yield and **329e** in 48% yield within 30 min (Scheme 4.104). Moreover, oximes **329a–e** are easily hydrolyzed using concentrated HCl in THF to give the corresponding coumarins **330a–e** in good yields.[150]

S_N2' type products **332** have been obtained in moderate to good yields from MBH adduct acetates, which were derived from aromatic aldehydes and cyclohexane-1,3-diones, within a short time under refluxing conditions (Scheme 4.105).[151] Interestingly, by prolonging the reaction time, 3-arylmethyl-7,8-dihydro-6*H*-chromene-2,5-diones **333** were obtained. After optimizing the reaction conditions, an efficient method to synthesize compounds **333** in 63–91% yields was established under solvent-free conditions. However, if MBH adducts derived from aliphatic aldehydes were examined under similar conditions, the unexpected compounds **334** were obtained in 31–37% yields.

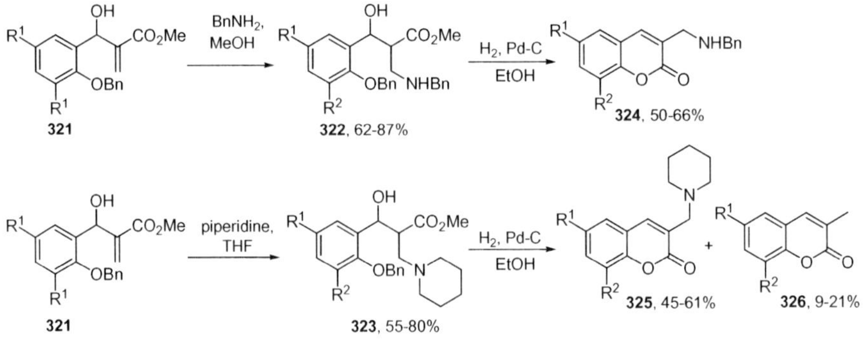

Scheme 4.103

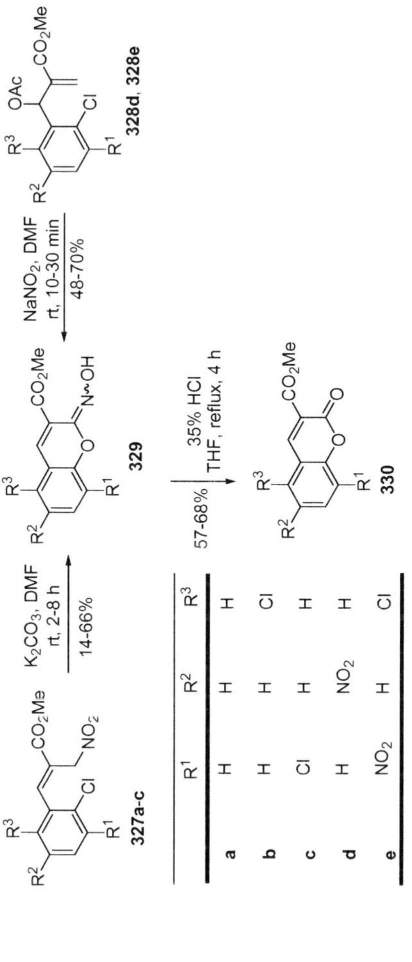

	R^1	R^2	R^3
a	H	H	H
b	H	H	Cl
c	Cl	H	H
d	H	NO_2	H
e	NO_2	H	Cl

Scheme 4.104

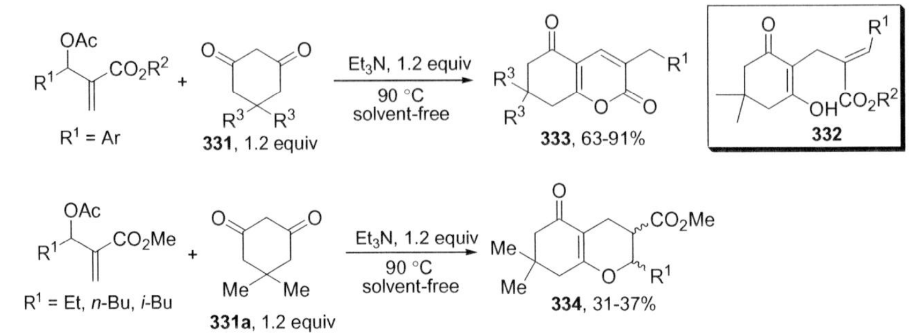

Scheme 4.105

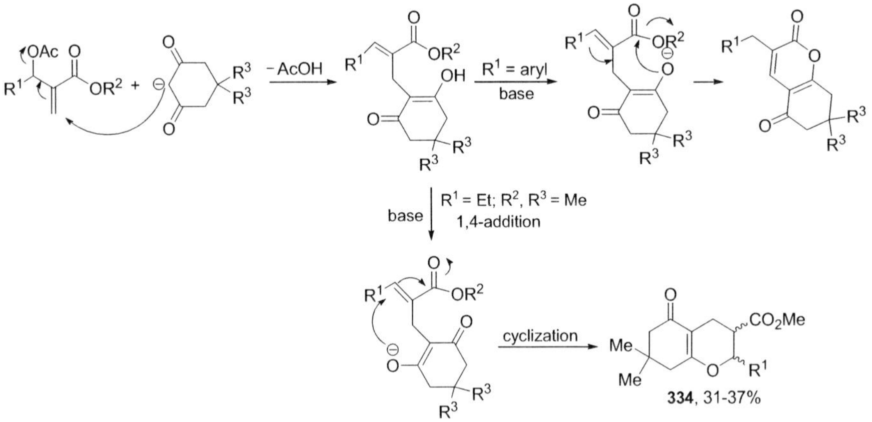

Scheme 4.106

A plausible mechanism for the formation of **333** and **334** has been proposed (Scheme 4.106).

Using chiral {(salen)Cr^{III}} complex **336** as catalyst, 2-silyloxyaryl carbinol **338** was obtained in 79% yield and 86% ee from the enantioselective addition of silyloxyallene **335** to aldehyde **337**. Deprotection of the silyl group of **338** at low temperature, by treatment with n-Bu$_4$NF in THF, delivered the 2,3-disubstituted chromene **339** in 61% yield with 68% ee (Scheme 4.107).[152]

Zhao and co-workers[153] have described a convenient and practical AlCl$_3$-promoted C–C coupling reaction between readily available α-hydroxyketene-*S,S*-acetals **340**[154] and various phenols, including 2-naphthalenols, to produce a series of bio- and pharmacologically important 3,4-disubstituted dihydrocoumarins in high yields. The reaction proceeds by a sequential Friedel–Crafts alkylation and intramolecular annulation under mild conditions (Scheme 4.108).

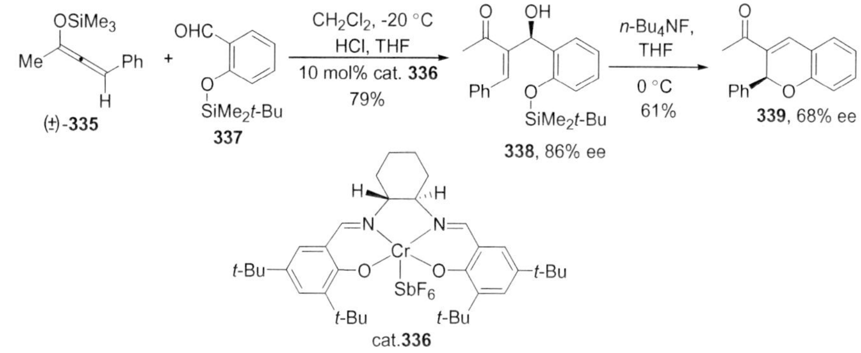

Scheme 4.107

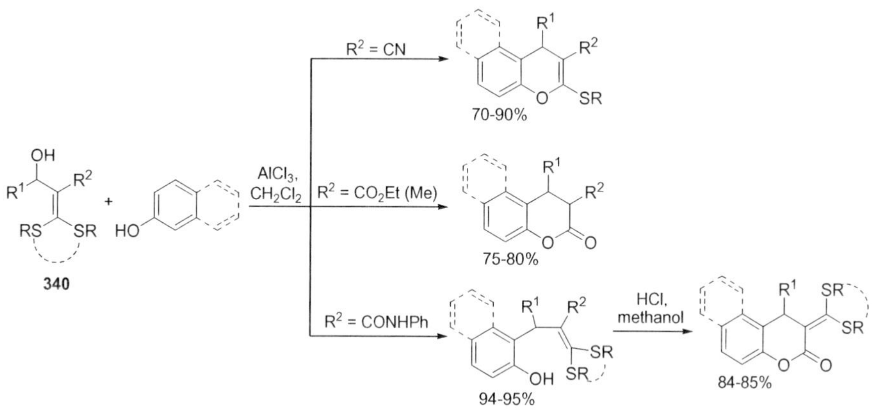

Scheme 4.108

An efficient protocol for the synthesis of highly substituted α-pyrones **346** has been developed by Kim and co-workers.[155] As shown in Scheme 4.109, α-pyrones **346** were synthesized *via* the sequential lactonization of **344** with TFAA and subsequent DBU-catalyzed isomerization of the intermediate **345**. The starting materials, δ-keto acids **344**, were easily synthesized from cinnamyl bromide **341** *via* S_N2 reaction and subsequent hydrolysis. This provides a synthetic route to multi-substituted α-pyrones, which have been used as important synthetic intermediates and are found in a wide variety of biologically active natural substances.[156]

4.3.8 Medium-sized-ring Oxygen-containing Heterocycles

Many natural products and synthetic derivatives bearing a 2-benzoxepine unit are found to be a class of medicinally important compounds, exhibiting

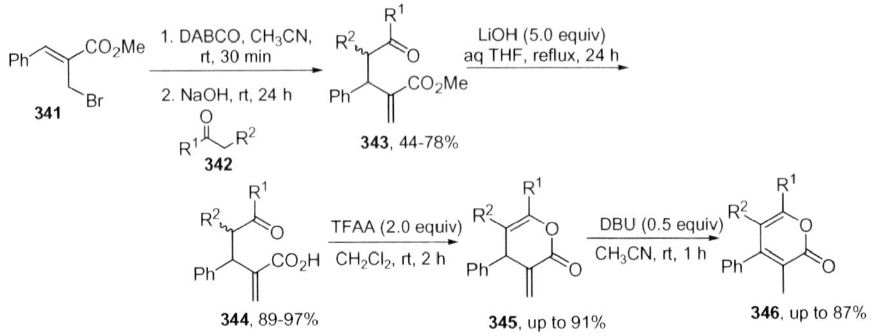

Scheme 4.109

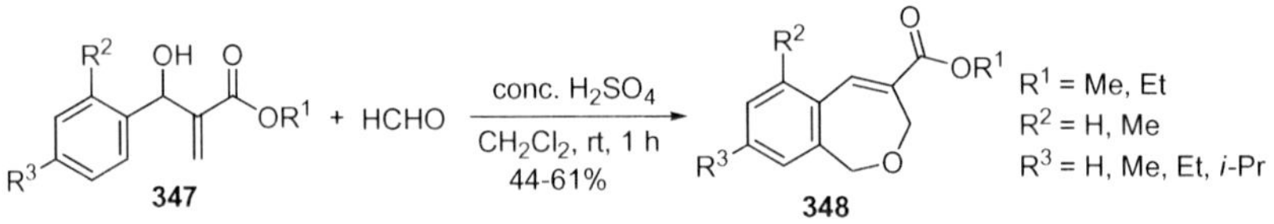

Scheme 4.110

antianaphylactic, oral hypotensive and antiulcer properties. Owing to their interesting and important biological properties, the development of simple, convenient methodologies for the synthesis of 2-benzoxepine derivatives is an attractive and interesting endeavor in synthetic organic chemistry and medicinal chemistry.[157]

Basavaiah and co-workers have developed a facile one-pot synthesis of 2-benzoxepines **348** using MBH adducts as starting materials.[158] For example, MBH adduct **347** (2.0 mmol) was treated with paraformaldehyde (2.0 mmol) in CH_2Cl_2 (4.0 mL) in the presence of concentrated H_2SO_4 (2.0 mmol) at room temperature for 1 h, providing 2-benzoxepines in 44–61% isolated yields (Scheme 4.110). Such a synthetic process involves a tandem construction of C–O and C–C bonds *via* Prins-type and Friedel–Crafts reactions. Subsequently, Das *et al.* also reported a novel catalytic synthetic method for the preparation of 2-benzoxepines **348**.[159] The reaction of MBH adducts **347** with paraformaldehyde was catalyzed by silica gel supported perchloric acid ($HClO_4 \cdot SiO_2$) or Amberlyst-15 in CH_2Cl_2 under reflux for a short period of time (1.5–2.5 h), and much higher yields of up to 82% of 2-benzoxepines were obtained. Moreover, these catalysts can be recovered and recycled three times with a minimum variation of the yields. However, disappointingly, these new benzoxepines showed no antibacterial activity against different Gram-positive (*Bacillus subtilis*, *Bacillus sphaericus* and *Staphylococcus aureus*) and Gram-negative (*Pseudomonas aeruginosa*, *Klebsiella aerogenes* and *Chromobacterium violaceum*) microorganisms.

n-Bu$_3$SnH/AIBN-mediated vinyl radical cyclization has also been applied to construct the oxepane skeleton (Scheme 4.111).[160] The key intermediates **349** were prepared by the *o*-alkylation of MBH adducts with homopropargyl alcohol or aryl substituted *o*-homopropargyl derivatives under clay-catalytic conditions in 34–48% yields. Radical cyclization of **349** was carried out with 1.5 equiv. of tri-*n*-butyltin hydride and a catalytic amount of AIBN in benzene under reflux for 12 h under an inert atmosphere to afford good yields of the crude stannylated compound **350**, which was protiodestannylated in dichloromethane with pyridinium *p*-toluenesulfonate (PPTS) to give tri- or tetra-substituted oxepane **351** in 48–63% yield.

The [4 + 3] cycloaddition reaction offers a convenient way to prepare relatively complex seven-membered rings from simple starting materials.[161] Using MBH adducts **352** as C3 component, a phosphine-catalyzed [4 + 3] annulation has been developed to construct the bicyclo[3.2.2]nonadiene skeleton (Scheme 4.112).[162] The reactions proceeds smoothly with various MBH adducts as well as various substituted pyrones **353** such as ester, nitrile and amide substituted pyrones, resulting in compounds **354** in moderate to good yields.

Given the successful applications of radical cyclization in the construction of seven-membered-ring compounds from MBH adducts, Chattopadhyay and co-workers have designed and synthesized several novel MBH adducts that

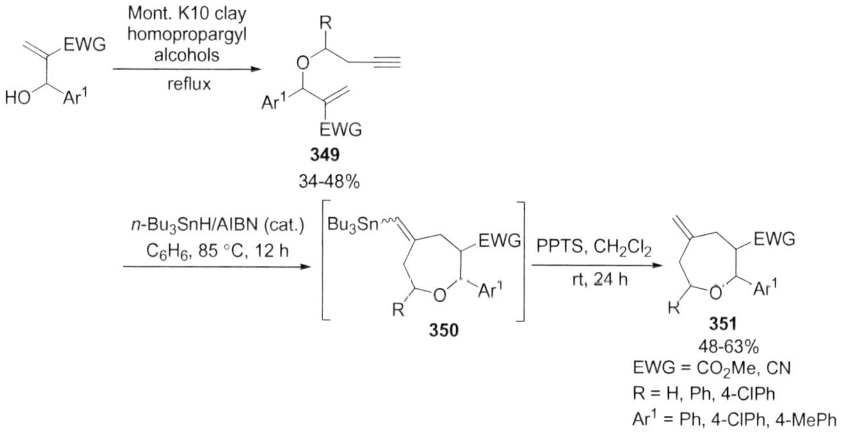

Scheme 4.111

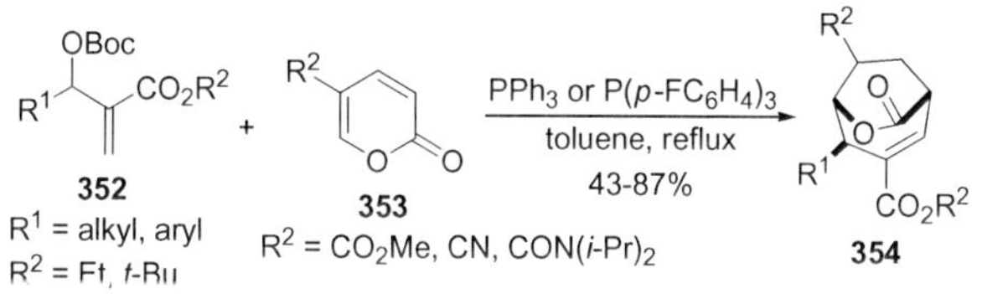

Scheme 4.112

R = CO$_2$Me, CO$_2$Et, CN
R^1 = R^2 = H, OMe, OCH$_2$O

Scheme 4.113

R^1 = R^2 = H, *t*-Bu; R^3 = H, OMe; R^4 = H, OMe,
OCH$_2$O; R^5 = H, OMe, OCH$_2$O; X = Br, I

Scheme 4.114

could be used to deliver furanose derivatives bearing a nine-membered-ring.[163] These dibenzo-heterocycles are structural fragments of many biologically important natural products[164] and continue to draw much attention from synthetic organic chemists. MBH adducts **355** and **356** containing chiral sugar units were easily prepared, and radical cyclization of these compounds in refluxing benzene with Bu$_3$SnH (1.5 equiv) and AIBN (5 mol.%) followed by separation of the tin compounds, acetylation and chromatography furnished the tricyclic ethers **357** and **358** in 53–64% and 63–71% yields, respectively (Scheme 4.113). The same synthetic strategy was further applied in the synthesis of dibenzo-heterocycles.[165] As shown in Scheme 4.114, good yields of dibenz-annulated ethers were acquired under the similar reaction conditions.

4.4 Synthesis of Nitrogen-containing Heterocyclic Compounds

4.4.1 Aziridines

Aziridination of olefins has received particular interest due to the enormous synthetic potential of aziridines.[166] Aziridines have a high strain energy

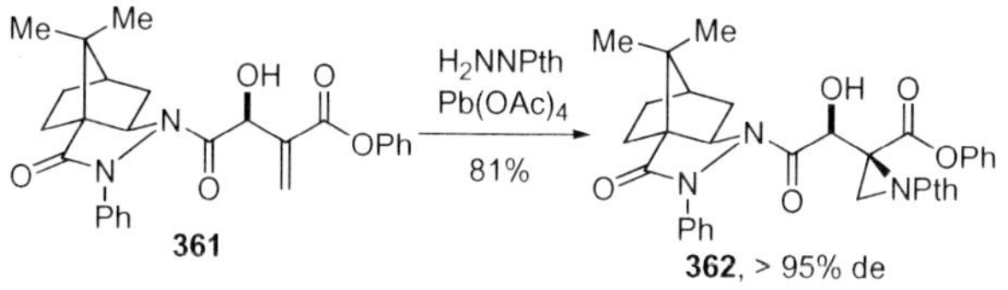

Scheme 4.115

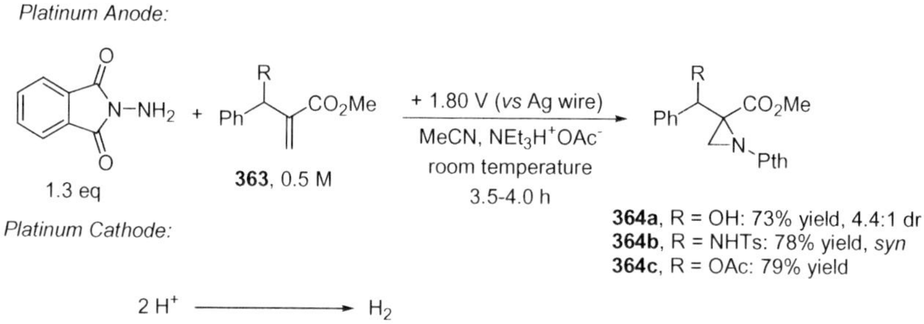

Scheme 4.116

$(28 \text{ kcal mol}^{-1})^{167}$ and are amenable to ring-opening reactions with a wide range of nucleophiles to provide molecules with valuable 1,2-heteroatom relationships commonly found in various natural products and in pharmaceuticals.[168]

MBH adduct **361**, containing the chiral auxiliary camphorpyrazolidinone,[169] has been reacted with *N*-aminophthalimide (H₂NNPth) in the presence of lead tetraacetate to give the corresponding *N*-phthalimidoaziridine **362** with 95% de in 85% yield (Scheme 4.115).[170]

Compared with metal-catalyzed olefin aziridination accompanied with metal based reagents, catalysts and stoichiometric oxidants,[171] electrochemical-catalyzed aziridination is a green catalytic method.[172] A simple combination of platinum electrodes, triethylamine and acetic acid led to a highly efficient formal nitrogen transfer from *N*-aminophthalimide to olefins; +1.80 V *versus* Ag wire produced the highest isolated yield of the aziridine from cyclohexene. Under the optimized reaction conditions, both electron-rich and electron-poor olefins were converted into aziridines with high efficiency. Using compounds **363** as substrates, 73–79% yields of compounds **364** were obtained with high selectivities (Scheme 4.116).

4.4.2 Azetidines

Azetidines, whose structures have been found in many natural products, constitute an important class of compounds, because of their interesting pharmacological activities and synthetic utility.[173] Whereas the strain associated with the azetidine ring system leads to difficulties in its synthesis, functionalization

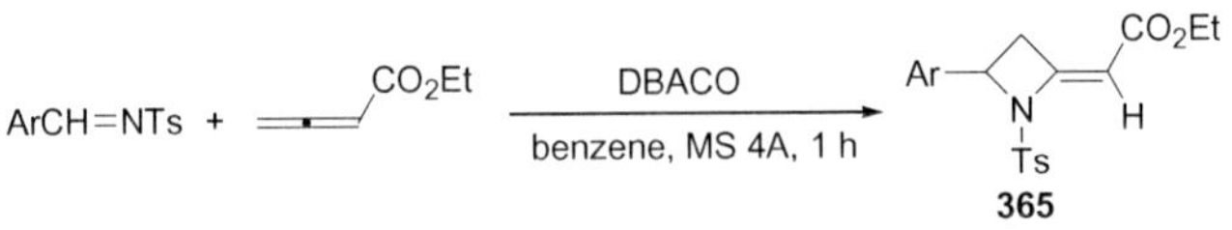

ArCH=NTs +
CO2Et
DBACO
benzene, MS 4A, 1 h
Ar
CO2Et
N
H
Ts
365

Scheme 4.117

OH
Ar2
EWG
1 equiv
(EtO)2PNHAr1
366 (1 equiv)
(i) 366, NaH (2 equiv)
benzene, 60 °C, 30 min
(ii) BH adduct, benzene,
rt, 3-5 h
EWG
Ar2
N
Ar1
367, 84-93% yield
(eq 1)
(EtO)2PNHAr1
366
NaH
(EtO)2PNAr1
acrylonitrile
or methyl acrylate
aldehyde
EWG
Ar2
N
Ar1
367, 21-30% yield
+ Ar1N=CHAr2
368, 66-74% yield
(eq 2)

Scheme 4.118

and modification, the preparation methods for functionalized azetidines have received great attention. The so-called "abnormal" aza-MBH reaction of *N*-tosylated imines with ethyl 2,3-butadienoate catalyzed by DABCO has been disclosed by Shi's group to give azetidines **365** in good yields (Scheme 4.117).[174]

MBH adducts as starting materials could also produce azetidines in high yields. In a one-pot procedure, diethyl *N*-arylphosphoramidates **366** have been treated with sodium hydride in dry benzene to generate an anion *in situ*, which underwent aza-Michael addition to MBH adducts followed by cyclization to afford the azetidine-3-carbonitriles/carboxylates **367** in 84–93% yields [Scheme 4.118, eqn (1)].[381] However, in their initial experiments, the aza-Michael addition of the anion of phosphoramidate **366** to acrylonitrile/methyl acrylate followed by addition–cyclization with aldehyde afforded azetidines **367** in only 21–30% yields, and the major products of the reactions were the Schiff bases **368** formed in 66–74% yields [Scheme 4.118, eqn (2)].

4.4.3 β-Lactams

Since the advent of penicillin, β-lactam antibiotics have occupied a central role against pathogenic bacteria.[175] Moreover, more and more novel β-lactam antibiotics (Figure 4.3) have been developed due to bacterial tolerance and resistance. Special impetus for research efforts on β-lactam chemistry has been provided by the introduction of the β-lactam synthon method, a term coined by Ojima over 15 years ago, according to which 2-azetidinones can be employed as useful intermediates in organic synthesis.[176]

Cyclization under Mitsunobu conditions has been reported to synthesize an azetidine core.[177] Michael additions to MBH adduct have provided 91% yield

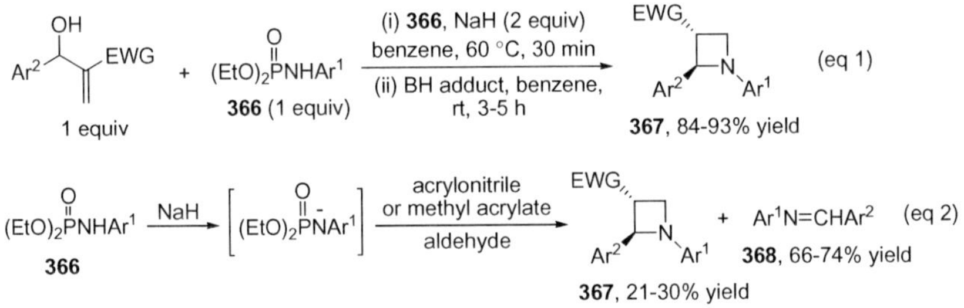

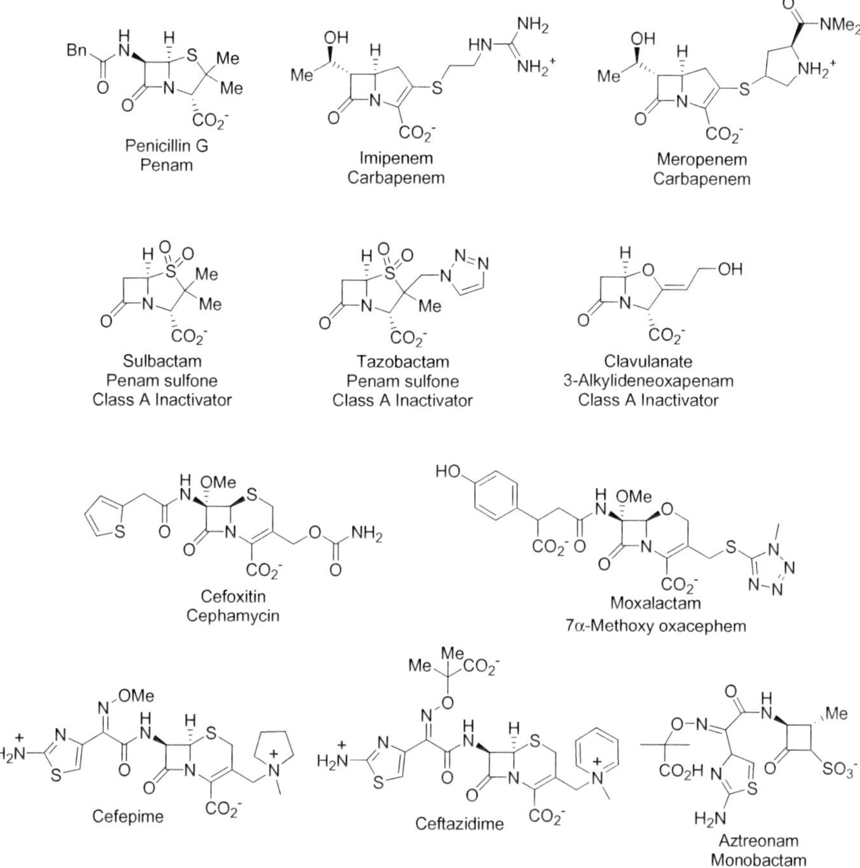

Figure 4.3 Examples of novel β-lactam antibiotics.

of compound **369**, which was successfully transformed into β-lactam **370** in moderate yield in the presence of (Me₃Si)₂NLi (Scheme 4.119).[178]

The S_N2' reaction product, β-aminoesters **372**, produced from the reaction of allyl bromide **371** and arylamine by treatment with triethylamine, could be converted into substituted β-lactams in the presence of base. Conventional bases such as ethylmagnesium bromide, *n*-butyllithium, potassium *tert*-butoxide, lithium hydroxide and potassium hydroxide used in such cyclizations failed to give β-lactam **373**. However, Sn[N(TMS)₂]₂ as a base in the cyclization of β-aminoesters[179] could successfully cyclize **372** to give β-lactam **373** (Scheme 4.120).[180] Under optimized cyclization conditions, it was found that 1.5 equiv. of Sn[N(TMS)₂]₂ in toluene under reflux for 6 h achieved the best yield (up to 83%.)

When (*S*)-**374** was used as starting material, the Pd-catalyzed allylic amination afforded dehydro-β-aminoesters **375**,[181] in which chirality was retained, but the reaction showed a strong solvent-dependent regiocontrol.[182]

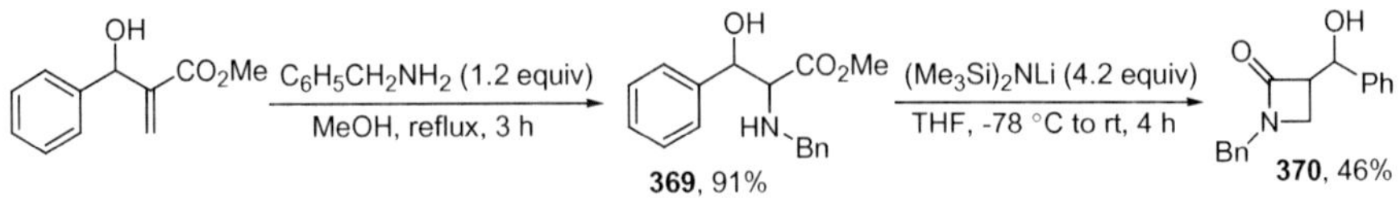

Scheme 4.119

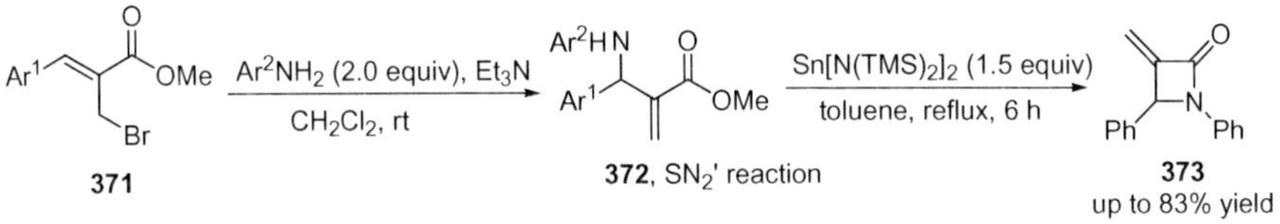

Scheme 4.120

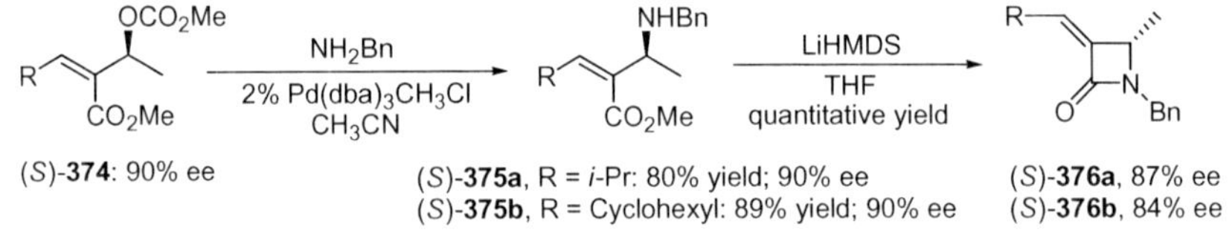

Scheme 4.121

Using MeCN as solvent, (*S*)-**374** reacted with benzylamine in the presence of Pd$_2$(dba)$_3$CHCl$_3$ catalyst to give (*S*)-**375** in 80–89% yields. These dehydro-β-aminoesters, (*S*)-**375a,b**, could be easily transformed into the corresponding (*Z*)-β-lactams (*S*)-**376a,b** in quantitative yields by treatment with LiHMDS in anhydrous THF at –20 °C (Scheme 4.121).

4-Acetoxyazetidinone **377** is the key intermediate for the synthesis of penems and carbapenems; it is available on an industrial scale[183] and is a considerable proportion of the cost of materials for these antibiotics. A cost-effective, scalable method for the stereoselective synthesis of **377** has been developed by Singh *et al.* (Scheme 4.122).[184] Using commercially available (–)-D-2,10-camphorsultam as chiral auxiliary, the enantioselective synthesis of MBH adduct **379** was achieved. Silylation of compound **378** in quantitative yield and diastereoselective Michael addition with benzyl amine afforded the product **380** in 91% yield. Hydrogenolysis of **380** with 20% Pd(OH)$_2$ on carbon in methanol at room temperature yielded amino compound **381** in quantitative yield. Using a catalytic amount of trimethylchlorosilane (TMCS), the silylation and cyclization of amino **381** with hexamethyldisilazane (HMDS) afforded β-lactam **382** in 54% yield. Subsequent oxidation using RuCl$_3$ as catalyst furnished the target molecule **377** in 72% yield according to the literature procedure.

Adam and co-workers have reported the synthesis of β-lactams from the cyclization of an amide.[185] As shown in Scheme 4.123, the MBH adducts **383**

378
(1) CH₃CHO, DABCO, CH₂Cl₂, 0-5 °C, 48 h, quantitative yield
(2) BF₃·Et₂O, MeOH, 0 °C-reflux, 52%
379
(1) TBDMSCl, imidazole, DMF, rt, 5 h, quantitative yield
(2) PhCH₂NH₂, MeOH, rt, 48 h, 91%
380
TBDMSO
N–Bn
H
20% Pd(OH)₂/C
H₂ (8 atm), rt, quantitative yield
381
TBDMSO
NH₂
(1) HMDS, TMCS, CH₂Cl₂, reflux
(2) t-BuMgCl, THF, rt, 2 h, 54%
382
OTBDMS
H
NH
O
RuCl₃·3H₂O, NaOAc, 40% CH₃CO₃H, 0-5 °C, CH₃CN, 72%
377
OTBDMS
H
OAc
NH
O

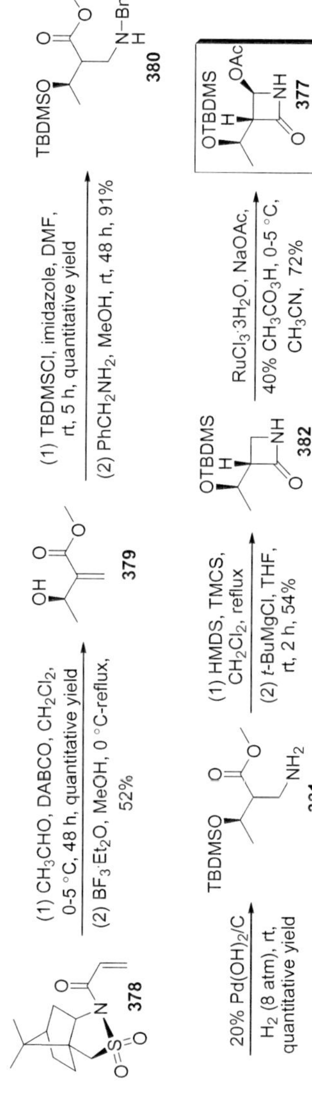

Scheme 4.122

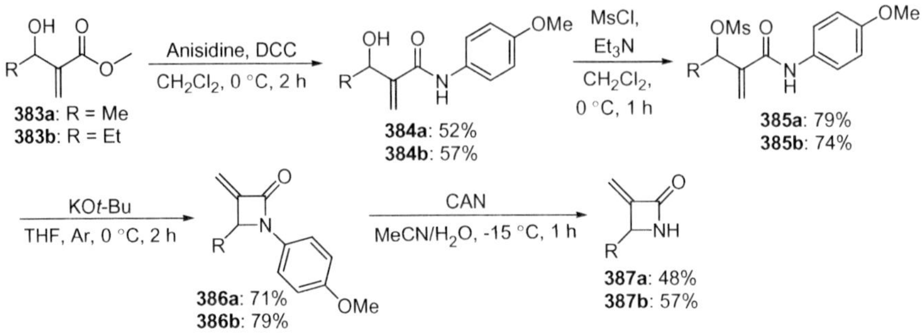

383a: R = Me
383b: R = Et

Anisidine, DCC
CH$_2$Cl$_2$, 0 °C, 2 h

384a: 52%
384b: 57%

MsCl, Et$_3$N
CH$_2$Cl$_2$, 0 °C, 1 h

385a: 79%
385b: 74%

KO*t*-Bu
THF, Ar, 0 °C, 2 h

386a: 71%
386b: 79%

CAN
MeCN/H$_2$O, -15 °C, 1 h

387a: 48%
387b: 57%

Scheme 4.123

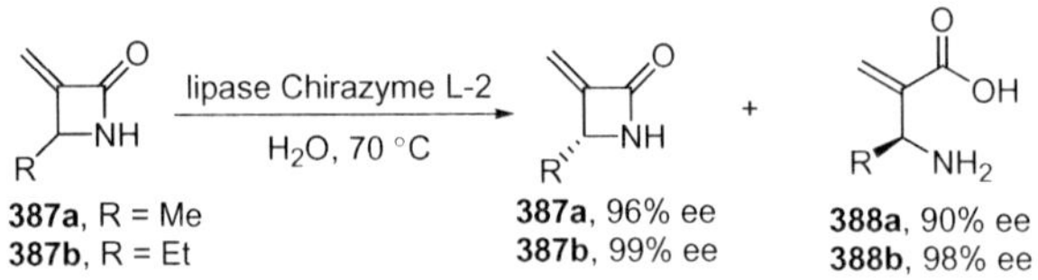

lipase Chirazyme L-2
H$_2$O, 70 °C

387a, R = Me
387b, R = Et

387a, 96% ee
387b, 99% ee

388a, 90% ee
388b, 98% ee

Scheme 4.124

react with *p*-anisidine in the presence of DCC to afford 3-hydroxy 2-methylene amides **384**, which are then converted into mesyl amides **385** by mesyl chloride using triethylamine as a base. Employing *t*-BuOK as a strong base in THF at 0 °C could achieve the cyclization of amides **385** in good yields. Despite the considerable ring strain, the oxidative deprotection of the aryl group with ceric ammonium nitrate (CAN) afforded β-methylene β-lactams **387** (48–57% yield), which may serve as synthons in organic synthesis. Lipase-catalyzed resolution of β-lactams **387** was further investigated by these researchers. From a screening of numerous enzymes, the lipase Chirazyme L-2 (from *Candida antartica*) emerged as the best enzyme employed in the resolution of β-lactams **387**, in up to 99% ee (Scheme 4.124), while it showed no activity for the resolution of N-substituted β-lactam derivatives **386**.

An aza-MBH reaction can provide β-amino acid esters that may be applied to construct valuable β-lactam derivatives. Hatakeyama *et al.* have proposed an asymmetric version of an aza-MBH reaction using β-isocupreidine (β-ICD) as catalyst and have synthesized a β-lactam derivative from an aza-MBH adduct (Scheme 4.125).[186] Adduct **389** could be prepared in 95% ee by the reaction of 1,1,1,3,3,3-hexafluoroisopropyl acrylate (HFIPA) with an activated imine bearing a diphenylphosphinoyl group. Upon acid hydrolysis of **389** in boiling hydrochloric acid, the diphenylphosphinoyl group was cleaved cleanly to give β-amino acid hydrochloride **390**. Treatment of **390** with BOPCl in the presence of triethylamine gave β-lactam **391** in 46% overall yield. This synthetic methodology presents an easy and effective route for the synthesis of β-lactams.

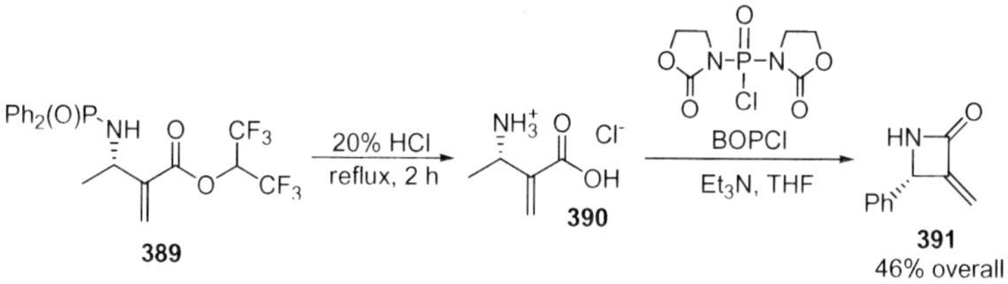

Scheme 4.125

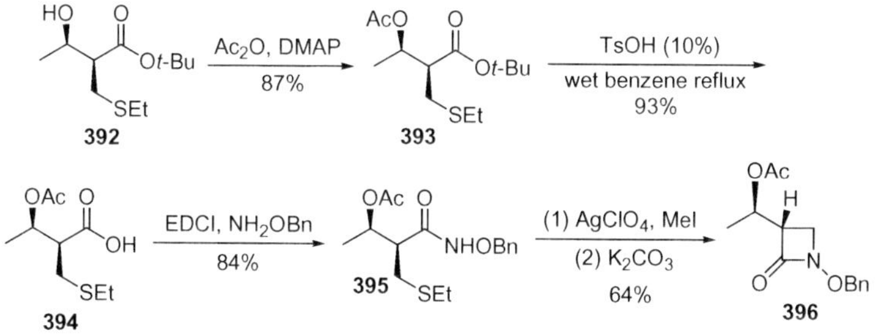

Scheme 4.126

The compounds **392**, prepared from Michael addition of aliphatic thiols to MBH adducts with high *syn*-selectivity (*syn/anti* up to 98%), have also been applied to the synthesis of β-lactams (Scheme 4.126).[187] The OH group of **392** was changed into an OAc group (**393**) in 87% yield. Subsequent acidic hydrolysis gave free carboxylic acid **394**, which was then converted into amide **395** in 84% yield. The intramolecular S_N2 reaction occurred in the presence of the sulfonium salt, affording β-lactam **396** in diastereomerically pure form. This new synthetic method disclosed that β-lactams can be prepared with easy and practical manipulations as well as sufficient prevention of epimerization problems.

4.4.4 Pyrrole Derivatives

It is well-known that the introduction of an amino group into MBH adducts having ester group is an effective protocol to construct N-heterocyclic compounds by intramolecular amidation. The Boc group has been subjected to deprotection from ester **397** in the presence of CF_3CO_2H to give the desired product in 68% yield, and the subsequent cyclization can produce lactam **399** in 82% yield (Scheme 4.127).[188]

Basavaiah and Rao first disclosed that a MBH adduct bearing a nitro group can furnish γ-lactams in good yields *via* a reductive cyclization.[189] Treatment of the starting material **400** (2.0 mmol) with nitroethane (8.0 mmol) in the presence

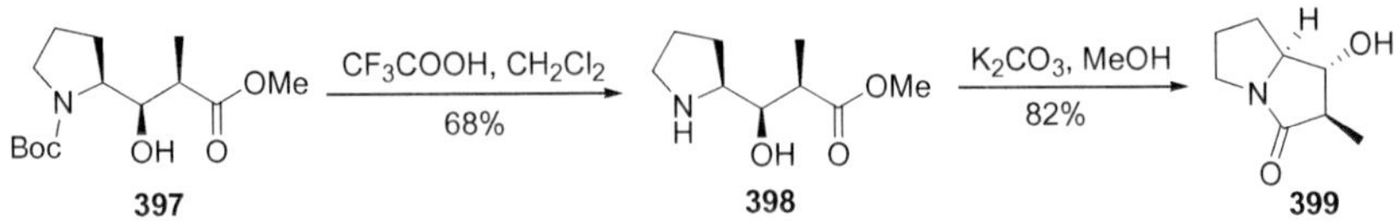

Scheme 4.127

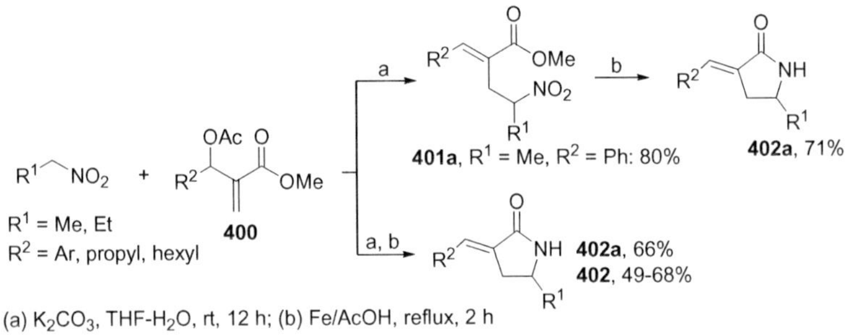

Scheme 4.128

(a) K_2CO_3, THF-H_2O, rt, 12 h; (b) Fe/AcOH, reflux, 2 h

of K_2CO_3 (8.0 mmol) in THF (5.0 mL) and water (0.1 mL) at room temperature for 12 h provided the S_N2'-type trisubstituted alkene **401a** in 80% yield. Subsequent treatment of **401a** with Fe/AcOH at 110 °C for 2 h provided the desired γ-lactam **402a** in 71% yield (57% overall) (Scheme 4.128). This reaction pathway can also be accomplished in a one-pot manner to give **402a** in 66% yield. With this convenient, operationally simple, one-pot procedure in hand, a series of substituted pyrrolidin-2-ones **402** were obtained in 49–68% yields.

3-Aryl-2-methylene-4-nitroalkanoates, obtained by S_N2 nucleophilic reaction of the acetate of MBH products with nitroalkanes, also offers opportunities to construct highly substituted pyrrolidin-2-ones by intramolecular reductive cyclization.[190] As shown in Scheme 4.129, the indium–HCl reductive system can successfully achieve the reductive cyclizations of adducts **403** to pyrrolidin-2-ones **404** and **405** in moderate yields. However, $SnCl_2 \cdot 2H_2O$-promoted reduction mainly produced oxime derivatives **407**, with one exception that formed **406**; the oxime derivatives **407** could be further converted into 1H-pyrroles **409**.

The reaction of MBH adduct with alkane-amines also provided an effective protocol to synthesize pyrrolidin-2-one derivatives *via* nucleophilic reaction and subsequent cyclization processes. Ayed *et al.* first reported the direct condensation of MBH adduct with primary amines in methanol to afford 3-hydroxyl pyrrolidin-2-ones **410** in good to excellent yields (Scheme 4.130).[191] The reaction of (*S*)-phenylethylamine with MBH adduct **411** was reported to generate an equimolar mixture of the 4,5-*cis*-disubstituted pyrrolidin-2-ones **412** and **413** (Scheme 4.131).[192] Interestingly, pyrrolidin-2-one **413** could be further used to synthesize the glycosidase inhibitor **414**.[193]

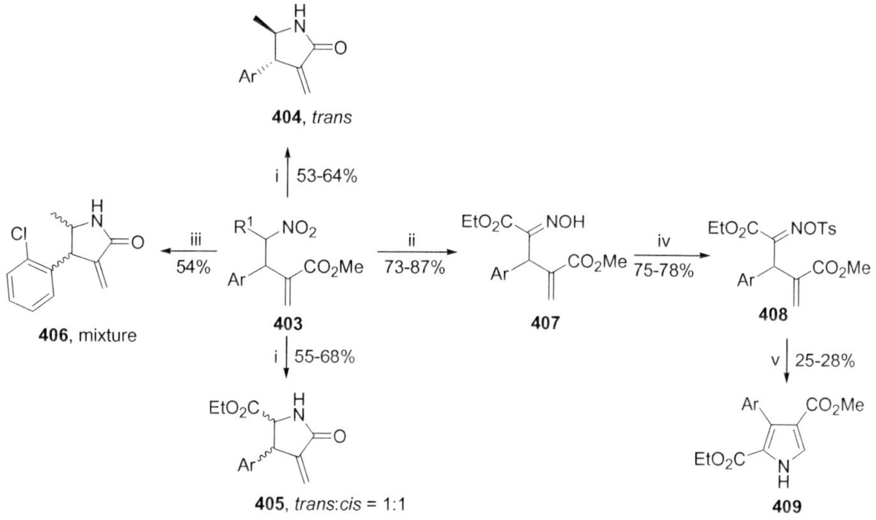

404, *trans*

i | 53-64%

406, mixture

iii | 54%

403

ii | 73-87%

407

iv | 75-78%

408

v | 25-28%

i | 55-68%

405, *trans:cis* = 1:1

409

Reagents and conditions: (i) In, HCl, THF-H_2O, rt, 2 h; (ii) $SnCl_2\cdot2H_2O$, MeOH, reflux, 1.5 h; (iii) $SnCl_2\cdot2H_2O$, MeOH, reflux, 24 h; (iv) TsCl, Et_3N, CH_2Cl_2, rt, 3 h; (v) DBU, CH_2Cl_2, rt, 330 min.

Scheme 4.129

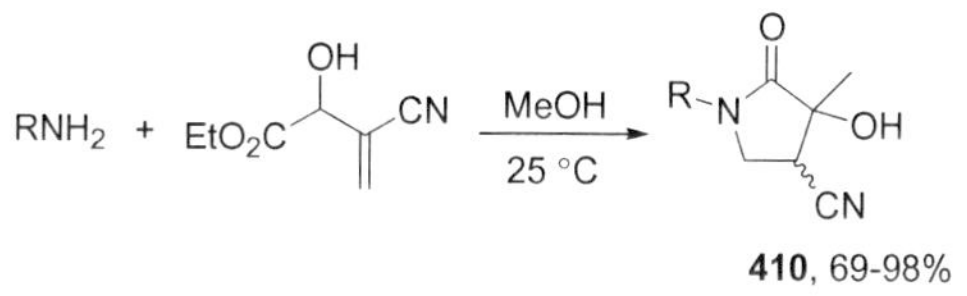

RNH_2 + [structure] $\xrightarrow[25\,°C]{MeOH}$

410, 69-98%

Scheme 4.130

411 + [structure] $\xrightarrow[\text{(a) MeOH, rt, 24 h}]{\text{(b) toluene, reflux, 4 h}}$

412, 40% + **413**, 38%

414

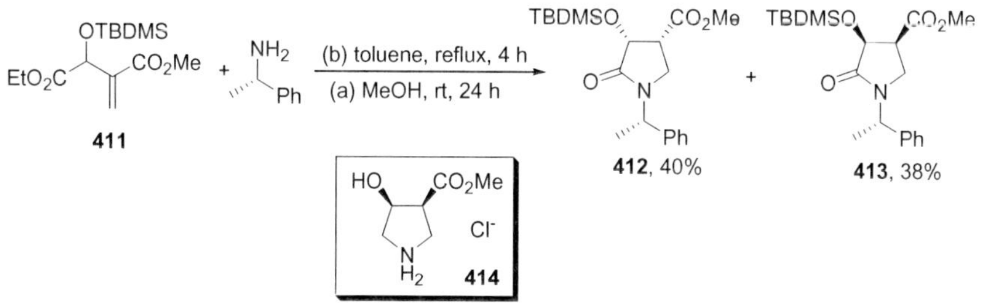

Scheme 4.131

3-Aryl-3-hydroxypyrrolidin-2-ones have been found in many natural products such as chimonamidine[194] and donaxaridine,[195] and an efficient synthetic route for the preparation of such derivatives has been developed by Kim's group.[196] Using isatins **415** as substrate, the reactions were carried out in the presence of benzylamine in MeOH at room temperature to give diastereomeric mixtures of **416** (*syn*-**416** and *anti*-**416**) in 67–88% yields (Scheme 4.132); in all cases, the *syn* isomers were obtained as the major products. When the cyano group in isatin **415** was replaced with an ester group, the reactions of isatin **417** with primary amines in methanol mainly produced tricyclic compounds **419**. This one-pot reaction took place *via* sequential Michael addition of amine to isatin **417**, intramolecular cyclization and concomitant ring opening of lactam of isatin moiety **418**. However, the reaction outcome using **417c** as substrate was different from those of **417a** and **417b**, which afforded a diastereomeric mixture of **418d**. Moreover, the *anti*-**418d** was obtained in 76% yield as the major component of the reaction mixtures.

Similarly, methyl 1-hexyl-4-hydroxy-5-oxopyrrolidine-3-carboxylate **421** has been synthesized from the reaction of MBH adduct **420** with 1.0 equiv. of *n*-hexylamine (Scheme 4.133).[197] Subsequent transformation of such a 2-pyrrolidinone **421** into acid **422** was achieved by hydrolysis.

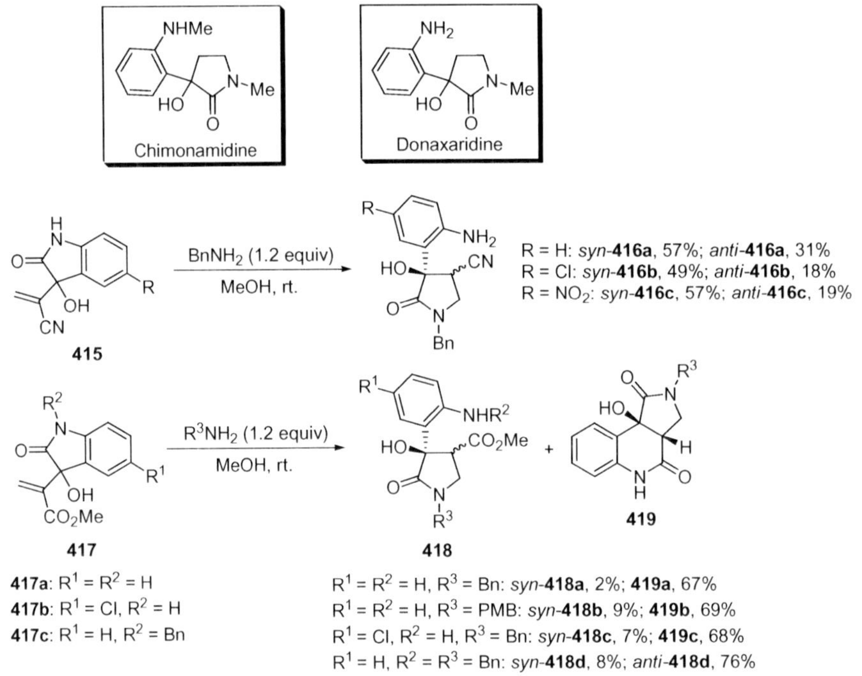

Scheme 4.132

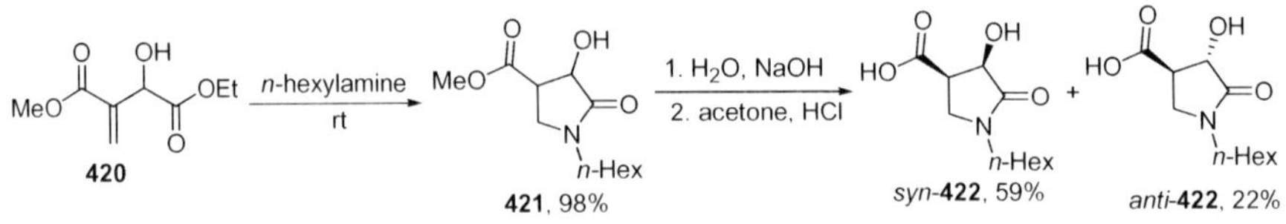

Scheme 4.133

Highly enantioselective Boc-protected MBH adducts **423** and **424** ($>98\%$ ee) – obtained from diastereoselective racemization-free MBH reaction using β-isocupreidine (β-ICD) as catalyst – have been treated with iodotrimethylsilane, generated *in situ* from chlorotrimethylsilane and NaI, in acetonitrile at $0\,^{\circ}C$ to give the highly functionalized pyrrolidinones **425** and **426** in quantitative yields *via* the cleavage of the Boc group and concomitant cyclization (Scheme 4.134).[198]

Orena *et al.* have further reported a synthetic method for the formation of 3,4-*trans*-disubstituted pyrrolidin-2-ones and their application in the synthesis of analogues of (*S*)-β-homoserine and (*S*)-aspartic acid.[199] *N*-Acylamino derivatives **428** were first reacted with chiral amine in MeOH at room temperature to 60 °C, and the resulting mixtures were further treated with DBU in toluene at room temperature to afford the *trans*-3,4-trans-disubstituted derivatives **429** and **430** exclusively and in good yields. These derivatives were then easily separated by flash chromatography on silica gel (Scheme 4.135). Using compounds **429** as starting materials, researchers have, notably, successfully synthesized several analogues of (*S*)-β-homoserine[200] and (*S*)-aspartic acid.[201]

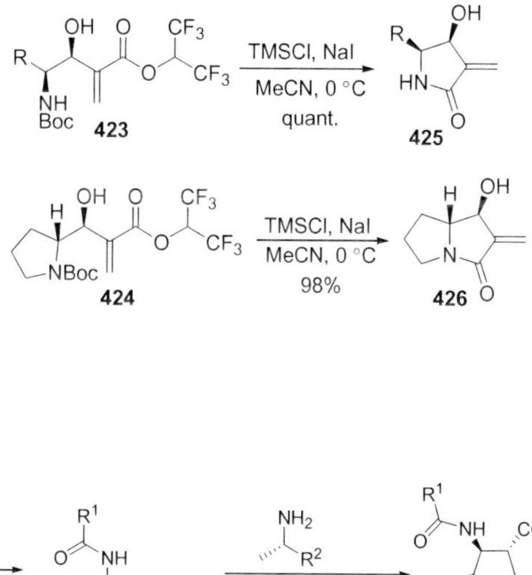

Scheme 4.134

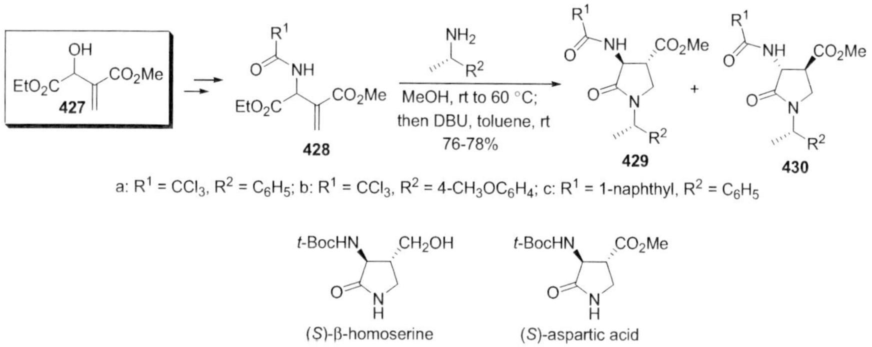

a: $R^1 = CCl_3$, $R^2 = C_6H_5$; b: $R^1 = CCl_3$, $R^2 = 4\text{-}CH_3OC_6H_4$; c: $R^1 = 1\text{-naphthyl}$, $R^2 = C_6H_5$

Scheme 4.135

Previous reports have disclosed that isoxazole derivatives could be utilized to synthesize 2-pyrrolidinone derivatives under hydrogenolysis[202] or Pd/C-promoted hydrogenation.[203] A series of isoxazole derivatives **431** and **432**, prepared from 3-isoxazolecarbaldehydes with acrylates, were investigated in intramolecular ring-closure reactions (Scheme 4.136).[204] These substrates (**431** or **432**) were subjected to hydrogenolysis in the presence of Raney-Ni to yield enaminones **433**, which could undergo intramolecular lactonization by NaH or DBU catalysis or silica gel treatment to give 2-pyrrolidinones **434**. However, Michael addition products **435** and **438** from compounds **431** afforded 1,5-dihydro-2-pyrrolones **437** and *N*-substituted pyrrolidines **440**, respectively, in good yields *via* intramolecular ring-closure reactions. Such results suggest that the elimination of a secondary base was favored in the ring closure of enaminones **436** but elimination of ammonia was favored in the ring closure reactions of compounds **439**.

With the development of the enantioselective alkylation of the Schiff base **441** with MBH acetates employing cinchonidine-derived catalysts under phase-transfer conditions,[205] optically active 4-alkylidenylglutamate (*S*)-**442** (92% ee) was hydrolyzed and subsequently lactonized to give 4-alkylidene pyroglutamate (*S*)-**443** in 97% yield in the presence of excess 10% aqueous citric acid; subsequent Pd-C hydrogenation of compound (*S*)-**443** produced 4-substituted pyroglutamate (2*S*,4*S*)-**444** in 82% yield (Scheme 4.137), which can be readily converted into 4-substituted glutamic acids.[206] This synthetic method presented a novel, general, and practical procedure for the asymmetric synthesis of 4-alkylidenyl-glutamic acid derivatives. According to Scheme 4.137, (*S*)-**442** and (*S*)-**443** are 4-benzylidene substituted compounds but not 4-phenylidene substituted compounds.

Allyl cyanide **445** have been oxidized to yield the corresponding amide derivative **446** in 82% yield, which then afforded the 3-benzylidene-pyrrolidine-2,5-dione **447** in 78% yield upon treatment with sodium hydride (Scheme 4.138).[207]

The ring-opening reaction of *N*-tosylaziridines with various nucleophiles has been widely used in organic synthesis. The reaction of *N*-tosylaziridine **448a** and aniline, as a representative external nucleophile, in CH_3CN in the presence of $LiClO_4$ (1.2 equiv) afforded 3-arylidenelactam **449a** in 75% yield as a single stereoisomer (*trans*-form) under refluxing conditions (Scheme 4.139).[208] When the same reaction was carried out at room temperature, vicinal diamine derivative **450a** was obtained in 85% yield as a single *anti*-stereoisomer. Upon continuous stirring in refluxing CH_3CN, the vicinal diamine **450a** was converted into **449a** in 78% yield in the presence of $LiClO_4$. These results suggest that the reaction might occur *via* an S_N1 type mechanism. By employing arylamines as nucleophiles, various 3-arylidenelactam derivatives **449** were synthesized from *N*-tosylaziridines **448** in good yields.

The introduction of an allyl group on the N-substituent of an aza-MBH adduct affords compounds that could be used as suitable starting materials for RCM reactions to construct pyrrole skeletons. Kim's group,[209] Adolfsson's group[210] and Lamaty's group[211] have each reported the RCM towards the synthesis of 2,5-dihydropyrrole skeletons from MBH adducts (Scheme 4.140). The RCM of adducts **451** by use of the commercially available second-generation

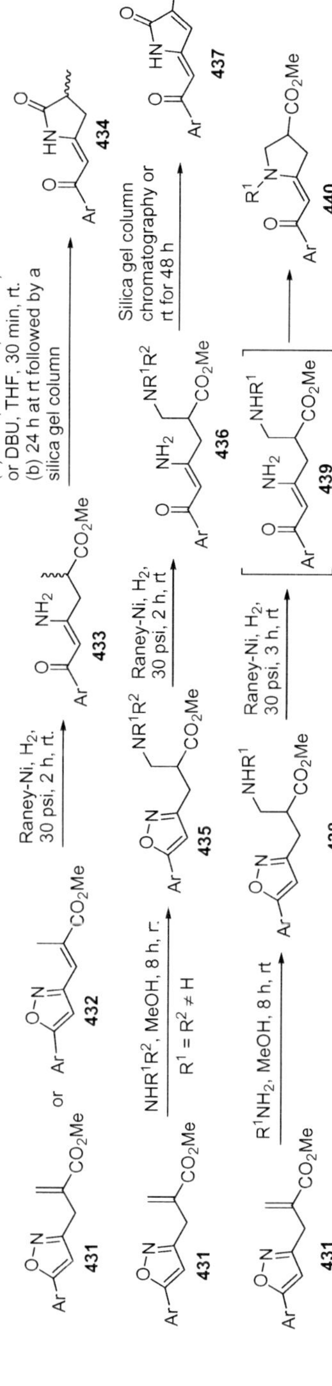

Scheme 4.136

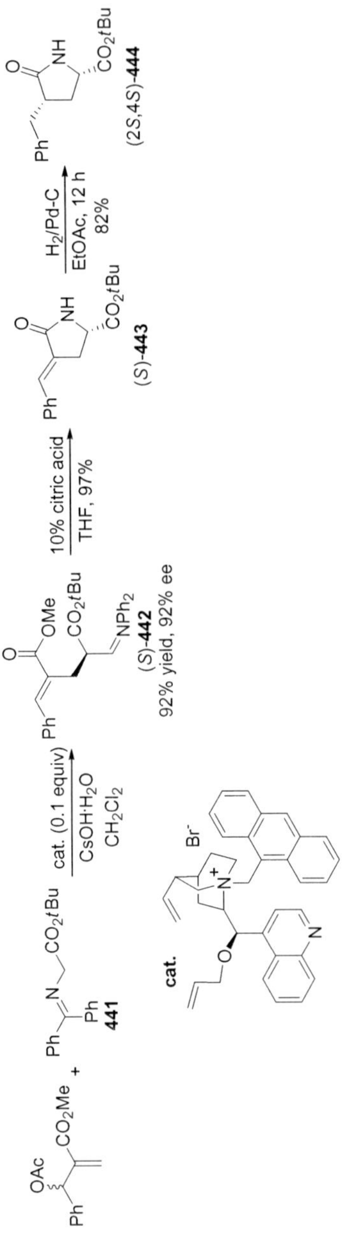
OAc
CO2Me
Ph
+
Ph
Ph
N
CO2tBu
441
cat. (0.1 equiv)
CsOH·H2O
CH2Cl2
Ph
O
OMe
CO2tBu
NPh2
(S)-442
92% yield, 92% ee
10% citric acid
THF, 97%
Ph
O
NH
CO2tBu
(S)-443
H2/Pd-C
EtOAc, 12 h
82%
Ph
O
NH
CO2tBu
(2S,4S)-444
cat.
Br-
N+
N
O

Scheme 4.137

CO$_2$Me
CN
Me
445
TFA:H$_2$SO$_4$, neat
rt, 4 h
CO$_2$Me
CONH$_2$
Me
446, 82%
NaH, toluene
rt, 30 min.
O
NH
O
Me
447, 78%

Scheme 4.138

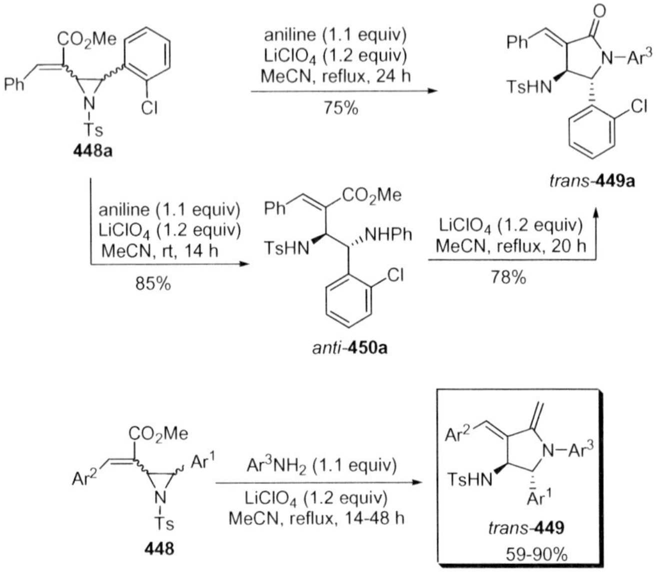

Scheme 4.139

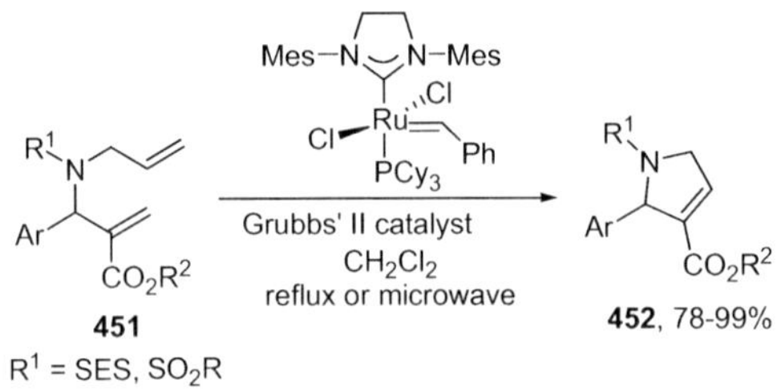

Scheme 4.140

of Grubbs catalyst in refluxing DCM for 4 h afforded the 2,5-dihydropyrroles **452** in high yields (78–99%). Using microwave irradiation could conveniently accelerate the RCM reaction rate to give the desired cyclic products in similar yields within 5 min.

However, using the simply modified MBH adducts **453** as substrates, 2,5-dihydropyrroles **454** were, unexpectedly, obtained in good yields with the elimination of styrene moiety (Scheme 4.141). Moreover, the configuration of the double bond in adducts **453** did not show any influence on the yields of compounds **454**.

2,5-Dihydropyrroles **455** containing an SES group, which is a valuable protecting group of amines in organic synthesis,[212] have been converted into the corresponding pyrroles **456** in high yields by cleavage of the SES group and aromatization in the presence of *t*-BuOK at room temperature (Scheme 4.142).

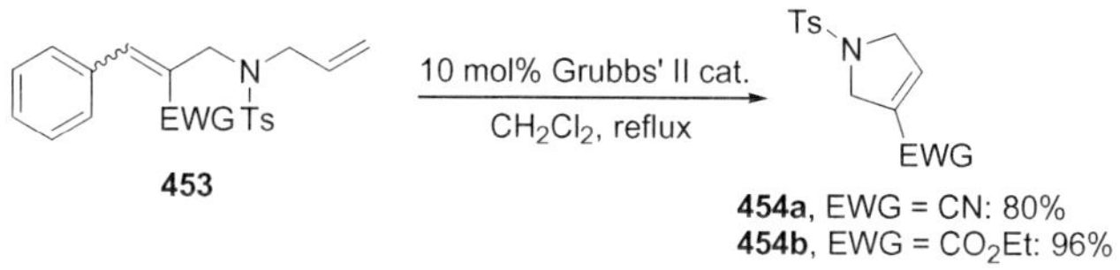
10 mol% Grubbs' II cat.
CH2Cl2, reflux
EWG Ts
453
Ts
EWG
454a, EWG = CN: 80%
454b, EWG = CO2Et: 96%

Scheme 4.141

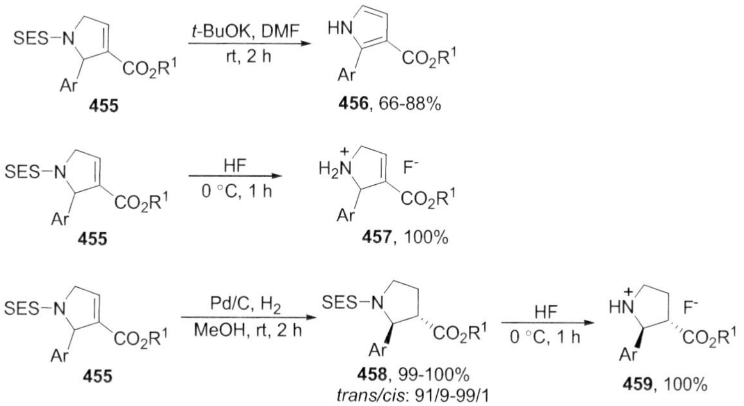
SES–N
CO2R1
Ar
455
t-BuOK, DMF
rt, 2 h
HN
456, 66-88%
HF
0 °C, 1 h
H2N
F-
457, 100%
Pd/C, H2
MeOH, rt, 2 h
458, 99-100%
trans/cis: 91/9-99/1
HN
459, 100%

Scheme 4.142

However, 2,5-dihydropyrroles **455** were completely deprotected to give the pyrrolines **457** as a hydrofluoride salt in quantitative yields, when they were used to react with neat hydrofluoric acid. Furthermore, hydrogenation of **455** with a Pd/C catalyst followed by HF-mediated deprotection of the resultant compound **458** yielded pyrrolidines **459** in excellent yields and good diastereomeric ratios.

Recently, the synthetic protocol employing the RCM of MBH adducts to construct pyrrole skeletons was applied in the synthesis of novel polyhydroxylated azasugars,[213] which are a family of important potent glycosidase inhibitors.[214] As shown in Scheme 4.143, novel iminosugars **468** and **470** were successfully prepared in good yields, and proved to be moderate inhibitors of β-galactosidase, α-glacto- and α-mannosidases.[215]

Kim *et al.* have reported an efficient method for the synthesis of polysubstituted pyrrole derivatives. Trisubstituted tetrahydropyrrole derivatives **475** were synthesized in 63–86% yields in the presence of K_2CO_3 in DMF from aza-MBH adducts **471** by sequential N-alkylation with phenacyl bromide (**472a**) and Michael addition at the conjugate vinyl moiety of the corresponding intermediate.[216] Using DBU to eliminate the tosyl group in MeCN, 2,3,5-trisubstituted pyrroles **476** were obtained in moderate yields (Scheme 4.144). However, using rearranged tosylamide derivatives **473** as reactants, which were easily synthesized from the corresponding MBH acetates and tosylamide in a

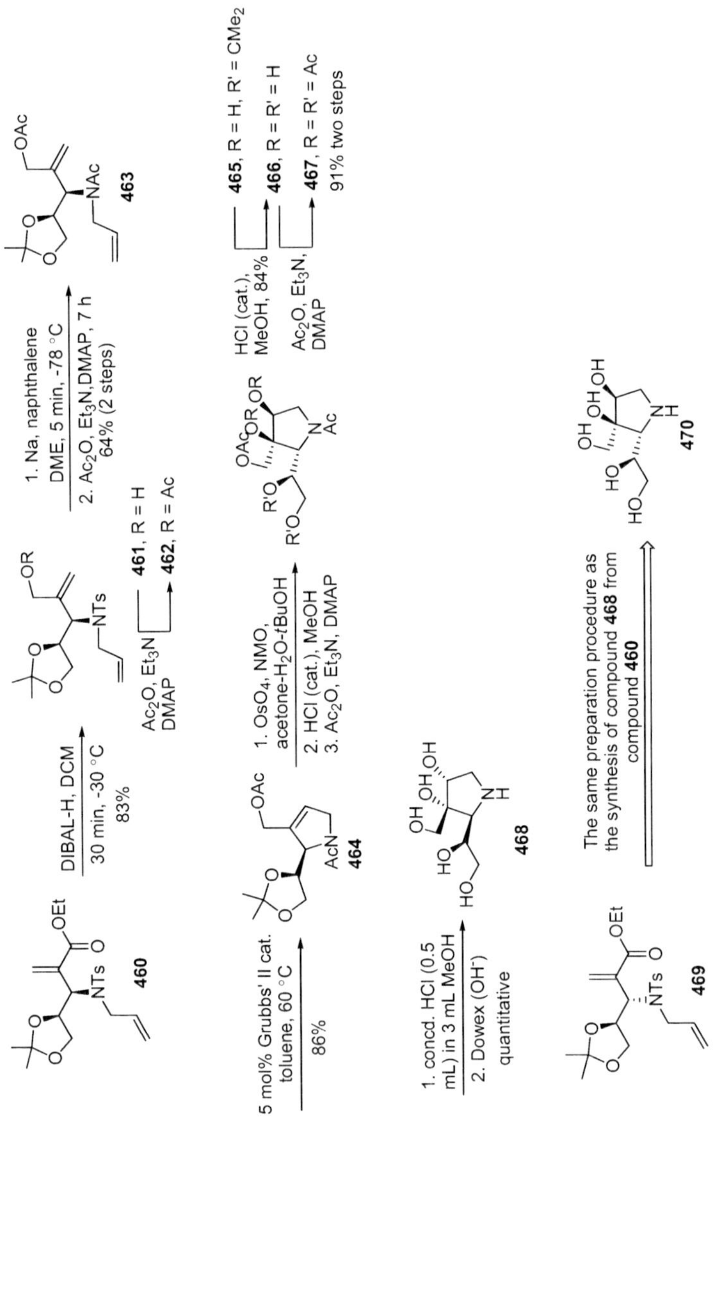

460
DIBAL-H, DCM
30 min, -30 °C
83%
461, R = H
462, R = Ac
Ac2O, Et3N
DMAP
1. Na, naphthalene
DME, 5 min, -78 °C
2. Ac2O, Et3N,DMAP, 7 h
64% (2 steps)
463
5 mol% Grubbs' II cat.
toluene, 60 °C
86%
464
1. OsO4, NMO,
acetone-H2O-tBuOH
2. HCl (cat.), MeOH
3. Ac2O, Et3N, DMAP
HCl (cat.),
MeOH, 84%
Ac2O, Et3N,
DMAP
465, R = H, R' = CMe2
466, R = R' = H
467, R = R' = Ac
91% two steps
1. concd. HCl (0.5
mL) in 3 mL MeOH
2. Dowex (OH-)
quantitative
468
469
The same preparation procedure as
the synthesis of compound 468 from
compound 460
470

Scheme 4.143

S_N2' manner, afforded 2,3,4-trisubstituted tetrahydropyrroles **477** and pyrroles **478** in 56–60% and 73–81% yields, respectively (Scheme 4.144). The products **479**, from **474**, were separated in 50–86% yields, which may undergo an intramolecular aldol process. Upon treatment with DBU, 2,3,4-trisubstituted pyrroles **480** were obtained in 51–71% yields (Scheme 4.144).

Radical cyclization of **482**, prepared from the alkylation of MBH adducts **481** with 1,4-dibromo-2-butene in the presence of K_2CO_3, has been utilized to synthesize 3,3,4-trisubstituted pyrrolidines **483** (Scheme 4.145).[217] As shown in

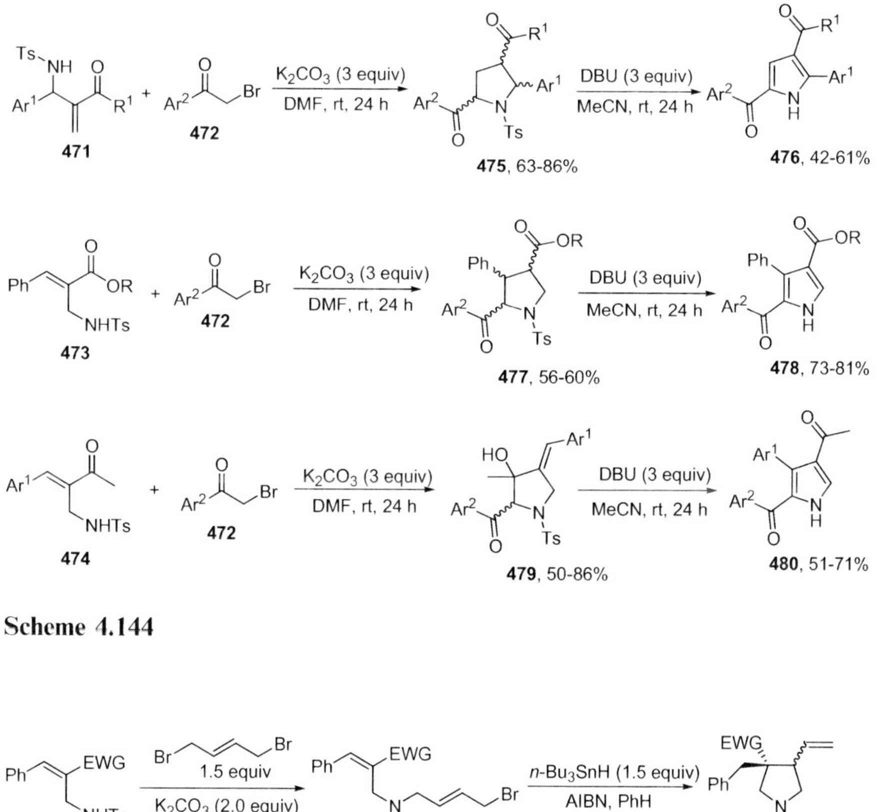

Scheme 4.144

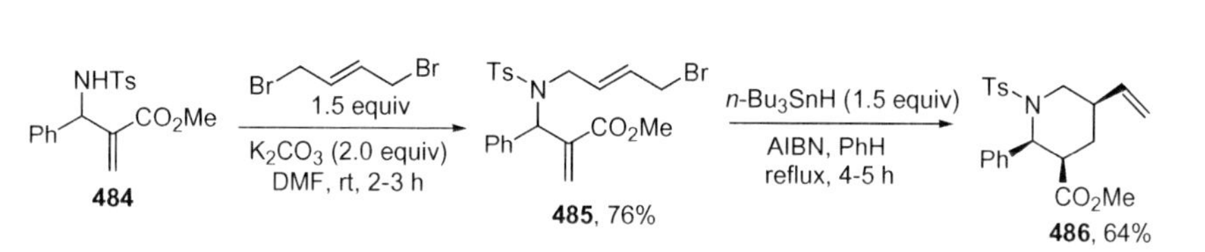

Scheme 4.145

Scheme 4.145, pyrrolidines **483** were obtained in moderate to good yields, in a poor *syn/anti* ratio, in the reaction of **482** with *n*-Bu$_3$SnH (1.5 equiv) using AIBN as an initiator. However, piperidine derivative **486** was isolated as the sole *anti*-product in 64% yield from the cyclization of **485**.

The titanium-mediated intramolecular radical vinylations of oxirane was first reported to yield alkylidene pyrrolidines **488** *via* the radical β-elimination of phosphinoyl radical.[218] The reactions of phosphine oxide precursors **487** with a stoichiometric amount of Cp$_2$TiCl$_2$ were carried out at room temperature using powdered manganese as reductant, from which pyrrolidines **488** were obtained in 57–82% yields (Scheme 4.146). The procedure for the synthesis of phosphine oxide precursors **487** is also shown in Scheme 4.146.

Llebaria *et al.* have developed a palladium-catalyzed intramolecular cyclization[219] of the sodium salts of substrates **489** containing a sulfinyl group as a chiral director to give the corresponding 2-[1-(*p*-tolylsulfinyl)ethenyl]pyrrolidines **490** in 53–88% yield (Scheme 4.147).[220] The stereo- and regiochemical outcomes were strongly dependent on the temperature and Pd catalyst, and up to 90% diastereomeric excess was achieved.

Sulfinamide **491**, upon treatment with *m*-CPBA in toluene, yielded in an unisolated mixture of diastereomeric epoxides;[221] subsequent electrophilic cyclization using a catalytic amount of camphor-10-sulfonic acid (CSA) afforded a 75 : 25 mixture of sulfonyl dihydropyrroles **492** in 73% yield (Scheme 4.148).[222] Treatment of the more acidic sulfinamide **494** with TBATB in the presence of K$_2$CO$_3$ gave cyclic products **495** in moderate yields, mainly as 2,5-*cis* isomers. Either further bromination of dihydropyrroles **492** or oxidation

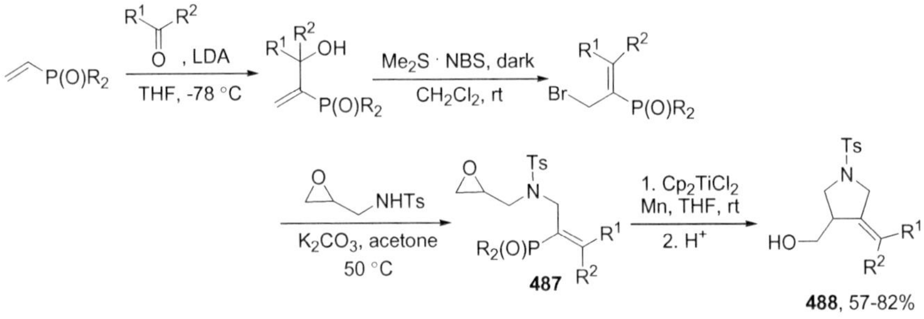

Scheme 4.146

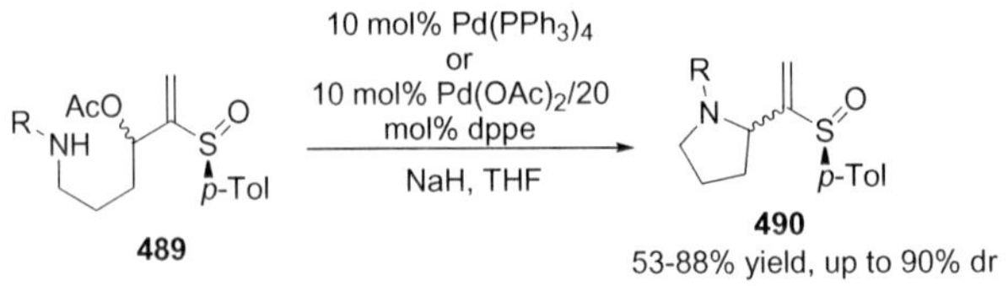

Scheme 4.147

Scheme 4.148

Scheme 4.149

of sulfinyl pyrroles **495** (R = Ph) could produce dihydropyrroles **493** in 51% and 83% yields, respectively.

Recently, Raghunathan *et al.* disclosed the synthesis of a series of novel spiropyrrolidines and polycyclic heterocycles by 1,3-dipolar cycloaddition reactions with MBH adducts.[223] The azomethine ylides, generated by the reaction of sarcosine with isatin, ninhydrin or acenaphthenequinone in boiling toluene for 48–78 h, reacted with MBH adducts to give the corresponding cycloadducts **496–498** as a single regioisomer in overall yields of 40–55% (Scheme 4.149). Interestingly, when the reactions were carried out in methanol, novel furo[3,4-*b*]pyrroles **499** and **500** were, respectively, obtained from ninhydrin and acenaphthenequinone. Such an unusual cyclization may be derived from a further hemiacetal cyclization of **497** and **498** (Figure 4.4).

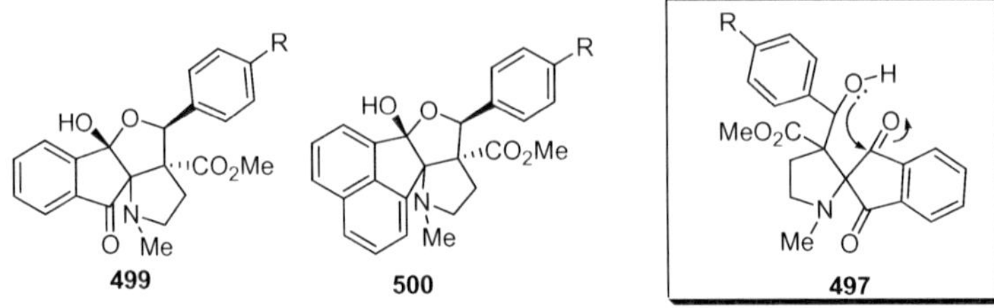

Figure 4.4   Novel spiropyrrolidines (**499** and **500**) and a proposed intermediate in their formation (**497**).

Scheme 4.150

4.4.5   Indoles and Indolizines

The indole ring system is probably the most ubiquitous heterocycle in a large number of biologically active natural products and pharmaceutical agents. Unsurprisingly, the synthesis and functionalization of indoles have been the subjects of much research and various well-established classical methods have become available since the structure of indole was elucidated by Baeyer and Knop in 1866.[224]

MBH adducts **501**, prepared from the reaction of pyridine-2-carbaldehydes, were first acetylated and then subjected to thermal cyclization to give 2-substituted indolizines in good yields.[225] This method offers a convenient and relatively efficient access to 2-carbonyl and 2-cyano-indolizines **503** (Scheme 4.150). In addition, this general approach appeared to be limited only by the availability of suitably substituted pyridine-2-carbaldehydes.

Compared with multi-steps used to form the desired 1-azabicyclo[4.3.0]nonane framework, Basavaiah and Rao have reported a novel, facile, convenient methodology for the synthesis of indolizines in a one-pot operation that was the first example of an electrophile-induced MBH reaction. The treatment of pyridine-2-carbaldehyde with various activated alkenes in the presence of

TMSOTf provided indolizines **504** in 38–55% yield (Scheme 4.151).[226] Subsequently, they also demonstrated a facile synthesis of indolizine-fused chromones **505** from the reaction between the MBH adducts of pyridine-2-carbaldehyde and 1-benzopyran-4(4*H*)-ones (Scheme 4.152).[227]

Another simple and efficient strategy to construct substituted indolizines under neutral conditions in a one-pot operation *via* MBH reaction was disclosed by Virieux *et al.*[228] In dry acetonitrile at 20 °C, pyrrolecarbaldehyde **506** was reacted with an excess of allene **507a** (2.5 equiv) catalyzed by Bu$_3$P for 5 h, affording the indolizine **508a** in 57% isolated yield. Similarly, electron-deficient alkyne **507d** was able to produce substituted indolizine **508d** in 50% isolated yield (Scheme 4.153). While for substrate **507b** and **507c**, only trace amounts of indolizines were isolated.

Following the synthesis of substituted indole *N*-oxides *via* a TiCl$_4$-mediated MBH reaction of α-oxo cyclic ketene-*S*,*S*-acetal with 2-nitrobenzaldehydes,[229] Dong and Liu and their co-workers further developed a novel method to rapidly synthesize indolizines from MBH adducts **509**, which in turn were prepared from the reaction of α-EWG ketene *S*,*S*-acetals with 2-pyridinecarbaldehyde

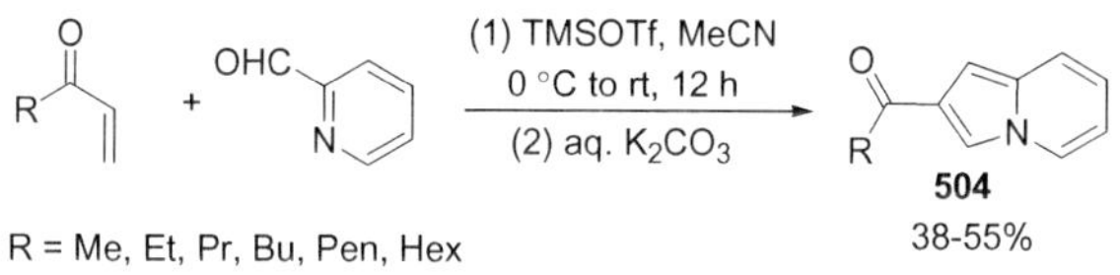

Scheme 4.151

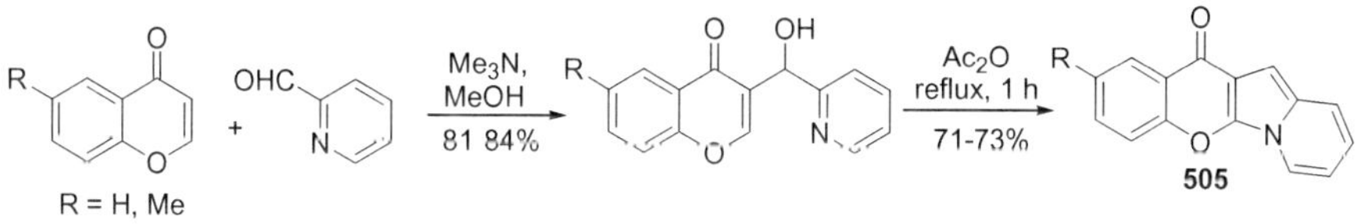

Scheme 4.152

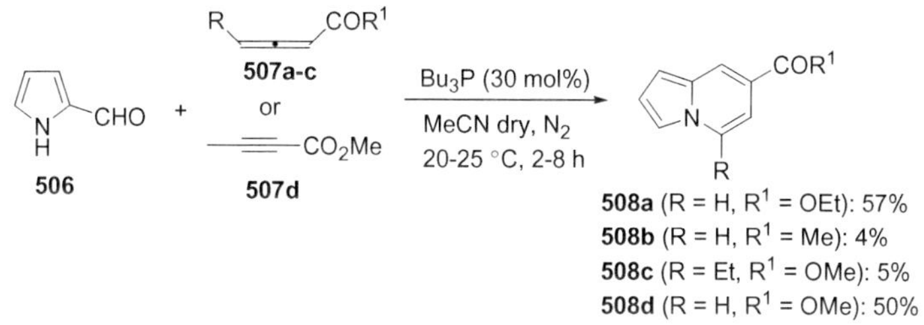

Scheme 4.153

(or 2-quinolinecarbaldehyde) using $TiCl_4$ as a catalyst.[230] Upon treatment with Ac_2O, the annulation of adducts **509** proceeded smoothly, affording 2,3-disubstituted and 1,2,3-trisubstituted indolizines **510** in good yields (Scheme 4.154). The utility of these annulation products in organic synthesis has been investigated further. The straightforward synthesis of bis(1-indolizinyl) methanes **511** has been achieved *via* the condensation of 2,3-disubstituted indolizines with various aldehydes or ketones in the presence of a catalytic amount of $BF_3 \cdot OEt_2$ (Scheme 4.155).

With the development of asymmetric allylic N-alkylation from MBH carbonates, a method for constructing novel tricycle compounds with indole ring was disclosed by Chen (Scheme 4.156).[231] A regioselective bromination was

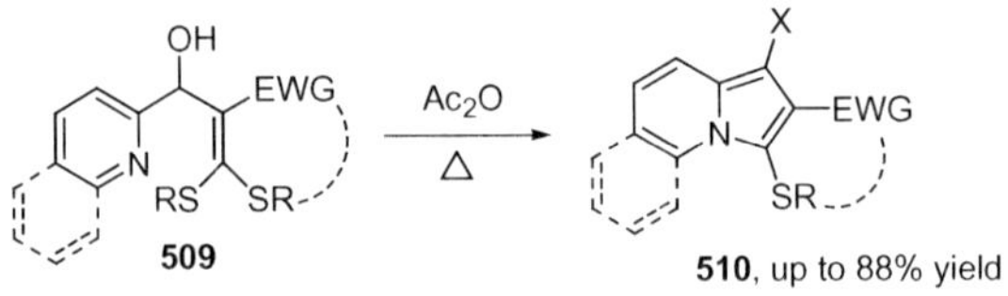

Scheme 4.154

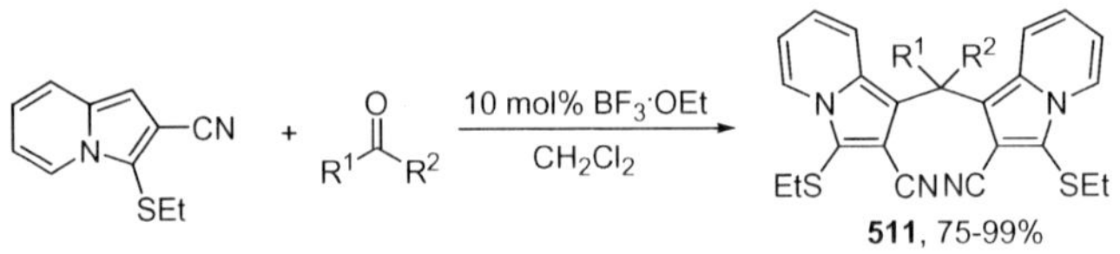

Scheme 4.155

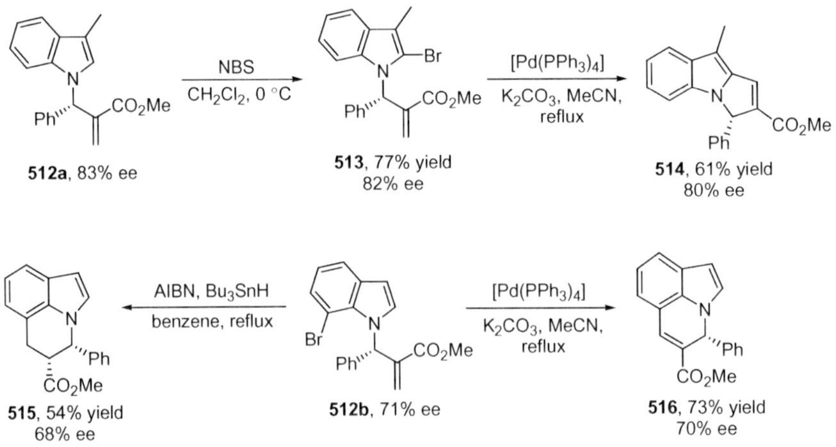

Scheme 4.156

easily realized by simply treating **512a** with NBS at 0 °C to give compound **513** in 77% yield (82% ee), and a subsequent palladium-catalyzed intramolecular Heck reaction of **513** afforded the desired pyrrolo[1,2-*a*]indole framework **514** (61% yield, 80% ee), which possesses the core structure of the natural product yuremamine.[232] For 7-bromine indole substrate **512b**, an intramolecular radical cyclization or Heck reaction could be conducted to deliver pyrrolo[3,2,1-*ij*]-quinoline derivatives **515** and **516**, respectively, in good yields. These derivatives might find valuable applications in medicinal chemistry.[233]

Reduction of the corresponding *o*-nitro compounds with concomitant cyclization is a well-established route to important heterocycles such as indoles, quinolones and quinolines.[234] Using 5 mol.% cyclopentadienyliron dicarbonyl dimer (Fp$_2$)[235] as catalyst, nitroenone **517** was heated in dioxane under CO (800 psi) to furnish 3-methylquinolin-4(1*H*)-one (**518**) in 76% yield. However, using the Fp$_2$–CO system, the reduction of MBH adducts **519** gave markedly different products, namely, indoles **520** and *N*-formylindolines **521**, and only a trace amount of quinoline or/and 2-quinolone derivatives (Scheme 4.157).[236]

The nucleophilic aromatic substitution reaction of MBH acetates in the presence of base provides a convenient and easy procedure for the synthesis of 3-nitroindole derivatives. MBH adduct **522** (R = 4-Cl) in DMF using three equivalents of KNO$_2$ at 0 °C gave the highest yield of 3-nitroindole **523** (R = 4-Cl) (77% isolated yield). With the optimal conditions in hand, various substituted MBH adducts could effectively provide 3-nitroindoles **523** in low to moderate yields (Scheme 4.158).[237]

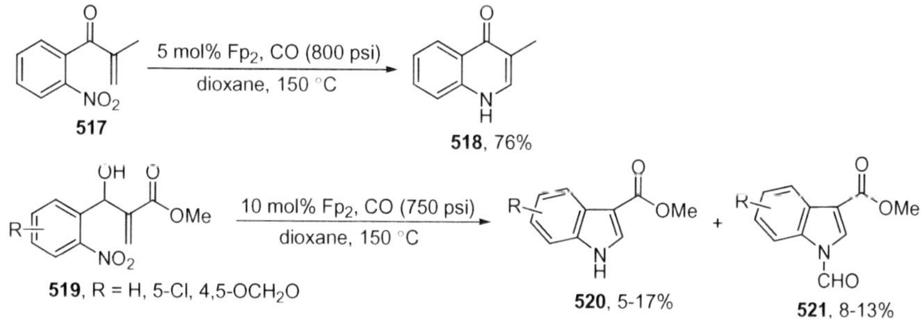

Scheme 4.157

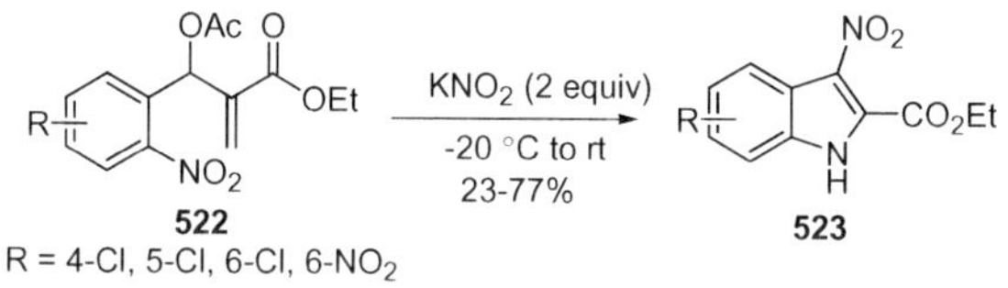

Scheme 4.158

4.4.6 Piperidines and Pyridines

Piperidine and pyridine ring systems are the key structural elements in a vast array of natural products as well as in a large class of biologically active natural products. They are also often embedded within scaffolds as privileged structures by medicinal chemists. Indeed, over 12 000 piperidine derivatives have been mentioned in clinical or preclinical studies during the last ten years.[238] The development of new methods for the synthesis of multi-functionalized piperidines and pyridines is therefore of considerable importance.[239]

Following the synthesis of dihydropyridines *via* the abnormal aza-MBH reaction of *N*-tosylated imines with ethyl 2,3-butandieonate using DMAP as catalyst,[240] a one-pot stepwise [3 + 3] annulation to yield piperidine-2,6-diones from MBH adducts was described by Chang *et al.*[241] After deprotonation of **524** with sodium hydride in tetrahydrofuran at room temperature, the resulting dianion reacted with MBH adducts **525** to afford piperidine-2,6-diones **527** and **528** at refluxing temperature in good yields (Scheme 4.159). However, interestingly, the sole products **527** were produced in modest yields when using **526** as the Michael acceptor. The same authors further investigated the application of piperidine-2,6-diones in natural product synthesis. As shown in Scheme 4.160, a convenient strategy for the formal synthesis of tacamonine (**532**), which possesses vasodilator and hypotensive activities and was isolated from *Tabernaemontana eglandulosa*,[242] has been developed from piperidine-2,6-dione **529**.

1,5-Dipentanoate derivatives **533**, prepared from the nucleophilic addition reaction of ethyl cyanoacetate with acetyl derivatives of the MBH adducts, serve as versatile precursor to construct piperidones (Scheme 4.161).[243] The reduction of compounds **533** in the presence of Raney-nickel under

Scheme 4.159

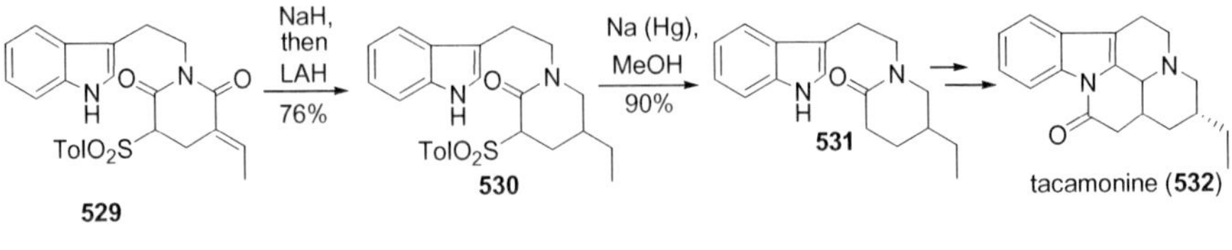

Scheme 4.160

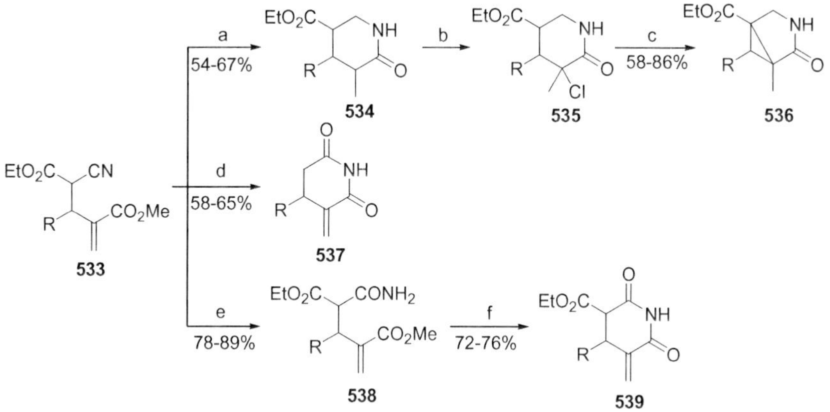

Reagents and conditions: (a) Raney-Ni, H_2, 40 psi, rt, 3 h; (b) $POCl_3$, PCl_3, reflux, 2 h; (c) DBU, MeCN, reflux, 14 h; (d) $FeCl_3 \cdot 6H_2O$, propionic acid, reflux, 2 h; (e) $TFA:H_2SO_4$, neat, rt, 3 h; (f) NaH, toluene, rt, 30 min.

Scheme 4.161

hydrogenation conditions yielded piperidine-2-ones **534** in 54–67% yields, and subsequent chlorination of **534** was attempted with a mixture of PCl_5 and $POCl_3$ to yield **535**, which underwent the further transformation in the presence of DBU to give 4-oxo-6-aryl-3-aza-bicyclo[3.1.0]hexane-1-carboxylates **536** in good yields. By employing an amide instead of a cyano group, intramolecular cyclization delivered piperidones.[244] Following this strategy, the treatment of **533** with 3 equiv. of $FeCl_3 \cdot 6H_2O$ in propionic acid under reflux for 2 h afforded piperidine-2,6-diones **537** in 58–65% yields. Another strategy involves hydrogenolysis of the cyano group and subsequent intramolecular nucleophilic substitution to give different substituted piperidine-2,6-diones **539** in good yields. However, hydrogenolysis of the cyano group in MBH adducts **540** and **541** with a 4 : 1 mixture of $TFA–H_2SO_4$ directly afforded piperidine-2-ones **542** and **543** in moderate yields (Scheme 4.162), respectively.[245] Moreover, compounds **543** have been employed as dipolarophiles in 1,3-dipolar cycloaddition reactions with nitrile oxides to produce good yields of spiroheterocyclic derivatives **544** containing 2-pyridone and isoxazoline rings.

In a one-pot reaction, Garrido's group has reported the hydrogenolysis of chiral γ-substituted δ-amino acid **548** and subsequent *in situ* lactamization, giving the piperidin-2-ones **549** in good yields. The starting materials **548** were synthesized, with high diastereoselectivities and enantioselectivities, from MBH adducts **546** and **547** using chiral lithium amide, in what is the first one-pot asymmetric Ireland–Claisen rearrangement/Michael addition domino reaction (Scheme 4.163).[246] Products **549a** and **549b** were obtained from **548a** and **548b** in 81% and 85% yields, respectively. Reduction of **549a** with LAH in THF and reduction of **549b** with $BH_3 \cdot THF$ could further furnish 2,3-disubstituted piperidines[247] in 72% and 89% isolated yields, respectively.

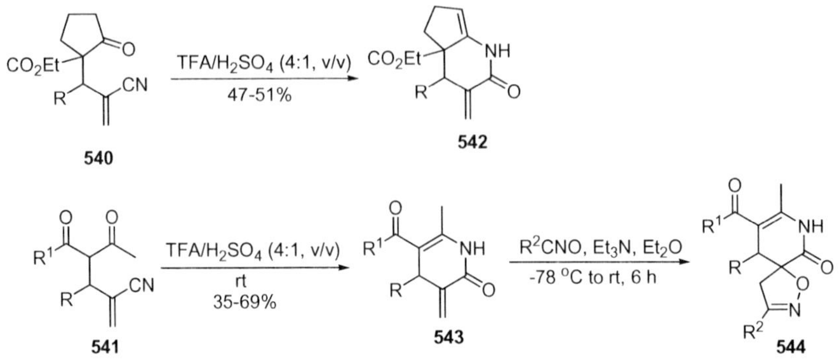

Scheme 4.162

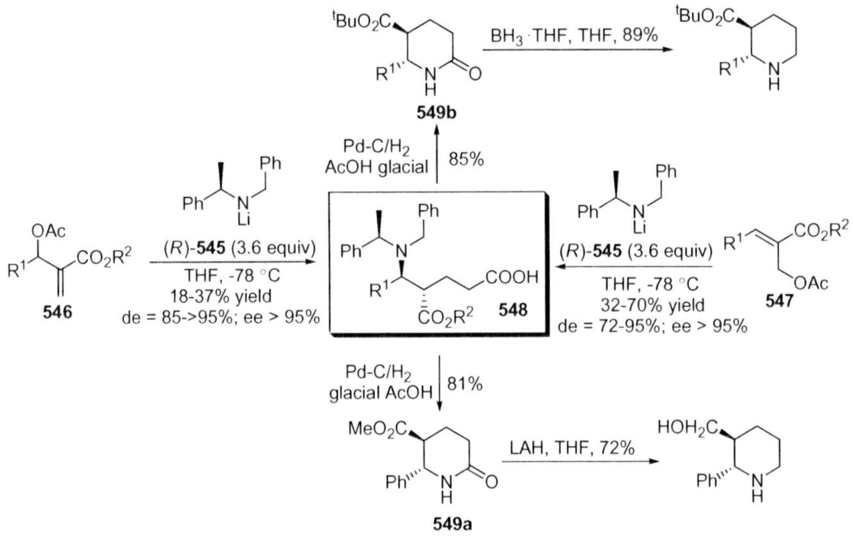

Scheme 4.163

The Barbier reaction of γ-cyanoesters **550** with allylindium reagent generated *in situ* from allyl bromide and indium powder in THF mainly afforded diallylated piperidine-2-ones **551** in 52–61% yields (Scheme 4.164).[248] However, the mono-allylated compounds **552** were isolated in low yields (12–16%), which may be due to a double Barbier reaction *via* the cyclic *N*-acylimine intermediate from **552**. Further synthetic application of diallylated piperidine-2-ones **551** was carried out, and the RCM (ring-closing metathesis) reaction of **551a** with second-generation Grubbs catalyst (3 mol.%) in toluene (50 °C, 10 h) produced novel spiro-cyclopentene compound **553** in 84% yield. RCM was also applied to construct the core tetrahydropyridine skeleton.[249] A series of functionalized

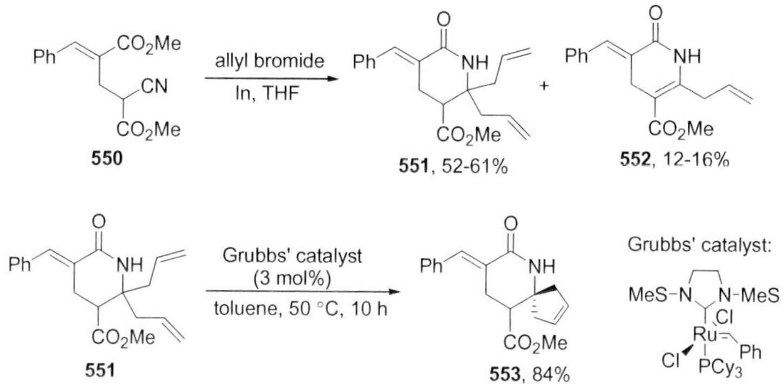

Scheme 4.164

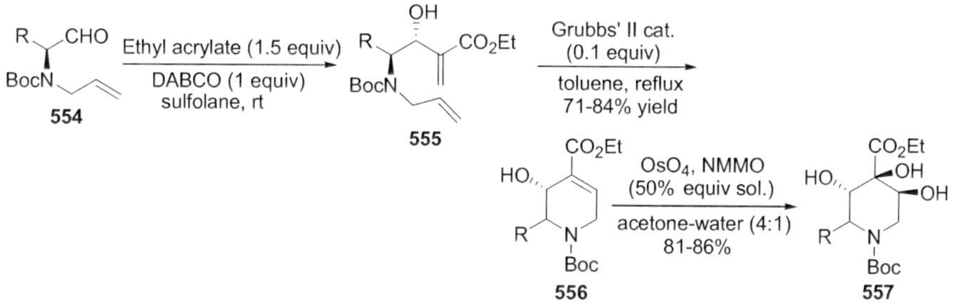

Scheme 4.165

chiral tetrahydropyridines **556** as potential analogues of isoguvacine[250] have been prepared in 71–84% yields using the second-generation Grubbs' catalyst; these compounds were further transformed into stereochemically highly diverse azasugars **557**[251] in 81–86% yields and $\geq$ 99 : 1 de *via cis*-dihydroxylation[252] (Scheme 4.165).

Using tosylamide **558** as starting material, a facile synthetic method to construct 3,4,5-trisubstituted pyridines **561** has been reported by Kim *et al* (Scheme 4.166).[253] Initially, in the presence of DBU, the reaction of **558** with various Michael acceptors **559** produced two major components at room temperature, which were very difficult to separate, and were presumed to be diastereomers **560** formed *via* Michael reaction and the following aldol cyclization reaction. After simple aqueous workup, crude mixtures **560** were subjected to dehydration in benzene with catalytic amounts of *p*-TsOH under refluxing conditions to successfully give piperidines **561**. Deprotection of the tosyl group in DMF using Cs_2CO_3 as base at 120–130 °C afforded pyridines **562** in moderate yields.

Piperidines **564** could also be synthesized *via* palladium-mediated Heck-type reactions, which have been presented as an efficient synthetic method for the

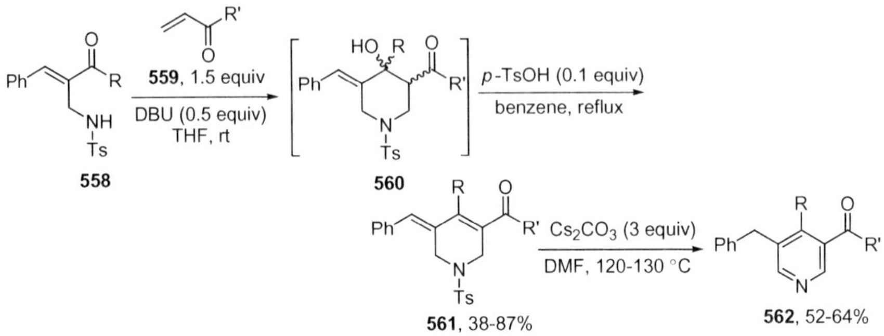

Scheme 4.166

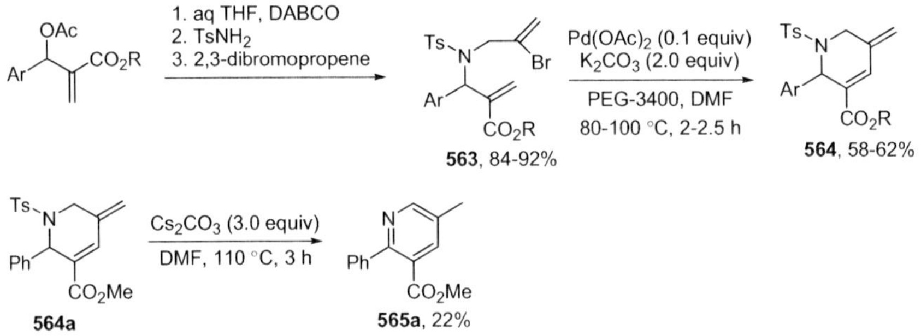

Scheme 4.167

construction of cyclic compounds.[254] With the introduction of tosylamide *via* the DABCO salt of the corresponding MBH acetates, and following alkylation with 2,3-dibromopropene, compounds **563** were prepared in high yields (84–92%). Heck-type cyclizations of **563** were conducted under the conditions of Pd(OAc)$_2$/K$_2$CO$_3$/PEG-3400/DMF/80–100 °C to produce *exo*-methylene tetrahydropyridine **564** in moderate yields (58–62%, Scheme 4.167). However, in the presence of 3.0 equiv. of Cs$_2$CO$_3$ in DMF at 110 °C, it was very difficult to convert **564a** into the corresponding 2,3,5-trisubstituted pyridine **565a**, which was produced in only 22% yield.

Kim's group have further developed a different annulation protocol, acid-catalyzed dehydration cyclization, to synthesize poly-substituted pyridines (Scheme 4.168).[255,256] Compounds **566** along with 3 equivalents of NH$_4$OAc were boiled under reflux in acetic acid for 1–15 h, giving the tetra-substituted pyridines **567** in moderate to good yields; only one exception produced abnormal tetra-substituted pyridines **568** in 54% yield. With 2-arylpyridines **567** in hand, regioselective *ortho*-hydroxylation of the aryl moiety was carried out *via* palladium-mediated C–H bond activations. Using PhI(OAc)$_2$ as oxidant, acetylated compound **569a** was obtained in 61% yield from **567a**; when the reaction was carried out in PEG-3400/*tert*-butanol as reaction medium at

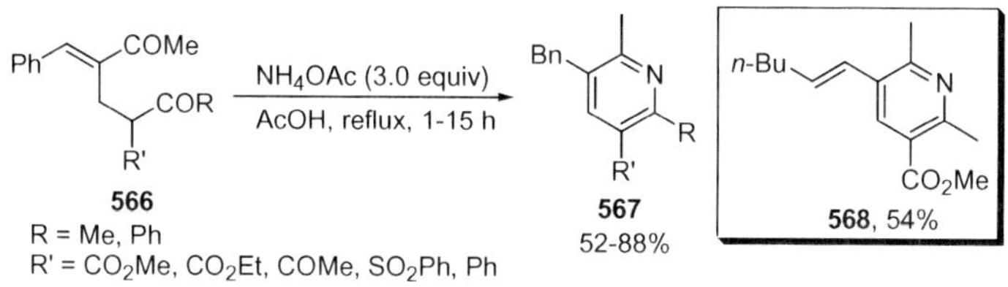

Scheme 4.168

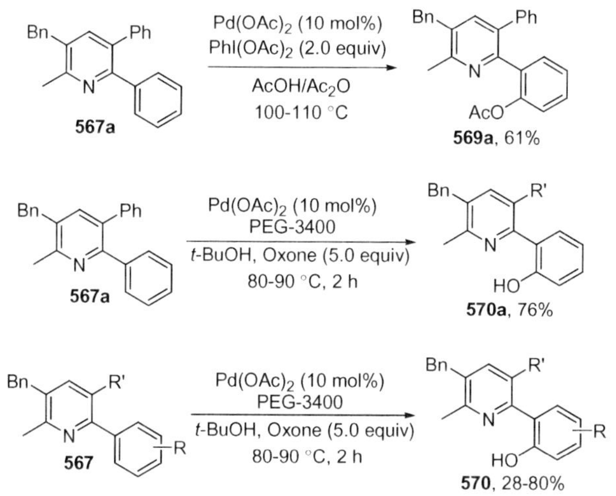

Scheme 4.169

80–90 °C using 5.0 equiv. of Oxone as oxidant, phenol derivative **570a** was formed in 76% yield (Scheme 4.169). A series of *ortho*-hydroxylations of the aryl moiety of 2-arylpyridines **570** was carried out, giving products in 28–80% yields.

Radical cyclizations have been used in the synthesis of various cyclic compounds from MBH adducts. Recently, Kim *et al.*[257] have reported a radical cyclization to construct tetrahydropyridine derivatives, which have been regarded as important synthetic intermediates for the synthesis of various important compounds.[258] Substrates **572**, which were prepared from compounds **571** with allyl bromide in the presence of K_2CO_3, have been examined in the radical cyclization reaction under the conditions of n-Bu₃SnH/AIBN in refluxing benzene (Scheme 4.170). When the N-substituted group was a tosyl group, 1,4,5,6-tetrahydropyridines **573** were obtained in good to moderate yields (56–82%) in a short time; when it was a phenyl or benzyl group, more n-Bu₃SnH was required for these cyclizations. The mechanism for radical cyclization may involve consecutive 1,5-hydrogen transfer and double bond isomerization processes.

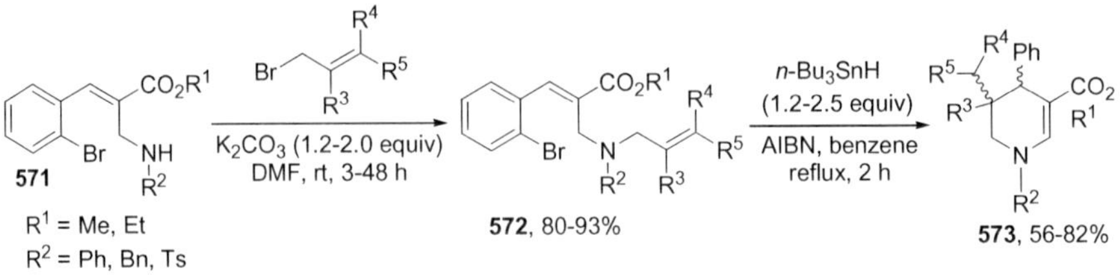

Scheme 4.170

4.4.7   Quinolines

Substituted quinolines are one of the oldest known classes of pharmaceutical agents and their relevance in chemotherapy, especially against malaria, is widely known.[259] Besides antimalarials, a spectrum of other pharmacological activities[260] has been the major reason for the development of novel and efficient syntheses of this heterocycle. As a result, the recent past has witnessed the publication of several simple and elegant syntheses of substituted quinolines[261] since the first synthesis of quinolines was reported by Familoni *et al.* in 1998.[262] MBH adducts or their derivatives have been illustrated as suitable starting materials for the synthesis of various quinoline systems.

1,2-Dihydroquinolines have received substantial attention due to their potential biological activities arising from their antioxidative properties[263] as well as their usefulness as precursors of some other biologically active compounds.[264] However, synthetic methods for the preparation of 1,2-dihydroquinolines are limited.[265] Using MBH adducts as a key intermediate offers an effective route to constructing the 1,2-dihydroquinoline backbone.

2-Nitrophenyl MBH adducts have been discovered to be highly effective and efficient precursors for constructing targeted quinoline derivatives *via* reductive cyclization. Familoni *et al.* first disclosed the catalytic hydrogenation, using a 10% Pd-C catalyst, of MBH adducts **573**, which underwent cyclization to the desired quinoline derivatives.[266,267] Catalytic hydrogenation of ketone precursors **573** ($R^1 = COR$; Scheme 4.171) yielded quinolines **574** and quinoline-*N*-oxides **575** *via* nucleophilic carbonyl addition and dehydration, whereas ester precursors **573** ($R^1 = CO_2R$) afforded quinolone derivatives **576–578** *via* acyl substitution. Ketone precursors **573** ($R^1 = COR$) thus tend to yield quinoline derivatives, while ester precursors **573** ($R^1 = CO_2R$) afford quinolone derivatives. The formation of quinoline-*N*-oxides is attributed to early cyclization of incompletely reduced, nucleophilic, *N*-oxygenated intermediates. Quinoline derivative **576a** could be further dehydrated to give 3-methyl-2-quinolinone **577a** under the prescence of TsOH in 70% yield in refluxing toluene (Scheme 4.172). Given the reactivity of the double bond of MBH products **573a** in reductive cyclization, Michael addition product **581** was reduced and then cyclized to afford exclusively the 2-quinolinone derivative **582** *via* acyl substitution.

The reduction of MBH adducts **573** with $SnCl_2 \cdot 2H_2O$ favored cyclization *via* conjugate addition to give 1,2-dihydropyridine derivatives, but the product

Scheme 4.171

Scheme 4.172

pattern also appeared to be substrate-dependent (Scheme 4.173). As shown in Scheme 4.173, ketone precursors **573** gave the *N*-hydroxydihydroquinolines **579** in very low yield. The ester precursors **573** were cyclized to give the dihydroquinoline derivatives **580** in low to moderate yields.

The SnCl$_2$-reduction system has also been applied in the reduction of S_N2 nucleophilic substitution products **583**, affording more functional quinolines, 4-(substituted vinyl)-quinolines **584**, in moderate yields, with several exclusions of the formation of dihydroquinoline derivative **585** (Scheme 4.174).[268] However, using compounds **586** as substrates without a ketone moiety, the ester group can also participate in the intramolecular cyclization, but the subsequent dehydrogenation does not occur and, therefore, tetrahydroquinolin-2-ones **587** were obtained in 51–62% yields (*trans* form only). From this study, the preference of the activated carbonyl group COR for cyclization has the order: R = Me > Ph > O-alkyl.

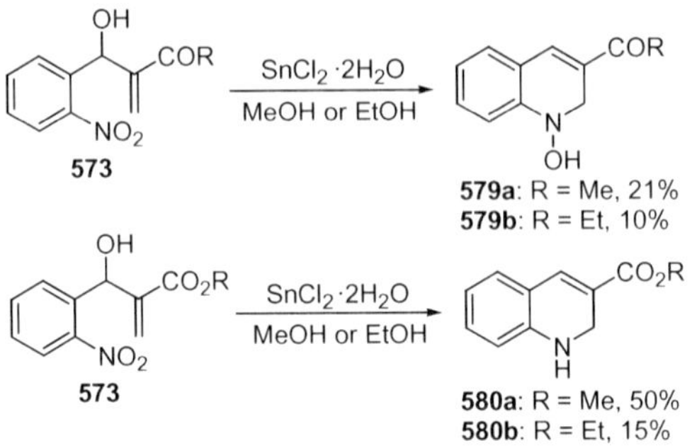

Scheme 4.173

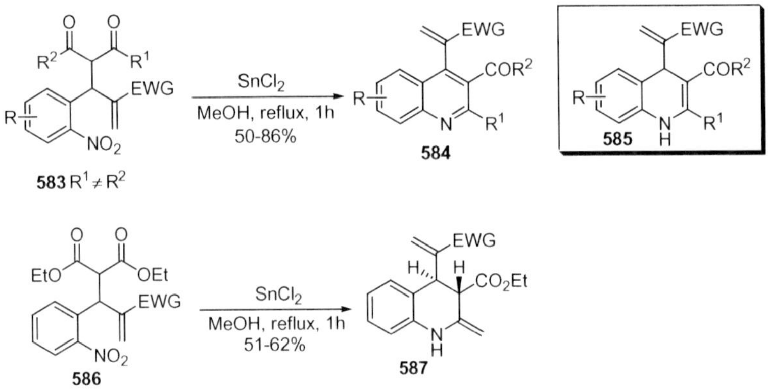

Scheme 4.174

The Fe–AcOH reduction system is another well-known reductive method for the conversion of nitrobenzene derivatives into aniline derivatives. Basavaiah *et al.* have examined the application of Fe/AcOH in the reductive cyclization of MBH adducts. As shown in Scheme 4.175, ester precursors **588** as substrates treated with Fe/AcOH at 110 °C for 30 min provide 3-acetoxymethyl-(1*H*)-quinol-2-one derivatives **589** in 72–89% isolated yields, while ketone precursors **591** give 3-acetoxymethylquinoline derivatives **592** in 56–83% isolated yields under the same conditions.[269] Precursors **594** as starting materials provide a simple, facile and one-pot synthesis of functionalized 1,2,3,4-tetra-hydroacridines **595** (*n* = 1) and cyclopenta[*b*]quinolines **595**(*n* = 0) in 61–82% yields.[270] Moreover, an easy, convenient and operationally simple one-pot procedure for the synthesis of 3-benzoylquinoline derivatives **597** in moderate yields from MBH alcohols **596** has been developed.[271]

In contrast to the stoichiometric Fe–AcOH reduction system, O'Dell and Nicholas have described another reductive cyclization of *o*-nitro-substituted MBH acetates by carbon monoxide, using [Cp*Fe(CO)$_2$]$_2$ as catalyst.[272]

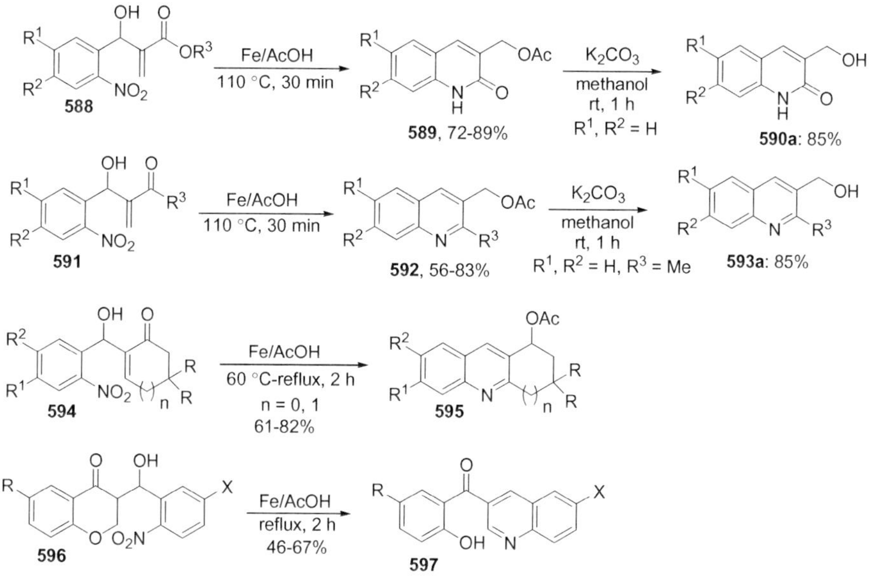

Scheme 4.175

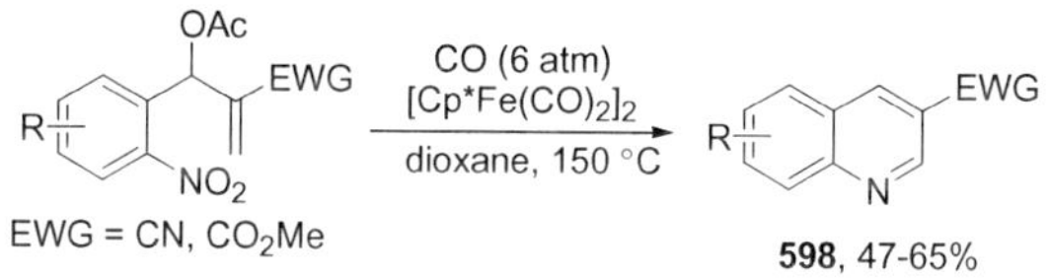

Scheme 4.176

Moderate to good yields of 3-substituted quinolines **598**[273] were achieved, and the reaction's tolerance for electronically diverse substituents on the aromatic ring promises to make this a general and preferred route to these quinolines (Scheme 4.176).

MBH acetates having an *ortho*-azide group have also been developed to construct a quinoline skeleton through the Staudinger reaction, an intramolecular S_N2' reaction and followed by the Michael addition (MA) rearrangement.[274] As shown in Scheme 4.177, the MBH acetate **599** (EWG = CO_2Me or CN) reacts with triethyl phosphite (TEP) at 0–5 °C to give the unstable iminophosphorane **603**, which was subjected to reflux in toluene without isolation, leading to 1,2-dihydroquinoline derivatives **600** in 57–91% yields. However, notably, quinolines **601** were obtained as by-products when using the acetates **599** as reactants with an electron-withdrawing group at the 5-position. With EWG as ketone, the intermediate iminophosphoranes **603** were also obtained, which then furnished acetoxymethylquinoline derivatives **602** in 60–76% yields.

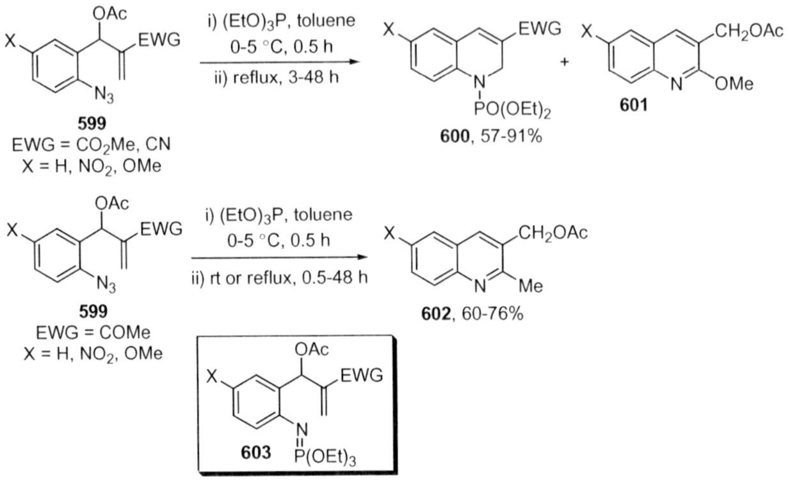

Scheme 4.177

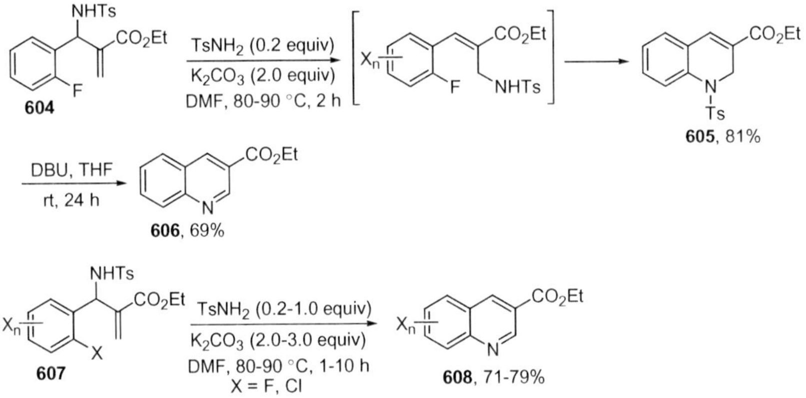

Scheme 4.178

If the aromatic moiety of a cinnamylamine derivative has an *ortho*-halogen substituent, 1,2-dihydroquinoline would be obtained *via* the subsequent S_NAr reaction. In the presence of catalytic amounts of tosylamide, MBH adduct **603** was rearranged to the thermodynamically more stable tosylamide derivative, which then could be easily subjected to nucleophilic aromatic substitution reaction at the *ortho* position, giving 1,2-dihydroquinoline **605** in 81% yield. Furthermore, using DBU as a base, elimination of *p*-toluenesulfinic acid afforded quinoline **606** in 69% yield (Scheme 4.178).[275] However, interestingly, X_n-substituted MBH adducts **607** were directly converted into quinolines **608** in a one-pot reaction in moderate yields. The discrepancy between **604** and **607**

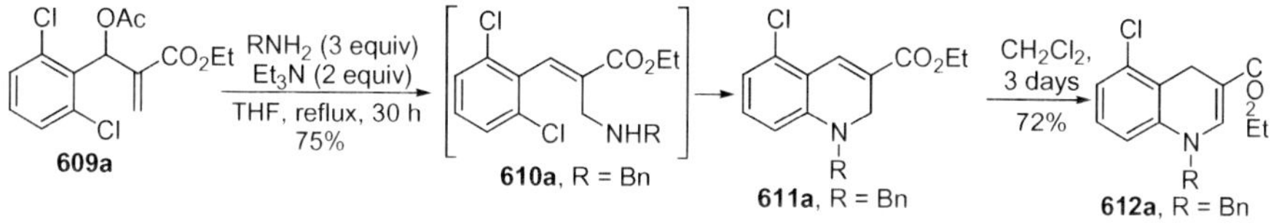

Scheme 4.179

might be due to the subtle difference in acidity of the proton at the 2-position of the corresponding dihydroquinolines.

Kim's group then developed an effective synthetic route for the synthesis of 1,2-dihydroquinoline and quinoline derivatives using MBH acetates **609** as starting materials.[276] The reaction of the MBH acetates **609a** and benzylamine gave the 1,2-dihydroquinoline **611a** as expected product in 75% yield by rapid column chromatography (Scheme 4.179), a reaction that might proceed *via* the initial allylic substitution (S_N2') of benzylamine[277] and subsequent S_NAr reaction of the cinnamylamine derivative **610a**. However, the yellow solid **611a** was very unstable, and slowly converted into the 1,4-dihydro analog **612a** even during the separation. Moreover, other MBH acetates **609** with *ortho*-halogen substituents gave products **611** that were difficult to separate in the pure state, and, therefore, were further converted into **612** in methylene chloride over 3 days at room temperature. In general, 1,4-dihydroquinolines **612** could be synthesized in 52–72% yields from MBH acetates **609**.

Cyclization of **613** takes place in the presence of sodium hydride in THF at room temperature after 30 min or at reflux temperature, giving methyl 4-oxo-1,4-dihydroquinoline-3-carboxylates **614** in 75–81% yields (Scheme 4.180).[278] The requirement for the presence of certain electron-withdrawing groups *para* to the leaving group does not promote the generality of this cyclization reaction, which works well with hydrogen or electron-donating methoxy substituted compounds. Esters **614**, upon heating with aqueous 10% HCl in MeOH for 8 h, give most of the known 4-oxo-1,4-dihydroquinoline-3-carboxylic acids **615** in 84–92% yields.[279] Repetition of this reaction sequence with 2,3,4,5-tetrafluorobenzaldehyde (**616**) and 2-aminopropan-1-ol afforded 9,10-difluoro-3-methyl-7-oxo-2,3-dihydro-7*H*-pyrido[1,2,3-*de*][1,4]benzoxazine-6-carboxylic acid **617** in five steps and 40% overall yield (Scheme 4.181).

MBH acetates have been conveniently transformed into multi-substituted quinolines and cyclopenta[*g*]quinolines by the reaction with nitroethane or ethyl cyanoacetate *via* a successive S_N2'-S_NAr elimination strategy.[280] As shown in Scheme 4.182, reaction of MBH acetates (**618** and **619**) in DMF, in the presence of potassium carbonate and nitroethane at 50–75 °C, afforded the desired substituted 8-methylquinolines **620** in 80–92% yields or substituted 9-methyl cyclopenta[*g*]quinoline-6-ones **622** in 85–94% yields in a short time (3–4 h). Similarly, substituted 8-cyanoquinolines **621** were obtained from the reaction of MBH acetates **618** with ethyl cyanoacetate at relatively higher temperatures (110–125 °C) and for longer times (~10 h) in 50–60% yields.

RNH$_2$, Et$_3$N
THF, 0 °C - rt
30 min
83-97%
OCH$_3$
F
X
Y
I
DMA
CH$_2$Cl$_2$, rt, 30 min
83-88%
10% HCl, MeOH
reflux, 8 h
84-92%
OH
615
OCH$_3$
614
—CO$_2$CH$_3$
Bu$_4$NI, ZrCl$_4$
CH$_2$Cl$_2$, 0 °C, 8 h
81-84%
OH
OCH$_3$
X
Y
F
I
NaH, THF, rt or
reflux, 30 min
75-81%
OCH$_3$
NHR
F
X
Y
613
H
O
F
X
Y

Scheme 4.180

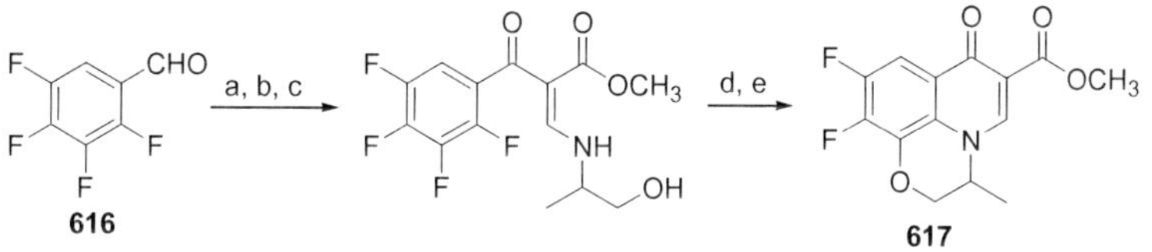

a) methyl propiolate, Bu$_4$NI/ZrCl$_4$, CH$_2$Cl$_2$, 0 °C, 8 h, 85%; b) Dess-Martin periodinane, CH$_2$Cl$_2$, rt, 30 min 85%; c) H$_2$NCH(Me)CH$_2$OH, Et$_3$N, THF, 30 min, 92%; d) NaH, dioxane, reflux, 4 h, 76%; e) 10% HCl, MeOH, reflux, 8 h, 80%

Scheme 4.181

Unlike the facile S_NAr reaction with the chlorinated or fluorinated aromatics as substrates, brominated substrates **624** must be treated with a strong base such as *t*-butyllithium *in situ* in ethyl ether at -78 °C to furnish the products **625** in overall yields of 40% from the acids **623**. Removal of the protecting groups with TBAF in THF at room temperature gave the 3,4-disubstituted dihydroisoquinolin-1(2*H*)-ones **626** in 76% yields (Scheme 4.183).[281] The same authors further attempted the Bischler–Napieralski reaction to construct the dihydroisoquinoline skeleton. When mixtures of the MBH adduct **627** with POCl$_3$ were refluxed for 1 h in the presence of pyridine, the isoquinolol **628** was afforded in 42% yield (Scheme 4.184).

Using MBH adducts **629** as starting materials, Heck-type cyclizations are supposed to synthesize dihydroquinoline **630**. However, employing the conditions of Lamaty's coupling reaction [Pd(OAc)$_2$/K$_2$CO$_3$/PEG-3400/DMF/ 80–90 °C],[282] 2-phenylquinoline-3-carboxylic acid derivatives **631** were obtained directly in moderate yields (53–69%), presumably *via* Pd-mediated aerobic oxidation[283] of the intermediate dihydroquinoline **630** (Scheme 4.185).[284]

The amine **632** undergoes cyclization upon treatment with palladium acetate and BINAP in toluene at 100 °C for 12 h to give a dihydroquinoline **633** in 82% yield (Scheme 4.186). The dihydroquinoline generated during the reaction seemed to be quite stable under the reaction conditions even after a long reaction time (48 h), based on thin-layer chromatography (TLC).[285]

In 2002 Kim *et al.* reported the synthesis of quinolines from MBH acetates *via* the oxidative cyclization of sulfonamidyl radical as the key step. In the presence of iodosobenzene diacetate and iodine, cinnamylamine derivative **634a** was converted into *N*-tosyldihydroquinoline **635a** in 51% yield along with de-tosylated product **636a** in 25% yield (Scheme 4.187).[286] The isolated dihydroquinoline **635a** could also be converted into **636a** quantitatively under the normal elimination reaction conditions (K$_2$CO$_3$, DMF, 120–130 °C) with the elimination of *p*-toluenesulfinic acid. In other cases, the dihydroquinolines **635** could not be isolated, and the next elimination step was performed directly after usual aqueous workup. Various substituted quinolines **636** were obtained in 52–80% yields.

o-Nitro derivatives **637** with various EWG groups and X$_n$ substituents on the phenyl ring provide a facile method for the formation of 3-ethoxycarbonyl-4-hydroxyquinoline *N*-oxide derivatives **638** (EWG = CO$_2$Et) upon catalysis by trifluoroacetic acid or trifluoroacetic acid in the presence of triflic acid

Scheme 4.182

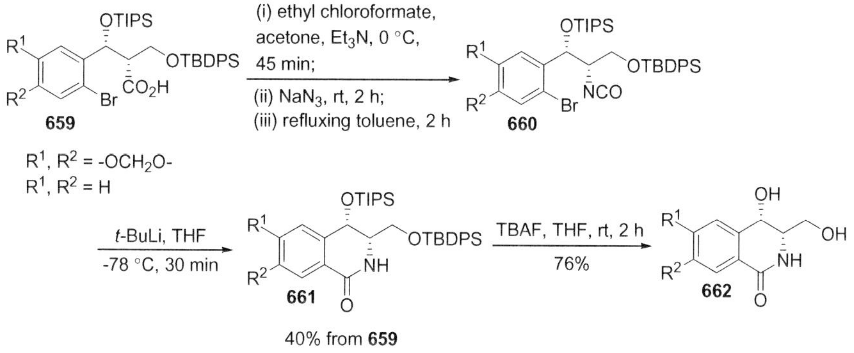

OTIPS

R^1 ... OTBDPS

R^2 ... Br CO$_2$H

659

(i) ethyl chloroformate, acetone, Et$_3$N, 0 °C, 45 min;
(ii) NaN$_3$, rt, 2 h;
(iii) refluxing toluene, 2 h

OTIPS

R^1 ... OTBDPS

R^2 ... Br NCO

660

R^1, R^2 = -OCH$_2$O-
R^1, R^2 = H

t-BuLi, THF
-78 °C, 30 min

OTIPS

R^1 ... OTBDPS

R^2 ... NH

661 O

40% from **659**

TBAF, THF, rt, 2 h
76%

OH

R^1 ... OH

R^2 ... NH

662 O

Scheme 4.183

OH

CO$_2$Me

PMB trichloroacetimidate, CSA, CH$_2$Cl$_2$, 12 h, rt
76%

OPMB

CO$_2$Me

(i) DIBAL-H, CH$_2$Cl$_2$, -78 °C, 2 h;
(ii) TBDPSCl, DMF, imidazole, rt, 14 h.

OPMB

OTBDPS

NHCO$_2$Me

627

POCl$_3$/Py, toluene, reflux, 1 h
42%

OTBDPS

N

628 OH

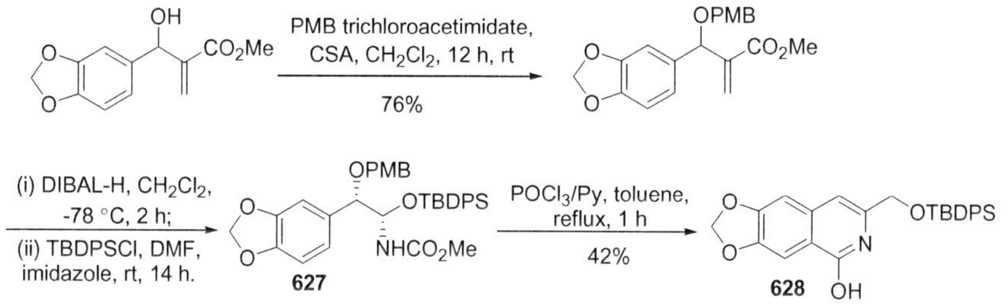

Scheme 4.184

(0.2 equiv) (Scheme 4.188).[287] However, the MBH adduct in which the EWG group was a cyano group did not form the quinoline ring. Using other Lewis acid such as acetic or formic acid did not promote this reaction. Possible reaction mechanisms have been proposed by Kim *et al.*[288] and Coelho *et al.*,[289] respectively. Deoxygenation of **638a** with triphenylphosphine in refluxing THF gave **639**[290] in 70% yield (Scheme 4.189).

Trifluoroacetic acid was also discovered to promote the tandem Claisen rearrangement and cyclization reaction to yield 3-arylmethylene-3,4-dihydro-1*H*-quinolin-2-ones **642** from **640** (Scheme 4.190).[291] Treatment of compounds **640** with neat trifluoroacetic acid at reflux temperature for 8–14 h yielded **642** in good yields. However, treatment of MBH adduct **641**, having a cyano group, with trifluoroacetic acid directly furnished 3-arylmethyl-2-amino-quinoline **643** *via* a tandem Claisen rearrangement, cyclization and isomerization in one step. A practical route for the synthesis of 3-arylmethyl-2-methoxy-quinolines **645** was further developed. Moreover, compound **645a** was useful starting material[293] for the synthesis of R207910,[292] which showed significant activity against the drug sensitive and drug-resistant *Mycobacterium tuberculosis*.

Polyphosphoric acid (PPA) can be used to promote the cyclization of MBH adducts. The reaction of **646** in PPA produced 3-benzylidene-3,4-dihydro-1*H*-quinolin-2-ones (**648**) in 73–80% yields (Scheme 4.191).[294] Such compounds

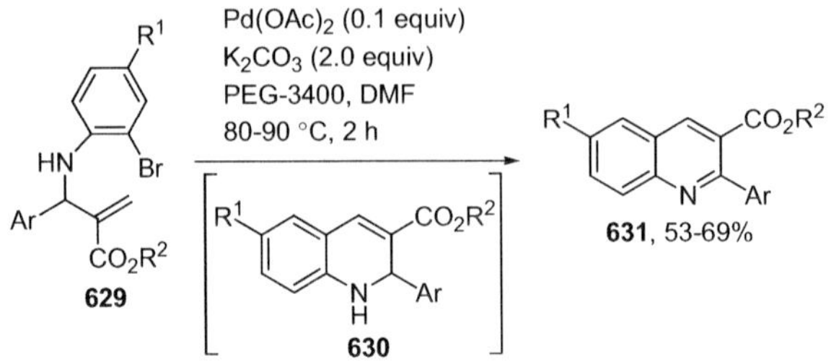

Pd(OAc)$_2$ (0.1 equiv)
K$_2$CO$_3$ (2.0 equiv)
PEG-3400, DMF
80-90 °C, 2 h

629

630

631, 53-69%

Scheme 4.185

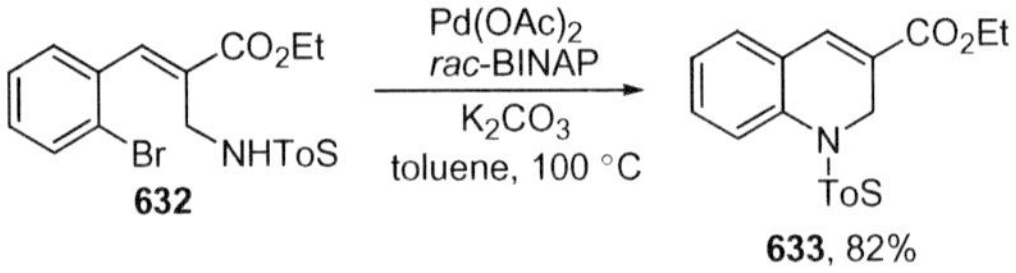

Pd(OAc)$_2$
rac-BINAP
K$_2$CO$_3$
toluene, 100 °C

632

633, 82%

Scheme 4.186

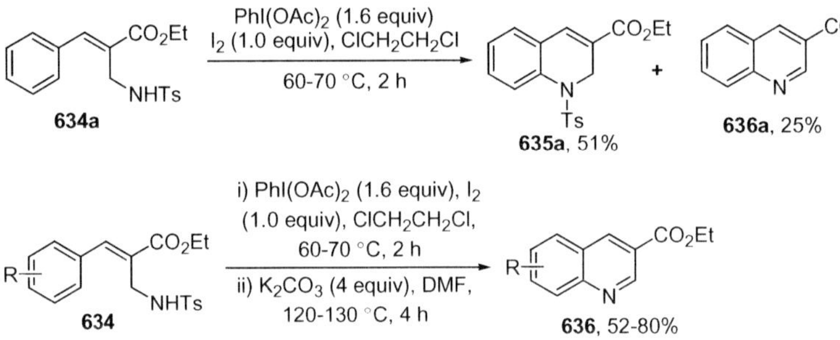

PhI(OAc)$_2$ (1.6 equiv)
I$_2$ (1.0 equiv), ClCH$_2$CH$_2$Cl
60-70 °C, 2 h

634a

635a, 51%   +   **636a**, 25%

i) PhI(OAc)$_2$ (1.6 equiv), I$_2$
(1.0 equiv), ClCH$_2$CH$_2$Cl,
60-70 °C, 2 h

ii) K$_2$CO$_3$ (4 equiv), DMF,
120-130 °C, 4 h

634

636, 52-80%

Scheme 4.187

CF$_3$CO$_2$H, 60-70 °C, 20 h

or

TFA, TfOH (0.2 equiv), 40-50 °C, 20 h

637
X$_n$ = Cl, OMe, OCH$_2$O
EWG: CO$_2$Et, COMe, SO$_2$Ph

638, 48-83%

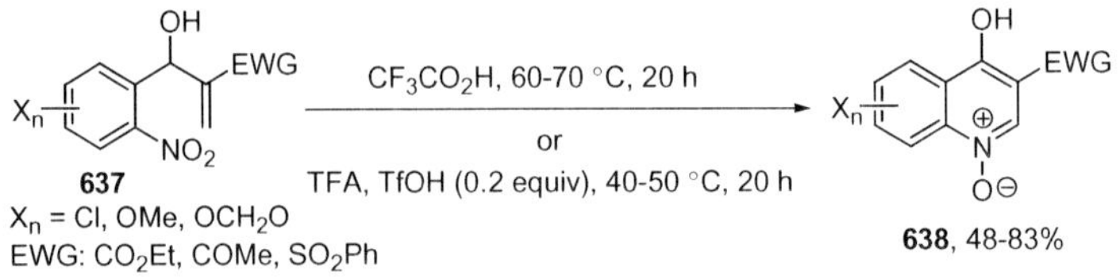

Scheme 4.188

might be generated *via* Claisen rearrangement and amide bond formation. Upon treatment of DBU, compounds **648** were converted into **649** effectively in THF at room temperature in a short time. However, when using **647**, which contain an electron-donating group in the aniline moiety, as the starting

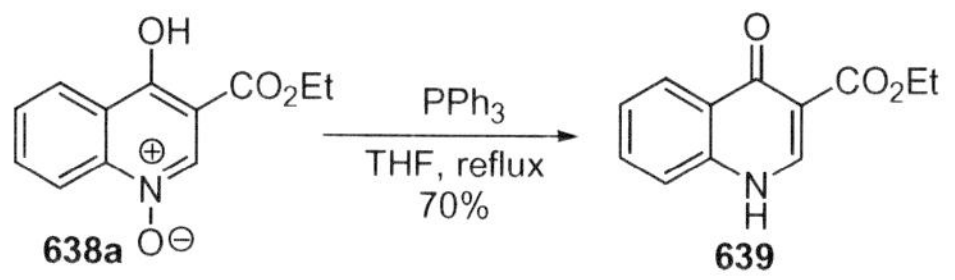

Scheme 4.189

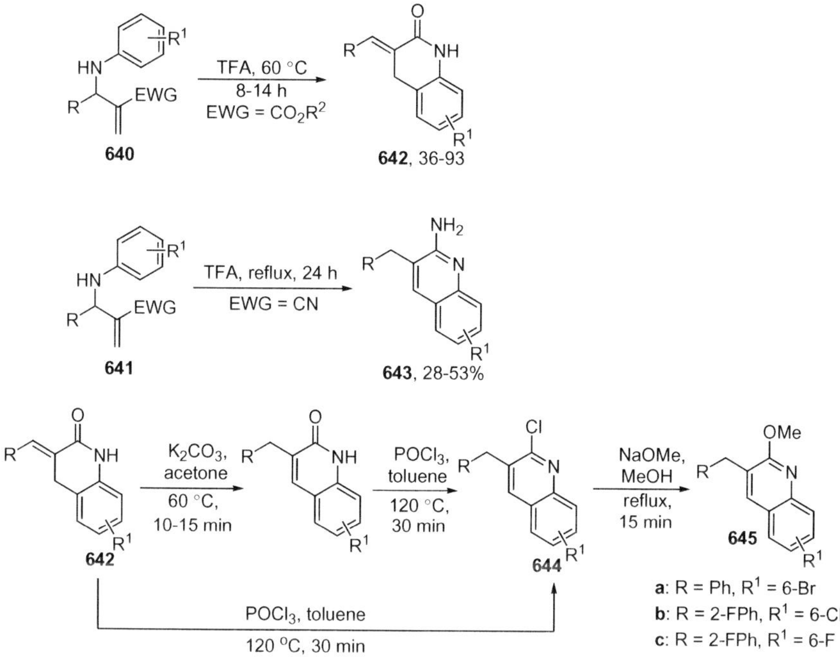

Scheme 4.190

materials, two types of major products (**650** and **651**) were isolated in 57–61% and 13–18% yields, respectively, but no appreciable amount of the analogue **648** was formed. It is interesting that the subtle difference in electron density at the aniline moiety caused such strikingly different results.

H_2SO_4-assisted intramolecular Friedel–Crafts cyclization of MBH derivatives **653** – which were synthesized from the reaction of acid **652** and aniline derivatives by using EDC in good to moderate yields (59–75%) – has been achieved to construct 2(1*H*)-quinolinone structures **654** [H_2SO_4 (3.0 equiv) in CH_2Cl_2 at reflux temperature in short time (20 min)] in moderate to good yields (43–91%, Scheme 4.192).[295] Further DBU-mediated isomerization of these *exo*-methylene compounds **654** in CH_3CN proceeded smoothly, affording their *endo*-isomers **655** in high yields (80–99%).

Hexahydroquinolizines have been synthesized by deprotection of an N-Boc protecting group, followed by a Michael addition, and a S_N2' reaction.[296]

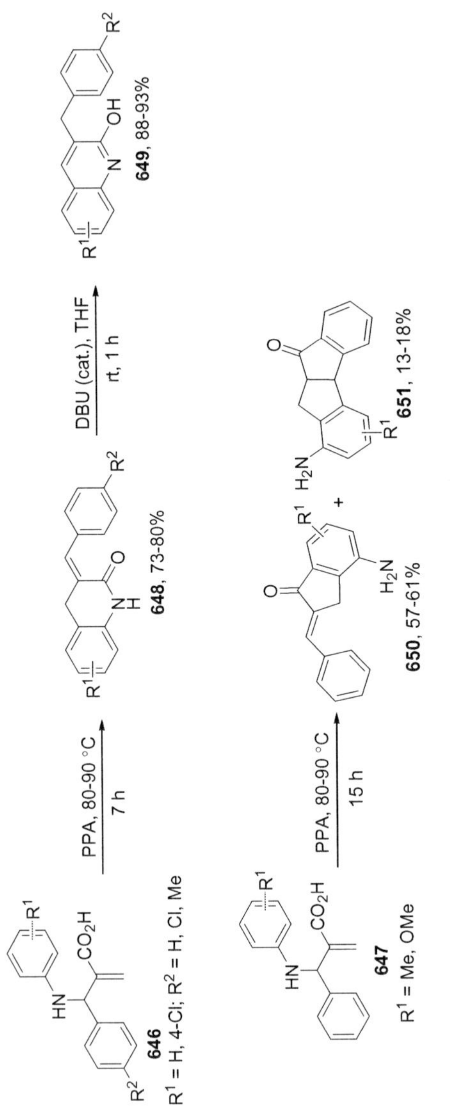

R²
OH
N
R¹
649, 88-93%
DBU (cat.), THF
rt, 1 h
R²
O
N
H
R¹
648, 73-80%
PPA, 80-90 °C
7 h
R¹
CO₂H
HN
646
R² = H, 4-Cl; R² = H, Cl, Me
O
R¹
H₂N
651, 13-18%
+
O
R¹
H₂N
650, 57-61%
PPA, 80-90 °C
15 h
R¹
CO₂H
HN
647
R¹ = Me, OMe

Scheme 4.191

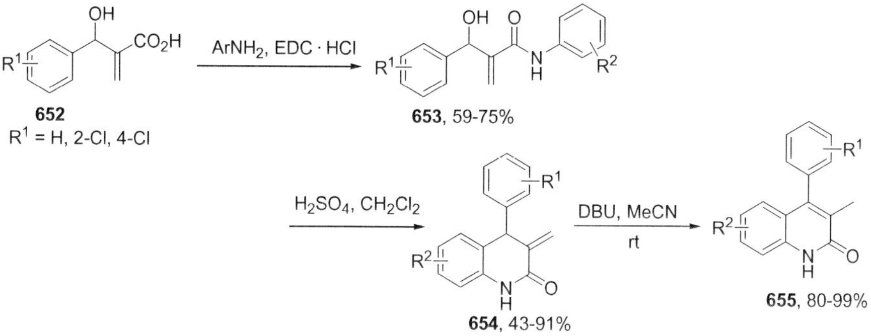

Scheme 4.192

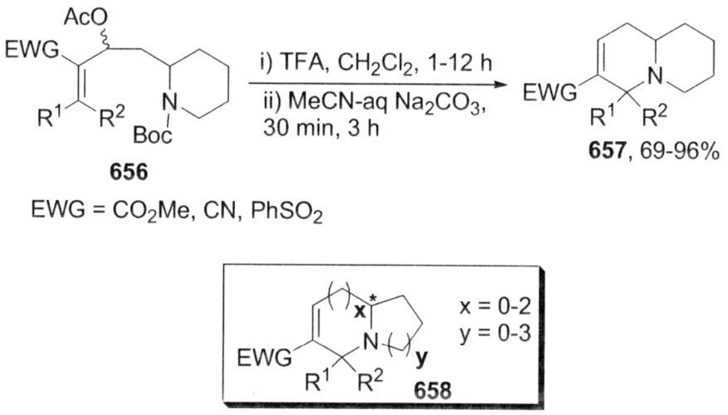

Scheme 4.193

For example, acetates **656** were treated with CF_3CO_2H to remove the Boc group, and the resulting products were then stirred in saturated aqueous Na_2CO_3 to afford the bicyclic amines **657**, often in moderate to over 90% yields (Scheme 4.193). The method of Scheme 4.193 represents a general way of making bicyclic amines **658** with nitrogen at a ring-fusion position, and has also been shown to work for a range of ring sizes, and the stereochemistry α to the nitrogen is preserved in the MBH step.

Recently, Su and co-workers have developed a simple, efficient method to convert MBH acetates into 3-substituted 7,8-dihydro-6*H*-chromene-2,5-diones **662** in excellent isolated yields under solvent-free conditions;[297] they further developed a one-pot, three-component process under solvent-free conditions to construct quinoline derivatives.[298] A series of 2-hydroxy-7,8-dihydroquinolin-5(6*H*)-ones **660** and 7,8-dihydroquinoline-2,5(1*H*,6*H*)-diones **661** have been synthesized in good to excellent yields from the reaction of MBH acetates, cyclohexane-1,3-diones **659** and ammonium acetate or primary amines (Scheme 4.194).

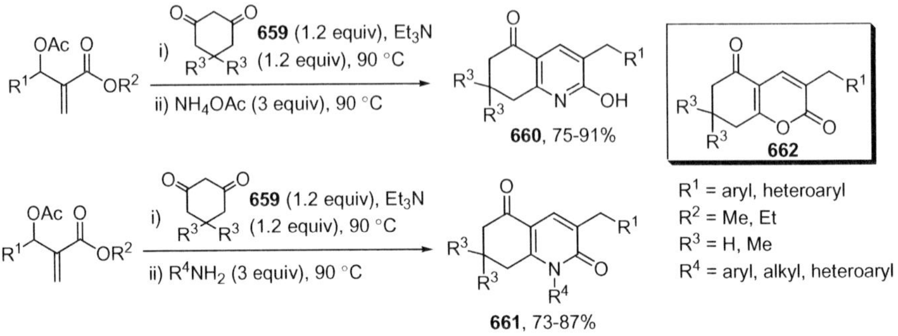

659 (1.2 equiv), Et$_3$N
(1.2 equiv), 90 °C
ii) NH$_4$OAc (3 equiv), 90 °C

660, 75-91%

662

R^1 = aryl, heteroaryl
R^2 = Me, Et
R^3 = H, Me
R^4 = aryl, alkyl, heteroaryl

659 (1.2 equiv), Et$_3$N
(1.2 equiv), 90 °C
ii) R^{4}NH$_2$ (3 equiv), 90 °C

661, 73-87%

Scheme 4.194

NaBH$_4$, HCl
MeOH
87%

663

664

5% CF$_3$SO$_3$H
CH$_2$Cl$_2$

665   55%

666   CO$_2$Me

667   CO$_2$Me

668   CO$_2$Me

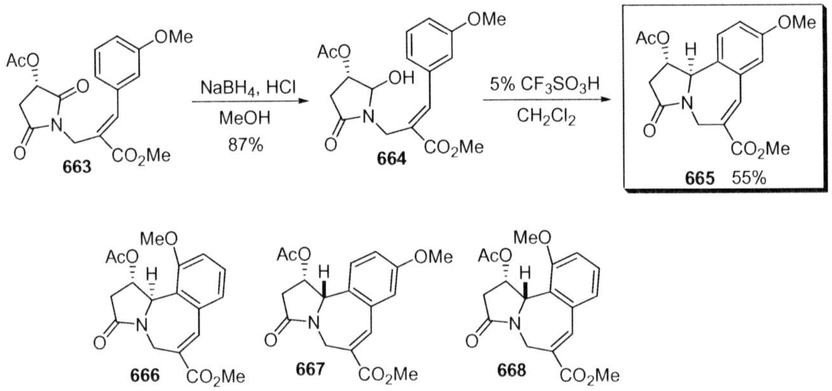

Scheme 4.195

4.4.8   Medium-sized-ring Nitrogen-containing Heterocycles

Imide **663**, prepared from the reaction of (3S)-3-acetoxysuccinimide and MBH bromide (K$_2$CO$_3$, DMF, 62% yield), has been reduced to afford a 1 : 2.5 mixture of epimers **664** in 87% yield. Intramolecular Friedel–Crafts reaction of **664** by treatment with 5% CF$_3$SO$_3$H in CH$_2$Cl$_2$ then furnished a mixture of tricyclic lactams **665–668**. Lactam **665** as main product was obtained in 55% yield (Scheme 4.195).[299] This enantioselective synthesis of the key lactam intermediate **665** established a convergent strategy for the synthesis of highly functionalized aza-analogues of natural products that incorporate an angularly fused 5-7-6 tricyclic system, including aza-analogues (X = N) of phorbol[300] and aconitine.[301] Friedel–Crafts cyclization of MBH adduct **669** takes place with 95% H$_2$SO$_4$ in CCl$_4$ at room temperature in 10 min, affording methyl 4H-pyrrolo[1,2-a]-benzazepine-5-carboxylates **670** in 70–96% yields (Scheme 4.196).[302]

The reaction of MBH adduct and alkyl nitrile to give (E)-allyl amides **671** in 72–83% yields *via* Ritter reaction (MeSO$_3$H, 110 °C, 5 h) has been achieved by

Scheme 4.196

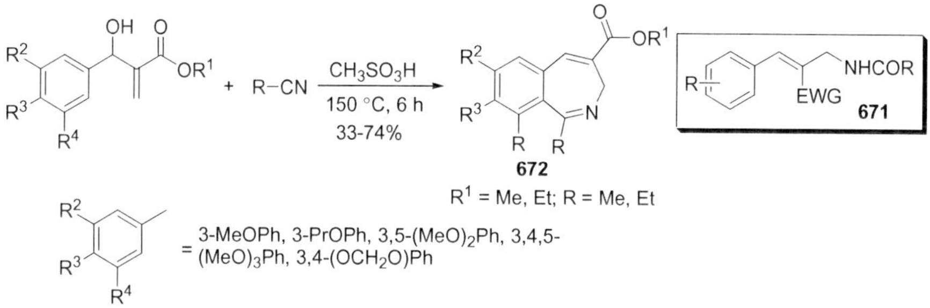

Scheme 4.197

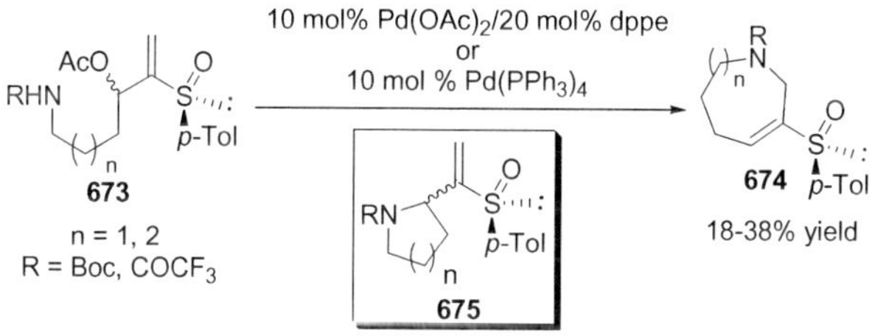

Scheme 4.198

Basavaiah *et al.* Subsequently, they elevated the reaction temperature to 150 °C, leading to the formation of 2-benzazepine derivatives **672** *via* consecutive Ritter and Houben–Hoesch reactions (Scheme 4.197).[303] Using this convenient one-pot procedure, 2-benzazepine derivatives **672** were obtained in 33–74% yields.

Intramolecular palladium-catalyzed allylation of nucleophiles has been developed to construct cyclic products. MBH acetates **673** were first investigated under the Pd(OAc)$_2$/dppe or Pd(PPh$_3$)$_4$ catalytic system, and seven- or eight-membered ring compounds **674** were obtained in low yields (18–38%), while pyrrolidines or piperidines **675** were mainly obtained in 38–88% yields (Scheme 4.198).[304] By optimizing the reaction temperature, a higher yield of the seven-membered ring compound (R = COCF$_3$) was obtained at lower temperature along with a lower yield of the pyrrolidine.

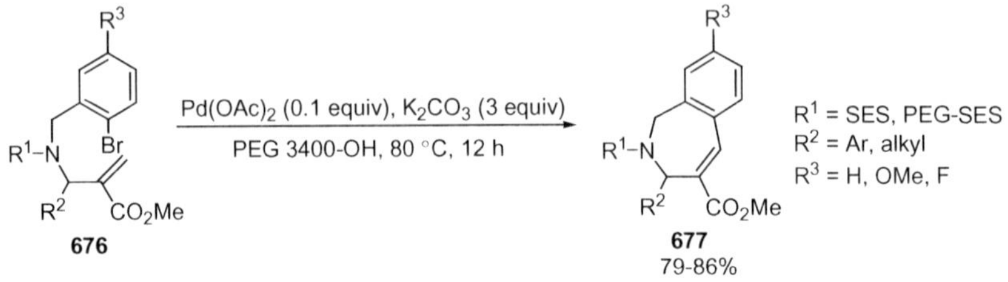

Scheme 4.199

A very efficient and selective Heck cyclization of the PEG polymer supported MBH adduct **676** has been developed for the preparation of novel heterocyclic structures (Scheme 4.199).[305] By carrying out the reaction with K_2CO_3 as the base and PEG 3400-OH as solvent at $80\,^{\circ}C$ for 12 h, benzazepines **677** were acquired with optimized yields. MBH **676** adducts derived from various aldehydes such as aromatic aldehyde, alkyl aldehyde, *etc* all gave the corresponding coupling products **677** in good yields. Compared with longer reaction times under conventional heating conditions, microwave activation has emerged as a powerful technique to accelerate most organic transformations such as the Heck reaction. Consequently, these reactions were also examined under microwave irradiation, furnishing benzazepines **677** in quantitative yields in a shorter time (30 min).[306]

MBH derivatives **678** – derived from the reaction of MBH acetates and indole derivatives (KOH, DMF, $0\,^{\circ}C$) in 45–95% yields – have been examined in terms of intramolecular palladium-catalyzed arylation (Scheme 4.200).[307] Seven-membered benzoazepino[1,2-*a*]indole derivatives **679**, which have also been synthesized from MBH derivatives *via* ring-closing metathesis (RCM) reaction,[308] were produced in good yields (65–82%) using MBH derivatives **678** ($R^1 = H$) as reactants. However, under identical reaction conditions, MBH derivatives **678** with $R^1 = Me$ formed exclusively eight-membered compounds **680** instead of compounds **679** in moderate yields (53–60%). The steric and electronic effects of the R^1 group may play a key role in the formation of the coupling product. Various MBH derivatives **681a–h** bearing isatin, benzimidazole and imidazole substituents were tested with such a Pd-catalyzed Heck reaction and showed the same reaction patterns as shown in Scheme 4.200.[309] Tetracyclic compounds containing an eight- or seven-membered ring were obtained in 36–55% yields.

Batra has reported the reactions of MBH acetates **682** with benzylamine *via* a S_N2' reaction, affording the corresponding allylic amines in good yields (Scheme 4.201).[310] However, the intermediate **684** bearing an ester group on the isoxazole ring was also facilely converted into cyclic product *via* intramolecular amidation. As expected, 5,6-dihydroisoxazolo[4,5-*c*]azepin-4-ones **683** were furnished in these reactions, but their formation was determined by the EWG group. From an examination of various MBH acetates **682**, it could be argued

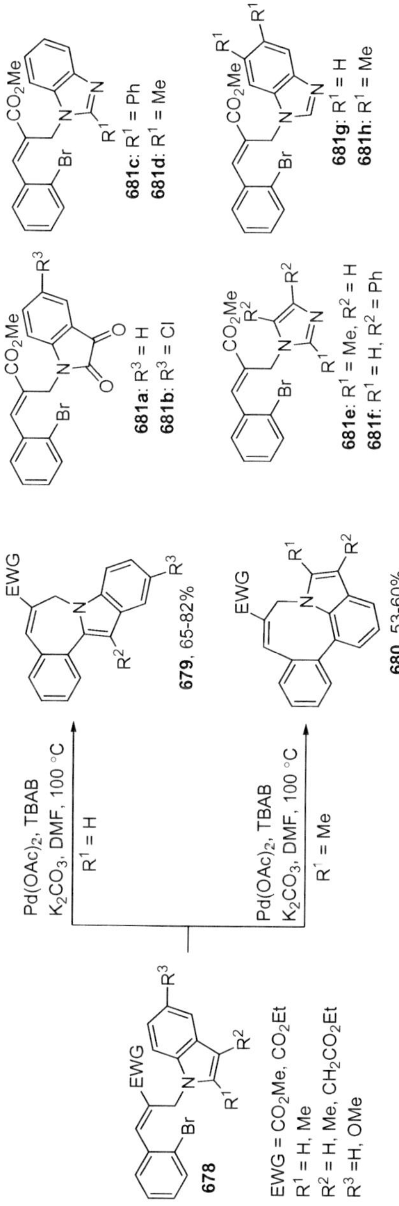

Scheme 4.200

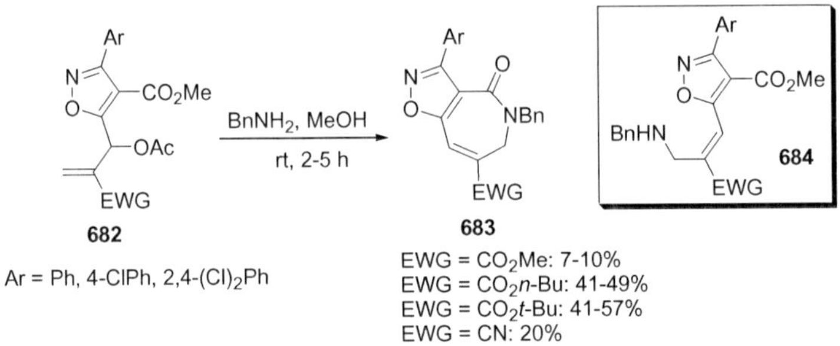

Scheme 4.201

682 (Ar, CO$_2$Me, OAc, EWG) → BnNH$_2$, MeOH, rt, 2-5 h → **683** + **684**

Ar = Ph, 4-ClPh, 2,4-(Cl)$_2$Ph

EWG = CO$_2$Me: 7-10%
EWG = CO$_2$*n*-Bu: 41-49%
EWG = CO$_2$*t*-Bu: 41-57%
EWG = CN: 20%

that the presence of a bulky group such as *n*-butyl or *t*-butyl in isoxazole ring facilitates the cyclization.

Later on, this group further developed a method for the reduction of a nitro group and then intramolecular amidation to construct benzazepines (Scheme 4.202).[311] 2-Nitro-4-(2-nitrobenzylidene)alkanoates **685** synthesized from the S_N2' reaction of ethyl nitroacetate and MBH acetate were treated with the $SnCl_2$-reduction system, giving the corresponding substituted 1*H*-1-benzazepines **686** in moderate yields. However, MBH derivatives **685** bearing electron-donating substituents on the phenyl ring were reduced to give the 3*H*-1-benzazepine derivatives **687** in 54–57% yields. To broaden the scope of MBH derivatives **685**, 4-nitro-2-(2-nitroalkylidene)alkanoates **688** were subjected to consecutive reduction and cyclization to give 3*H*-1-benzazepines **689** in 51–58% yields. In other work, the Fe/AcOH reduction system has been employed in a facile synthesis of tri-/tetracyclic heterocyclic products **690** and **691**, which contain an important azocine moiety. An alkylation, reduction and cyclization sequence was involved in this one-pot multistep protocol from MBH acetates (Scheme 4.203).[312] The products **694**, synthesized from the highly α-regioselective nucleophilic substitution of MBH acetates **693** with indoles **692** catalyzed by AgOTf (89–99% yields), were reduced in the presence of 10% Pd/C under a hydrogen atmosphere at room temperature to afford azepinoindoles **695a–c** in good yields (63–93%) (Scheme 4.204).[313] The reaction is likely to proceed through the reduction of **694** and an *in situ* aza-Michael addition, followed by cleavage of a hemiaminal in one-pot manner to provide the azepino[4,3,2-*cd*]indoles **695**.

Reduction of the nitro group of 2-(cyanomethyl)-3-(2-nitrophenyl)-propenoates **696** with the Fe/AcOH reduction system has been developed to synthesize 2-aminobenzazepines **697**, which may be used in several clinical applications, such as in methods for abating conditions involving unwanted activity, including inflammatory, autoimmune disorders and the treatment of cancer.[314] However, the key intermediate **696** could not be prepared from MBH acetate or MBH bromide. Lee and co-workers, however, have presented an effective synthetic strategy to construct the 2-aminobenzazepine framework from MBH acetate **698** (Scheme 4.205).[315] With the introduction of a cyano

CO2Me
CO2RMe
687, 54-57%
R = 3,4-OCH2O, 3,4-(OMe)2
CO2R1
CO2R2
686, 45-62%
R1 = Me, Et, t-Bu
R2 = Me, Et
R = H, 5-Cl
CO2R1
Me
689, 51-58%
R1 = Me, Et
R = H, 3,4-OCH2O, 3,4-(OMe)2
SnCl2 · 2H2O
MeOH or EtOH
N2, reflux, 2 h
CO2R1
CO2Et
NO2
NO2
685
SnCl2 · 2H2O
MeOH or EtOH
N2, reflux, 2 h
CO2R1
Me
NO2
NO2
688

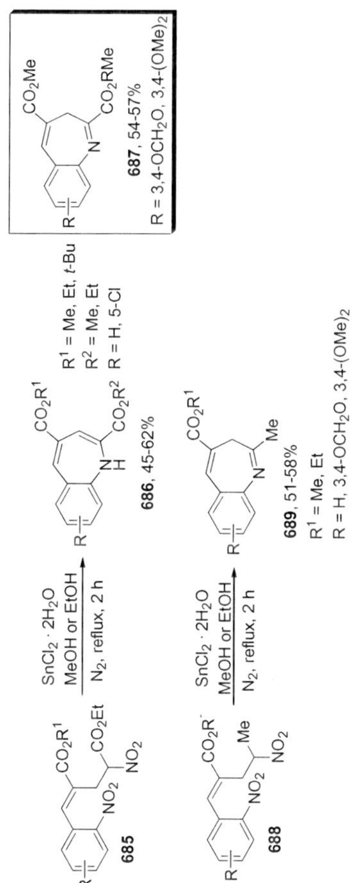

Scheme 4.202

CO2R3
O
R
R
HN
R1
R2
690, 55-77%
R1 = H, Br, Cl, OMe
R2 = H, OMe
R3 = Me, Et
OAc
CO2R3
NO2
R1
R2
1) K2CO3 / THF or DMF
rt or 80 °C, 2-6 h
2) Fe / AcOH
reflux, 1.5 h
O
O
R
R
n = 0, 1
R = H, Me
OAc
CO2R3
NO2
1) K2CO3 / THF
rt, 4 h
2) Fe / AcOH
reflux, 1.5 h
CO2R3
O
R
R
HN
691, 33-35%
R3 = Me, Et

Scheme 4.203

OAc
693
R
10 mol%
AgOTf
CH₂Cl₂,
reflux
NO₂
692
NO₂
694
R
H₂, 10% Pd/C
MeOH, rt
overnight
HO
R
695a, R = p-Cl: 96%
695b, R = o-F: 63%
695c, R = o-CF₃: 72%

Scheme 4.204

Scheme 4.205

group, the intermediate **699** (R = H) was treated with sodium methoxide in methanol at room temperature, and 3*H*-1-benzazepine derivatives **700** were obtained *via* base-promoted hydrolysis and subsequent formation of the amidine functionality between the amino and cyano groups. Interestingly, the reaction of intermediate **699** (R = Me or Et) under the same conditions proceeded smoothly to furnish methyl 2-(cyanomethyl)-2,3-dihydro-1*H*-indole-2-carboxylates **701** in 41–65% yields.

β-Lactamases were discovered before penicillin was introduced into medical use.[316] The increased resistance of bacteria to common β-lactam antibiotics has been estimated to lead to an annual economic burden of $30 billion in the USA.[317] Developing new structural types of β-lactamases has received wide attention from both academia and industry. The ever-growing new applications of azetidin-2-ones – in fields ranging from enzyme inhibition[318] to the use of these products as starting materials to develop new synthetic methodologies – have triggered renewed interest in building new bi- and polycyclic β-lactam systems in an attempt to move away from the classical β-lactam antibiotic structures.[319]

When MBH adduct **702a** was heated in toluene or *p*-xylene in a sealed tube at 210 °C the bicyclic products **703a** or **703b** formed *via* radical cyclization could be isolated in 37–45% yields. On broadening the MBH adduct **702a** to MBH enynes **702b,c**, the reactions proceeded in the same manner, and moderate yields of the corresponding bicyclic products were obtained with the exception of only a 20% yield of **703d** for enyne **702c** (Scheme 4.206).[320] Furthermore, enynes **702b–d** have also been treated with more electrophilic radicals such as PhS• and Ph₃Sn• instead of the benzylic radical in the presence of AIBN. In which case, the bicyclic β-lactams **704** were formed as the exclusive

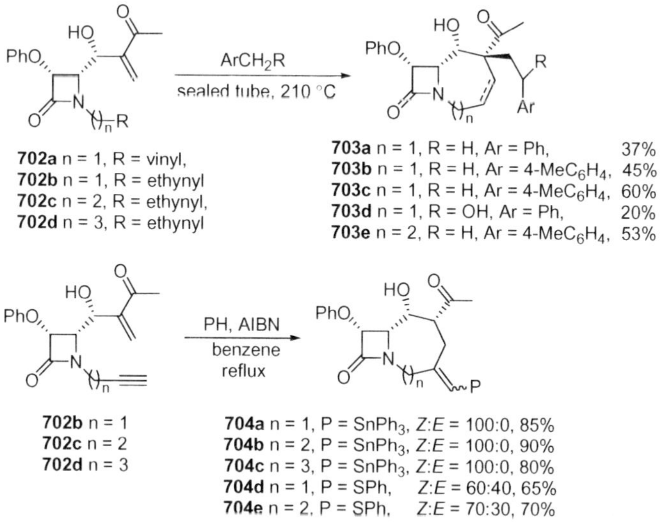

Scheme 4.206

products from **702** using Ph$_3$SnH but β-lactams **704** were obtained in relatively low stereoselectivities for PhSH. With this novel synthetic tool for the asymmetric synthesis of densely functionalized monocyclic β-lactams in hand, several unusual bicyclic β-lactams with medium-sized rings have been prepared by changing the substituents in readily available 4-oxoazetidine-2-carbaldehydes, activated alkenes or radical precursors.[321]

4.4.9 Construction of Nitrogen Heterocycles Containing Two Nitrogen Atoms

Since the development of the reaction of 2-aminopyridine with alkyl acrylates or methyl propiolate to construct a pyrimidine ring,[322] the reaction of MBH acetates with 2-aminopyridine has been investigated for the synthesis of substituted fused pyrimidine derivatives.[323] The reactions were conducted in H$_2$O/MeOH (1 : 1) at room temperature for 6 h, furnishing 3-aryl-substituted-1,5-diazabicyclo[4.4.0]deca-2,5,7,9-tetraen-4-ones **707** in moderate to good yields (Scheme 4.207). However, MBH acetates **705** (R = *n*-pentyl) derived from an alkyl aldehyde gave a low yield of the corresponding product **707**. As well as 2-aminopyridine **706**, thiazol-2-amines **708** have also been developed to synthesize 5*H*-thiazolo[3,2-*a*]pyrimidin-5-ones **711**, which are very important intermediates and widely used in the pharmaceutical chemistry.[324] As shown in Scheme 4.208,[325] a range of aryl- and hetaryl-substituted substrates (**709**) and thiazol-2-amines (**708**) have been reacted at room temperature, affording 6-benzylidene-6,7-dihydro-5*H*-thiazolo[3,2-*a*]pyrimidin-5-ones **710** in good to excellent yields with high regioselectivity *via* nucleophilic addition and subsequent cyclization (Scheme 4.208). Upon elevating the reaction temperature to 130 °C, a thermo-induced sigmatropic shift procedure took place to transform compound **710** into 5*H*-thiazolo[3,2-*a*]pyrimidin-5-one **711** in quantitative yield. Therefore, various substituted compounds **711** were easily prepared at 130 °C in a one-pot procedure in good to excellent yields. However, only a 30% yield of product **712** was obtained from MBH adduct **709** derived from an alkyl aldehyde.

MBH adducts **713**, synthesized from 2-cyclohexen-1-one, have been converted into the corresponding pyrazole derivatives **714** in 48–57% yields.[326] The oxidation of **714** with DDQ (2.0 equiv) in benzene at reflux temperature gave

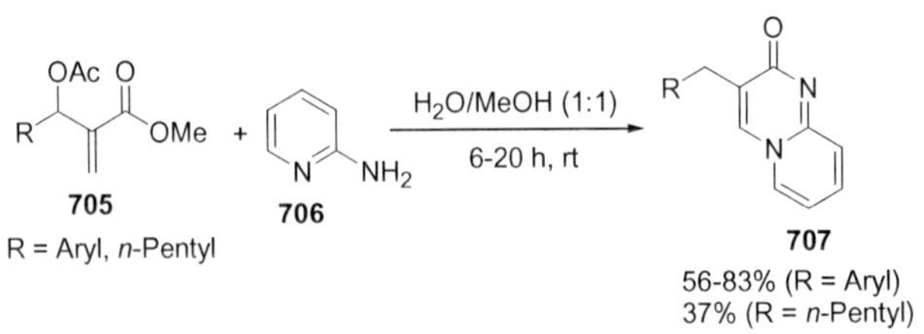

Scheme 4.207

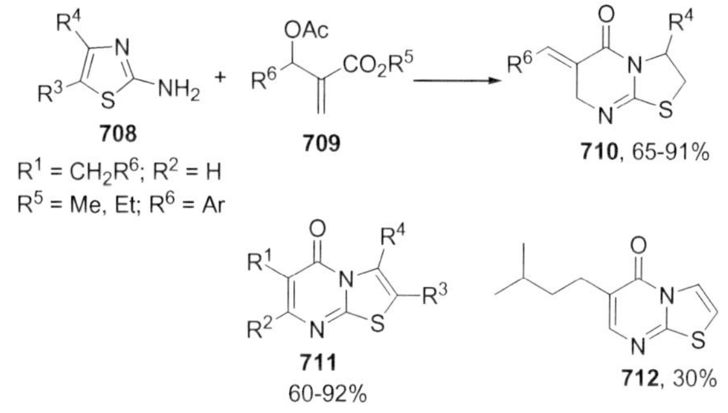
R4
N
R3
S
NH2
708
R1 = CH2R6; R2 = H
R5 = Me, Et; R6 = Ar
+
OAc
R6
CO2R5
709
O
R4
R6
N
N
S
710, 65-91%
O
R4
R1
N
R3
R2
N
S
711
60-92%
O
N
N
S
712, 30%

Scheme 4.208

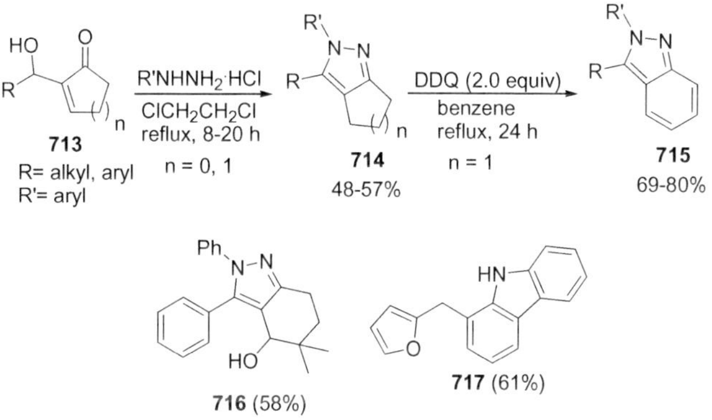
HO
O
R
n
713
R= alkyl, aryl
R'= aryl
R'NHNH2·HCl
ClCH2CH2Cl
reflux, 8-20 h
n = 0, 1
R'
N-N
R
n
714
48-57%
DDQ (2.0 equiv)
benzene
reflux, 24 h
n = 1
R'
N-N
R
715
69-80%
Ph
N-N
HO
716 (58%)
HN
O
717 (61%)

Scheme 4.209

the desired 2*H*-indazole derivatives **715** in moderate to good yields (69–80%, Scheme 4.209).[327] During the DDQ oxidation of pyrazole derived from 4,4-dimethylcyclohex-2-en-1-one in refluxing benzene, the hydroxylated compound **716** was obtained in 58% yield. Interestingly, when the MBH adduct derived from furan-2-carbaldehyde was treated with phenylhydrazine hydrochloride the aromatized compound **717** was obtained directly in 61% yield without isolation of the pyrazole intermediate.

To develop the new cyclic ureide analogs and evaluate their antibacterial activity, several types of tetrahydropyrimidin-2-ones have been synthesized using MBH adducts as the key starting materials.[328] As shown in Scheme 4.210, 1-(2-mcyano-3-aryl-allyl)-3-aryl-urea (thiourea) derivatives **718**, **720** and **722** could be easily cyclized to the corresponding product in moderate yields in the presence of K_2CO_3 or NaH. On the basis of testing their antibacterial activity against susceptible Gram-positive and the Gram-negative bacteria, including *Staphylococcus aureus*, *Streptococcus faecalis*, *Klebsiella pneumoniae*, *Escherichia coli* and *Pseudomonas aeruginosa* strains, few of these cyclized compounds

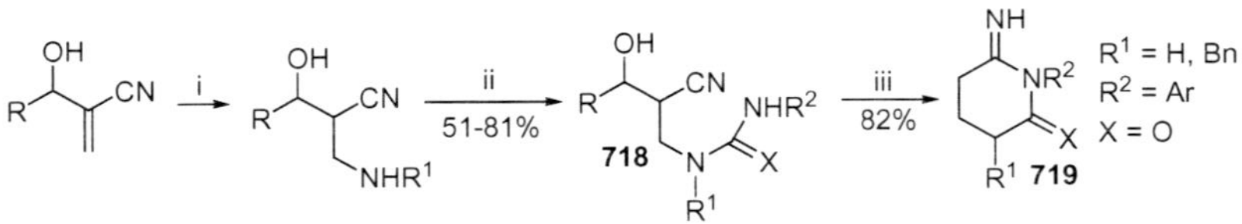

(i) methanolic ammonia, rt, 8 h or BnNH$_2$, MeOH, rt, 8 h; (ii) R$_2$NCO, THF, rt, 1.5-2 h; (iii) K$_2$CO$_3$, MeOH, reflux, 8-9 h or NaH, toluene, reflux, 8-9 h.

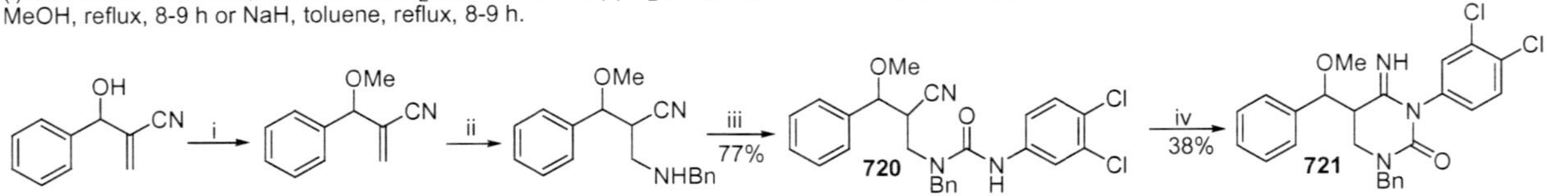

(i) Ag$_2$O, MeI, CH$_2$Cl$_2$, rt, 5 h; (ii) BnNH$_2$, MeOH, rt, 6 h; (iii) 3,4-(Cl)$_2$-C$_6$H$_3$NCO, THF, 45 min; (iv) NaH, toluene, reflux, 8 h.

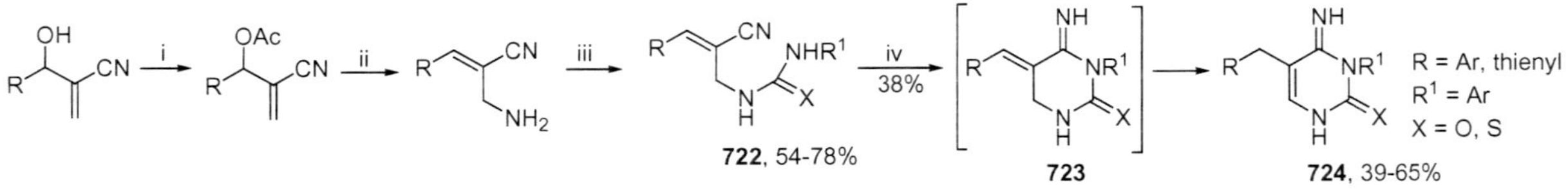

(i) AcCl, pyridine, CH$_2$Cl$_2$, rt, 3 h; (ii) methanolic ammonia, rt, 1 h; (iii) R^{1}NCO, THF, rt, 1 h; (iv) K$_2$CO$_3$, MeOH, reflux, 8-9 h or NaH, toluene, reflux, 8-9 h.

Scheme 4.210

showed superior activity or were equipotent to the standard antibacterial agents. At a later date, the same group demonstrated another practical and convenient synthesis of 1,5-disubstituted uracil derivatives from the corresponding cyanamides derived from the MBH adducts (Scheme 4.211).[329]

Benzo[*b*][1,4]diazepin-2-ones constitute unique structures that exhibit a spectrum of biological activities such as interleukin 1β enzyme inhibition and potassium current blocking.[330] The diamino esters **725** and **726**, derived from S_N2 nucleophilic substitution and S_N2' nucleophilic substitution of MBH acetates, respectively, were easily cyclized by treatment with sodium hydride in toluene at 80 °C to afford the desired diazepinones **727** and **728** in moderate to good yields (Scheme 4.212).[331]

The intramolecular alkylation of MBH derivatives in the presence of NaH[332] and intramolecular condensation in the presence of EDCI·HCl[333] were, respectively, designed to give 1,4-diazepane-2,5-diones **731** and **732** (Scheme 4.213). These compounds were then used as intermediates for further conversion into products designed to contain the 1,4-diazepane-2,5-dione scaffold, products that showed potent and selective inhibition of human chymase.

4.4.10 Polyheterocyclic Compounds Containing a Nitrogen Atom

Indenoquinoline derivatives, with a 1,4-DHP parent nucleus, have shown a diverse range of biological properties such as 5-HT-receptor binding[334] and anti-inflammatory activities,[335] and also act as antitumor agents.[336] Consequently, these compounds have distinguished themselves as heterocycles of profound chemical and biological significance. Thus the synthesis of these molecules has attracted considerable attention.[337] Using S_N2' type aniline-substituted MBH derivatives **733** as starting materials, an efficient synthetic method for indenoquinoline skeletons has been developed by Kim *et al.*[338] As shown in Scheme 4.214, compound **733a** was heated in polyphosphoric acid (PPA) at around 120 °C. After careful isolation, **734a** (3%), 4b,5,10a,11-tetra-hydroindeno[1,2-*b*]quinolin-10-one (**735a**) (62%) and 7*H*-indeno[2,1-*c*]quinoline (**736a**) (4%) were obtained and confirmed by analysis of their spectroscopic data. Various *para*-substituted arenes **733** both at the MBH moiety and at the aniline moiety also gave the corresponding products under identical reaction conditions. To examine the postulated reaction mechanism, a series of control experiments was conducted. It was found that intermediate **734a** was obtained as major product in 73% yield at 90 °C after 8 h, and could be further converted into **735a** at elevated temperature (120 °C) in 85% yield.

With the development of a facile route for the synthesis of novel 5-substituted-2-amino-1,4,5,6-tetrahydropyrimidines **740** from MBH adducts **737**, pyrimidines **740** were treated with ethyl bromoacetate in the presence of potassium carbonate to give imidazo[1,2-*a*]pyrimidine derivatives **741** in high yields (Scheme 4.215).[339] Subsequently, three additional convenient and

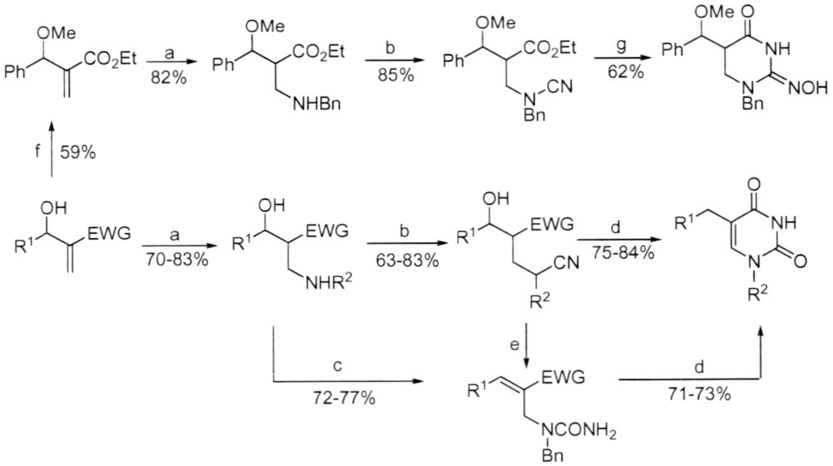

(a) R^2NH_2, EtOH-MeOH, rt, 8-12 h; (b) BrCN, NaHCO$_3$, benzene, rt, 15 min; (c) BrCN, NaHCO$_3$, benzene, rt, 12 h; (d) NaH, toluene, rt, 45 min; (e) NaHCO$_3$, benzene, rt, 12 h; (f) MeI, Ag$_2$O, CH$_2$Cl$_2$, rt, 24 h; (g) NH$_2$OH·HCl, K$_2$CO$_3$, EtOH-H$_2$O (3:2), rt, 5 h.

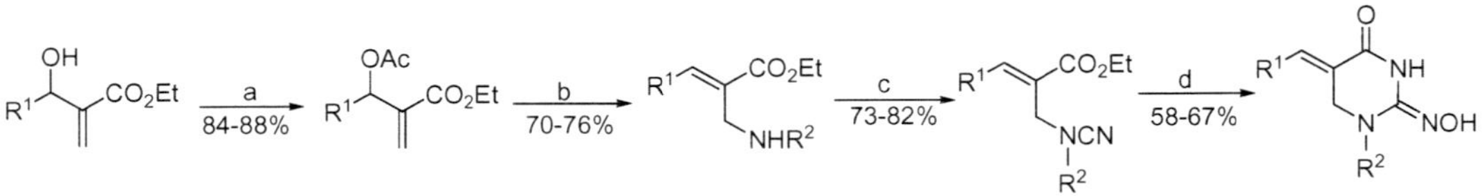

(a) AcCl, py, CH$_2$Cl$_2$, rt, 3-5 h; (b) R^2NH_2, EtOH, rt, 1-1.5 h; (c) BrCN, NaHCO$_3$, benzene, rt, 15 min; (d) NH$_2$OH·HCl, K$_2$CO$_3$, EtOH-H$_2$O (3:2), rt, 14-16 h.

Scheme 4.211

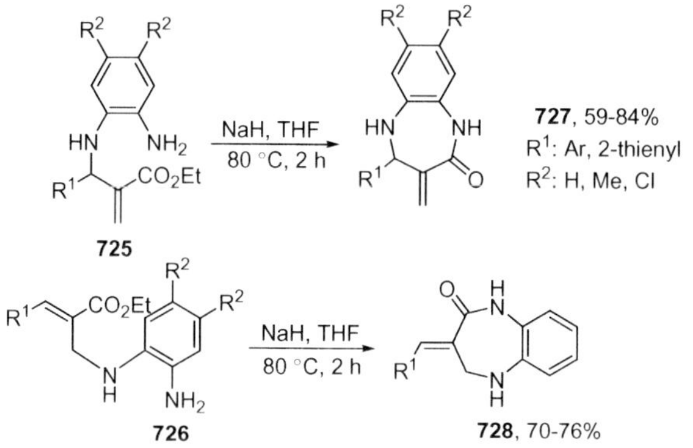
R2 R2
HN NH2
R1 CO2Et
NaH, THF
80 °C, 2 h
725
727, 59-84%
R1: Ar, 2-thienyl
R2: H, Me, Cl
R1 CO2Et
R2 R2
N H
NH2
726
NaH, THF
80 °C, 2 h
728, 70-76%

Scheme 4.212

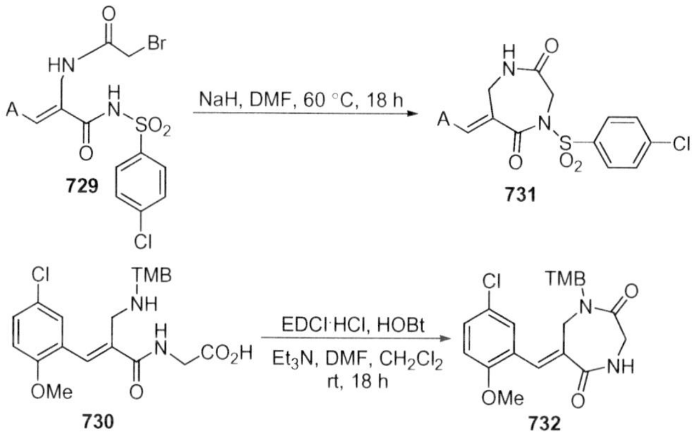
Br
HN
A
SO2
729
Cl
NaH, DMF, 60 °C, 18 h
731
Cl
TMB
NH
CO2H
OMe
730
EDCI·HCl, HOBt
Et3N, DMF, CH2Cl2
rt, 18 h
732

Scheme 4.213

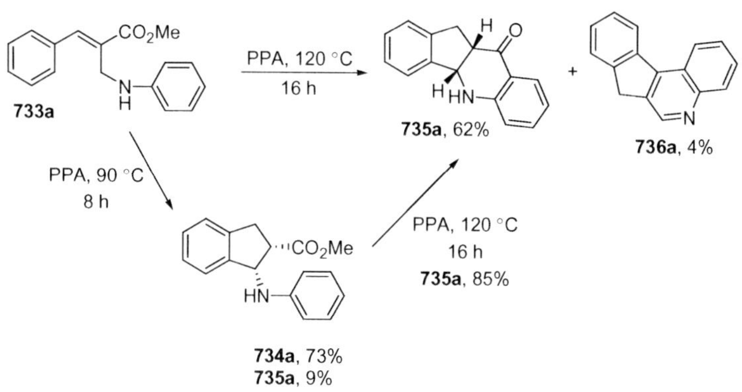
CO2Me
733a
PPA, 120 °C
16 h
735a, 62%
736a, 4%
PPA, 90 °C
8 h
CO2Me
HN
PPA, 120 °C
16 h
735a, 85%
734a, 73%
735a, 9%

Scheme 4.214

Scheme 4.215

practical strategies (Schemes 4.216–4.218) for the synthesis of differently substituted annulated 5,6,7,8-tetrahydro-imidazo[1,2-*a*]pyrimidine-2-ones and 3,4,6,7,8,9-hexahydro-pyrimido[1,2-a]pyrimidin-2-ones from MBH adduct were further developed.[340]

More recently, to find more effective methods to construct bis(heterocycle)s containing a pyrimidine ring from MBH derivatives, the solid-phase parallel synthesis of new annulated pyrimidinone derivatives was disclosed (Scheme 4.219).[341] The resin-bound allyl amine derivatives **742** were treated with cyanogen bromide to yield the pyrazole derivatives **743** in good yields under standard conditions. The subsequent lactonization with cleavage of the resin afforded the pyrimidinone derivative **744** in good yields in the presence of 20% triethylamine in chloroform under reflux.

Imidazo[1,2-*a*]pyrimidines have been attractive targets for synthetic chemists due to their interesting biological activities,[342] and some of them have been reached the market, such as the drug divaplon.[343] Encouraged by the above synthetic methods, MBH acetates **745** as starting materials were used to prepare 6-arylmethylimidazo[1,2-*a*]pyrimidin-7-ylamine derivatives **748** (Scheme 4.220).[344] Products **748** could also be obtained by a one-pot procedure from allylamines **746**

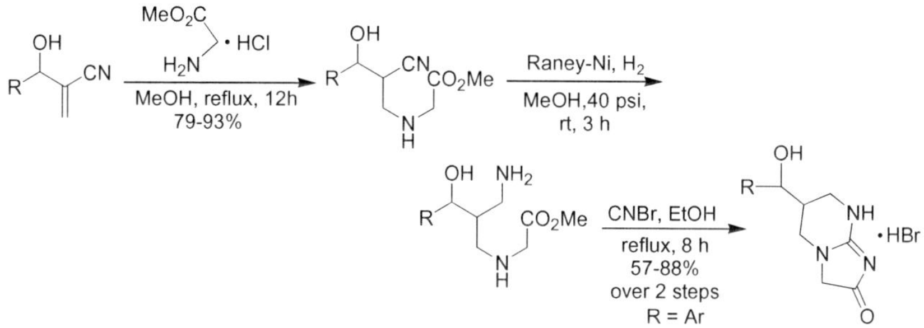

Scheme 4.216

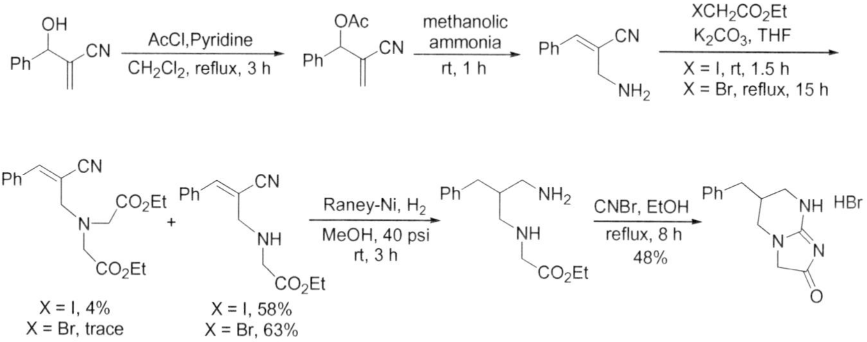

Scheme 4.217

Scheme 4.218

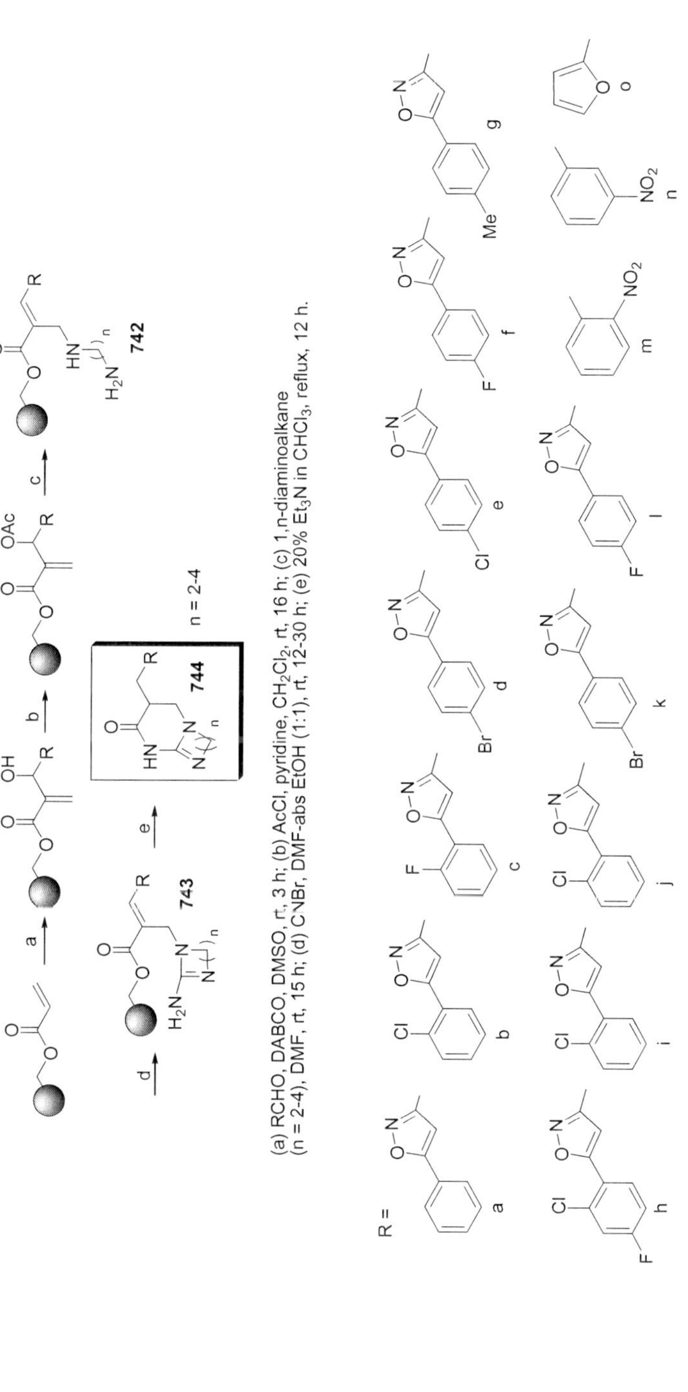

(a) RCHO, DABCO, DMSO, rt, 3 h; (b) AcCl, pyridine, CH$_2$Cl$_2$, rt, 16 h; (c) 1,n-diaminoalkane (n = 2-4), DMF, rt, 15 h; (d) CNBr, DMF-abs EtOH (1:1), rt, 12-30 h; (e) 20% Et$_3$N in CHCl$_3$, reflux, 12 h.

Scheme 4.219

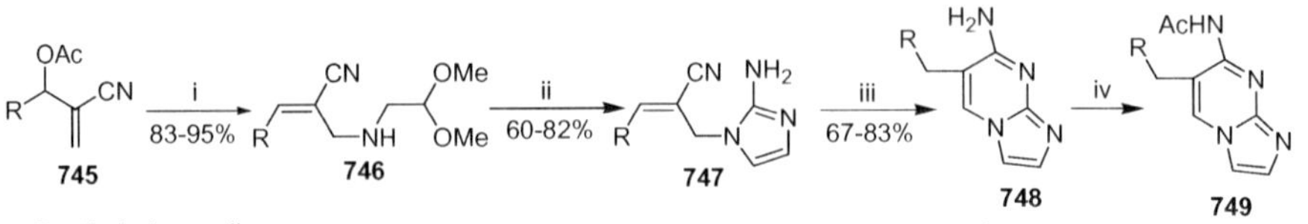

745 → **746** → **747** → **748** → **749**

R = Ar, heterocyclic group

(i) 2,2-dimethoxyethylamine, MeOH, rt, 1 h; (ii) (1) NH_2CN, $AcOH/H_2O$, 90-100 °C, 2 h, (2) HCl (conc.), 90-100°C, 5 min; (iii) NaOMe, MeOH, rt., 1 h; (iv) Ac_2O, pyridine, rt, 3 h

Scheme 4.220

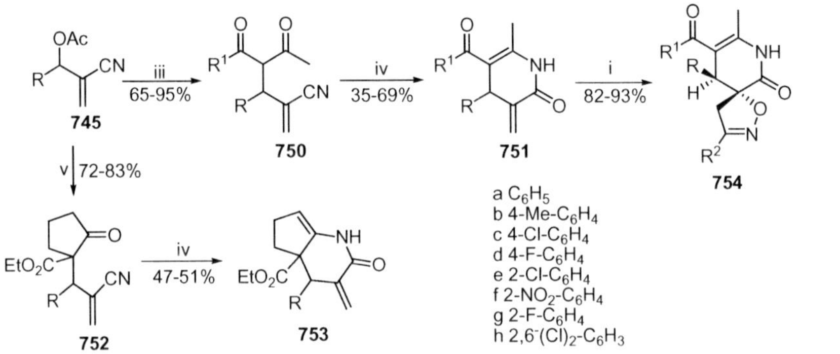

745 → **750** → **751** → **754**

745 → **752** → **753**

a C_6H_5
b 4-Me-C_6H_4
c 4-Cl-C_6H_4
d 4-F-C_6H_4
e 2-Cl-C_6H_4
f 2-NO_2-C_6H_4
g 2-F-C_6H_4
h 2,6-$(Cl)_2$-C_6H_3

(i) DABCO, neat, rt, 0.5-15 h; (ii) AcCl, pyridine, CH_2Cl_2, 0 °C to rt, 3 h; (iii) DABCO, acetyl acetone or ethylacetoacetate, THF/H_2O (1:1, v/v), rt, 2 h; (iv) TFA/H_2SO_4 (4:1, v/v), rt, 5-7 min; (v) DABCO, ethylcyclopentanone-2-carboxylate, THF/H_2O (1:1, v/v), rt, 2 h. (iv) R^2CNO, Et_3N, dry Et_2O, -78 °C to rt, 6 h.

Scheme 4.221

in 48–79% yields. All MBH acetates **745** derived from aromatic aldehydes as well as heterocyclic aldehydes successfully gave the corresponding products **748** in good yields. Using MBH acetates **745** as starting materials, Batra and co-workers also demonstrated a facile approach to the synthesis of substituted 3-methylene-2-pyridones **751** and **753** (Scheme 4.221).[345] The utility of pyridone derivatives **751** for the synthesis of new spiroisoxazolines **754** in highly regio- and stereoselective fashion was also illustrated.

The pyrrolo[3,2-*c*]quinoline and its analogue pyrroloquinoxaline skeletons possessing a wide spectrum of biological activities have been known for several years as one of the most widely used motifs in medicinal chemistry.[346] To introduce successfully different substituents at the 4-position of the pyrrolo[3,2-*c*]-quinoline skeleton, imidoyl chloride **762** was designed and synthesized by the ring-closing metathesis with the formation of pyrroline and microwave-promoted lactonization as the key steps in the synthetic strategy (Scheme 4.222).[347] Under the microwave irradiation, electron-rich alkylamines as nucleophiles reacted with imidoyl chloride **762** to afford the corresponding ammoniated products, 4-amino-substituted pyrrolo[3,2-*c*]quinoline **763**, in good yields. The

Scheme 4.222

Pd-catalyzed and microwave-enhanced Heck reaction of an imidoyl chloride derivative with boronic acid derivatives also furnished 4-aryl-substituted pyrrolo[3,2-*c*]quinoline **764** in a straightforward manner and in good yield. Subsequently, this group further developed a similar synthetic method to construct 4-alkyl-substituted pyrrolo[3,2-*c*]quinoline **766** (Scheme 4.223) using penten-2-one **765** as starting material.[348]

The azido-β-amino esters **768**, obtained from conjugate addition of aza-MBH adduct with HN_3 in high yields as a mixture of *anti* and *syn* diastereoisomers, have been subjected to alkylation with propargyl bromide in the presence of Cs_2CO_3, furnishing compounds **769** as an inseparable mixture of *anti* and *syn* isomers in excellent yields. Subsequent 1,3-dipolar cycloaddition produced the new bicyclic triazoles **770** in *trans*-form with extremely high diastereomeric ratios (Scheme 4.224).[349] Deprotection of the SES group by anhydrous HF and neutralization of the hydrofluoride salt with $NaHCO_3$ could quantitatively provide the deprotected bicyclic triazoles.

The MBH derivative enamide **771a** has, under typical Heck cyclization reaction conditions, smoothly delivered benzoazepino[2,1-*a*]isoindole derivative

Scheme 4.223

Scheme 4.224

772a instead of forming seven- or eight-membered ring compounds.[350] Optimization of the reaction conditions revealed that the use of Pd(OAc)$_2$/n-Bu$_4$NBr/NaHCO$_3$/DMF/80 °C gave the best results for the formation of **772a**, in 55% yield (Scheme 4.225). Since the benzoazepino[2,1-a]isoindole skeleton is found abundance in natural products with interesting biological activities, such compounds and related compounds have been studied extensively.[351] Other enamides bearing different substituents on the MBH derivative or isoindolin-1-one have been subjected to Heck reactions under the optimized conditions. As expected, pentacyclic benzoazepino[2,1-a]isoindole compounds were obtained in 46–55% yields. However, seven-membered ring compound **773** was obtained in 57% yield when R^3 was an ethyl group. In addition, (Z)-nitrile derivative **774** showed low reaction activity; only a trace amount of product **772** was obtained.

More recently, S_N2'-substituted MBH derivatives **775** were also subjected to the Pd-catalyzed reaction, delivering novel 1-phenyl-1,6a-dihydro-6-oxacyclopropa[a]indene-1a-carboxylic acid derivatives (Scheme 4.226).[352] In this case, compounds **776** were isolated as the major products in 40–57% yields and **777** as the minor products in 14–23% yields. However, for the more sterically

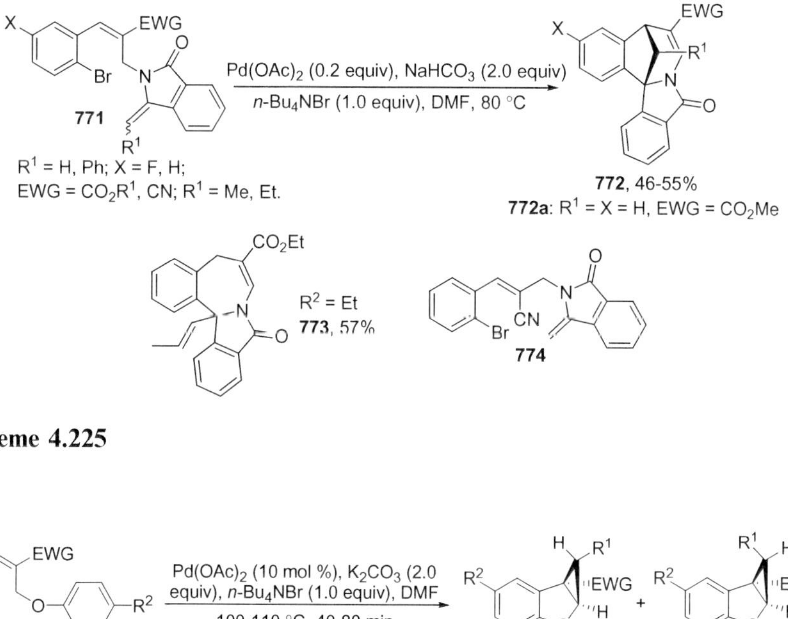

Scheme 4.225

Scheme 4.226

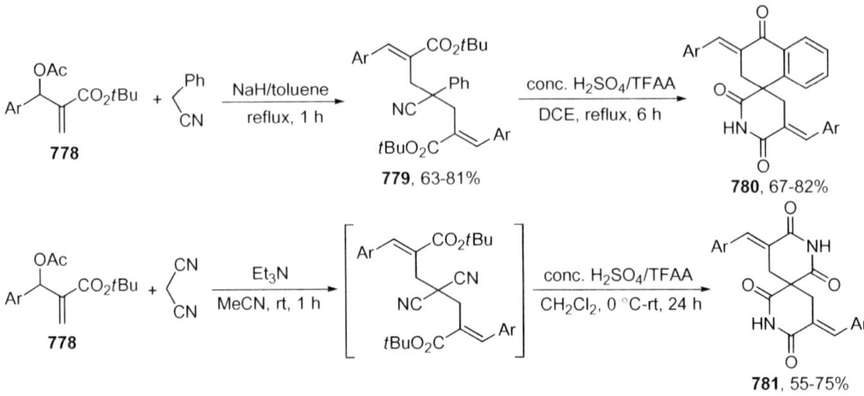

Scheme 4.227

hindered substrate **775** bearing a 2-naphthyl group at R^1, the corresponding products were formed in 13% and 26% yields, respectively.

MBH acetates **778** as alkylating agents (5 mmol) were used to react with benzyl cyanide (2 mmol) in the presence of excess NaH (10 mmol) in anhydrous toluene under reflux for 1 h to provide the desired bis-adducts **779** in good yields. Such compounds **779** were further subjected to an intramolecular Friedel–Crafts reaction to achieve a bis-cyclization strategy involving facile C–C and C–N bond formation: treatment with conc. H_2SO_4 and TFAA furnished di(E)-arylidene-tetralone-spiro-glutarimides **780** in 67–82% yields (Scheme 4.227).[353] Moreover, using malononitrile instead of benzyl cyanide, a one-pot multistep transformation of the MBH acetates **778** into di(E)-arylidene-spiro-bis-glutarimides **781** in 55–75% yields was also disclosed by these researchers.

MBH derivatives **782**, prepared from the reaction of MBH acetate with pyrrole-2-carbaldehyde or the reaction of MBH bromide with MBH acetate in the presence of K_2CO_3, were procured to examine the intramolecular [3 + 2] cycloaddition. The condensation of **782** with sarcosine (**783**) in refluxing toluene under Dean-Stark conditions produced *cis*-adducts **784** in moderate yields (Scheme 4.228).[354] Further broadening of the scope of amino acids to proline and thiazolidine revealed that all of the reactions proceeded very well – the corresponding pyrrolizidine and thiopyrrolizidine heterocycles were obtained in similar yields. To improve the yield, the same reactions were also carried out under microwave irradiation, affording adducts **784** and **787** in higher yields (78–87%).

4.5 Synthesis of Other Heterocyclic Compounds

The 1,3-dipolar cycloaddition of alkenes to nitrile oxides is a fundamental reactions because the resulting isoxazolines are very useful "building blocks"

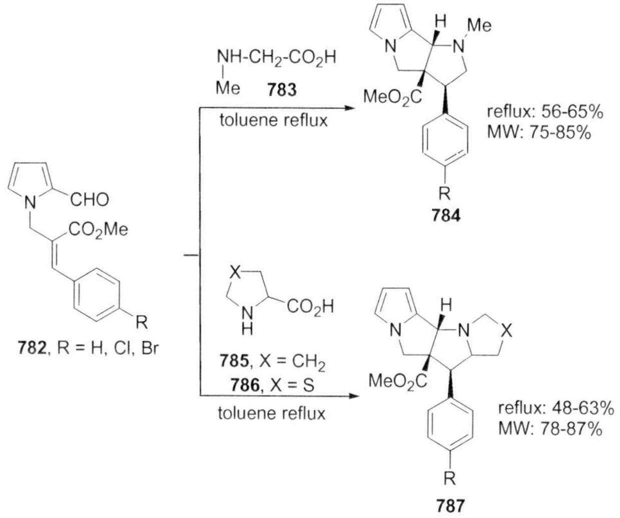

Scheme 4.228

in organic synthesis.[355] The presence of a double bond in the MBH adduct makes it a suitable substrate for a cycloaddition reaction. Since the initial report of 1,3-dipolar cycloaddition of a MBH adduct with nitrile oxides to give isoxazolines in good yields,[356] cycloaddition reactions of these substrates with MBH adducts have been reported extensively.

1-Pyrazolines **788a**,[357] easily prepared by the 1,3-dipolar cycloaddition reaction of MBH adducts with diazomethane, have been reacted with butyraldehyde to give cyclic emiaminal **789** in 30% yield in the presence of acetyl chloride and pyridine (Scheme 4.229).[358] This reaction suggests that **788a** might undergo an isomerization of the pyrazolinic ring. The carboxylic ester of **788b** could be reduced by DIBAL-H to give the corresponding alcoholic product **790** in 44% yield. After treating with 30% HF, 2-pyrazoline **791** was formed in which the primary alcohol was protected by TBDMS and the pyrazolinic ring was isomerized. Based on this observation, in a one-pot treatment of **790** with 2,2-dimethoxypropane and HF, the desired emiaminal **792** was successfully obtained in 66% yield.

Some spiroisoxazolines occur naturally and have significant biological activities. For example, araplysillins are inhibitors of ATPase.[359] Isoxazoline **793**, prepared from the reaction of mesitonitrile oxide with MBH adducts, can be lactonized in 70% aqueous acetic acid, affording spiroisoxazoline **794** in excellent yield (Scheme 4.230).[360]

Swern oxidation of isoxazolines **795** yields β-ketoesters **796** in moderate yields. By treatment with hydrazines, β-ketoesters **796** can be condensed to give the corresponding hydrazones **797**, which concomitantly undergo elimination to give novel spiro-fused isoxazolinopyrazolones **798** (3,7,9-substituted-1-oxa-2,7,8-triazaspiro[4.4]nona-2,8-dien-6-one) in moderate to good yields (Scheme 4.231).[361]

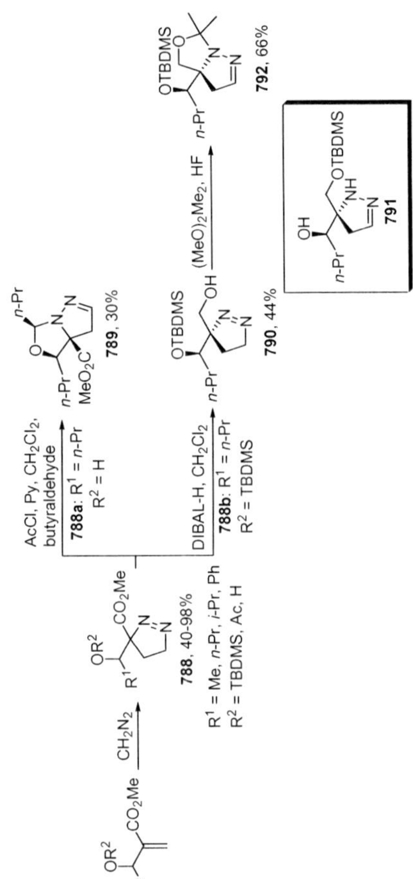

Scheme 4.229

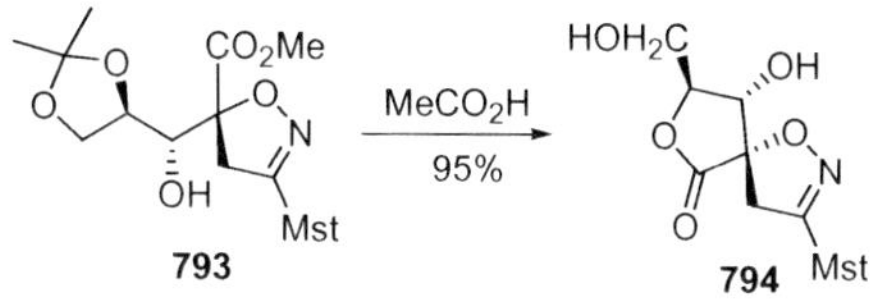

Scheme 4.230

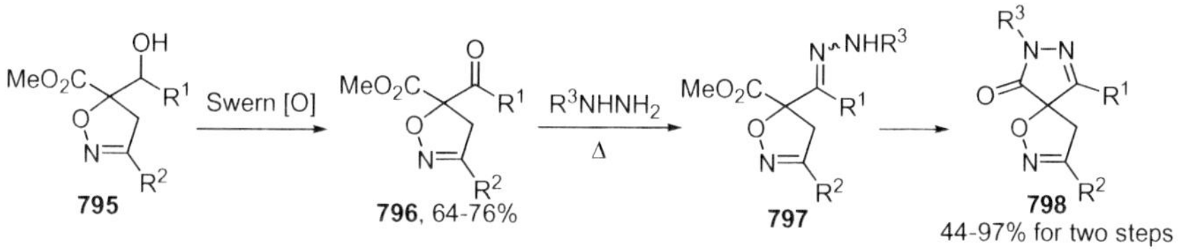

Scheme 4.231

In this one-pot reaction, it was found that hydrazone formation was the rate-determining step and the yield of compounds **798** mainly depended on the steric hindrance of these ketones.

In contrast with the isoxazoline obtained from the reaction of MBH adduct with nitrile oxide, isoxazolidine[362] was formed in such a 1,3-dipolar cycloaddition of MBH adduct with nitrone. To establish the relative configuration of the cycloadducts, cycloaddition product **800** was further converted into acetonide **802** *via* the corresponding diol **801** (Scheme 4.232). NMR analysis of this rigid, bicyclic compound (**802**) revealed its relative configuration.[363]

A facile approach to synthesize polyfunctionalized 4,5-dihydro-1,3-oxazoles has been disclosed by Orena and co-workers.[364] Trichloroacetimidates **803**, prepared in excellent yields from the reactions of MBH adducts and CCl$_3$CN for 1 h at temperatures ranging from –40 °C to room temperature, were treated with NIS in CHCl$_3$, affording the diastereomeric 4,5-dihydro-1,3-oxazoles **804a** and **804b** in 88% total yield, with the *cis*-isomer as the major component (Scheme 4.233). However, *cis*-4,5-dihydro-1,3-oxazole **806** was obtained exclusively and in good yield from trichloroacetamide **805a**, which, in turn, was synthesized from trichloroacetimidate **803** in the presence of DABCO.

MBH adduct as starting material has also been used to synthesize oxazolidin-2-ones that showed various pharmacological activities.[365] As shown in Scheme 4.234, an approach for the preparation of functionalized oxazolidin-2-ones **813** from the MBH adduct in eight steps in an overall yield of 26% has been disclosed by Coelho and Rossi.[366] In such a synthetic method, hydroboration of the double bond, Curtius rearrangement of hydroxy-acid **811** and oxidation of the primary alcohol to the carboxylic acid were the key steps.

Seven-membered heterocycles with two heteroatoms in a 1,4-configuration are known to possess manifold biological activities,[367] and are also used as

Scheme 4.232

Scheme 4.233

valuable chiral templates for stereoselective syntheses.[368] A method for the synthesis of [1,4]oxazepin-7-ones **819** from readily available aldehydes and α-amino alcohols has been developed using the MBH reaction as the key step. As shown in Scheme 4.235, the [1,4]oxazepine-7-ones could be obtained in overall yields of 14–41% starting from **815** using 2-piperidinylmethanol as the α-amino alcohol *via* a split synthesis approach on soluble polymer-supported MeOPEG **814**.[369] This synthetic method proved to be a more efficient way to rapidly screen the substrate spectrum in a multistep reaction sequence compared with a parallel synthesis approach.

A concise α-amino acid-based synthetic approach *via* sequential intermolecular nucleophilic substitution and intramolecular Michael addition reactions starting from MBH acetates has also been described to synthesize disubstituted [1-4]oxazepin-2-ones **821**.[370] Such a synthetic method was operationally simple under ambient conditions, and gave 81–93% yields of the target [1,4]oxazepin-2-ones (Scheme 4.236).

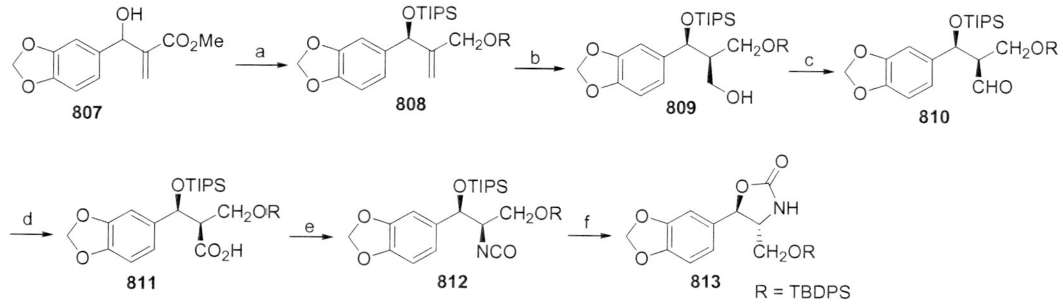

(a) (i) TBDPSCl, imidazole, DMF, rt, 14 h, 90%; (ii) DIBAL-H, CH_2Cl_2, -78 °C, 2 h, 91%; (iii) TBDPSCl, imidazole, DMF, rt, 14 h, 90%. (b) (i) $BH_3 \cdot (CH_3)_2$ or 9-BEN, THF, 0 °C to rt, 16 h; (ii) NaOH (3 M), H_2O_2 (30%), 0 °C to rt, 1.5 h. (c) TPAP, NMO, CH_2Cl_2, MS 4 A, 15 min, rt, 96%. (d) $NaClO_2$, NaH_2PO_4, t-BuOH, H_2O, 2-methyl-but-2-ene, rt, 14 h, 90%, chromatographic separation; (e) (i) $ClCO_2Et$, NEt_3, 0 °C, 40 min; (ii) NaN_3, H_2O, 0 °C, 2 h; (iii) reflux in toluene. (f) $SnCl_4$, CH_2Cl_2, rt, 16 h, 30% overall yield (five steps).

Scheme 4.234

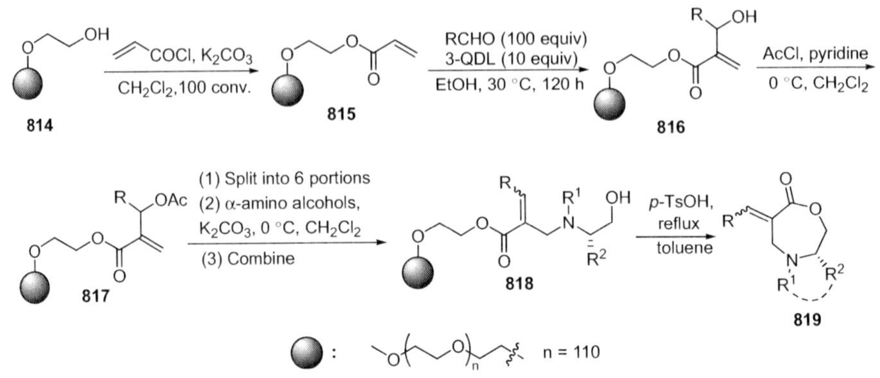

Scheme 4.235

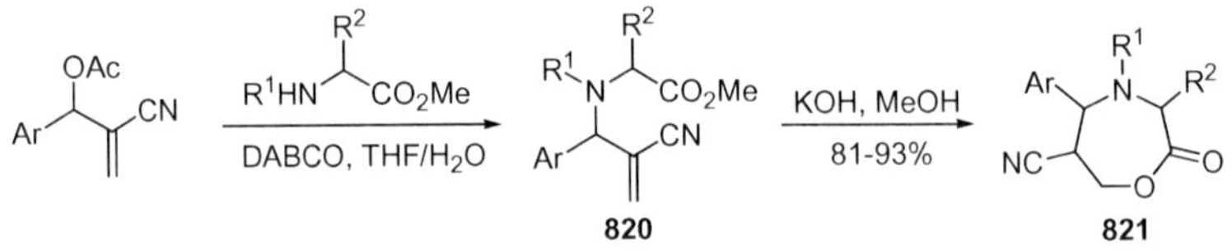

Scheme 4.236

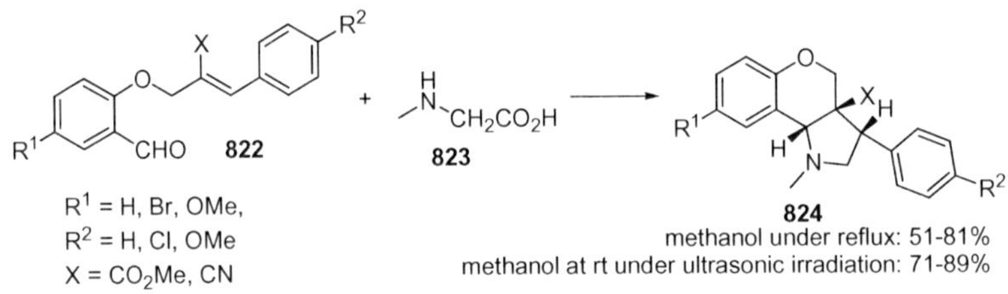

Scheme 4.237

Raghunathan has reported that, using substrates prepared from MBH adducts or MBH bromides, a series of chromano[4,3-*b*]pyrroles **824** can be synthesized through intramolecular 1,3-dipolar cycloaddition (Scheme 4.237).[371] To improve the yield, the reactions carried out under ultrasonic irradiation were further studied in methanol at room temperature. A dramatic increase in the yields of **824** along with a decrease in reaction time was achieved under ultrasonic irradiation.

MBH adduct **825**, obtained from *N*-Boc-α-amino aldehydes in the presence of DABCO upon ultrasound radiation at room temperature, has been treated with 2,2-dimethoxypropane in the presence of a catalytic amount of camphorsulfonic acid to successfully furnish the corresponding oxazolidine **826** in

91% yield (Scheme 4.238).[372] Chiral oxazolidines have been widely used as chiral auxiliaries in many stereoselective syntheses. They can also be employed in effectively directing the stereochemical course of a great variety of MBH reactions.[373]

With the development of the application of MBH adducts in the synthesis of cyclic products, other heteroatoms such as an S-atom have also been introduced into cyclic products. As by-product of a MBH reaction, 3-(α-hydroxybenzyl)thiochromanone **827** and 3-(α-hydroxybenzyl)selenochromanone **828** were obtained, respectively, in the chalcogeno MBH reaction of 2-(methylchalcogeno)phenyl vinyl ketones with aldehydes (Scheme 4.239).[374] More recently, MBH bromides **829** have been treated with thiourea in the presence of a base such as $NaHCO_3$ to furnish 1,3-thiazin-4-ones **830** in 75–91% yields (Scheme 4.240).[375]

Subsequently, a one-pot three-component highly stereoselective synthesis of hitherto unknown trisubstituted (*E*)- and (*Z*)-allyl dithiocarbamates **831** and **832** in water at room temperature without using a catalyst was disclosed (Scheme 4.241).[376] This one-pot procedure was performed simply by stirring a mixture of an amine, CS_2, and an acetate of acrylonitrile/acrylate ester-derived MBH adduct in water at room temperature for 6–10 h to afford the corresponding allyl dithiocarbamates **831** and **832** in 83–94% yields along with

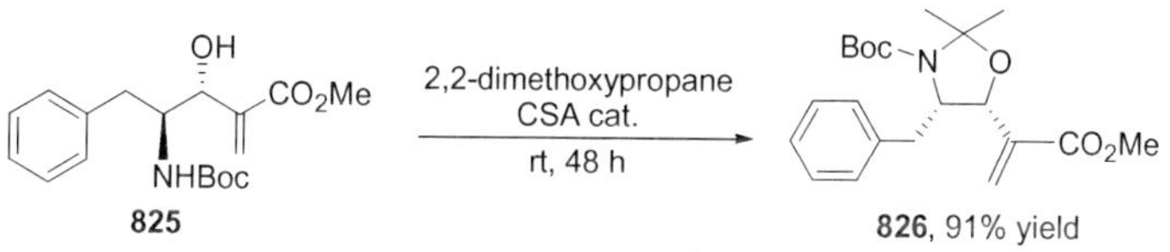

Scheme 4.238

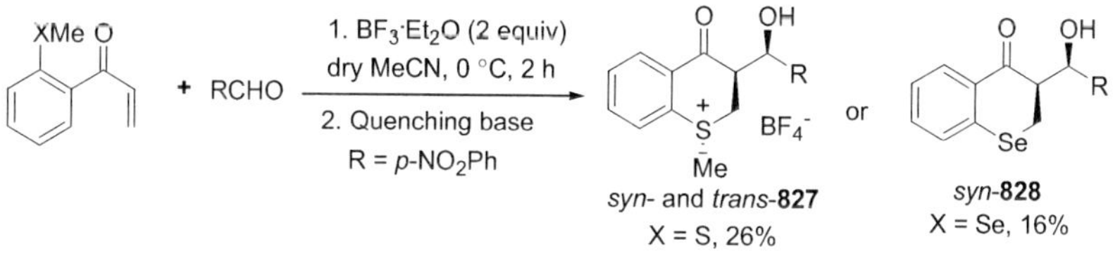

Scheme 4.239

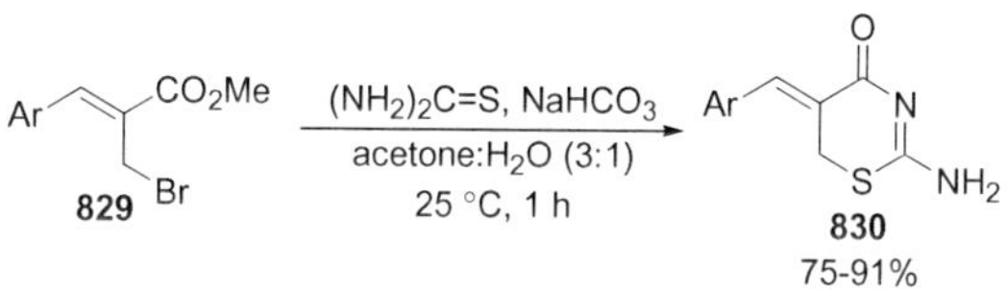

Scheme 4.240

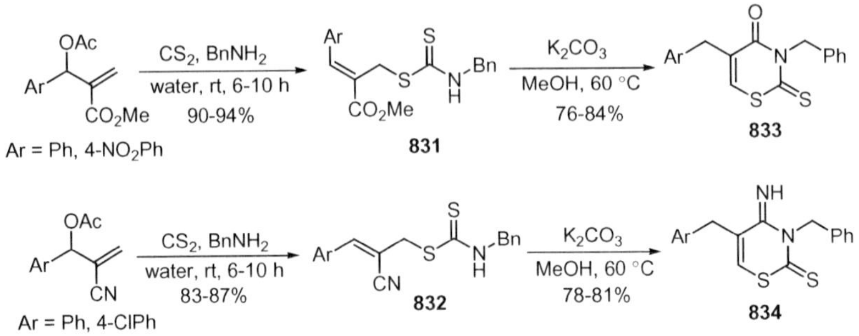

Scheme 4.241

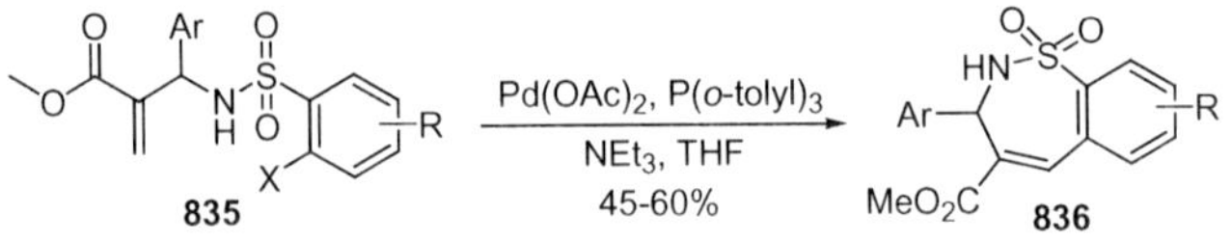

Scheme 4.242

sole diastereoselectivities. These allyl dithiocarbamates (**831** and **832**) can undergo an intramolecular cyclization reaction in the presence of K_2CO_3 in methanol upon heating the reaction mixture at 60 °C to yield the chemically and pharmaceutically interesting entities 3,5-dibenzyl-1,3-thiazines **833** and **834**.

An intramolecular Heck coupling reaction of aza-MBH products obtained from *ortho*-halobenzaldehydes has been conducted in the presence of the $Pd(OAc)_2/P(o\text{-tolyl})_3$ catalytic system with triethylamine, giving conformationally constrained scaffolds **836** in 45–60% yields (Scheme 2.242).[377]

Bis-MBH adduct **837**, based on a ferrocene framework, has been heated with an excess of propargyl alcohol and 100% w/w montmorillonite K10 clay under reflux for 12 h to afford the corresponding bis-isomerized adduct **838** in 68% yield. A subsequent Eglinton coupling reaction of propargyl derivative **838** gave the highly functionalized 36-membered macrocycle diyne ether derivative **839** in 43% yield (Scheme 4.243).[378] Moreover, the bis-allyl derivative **840** of ferrocene underwent a ring closing metathesis with Grubbs II generation catalyst to form ferrocenophane **841** in 40% yield.

MBH adduct **842** has subjected to epoxidation with *m*-CPBA followed by a Swern oxidation to afford aldehyde **844** (Scheme 4.244).[379] Wittig methodology using (iodomethylene)triphenylphosphorane gave iodide **845** in 34% overall yield. Iodide **845** was subsequently treated with sodium benzylselenoate to afford **846** and then free-radical mediated ring closure as well as ester hydrolysis afforded selenophene-3-carboxylic acid **847** in 26% yield within two steps. Acid

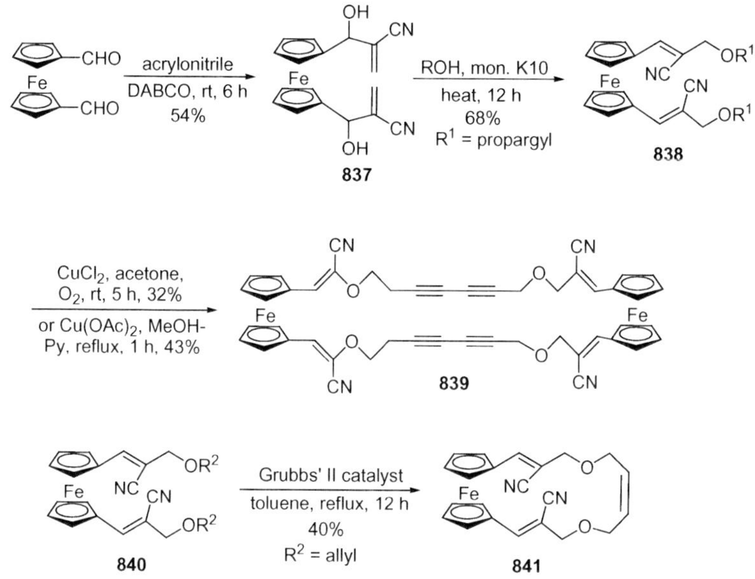

Scheme 4.243

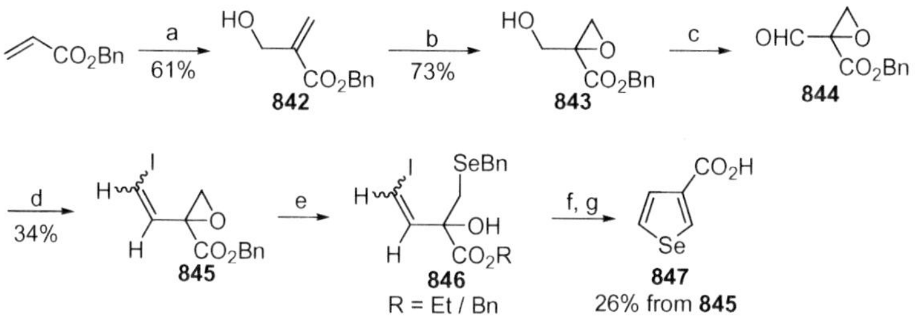

(a) $(CHO)_n$, NMe_3 (aq), H_2O, 60 °C; (b) *m*-CPBA, 3-tert-butyl-4-hydroxy-5-methylphenyl sulfide, CCl_4, reflux, 73%; (c) $(COCl)_2$, DMSO, NEt_3, CH_2Cl_2, -78 °C; (d) $(Ph_3PCH_2I)^+I$, NaHMDS, HMPA, THF, -78 °C to rt; (e) $(SeBn)_2$, $NaBH_4$, EtOH, rt; (f) TTMSS, AIBN, benzene, reflux; (g) LiOH, THF, H_2O.

Scheme 4.244

847 has been used as a useful synthetic intermediate in the preparation of the antitumor agent selenophenfurin[380] and the procedure described above represents a novel synthesis of this useful compound.

References

1. (a) M. G. P. Buffat, *Tetrahedron*, 2004, **60**, 1701; (b) J. P. Michael, *Nat. Prod. Rep.*, 2004, **21**, 625; (c) J. Christoffers and A. Mann, *Angew. Chem. Int. Ed.*, 2001, **40**, 4591; (d) D. O'Hagan, *Nat. Prod. Rep.*, 2000, **17**,

435; (e) E. J. Corey and A. Guzman-Perez, *Angew. Chem. Int. Ed.*, 1998, **37**, 388; (f) K. Fuji, *Chem. Rev.*, 1993, **93**, 2037.

2. (a) R. Faust, *Angew. Chem. Int. Ed.*, 2001, **40**, 2251; (b) M. P. Doyle, *Chem. Rev.*, 1986, **86**, 919; (c) V. K. Singh, A. DattaGupta and G. Sekar, *Synthesis*, 1997, 137; (d) W. A. Donaldson, *Tetrahedron*, 2001, **57**, 8589; (e) J. Salaun, *Chem. Rev.*, 1989, **89**, 1247; (f) P. Helquist, in *Comprehensive Organic Synthesis: Stereoselectivity, Strategy, and Efficiency in Modern Organic Chemistry*, ed. B. M. Trost and I. Felming, Pergamon Press, New York, 1991, **vol. 3**, ch. 4.6; (g) R. E. Taylor, F. C. Engelhardt and M. J. Schmitt, *Tetrahedron*, 2003, **59**, 5623; (h) H. Lebel, J.-F. Marcoux, C. Molinaro and A. B. Charette, *Chem. Rev.*, 2003, **103**, 977.

3. K. Miura, D. Wang and A. Hosomi, *J. Am. Chem. Soc.*, 2005, **127**, 9366.

4. (a) Y. Ueno, M. Ohta and M. Okawara, *Tetrahedron Lett.*, 1982, **23**, 2577; (b) N. Isono and M. Mori, *J. Org. Chem.*, 1996, **61**, 7867; (c) H. G. Kuivila and N. M. Scarpa, *J. Am. Chem. Soc.*, 1970, **92**, 6990; (d) D. D. Davis, R. L. Chambers and H. T. Johnson, *J. Organomet. Chem.*, 1970, **25**, C13–C16.

5. (a) R. Grigg, B. Putnicovic and C. J. Urch, *Tetrahedron Lett.*, 1996, **37**, 695; (b) K. Okada, M. Kondo, H. Tanino, H. Kakoi and S. Inoue, *Heterocycles*, 1992, 589; (c) E. Wenkert and S. Liu, *Synthesis*, 1992, 323; (d) D. J. Hart and S. C. Wu, *Tetrahedron Lett.*, 1991, **32**, 4099; (e) C. J. Flann, L. E. Overman and A. K. Sarkar, *Tetrahedron Lett.*, 1991, **32**, 6993; (f) R. Grigg, P. Stevenson and T. Worakun, *Tetrahedron*, 1988, **44**, 2049; (g) I. Fleming, M. A. Loreto, I. H. M. Wallace and J. P. Michael, *J. Chem. Soc., Perkin Trans. 1*, 1986, 349; (h) K. Jones, M. Thompson and C. Wright, *J. Chem. Soc., Chem. Commun.*, 1986, 115; (i) K. Joshi, R. Jain and P. Chand, *Heterocyles*, 1985, **23**, 957; (j) C. G. Richael and D. E. Thurston, *Tetrahedron*, 1983, **39**, 1817; (k) K. Okada, H. Sakuma, M. Kondo and S. Inoue, *Chem. Lett.*, 1979, 213; (l) K. Kieslich, *Justus Liebigs Ann. Chem.*, 1978, **5**, 717.

6. P. Shanmugam, V. Vaithiyanathan and B. Viswambharan, *Tetrahedron*, 2006, **62**, 4342.

7. S. Gessler, S. Randl and S. Blechert, *Tetrahedron Lett.*, 2000, **41**, 9973.

8. B. M. Trost and M. K. Brennan, *Org. Lett.*, 2007, **9**, 3961.

9. (a) O. I. Kolodiazhnyi, *Phosphorus Ylides: Chemistry and Application in Organic Synthesis*, Wiley-VCH Verlag GmbH, Weinheim, 1999; (b) B. E. Maryanoff and A. B. Reitz, *Chem. Rev.*, 1989, **89**, 863; (c) A.-H. Li, L.-X. Dai and V. K. Aggarwal, *Chem. Rev.*, 1997, **97**, 2341.

10. (a) G. Büchi and H. Wüest, *Helv. Chim. Acta*, 1971, **54**, 1767; (b) F. Bohlmann and C. Zdero, *Chem. Ber.*, 1973, **106**, 3779; (c) W. G. Dauben and J. Ipaktschi, *J. Am. Chem. Soc.*, 1973, **95**, 5088; (d) A. Padwa and L. Brodsky, *J. Org. Chem.*, 1974, **39**, 1318; (e) D. F. Schneider and A. C. Venter, *Synth. Commun.*, 1999, **29**, 1303.

11. (a) W. G. Dauben, D. J. Hart, J. Ipaktschi and A. P. Kozikowski, *Tetrahedron Lett.*, 1973, **14**, 4425; (b) W. G. Dauben and A. P. Kozikowski, *Tetrahedron Lett.*, 1973, **14**, 3711.

12. Y. Du, X. Lu and C. Zhang, *Angew. Chem. Int. Ed.*, 2003, **42**, 1035.

13. (a) X. Lu, C. Zhang and Z. Xu, *Acc. Chem. Res.*, 2001, **34**, 535; (b) B. M. Trost and U. Kazmaier, *J. Am. Chem. Soc.*, 1992, **114**, 7933; (c) C. Guo and X. Lu, *J. Chem. Soc., Perkin. Trans.* I, 1993, 1921; (d) J. Inanaga, Y. Baba and T. Hanamoto, *Chem. Lett.*, 1993, 241; (e) B. M. Trost and C.-J. Li, *J. Am. Chem. Soc.*, 1994, **116**, 3167; (f) B. M. Trost and C.-J. Li, *J. Am. Chem. Soc.*, 1994, **116**, 10819; (g) C. Zhang and X. Lu, *J. Org. Chem.*, 1995, **60**, 2906; (h) B. M. Trost and G. R. Dake, *J. Am. Chem. Soc.*, 1997, **119**, 7595; (i) B. M. Trost and G. R. Dake, *J. Org. Chem.*, 1997, **62**, 5670; (j) Z. Xu and X. Lu, *J. Org. Chem.*, 1998, **63**, 5031.

14. L.-B. Han, F. Mirzaei and M. Tanaka, *Organometallics*, 2000, **19**, 722.

15. J. Feng, X. Lu, A. Kong and X. Han, *Tetrahedron*, 2007, **63**, 6035.

16. J. Hartung, M. E. Pulling, D. M. Smith, D. X. Yang and J. R. Norton, *Tetrahedron*, 2008, **64**, 11822.

17. D. Basavaiah and R. M. Reddy, *Tetrahedron Lett.*, 2001, **42**, 3025.

18. (a) J. N. Kim, H. J. Lee, K. Y. Lee and J. H. Gong, *Synlett*, 2002, **173**; (b) J. H. Gong, H. R. Kim, E. K. Ryu and J. N. Kim, *Bull. Korean Chem. Soc.*, 2002, **23**, 789; (c) J. M. Kim, K. Y. Lee and J. N. Kim, *Bull. Korean Chem. Soc.*, 2004, **25**, 328; (d) Y. M. Chung, J. H. Gong, T. H. Kim and J. N. Kim, *Tetrahedron Lett.*, 2001, **42**, 9023; (e) Y. J. Im, J. M. Kim, J. H. Mun and J. N. Kim, *Bull. Korean Chem. Soc.*, 2001, **22**, 349.

19. C. G. Lee, K. Y. Lee, S. Gowrisankar and J. N. Kim, *Tetrahedron Lett.*, 2004, **45**, 7409.

20. C. G. Lee, K. Y. Lee, S. Lee and J. N. Kim, *Tetrahedron*, 2005, **61**, 1493.

21. H. N. Lim, S.-H. Ji and K.-J. Lee, *Synthsis*, 2007, 2454.

22. (a) A. S. Carlstroem and T. Frejd, *J. Chem. Soc., Chem. Commun.*, 1991, **17**, 1216; (b) K. Akaji and Y. Kiso, *Tetrahedron Lett.*, 1997, **38**, 5185; (c) N. Pitt and D. Gani, *Tetrahedron Lett.*, 1999, **40**, 3811; (d) R. P. Rajamohan, V. Balraju, G. R. Madhavan, B. Biswadip and J. Iqbal, *Tetrahedron Lett.*, 2002, **44**, 353.

23. A. Vasudevan, P.-S. Tseng and S. W. Djuric, *Tetrahedron Lett.*, 2006, **47**, 8591.

24. D. Seomoon, K. Lee, H. Kim and P. H. Lee, *Chem. Eur. J.*, 2007, **13**, 5197.

25. (a) P. Nilsson, M. Larhed and A. Hallberg, *J. Am. Chem. Soc.*, 2001, **123**, 8217; (b) A. Stadler, B. H. Yousefi, D. Dallinger, P. Walla, E. Van Der Eycken, N. Kaval and C. O. Kappe, *Org. Proc. Res. Dev.*, 2003, **7**, 707.

26. T. E. Reynolds and K. A. Scheidt, *Angew. Chem. Int. Ed.*, 2007, **46**, 7806.

27. A. M. Zawisza, B. Ganchegui, I. González, S. Bouquillon, A. Roglans, F. Hénin and J. Muzart, *J. Mol. Catal. Chem.*, 2008, **283**, 140.

28. S. Tu, L.-H. Xu and C.-R. Yu, *Synth. Commun.*, 2008, **38**, 2662.

29. C. Grundke and H. M. R. Hoffmann, *Chem. Ber.*, 1987, **120**, 1461.

30. D. Basavaiah, T. K. Bharathi and V. V. L. Gowriswari, *Tetrahedron Lett.*, 1987, **28**, 4351.

31. P. V. Ramachandran, M. T. Rudd, T. E. Burghardt and M. V. R. Reddy, *J. Org. Chem.*, 2003, **68**, 9310.

32. S. Kobbelgaard, S. Brandes and K. A. Jørgensen, *Chem. Eur. J.*, 2008, **14**, 1464.
33. A. Chamakh and H. Amri, *Tetrahedron Lett.*, 1998, **39**, 375.
34. Y. J. Im, C. G. Lee, H. R. Kim and J. N. Kim, *Tetrahedron Lett.*, 2003, **44**, 2987.
35. X. Liu, J. Zhao, G. Jin, G. Zhao, S. Zhu and S. Wang, *Tetrahedron*, 2005, **61**, 3841.
36. M. J. Lee, D. Y. Park, K. Y. Lee and J. N. Kim, *Tetrahedron Lett.*, 2006, **47**, 1833.
37. S. Zheng and X. Lu, *Tetrahedron Lett.*, 2009, **50**, 4532.
38. S. GowriSankar, C. G. Lee and J. N. Kim, *Tetrahedron Lett.*, 2004, **45**, 6949.
39. H.-L. Cui, J. Peng, X. Feng, W. Du, K. Jiang and Y.-C. Chen, *Chem. Eur. J.*, 2009, **15**, 1574.
40. P. Wasnaire, M. Wiaux, R. Touillaux and I. E. Markó, *Tetrahedron Lett.*, 2006, **47**, 985.
41. (a) N. R. Andersen, P. R. Rasmussen, C. P. Falshaw and T. J. King, *Tetrahedron Lett.*, 1984, **25**, 469; (b) T. Tokoroyama, *Synthesis*, 2000, 611.
42. P. Wasnaire, T. de Merode and I. E. Markó, *Chem. Commun.*, 2007, 4755.
43. L. C. Blaszczak and J. E. McMurry, *J. Org. Chem.*, 1974, **39**, 258.
44. R. Manchanayakage, D. Omune, C. Hayes and S. T. Handy, *Tetrahedron*, 2007, **63**, 9691.
45. J. N. Kim, Y. J. Im, J. H. Gong and K. Y. Lee, *Tetrahedron Lett.*, 2001, **42**, 4195.
46. P. Narender, U. Srinivas, M. Ravinder, B. Ananda Rao, Ch. Ramesh, K. Harakishore, B. Gangadasu, U. S. N. Murthy and V. Jayathirtha Rao, *Bioorg. Med. Chem.*, 2006, **14**, 4600.
47. X. Bi, D. Dong, Q. Liu, W. Pan, L. Zhao and B. Li, *J. Am. Chem. Soc.*, 2005, **127**, 4578.
48. Q. Zhang, S. Sun, J. Hu, Q. Liu and J. Tan, *J. Org. Chem.*, 2007, **72**, 139.
49. (a) M. J. Lee, K. Y. Lee, S. Gowrisankar and J. N. Kim, *Tetrahedron Lett.*, 2006, **47**, 1355; (b) M. J. Lee, K. Y. Lee, D. Y. Park and J. N. Kim, *Tetrahedron*, 2006, **62**, 3128.
50. S. C. Kim, H. S. Lee, Y. J. Lee and J. N. Kim, *Tetrahedron Lett.*, 2006, **47**, 5681.
51. J. N. Kim, Y. J. Im and J. M. Kim, *Tetrahedron Lett.*, 2002, **43**, 6597.
52. D. Y. Park, S. J. Kim, T. H. Kim and J. N. Kim, *Tetrahedron Lett.*, 2006, **47**, 6315.
53. D. Y. Park, K. Y. Lee and J. N. Kim, *Tetrahedron Lett.*, 2007, **48**, 1633.
54. E. S. Kim, K. H. Kim, S. H. Kim and J. N. Kim, *Tetrahedron Lett.*, 2009, **50**, 5098.
55. (a) A. Citterio, D. Fancelli, C. Finzi, L. Pesce and R. Santi, *J. Org. Chem.*, 1989, **54**, 2713; (b) A. Citterio, R. Sebastiano, A. Marion and R. Santi, *J. Org. Chem.*, 1991, **56**, 5328; (c) B. B. Snider, *Chem. Rev.*, 1996, **96**, 339.
56. Y. J. Im, K. Y. Lee, T. H. Kim and J. N. Kim, *Tetrahedron Lett.*, 2002, **43**, 4675.

57. K. Y. Lee, S. Gowrisankar, Y. J. Lee and J. N. Kim, *Tetrahedron*, 2006, **62**, 8798.

58. (a) J. S. Yadav, B. V. S. Reddy, S. Abraham and G. Sabitha, *Tetrahedron Lett.*, 2002, **43**, 1565; (b) Y. L. Bennani, G.-D. Zhu and J. C. Freeman, *Synlett*, 1998, 754; (c) L. Dubois, A. Mehta, E. Tourette and R. H. J. Dodd, *Org. Chem.*, 1994, **59**, 434; (d) H. Stamm, A. Onistschenko, B. Buchholz and T. Mall, *J. Org. Chem.*, 1989, **54**, 193; (e) E. Vedejs, A. Klapars, D. L. Warner and A. H. Weiss, *J. Org. Chem.*, 2001, **66**, 7542.

59. K. Y. Lee, S. C. Kim and J. N. Kim, *Tetrahedron Lett.*, 2006, **47**, 977.

60. P.-Y. Chen, H.-M. Chen, L.-Y. Chen, J.-Y. Tzeng, J.-C. Tsai, P.-C. Chi, S.-R. Li and E.-C. Wang, *Tetrahedron*, 2007, **63**, 2824.

61. A. Weichert and H. M. R. Hoffman, *J. Org. Chem.*, 1991, **56**, 4098.

62. (a) F. Sato, H. Urabe and S. Okamoto, *Chem. Rev.*, 2000, **100**, 2835; (b) O. G. Kulinkovich and A. de Meijere, *Chem. Rev.*, 2000, **100**, 2789.

63. S. Okamoto, H. Ito, S. Tanaka and F. Sato, *Tetrahedron Lett.*, 2006, **47**, 7537.

64. S. Mix and S. Blechert, *Org. Lett.*, 2005, **7**, 2015.

65. R. Takagi, Y. Miwa, T. Nerio, Y. Inoue, S. Matsumura and K. Ohkata, *Org. Biomol. Chem.*, 2007, **5**, 286.

66. (a) G. Biggi, F. Deloma and F. Pietra, *J. Am. Chem. Soc.*, 1973, **95**, 7101; (b) F. Pietra, *Acc. Chem. Res.*, 1979, **12**, 132.

67. (a) K. N. Houk and N. G. Rondan, *J. Am. Chem. Soc.*, 1984, **106**, 3882; (b) L. A. Paquette, S. J. Hathaway and P. F. T. Schirch, *J. Org. Chem.*, 1985, **50**, 4199; (c) R. L. Funk and G. L. Bolton, *J. Am. Chem. Soc.*, 1986, **108**, 4655; (d) J. H. Rigby, *Org. React.*, 1997, **49**, 331; (e) K. Saito, S. Ando and Y. Kondo, *Heterocycles*, 2000, **53**, 2601; (f) L. Isakovic, J. A. Ashenhurst and J. L. Gleason, *Org. Lett.*, 2001, **3**, 4189; (g) R. Gompper and U. Wolf, *Liebigs Ann. Chem.*, 1979, 1388; (h) K. Hayakawa, H. Nishiyama and K. Kanematsu, *J. Org. Chem.*, 1985, **50**, 512; (i) B. M. Trost and P. R. Seoane, *J. Am. Chem. Soc.*, 1987, **109**, 615; (j) K. Kumar, A. Kapur and M. P. S. Ishar, *Org. Lett.*, 2000, **2**, 787; (k) R. P. Gandhi and M. P. S. Ishar, *Chem. Lett.*, 1989, 101; (l) M. P. S. Ishar and R. P. Gandhi, *Tetrahedron*, 1993, **49**, 6729.

68. Y. Du, J. Feng and X. Lu, *Org. Lett.*, 2005, **7**, 1987.

69. L.-W. Ye, X.-L. Sun, Q.-G. Wang and Y. Tang, *Angew. Chem. Int. Ed.*, 2007, **46**, 5951.

70. (a) J.-C. Wang, S.-S. Ng and M. J. Krische, *J. Am. Chem. Soc.*, 2003, **125**, 3682; (b) J.-C. Wang and M. J. Krische, *Angew. Chem.*, 2003, **115**, 6035; *Angew. Chem. Int. Ed.*, 2003, **42**, 5855.

71. L.-W. Ye, X. Han, X.-L. Sun and Y. Tang, *Tetrahedron*, 2008, **64**, 1487.

72. Y. G. Gololobov, A. N. Nesmeyanov, V. P. Lysenko and I. E. Boldeskul, *Tetrahedron*, 1987, **43**, 2609.

73. H. Akiyama, T. Fujimoto, K. Ohshima, K. Hoshino and I. Yamamoto, *Org. Lett.*, 1999, **1**, 427.

74. H. Akiyama, T. Fujimoto, K. Ohshima, K. Hoshino, Y. Saito, A. Okamoto, I. Yamamoto, A. Kakehi and R. Iriye, *Eur. J. Org. Chem.*, 2001, 2265.

75. S. GowriSankar, K. Y. Lee, C. G. Lee and J. N. Kim, *Tetrahedron Lett.*, 2004, **45**, 6141.

76. V. S. Singh and S. Batra, *Tetrahedron Lett.*, 2006, **47**, 7043.

77. (a) D. L. J. Clive, M. Yu and Z. Li, *Chem. Commun.*, 2005, 906; (b) D. L. J. Clive, Z. Li and M. Yu, *J. Org. Chem.*, 2007, **72**, 5608.

78. B. Prabhudas and D. L. J. Clive, *Angew. Chem. Int. Ed.*, 2007, **46**, 9295.

79. D. A. Spiegel, J. T. Njardarson, I. M. McDonald and J. L. Wood, *Chem. Rev.*, 2003, **103**, 2691.

80. J. Méndez-Andino and L. A. Paquette, *Adv. Synth. Catal.*, 2002, **344**, 303.

81. L. Banfi, A. Bernardi, L. Colombo, C. Gennari and C. Scolastico, *J. Org. Chem.*, 1984, **49**, 3784.

82. E. V. Ramachandran and M. E. Krzeminski, *Tetrahedron Lett.*, 1999, **40**, 7879.

83. L. D. S. Yadav and C. Awasthi, *Tetrahedron Lett.*, 2009, **50**, 715.

84. (a) T. Katsuki, in *Comprehensive Asymmetric Catalysis*, ed. E. N. Jacobsen, A. Pfaltz and H. Yamamoto, Springer, New York, 1999, ch. 18.1; (b) T. Katsuki and V. S. Martín, *Org. React.*, 1996, **48**, 1; (c) *Catalytic Asymmetric Synthesis*, ed. I. Ojima , Wiley-VCH Verlag GmbH, Weinheim, 2000; (d) E. N. Jacobsen and M. H. Wu, in *Comprehensive Asymmetric Catalysis*, ed. E. N. Jacobsen, A. Pfaltz and H. Yamamoto, Springer, New York, 1999, ch. 18.2; (e) M. Frohn and Y. Shi, *Synthesis*, 2000, 1979.

85. (a) J. P Marino, L. J. Anna, R. F. de la Pradilla, M. V. Martínez, C. Montero and A. Viso, *J. Org. Chem.*, 2000, **65**, 6462; (b) R. F. de la Pradilla, M. V. Buergo, M. V. Martínez, C. Montero, M. Tortosa and A. Viso, *J. Org. Chem.*, 2004, **69**, 1978.

86. R. F. de la Pradilla, J. Fernández, P. Manzano, P. Méndez, J. Priego, M. Tortosa, A. Viso, M. Martínez-Ripoll and A. Rodríguez, *J. Org. Chem.*, 2002, **67**, 8166.

87. R. F. de la Pradilla, A. Castellanos, J. Fernández, M. Lorenzo, P. Manzano, P. Méndez, J. Priego and A. Viso, *J. Org. Chem.*, 2006, **71**, 1569.

88. Y. Iwabuchi, M. Furukawa, T. Esumi and S. Hatakeiyama, *Chem. Commun.*, 2001, 2030.

89. R. S. Porto, M. L. A. A. Vasconcellos, E. Ventura and F. Coelho, *Synthesis*, 2005, 2297.

90. J. Švenda and A. G. Myers, *Org. Lett.*, 2009, **11**, 2437.

91. B. Das, H. Holla, K. Venkateswarlu and A. Majhi, *Tetrahedron Lett.*, 2005, **46**, 8895.

92. I. Martínez, A. E. Andrews, J. D. Emch, A. J. Ndakala, J. Wang and A. R. Howell, *Org. Lett.*, 2003, **5**, 399.

93. (a) Y. S. Rao, *Chem. Rev.*, 1976, **76**, 625; (b) H. M. R. Hoffmann and J. Rabe, *Angew. Chem., Int. Ed. Engl.*, 1985, **24**, 94; (c) P. A. Grieco, *Synthesis*, 1975, **67**; (d) Y. Ohfune, P. A. Grieco, C. L. J. Wang and

G. Maetich, *J. Am. Chem. Soc.*, 1978, **100**, 5946; (e) J. Banerji and B. Das, *Heterocycles*, 1985, **23**, 661; (f) E. Lee, C. U. Hur, Y. C. Geong, Y. H. Rhee and M. H. Chang, *J. Chem. Soc., Chem. Commun.*, 1991, **9**, 1314; (g) G. L. Larson and M. B. Betancourtde Perez, *J. Org. Chem.*, 1985, **50**, 5257; (h) M. Tanaka, C. Mukaiyama, H. Mitsuhashi and T. Wakamatsu, *Tetrahedron Lett.*, 1992, **33**, 4165; (i) M. Tanaka, H. Mitsuhashi and T. Wakamatsu, *Tetrahedron Lett.*, 1992, **33**, 4161; (j) D. B. Berkowitz, S. Choi and J.-H. Maeng, *J. Org. Chem.*, 2000, **65**, 847; (k) M.-J. Chen, C.-Y. Lo, C.-C. Chin and R.-S. Liu, *J. Org. Chem.*, 2000, **65**, 6362.

94. (a) R. Grigg and V. Savic, *Chem. Commun.*, 2000, 2381; (b) D. Johnston, N. Francon, D. J. Edmonds and D. J. Procter, *Org. Lett.*, 2001, **3**, 2001; (c) S. V. Gagnier and R. C. Larock, *J. Org. Chem.*, 2000, **65**, 1525; (d) R. Rossi, F. Bellina, C. Bechini, L. Mannina and P. Vergamini, *Tetrahedron*, 1998, **54**, 135; (e) R. Ballini, E. Marcantoni and S. Perella, *J. Org. Chem.*, 1999, **64**, 2954; (f) K.-W. Liang, W.-T. Li, S.-M. Peng, S.-L. Wang and R.-S. Liu, *J. Am. Chem. Soc.*, 1997, **119**, 4404; (g) K. Lee, J. A. Jackson and D. F. Wiemer, *J. Org. Chem.*, 1993, **58**, 5967; (h) L. D. Martin and J. K. Stille, *J. Org. Chem.*, 1982, **47**, 3630; (i) T. Minami, I. Niki and T. Agawa, *J. Org. Chem.*, 1974, **39**, 3236; (j) H. Zimmer and J. Rothe, *J. Org. Chem.*, 1959, **24**, 28; (k) B. Liu, M.-J. Chen, C.-Y. Lo and R.-S. Liu, *Tetrahedron Lett.*, 2001, **42**, 2533; (l) M.-J. Chen, C.-Y. Lo and R.-S. Liu, *Synlett*, 2000, 1205; (m) Y. Masuyama, Y. Nimura and Y. Kurusu, *Tetrahedron Lett.*, 1991, **32**, 225.

95. A. Bernardi, M. G. Beretta, L. Colombo, C. Gennari, G. Poli and C. Scolastico, *J. Org. Chem.*, 1985, **50**, 4442.

96. P. R. Krishna, V. Kannan and G. V. M. Sharma, *J. Org. Chem.*, 2004, **69**, 6467.

97. S. GowriSankar, C. G. Lee and J. N. Kim, *Tetrahedron Lett.*, 2004, **45**, 6949.

98. (a) L. J. Brzezinski, S. Rafel and J. W. Leahy, *J. Am. Chem. Soc.*, 1997, **119**, 4317; (b) P. V. Ramachandran, M. T. Rudd, T. E. Burghardt and M. V. R. Reddy, *J. Org. Chem.*, 2003, **68**, 9310; (c) R. S. Porto and F. Coelho, *Synth. Commun.*, 2004, **34**, 3037.

99. M. Shimizu, M. Kimura, S. Tanaka and Y. Tamaru, *Tetrahedron Lett.*, 1998, **39**, 609.

100. A.-C. L. Lamer, N. Gouault, M. David, J. Boustie and P. Uriac, *J. Comb. Chem.*, 2006, **8**, 643.

101. A. G. González, M. H. Silva, J. I. Padrón, F. León, E. Reyes, M. Álvarez-Mon, J. P. Pivel, J. Quintana, F. Estévez and J. Bermejo, *J. Med. Chem.*, 2002, **45**, 2358.

102. P. V. Ramachandran, D. Pratihar, D. Biswas, A. Srivastava and M. V. R. Reddy, *Org. Lett.*, 2004, **6**, 481.

103. S. H. Yu, M. J. Fergusaon, R. McDonald and D. G. Hall, *J. Am. Chem. Soc.*, 2005, **127**, 12808.

104. (a) A. R. Carroll and W. C. Taylor, *Aust. J. Chem.*, 1991, **44**, 1615; (b) A. R. Carroll and W. C. Taylor, *Aust. J. Chem.*, 1991, **44**, 1705.

105. G. W. Kabalka, B. Venkataiah and C. Chen, *Tetrahedron Lett.*, 2006, **47**, 4187.

106. A. Masunari, E. Ishida, G. Trazzi, W. P. Almeida and F. Coelho, *Synth. Commun.*, 2001, **31**, 2127.

107. (a) R. S. Ward, *Tetrahedron*, 1990, **46**, 5029; *Synthesis*. 1992, 719; (b) D. A. Whiting, *Nat. Prod. Rep.*, 1990, **7**, 349; (c) J. L. Charlton and G.-L. Chee, *Can. J. Chem.*, 1997, **75**, 1076; (d) S. V. Bhat, R. P. Tanpure and S. B. Hadimani, *Tetrahedron Lett.*, 1996, **37**, 4791; (e) H. Ishibashi, K. Ito, T. Hirano, M. Tabuchi and M. Ikeda, *Tetrahedron*, 1993, **49**, 4173.

108. (a) I. Jardine, in *Anticancer Agents Based on Natural Products*, ed. J. M. Cassady and J. D. Douros, Academic Press: New York, 1980, ch. 9; (b) S. G. Weiss, M. Tin-Wa, R. E. Perdue and N. R. Farnsworth, *J. Pharm. Sci.*, 1975, **64**, 95; (c) C. Keller-Juslen, M. Kuhn, A. von Wartburg and H. Stahelin, *J. Med. Chem.*, 1971, **14**, 936; (d) R. S. Ward, *Nat. Prod. Rep.*, 1999, **16**, 75.

109. F. Coelho, D. Verones, C. H. Pavam, W. I. de Paula and R. Buffon, *Tetrahedron*, 2005, **68**, 4563.

110. L. K. Kohn, C. H. Pavam, D. Veronese, F. Coelho, J. E. D. Carvalho and W. P. Almeida, *Eur. J. Med. Chem.*, 2006, **41**, 738.

111. P. R. Krishna and M. Narsingam, *J. Comb. Chem.*, 2007, **9**, 62.

112. A. K. Chatterjee, J. P. Morgan, M. Scholl and R. H. Grubbs, *J. Am. Chem. Soc.*, 2000, **122**, 3783.

113. J. M. Kim, K. Y. Lee, S. Lee and J. N. Kim, *Tetrahedron Lett.*, 2004, **45**, 2805.

114. (a) T. J. Donohoe, A. J. Orr, K. Gosby and M. Bingham, *Eur. J. Org. Chem.*, 2005, **1969**; (b) T. J. Donohoe, N. M. Kershaw, A. J. Orr, K. M. P. Wheelhouse, L. P. Fishlock, A. R. Lacy, M. Bingham and P. A. Procopiou, *Tetrahedron*, 2008, **64**, 809.

115. (a) B. Alcaide, P. Almendros and C. Aragoncillo, *J. Org. Chem.*, 2001, **66**, 1612; (b) B. Alcaide, P. Almendros and C. Aragoncillo, *Chem. Commun.*, 1999, **1913**; (c) E. P. Kundig, L. He-Xu and P. Romanens, *Tetrahedron Lett.*, 1995, **36**, 4047.

116. P. Shanmugam and P. Rajasingh, *Tetrahedron*, 2004, **60**, 9283.

117. P. Shanmugam, P. Rajasingh, B. Viswambharan and V. Vaithiyanathan, *Synth. Commun.*, 2007, **37**, 2291.

118. S. Gowrisankar, K. Y. Lee and J. N. Kim, *Tetrahedron Lett.*, 2005, **46**, 4859.

119. S. Gowrisankar, K. Y. Lee and J. N. Kim, *Tetrahedron*, 2006, **62**, 4052.

120. M. Frezza, D. Balestrino, L. Soulère, S. Reverchon, Y. Queneau, C. Forestier and A. Doutheau, *Eur. J. Org. Chem.*, 2006, 4731.

121. B. M. Trost, H.-C. Tsui and F. D. Toste, *J. Am. Chem. Soc.*, 2000, **122**, 3534.

122. (a) R. Fernández de la Pradilla, C. Montero, J. Priego and L. Martínez-Cruz, *A. J. Org. Chem.*, 1998, **63**, 9612; (b) R. Fernández de la Pradilla, P. Manzano, C. Montero, J. Priego, M. Martínez-Ripoll and L. A. Martínez -Cruz, *J. Org. Chem.*, 2003, **68**, 7755; (c) R. Fernández de la

Pradilla, A. Castellanos, J. Fernández, M. Lorenzo, P. Manzano, P. Méndez, J. Priego and A. Viso, *J. Org. Chem.*, 2006, **71**, 1569; (d) R. Fernández de la Pradilla, J. Fernández, A. Viso, J. Fernández and A. Gómez, *Heterocycles*, 2006, **68**, 1579.

123. R. Fernández de la Pradilla, P. Manzano, C. Montero, J. Priego, M. Martínez-Ripoll and L. A. Martínez-Cruz, *J. Org. Chem.*, 2003, **68**, 7755.

124. (a) R. Fernández de la Pradilla and A. Castellanos, *Tetrahedron Lett.*, 2007, **48**, 6500; (b) R. Fernández de la Pradilla, A. Castellanos, I. Osante, I. Colomer and M. I. Sánchez, *J. Org. Chem.*, 2009, **74**, 170.

125. B. Alcaide, P. Almendros, T. M. del Campo and M. T. Quirós, *Chem. Eur. J.*, 2009, **15**, 3344.

126. A. Weichert and H. M. R. Hoffman, *J. Org. Chem.*, 1991, **56**, 4098.

127. (a) Q. Cheng and J. K. Snyder, *J. Org. Chem.*, 1988, **53**, 4562; (b) R. G. Salomon, K. Lal, S. M. Mazza, E. A. Zarate and W. J. Youngs, *J. Am. Chem. Soc.*, 1988, **110**, 5213; (c) M. Krause and H. M. R. Hoffmann, *Tetrahedron Lett.*, 1990, **31**, 6629.

128. (a) T. D. J. DSilva and D. W. Peck, *J. Org. Chem.*, 1972, **37**, 1828; (b) B. P. Mundy, R. D. Otzenberger and A. R. DeBernadis, *J. Org. Chem.*, 1971, **36**, 2390.

129. P. Hayes and C. Maignan, *Tetrahedron: Asymmetry*, 1999, **10**, 1041.

130. (a) D. Basavaiah, M. Bakthadoss and S. Pandiaraju, *Chem. Commun.*, 1998, 1839; (b) A. Marx, R. Suresh, C. C. Kanakam, V. Manivannan and C. S. Vasam, *Acta Crystallogr., Sect. E*, 2008, **64**, 27.

131. (a) D. D. McPherson, G. A. Cordell, D. D. Soejarto, J. M. Pezzuto and H. H. S. Fong, *Phytochemistry*, 1983, **22**, 2835; (b) W. T. L. Sidwell and C. Tamm, *Tetrahedron Lett.*, 1970, **11**, 475; (c) P. Bohler and C. Tamm, *Tetrahedron Lett.*, 1967, **8**, 3479.

132. T. Al-Nakib, V. Benjak, M. J. Meegan and R. Chandy, *Eur. J. Med. Chem.*, 1990, **25**, 455; *Chem. Abstr.*, 1990, **113**, 231160v.

133. P. Shanmugam and P. Rajasingh, *Synlett*, 2005, 939.

134. S. Gowrisankar, K. Y. Lee and J. N. Kim, *Tetrahedron Lett.*, 2005, **46**, 4859.

135. S. Gowrisankar, K. Y. Lee, T. H. Kim and J. N. Kim, *Tetrahedron Lett.*, 2006, **47**, 5785.

136. J. S. Yadav, B. V. Subba Reddy, S. S. Mandal, A. K. Basak, C. Madavi and A. C. Kunwar, *Synlett*, 2008, 1175.

137. B. M. Trost, M. R. Machacek and H. C. Tsui, *J. Am. Chem. Soc.*, 2005, **127**, 7014.

138. (a) P. A. Grieco, *Synthesis*, 1975, 67; (b) G. M. Ksander, J. E. McMurry and M. Johnson, *J. Org. Chem.*, 1977, **42**, 1180; (c) E. Ghera, T. Yechezkel and A. Hassner, *J. Org. Chem.*, 1990, **55**, 5977; (d) I. Paterson and I. Fleming, *Tetrahedron Lett.*, 1979, **20**, 993; (e) M. Ochiai, E. Fujita, M. Arimoto and H. Yamaguchi, *Tetrahedron Lett.*, 1983, **24**, 777; (f) M. R. Carlson and L. L. White, *Synth. Commun.*, 1983, **13**, 237; (g) M. Mori, Y. Washioka, T. Urayama, K. Yoshiura, K. Chiba and Y. Ban, *J. Org. Chem.*, 1983, **48**, 4058; (h) E. Lee, C. U. Hur and C. M. Park, *Tetrahedron*

Lett., 1990, **35**, 5039; (i) K. Nishitani, Y. Harada, Y. Nakamura, K. Yokoo and K. Yamakawa, *Tetrahedron Lett.*, 1994, **35**, 7809; (j) A. Gupta and Y. D. Vankar, *Tetrahedron*, 2000, **56**, 8525; (k) H. Krawczyk and M. Sliwinski, *Tetrahedron*, 2003, **59**, 9199; (l) H. Krawczyk, M. Sliwinski, W. M. Wolf and R. Bodalski, *Synlett*, 2004, 1995.

139. (a) S. M. Kupchan, R. J. Hemingway, D. Werner, A. Karim, A. T. McPhail and G. A. Sim, *J. Am. Chem. Soc.*, 1968, **90**, 3596; (b) X.-B. Liao, J.-Y. Han and Y. Li, *Tetrahedron Lett.*, 2001, **42**, 2843; (c) S. Ekthawatchai, S. Kamchonwongpaisan, P. Kongsaeree, B. Tarnchompoo, Y. Thebtaranonth and Y. Yuthavong, *J. Med. Chem.*, 2001, **44**, 4688; (d) M. A. Avery, M. Alvim-Gaston, J. A. Vroman, B. Wu, A. Ager, W. Peters, B. L. Robinson and W. Charman, *J. Med. Chem.*, 2002, **45**, 4321; (e) D. E. Cane and T. Rossi, *Tetrahedron Lett.*, 1979, **20**, 2973; (f) A. Nangia, G. Prasuna and P. B. Rao, *Tetrahedron*, 1997, **53**, 14507; (g) J. E. McMurry and R. G. Dushin, *J. Am. Chem. Soc.*, 1990, **112**, 6942.

140. J.-H. Cho and S. Y. Ko, *Helv. Chim. Acta*, 2002, **85**, 3994.

141. E. J. Kim and S. Y. Ko, *Bioorg. Med. Chem.*, 2005, **13**, 4103.

142. V. Singh and S. Batra, *Synthesis*, 2006, 63.

143. V. Singh, S. Madapa and S. Batra, *Synth. Commun.*, 2008, **38**, 2113.

144. S. K. Mandal, M. Paira and S. C. Roy, *J. Org. Chem.*, 2008, **73**, 3823.

145. S. J. Kim, H. S. Lee and J. N. Kim, *Tetrahedron Lett.*, 2007, **48**, 1069.

146. D. Basavaiah and T. Satyanarayana, *Org. Lett.*, 2001, **3**, 3619.

147. P. T. Kaye and M. A. Musa, *Synth. Commun.*, 2003, **33**, 1755.

148. R. D. H. Murray, in *Progress in the Chemistry of Natural Products*, ed. W. Herz, H. Grisebach and G. W. Kirby, Springer-Verlag, Wein–New York, 1978, p. 199.

149. W. P. Hong and K.-J. Lee, *Synthesis*, 2005, 33.

150. T. Keumi, K. Matsuura, N. Nakayama, T. Tsubota, T. Morita, I. Takahashi and H. Kitajima, *Tetrahedron*, 1993, **49**, 537.

151. W. Zhong, Y. Zhao and W. Su, *Tetrahedron*, 2008, **64**, 5491.

152. T. E. Reynolds and K. A. Scheidt, *Angew. Chem. Int. Ed.*, 2007, **46**, 7806.

153. C.-R. Piao, Y.-L. Zhao, X.-D. Han and Q. Liu, *J. Org. Chem.*, 2008, **73**, 2264.

154. (a) Y. Yin, M. Wang, Q. Liu, J. Hu, S. Sun and J. Kang, *Tetrahedron Lett.*, 2005, **46**, 4399; (b) Q. Zhang, Y. Liu, M. Wang, Q. Liu, J. Hu and Y. Yin, *Synthesis*, 2006, 3009; (c) S. Sun, Q. Zhang, Q. Liu, J. Kang, Y. Yin, D. Li and D. Dong, *Tetrahedron Lett.*, 2005, **46**, 6271.

155. E. S. Kim, K. H. Kim, S. H. Kim and J. N. Kim, *Tetrahedron Lett.*, 2009, **50**, 5098.

156. (a) D.-C. Oh, E. A. Gontang, C. A. Kauffman, P. R. Jensen and W. Fenical, *J. Nat. Prod.*, 2008, **71**, 570; (b) K. Trisuwan, V. Rukachaisirikul, Y. Sukpondma, S. Preedanon, S. Phongpaichit, N. Rungjindamai and J. Sakayaroj, *J. Nat. Prod.*, 2008, **71**, 1323; (c) A. H. Aly, R. Edrada-Ebel, V. Wray, W. E. G. Muller, S. Kozytska, U. Hentschel, P. Proksch and R. Ebel, *Phytochemistry*, 2008, **69**, 1716; (d) A. Cutignano, A. Fontana, L. Renzulli and G. Cimino, *J. Nat. Prod.*, 2003, **66**, 1399; (e) N. Sata,

H. Abinsay, W. Y. Yoshida, F. D. Horgen, N. Sitachitta, M. Kelly and P. J. Scheuer, *J. Nat. Prod.*, 2005, **68**, 1400.

157. (a) M. Yus, T. Soler and F. Foubelo, *Tetrahedron*, 2002, **58**, 7009; (b) C. A. Maier and B. Wunsch, *Eur. J. Org. Chem.*, 2003, **714**; (c) R. Grigg, V. Savic, V. Sridharan and C. Terrier, *Tetrahedron*, 2002, **58**, 8613.

158. D. Basavaiah, D. S. Sharada and V. Veerendhar, *Tetrahedron Lett.*, 2004, **45**, 3081.

159. B. Das, A. Majhi, J. Banerjee, N. Chowdhury, H. Holla, K. Harakishore and U. S. Murty, *Chem. Pharm. Bull.*, 2006, **54**, 403.

160. P. Shanmugam and P. Rajasingh, *Tetrahedron Lett.*, 2005, **46**, 3369.

161. (a) M. Harmata, *Adv. Synth. Catal.*, 2006, **348**, 2297; (b) M. Harmata, *Acc. Chem. Res.*, 2001, **34**, 595.

162. S. Zheng and X. Lu, *Org. Lett.*, 2009, **11**, 3978.

163. T. P. Majhi, A. Neogi, S. Ghosh, A. K. Mukherjee and P. Chattopadhyay, *Tetrahedron*, 2006, **62**, 12003.

164. (a) H. Uprety and D. S. Bhakuni, *Tetrahedron Lett.*, 1975, **16**, 1201; (b) S. Brandt, A. Marfat and P. Helquist, *Tetrahedron Lett.*, 1979, **20**, 2193; (c) H. G. Theuns, H. B. N. Lenting, C. A. Salemink, H. Tanaka, M. Shibata, K. Ito and R. J. J. C. Lousberg, *Phytochemistry*, 1984, **23**, 1157; (d) J. B. Bremner, W. Jaturonrusmee, L. M. Engelhardt and A. H. White, *Tetrahedron Lett.*, 1989, **30**, 3213.

165. T. P. Majhi, A. Neogi, S. Ghosh, A. K. Mukherjee, M. Helliwell and P. Chattopadhyay, *Synthesis*, 2008, 94.

166. W. McCoull and F. A. Davis, *Synthesis*, 2000, **10**, 1347.

167. A. Skancke, D. van Vechten, J. F. Liebman and P. N. Skancke, *J. Mol. Struct.*, 1996, **376**, 461.

168. H. C. Kolb, M. G. Finn and K. B. Sharpless, *Angew. Chem., Int. Ed.*, 2001, **40**, 2004.

169. (a) K.-S. Yang and K. Chen, *Org. Lett.*, 2000, **2**, 729; (b) C. L. Fang, W.-D. Lee, N. W. Teng, Y.-C. Sun and K. Chen, *J. Org. Chem.*, 2003, **68**, 9816; (c) K. S. Yang and K. Chen, *J. Org. Chem.*, 2001, **66**, 1676; (d) K. S. Yang, J. C. Lain, C. H. Lin and K. Chen, *Tetrahedron Lett.*, 2000, **41**, 1453; (e) C. H. Lin, K. S. Yang, J. F. Pan and K. Chen, *Tetrahedron Lett.*, 2000, **41**, 6815.

170. J.-F. Pan and K. Chen, *Tetrahedron Lett.*, 2004, **45**, 2541.

171. (a) P. Muller, *Adv. Catal. Processes*, 1997, **2**, 113; (b) Z. Li, K. R. Conser and E. N. Jacobsen, *J. Am. Chem. Soc.*, 1993, **115**, 5326; (c) D. A. Evans, M. T. Bilodeau and M. M. Faul, *J. Am. Chem. Soc.*, 1994, **116**, 2742; (d) K. Noda, N. Hosoya, R. Irie, Y. Ito and T. Katsuki, *Synlett*, 1993, **7**, 469; (e) P. Muller, C. Baud and Y. Jacquier, *Tetrahedron*, 1996, **52**, 1543; (f) H. J. Jeon and S. T. Nguyen, *Chem. Commun.*, 2001, **235**.

172. (a) T. Siu and A. K. Yudin, *J. Am. Chem. Soc.*, 2002, **124**, 530; (b) T. Siu, C. J. Picard and A. K. Yudin, *J. Org. Chem.*, 2005, **70**, 932.

173. (a) L. Fowden, *Biochem. J.*, 1956, **64**, 323; (b) T. Takemoto, K. Nomoto, S. Fushiya, R. Ouchi, G. Kusano, H. Hikino, S. Takagi, Y. Matsuura and M. Kakudo, *Proc. Jpn. Acad., Ser. B*, 1978, **54**,

469; (c) F. Matsuura, Y. Hamada and T. Shioiri, *Tetrahedron*, 1994, **50**, 265; (d) T. Shioiri, Y. Hamada and F. Matsuura, *Tetrahedron*, 1995, **51**, 3939; (e) S. Singh, G. Crossley, S. Ghosal, Y. Lefievre and M. W. Pennington, *Tetrahedron Lett.*, 2005, **46**, 1419; (f) E. Kinoshita, J. Yamakoshi and M. Kikuchi, *Biosci. Biotechnol. Biochem.*, 1993, **57**, 1107; (g) S. Fushiya, T. Tamura, T. Tashiro and S. Nozoe, *Heterocycles*, 1984, **22**, 1039; (h) K. Isono, K. Asahi and S. Suzuki, *J. Am. Chem. Soc.*, 1969, **91**, 7490; (i) T. Akihisa, S. Mafune, M. Ukiya, Y. Kimura, K. Yasukawa, T. Suzuki, H. Tokuda, N. Tanabe and T. Fukuoka, *J. Nat. Prod.*, 2004, **67**, 479; (j) B. I. Ericksson, S. Carlsson, M. Halvarsson, B. Risberg and C. Mattson, *Thromb. Haemostasis.*, 1997, **78**, 1404; (k) AstraZeneca; *PCT Int. Appl.*, WO 00/41716, 2000; *Chem. Abstr.*, 2000, **133**, 99559.

174. (a) G. L. Zhao, J.-W. Huang and M. Shi, *Org. Lett.*, 2003, **5**, 4737; (b) G. L. Zhao and M. Shi, *J. Org. Chem.*, 2005, **70**, 9975.

175. (a) E. L. Setti and R. G. Micetich, *Curr. Med. Chem.*, 1998, **5**, 101; (b) *The Organic Chemistry of β-Lactams*, ed. G. I. Georg, VCH, New York, 1993; (c) F. C. Neuhaus and N. H. Georgeopapadakou, in *Emerging Targets in Antibacterial and Antifungal Chemoterapy*, ed. J. Sutcliffe and N. H. Georgeopapadakou, Chapman & Hall, New York, 1992; (d) P. G. Sammes, *Chem. Rev.*, 1976, **76**, 113; (e) B. Alcaide, P. Almendros and C. Aragoncillo, *Chem. Rev.*, 2007, **107**, 4437; (f) J. F. Fisher, S. O. Meroueh and S. Mobashery, *Chem. Rev.*, 2005, **105**, 395.

176. (a) I. Ojima, *Adv. Asym. Synth.*, 1995, **1**, 95; R. Noyori, T. Ikeda, T. Ohkuma, M. Widhalm, M. Kitamura, H. Takaya, S. Akutagawa, N. Sayo, T. Saito, T. Taketomi and H. Kumobayashi, *J. Am. Chem. Soc.*, 1989, **111**, 9134; (b) S. Murahashi, T. Naota, T. Kuwabara, T. Saito, H. Kumobayashi and S. Akutagawa, *J. Am. Chem. Soc.*, 1990, **112**, 7820; (c) N. Sayo, T. Saito, Y. Okeda, H. Nagashima and H. Kumobayashi, *US Pat.* 4981992, Jan. 1, 1991; (d) T. Saito and H. Kumobayashi, US Pat. 5081239, Jan. 14, 1992; (e) T. Saito, H. Kumobayashi and S. Murahashi, *US Pat.* 5288862, Feb. 22, 1994; (f) Other-Ref: T. Saito, H. Kumobayashi and S. Murahashi, *US Pat.* 5191076, Mar. 2, 1993; (g) T. Nakatsuka, H. Iwata, R. Tanaka, S. Imajo and M. Ishiguro, *J. Chem. Soc., Chem. Commun.*, 1991, 662; (h) M. Ishiguro, H. Iwata, T. Nakatsuka and Y. Yamada, *US Pat.* 5026844, Jun. 25, 1991.

177. (a) C. Gennari, I. Venturini, G. Gislon and G. Schimperna, *Tetrahedron Lett.*, 1987, **28**, 227; (b) G. Guanti, E. Narisano and L. Banfi, *Tetrahedron Lett.*, 1987, **28**, 4331; (c) T. Matsumoto, T. Murayama and M. Takashi, *Eur. Pat. EP* 742223, 1996.

178. B. A. Kulkarni and A. Ganesan, *J. Comb. Chem.*, 1999, **1**, 373.

179. W.-B. Wang and E. J. Roskamp, *J. Am. Chem. Soc.*, 1993, **115**, 9417.

180. H.-Y. Chen, L. N. Patkar, S.-H. Ueng, C.-C. Lin and A. S.-Y. Lee, *Synlett*, 2005, **13**, 2035.

181. F. Benfatti, G. Cardillo, L. Gentilucci, E. Mosconi and A. Tolomelli, *Org. Lett.*, 2008, **10**, 2425.

182. (a) S. Rajesh, B. Banerij and J. Iqbal, *J. Org. Chem.*, 2002, **67**, 7852; (b) I. G. Watson and A. K. Yudin, *J. Am. Chem. Soc.*, 2005, **127**, 17516.

183. (a) R. Noyori, T. Ikeda, T. Ohkuma, M. Widhalm, M. Kitamura, H. Takaya, S. Akutagawa, N. Sayo, T. Saito, T. Taketomi and H. Kumobayashi, *J. Am. Chem. Soc.*, 1989, **111**, 9134; (b) S. Murahashi, T. Naota, T. Kuwabara, T. Saito, H. Kumobayashi and S. Akutagawa, *J. Am. Chem. Soc.*, 1990, **112**, 7820; (c) N. Sayo, T. Saito, Y. Okeda, H. Nagashima and H. Kumobayashi, *US Pat.* 4981992, Jan. 1, 1991; (d) T. Saito and H. Kumobayashi, *US Pat.* 5081239, Jan. 14, 1992; (e) T. Saito, H. Kumobayashi and S. Murahashi, *US Pat.* 5288862, Feb. 22, 1994; (f) T. Saito, H. Kumobayashi and S. Murahashi, *US Pat.* 5191076, Mar. 2, 1993; (g) T. Nakatsuka, H. Iwata, R. Tanaka, S. Imajo and M. Ishiguro, *J. Chem. Soc., Chem. Commun.*, 1991, 662; (h) M. Ishiguro, H. Iwata, T. Nakatsuka and Y. Yamada, *US Pat.* 5026844, Jun. 25, 1991.

184. S. K. Singh, G. B. Singh, V. K. Byri, B. Satish, R. Dhamjewar and B. Gopalan, *Synth. Commun.*, 2008, **38**, 456.

185. W. Adam, P. Groer, H.-U. Humpf and C. R. Saha-Möller, *J. Org. Chem.*, 2000, **65**, 4919.

186. S. Kawahara, A. Nakano, T. Esumi, Y. Iwabuchi and S. Hatakeyama, *Org. Lett.*, 2003, **5**, 3103.

187. (a) A. Kamimura, R. Morita, K. Matsuura, Y. Omata and M. Shirai, *Tetrahedron Lett.*, 2002, **43**, 6189; (b) A. Kamimura, R. Morita, K. Matsuura, H. Mitsudera and M. Shirai, *Tetrahedron*, 2003, **43**, 9931.

188. W. P. Almeida and F. Coelho, *Tetrahedron Lett.*, 2003, **44**, 937.

189. D. Basavaiah and J. S. Rao, *Tetrahedron Lett.*, 2004, **45**, 1621.

190. V. Singh, S. Kanojiya and S. Batra, *Tetrahedron*, 2006, **62**, 10100.

191. (a) H. Amri, M. M. El Gaied, T. Ben Ayed and J. Villieras, *Tetrahedron Lett.*, 1992, **33**, 7345; (b) T. Ben Ayed, H. Amri, M. M. El Gaied and J. Villieras, *Tetrahedron*, 1995, **51**, 9633.

192. R. Galeazzi, G. Martelli, G. Mobbili, M. Orena and S. Rinaldi, *Tetrahedron: Asymmetry*, 2004, **15**, 3249.

193. (a) K. Makino and Y. Ichikawa, *Tetrahedron Lett.*, 1998, **39**, 8245; (b) S. Karlsson and H.-E. Högberg, *Tetrahedron: Asymmetry*, 2001, **12**, 1977; (c) M. D. Sorensen, N. M. Khalifa and E. B. Pedersen, *Synthesis*, 1999, **8**, 1937–1943; (d) M. Godskesen and I. Lundt, *Tetrahedron Lett.*, 1998, **39**, 5841.

194. H. Takayama, Y. Matsuda, K. Masubuchi, A. Ishida, M. Kitajima and N. Aimi, *Tetrahedron*, 2004, **60**, 893.

195. V. U. Khuzhaev, *Chem. Nat. Compd.*, 2004, **40**, 516.

196. S. C. Kim, S. Gowrisankar and J. N. Kim, *Tetrahedron Lett.*, 2006, **47**, 3463.

197. J.-F. Morizur and L. J. Mathias, *Tetrahedron Lett.*, 2007, **48**, 5555.

198. A. Nakano, K. Takahashi, J. Ishihara and S. Hatakeyama, *Org. Lett.*, 2006, **8**, 5357.

199. R. Galeazzi, G. Martelli, M. Orena, S. Rinaldi and P. Sabatino, *Tetrahedron*, 2005, **61**, 5465.

200. (a) F. Barclay, E. Chrystal and D. Gani, *J. Chem. Soc. Chem. Commun.*, 1994, **1**, 815; (b) F. Barclay, E. Chrystal and D. Gani, *J. Chem. Soc., Perkin Trans. 1*, 1996, 683.
201. (a) A. R. Chamberlin, H. P. Koch and R. J. Bridges, *Methods Enzymol.*, 1998, **296**, 175; (b) P. Stefanic and M. S. Dolenc, *Curr. Med. Chem.*, 2004, **11**, 945.
202. R. V. Stevens, *Tetrahedron*, 1976, **32**, 1599.
203. R. C. F. Jones, S. H. Dunn and K. A. M. Duller, *J. Chem. Soc., Perkin Trans. 1*, 1996, 1319.
204. V. Singh, R. Saxena and S. Batra, *J. Org. Chem.*, 2005, **70**, 353.
205. P. V. Ramachandran, S. Madhi, L. Bland-Berry, M. V. R. Reddy and M. J. O'Donnell, *J. Am. Chem. Soc.*, 2005, **127**, 13450.
206. J. Ezquerra, C. Pedregal, B. Yruretagoyena, A. Rubio, M. C. Carreno, A. Escribano and J. L. G. Ruano, *J. Org. Chem.*, 1995, **60**, 2925.
207. V. Singh, R. Pathak and S. Batra, *Catal. Commun.*, 2007, **8**, 2048.
208. K. Y. Lee, H. S. Lee and J. N. Kim, *Tetrahedron Lett.*, 2007, **48**, 2007.
209. J. M. Kim, K. Y. Lee, S. Lee and J. N. Kim, *Tetrahedron Lett.*, 2004, **45**, 2805.
210. D. Balan and H. Adolfsson, *Tetrahedron Lett.*, 2004, **45**, 3089.
211. (a) V. Declerck, P. Ribière, J. Martinez and F. Lamaty, *J. Org. Chem.*, 2004, **69**, 8372; (b) V. Declerck, H. Allouchi and J. Martinez and F. Lamaty, *J. Org. Chem.*, 2007, **72**, 1518.
212. (a) S. M. Weinreb and J. L. Ralbovsky, in *Handbook of Reagents for Organic Synthesis, Activating Agents and Protecting Groups*, ed. A. J. Pearson and W. J. Roush, John Wiley & Sons, Ltd., Chichester, UK, 1999, p. 425; (b) T. W. Greene and P. G. M. Wuts, *Protective Groups in Organic Synthesis*, John Wiley & Sons, Inc., New York, 1999, p. 612; (c) P. J. Kocienski, in *Protecting Groups*, ed. D. Enders, R. Noyori and B. M. Trost, Thieme, New York, 2000, p. 215.
213. For review articles, see: (a) C.-H. Wong, R. L. Halcomb, Y. Ichikawa and T. Kajimoto, *Angew. Chem. Int. Ed.*, 1995, 34, 412; (b) A. B. Hughes and A. J. Rudge, *Nat. Prod. Rep.*, 1994, **11**, 135; (c) H. Paulsen and K. Todt, *Adv. Carbohydr. Chem. Biochem.*, 1968, **23**, 115; (d) S. Picasso, *Chimia*, 1996, **50**, 648; (e) L. A. G. M. van den Broek, in *Carbohydrates in Drug Design*, ed. Z. J. Witczak and K. A. Nieforth, Dekker, New York, 1997, p. 471; and (f) Z. J. Witczak, in *Carbohydrates in Drug Design*, ed. Z. J. Witczak and K. A. Nieforth, Dekker, New York, 1997, p. 1
214. (a) S. Cren, C. Wilson and N. R. Thomas, *Org. Lett.*, 2005, **7**, 3521; (b) M. L. Ulf, D. Rui and H. Olle, *Chem. Commun.*, 2005, **13**, 1773; (c) T. Ayad, Y. Genisson, S. Broussy, M. Baltas and L. Gorrichon, *Eur. J. Org. Chem.*, 2003, **15**, 2903; (d) A. T. Carmona, J. Fuentes, I. Robina, G. E. Rodriguez, D. Raynald, P. Vogel and A. L. Winters, *J. Org. Chem.*, 2003, **68**, 3874; (e) T. J. Donohoe, C. E. Headley, R. P. C. Cousins and A. Cowley, *Org. Lett.*, 2003, **5**, 999; (f) A. T. Carmona, J. Fuentes and I. Robina, *J. Org. Chem.*, 2003, **68**, 3874.
215. V. R. Doddi and Y. D. Vankar, *Eur. J. Org. Chem.*, 2007, 5583.

216. H. S. Lee, J. M. Kim and J. N. Kim, *Tetrahedron Lett.*, 2007, **48**, 4119.

217. H. S. Lee, H. S. Kim, J. M. Kim and J. N. Kim, *Tetrahedron*, 2008, **64**, 2397.

218. D. Leca, K. Song, M. Albert, M. G. Gonçalves, L. Fensterbank and E. Lacôte, *Synthesis*, 2005, 1405.

219. (a) B. M. Trost and D. L. Van Vranken, *Chem. Rev.*, 1996, **96**, 395; (b) B. M. Trost, *Acc. Chem. Res.*, 1996, **29**, 355; (c) P. Metz, *Methoden Org. Chem; (Houben-Weyl)*, Vol. **E21**, 5643.

220. M. Henrich, A. Delgado, E. Molins, A. Roig and A. Llebaria, *Tetrahedron Lett.*, 1999, **40**, 4259.

221. (a) R. Fernández de la Pradilla, P. Manzano, C. Montero, J. Priego, M. Martínez-Ripoll and L. A. Martíez- Cruz, *J. Org. Chem.*, 2003, **68**, 7755; (b) R. Fernández de la Pradilla and A. Castellanos, *Tetrahedron Lett.*, 2007, **48**, 6500.

222. A. Viso, R. Fernández de la Pradilla, M. Ureña and I. Colomer, *Org. Lett.*, 2008, **10**, 4775.

223. J. Jayashankaran, R. D. R. S. Manian, M. Sivaguru and R. Raghunathan, *Tetrahedron Lett.*, 2006, **47**, 5535.

224. (a) R. J. Sundberg, *The Chemistry of Indoles*, Academic Press, New York, 1970; (b) R. J. Sundberg, *Best Synthetic Methods, Indoles*, Academic Press, New York, 1996, pp. 7–11; (c) J. A. Joule, Indole and its derivatives, in *Science of Synthesis: Houben-Weyl Methods of Molecular Transformations*, ed E. J. Thomas, George Thieme Verlag, Stuttgart, 2000, category 2, vol. 10, ch 10.13; (d) R. K. Brown, in *Indoles*, ed. W. J. Houlihan, Wiley-Interscience, New York, 1972; (e) *Indoles*; ed. R. J. Sundberg, Academic Press, London, 1996; (f) A. Kleeman, J. Engel, B. Kutscher and D. Reichert, *Pharmaceutical Substances*, 4th edn., Thieme, New York, 2001; (g) S. Cacchi and G. Fabrizi, *Chem. Rev.*, 2005, **105**, 2873; (h) G. R. Humphrey and J. T. Kuethe, *Chem. Rev.*, 2006, **106**, 2875; (i) G. Zeni and R. C. Larock, *Chem. Rev.*, 2006, **106**, 4644.

225. (a) M. L. Bode and P. T. Kaye, *J. Chem. Soc., Perkin Trans. 1*, 1993, 1809; (b) M. L. Bode and P. T. Kaye, *J. Chem. Soc., Perkin Trans. 1*, 1990, 2612.

226. D. Basavaiah and A. J. Rao, *Chem. Commun.*, 2003, 604.

227. D. Basavaiah and A. J. Rao, *Tetrahedron Lett.*, 2003, **44**, 4365.

228. D. Virieux, A.-F. Guillouzic and H.-J. Cristau, *Tetrahedron*, 2006, **62**, 3710.

229. W. Pan, D. Dong, S. Sun and Q. Liu, *Synlett*, 2006, **7**, 1090.

230. S. Sun, M. Wang, H. Deng and Q. Liu, *Synthesis*, 2008, 573.

231. H.-L. Cui, X. Feng, J. Peng, J. Lei, K. Jiang and Y.-Y. Chen, *Angew. Chem. Int. Ed.*, 2009, **48**, 5737.

232. (a) M. B. Johansen and M. A. Kerr, *Org. Lett.*, 2008, **10**, 3497; (b) H. Kusama, J. Takaya and N. Iwasawa, *J. Am. Chem. Soc.*, 2002, **124**, 11592.

233. D. Paris, M. Cottin, P. Demonchaux, G. Augert, P. Dupassieux, P. Lenoir, M. J. Peck and D. Jasserand, *J. Med. Chem.*, 1995, **38**, 669.

234. (a) J. J. Li and G. W. E. Gribble, *Palladium in Heterocyclic Chemistry: A Guide for the Synthetic Chemist*, Pergamon, New York, 2000; (b) L. S. Hegedus, *Angew. Chem.*, 1988, **100**, 1147.

235. R. S. Srivastava and K. M. Nicholas, *Chem. Commun.*, 1998, **24**, 2705.

236. D. K. O'Dell and K. M. Nicholas, *Tetrahedron*, 2003, **59**, 747.

237. C. R. Horn and M. Perez, *Synlett*, 2005, **9**, 1480.

238. P. S. Watson, B. Jiang and B. Scott, *Org. Lett.*, 2000, **2**, 3679.

239. (a) P. D. Bailey, P. A. Millwood and P. D. Smith, *Chem. Commun.*, 1998, 633; (b) S. Laschat and T. Dickner, *Synthesis*, 2000, 1781; (c) V. Baliah, *Chem. Rev.*, 1983, **83**, 379; (d) C. D. Risi, G. Fanton, G. P. Pollini, C. Trapella, F. Valente and V. Zanirato, *Tetrahedron: Asymmetry*, 2008, **19**, 131; (e) F.-X. Felpin and J. Lebreton, *Eur. J. Org. Chem.*, 2003, 3693; (f) J. A. Varela and C. Saá, *Chem. Rev.*, 2003, **103**, 3787; (g) G. R. Newkome, J. D. Sauer, J. M. Roper and D. C. Hager, *Chem. Rev.*, 1977, **77**, 513.

240. (a) G.-L. Zhao, J.-W. Huang and M. Shi, *Org. Lett.*, 2003, **5**, 4737; (b) G.-L. Zhao and M. Shi, *J. Org. Chem.*, 2005, **70**, 9975.

241. C.-Y. Chen, M.-Y. Chang, R.-T. Hsu, S.-T. Chen and N.-C. Chang, *Tetrahedron Lett.*, 2003, **44**, 8627.

242. (a) T. A. Van Beek, R. Verpoorte and A. Baerheim Svendsen, *Tetrahedron*, 1984, **40**, 737; (b) G. Massiot, F. Sousa Oliverira, T. A. Van Beek, R. Verpoorte and A. Baerheim Svendsen, *Tetrahedron*, 1984, **40**, 737 J. Lévy, *Bull. Soc. Chim. Fr. II*, 1982, 185; (c) B. Danieli, G. Lesma, S. Macecchini, D. Passarella, T. A. Van Beek, R. Verpoorte and A. Baerheim Svendsen, *Tetrahedron*, 1984, **40**, 737; (d) A. Silvani, *Tetrahedron: Asymmetry*, 1999, **10**, 4057.

243. V. Singh, G. P. Yadav, P. R. Maulik and S. Batra, *Tetrahedron*, 2006, **62**, 8731.

244. S. Wang, T. Tan, J. Li and H. Hu, *Synlett*, 2005, 2658.

245. V. Singh, G. P. Yadav and P. R. Maulik and S. Batra, *Tetrahedron*, 2008, **64**, 2979.

246. N. M. Garrido, M. García, D. Díez, M. R. Sánchez, F. Sanz and J. G. Urones, *Org. Lett.*, 2008, **10**, 1687.

247. (a) P. Raubo, J. J. Kulagowski and G. G. Chicchi, *Synlett*, 2006, 271; (b) E. M. Seward and C. J. Swain, *Expert Opin. Therap. Pat.*, 1999, **9**, 571; (c) J. P. Michael, C. B. Koning and D. P. Pienaar, *Synlett*, 2006, 383; (d) J. B. Koepfli, J. A. Brockman Jr. and J. Moffat, *J. Am. Chem. Soc.*, 1950, **72**, 3323.

248. S. H. Kim, H. S. Lee, K. H. Kim and J. N. Kim, *Tetrahedron Lett.*, 2009, **50**, 1696.

249. P. R. Krishna and P. S. Reddy, *J. Comb. Chem.*, 2008, **10**, 426.

250. P. Krogsgaard-Larsen, H. Hjeds, D. R. Curtis, J. D. Leah and M. J. Peet, *J. Neurochem.*, 1982, **39**, 1319.

251. (a) C. R. Bertozzi and L. Kiessling, *Science.*, 2001, **291**, 2357; (b) T. D. Heightman and A. T. Vasella, *Angew Chem., Int. Ed.*, 1999, **38**, 750; (c) N. Asano, *Curr. Top. Med. Chem.*, 2003, **3**, 471; (d) P. Compain and O. R. Martin, *Curr. Top. Med. Chem.*, 2003, **3**, 541.

252. J. K. Cha and N.-S. Kim, *Chem. Rev.*, 1995, **95**, 1761.

253. D. Y. Park, M. J. Lee, T. H. Kim and J. N. Kim, *Tetrahedron Lett.*, 2005, **46**, 8799.

254. S. Gowrisankar, H. S. Lee, J. M. Kim and J. N. Kim, *Tetrahedron Lett.*, 2008, **49**, 1670.

255. S. H. Kim, K. H. Kim, H. S. Kim and J. N. Kim, *Tetrahedron Lett.*, 2008, **49**, 1948.

256. S. H. Kim, H. S. Lee, S. H. Kim and J. N. Kim, *Tetrahedron Lett.*, 2008, **49**, 5863.

257. H. S. Lee, E. S. Kim, S. H. Kim and J. N. Kim, *Tetrahedron Lett.*, 2009, **50**, 2274.

258. (a) M. G. Kim, E. T. Bodor, C. Wang, T. K. Harden and H. Kohn, *J. Med. Chem.*, 2003, **46**, 2216; (b) M. Tingoli, M. Tiecco, L. Testaferri, R. Andrenacci and R. Balducci, *J. Org. Chem.*, 1993, **58**, 6097; (c) I. Carranco, J. L. Diaz, O. Jimenez and R. Lavilla, *Tetrahedron Lett.*, 2003, **44**, 8449; (d) J. Jiang, J. Yu, X.-X. Sun, Q.-Q. Rao and L.-Z. Gong, *Angew. Chem., Int. Ed.*, 2008, **47**, 2458.

259. (a) P. J. Rosenthal, *Antimalarial Chemotherapy: Mechanisms of Action, Resistance, and New Directions in Drug Discovery*, Humana Press, Inc., Totowa, NJ, 2001; (b) A. K. Bhattacharjee and J. M. Karle, *J. Med. Chem.*, 1996, **39**, 4622; (c) D. De, F. M. Krogstad, L. D. Byers and D. J. Krogstad, *J. Med. Chem.*, 1998, **41**, 4918; (d) P. A. Stocks, K. J. Raynes, P. G. Bray, B. K. Park, P. M. O'Neill and S. A. Ward, *J. Med. Chem.*, 2002, **45**, 4975; (e) J. L. Vennerstrom, A. L. Ager Jr, A. Dorn, S. L. Andersen, L. Gerena, R. G. Ridley and W. K. Milhous, *J. Med. Chem.*, 1998, **41**, 4360; (f) C. H. Kaschula, T. J. Egan, R. Hunter, N. Basilico, S. Parapini, D. Taramelli, E. Pasini and D. Monti, *J. Med. Chem.*, 2002, **45**, 3531; (g) S. Delarue, S. Girault, L. Maes, M.-A. Debreu-Fontaine, M. Labaeid, P. Grellier and C. Sergheraert, *J. Med. Chem.*, 2001, **44**, 2827.

260. (a) J. A. Joule and K. Mills, *Heterocyclic Chemistry*, 4th edn, Blackwell Science, Oxford, 2000, p. 121; (b) H. J. Roth and H. Fenner, *Arzneistoffe*, 3rd edn, Deutscher Apotheker, Stuttgart, 2000, p. 51; (c) M. Balasubramanian and J. G. Keay, in *Comprehensive Heterocyclic Chemistry II*, ed. A. R. Katritzky, C. W. Rees and E. F. V. Scriven, Pergamon, Oxford, 1996, **vol. 5**, p. 245; (d) Y. L. Chen, K. C. Fang, J. Y. Sheu, S. L. Hsu and C. C. Tzeng, *J. Med. Chem.*, 2001, **44**, 2374; (e) G. Roma, M. D. Braccio, G. Grossi, F. Mattioli and M. Ghia, *Eur. J. Med. Chem.*, 2000, **35**, 1021; (f) H. Shinkai, T. Ito, T. Iida, Y. Kitao, H. Yamada and I. Uchida, *J. Med. Chem.*, 2000, **43**, 4667; (g) N. C. R. van Straten, T. H. J. van Berkel, D. R. Demont, W.-J. F. Karstens, R. Merkx, J. Oosterom, J. Schulz, R. G. van Someren, C. M. Timmers and P. M. van Zandvoort, *J. Med. Chem.*, 2005, **48**, 1697; (h) D. H. Boschelli, Y. D. Wang, S. Johnson, B. Wu, F. Ye, A. C. Barrios Sosa, J. M. Golas and F. Boschelli, *J. Med. Chem.*, 2004, **47**, 1599.

261. (a) X.-F. Lin, S.-L. Cui and Y.-G. Wang, *Tetrahedron Lett.*, 2006, **47**, 3127; (b) N. Sakai, D. Aoki, T. Hamajima and T. Konakahara,

Tetrahedron Lett., 2006, **47**, 1261; (c) N. Sakai, K. Annaka and T. Konakahara, *J. Org. Chem.*, 2006, **71**, 3653; (d) S.-Y. Tanaka, M. Yasuda and A. Baba, *J. Org. Chem.*, 2006, **71**, 800; (e) G.-W. Wang, C.-S. Jia and Y.-W. Dong, *Tetrahedron Lett.*, 2006, **47**, 1059; (f) X. Wang, S. Dixon, M. J. Kurth and K. S. Lam, *Tetrahedron Lett.*, 2005, **46**, 5361; (g) S. K. De and R. A. Gibbs, *Tetrahedron Lett.*, 2005, **46**, 1647; (h) K. Taguchi, S. Sakaguchi and Y. Ishii, *Tetrahedron Lett.*, 2005, **46**, 4539; (i) K. Kobayashi, K. Yoneda, K. Miyamoto, O. Morikawa and H. Konishi, *Tetrahedron*, 2004, **60**, 11639; (j) X. Zhang, M. A. Campo, T. Yao and R. C. Larock, *Org. Lett.*, 2004, **7**, 763.

262. O. B. Familoni, P. T. Kaye and P. J. Klaas, *Chem. Commun.*, 1998, 2563.

263. (a) A. Tetsuya, D. Takayuki and F. Koji, *Japan Pat.* JP 1992, 04282370; *Chem. Abstr.*, 1993, **118**, 147472u; (b) M. Uchida, M. Chihiro, S. Morita, H. Yamashita, K. Yamasaki, T. Kanbe, Y. Yabuuchi and K. Nakagawa, *Chem. Pharm. Bull.*, 1990, **38**, 534; (c) S. S. Epstein, I. B. Saporoschetz, M. Small, W. Park and N. Mantel, *Nature*, 1965, **208**, 655.

264. (a) T. A. Engler, K. O. Lynch Jr, W. Chai and S. P. Meduna, *Tetrahedron Lett.*, 1995, **36**, 2713; (b) T. A. Engler, K. O. La Tessa, Iyenger, R. W. Chai and K. Agrios, *Bioorg. Med. Chem.*, 1996, **4**, 1755.

265. (a) I. Yavari, A. Ramazani and A. A. Esmaili, *J. Chem. Res. (S)*, 1997, 208; (b) N. M. Williamson, D. R. March and A. D. Ward, *Tetrahedron Lett.*, 1995, **36**, 7721; (c) K. Kobayashi, R. Nakahashi, A. Shimizu, T. Kitamura, O. Morikawa and H. Konishi, *J. Chem. Soc., Perkin. Trans.*, 1999, 1547; (d) I. Yavari, A. A. Esmaili, A. Ramazani and A. R. Bolbol-Amiri, *Monatsh. Chem.*, 1997, **128**, 927; (e) Y. Kikugawa, M. Kuramoto, I. Saito and S.-I. Yamada, *Chem. Pharm. Bull.*, 1973, **21**, 1914; (f) W. S. Johnson and B. G. Buell, *J. Am. Chem. Soc.*, 1952, **74**, 4517; (g) K. Kobayashi, S. Nagato, M. Kawakita, O. Morikawa and H. Konishi, *Chem. Lett.*, 1995, 575; (h) V. Kouznetsov, A. Palma, C. Ewert and A. Varlamov, *J. Heterocyclic Chem.*, 1998, **35**, 761; (i) M. Grignon-Dubois, F. Diaba and M.-C. Grellier-Marly, *Synthesis*, 1994, 800; (j) F. Diaba, C. L. Houerou, M. Grignon-Dubois and P. Gerval, *J. Org. Chem.*, 2000, **65**, 907; (k) F. Diaba, I. Lewis, M. Grignon-Dubois and S. Navarre, *J. Org. Chem.*, 1996, **61**, 4830; (l) M. Grignon-Dubois and F. Diaba, *J. Chem. Res; (S).*, 1998, 660; (m) M. Maeda, *Chem. Pharm. Bull.*, 1990, **38**, 2577; (n) N. S. Mani, P. Chen and T. K. Jones, *J. Org. Chem.*, 1999, **64**, 6911.

266. O. B. Familoni, P. T. Kaye and P. J. Klaas, *Chem. Commun.*, 1998, 2563.

267. O. B. Familoni, P. J. Klaas, K. A. Lobb, V. E. Pakade and P. T. Kaye, *Org. Biomol. Chem.*, 2006, **4**, 3960.

268. S. Madapa, V. Singh and S. Batra, *Tetrahedron*, 2006, **62**, 8740.

269. D. Basavaiah, R. M. Reddy, N. Kumaragurubaran and D. S. Sharada, *Tetrahedron*, 2002, **58**, 3693.

270. D. Basavaiah, J. S. Rao and R. J. Reddy, *J. Org. Chem.*, 2004, **69**, 7379.

271. D. Basavaiah, R. M. Reddy and J. S. Rao, *Tetrahedron Lett.*, 2006, **47**, 73.

272. D. K. O'Dell and K. M. Nicholas, *J. Org. Chem.*, 2003, **68**, 6427.

273. C. J. Ohnmacht, F. Davis and R. E. Lutz, *J. Med. Chem.*, 1971, **14**, 17.
274. H.-W. Yi, H. W. Park, Y. S. Song and K.-J. Lee, *Synthesis*, 2006, 1953.
275. J. N. Kim, H. J. Lee, K. Y. Lee and H. S. Kim, *Tetrahedron Lett.*, 2001, **42**, 3737.
276. J. N. Kim, H. S. Kim, J. H. Gong and Y. M. Chung, *Tetrahedron Lett.*, 2001, **42**, 8341.
277. A. Foucaud and F. El Guemmout, *Bull. Soc. Chim. Fr.*, 1989, 403.
278. W. P. Hong and K.-J. Lee, *Synthesis*, 2006, 963.
279. (a) K. Grohe and H. Heitzer, *Liebigs Ann. Chem.*, 1987, 29; (b) H. Egawa, T. Miyamoto and J.-I. Matsumoto, *Chem. Pharm. Bull.*, 1986, **34**, 4098.
280. P. Narender, U. Srinivas, M. Ravinder, B. A. Rao, C. Ramesh, K. Harakishore, B. Gangadasu, U. S. N. Murthy and V. J. Rao, *Bioorg. Med. Chem.*, 2006, **14**, 4600.
281. F. Coelho, D. Veronese, E. C. S. Lopes and R. C. Rossi, *Tetrahedron Lett.*, 2003, **44**, 5731.
282. (a) P. Ribiere, V. Declerck, Y. Nedellec, N. Yadav-Bhatnagar, J. Martinez and F. Lamaty, *Tetrahedron*, 2006, **62**, 10456; (b) V. Declerck, P. Ribiere, Y. Nedellec, H. Allouchi, J. Martinez and F. Lamaty, *Eur. J. Org. Chem.*, 2007, 201; (c) M. Szlosek-Pinaud, P. Diaz, J. Martinez and F. Lamaty, *Tetrahedron*, 2007, **63**, 3340.
283. (a) J.-R. Wang, Y. Fu, B.-B. Zhang, X. Cui, L. Liu and Q.-X. Guo, *Tetrahedron Lett.*, 2006, **47**, 8293; (b) J.-R. Wang, C.-T. Yang, L. Liu and Q.-X. Guo, *Tetrahedron Lett.*, 2007, **48**, 5449.
284. S. Gowrisankar, H. S. Lee, J. M. Kim and J. N. Kim, *Tetrahedron Lett.*, 2008, **49**, 1670.
285. Y. S. Park, M. Y. Cho, Y. B. Kwon, B. W. Yoo and C. M. Yoon, *Synth. Commun.*, 2007, **37**, 2677.
286. J. N. Kim, Y. M. Chung and Y. J. Im, *Tetrahedron Lett.*, 2002, **43**, 6209.
287. J. N. Kim, K. Y. Lee, H. S. Kim and T. Y. Kim, *Org. Lett.*, 2000, **2**, 343.
288. K. Y. Lee, J. M. Kim and J. N. Kim, *Tetrahedron*, 2003, **59**, 385.
289. G. W. Amarante, M. Benassi, A. A. Sabino, P. M. Esteves, F. Coelho and M. N. Eberlin, *Tetrahedron Lett.*, 2006, **47**, 8427.
290. (a) J. C. Carretero, J. L. Garcia Ruano and M. Vicioso, *Tetrahedron*, 1992, **48**, 7373; (b) G. L. Anderson, *J. Heterocycl. Chem.*, 1985, **22**, 1469; (c) D. Kaminsky, *Fr. Demande* 2002888; *Chem. Abstr.*, 1970, **72**, 90322v.
291. R. Pathak, S. Madapa and S. Batra, *Tetrahedron*, 2007, **63**, 451.
292. (a) K. J. Andries and J. F. E. Van Gestel, WO 2005117875, 2005; (b) J. F. E. Van Gestel, J. E. G. Guillemont, M. G. Venet, H. J. J. Poignet, L. F. B. Decrane, D. F. J. Vernier and F. C. Odds, *US Pat.* 2005148581 A1, 2005.
293. (a) S. T. Cole and P. M. Alzari, *Science*, 2005, 306; (b) E. J. N. Rubin, *Engl. J. Med.*, 2005, **352**, 933; (c) B. Ji and V. Jarlier, *Antimicrob. Agents Chemother.*, 2006, **50**, 1558.
294. C. G. Lee, K. Y. Lee, S. Lee and J. N. Kim, *Tetrahedron*, 2005, **61**, 1493.
295. K. H. Kim, H. S. Lee and J. N. Kim, *Tetrahedron Lett.*, 2009, **50**, 1249.

296. (a) D. L. J. Clive, M. Yu and Z. Li, *Chem. Commun*, 2005, 906; (b) D. L. J. Clive, Z. Li and M. Yu, *J. Org. Chem.*, 2007, **72**, 5608.

297. W. Zhong, Y. Zhao and W. Su, *Tetrahedron*, 2008, **64**, 5491.

298. W. Zhong, F. Lin, R. Chen and W. Su, *Synthesis*, 2008, 2561.

299. C. M. Marson and J. H. Pink, *Tetrahedron Lett.*, 1995, **36**, 8107.

300. P. R. Ocken, *J. Lipid. Res.*, 1969, **10**, 460.

301. (a) M. H. Benn and J. M. Jacyno, in *Alkaloids – Chemical and Biological Perspectives*, ed. S. W. Pelletier, John Wiley & Sonsm, Inc., New York, 1983, **vol. 1**, p. 153; (b) K. Wiesner, *Pure Appl. Chem.*, 1979, **51**, 689.

302. S. P. Park, Y. S. Song and K.-J. Lee, *Tetrahedron*, 2009, **65**, 4703.

303. D. Basavaiah and T. Satyanarayana, *Chem. Commun.*, 2004, 32.

304. M. Henrich, A. Delgado, E. Molins, A. Roig and A. Llebaria, *Tetrahedron Lett.*, 1999, **40**, 4259.

305. P. Ribière, V. Declerck, Y. Nédellec, N. Yadav-Bhatnagar, J. Martinez and F. Lamaty, *Tetrahedron*, 2006, **62**, 10456.

306. V. Declerck, P. Ribière, Y. Nédellec, H. Allouchi, J. Martinez and F. Lamaty, *Eur. J. Org. Chem.*, 2007, 201.

307. H. S. Lee, S. H. Kim, T. H. Kim and J. N. Kim, *Tetrahedron Lett.*, 2008, **49**, 1773.

308. (a) T. W. Hudyma, X. Zheng, F. He, M. Ding, C. P. Bergstrom, P. Hewawasam, S. W. Martin and R. G. Gentles, WO 2007092000, 2007; *Chem. Abstr.,* 2007, **147**, 277782; (b) N. A. Meanwell, R. G. Gentles, M. Ding, J. A. Bender, J. F. Kadow, P. Hewawasam, T.W. Hudyma and X. Zheng, *US Pat.* US 2007,184,024, 2007; *Chem. Abstr.*, 2007, **147**, 257667; (c) C. P. Bergstrom, J. A. Bender, R. G. Gentles, P. Hewawasam, T. W. Hudyma, J. F. Kadow, S.W. Martin, A. Regueiro-Ren, K.-S. Yeung, Y. Tu, K. A. Grant-Young and X. Zheng, WO 2007033175, 2007; *Chem. Abstr.*, 2007, **146**, 358820; (d) C. P. Bergstrom, S. W. Martin and T. W. Hudyma, *US Pat.* US 2007,185,083, 2007; *Chem. Abstr.*, 2007, **147**, 235031; (e) R. G. Gentles, M. Ding and P. Hewawasam, *US Pat.* US 2007,275,930, 2007; *Chem. Abstr.*, 2007, **148**, 11085.

309. H. S. Lee, S. H. Kim, S. Gowrisankar and J. N. Kim, *Tetrahedron*, 2008, **64**, 7183.

310. S. Batra and A. K. Roy, *Synthesis*, 2004, 2550.

311. V. Singh and S. Batra, *Eur. J. Org. Chem.*, 2007, 2970.

312. D. Basavaiah and K. Aravindu, *Org. Lett.*, 2007, **9**, 2453.

313. Z. Shafiq, L. Liu, Z. Liu, D. Wang and Y.-J. Chen, *Org. Lett.*, 2007, **9**, 2525.

314. (a) C. B. Cooper, J. P. Lyssikatos, D. W. Mann and S. J. Hecker, EP 825,186, 1998; *Chem. Abstr.*, 1998, **128**, 192567; (b) G. A. Doherty, T. C. Eary, R. D. Groneberg and Z. Jones, WO 2007,024,612, 2007; *Chem. Abstr.*, 2007, **146**, 274240.

315. H. L. Lim, Y. S. Song and K.-J. Lee, *Synthesis*, 2007, 3376.

316. E. P. Abraham and E. Chain, *Nature*, 1940, **146**, 837.

317. (a) M. I. Page, *Acc. Chem. Res.*, 1984, **17**, 144; (b) R. D. G. Cooper, L. D. Hatfield and D. O. Spry, *Acc. Chem. Res.*, 1973, **6**, 32; (c) P. G. Sammes,

Chem. Rev., 1976, **76**, 113; (d) M. J. Miller, *Acc. Chem. Res.*, 1986, **19**, 49; (e) I. Massova and S. Mobashery, *Acc. Chem. Res.*, 1997, **30**, 162; (f) J. F. Fisher, S. O. Meroueh and S. Mobashery, *Chem. Rev.*, 2005, **105**, 395.

318. (a) O. A. Mascaretti, C. E. Boschetti, G. O. Danelon, E. G. Mata and O. A. Roveri, *Curr. Med. Chem.*, 1995, **1**, 441; (b) P. D. Edwards and P. R. Bernstein, *Med. Res. Rev.*, 1994, **14**, 127.

319. J. Kant and D. G. Walker, in *The Organic Chemistry of β-Lactams*, ed. G. I. Georg, VCH, New York, 1993, ch. 3.

320. B. Alcaide, P. Almendros and C. Aragoncillo, *Chem. Commun.*, 1999, 1913.

321. B. Alcaide, P. Almendros and C. Aragoncillo, *J. Org. Chem.*, 2001, **66**, 1612.

322. (a) G. R. Lappin, *J. Org. Chem.*, 1958, **23**, 1358; (b) G. R. Lappin, *J. Org. Chem.*, 1961, **26**, 2350.

323. D. Basavaiah and T. Satyanarayana, *Tetrahedron Lett.*, 2002, **43**, 4301.

324. (a) G. Meinhardt, E. Eppinger and R. Schmidmaier, *Anti-Cancer Drugs*, 2002, **13**, 725; (b) D. E. Jones, J. A. V. Coates, D. I. Rhodes, J. J. Deadman, N. A. Vandegrafe, L. J. Winfield, N. Thienthong, W. Issa, N. Choi and K. Macfarlane, *WO Pat.* 77188, 2008; *Chem. Abstr.*, 2008, **149**, 128851; (c) P. A. Brough, S. C. Cheetham, F. Kerrigan and J. P. Watts, *WO Pat.* 71549, 2000; *Chem. Abstr.*, 2000, **134**, 29416.

325. W. Zhong, B. Guo, F. Lin, Y. Liu and W. Su, *Synthesis*, 2009, 1615.

326. K. Y. Lee, J. M. Kim and J. N. Kim, *Tetrahedron Lett.*, 2003, **44**, 6737.

327. K. Y. Lee, S. Gowrisankar and J. N. Kim, *Tetrahedron Lett.*, 2005, **46**, 5387.

328. S. Nag, R. Pathak, M. Kumar, P. K. Shukla and S. Batra, *Bioorg. Med. Chem. Lett.*, 2006, **16**, 3824.

329. S. Nag, G. P. Yadav, P. R. Maulik and S. Batra, *Synthesis*, 2007, 911.

330. D. A. Horton, G. T. Bourne and M. L. Smythe, *Chem. Rev.*, 2003, **103**, 893.

331. R. Pathak, S. Nag and S. Batra, *Sythesis*, 2006, 4205.

332. T. Tanaka, T. Muto, H. Maruoka, S. Imajo, H. Fukami, Y. Tomimori, Y. Fukuda and T. Nakatsuka, *Bioorg. Med. Chem. Lett.*, 2007, **17**, 3431.

333. H. Maruoka, T. Muto, T. Tanaka, S. Imajo, Y. Tomimori, Y. Fukuda and T. Nakatsuka, *Bioorg. Med. Chem. Lett.*, 2007, **17**, 3435.

334. M. Anzini, A. Cappelli, S. Vomero, A. Cagnotto and M. Skorupska, *Med. Chem. Res.*, 1993, **3**, 44.

335. M. A. Quraishi, V. R. Thakur and S. N. Dhawan, *Indian J. Chem., Sect. B: Org. Chem. Incl. Med. Chem.*, 1989, **28**, 891.

336. (a) M. Yamato, Y. Takeuchi, K. Hashigaki, Y. Ikeda, M. C. Chang, K. Takeuchi, M. Matsushima, T. Tsuruo, T. Tashiro, S. Tsukagoshi, Y. Yamashita and H. Nakano, *J. Med. Chem.*, 1989, **32**, 1295; (b) L. W. Deady, J. Desneves, A. J. Kaye, G. J. Finlay, B. C. Baguley and W. A. Denny, *Bioorg. Med. Chem.*, 2000, **8**, 977.

337. (a) L. W. Deady, J. Desneves, A. J. Kaye, G. J. Finlay, B. C. Baguley and W. A. Denny, *Bioorg. Med. Chem.*, 2001, **9**, 445; (b) X. Bu and L. W. Deady, *Synth. Commun.*, 1999, **29**, 4223; (c) X. L. Lu, J. L. Petersen and K. K. Wang, *Org. Lett.*, 2003, **5**, 3277; (d) G. Babu and P. T. Perumal, *Tetrahedron Lett.*, 1998, **39**, 3225; (e) J. S. Yadav, B. V. Subba Reddy, R. Srinivas, Ch. Madhuri and T. Ramalingam, *Synlett*, 2001, 240; (f) S.-J. Tu, B. Jiang, R.-H. Jia, J.-Y. Zhang, Y. Zhang, C.-S. Yao and F. Shi, *Org. Biomol. Chem.*, 2006, **4**, 3664.

338. C. G. Lee, K. Y. Lee, S. GowriSankar and J. N. Kim, *Tetrahedron Lett.*, 2004, **45**, 7409.

339. R. Pathak, A. K. Roy and S. Batra, *Synlett*, 2005, 848.

340. R. Pathak and S. Batra, *Tetrahedron*, 2007, **63**, 9448.

341. R. Pathak, A. K. Roy, S. Kanojiya and S. Batra, *Tetrahedron Lett.*, 2005, **46**, 5289.

342. (a) W. R. Tully, C. R. Gardner, R. J. Gillespie and R. Westwood, *J. Med. Chem.*, 1991, **34**, 2060; (b) A. F. Almansa, F. L. Cavalcanti, L. A. Gomez, A. Miralles, M. Merlos, J. Garcia-Rafanell and J. Forn, *J. Med. Chem.*, 2001, **44**, 350; (c) G. R. Revankar, T. R. Matthews and R. K. Robins, *J. Med. Chem.*, 1975, **18**, 1253; (d) Y. Rival, G. Grassy and G. Michel, *Chem. Pharm. Bull.*, 1992, **40**, 1170; (e) Y. Rival, G. Grassy, A. Taudou and R. Ecalle, *Eur. J. Med. Chem.*, 1991, **26**, 13.

343. S. Clements-Jewery, G. Danswan, C. R. Gardner, S. S. Matharu, R. Murdoch, W. R. Tully and W. Westwood, *J. Med. Chem.*, 1988, **31**, 1220.

344. S. Nag, A. Mishra and S. Batra, *Eur. J. Org. Chem.*, 2008, 4334.

345. V. Singh, G. P. Yadav, P. R. Maulik and S. Batra, *Tetrahedron*, 2008, **64**, 2979.

346. (a) P. Helissey, H. Parrot-Lopez, J. Renault and S. Cros, *Eur. J. Med. Chem.*, 1987, **22**, 366; (b) G. C. Wright, E. J. Watson, F. F. Ebetino, G. Lougheed, B. F. Stevenson, A. Winterstein, R. K. Bickerton, R. P. Halliday and D. T. Pals, *J. Med. Chem.*, 1971, **14**, 1060; (c) C. A. Leach, T. H. Brown, R. J. Ife, D. J. Keeling, S. M. Laing, M. E. Parsons, C. A. Price and K. J. Wiggall, *J. Med. Chem.*, 1992, **35**, 1845; (d) C. Escolano and K. Jones, *Tetrahedron Lett.*, 2000, **41**, 8951; (e) T. H. Brown, R. J. Ife, D. J. Keeling, S. M. Laing, C. A. Leach, M. E. Parsons, C. A. Price, D. R. Reavill and K. J. Wiggall, *J. Med. Chem.*, 1990, **33**, 527; (f) M. L. Testa, L. Lamartina and F. Mingoia, *Tetrahedron*, 2004, **60**, 5873.

347. H. Benakki, E. Colacino, C. André, F. Guenoun, J. Martinez and F. Lamaty, *Tetrahedron*, 2008, **64**, 5949.

348. E. Colacino, C. André, J. Martinez and F. Lamaty, *Tetrahedron Lett.*, 2008, **49**, 4953.

349. V. Declerck, L. Toupet, J. Martinez and F. Lamaty, *J. Org. Chem.*, 2009, **74**, 2004.

350. S. Gowrisankar, H. S. Lee, K. Y. Lee, J.-E. Lee and J. N. Kim, *Tetrahedron Lett.*, 2007, **48**, 8619.

351. (a) Y. S. Lee, B. J. Min, Y. K. Park, J. Y. Lee, S. J. Lee and H. Park, *Tetrahedron Lett.*, 1999, **40**, 5569; (b) P. Pigeon and B. Decroix,

Tetrahedron Lett., 1997, **38**, 1041; (c) P. Pigeon and B. Decroix, *Synth. Commun.*, 1998, **28**, 2507; (d) G. Hilt, F. Galbiati and K. Harms, *Synthesis*, 2006, 3575; (e) G. N. Walker, A. R. Engle and R. J. Kempton, *J. Org. Chem.*, 1972, **37**, 3755; (f) H. Heaney and K. F. Shuhaibar, *Tetrahedron Lett.*, 1994, **35**, 2751; (g) H. Heaney and K. F. Shuhaibar, *Synlett*, 1995, 47; (h) M. T. El Gihani, H. Heaney and K. F. Shuhaibar, *Synlett*, 1996, 871; (i) Y. Ishihara, T. Tanaka, S. Miwatashi, A. Fujishima and G. Goto, *J. Chem. Soc., Perkin Trans.*, 1994, 2993; (j) S. Gowrisankar, K. Y. Lee and J. N. Kim, *Bull. Korean Chem. Soc.*, 2005, **26**, 1112.

352. H. S. Kim, S. Gowrisankar, S. H. Kim and J. N. Kim, *Tetrahedron Lett.*, 2008, **49**, 3858.

353. D. Basavaiah and R. J. Reddy, *Org. Biomol. Chem.*, 2008, **6**, 1034.

354. S. Kathiravan, E. Ramesh and R. Raghunathan, *Tetrahedron Lett.*, 2009, **50**, 2389.

355. (a) A. P. Kozikowski, *Acc. Chem. Res.*, 1984, **17**, 410; (b) B. B. Shankar, D. Y. Yang, S. Girton and A. K. Ganguly, *Tetrahedron Lett.*, 1998, **39**, 2447.

356. R. Annunziata, M. Benaglia, M. Cinquini, F. Cozzi and L. Raimondi, *J. Org. Chem.*, 1995, **60**, 4697.

357. (a) I. T. Raheem and E. N. Jacobsen, *Adv. Synth. Catal.*, 2005, **347**, 1701; (b) D. Bhuniya, S. Gujjary and S. Sengupta, *Synth. Commun.*, 2006, **36**, 151; (c) K. Jiang, J. Peng, H.-L. Cui and Y.-C. Chen, *Chem. Commun.*, 2009, 3955; (d) B. Das, G. Mahender, H. Holla and J. Banerjee, *Arkivoc*, 2005, 27; (e) J.-M. Becht, S. De Lamo Marin, M. Maruani, A. Wagner and C. Mioskowski, *Tetrahedron*, 2006, **62**, 4430.

358. R. Annunziata, M. Benaglia, M. Cinquini, F. Cozzi and L. Raimondi, *J. Org. Chem.*, 1995, **60**, 4697.

359. A. Longeon, M. Guoyot and J. Vacelet, *Experimentia*, 1990, **46**, 548.

360. (a) P. Mičúch, L'. Fišera, M. K. Cyrański and T. M. Krygowski, *Tetrahedron Lett.*, 1999, **40**, 167; (b) P. Mičúch, L'. Fišera, M. K. Cyrański, T. M. Krygowski and J. Krajèik, *Tetrahedron*, 2000, **56**, 5465.

361. R. E. Sammelson, C. D. Gurusinghe, J. M. Kurth, M. M. Olmstead and M. J. Kurth, *J. Org. Chem.*, 2002, **67**, 876.

362. (a) B. Dugovič, L. Fišera, C. Hametner and N. Prónayová, *Arkivoc*, 2003, 162; (b) B. Dugovič, L. Fišera, C. Hametner, M. K. Cyrański and N. Prónayová, *Monatsh. Chem.*, 2004, **135**, 685; (c) L. Bernardi, B. F. Bonini, M. Comes-Franchini, M. Fochi, M. Folegatti, S. Grilli, A. Mazzanti and A. Ricci, *Tetrahedron: Asymmetry*, 2004, **15**, 245.

363. R. Annunziata, M. Benaglia, M. Cinquini, F. Cozzi and L. Raimondi, *Eur. J. Org. Chem.*, 1998, 1823.

364. R. Galeazzi, G. Martelli, M. Orena and S. Rinaldi, *Synthesis*, 2004, 2560.

365. (a) K. Danielmeiera, R. Galeazzi, G. Martelli, M. Orena and S. Rinaldi, *Synthesis*, 2004, 2560; (b) E. Steckhan, *Tetrahedron: Asymmetry*, 1995, **6**, 1181.

366. (a) F. Coelho, R. Galeazzi, G. Martelli, M. Orena and S. Rinaldi, *Synthesis*, 2004, 2560; (b) R. C. Rossi, *Tetrahedron Lett.*, 2002, **43**, 2797.

367. (a) For a review on 1,4-oxazepines and 1,4-thiazepines, see: I. Ninomiya, T. Naito and O. Miyata, *Comp. Heterocycl. Chem., II* 1996, 217; (b) R. M. Julien, *Drogen und Psychopharmaka*, Spektrum Akademischer Verlag, Heidelberg, 1997.

368. (a) T. Mukaiyama, T. Takeda and M. Osaki, *Chem. Lett.*, 1977, 1165; (b) L. F. Tietze, S. Brand, T. Pfeiffer, J. Antel, K. Harms and G. M. Sheldrick, *J. Am. Chem. Soc.*, 1987, **109**, 921.

369. R. Räcker, K. Döring and O. Reiser, *J. Org. Chem.*, 2000, **65**, 6932.

370. L. D. S. Yadav, V. P. Srivastava and R. Patel, *Tetrahedron Lett.*, 2009, **50**, 1423.

371. E. Ramesh and R. Raghunathan, *Tetrahedron Lett.*, 2008, **49**, 1125.

372. F. Coelho, G. Diaz, C. A. M. Abella and W. P. Almedia, *Synlett*, 2006, 435.

373. E. D. Bergmann, *Chem. Rev.*, 1953, **53**, 309.

374. H. Kinoshita, S. Kinoshita, Y. Munechika, T. Iwamura, S. Watanabe and T. Kataoka, *Eur. J. Org. Chem.*, 2003, 4852.

375. M. M. Sá, L. Fernandes, M. Ferreira and A. J. Bortoluzzi, *Tetrahedron Lett.*, 2008, **49**, 1228.

376. L. D. S. Yadav, R. Patel and V. P. Srivastava, *Tetrahedron Lett.*, 2009, **50**, 1335.

377. A. Vasudevan, P.-S. Tseng and S. W. Djuric, *Tetrahedron Lett.*, 2006, **47**, 8591.

378. P. Shanmugam, S. Madhavan, K. Selvakumar, V. Vaithiyanathan and B. Viswambharan, *Tetrahedron Lett.*, 2009, **50**, 2213.

379. R. L. Grange, J. Ziogas, A. J. North, J. A. Angus and C. H. Schiesser, *Bioorg. Med. Chem. Lett.*, 2008, **18**, 1241.

380. P. Franchetti, L. Cappellacci, G. Abu Sheikha, H. N. Jayaraman, V. V. Gurudutt, T. Sint, B. P. Schneider, W. D. Jones, B. M. Goldstein, G. Perra, A. De Montis, A. G. Loi, P. La Colla and M. Grifantini, *J. Med. Chem.*, 1997, **40**, 1731.

381. S. Lal Dhar, S. Yadav, V. P. Srivastava and R. Patel, *Tetrahedron Lett.*, 2008, **49**, 5652.

Application of Morita–Baylis–Hillman Reaction for the Synthesis of Natural Products

FEI-JUN WANG, YIN WEI AND MIN SHI

5.1 Introduction

Highly functionalized MBH adducts and their derivatives have afforded access to structurally complex and diverse molecules. The synthetic applications of these compounds have clearly established the Morita–Baylis–Hillman (MBH) reaction as a standard synthetic methodology in the arsenal of organic chemists. This chapter describes the general application of the MBH reaction in the synthesis of a series of natural products and their analogues.

5.2 Acaterin

Acaterin (**1**), isolated from a culture broth of *Pseudomonas* sp. A92 by Endo and co-workers,[1] is one of the acyl-CoA cholesterol acyltransferase (ACAT) inhibitors that are expected to be effective for the treatment of atherosclerosis and hypercholesterolemia, and also has remarkable antitumor activity. Since the total synthesis of acaterin reported by Kitahara and co-workers,[2] synthetic strategies based on the MBH reaction have more recently been reported for this molecule almost simultaneously by three different research groups. Franck and Figadère first reported the synthesis of racemic acaterin **1** by condensation of α,β-unsaturated γ-lactone **2** (99% ee) and octanal *via* a DABCO-mediated MBH reaction (Scheme 5.1).[3] Unfortunately, using chiral (*S*)-**2** (99% ee) as a reactant,

---

RSC Catalysis Series No. 8
The Chemistry of the Morita–Baylis–Hillman Reaction
By Min Shi, Fei-Jun Wang, Mei-Xin Zhao and Yin Wei
© Min Shi, Fei-Jun Wang, Mei-Xin Zhao and Yin Wei 2011
Published by the Royal Society of Chemistry, www.rsc.org

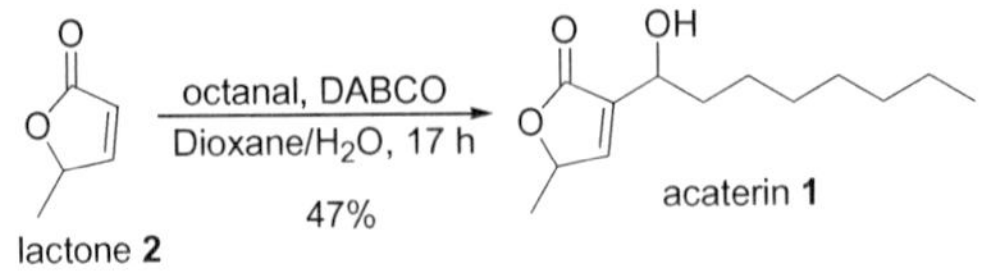

Scheme 5.1

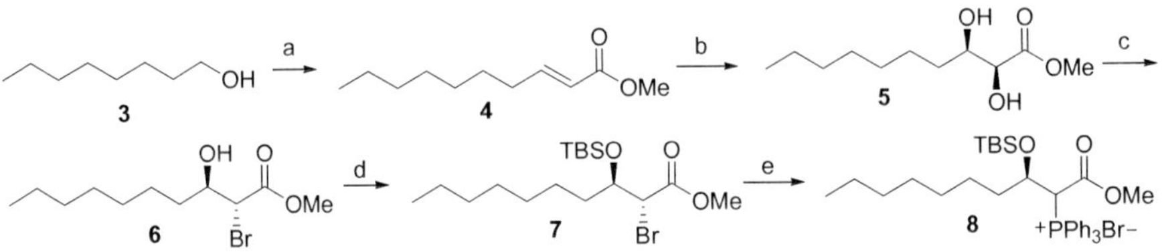

(a) (i) P$_2$O$_5$, DMSO, CH$_2$Cl$_2$, Et$_3$N, 0 °C, 4 h, 96%; (ii) Ph$_3$P=CHCO$_2$Me, THF, reflux, 12 h, 93%; (b) (DHQD)$_2$PHAL, OsO$_4$, CH$_3$SO$_2$NH$_2$, K$_3$Fe(CN)$_6$, K$_2$CO$_3$, *t*-BuOH:H$_2$O (1:1), 24 h, 0 °C, 98%; (c) HBr/AcOH, dry MeOH, 40 °C, 24 h, 83%; (d) TBDMSCl, imidazole, DMAP (cat.), CH$_2$Cl$_2$, 36 h, 92%; (e) PPh$_3$, CH$_3$CN, reflux, 12 h, 66%.

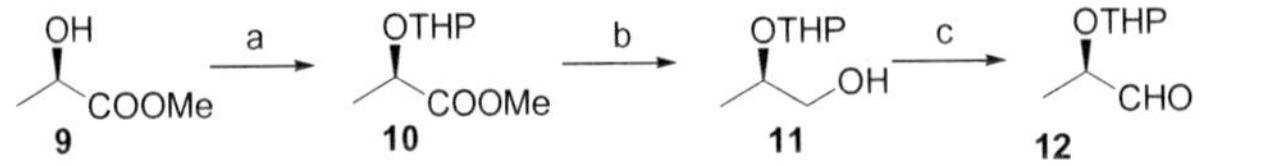

(a) DHP, *p*-TsOH (cat.), dry CH$_2$Cl$_2$, 96%; (b) LiAlH$_4$, dry THF, 0 °C to rt, 12 h, 92%;
(c) PCC, anhydrous CH$_3$CO$_2$Na, Celite, 4 h.

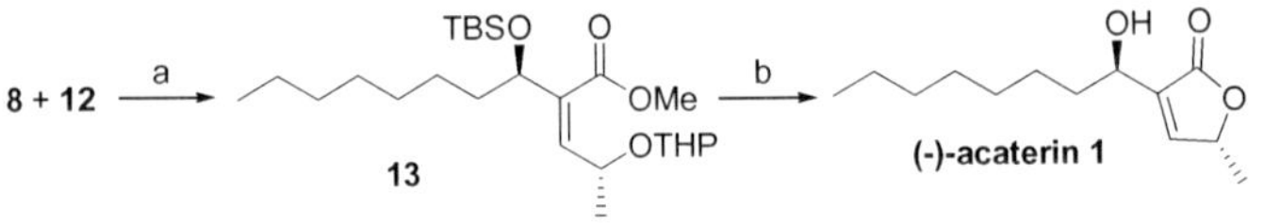

(a) LHMDS, dry THF, -78 °C, 30 min then **12**, 10 h, 73%; (b) cat. *p*-TsOH, MeOH, overnight, 68%.

Scheme 5.2

nearly racemic acaterin **1** only was obtained with low enantioselectivity ($\sim$15% ee). It was unclear whether racemization occurred after the MBH reaction or during the reaction involving enolization followed by α-aldolization.

Shortly after this, Kandula and Kumar[4] developed a synthetic route to (−)-acaterin by employing Sharpless asymmetric dihydroxylation and Wittig olefination as key steps (Scheme 5.2). Singh *et al.* have reported another synthetic route, in which a ring-closing metathesis (RCM) reaction as a key step was employed (Scheme 5.3), with the synthesis of natural (−)-acaterin **1** being accomplished in an overall yield of 22%.[5]

5.3 Analogues of *N*-Benzoyl-*syn*-phenylisoserine, (*S*)-β-Homoserine and (*S*)-Aspartic Acid

β-Amino-α-hydroxy acids have demonstrated important biological activities.[6] In particular, *N*-benzoyl-*syn*-phenylisoserine (**21**), the C13 side chain of

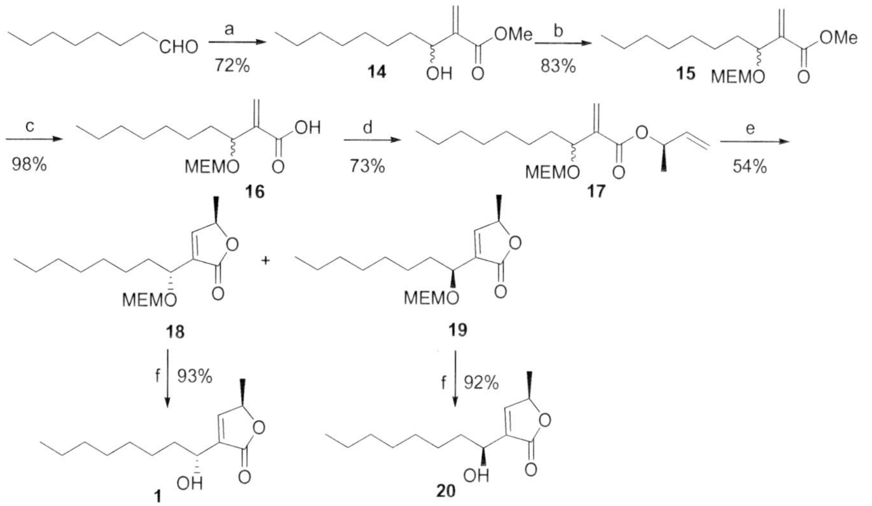

(a) Methyl acrylate, quinuclidine, 48 h; (b) 2-methoxyethoxymethyl chloride, N-ethyl diisopropylamine, DCM, 6 h; (c) 1.0 N aq. LiOH, THF/water (2:1), 24 h; (d) *R*-(-)-3-buten-2-ol, DCC, DMAP, DCM, 24 h; (e) Grubbs' catalyst (30 mol%), DCM, reflux, 48 h; (f) TiCl$_4$, DCM, 8 h.

Scheme 5.3

paclitaxel (Taxol®, **22**), has been the target of numerous synthetic efforts to develop new products with broader activity and lower toxicity than the parent drug. Galeazzi and co-workers[7] have described an efficient method to construct dihydro-1,3-oxazole derivatives with high stereoselectivities starting from the MBH adducts **23**, which easily leads to derivatives of racemic *syn*-2-substituted 2-hydroxy-3-amino acids **28–31**, which are analogues of **21**. As shown in Scheme 5.4, the reaction of isocyanates with MBH adducts **23** quantitatively yielded esters **25**, which in turn were treated with DABCO in DCM at room temperature to afford 3-acylamino-2-methylene-3-arylpropanoates **26** in 71–94% yields. Compounds **26** were then subjected to an iodocyclization reaction to give the corresponding dihydro-1,3-oxazoles **27** with *cis* configuration. These substituted dihydro-1,3-oxazoles could be easily converted into β-amino-α-hydroxy acids **28–31** with 3.0 M HCl in methanol.

Subsequently, Orena *et al.* further described an effective method to synthesize chiral 3,4-*trans*-disubstituted pyrrolidin-2-ones **33** in 76–78% yields from compounds **32** (Scheme 5.5).[8] These pyrrolidin-2-ones **33** could be easily transformed to furnish the constrained analogs of (*S*)-β-homoserine **34** and (*S*)-aspartic acid **35**, respectively.

5.4   Asmarines A and B

Asmarines A and B,[9] members of the clerodane class of natural products,[10] exhibit potent antiproliferative activity against several types of human-cancer-cell

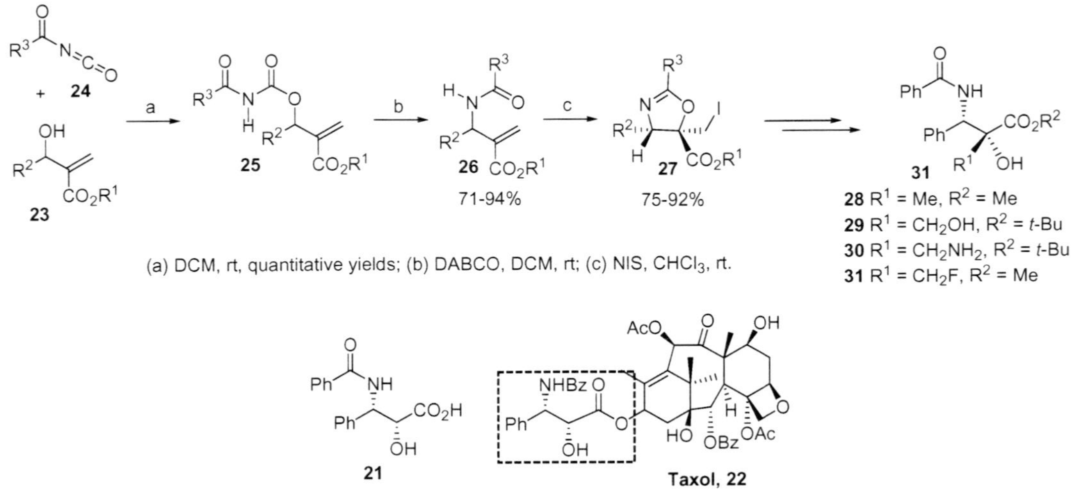

(a) DCM, rt, quantitative yields; (b) DABCO, DCM, rt; (c) NIS, CHCl$_3$, rt.

28 R^1 = Me, R^2 = Me
29 R^1 = CH$_2$OH, R^2 = t-Bu
30 R^1 = CH$_2$NH$_2$, R^2 = t-Bu
31 R^1 = CH$_2$F, R^2 = Me

Taxol, **22**

Scheme 5.4

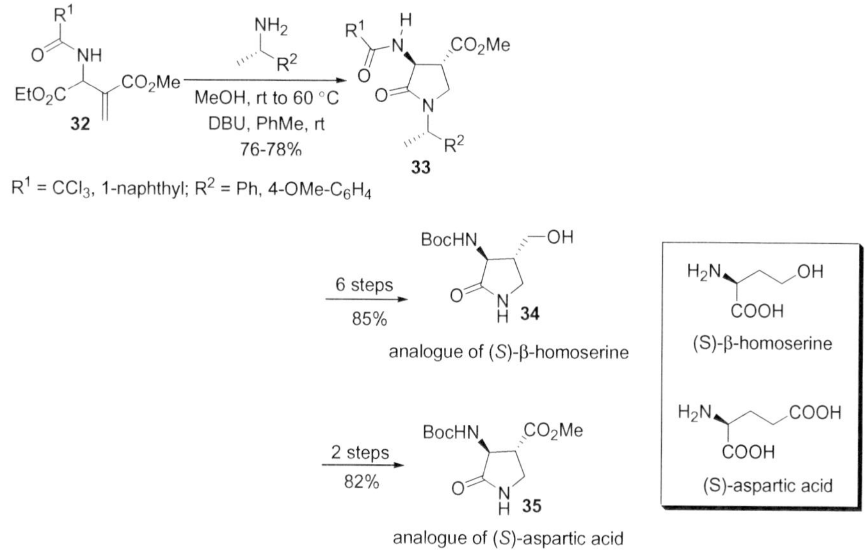

Scheme 5.5

lines.[11] A general and efficient strategy towards the core structure of clerodane was, therefore, attractive. Many approaches to construct substituted decalin structures have been reported, but these mainly gave racemic products.[12] Rodgen and Schaus have reported an efficient enantioselective route for construction of the clerodane decalin core (**36**) of asmarines A and B through an asymmetric MBH reaction followed by a Lewis acid-mediated ring-annulation strategy.[13] As shown in Scheme 5.6, the asymmetric MBH reaction of aldehyde **37** with cyclohexenone using the Brønsted acid catalyst (*R*)-**38** afforded alcohol **39**, in 86% isolated yield with 99% de, which in turn underwent an intramolecular Hosomi–Sakuari reaction in the presence of $BF_3 \cdot OEt_2$ in dichloromethane at $-78\,^{\circ}C$, leading to the clean formation of the desired clerodane core structure **36** in 81% isolated yield with 98% de.

5.5 Aza Analogues of the Tricyclic Skeleton of Daphnane and the ABC Ring System of Phorbol

The 5-7-6-tricyclic framework exists in many natural products, which often shows biological activities. For example, phorbol **40** (Figure 5.1) is extensively used in studies of tumor promotion because of its ability to activate protein kinase C.[14] The related 5-7-6 daphnane skeleton is exemplified by the antileukemic agent gnidilatin (**41**) and by the irritant resiniferatoxin, which also possesses analgesic properties.[15] However, the stereocontrolled construction of such 5-7-6-tricyclic systems is still a challenge.[16]

Scheme 5.6

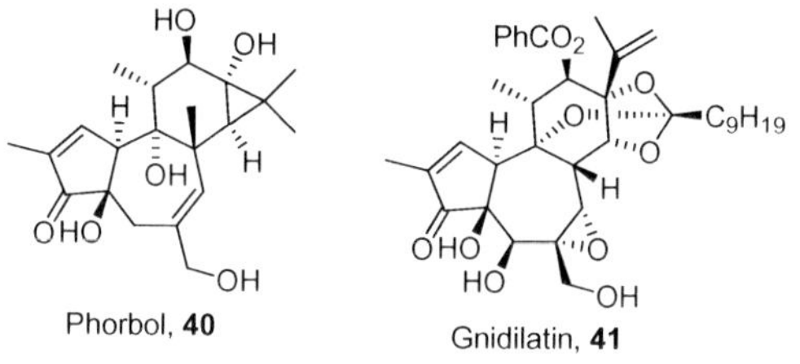

Figure 5.1 Chemical structure of phorbol (**40**) and of gnidilatin (**41**).

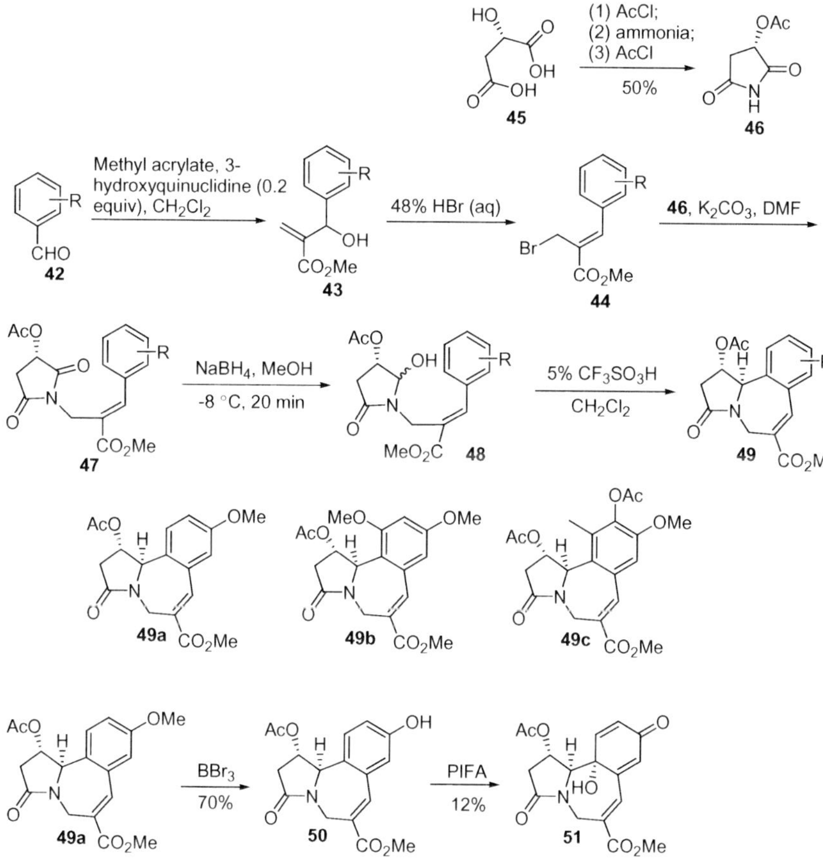

Scheme 5.7

Marson and co-workers have developed an elegant asymmetric synthetic
route to prepare aza analogues of the **ABC** ring system of phorbol and other
related compounds containing the 5-7-6-fused framework of daphnane.[17]
As shown in Scheme 5.7, using chiral succinimide **46** and MBH adducts **43**

as key intermediates, the general synthetic route to the enantioselective syntheses of the tricyclic lactams **49** was achieved in three steps. As shown in Scheme 5.7, compounds **49a–c** and the tricyclic carbinol **51** have been prepared. Herein, the construction of the central seven-membered ring was achieved by a regioselective reduction of a chiral imide and cyclization with trifluoro-methanesulfonic acid.

5.6   *N*-Boc-dolaproine

The unusual chiral β-methoxy-γ-amino acid dolaproine (Dap) is the most complex unit of dolastatin **52**, which has a remarkable antineoplastic activity[18] and is now in Phase II human cancer clinical trials.[19] Many synthetic strategies such as aldol condensation and a cobalt-catalyzed Reformatsky reaction have been employed in its synthesis. Almeida and Coelho have demonstrated a stereoselective synthetic method for *N*-Boc-dolaproine (**53**) through a sequence of MBH reaction, a diastereoselective double bond hydrogenation and hydrolysis of the ester functional group (Scheme 5.8).[20]

5.7   **Borrelidin**

Borrelidin is a structurally unique macrolide first isolated from *Streptomyces rochei* in 1949 by Berger and co-workers. It has emerged as a potent angio-genesis inhibitor from its recently described CDK inhibitory activity, with possible implications for anticancer therapy.[21] Its innovative structure and corresponding biological activity present an exciting challenge for chemical synthesis.[22] A concise synthetic route has been developed to construct the macrolide **58a** containing the (*Z,E*) C11–C15 conjugated cyanodiene fragment of borrelidin, which was the first example of a ring-closing metathesis (RCM) with a nitrile functionality on a diene.[23] Such a strategy can be applied in the total synthesis of the borrelidin. Substrate **57a** for RCM was obtained *via* a MBH analogous reaction. As shown in Scheme 5.9, the addition of undecylenic

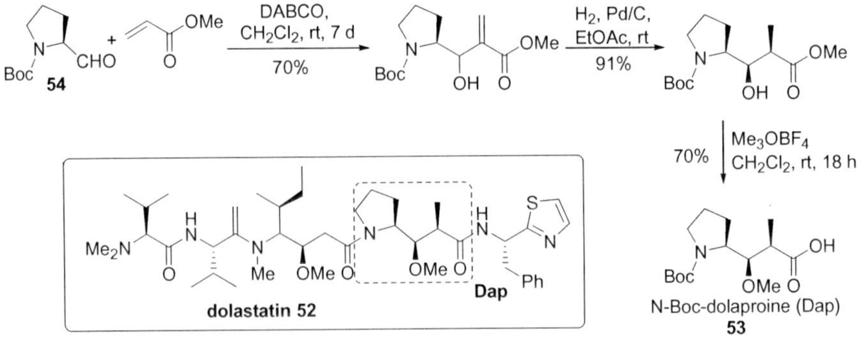

Scheme 5.8

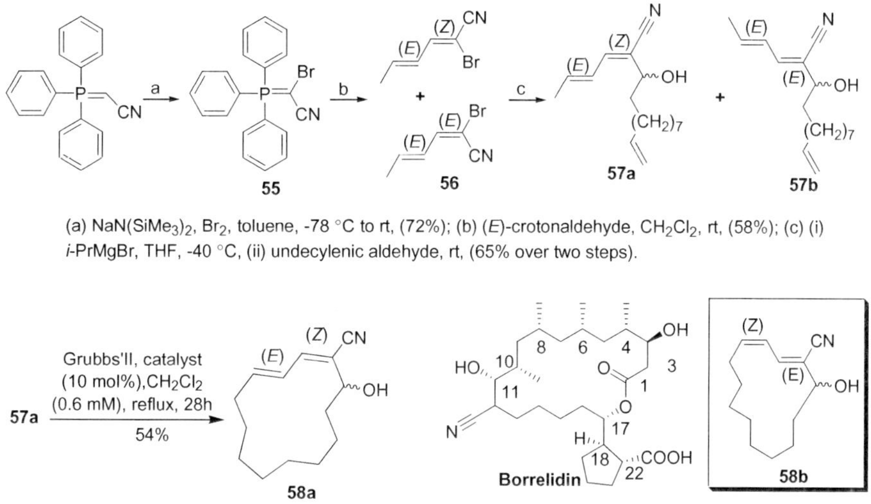

(a) NaN(SiMe$_3$)$_2$, Br$_2$, toluene, -78 °C to rt, (72%); (b) (*E*)-crotonaldehyde, CH$_2$Cl$_2$, rt, (58%); (c) (i) *i*-PrMgBr, THF, -40 °C, (ii) undecylenic aldehyde, rt, (65% over two steps).

Scheme 5.9

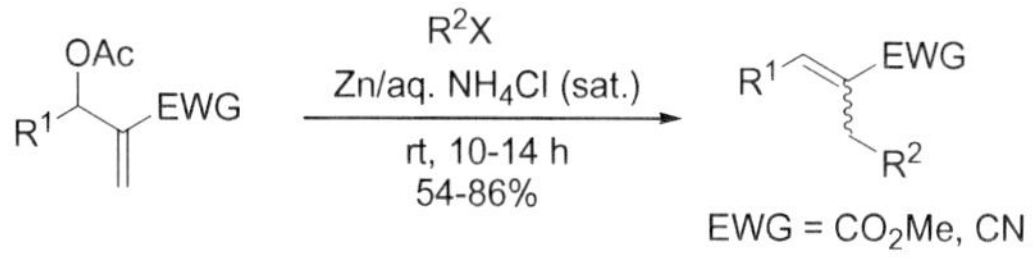

Scheme 5.10

aldehyde to the resulting Grignard product derived from the reaction of **56** and isopropylmagnesium bromide gave a mixture of (*E,Z*)-**57a** and (*E,E*)-**57b** isomers in 65% yield.

5.8 (2*E*)-2-Butyloct-2-enal

Since the first example of the construction of (*E*)- and (*Z*)-selective trisubstituted alkenes was reported using activated MBH adducts as starting materials with Grignard reagents,[24] an efficient alkylation of activated MBH adducts under mild conditions was further developed.[25] As shown in Scheme 5.10, (*E*)- and (*Z*)-trisubstituted alkenes were obtained in 54–86% yields and excellent stereoselectivities from the reaction of simple alkyl halides with activated MBH adducts in the presence of zinc and saturated aqueous NH$_4$Cl.

A trisubstituted alkene moiety, widely found in various natural bioactive molecules including several important pheromones and antibiotics, often has certain biological activities.[26] For example, the following methodology has been employed to synthesize (2*E*)-2-butyloct-2-enal (**62**),[27] an alarm pheromone component of the African weaver ant, *Oecophylla longinoda* (Scheme 5.11).

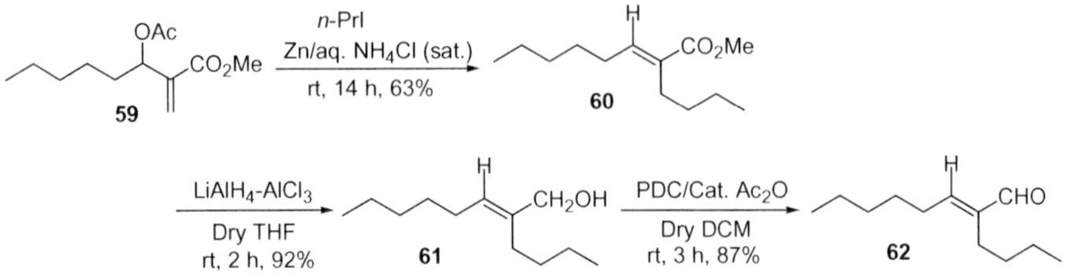

Scheme 5.11

Reduction of methyl (2*E*)-2-butyl-3-pentyl-propenoate (**60**) yielded the corresponding alcohol **61** in 92% yield. Upon oxidation of alcohol **61**, the product **62** was furnished in 87% yield.

5.9   Caribenolide I

Caribenolide I, isolated from dried cells of *Amphidinium* sp., is a 26-membered macrolactone of amphidinolides[28] and possesses important *in vivo* activity against P388 tumor grafted mice and important *in vitro* cytotoxicity against human colon tumor cells of wild type as well as against those that have shown a multi-drug resistance phenotype.[29] Although a few total syntheses of amphidinolides have appeared in the literatures,[30] no paper concerning the total or partial synthesis of caribenolide I emerged until the report by Franck *et al.* They successfully synthesized the C1–C11 fragment **63** of caribenolide I (Scheme 5.12) starting from intermediate **64**, which in turn was obtained in 62% yield from the reaction of 3-*para*-methoxybenzyloxypropanal with methyl acrylate in the presence of 3-hydroxyquinuclidine (3-HQD).[31]

5.10   Clusianone and Polycyclic Polyprenylated Acylphloroglucinols (PPAPs)

Clusianone **65**,[32] bearing a densely functionalized bicyclo[3.3.1]nonane-1,3,5-trione core structure and involved in other natural products such as nemorosone and hyperibone K (Figure 5.2), has been isolated from the floral resins of *Clusia* species.[33] A synthetic method for the construction of such a core structure of these compounds, which hold the promise of biological activities, remains a challenge. Using MBH adducts as starting materials, an effective approach employed an alkylative dearomatization–annulation to construct the bicyclo[3.3.1]nonane core has been developed by Porco *et al.* (Scheme 5.13).[34] Treatment of phloroglucinol **66** bearing an alkyl-aryl ketone ($R^1 = i$-Pr) with LHMDS (3 equiv.) followed by the addition of α-acetoxymethyl acrylate **67** (2 equiv.) at 0 °C led to an efficient, highly diastereoselective dearomatization–annulation process in which an additional Michael-elimination event had unexpectedly occurred to afford **68a** in 84% yield. For substrate **66** bearing

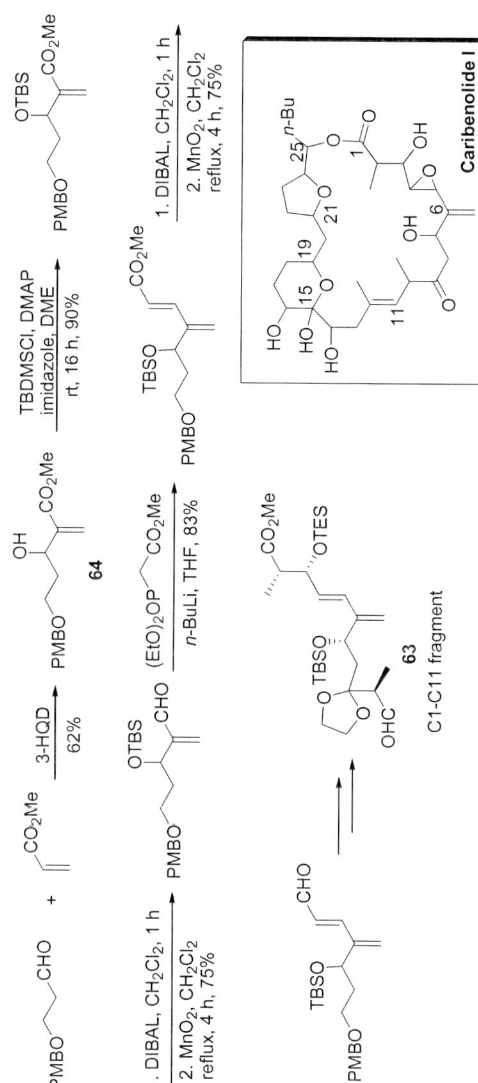

Scheme 5.12

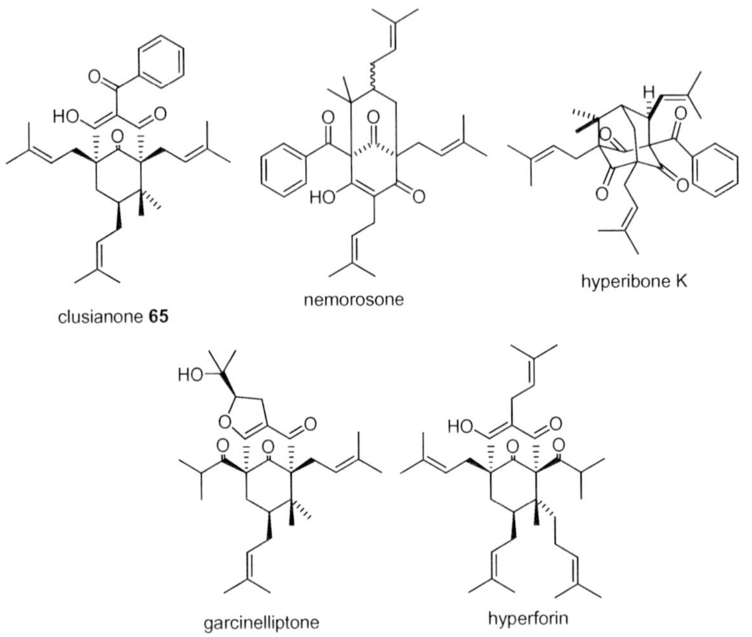

Figure 5.2 Examples of natural products with a functionalized bicyclo[3.3.1]nonane-1,3,5-trione core structure.

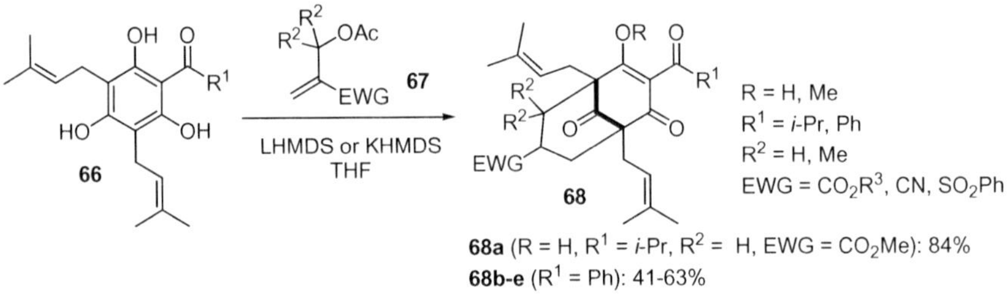

Scheme 5.13

aryl-aryl ketone ($R^1 = $ Ph) the reactions can proceed smoothly with various Michael acceptors having an EWG such as ester, acrylonitrile and sulfone in the presence of stronger base (KHMDS), but the bicyclo[3.3.1]nonane cores **68b–e** are afforded in lower yields (41–63%). By applying this synthetic strategy in the total synthesis of racemic natural products bearing densely functiona-lized bicyclo[3.3.1]nonane-1,3,5-trione core structures, clusianone **65** has been synthesized in five steps with 27% overall yield from **69** (Scheme 5.14).

More recently, Takagi and co-workers also developed an effective synthetic route to construct the bicyclo[3.3.1]nonane core, using **MBH** adduct **74** as starting material.[35] Functionalized adamantane **75** is the core of polycyclic

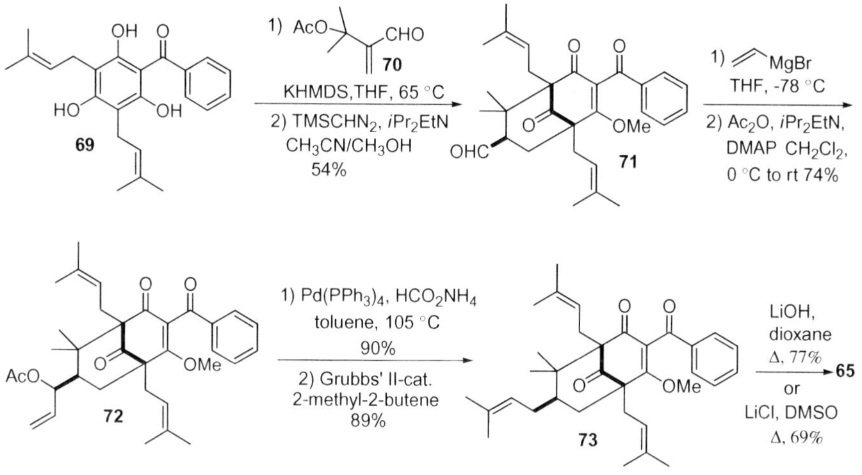

Scheme 5.14

polyprenylated acylphloroglucinols (PPAPs), which have been isolated from guttiferous plants. An efficient method for the construction of the adamantane core of plukenetione-type is described in Scheme 5.15. The method features two types of reactions: successive Michael reaction and acid-catalyzed cyclization.

5.11 Cyclic Peptides

Cyclic peptides are found everywhere in nature and have a range from just a few amino acids in length to hundreds. Their interesting bio-activities have also made them a hot topic in the drug discovery field. Most cyclic peptides can be synthesized routinely by the split-and-pool method. Iqbal and co-workers have employed RCM as a key step to construct a cyclic peptide structure (Scheme 5.16).[36] The α-dehydro β-amino acid derivatives **85**, prepared from Pd-catalyzed amination of MBH adduct in 75–80% yields, are very good nucleophiles as they cleaves epoxides in the presence of a catalytic amount of cobalt(II) chloride. In the presence of CoCl$_2$ catalyst, β-turn mimics containing an α-hydroxy β-amino residue (**86**) as the *trans* diastereomer have been obtained in 45–50% yields *via* a Co-catalyzed opening reaction of epoxy peptides **84**. The introduction of an allyl group into peptide **86** furnished the diallylated precursor **87**, which in turn was subjected to a RCM reaction using Grubbs catalyst to afford the cyclic peptide **88**, in 40–45% yields, as a constrained mimic of a β-turn [as a mixture of $E : Z$ (3:1) isomers].

5.12 (−)-(Z)-Deoxypukalide

(+)-(Z)-Deoxypukalide **89**, recently isolated from Pacific octocoral *Leptogorgia* sp.,[37] belongs to a family of furan cembranolides, which are an intriguing class

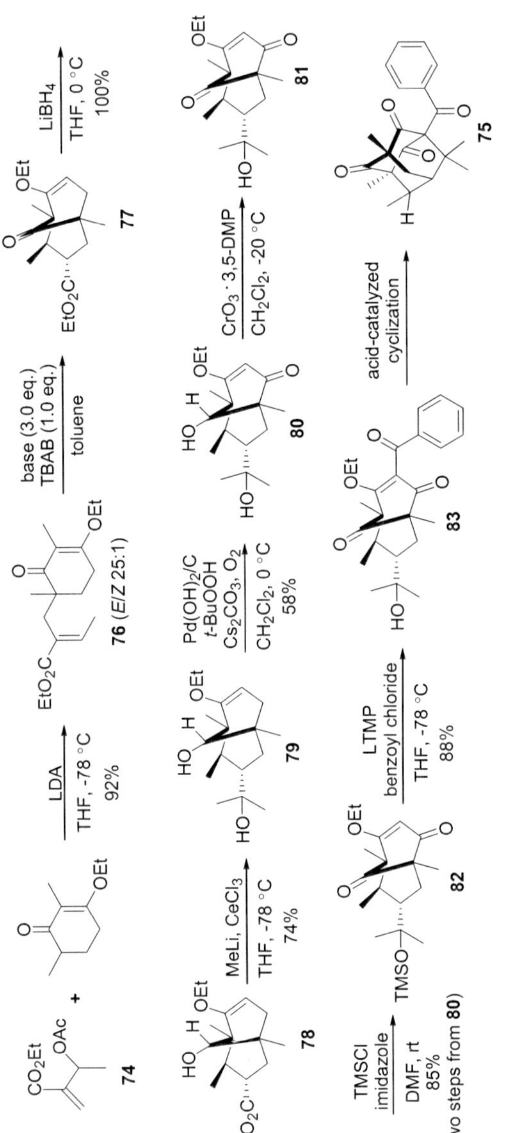
CO2Et
OAc
74
+
OEt
LDA
THF, -78 °C
92%
EtO2C
OEt
76 (E/Z 25:1)
base (3.0 eq.)
TBAB (1.0 eq.)
toluene
EtO2C
OEt
77
LiBH4
THF, 0 °C
100%
EtO2C
HO
H
OEt
78
MeLi, CeCl3
THF, -78 °C
74%
HO
HO
H
OEt
79
Pd(OH)2/C
t-BuOOH
Cs2CO3, O2
CH2Cl2, 0 °C
58%
HO
HO
H
OEt
80
CrO3 · 3,5-DMP
CH2Cl2, -20 °C
HO
OEt
81
TMSCl
imidazole
DMF, rt
85%
(two steps from 80)
TMSO
OEt
82
LTMP
benzoyl chloride
THF, -78 °C
88%
HO
OEt
83
acid-catalyzed
cyclization
H
75

Scheme 5.15

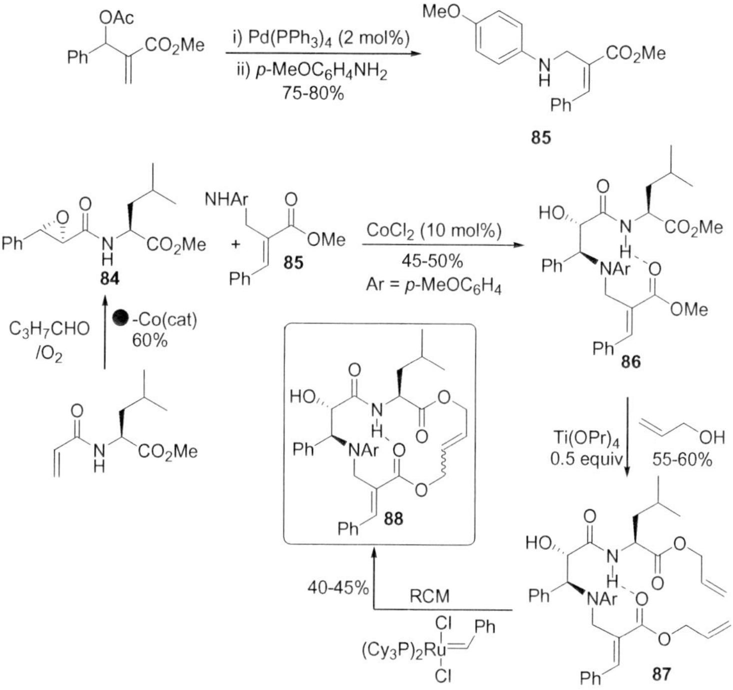

Scheme 5.16

of macrocyclic marine natural products that exhibit a wide range of biological activities from neurotoxicity to anti-inflammatory and antifeedant properties. After the development of a total synthesis of (+)-**89** by Marshall and Van Devender (through a 28-step synthesis),[38] (−)-(Z)-deoxypukalide (−)-**89** was synthesized in an overall yield of 15% with just 12 linear steps from easily available starting materials.[39] This elegant synthetic route featured the RCM/ aromatization protocol to prepare a disubstituted furan and a regioselective Negishi cross-coupling reaction to construct the fully substituted aromatic core of the target (Scheme 5.17). To prepare key intermediate **92**, a MBH analogous reaction was realized in the formation of allylic alcohol **91** from **90** and a vinylalane derived from the reaction of DIBAL-H with methyl propiolate, as an inconsequential (1 : 1) mixture of diastereoisomers which couldn't be obtained under MBH reaction conditions.

5.13 Diversonol

Diversonol (**97**), a fungal metabolite, was isolated by Turner from *Penicillium diversum* and its structural motif has also been found in some mycotoxins such as secalonic acids with interesting biological activity. Therefore, the total synthesis of diversonol will also promote the total synthesis of these

(S)-perillyl alcohol

a) TIPSCl, imid
b) O₃, py (1 eq.), isoprene

90

DIBAL-H, HMPA
methylpropiolate
48% (3 steps)

EtO OEt
PPTS

91

a) MePPh₃·Br, *n*-BuLi
b) TBAF
c) TEMPO, NaClO₂, NaOCl
90% (3 steps)

Grubbs II cat.
CH₂Cl₂, Δ
then add PPTS
85% (2 steps)

92

93

a) LDA (2.3 equiv), ZnBr₂, then PdCl₂(dppf)] (cat.)
b) TBAF 78%

94

MNBA, Et₃N, DMAP
73%

95

Grubbs' II cat.
toluene, Δ, 72%

(-)-(Z)-deoxypukalide (**89**)
12 liner steps and 15% overall yield

Scheme 5.17

compounds. After the development of novel synthetic strategy to construct tetrahydroxanthenone from salicylic aldehydes and cyclohexenones,[40] the first total synthesis of diversonol in racemic form through a sequence of 14 synthetic steps was achieved by Bräse *et al.* (Scheme 5.18).[41] In their synthetic strategy, a flexible route to constructing tetrahydroxanthenone mycotoxin framework was developed, using a domino oxa-Michael-aldol condensation as the key step. Tetrahydroxanthenone **96** was isolated in 61% yield and as a 1.5 : 1 mixture of the two possible diastereoisomers using imidazole as the base. Using DABCO and K_2CO_3 only led to decomposition of the cyclohexenone.[42] Such a synthetic approach has also been applied in the total synthesis of blennolide C as well as to several advanced synthetic intermediates[43] for the preparation of secalonic acids (Figure 5.3).[44]

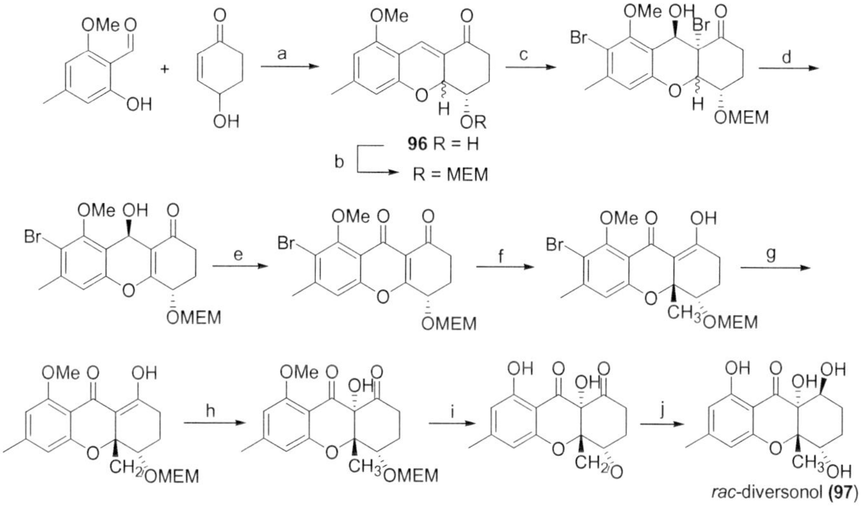

a) Imidazole, dioxane/H_2O, sonication, 1 d, 61%; b) MEMCl, *i*-Pr$_2$NEt, CH$_2$Cl$_2$, rt, 3 h, 75%; c) tetrabutylammonium tribromide, THF/H_2O, rt, 5 h, 52%; d) DABCO, dioxane, rt, 16 h, 53%; e) TPAP, NMO, CH$_2$Cl$_2$/CH$_3$CN, sonication, 40%; f) MeLi, CuCN, Et$_2$O, -78 °C, 5 h, 52%; g) *t*-BuLi, THF, -78 °C, NaHCO$_3$, 4 h, 93%; h) magnesium monoperoxophthalate, EtOH, rt, 5 h, 57%; i) BBr$_3$, CH$_2$Cl$_2$, rt, 7 h, 40%; j) NaBH$_4$, MeOH, -78 °C, 20 min, 66%.

Scheme 5.18

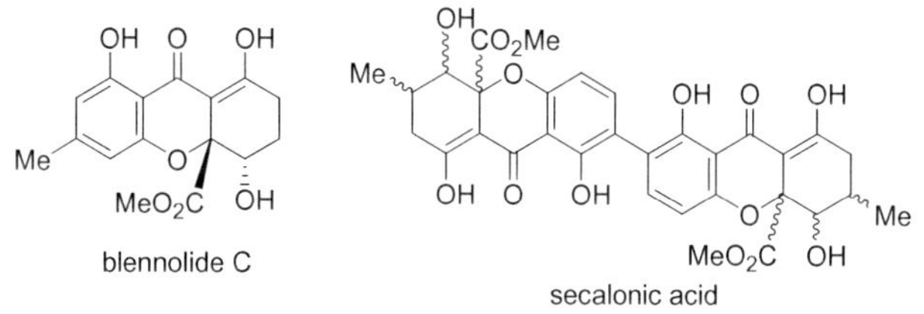

Figure 5.3 Chemical structures of blennolide C and secalonic acid.

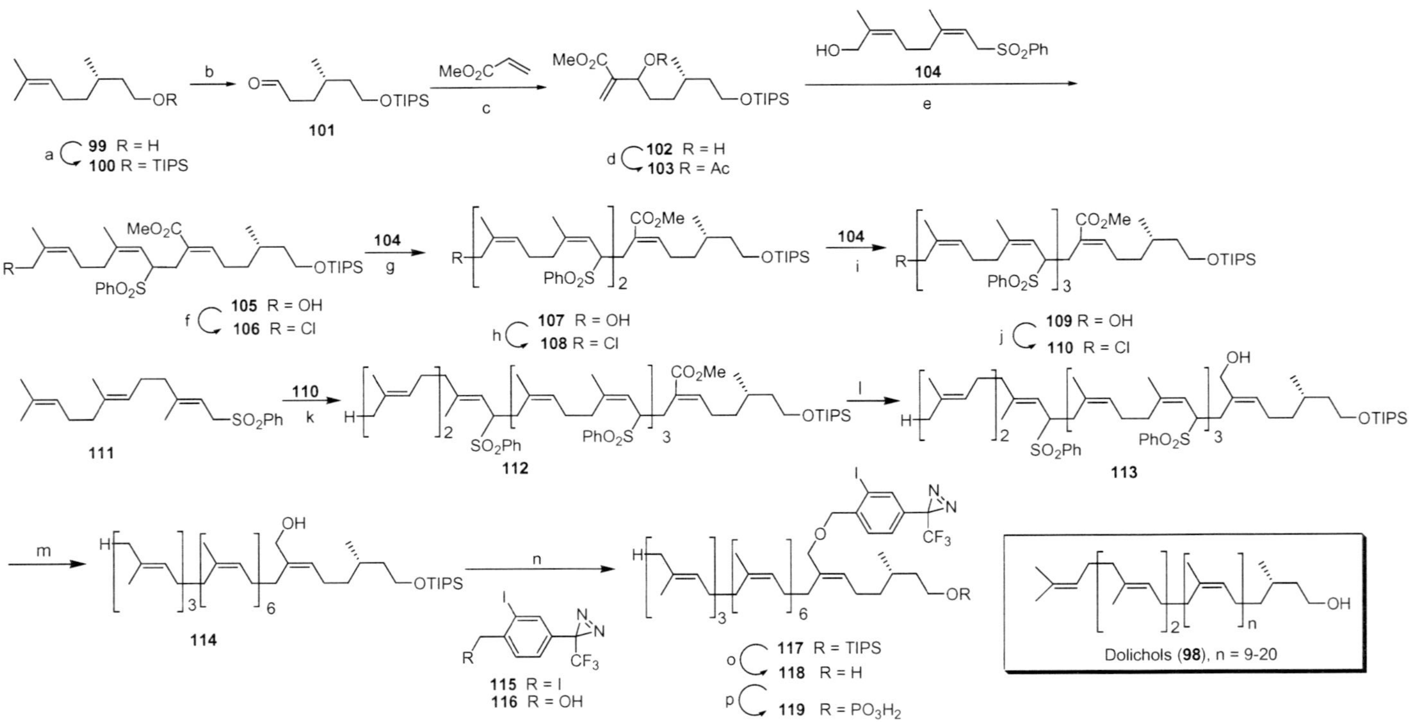

(a) TIPSCl, imidazole, CH$_2$Cl$_2$; (b) i). O$_3$, CH$_2$Cl$_2$, -78 °C; ii). Me$_2$S (65-83% from **99**); (c) DABCO (62-79%); (d) Ac$_2$O, pyridine, DMAP,CH$_2$Cl$_2$ (90-95%); (e) *n*-BuLi, THF, DMTP, -78 °C (69%); (f) MsCl, LiCl, collidine, DMF, 0 °C; (g) *n*-BuLi, THF, DMTP, -78 °C (52% from **105**); (h) MsCl, LiCl, collidine, DMF, 0 °C (92%); (i) *n*-BuLi, THF, DMTP, -78 °C (76%); (j) MsCl, LiCl, collidine, 0 °C (89%); (k) *n*-BuLi, THF, DMTP, -78 °C (66%); (l) DIBAH, CH$_2$Cl$_2$, -78 °C (78%); (m) LiEt$_3$BH, [PdCl$_2$(dppp)], THF, 0 °C (43-51%); (n) **115**, benzene, NaOH aq, TBAI; (o) THF, TBAF (56% from **114**); (p) POCl$_3$, NEt$_3$, hexane (68%)

Scheme 5.19

5.14 Dolichol Analogues

Dolichols (**98**), a family of polyisoprenoid alcohols, have isolated from all eukaryotic cells or archaebacteria.[45] They play a role in the co-translational modification of proteins and are known in *N*-glycosylation in the form of dolichol phosphate.[46] The synthesis of the photochemical probes **118** and **119** bearing a photoreactive group [3-(trifluoromethyl)-3-aryldiazirine], analogues of dolichol and dolichol phosphate, is described in Scheme 5.19.[47] The synthetic strategy involves the sequential alkylation of a monoterpenoid hydroxysulfonyl dianion with allyl chlorides, and MBH adduct **102** as a starting material.

5.15 Dykellic Acid and Gelastatin Analogues

Dykellic acid is a novel microbial metabolite isolated from the broth of *Westerdykella multispora* F 50733.[48] Investigations on the molecular function of dykellic acid revealed that this compound partially inhibited calcium influx, resulting in a decrease in Ca^{2+}-dependent endonuclease activation and DNA fragmentation induced by camptothecin. Other interesting biological properties of dykellic acid were also described, such as its ability to inhibit cell migration[49] and interfere with NF-*k*B transcriptional activity.[50] Hergenrother and co-workers first described the total synthesis of dykellic acid and its derivatives, along with the biological evaluation of these compounds.[51] From a retrosynthetic analysis, two key carbon–carbon bond forming reactions were envisioned: a Horner–Wadsworth–Emmons (HWE) reaction and a MBH reaction (Scheme 5.20). 2,4-Hexadienal **120** was reacted with methyl

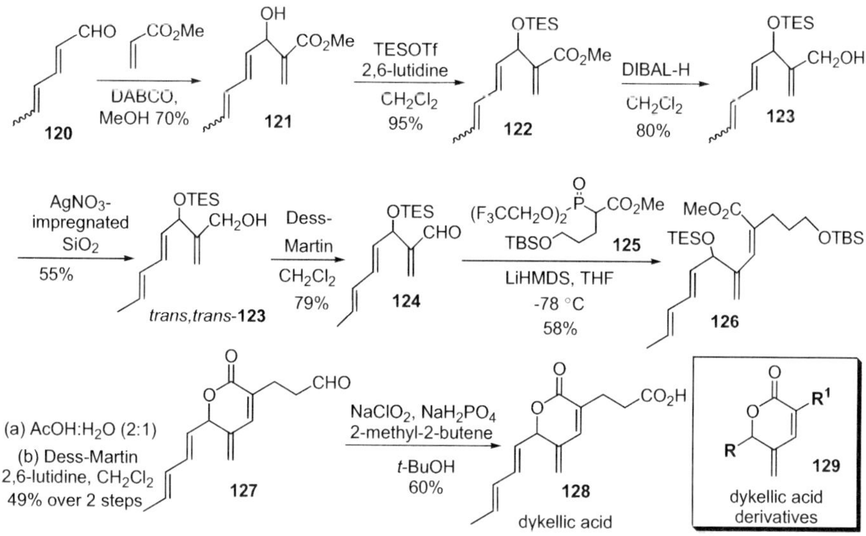

Scheme 5.20

acrylate at 0 °C for 3 days in the presence of DABCO, generating alcohol **121** in 70% yield. However, notably, even using pure *trans,trans*-2,4-hexadienal **120** also provided alcohol **121** as an isomeric mixture, presumably because of unproductive 1,6-addition of DABCO to the aldehyde. The TES-protected ether **122** was reduced by DIBAL-H to give alcohol **123** in 80% yield, which was then was subjected to chromatography separation with 25% AgNO$_3$-impregnated silica gel to afford pure isomer *trans,trans*-**123** in 55% yield. After Dess–Martin oxidation of *trans,trans*-**123** and subsequent HWE reaction of phosphonate carbanion **125** with aldehydes **124**, the trisubstituted olefin **126** was obtained with as a Z/E mixture. Two silyl-ether groups were deprotected with an AcOH–H$_2$O mixture, and spontaneous cyclization then gave an unstable lactone that was immediately oxidized to aldehyde **127** using Dess–Martin periodinane. Treatment of aldehyde **127** with sodium chlorite furnished dykellic acid (**128**) in a 60% yield. The various dykellic acid derivatives **129** have been synthesized according to the same procedure as that for dykellic acid (**128**).

Gelastatin A and B, isolated from culture broth of *Westerdykella multispora* F 50733, also contain a 5-methylene-5,6-dihydro-2*H*-pyran-2-one framework, and have been reported to exhibit MMP-inhibitory activities at the sub-micromolar level.[52] Cho and Ko first reported the synthesis of gelastatin analogues **130** (Scheme 5.21), in which the benzylidene group replaced the triene unit.[53] The MBH reaction between the aldehyde **131** and methyl acrylate was performed in the presence of DABCO/triethanolamine/[La(OTf)$_3$] or Bu$_3$P to give, respectively, (*E*)- or (*Z*)-MBH adduct **132**. Desilylation of adducts **132**

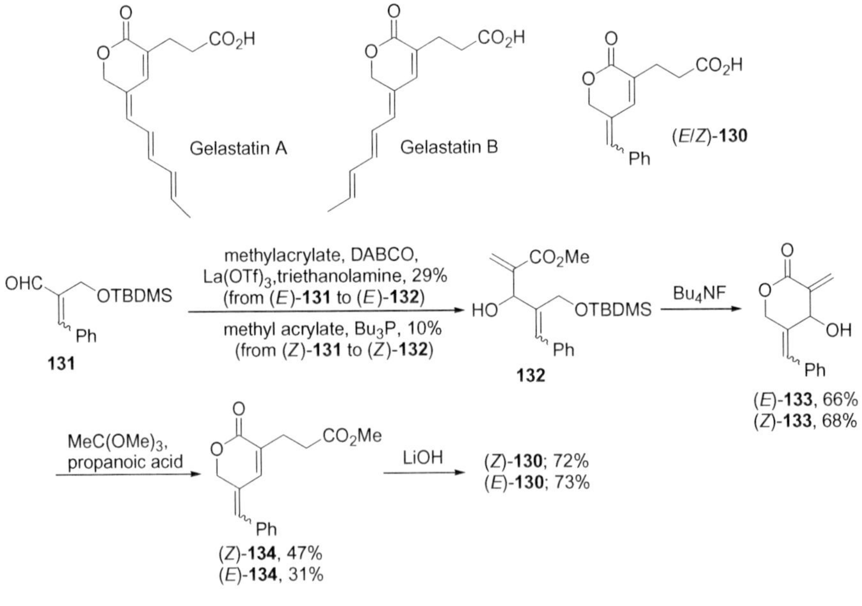

Scheme 5.21

using Bu_4NF resulted in a concurrent lactonization to give **133**, which following *ortho*-ester Claisen rearrangement produced the desired skeleton **134**. Final hydrolysis of each of the (*E/Z*)-isomers yielded the desired benzylidene-substituted gelastatin **130**.

5.16 (+)-Efaroxan

(+)-Efaroxan,[54] an α2 adrenoreceptor antagonist, could be used for the treatment of neurodegenerative diseases (Alzheimer and Parkinson), migraine and type II diabetes. Therefore, the total syntheses of (+)-efaroxan and their derivatives have drawn much attention.[55] The chiral 2,3-dihydrobenzofuran carboxylic acid **135**, the direct precursor of (+)-efaroxan, was obtained mainly from the resolution of racemic **135**.[56] Coelho *et al.* have reported a straightforward enantioselective synthesis of *R*-(+)-2-ethyl-2,3-dihydrofuran-2-carboxylic acid (**135**) achieved by a Sharpless–Katsuki asymmetric epoxidation reaction (Scheme 5.22).[57] The dihydrobenzofuran acid **135** was obtained in seven steps from MBH adduct **136** in an overall yield of 17%.

5.17 Epopromycin B

Epopromycin B is a novel plant cell wall synthesis inhibitor isolated from the culture broth of *Streptomyces* sp. NK0400.[58] This densely functionalized structural motif, bearing an epoxy-β-aminoketone moiety (Figure 5.4), is also found in the proteasome inhibitors TMC-86 and TMC-96[59] as well as angiostatic natural product eponemycin,[60] and thus has attracted considerable attention as a promising pharmacophore.[61] A facile, effective, enantio- and stereocontrolled route to synthesize the epoxy-β-aminoketone pharmacophore **141**, Dobler's key precursor of epopromycin B,[62] has been developed based on the cinchona alkaloid-catalyzed MBH reaction of *N*-Fmoc-leucinal, starting from (*S*)-*N*-Fmoc-leucinal (*S*-**137**), in six steps in 29% overall yield (Scheme 5.23).[63]

5.18 Eupomatilone 2

Eupomatilones 1–7 (**142a–g**, Figure 5.5), isolated by Carroll and Taylor in 1991 from the Australian shrub *Eupomatia bennettii*,[64] bear an α-methylene-γ-lactone fragment that has been considered to produce antigenic compounds *via* forming covalent bonds to cellular proteins and may be the cause chronic actinic dermatitis (CAD).[65] A short and efficient strategy for the total syntheses of eupomatilones 2 (**142b**) and 5 (**142e**) was first accomplished using MBH adducts as synthetic precursors.[66] As shown in Scheme 5.24, Suzuki coupling and allylation reactions as the key steps were involved in this route. In addition, allyl-metal reagents such as allylindium reagents obtained *in situ* from the reaction of allyl bromide **143** with In metal were utilized to prepare the lactone precursor **144**. Alcohols **144** were easily cyclized under mild acidic conditions (PTSA, CH_2Cl_2).

Scheme 5.22

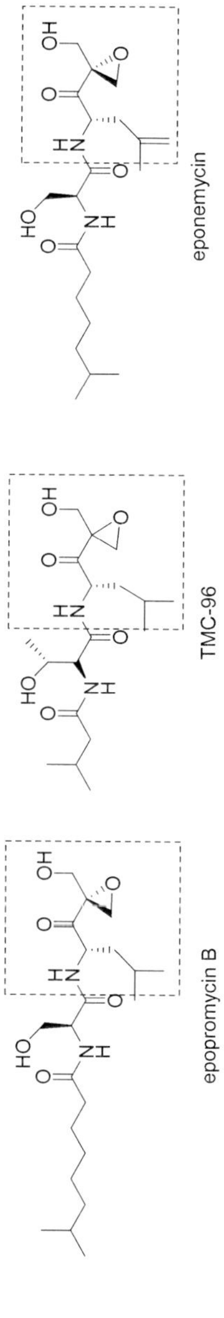

Figure 5.4 Compounds containing an epoxy-β-aminoketone moiety.

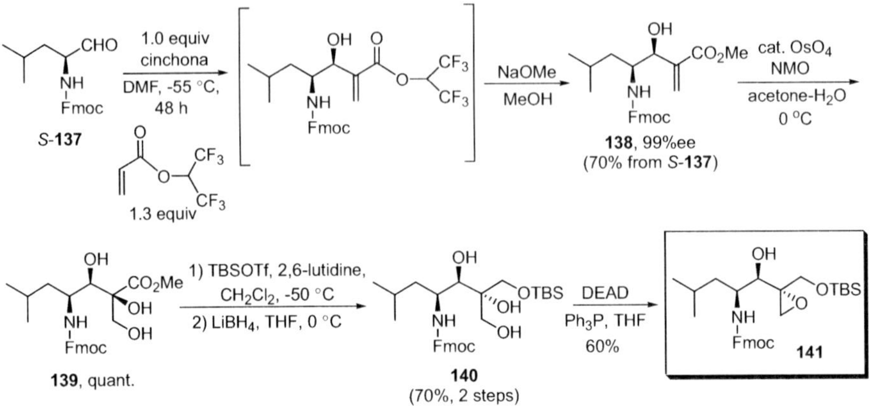

Scheme 5.23

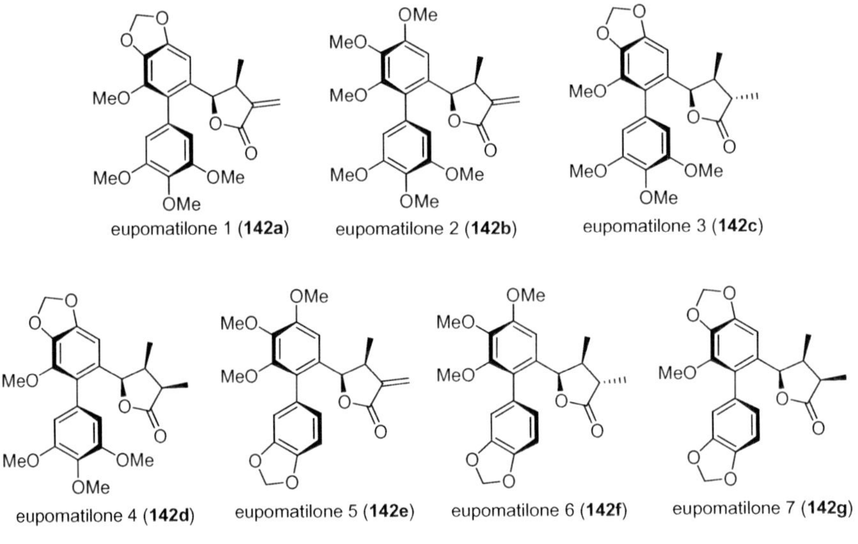

eupomatilone 1 (**142a**)

eupomatilone 2 (**142b**)

eupomatilone 3 (**142c**)

eupomatilone 4 (**142d**)

eupomatilone 5 (**142e**)

eupomatilone 6 (**142f**)

eupomatilone 7 (**142g**)

Figure 5.5   Eupomatilones 1–7.

More recently, an asymmetric crotyl-boration strategy to directly assemble the α-methylene-γ-lactone moiety with *syn*-stereochemistry at the C4 and C5 positions of the lactone ring was developed by Coleman and co-workers, and applied in the asymmetric total synthesis of eupomatilones (Scheme 5.25).[67] Eupomatilones 2 (**142b**) and 5 (**142e**) were synthesized in up to 74% yield with 88 : 12 er (enantiomeric ratio) from borane reagent **147**, which was prepared from the reaction of MBH acetates **145** with **146**, while eupomatilone 1 was obtained only in 30% yield and 3 : 1 er under similar conditions. This indicates that the steric interaction between lactone ring and aromatic ring may be the

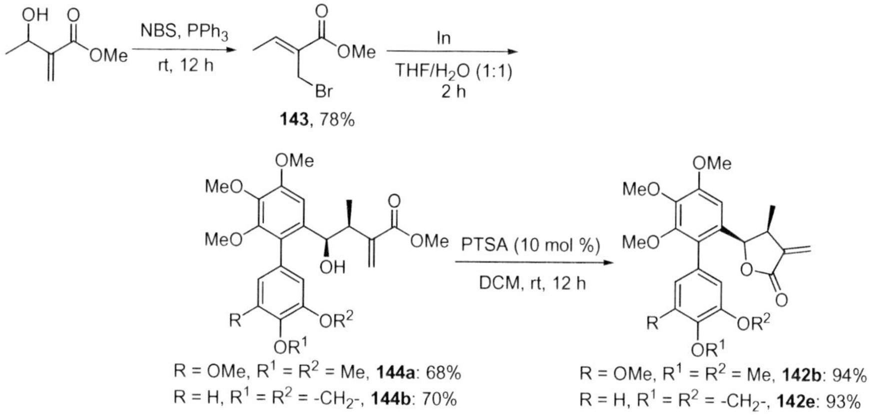

Scheme 5.24

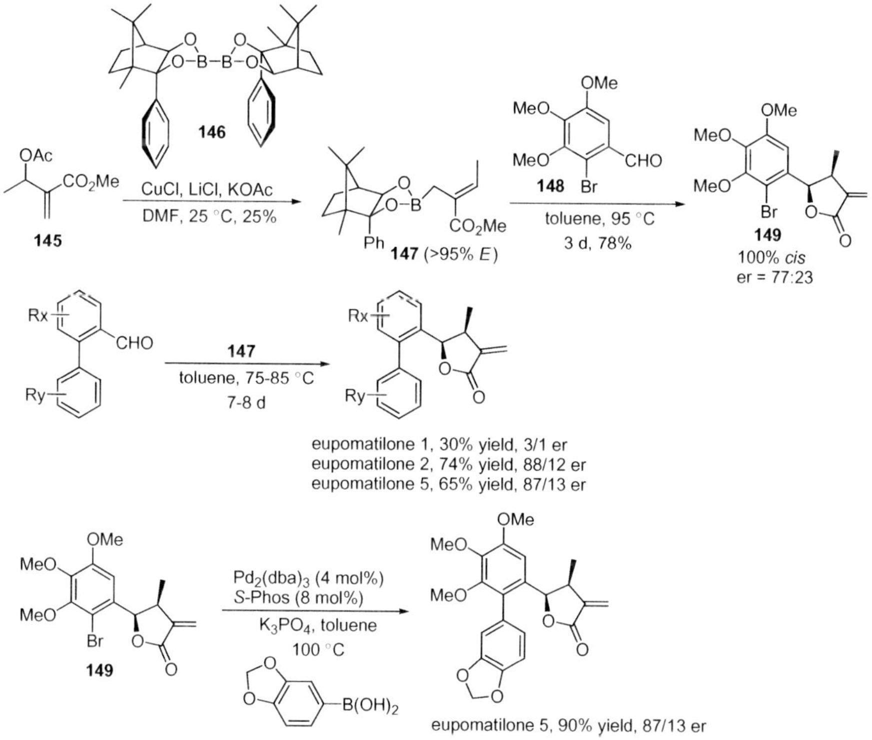

Scheme 5.25

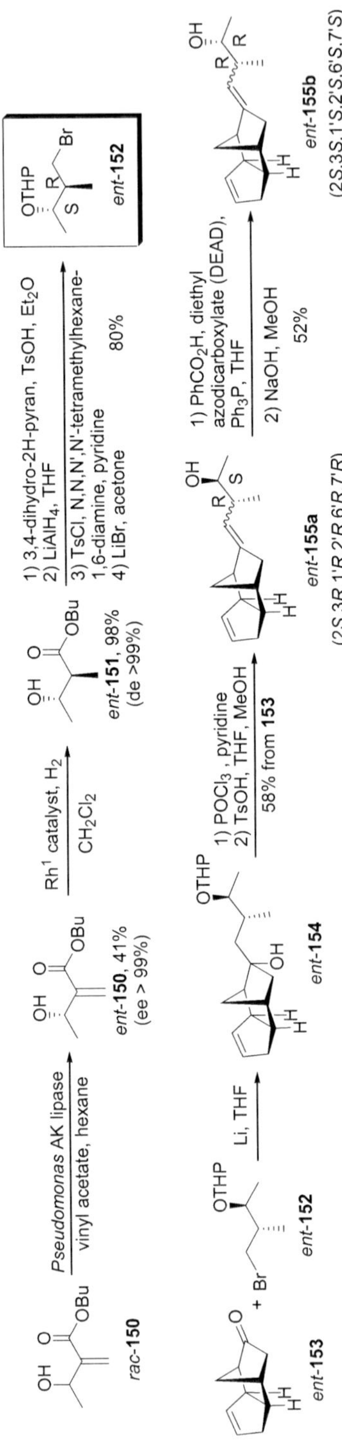

Scheme 5.26

key factor affecting the reaction results. The same authors also developed another asymmetric synthetic route of eupomatilone 5 in 90% yield and 74% ee *via* a Suzuki–Miyaura biaryl cross-coupling reaction.

5.19   Fleursandol®

Despite the large variety of synthetic sandalwood odorants commercially available, the search for new aroma chemicals possessing an even more natural, sandalwood-like odor character and/or a different C-skeleton is ongoing.[68] East-Indian sandalwood oil (*Santalum album* L.) belongs to the oldest known perfumery ingredients and is still a highly appreciated raw material for fragrances.[69] The enantioselective total synthesis of Fleursandol®,[70] a new class of sandalwood odorants, was first accomplished by Hölscher *et al.* (Scheme 5.26).[71] Using MBH adduct *rac*-150 as a starting material, a key chiral synthon (*ent*-150) has been prepared through a lipase-mediated resolution protocol. (–)-(*S*)-ester *ent*-150 was obtained in 41% yield and up to ≥ 99% ee in the presence of *Pseudomonas* AK lipase. *Ent*-150 could be easily converted into key

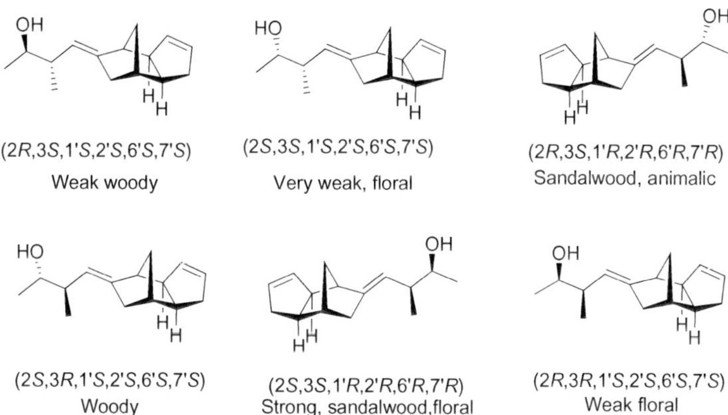

Figure 5.6   Other synthesized stereoisomers of Fleursandol®.

Figure 5.7   Frondosins A–E.

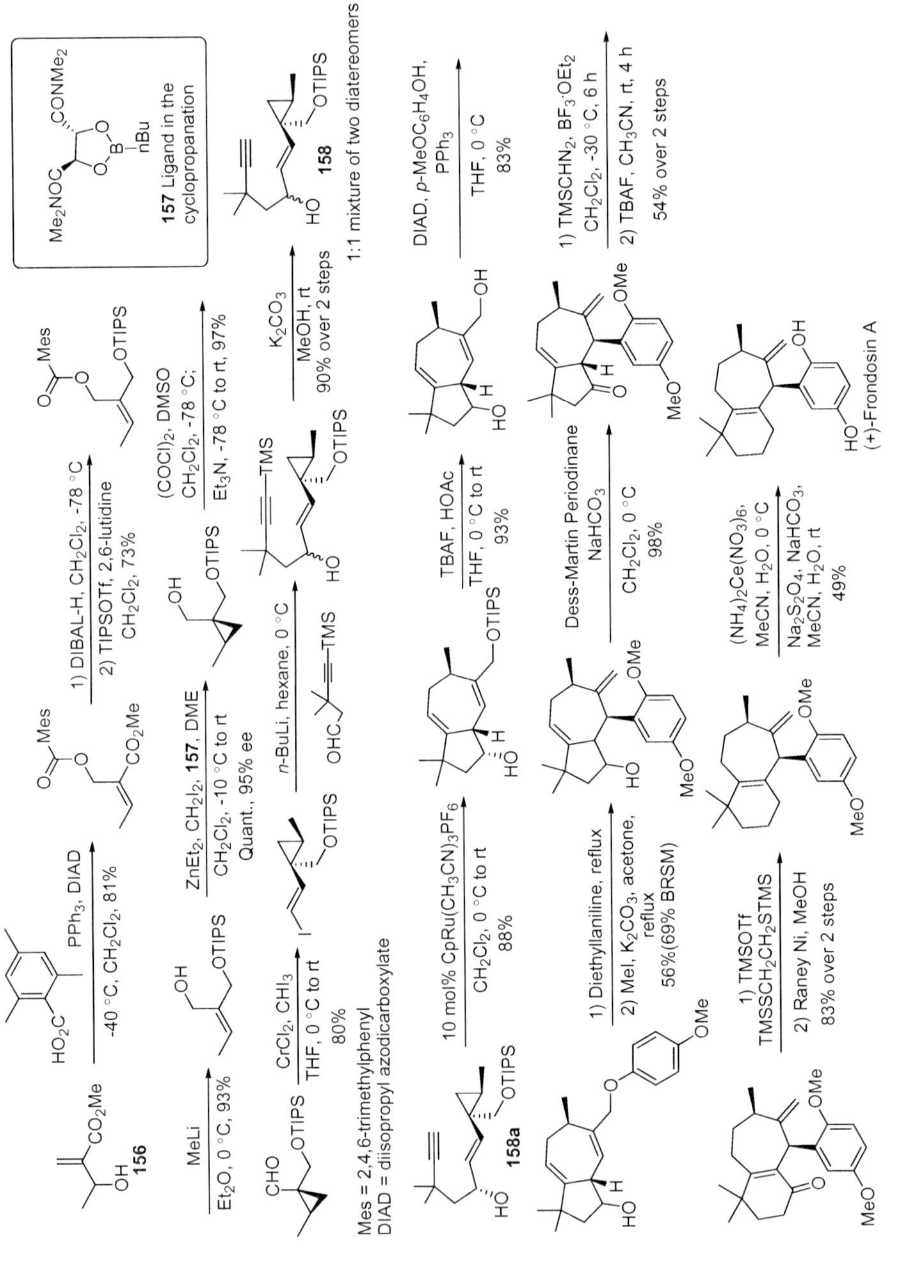

156
PPh3, DIAD
-40 °C, CH2Cl2, 81%
1) DIBAL-H, CH2Cl2, -78 °C
2) TIPSOTf, 2,6-lutidine CH2Cl2, 73%
Mes
CO2Me
OTIPS
Me2NOC
CONMe2
nBu
157 Ligand in the cyclopropanation
MeLi
Et2O, 0 °C, 93%
ZnEt2, CH2I2, 157, DME
CH2Cl2, -10 °C to rt
Quant., 95% ee
(COCl)2, DMSO
CH2Cl2, -78 °C;
Et3N, -78 °C to rt, 97%
CrCl2, CHI3
THF, 0 °C to rt
80%
n-BuLi, hexane, 0 °C
TMS
K2CO3
MeOH, rt
90% over 2 steps
158
1:1 mixture of two diatereomers
Mes = 2,4,6-trimethylphenyl
DIAD = diisopropyl azodicarboxylate
158a
10 mol% CpRu(CH3CN)3PF6
CH2Cl2, 0 °C to rt
88%
TBAF, HOAc
THF, 0 °C to rt
93%
DIAD, p-MeOC6H4OH, PPh3
THF, 0 °C
83%
1) Diethyllaniline, reflux
2) MeI, K2CO3, acetone, reflux
56%(69% BRSM)
Dess-Martin Periodinane
NaHCO3
CH2Cl2, 0 °C
98%
1) TMSCHN2, BF3·OEt2
CH2Cl2, -30 °C, 6 h
2) TBAF, CH3CN, rt, 4 h
54% over 2 steps
1) TMSOTf
TMSSCH2CH2STMS
2) Raney Ni, MeOH
83% over 2 steps
(NH4)2Ce(NO3)6,
MeCN, H2O, 0 °C
Na2S2O4, NaHCO3,
MeCN, H2O, rt
49%
(+)-Frondosin A

Scheme 5.27

intermediate *ent*-**152**, which was reacted with another chiral synthon **153** and could be further converted into chiral Fleursandols, *ent*-**155a** and *ent*-**155b**. Moreover, by using this synthetic strategy, other possible stereoisomers were synthesized (Figure 5.6); however, only four of 16 possible stereoisomers of *rac*-**155** possess the typical, very pleasant, long-lasting sandalwood odor.

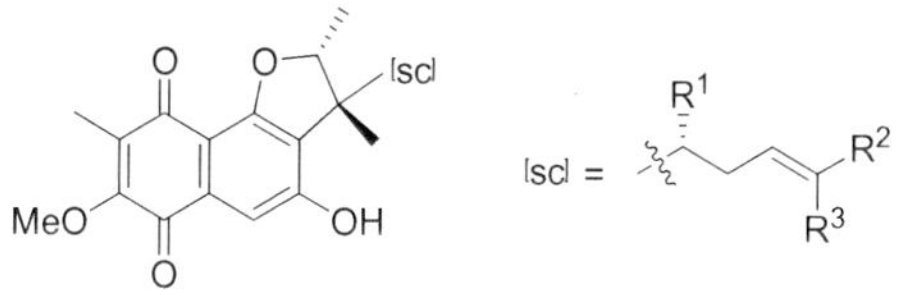

Furaquinocin	R^1	R^2	R^3
A	OH	CH_3	CH_2OH
B	OH	CH_2OH	CH_3
C	H	CH_3	CH_3
D	OH	CH_3	CH_3
F	H	CH_2OH	CH_3
H	OH	CH_2OH	CH_2OH

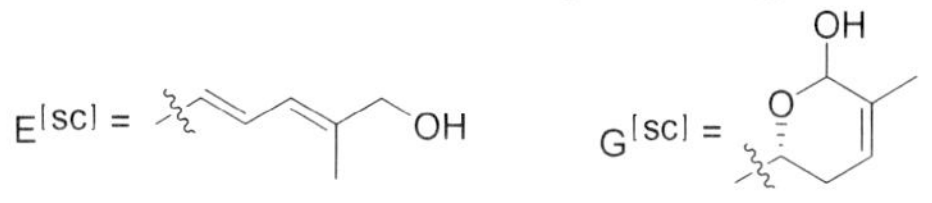

Figure 5.8 Furaquinocins A–H.

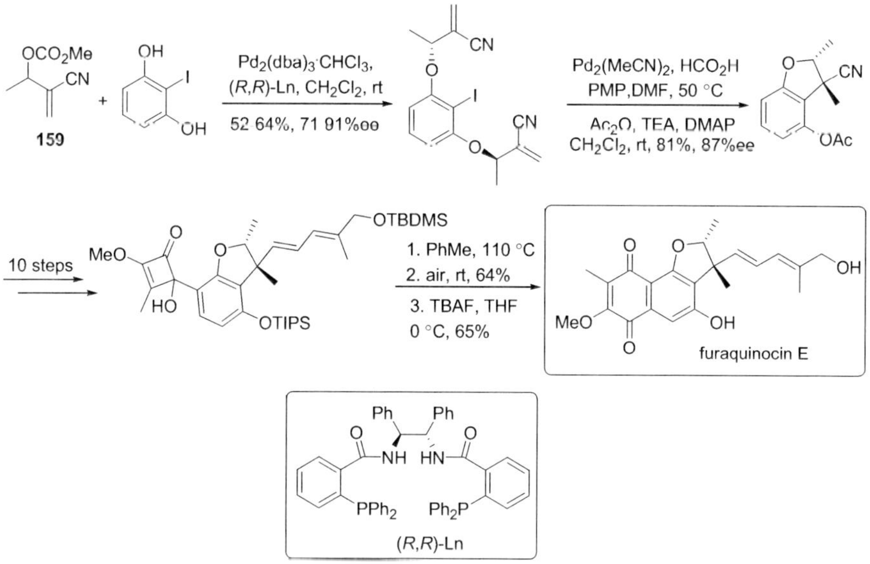

Scheme 5.28

5.20 (+)-Frondosin A

Frondosins A–E (Figure 5.7) are a family of five norsesquiterpenoid (14-carbon) natural products that were originally extracted from the marine sponge *Dysidea frondosa*[72] and later isolated from the HIV-inhibitory extract of *Euryspongia* sp.[73] These compounds have shown IL-8α, IL-8β, protein kinase C (PKC) activities and *anti*-HIV activity. Trost *et al.* first achieved the total synthesis of frondosin

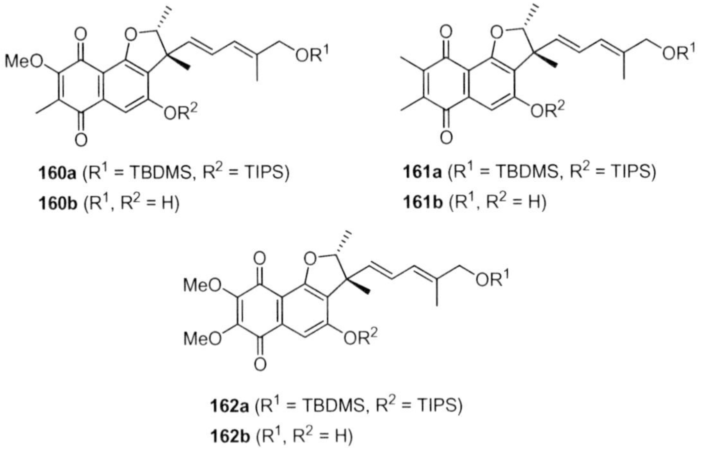

160a (R^1 = TBDMS, R^2 = TIPS)
160b (R^1, R^2 = H)

161a (R^1 = TBDMS, R^2 = TIPS)
161b (R^1, R^2 = H)

162a (R^1 = TBDMS, R^2 = TIPS)
162b (R^1, R^2 = H)

Figure 5.9 Analogs of furaquinocin E.

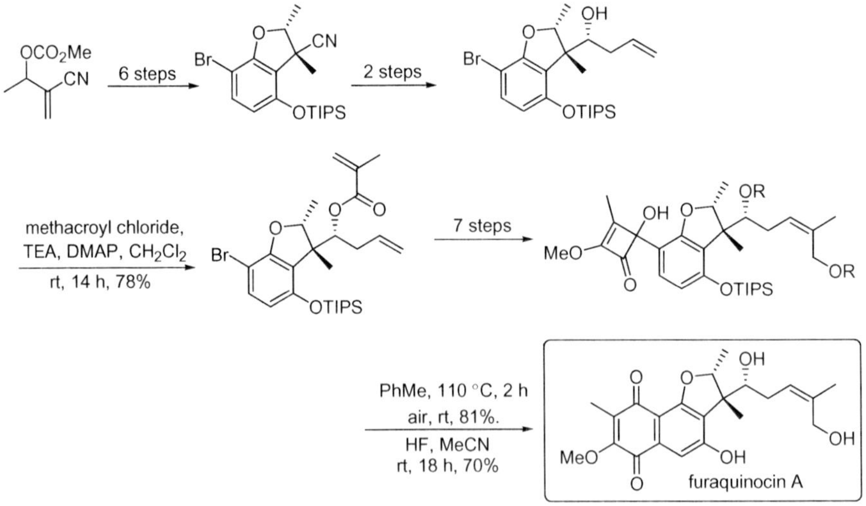

Scheme 5.29

A, which was the compound with greatest biological potential in this class,[74] though there had been several reports of syntheses of frondosin B and frondosin C.[75] Using MBH adduct **156** as a starting material, the total synthesis of (+)-frondosin A was accomplished in 7% overall yield through 19 longer linear and 21 total steps (Scheme 5.27). For this total synthesis strategy, a Ru-catalyzed [5 + 2] cycloaddition, a Claisen rearrangement, and a ring expansion were utilized to construct the core of frondosin A. In particular, a Ru-catalyzed [5 + 2] cycloaddition was the key step in constructing the bicyclo[5.3.0] ring system with high regio- and diastereoselectivity from an enantioenriched cyclopropyl enyne.

5.21 Furaquinocins

The furaquinocins (A–H, Figure 5.8) are a class of antibiotics, isolated from the fermentation broth of *Streptomyces* sp. KO-3998,[76] that showed a wide range of biological effects such as *in vitro* cytotoxicity against HeLa S3 and B16 melamona cells, antihypertensive activity and inhibition of platelet aggregation and coagulation. These biological activities make this class of compounds highly interesting synthetic targets.[77]

With their efficient procedure for deracemization of MBH adducts,[78] Trost and coworkers have applied dynamic asymmetric kinetic transformation (DYKAT) to the total synthesis of furaquinocin E.[79] As shown in Scheme 5.28, the asymmetric palladium-catalyzed alkylation of phenols combined with a reductive Heck reaction delivered an efficient approach to the synthesis of the key synthon, which is the core structure of the furaquinocins. A general synthetic route to furaquinocin E was established in 14 steps from MBH adduct **159**. Their work highlighted the ability to use racemic MBH adducts for asymmetric synthesis. They further extended the scope of their strategy by developing the synthesis of three more analogs of

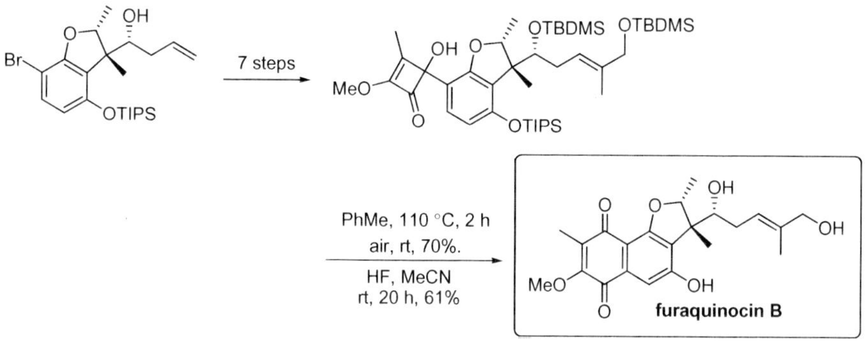

Scheme 5.30

163
164
165
166
TBDPSO
MeO2C
OH
TBDPSO
O
10 steps
methyl propiolate
HMPA, DIBAL-H
82%
1. AcCl, Py, CH2Cl2, 0 °C
2. TBAF, AcOH, 75%
MeO2C
OAc
HO
1. [Pd(π-allyl)Cl]2, (R,R)-Ln
N(Hex)4Cl, dioxane
2. LiOH, THF/H2O
Ph
Ph
HN
NH
O
O
PPh2
Ph2P
(R,R)-Ln
HO2C
O
(+)-hippospongic acid A
OAc

Scheme 5.31

furaquinocin E (**160**–**162**, Figure 5.9), furaquinocins A (Scheme 5.29) and B (Scheme 5.30).[80]

Later, they also successfully employed aliphatic alcohols **166** as competent nucleophiles in the Pd-catalyzed DYKAT reaction, the utility of which was demonstrated by a concise total synthesis of the gastrulation inhibitor (+)-hippospongic acid A (Scheme 5.31).[81] Key intermediate **165** was prepared in good yield from aldehyde **164** *via* a MBH analogous reaction in the presence of HMPA and DIBAL-H.

5.22 (+)-Heliotridine and (–)-Retronecine

Recently, a novel methodology has been developed to prepare densely functionalized heterocycles from the coupling of various Michael acceptors with readily available iminium ions (masked as N,O-acetals) *via* an intermolecular MBH-type reaction.[82] An intramolecular MBH-type reaction was developed to construct a bicyclic pyrrolizidine ring. As shown Scheme 5.32, the total synthesis of (+)-heliotridine, as this basic structure unit existed in many plants, was achieved.[83]

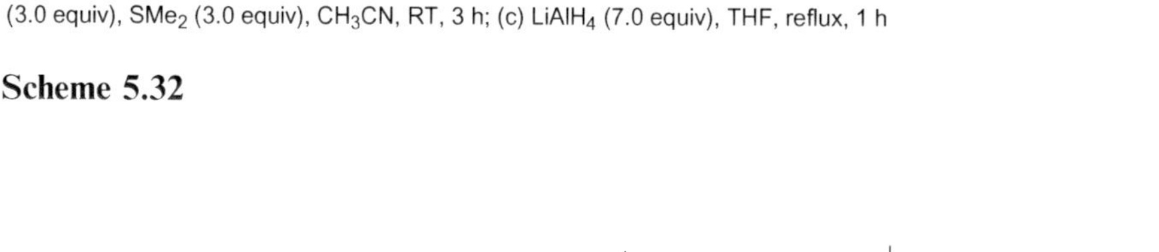

OAc
a
b
c
OEt
CHO
H
OH
+
167
168, 74%
169, 65% yield
3:1 trans:cis
170
(+)-heliotridine
38%
171
(-)-retronecine
12%

(a) acrolein (3.0 equiv), Grubbs-Hoveyda cat. (2.5 mol%), CH_2Cl_2, RT, 12 h; (b) TMSOTf (3.0 equiv), $BF_3{\cdot}OEt_2$ (3.0 equiv), SMe_2 (3.0 equiv), CH_3CN, RT, 3 h; (c) $LiAlH_4$ (7.0 equiv), THF, reflux, 1 h

Scheme 5.32

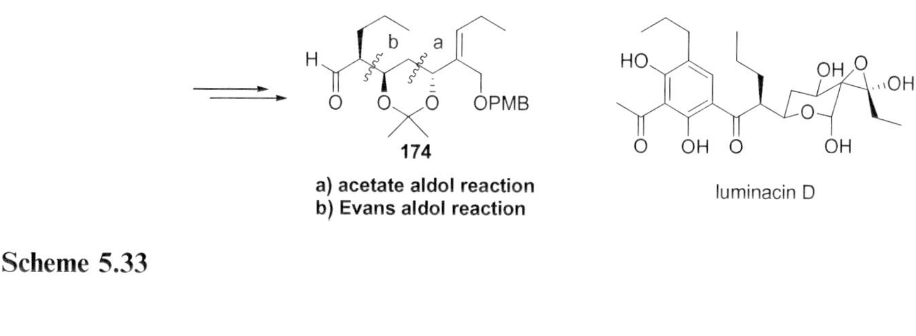

CHO
CO2Me
DABCO, rt, 5 d
HO
MeO2C
172
H
OPMB
173
174
a) acetate aldol reaction
b) Evans aldol reaction
luminacin D

Scheme 5.33

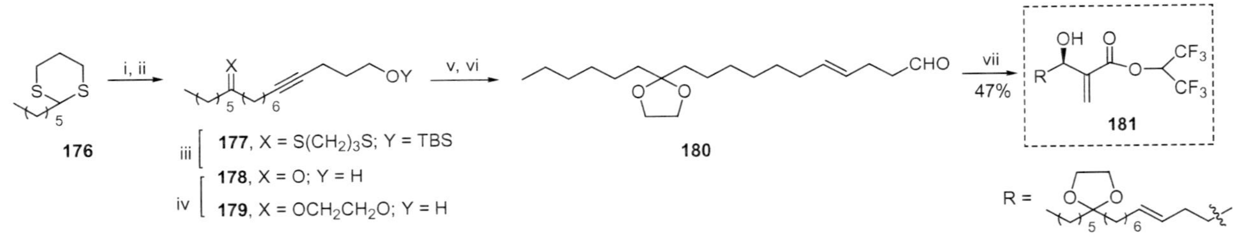

177, X = S(CH$_2$)$_3$S; Y = TBS
178, X = O; Y = H
179, X = OCH$_2$CH$_2$O; Y = H

176

180

181

R =

i, *n*-BuLi, THF, -30 °C, then 1,6-diiodohexane, 81%; ii, LiCC(CH$_2$)$_3$OTBS, THF-HMPA, -20 °C, 87%; iii, MeI, Na$_2$CO$_3$, THF-MeCN, sealed tube 70 °C, 91%; iv, ethylene glycol, cat. *p*-TsOH, benzene, 80 °C, 98%; v, LiAlH4, diglyme, 130 °C, 87%; vi, DMSO, (COCl)$_2$, Et$_3$N, CH$_2$Cl$_2$, -70 °C to 0 °C, quant; vii, cinchona alkaloid (0.2 equiv), 1,1,1,3,3,3-hexafluoroisopropyl acrylate (1.5 equiv), DMF-CH$_2$Cl$_2$ (1:1), -55 °C, 24 h

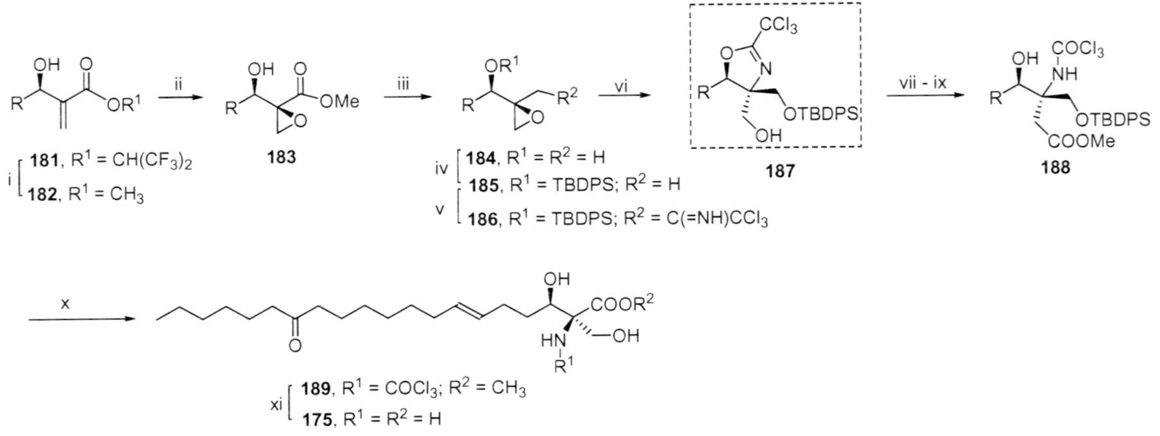

181, R^1 = CH(CF$_3$)$_2$
182, R^1 = CH$_3$

183

184, R^1 = R^2 = H
185, R^1 = TBDPS; R^2 = H
186, R^1 = TBDPS; R^2 = C(=NH)CCl$_3$

187

188

189, R^1 = COCl$_3$; R^2 = CH$_3$
175, R^1 = R^2 = H

i, cat. NaOMe, MeOH, rt, then Dowex- 50 H$^+$ form, 95%; ii, Ti(O*i*-Pr)$_4$, *t*-BuOOH, 4 Å molecular sieves, CH$_2$Cl$_2$, 220 °C, 73%; iii, NaBH$_4$, THF–MeOH (1:1), 0 °C, 79%; iv, TBDPSCl, DMAP, Et$_3$N, CH$_2$Cl$_2$, 0 °C, 95%; v, DBU, CCl$_3$CN, CH$_2$Cl$_2$, 0 °C, 81%; vi, BF$_3$·OEt$_2$, CH$_2$Cl$_2$, 223 °C, 71%; vii, DMSO, DCC, TFA, pyridine, benzene, 0 °C; viii, NaClO$_2$, NaH$_2$PO$_4$, 2-methylbut-2-ene, *t*-BuOH–H$_2$O (4:1); ix, CH$_2$N$_2$, THF, 55% (3 steps); x, 47% HF, MeCN, 60 °C, 87%; xi, 10% NaOH, MeOH, 80 °C, 70%.

Scheme 5.34

Cross-metathesis of **167** with acrolein was achieved using the Grubbs–Hoveyda catalyst to give intermediate **168** in 74% yield, and subsequent MBH type ring closure furnished compound **169** in 65% yield in the presence of TMSOTf, $BF_3 \cdot OEt_2$ and SMe_2. A solution of **169** in THF with $LiAlH_4$ was boiled under reflux to give (+)-heliotridine (**170**) in 38% yield. The isomer (–)-retronecine (**171**), which was the most common compound of the necine bases of the *Senecio* alkaloids,[84] was also isolated in 12% yield.

5.23 Luminacin D

Luminacin D, isolated from the fermentation broth of a soil bacterium, represented promising structural lead of angiogenesis inhibitors,[85] and also showed activity in a rat aorta matrix culture model.[86] A practical route has been developed by Jogireddy and Maier for its synthesis (Scheme 5.33).[87] This route began with the known MBH adduct **172**, followed by OH transposition to give the key intermediate aldehyde **173**, which was further converted into the carbohydrate sector **174** by two highly stereoselective asymmetric aldol reactions, Evans aldol reaction and acetate aldol reaction. This overall strategy was concise and convergent, which should make it suitable for the synthesis of other luminacins and its analogues.

5.24 Mycestericin E

(–)-Mycestericin E (**175**), isolated from the culture broth of the fungus *Mycelia sterilia* ATCC 20349, showed the activity in suppressing the proliferation of lymphocytes in the mouse allogeneic mixed lymphocyte reaction.[88] An enantio- and stereocontrolled synthesis[89] of (–)-mycestericin E (**175**) has been reported employing a new strategy.[90] As shown in Scheme 5.34, the asymmetric MBH reaction as the key step was utilized to synthesize the key chiral intermediate **181**. In addition, Lewis acid-promoted cyclization was achieved to stereoselectively construct the substituted oxazolidine **187** with a quaternary centre from an epoxytrichloroacetimidate **186**.

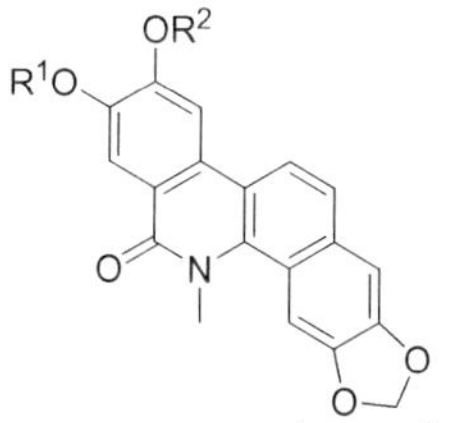

oxyterihanine **190a**, R[1] = H, R[2] = Me
oxyisoterihanine **190b**, R[1] = Me, R[2] = H

Figure 5.10 Structures of oxyterihanine and oxyisoterihanine.

Scheme 5.35

5.25 Oxyisoterihanine

Oxyterihanine **190a**, a phenolic benzo[*c*]phenanthridine, was isolated from *Xanthoxylum nitidum* (Roxb.) D. C. (*Fagara nitida* Roxb.) in 1984.[91] The structures of **190a** and **190b** were not confirmed until in 1987 by Ishii *et al.* (Figure 5.10).[92] More recently, Chang *et al.* achieved the total synthesis of oxyisoterihanine **190b** using MBH adduct **191** as a starting material.[93] As depicted in Scheme 5.35, key steps such as a [3 + 3] annulation, Friedel–Crafts reaction, aromatization and oxidation are involved in their synthetic route.

5.26 Phaseolinic Acid

Paraconic acids, isolated from different species of moss, lichens, fungi and cultures of *Penicillium* sp. (Figure 5.11),[94] are a group of highly substituted γ-butyrolactones and exhibit interesting biological activities such as antitumor,

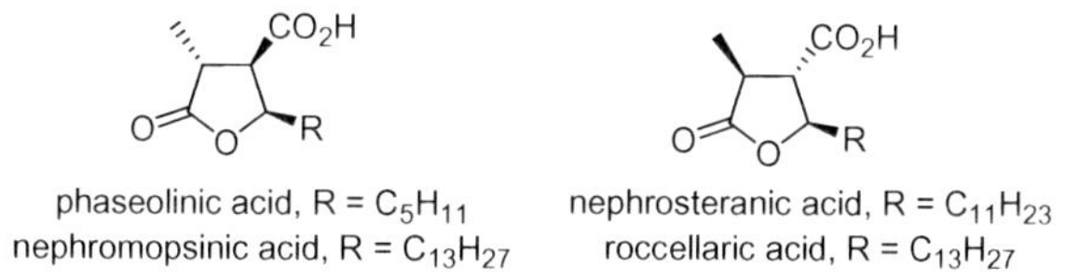

phaseolinic acid, R = C_5H_{11}
nephromopsinic acid, R = $C_{13}H_{27}$

nephrosteranic acid, R = $C_{11}H_{23}$
roccellaric acid, R = $C_{13}H_{27}$

Figure 5.11 Some paraconic acids.

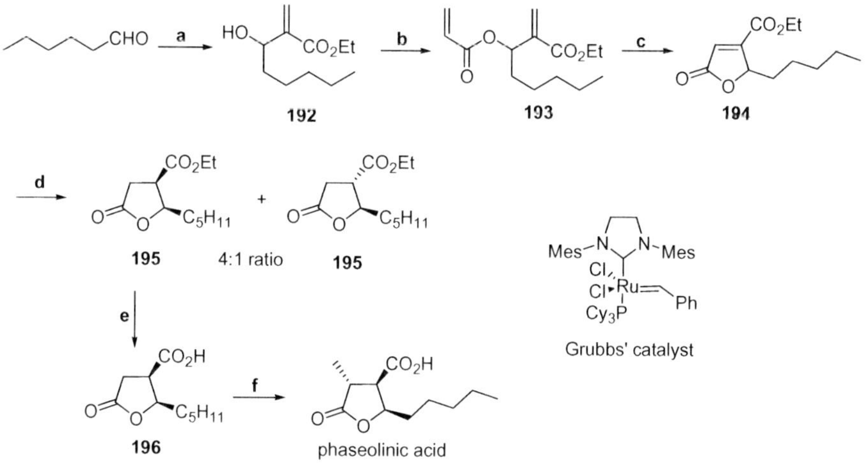

Grubbs' catalyst

(a) ethyl acrylate, DABCO, rt; (b) acryloyl chloride, Et_3N, CH_2Cl_2, 0 °C to rt, 95%; (c) Grubbs' catalyst, $Ti(OiPr)_4$, 50 °C, 87%; (d) 10% Pd-C, H_2, EtOAc, 84% (combined yield, 1:2 ratio) or 10% Pd-C, ammonium formate, MeOH, reflux, 83% (combined yield, 4:1 ratio); (e) 6 N HCl, dioxane, reflux, 91%; (f) $NaN(TMS)_2$, MeI, THF, -78 °C.

Scheme 5.36

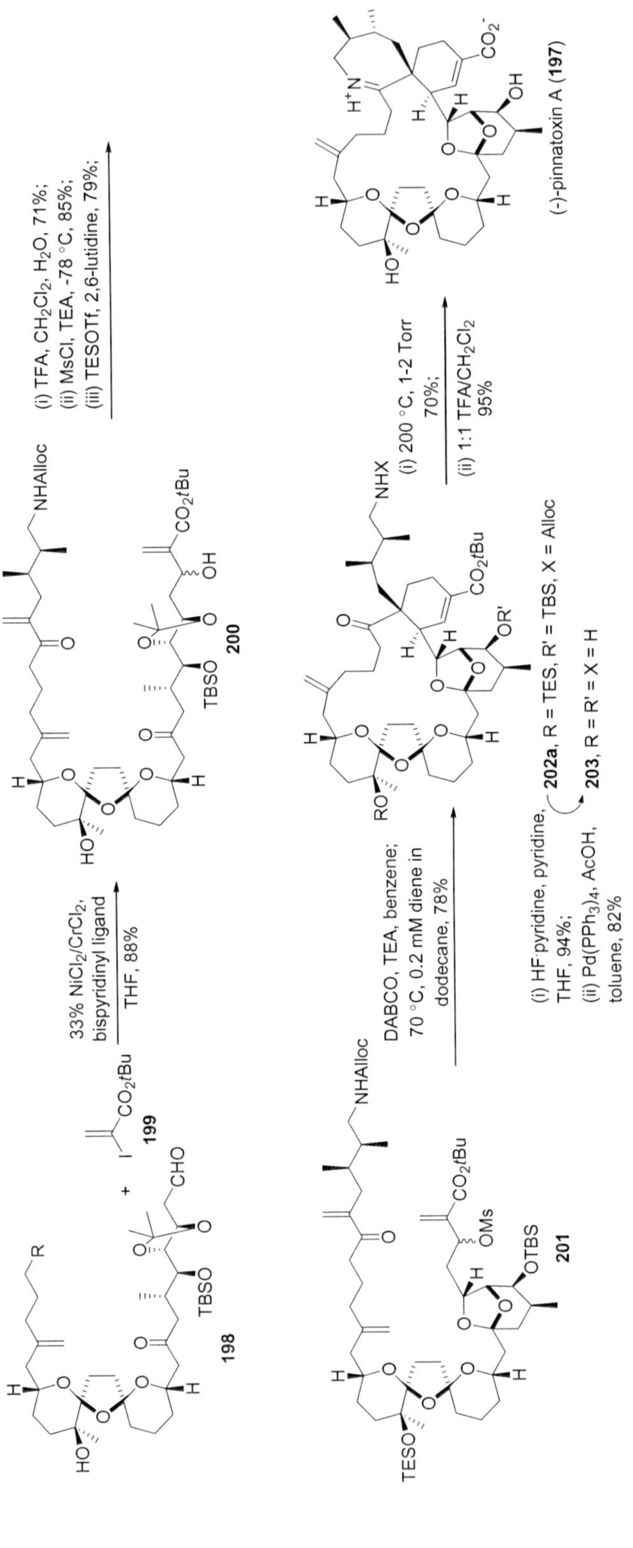

198
199
CHO
TBSO
CO2tBu
33% NiCl2/CrCl2, bispyridinyl ligand
THF, 88%
200
NHAlloc
(i) TFA, CH2Cl2, H2O, 71%;
(ii) MsCl, TEA, -78 °C, 85%;
(iii) TESOTf, 2,6-lutidine, 79%;
201
OMs
OTBS
TESO
DABCO, TEA, benzene; 70 °C, 0.2 mM diene in dodecane, 78%
NHX
OR'
(i) HF·pyridine, pyridine, THF, 94%;
(ii) Pd(PPh3)4, AcOH, toluene, 82%
202a, R = TES, R' = TBS, X = Alloc
203, R = R' = X = H
(i) 200 °C, 1-2 Torr 70%;
(ii) 1:1 TFA/CH2Cl2 95%
H+N
CO2-
(-)-pinnatoxin A (197)

Scheme 5.37

antifungal and antibacterial.[95] Consequently, the synthesis of paraconic acids has attracted widespread attention from synthetic chemists. Using the known MBH adduct **192** as starting material, the total synthesis of (±)-phaseolinic acid has been disclosed by Selvakumar *et al.* (Scheme 5.36).[96] They employed the RCM reaction of electron-deficient dienes to construct a butenolide skeleton. Consequently, butenolide **194**, a key intermediate for the synthesis of phaseolinic acid, was prepared in 87% yield from **193** using the second-generation Grubbs' catalyst in the presence of Ti(OPr^i)$_4$.

5.27 Pinnatoxin A

Pinnatoxin A (**197**), isolated from the shellfish *Pinna muricata*, is an important toxic principle in *Pinna* shellfish intoxication outbreaks in China and Japan.[97] Its unique molecular architecture, accompanied by its pronounced biological activity as a Ca^{2+}-channel activator, makes pinnatoxin an intriguing synthetic target.[98] A MBH analogous reaction, namely, Ni(II)/Cr(II)-mediated coupling between aldehyde **198** and 2-iodoacrylic acid derivative **199** to give MBH adduct **200** in 88% yield, was utilized as a key step in the total synthesis of (–)-pinnatoxin A (Scheme 5.37).[99] As part of the synthetic strategy, a biomimetic intramolecular Diels–Alder reaction was developed to construct the macrocyclic structure of (–)-pinnatoxin A.

5.28 Salinosporamide A and its Analogues

Salinosporamide A, discovered by Fenical and his group from a marine microorganism,[100] is a proteasome inhibitor that is more effective than omuralide[101] (Figure 5.12). In addition, it displays significantly high *in vitro* cytotoxic activity against many tumor cell lines (IC$_{50}$ of *ca.*10 nM). Corey *et al.* reported the first enantiospecific total synthesis of salinosporamide A from (*S*)-threonine, as outlined in Scheme 5.38.[102] The intramolecular MBH reaction was the key step in achieving the cyclization needed to furnish a highly substituted γ-lactam **205** from intermediate **204** (90% yield). γ-Lactam **205** served as the key intermediate in the enantioselective total synthesis of salinosporamide A. However, this intramolecular MBH reaction was time-consuming, along with 9 : 1 diastereoselectivity. Subsequently, they further developed a more effective method to synthesize γ-lactam **205**. As shown in Scheme 5.39, the MBH substrate **204** was treated

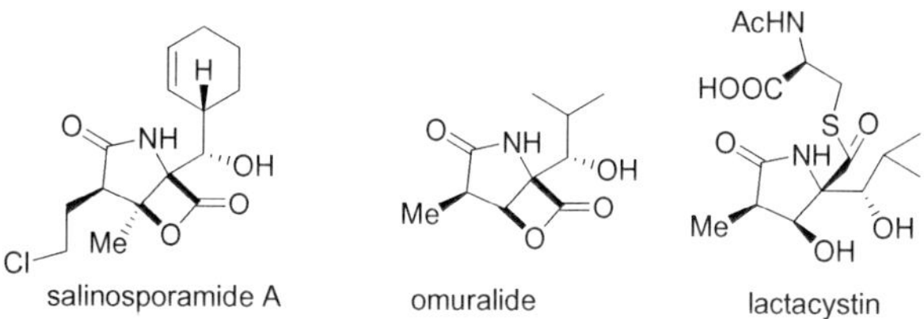

Figure 5.12 Proteasome inhibitors salinosporamide and omuralide and lactacystin.

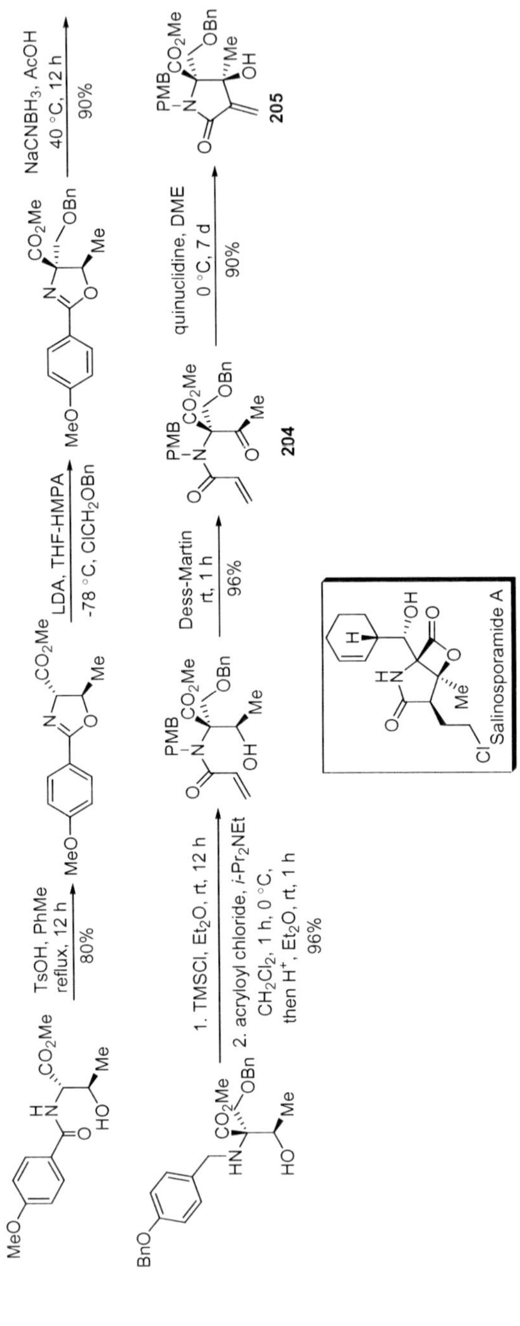

NaCNBH$_3$, AcOH
40 °C, 12 h
90%

LDA, THF–HMPA
–78 °C, ClCH$_2$OBn

TsOH, PhMe
reflux, 12 h
80%

Dess–Martin
rt, 1 h
96%

quinuclidine, DME
0 °C, 7 d
90%

1. TMSCl, Et$_2$O, rt, 12 h
2. acryloyl chloride, *i*-Pr$_2$NEt
CH$_2$Cl$_2$, 1 h, 0 °C,
then H$^+$, Et$_2$O, rt, 1 h
96%

204

205

Salinosporamide A

Scheme 5.38

sequentially with the Kulinkovich reagent, I_2 and Et_3N to afford γ-lactam **205** in 83% overall yield along with >99% dr.[103] In addition, omuralide analogue **207** and salinosporamide A analogue **206** were also synthesized from the γ-lactam **205**.

5.29   Selenomilfasartan

Milfasartan **208**, a thiophene-containing selective AT1 receptor antagonist (sartans), reached Phase I clinical trials;[104] therefore, the development of its novel analogues have drawn much attention. Selenium analogue **209** was first developed to test the AT1 receptor antagonist properties.[105] The synthetic route to compound **209** was established (Scheme 5.40), and selenophene **211** as key intermediate was prepared using MBH adduct **210** as a starting material. The result of tests for AT1 receptor antagonist properties showed that seleno-milfasartan (**209**) is as effective as milfasartan (**208**) at blocking AT1 receptor mediated responses, from pK_B estimates.

5.30   Solanoeclepin A

Solanoeclepin A (**212**) is the most active natural hatching agent of potato cyst nematodes (PCN), showing activity at nanomolar concentration.[106] However, the unavailability of the natural product in useful quantities from natural sources and the unique structure needed to develop an environmentally benign way to combat the nematodes, which cause serious losses in potato production, rendered solanoeclepin A a challenging synthetic target.

An intramolecular [2 + 2] photocycloaddition reaction was reported as the key step in constructing the tricyclic core **218** of solanoeclepin A, which includes an intricate bicycle[2.1.1]hexanone moiety.[107] Allene butenolide **217** as

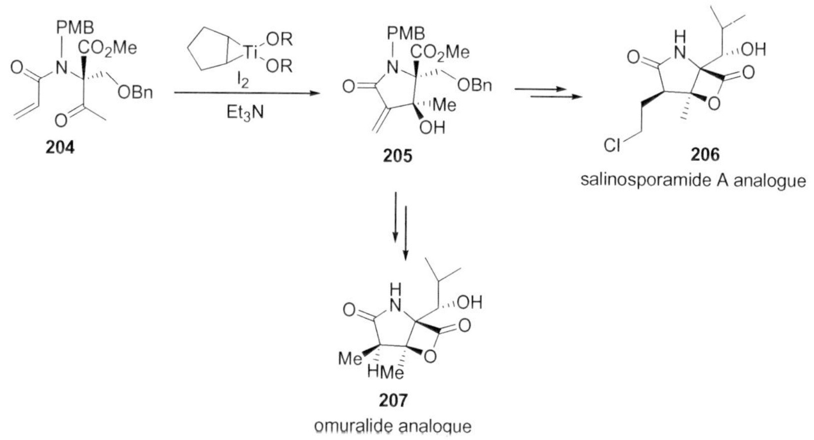

204

205

206
salinosporamide A analogue

207
omuralide analogue

Scheme 5.39

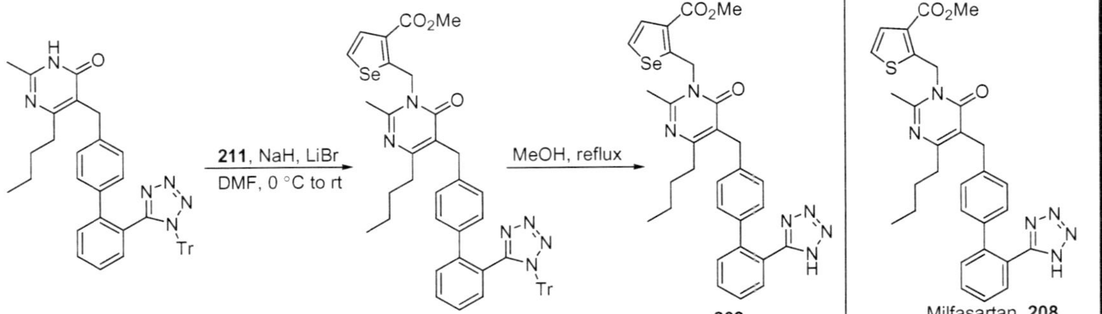

(a) (CHO)$_n$, NMe$_3$ (aq.), H$_2$O, 60 °C, 61%; (b) m-CPBA, 3-t-butyl-4-hydroxy-5-methylphenyl sulfide, CCl$_4$, reflux, 73%; (c) (COCl)$_2$, DMSO, NEt$_3$, CH$_2$Cl$_2$, -78 °C; (d) (Ph$_3$PCH$_2$I)$^+$I$^-$, NaHMDS, HMPA, THF, -78 °C to rt, 34%; (e) (SeBn)$_2$, NaBH$_4$, EtOH, rt; (f) TTMSS, AIBN, benzene, reflux; (g) LiOH, THF, H$_2$O, 26% from 16; (h) s-BuLi, -78 °C, THF; (i) MeI, rt, 90%; (j) K$_2$CO$_3$, MeI, acetone, rt, 87%; (k) NBS, AIBN, CCl$_4$, reflux, 45% (NMR yield).

Scheme 5.40

photosubstrate, prepared through silver-mediated coupling of silyloxy-furan **216** and allenic bromide **215**, gave **218** in 60% yield after irradiation at 300 nm for 1 h (Scheme 5.41). Bromide **215** was obtained in a five-step sequence starting from the MBH reaction between benzyl butadienolate and paraformaldehyde.

5.31 Sordarin Core

Sordarins, a family of natural products isolated from *Sordaria araneosa*, have captured significant attention due to their unique mode of action in antifungal activity.[108] Sordarin (**219**) and its congeners block protein synthesis by inhibiting the fungal elongation factor.[109] Ciufolini *et al.* have described an effective synthetic route to construct the sordarin cores, which were useful building blocks for the preparation of analogs of **219**.[110] As shown in Scheme 5.42, the MBH adduct **220** as key intermediate underwent a spontaneous intramolecular Diels–Alder reaction to furnish the core of sordarin (**221**) in moderate yields *via* trialkylsilyl triflate/Et$_3$N-catalyzed cyclization. For the MBH derivatives **223**, cyclization to give product **224** as a single regioisomer in 77% yield was also achieved, by heating to 140 °C in toluene.

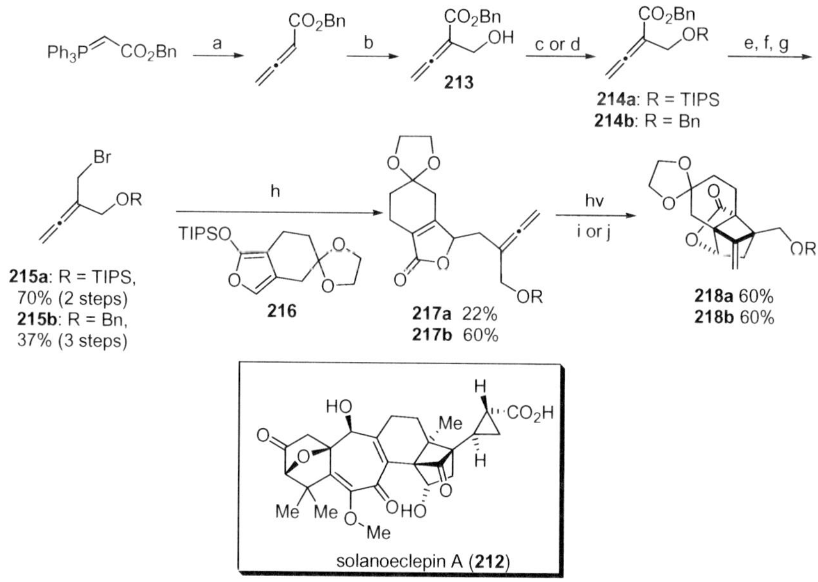

(a) MeCOCl, Et$_3$N, CH$_2$Cl$_2$, 82%; (b) DABCO (cat.), (CH$_2$O)$_n$, THF, 60%; (c) TIPSOTf, Et$_3$N, 79%; (d) PhCH$_2$OC(NH)CCl$_3$, TMSOTf (cat.), 67%; (e) DIBAL-H, CH$_2$Cl$_2$, -78 °C; (f) MsCl, Et$_3$N; (g) LiBr, acetone; (h) AgOCOCF$_3$, CH$_2$Cl$_2$, -78 °C to room temp.; (i) MeCN/acetone (9:1 v/v); (j) benzene/acetone (9:1 v/v).

Scheme 5.41

Scheme 5.42

NMe₂

(-)-spinosyn A (226), R = H
(-)-spinosyn D, R = Me

RhamO
TBSO
COMe
CO₂Me
Br
227

MeAlCl₂
CH₂Cl₂
-78 to 0 °C
93%
> 30 : 1 dr

tert-amyl-OH
40 °C, 67 h,
then PMe₃,
23 °C, 9 h
80%
96 : 4 selectivity

RhamO
TBSO
Br
H
H
H
COMe
CO₂Me
228

RhamO
TBSO
Br
H
H
H
COMe
CO₂Me
229

PMe₃
tert-amyl-OH
23 °C, 5 h
88%

RhamO
TBSO
Br
H
H
H
COMe
CO₂Me
228 (96%)

(a) HF·pyr, THF, 96%;
(b) 1,1'-thiocarbonyldiimidazole,
DMAP, PhCH₃, 65 °C, 95%;
(c) (TMS)₃SiH, AIBN, dioxane, 80 °C, 73%

Me
O
Me
O
Me
O
H
H
H
COMe
CO₂Me
230

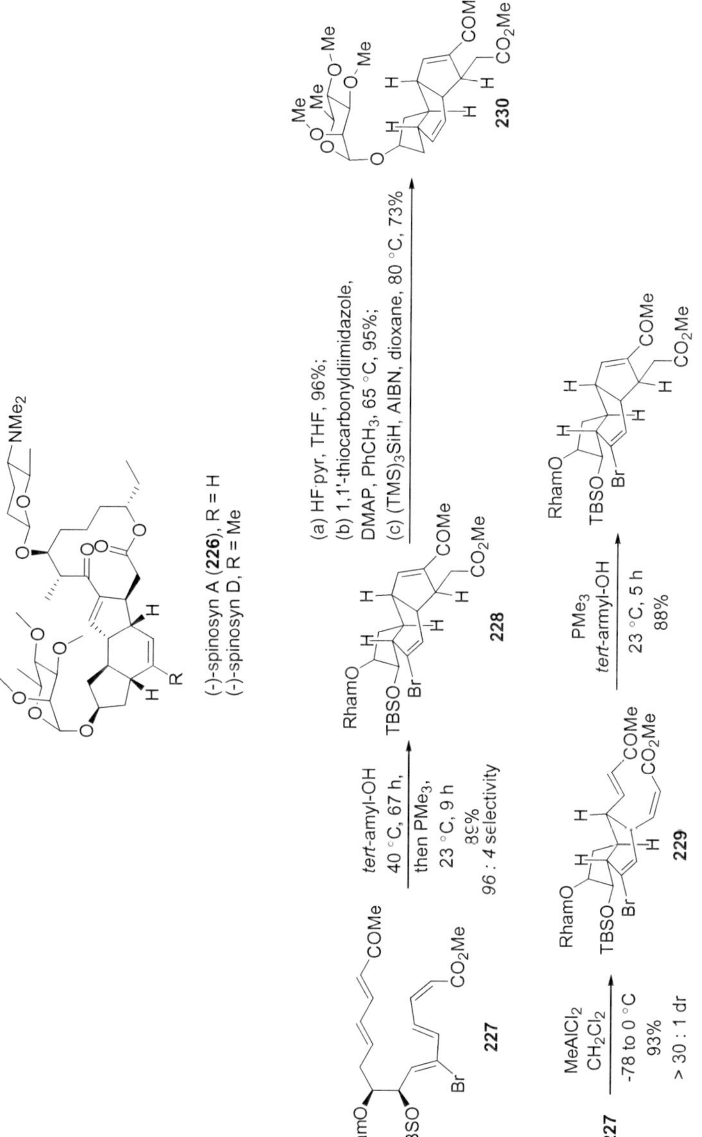

Scheme 5.43

5.32   (–)-Spinosyn A

(–)-Spinosyn A, a polyketide natural product, exhibits extraordinary insecticidal activity.[111] Its biosynthetic mixture, which consists of spinosyn A **226** (*ca.* 85%) and spinosyn D (*ca.* 10–15%) generated by *Saccharopolyspora spinosa*, has been marketed as an insecticide spinosad against various insects.[112] The total synthesis of spinosyn A has been reported by Evans and Black[113] and by Paquette *et al.*[114] An effective synthetic strategy – the intramolecular Diels–Alder reaction and MBH cyclization sequence to construct the spinosyn A tricycle **230** skeleton with exceptional stereocontrol – has been described by Franck *et al.* As shown in Scheme 5.43,[115] treatment of **227** with MeAlCl$_2$ in CH$_2$Cl$_2$ at –78 °C generated the Diels–Alder product **229** in 93% yield with a diastereomeric ratio >30 : 1 (Scheme 5.43). A solution of intermediate **229** in *tert*-amyl alcohol was treated with Me$_3$P (0.5 equiv.) to afford the desired tricyclic precursor **228** as a 96 : 4 mixture of regioisomers in 88% yield. Moreover, the precursor **228** could be also obtained in one pot. Heating of **227** at 40 °C in *tert*-amyl alcohol for 67 h and subsequent addition of Me$_3$P (0.6 equiv.) at room temperature yielded tricycle **228** in 89% yield as a 96 : 4 mixture of regioisomers. The subsequent application of this strategy in the transannular manifold to a total synthesis of (–)-spinosyn A (**235**) was accomplished in 3% overall yield, involving 23 steps in the longest linear sequence (Scheme 5.44).[116]

5.33   (–)-Spirotryprostatin B

(–)-Spirotryprostatin B is the most complex diketopiperazine alkaloid, isolated from the fungus *Aspergillus fumigatus* by Osada and co-workers,[117] and can inhibit G2/M phase progression of the mammalian cell cycle at micromolar

Scheme 5.44

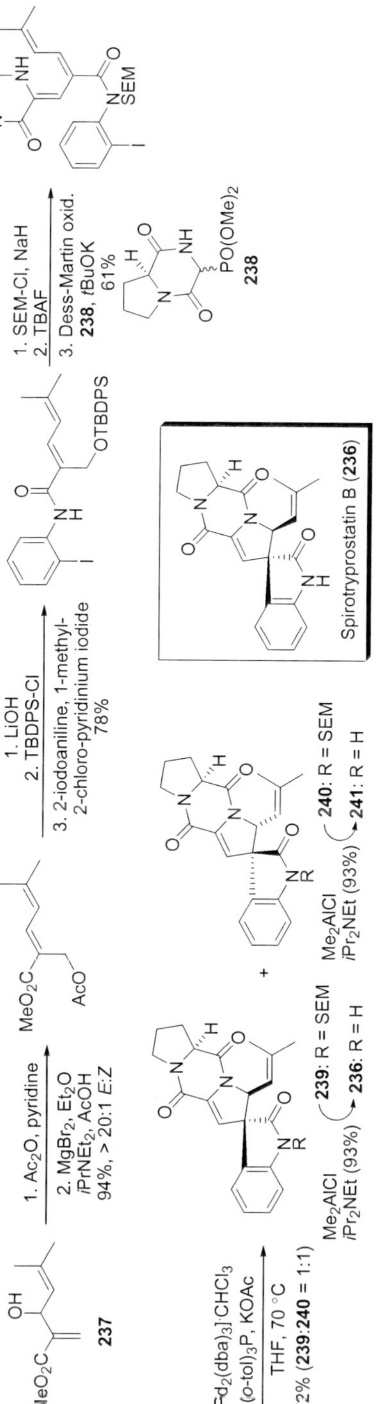

237
1. Ac2O, pyridine
2. MgBr2, Et2O
iPrNEt2, AcOH
94%, > 20:1 E:Z
MeO2C
AcO
1. LiOH
2. TBDPS-Cl
3. 2-iodoaniline, 1-methyl-
2-chloro-pyridinium iodide
78%
OTBDPS
1. SEM-Cl, NaH
2. TBAF
3. Dess-Martin oxid.
238, tBuOK
61%
PO(OMe)2
238
SEM
[Pd2(dba)3]·CHCl3
(o-tol)3P, KOAc
THF, 70 °C
72% (239:240 = 1:1)
Me2AlCl
iPr2NEt (93%)
239: R = SEM
236: R = H
240: R = SEM
241: R = H
Spirotryprostatin B (236)

Scheme 5.45

concentrations. The novelty of such a structure and the promising bioactivity as a cell-cycle inhibitor drew much attention in the field of organic synthesis, but the construction of the stereochemistry of the quaternary spiro carbon to the adjacent stereocenter bearing the 2-methylpropeny side chain is still a challenge.[118] (–)-Spirotryprostatin B (**236**) was synthesized in ten steps in overall 9% yield from MBH adduct **237**, which could be easily prepared from methyl acrylate and 3-methyl-2-butenal (Scheme 5.45).[119] The asymmetric Pd-catalyzed intramolecular Heck insertion of conjugated trienes as a key step was involved in this synthetic strategy. In this asymmetric Heck insertion, η^3-allylpalladium intermediates were trapped by the nitrogen of a tethered diketopiperazine.

5.34 Syributins and Syringolides, Spyhydrofurans and Secosyrins

The syributins along with secosyrins (Figure 5.13) were isolated by Sims *et al.*[120] as the co-isolates of syringolide elicitors from *Pseudomonas syringea pv. tomato* expressing virulence gene D (*avr*D-genes). Syringolides are of interest due to their unusual response to resistant soybean plants. Furthermore, syributins and secosyrins have gained attention because of their interesting structural features and their potential ability to provide vital clues to the biosynthesis of syringolides.[121] Krishna *et al.* have described the total synthesis of syributins 1 (**244**) and 2 (**245**).[122] As shown in Scheme 5.46, MBH reaction and RCM were the key procedures in this total synthetic method, using 2,3-*O*-isopropylidene-(*R*)-glyceraldehyde **248** as the starting material. Moreover, the lactone **250** was also an important advanced intermediate used in the total synthesis of several natural products such as sphydrofurans (Scheme 5.47) and secosyrins (Scheme 5.48).[123]

5.35 Tacamonine

As one of the few indole alkaloids of the tacamane type, tacamonine (**262**) was isolated by Beek and co-workers from *Tabernaemontana eglandulosa*;[124] it possesses vasodilator and hypotensive activities. With the development of

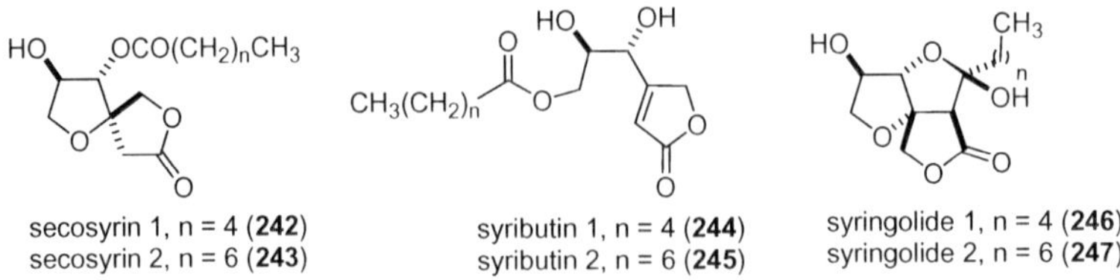

secosyrin 1, n = 4 (**242**)
secosyrin 2, n = 6 (**243**)

syributin 1, n = 4 (**244**)
syributin 2, n = 6 (**245**)

syringolide 1, n = 4 (**246**)
syringolide 2, n = 6 (**247**)

Figure 5.13   Secosyrins and syributins, the co-isolates of syringolides.

Scheme 5.46

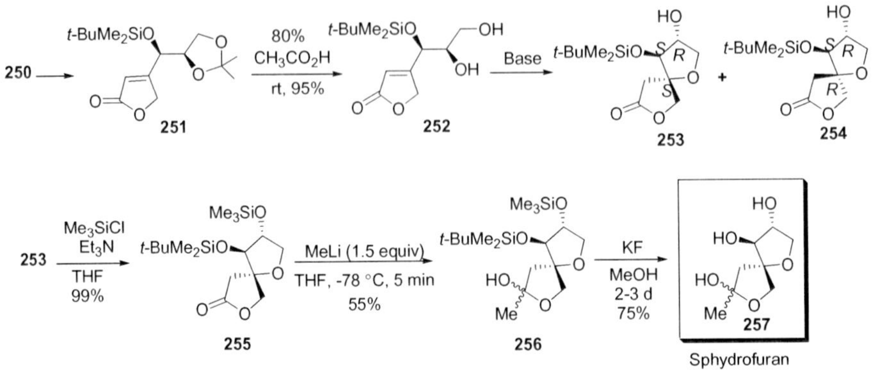

Scheme 5.47

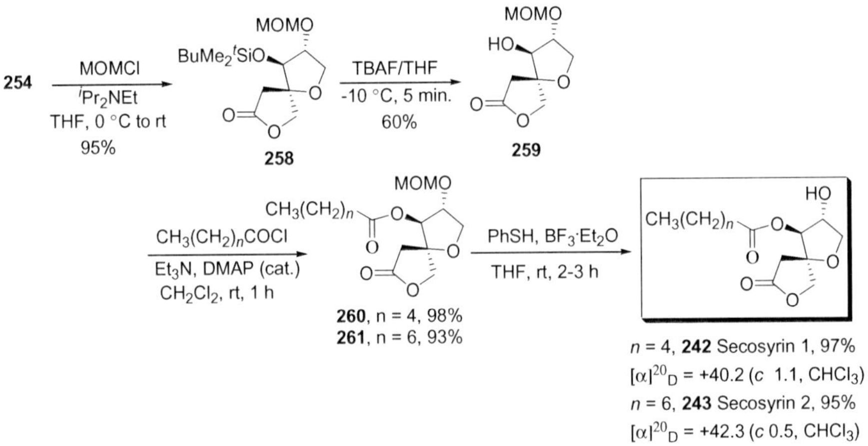

Scheme 5.48

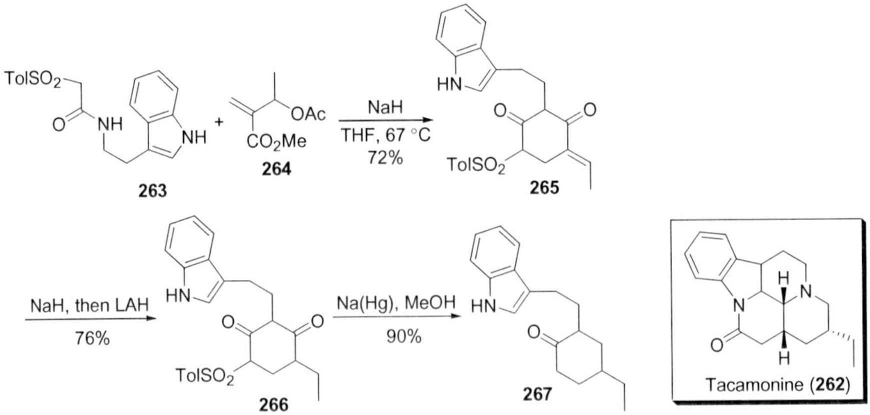

Scheme 5.49

Scheme 5.50

an efficient formal [3 + 3] reaction strategy to construct *N*-alkyl piperidin-2,6-diones from MBH adduct, the formal synthesis of tacamonine **262** has been described with application of such synthetic strategy (Scheme 5.49).[125] After deprotonation of **263** with NaH in tetrahydrofuran (THF) at room temperature, the resulting dianion reacted with four α,β-unsaturated esters **264** to furnish the corresponding piperidin-2,6-dione **265** in 72% yield at refluxing temperature. Upon treatment with NaH at room temperature, followed by addition of LAH,

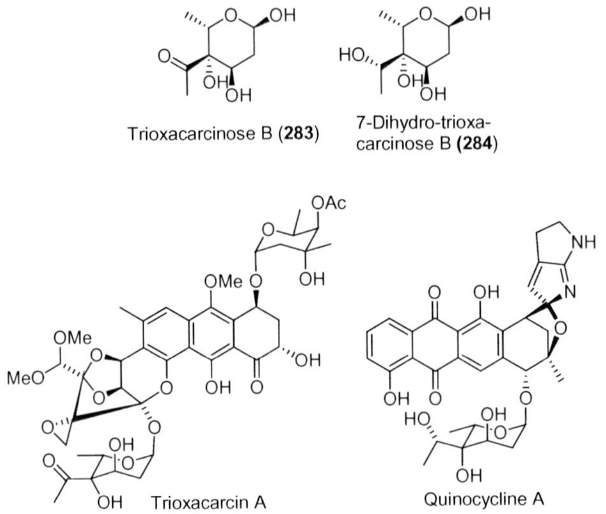

Figure 5.14 Trioxacarcinosides B are a key subunit in many natural products.

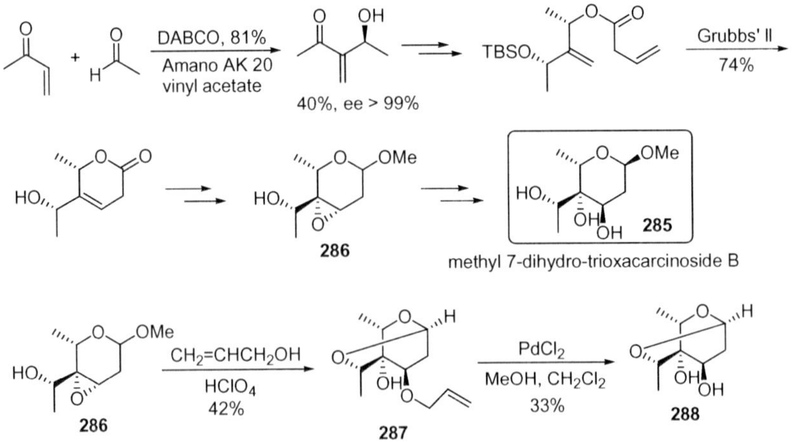

Scheme 5.51

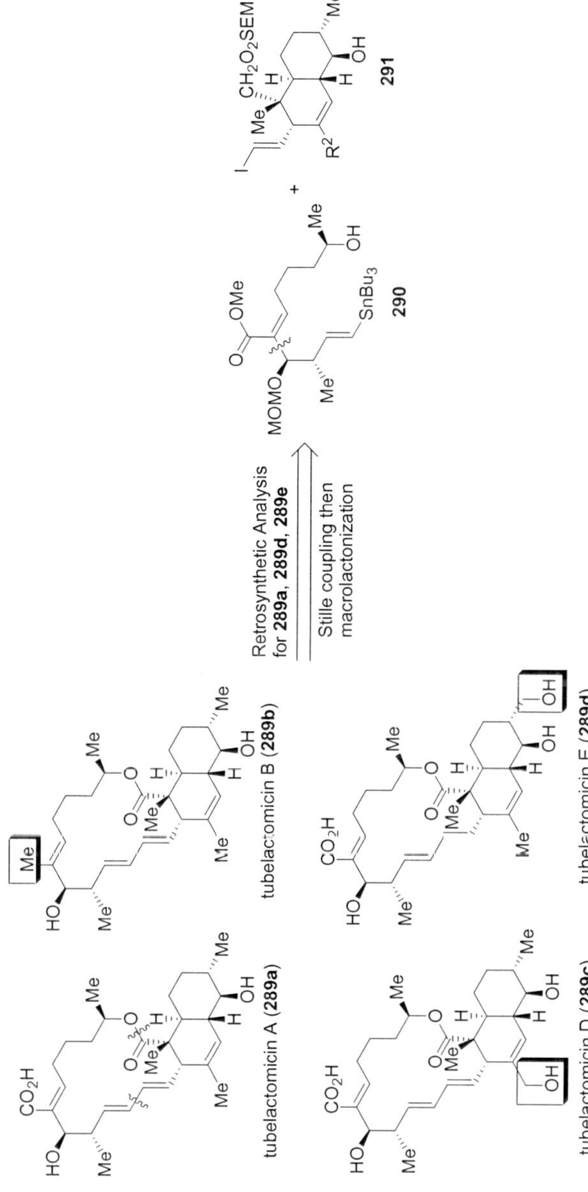

Figure 5.15 Tubelactomicins A, B, D and E.

piperidin-2,6-dione **265** was further converted into δ-lactam **266** in 76% yield. Reductive desulfonation of **266** with Na(Hg) in methanol furnished **267** in 90% yield, which could be readily converted into tacamonine **262** in three steps *via* Bischler–Napieralski cyclization.[126]

5.36   Core of Eleutherobin

Eleutherobin (**268**), an intricate diterpene glycoside first isolated in low yield (0.01–0.02%) from a rare alcyonacean coral *Eleutherobia* sp.,[127] has several distinct advantages compared with Taxol[®], the most widely used cancer chemotherapeutic agent in the United States. The structural complexity, biological significance and limited availability of **268** have fostered the development of novel synthetic methods for its total synthesis.[128] Winkler and co-workers have disclosed a brief way to construct the eleutherobin framework **282** in 15 linear steps from readily available starting materials (Scheme 5.50).[129] The tandem Diels–Alder reaction of **273** and **272** to produce **274a**, in which three new rings and six new stereogenic centers were formed, was a key step in this synthetic method. Among these intermediates, **271** was prepared *via* a MBH analogous reaction of propargyl bromide **269** and aldehyde **270** in the presence of $SnCl_2$ and NaI.

5.37   Trioxacarcinosides B

Trioxacarcinosides B, a family of octoses, exist as a key subunit in many natural products, such as quinocyclines (Figure 5.14).[130] The antibiotic and cytotoxic activities of natural products containing the trioxacarcinosides B (**283** or **284**) make these rare sugars attractive to synthetic chemists. So far, the synthetic methods for **283** or **284** make the use of a chiral-pool approach starting from sugar building blocks. Koert *et al.* have utilized the biocatalytic resolution of a MBH adduct and a subsequent ring-closing metathesis to

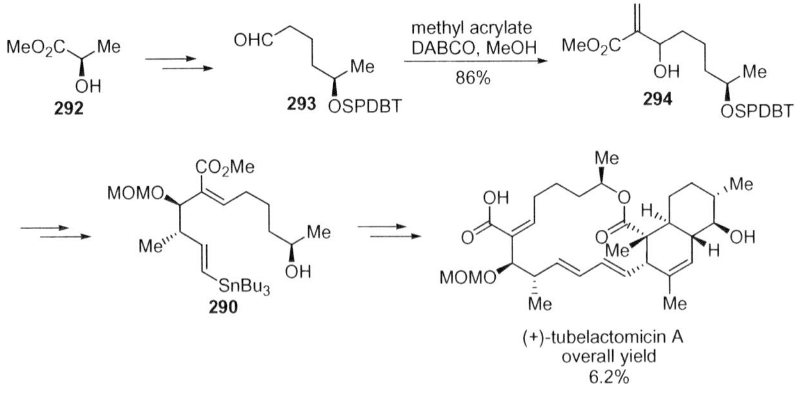

Scheme 5.52

MBH adduct
OR
CO2Me
Me
OH
R1
SnBu3
MOMO
Me
290, R1 = CO2Me
CO2SEM
H
Me
OH
Me
H
R2
291, R2 = CH2OMOM
Pd2(dba)3, AsPh3,
CuI, DMF, 60 °C
63%
MOMO
Me
R1
Me
HO
Me
CO2SEM
H
Me
OH
Me
H
R2
295, R1 = CO2Me, R2 = CH2OMOM
tubelactomicin D (289c)

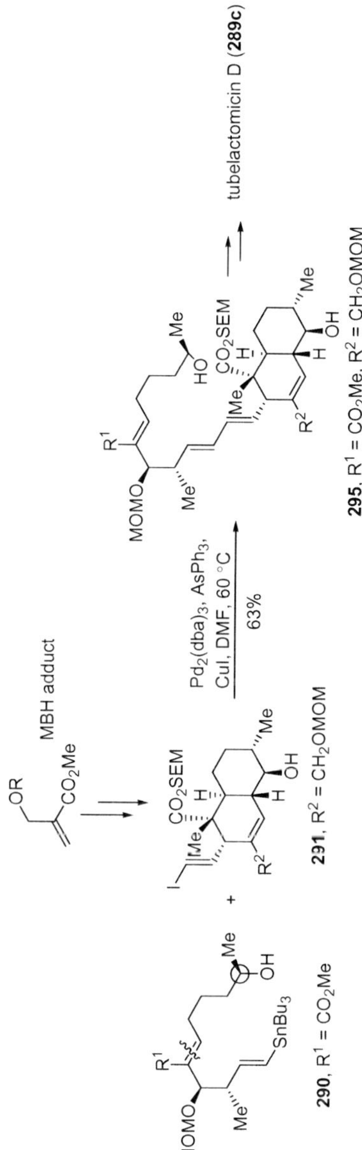

Scheme 5.53

assemble the chiral molecular framework (Scheme 5.51).[131] Following key steps such as a substrate-controlled epoxidation, stereo- and regiocontrolled epoxide opening by allyl alcohol were achieved for the stereoselective synthesis of methyl 7-dihydro-trioxacacarcinose B (**285**). Using key intermediate **286** as starting material, Koert *et al.* also accomplished the synthesis of 1,7-anhydrosugar **288**, whose spectral data and specific rotation were identical in all respects with those of the natural product reported by Webb *et al.*[132]

5.38 Tubelactomicin

Since the isolation from a culture broth of an actinomycete strain designated MK703-102F1 and the structure determination of (+)-tubelactomicin A (**289a**) were reported in 2000,[133] other structurally similar tubelactomicins (Figure 5.15),[134] having the tricyclic 16-membered macrolide, *i.e.*, (+)-tubelactomicin B (**289b**), (+)-tubelactomicin D (**289c**), and (+)-tubelactomicin E (**289d**), have been isolated from the same microorganism by the same group, and showed potential antimicrobial activity against acid-fast bacteria, including drug-resistant strains.

Based on a retrosynthetic analysis of tubelactomicins A, D and E, a MBH reaction could be employed for the synthesis of these compounds. Tadano *et al.* first accomplished the total synthesis of natural (+)-tubelactomicin A (Scheme 5.52),[135] which consisted of 54 total steps from methyl (*R*)-lactate (**292**) in 6.2% overall yield. As shown in Scheme 5.52, compound **294**, the key intermediate for the synthesis of upper-half segment **290**, was synthesized in 86% yield by the reaction of aldehyde **283** with methyl acrylate in the presence of DABCO for 5 days. With the upper-half segment **290** in hand, Tadano *et al.* subsequently

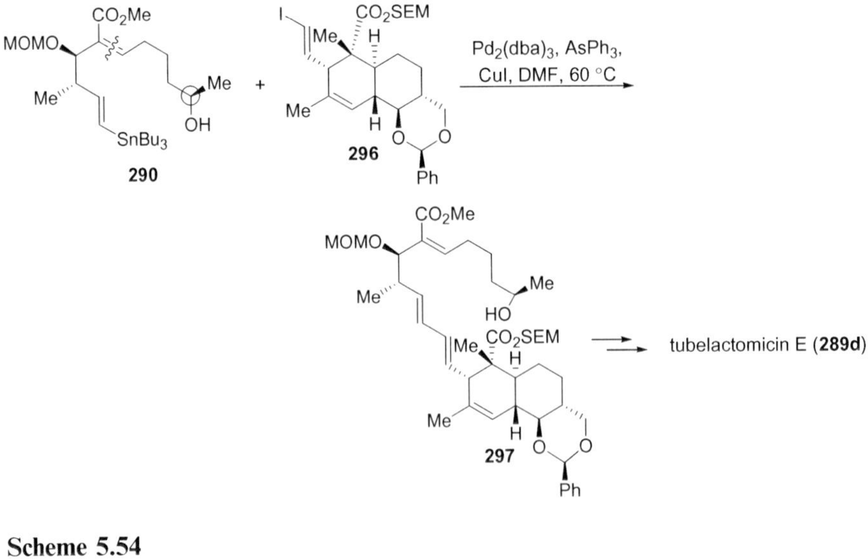

Scheme 5.54

Scheme 5.55

developed total synthetic approaches for the synthesis of (+)-tubelactomicin D (Scheme 5.53), and (+)-tubelactomicin E (Scheme 5.54).[136,137] Moreover, the lower-half segment of (+)-tubelactomicin D was also synthesized from MBH adduct.

5.39   6-Tuliposide B

6-Tuliposide B (**298**) often exists in tulip cultivars, and has potential anti-microbial activity against Gram-positive, Gram-negative, and certain fungicide-tolerant strains of bacteria, but it is not active against yeast. There are several existing synthetic methods to produce tulipalin B (**299**), including the non-sugar groups released by enzymatic or spontaneous hydrolysis of tuliposides under non-alkaline conditions.[138] However, the complete synthesis of **298** has not been

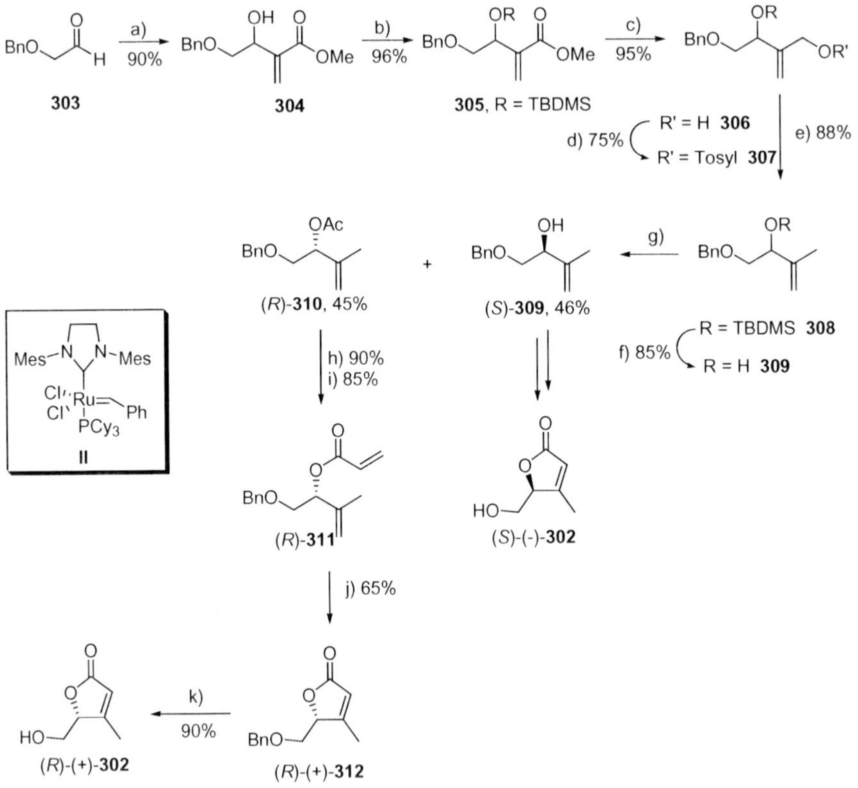

(a) Methyl acrylate, quinuclidine, MeOH, 8 h; (b) TBDMSCl, imidazole, CH_2Cl_2, DMAP, 0 °C to rt, 6 h; (c) DIBAL-H, CH_2Cl_2, -78 °C, 1 h; (d) TsCl, Et_3N, CH_2Cl_2, DMAP, rt, 4 h; (e) $LiAlH_4$, THF, 1 h; (f) TBAF, THF, 0 °C to rt, 2 h; (g) lipase PS-C, vinyl acetate, hexane, 6 h; (h) K_2CO_3, MeOH, 0 °C to rt, 1 h; (i) acryloyl chloride, Et_3N, CH_2Cl_2, MAP, 0 °C to rt, 30 min; (j) **II**, CH_2Cl_2, 35 °C, 48 h; (k) $TiCl_4$, CH_2Cl_2, 0 °C to rt, 10 min.

Scheme 5.56

reported, though there have been three unsuccessful attempts.[139] The complete synthesis of (+)-6-tuliposide B (**298**) was first achieved in nine steps from D-glucose *via* the MBH reaction of 2-(*tert*-butyldimethylsilyloxy)-acetaldehyde with 6-*O*-acryloyl-1-*O*-(2-trimethylsilylethyl)-β-D-glucopyranoside, followed by a mild deprotection procedure using TFA in CH_2Cl_2 (Scheme 5.55).[140]

5.40 (*R*)-Umbelactone

(*R*)-Umbelactone (**302**), one example of a naturally occurring γ-(hydroxymethyl)-α,β-butenolide, has been isolated from *Memycelon umbelatum Brum*, whose crude extracts have shown various biological activities, including antiviral (activity against Ranikhet disease virus), antiamphetory and spasmolytic.[141] Various synthetic approaches for umbelactones have been reported.[142] An efficient and convenient strategy for the enantioselective synthesis of enantiomerically enriched umbelactones, utilizing a lipase-mediated resolution protocol, has been reported by Kamal *et al.*, in which a MBH reaction and ring-closing metathesis were employed as key steps (Scheme 5.56).[143] The lipase-resolution of racemic **309** was carried out using several lipases, and about 50% conversion of **309** and up to > 99% ee were obtained using *Burkholderia cepacia* (PS-C) as lipase in hexane.

References

1. S. Naganuma, K. Sakai and A. Endo, *J. Antibiot.*, 1992, **45**, 1216.
2. K. Ishigami and T. Kitahara, *Tetrahedron*, 1995, **51**, 6431.
3. X. Franck and B. Figadère, *Tetrahedron Lett.*, 2002, **43**, 1449.
4. S. R. V. Kandula and P. Kumar, *Tetrahedron Lett.*, 2003, **44**, 6149.
5. R. V. Anand, S. Baktharaman and V. K. Singh, *Tetrahedron Lett.*, 2002, **43**, 5393.
6. (a) G. Cardillo and C. Tomasini, *Chem. Soc. Rev.*, 1996, **25**, 117; (b) *Enantioselective Synthesis of β-Amino Acids*, 2nd edn, ed. E. Juaristi and V. A. Soloshonok, Wiley-VCH, New York, 2005; (c) D. Seebach and J. L. Mattews, *J. Chem. Soc., Chem. Commun.*, 1997, 2015; (d) S. H. Gellman, *Acc. Chem. Res.*, 1998, **31**, 173.
7. R. Galeazzi, G. Martelli, G. Mobbili, M. Orena and S. Rinaldi, *Org. Lett.*, 2004, **6**, 25714.
8. R. Galeazzi, G. Martelli, M. Orena, S. Rinaldia and P. Sabatino, *Tetrahedron*, 2005, **61**, 5465.
9. (a) Y. Tesfamariam, A. Rudi, S. Zafra, I. Goldberg, G. Gravalos, M. Chleyer and Y. Kashman, *Tetrahedron Lett.*, 1998, **39**, 3323; (b) T. Yosief, A. Rudi and Y. Kashman, *J. Nat. Prod.*, 2000, **63**, 299.
10. (a) A. T. Merritt and S. V. Ley, *Nat. Prod. Rep.*, 1992, **9**, 243; (b) T. J. Tokoroyama, *Synth. Org. Chem. Jpn.*, 1993, **51**, 1164; (c) J. R. Hanson, *Nat. Prod. Rep.*, 2002, **19**, 125.
11. D. Pappo, S. Shimony and Y. Kashman, *J. Org. Chem.*, 2005, **70**, 199.

12. (a) T. Tokoroyama, M. Asada and H. Iio, *Tetrahedron Lett.*, 1984, **25**, 5068; (b) M. Tsukamoto, T. Asada, H. Iio and T. Tokoroyama, *Tetrahedron Lett.*, 1987, **28**, 6645; (c) T. Tokoroyama, H. Iio and K. Okada, *J. Chem. Soc. Chem. Commun.*, 1989, **1**, 1572; (d) T. Aoto and T. Tokoroyama, *J. Org. Chem.*, 1998, **63**, 4151; (e) P. Wasnaire, M. Wiaux, R. Touillaux and I. E. Markó, *Tetrahedron Lett.*, 2006, **47**, 985.

13. S. A. Rodgen and S. E. Schaus, *Angew. Chem. Int. Ed.*, 2006, **45**, 4929.

14. (a) A. M. Jeffrey and R. M. J. Liskamp, *Proc. Natl. Acad. Sci. U. S. A.*, 1986, **83**, 241; (b) P. A. Wender, K. F. Koehler, N. A. Sharkey, M. L. Dell' Aquila and P. M. Blumberg, *Proc. Natl. Acad. Sci. USA*, 1986, **83**, 4214; (c) P. A. Wender, C. M. Cribbs, K. F. Koehler, N. A. Sharkey, C. L. Herald, Y. Kamano, G. R. Pettit and P. M. Blumberg, *Proc. Natl. Acad. Sci. USA*, 1988, **85**, 7197; (d) Y. Nishizuka, *Nature*, 1984, **308**, 693.

15. S. M. Kupchan, Y. Shizuri, W. C. Sumner Jr, H. R. Haynes, A. P. Leighton and B. R. Sickles, *J. Org. Chem.*, 1976, **41**, 3850.

16. (a) C. M. Marson, D. W. M. Benzies, A. D. Hobson, H. Adams and N. A. Bailey, *J. Chem. Soc., Chem. Commun.*, 1990, 15168; (b) C. M. Marson, S. Harper, A. J. Walker, J. Pickering, J. Campbell, R. Wrigglesworth and S. J. Edge, *Tetrahedron*, 1993, **45**, 10339; (c) F. J. Evans and S. E. Taylor, *Prog. Chem. Org. Nat. Prod.*, 1983, **44**, 1.

17. C. M. Marson, J. H. Pink, D. Hall, M. B. Hursthouse, A. Malik and C. Smith, *J. Org. Chem.*, 2003, **68**, 792.

18. G. R. Pettit, Y. Kamano, C. L. Herald, A. A. Tuinman, F. E. Boettner, H. Kizu, J. M. Schmidt, L. Baczynskyz, K. B. Tomer and R. J. Bontems, *J. Am. Chem. Soc.*, 1987, **109**, 6883.

19. T. Madden, H. T. Tran, D. Beck, R. Huie, R. A. Newman, L. Pusztai, J. J. Wright and J. L. Abbruzzese, *Clin. Cancer. Res.*, 2000, **6**, 1293.

20. W. P. Almeida and F. Coelho, *Tetrahedron Lett.*, 2003, **44**, 937.

21. E. Tsuchiya, M. Yukawa, T. Miyakawa, K. Kimura and H. Takahashi, *J. Antibiot.*, 2001, **54**, 84.

22. For its total syntheses, see: (a) M. O. Duffey, A. LeTiran and J. P. Morken, *J. Am. Chem. Soc.*, 2003, **125**, 1458; (b) S. Hanessian, Y. Yang, S. Giroux, V. Mascitti, J. Ma and F. Raeppel, *J. Am. Chem. Soc.*, 2003, **125**, 13784; (c) B. G. Vong, S. H. Kim, S. Abraham and E. A. Theodorakis, *Angew. Chem. Int. Ed.*, 2004, **43**, 3947; and (d) T. Nagamitsu, D. Takano, T. Fukuda, K. Otoguro, I. Kuwajima, Y. Harigaya and S. Omura, *Org. Lett.*, 2004, **11**, 1865

23. C. V. Krishna, S. Maitra, R. V. Dev, K. Mukkanti and J. Iqbal, *Tetrahedron Lett.*, 2006, **47**, 6103.

24. D. Basavaiah, P. K. S. Sarma and A. K. D. Bhavani, *J. Chem. Soc., Chem. Commun.*, 1994, 1091.

25. B. Das, J. Banerjee, G. Mahender and A. Majhi, *Org. Lett.*, 2004, **6**, 3349.

26. (a) J. W. S. Bradshaw, R. Baker and P. E. Howse, *Nature*, 1975, **258**, 230; (b) R. Rossi, A. Carpita and T. Messari, *Synth. Commun.*, 1992, **22**, 603; (c) M. J. Garson, *Chem. Rev.*, 1993, **93**, 1699.

27. (a) J. W. S. Bradshaw, R. Baker and P. E. Howse, *Nature*, 1975, **258**, 230; (b) D. Basavaiah, P. K. S. Sarma and A. K. D. Bhavani, *J. Chem. Soc., Chem. Commun.*, 1994, 1091.

28. J. Kobayashi, M. Ishibashi, H. Nakamura, Y. Ohizumi, T. Yamasu, T. Sasaki and Y. Hirata, *Tetrahedron Lett.*, 1986, **27**, 5755.

29. I. Bauer, L. Maranda, K. A. Young, Y. Shimizu, C. Fairchild, L. Cornell, J. MacBeth and S. Huang, *J. Org. Chem.*, 1995, **60**, 1084.

30. For an overview of synthetic efforts, see: C. Aïssa, R. Riveiros, J. Ragot and A. Fürstner, *J. Am. Chem. Soc.*, 2003, **125**, 15512.

31. M. Seck, X. Franck, B. Seon-Meniel, R. Hocquemiller and B. Figadere, *Tetrahedron Lett.*, 2006, **47**, 4175.

32. (a) V. Rodeschini, N. S. Simpkins and C. Wilson, *J. Org. Chem.*, 2007, **72**, 4265; (b) P. Nuhant, M. David, T. Pouplin, B. Delpech and C. Marazano, *Org. Lett.*, 2007, **9**, 287; (c) N. M. Ahmad, V. Rodeschini, N. S. Simpkins, S. E. Ward and A. J. Blake, *J. Org. Chem.*, 2007, **72**, 4803.

33. L. E. Mccandlish, J. C. Hanson and G. H. Stout, *Acta Crystallogr., Sect. B*, 1976, **32**, 1793.

34. J. Qi and J. A. Porco, Jr., *J. Am. Chem. Soc.*, 2007, **129**, 12682.

35. R. Takagi, Y. Inoue and K. Ohkata, *J. Org. Chem.*, 2008, **73**, 9320.

36. S. Rajesh, J. Srivastava, B. Bannerji and J. Iqbal, *Arkivoc*, 2001, **1**, 20.

37. E. Dorta, A. R. Díaz-Marrero, I. Brito, M. Cueto, L. D'Croz and J. Darias, *Tetrahedron*, 2007, **63**, 9057.

38. J. A. Marshall and E. A. Van Devender, *J. Org. Chem.*, 2001, **66**, 8037.

39. T. J. Donohoe, A. Ironmonger and N. M. Kershaw, *Angew. Chem. Int. Ed.*, 2008, **47**, 7314.

40. C. F. Nising, U. K. Ohnemüller, A. Friedrich, B. Lesch, J. Steiner, H. Schnöckel, M. Nieger and S. Bräse, *Chem. Eur. J.*, 2006, **12**, 3647.

41. C. F. Nising, U. K. Ohhemuller and S. Bräse, *Angew. Chem.* 2006, **118**, 313; *Angew. Chem. Int. Ed.*, 2006, **45**, 307.

42. (a) B. Lesch and S. Bräse, *Angew. Chem.* 2004, **116**, 118; *Angew. Chem. Int. Ed.*, 2004, **43**, 115; (b) B. Lesch, J. Toräng, S. Vanderheiden and S. Bräse, *Adv. Synth. Cat.*, 2005, **4**, 555; (c) K. Y. Lee, J. M. Kim and J. N. Kim, *Bull. Korean Chem. Soc.*, 2003, **24**, 17.

43. U. K. Ohnemüller, C. F. Nising, M. Nieger and S. Bräse, *Eur. J. Org. Chem.*, 2006, 1535.

44. E. M. C. Gérard and S. Bräse, *Chem. Eur. J.*, 2008, **14**, 8086.

45. J. Lechner, F. Wieland and M. Sumper, *J. Biol. Chem.*, 1985, **260**, 8984.

46. R. Kornfeld and S. Kornfeld, *Annu. Rev. Biochem.*, 1985, **54**, 631.

47. D. Grassi, V. Lippuner, M. Aebi, J. Brunner and A. Vasella, *J. Am. Chem. Soc.*, 1997, **119**, 10992.

48. (a) H. J. Lee, C. H. Lee, M. C. Chung, H. K. Chun, J. S. Rhee and Y. H. Kho, *Tetrahedron Lett.*, 1999, **40**, 6949; (b) H. J. Lee, H. K. Chun, M. C. Chung, C. H. Lee, J. S. Rhee and Y. H. Kho, *J. Antibiot.*, 2000, **53**, 78; (c) S. B. Han, H. J. Lee, Y. H. Kho, Y. J. Jeon, S. H. Lee, H. C. Kim and H. M. Kim, *J. Antibiot.*, 2001, **54**, 840.

49. J. C. Heo, J. Y. Park, S. U. Woo, J. R. Rho, H. J. Lee, S. U. Kim, Y. H. Kho and S. H. Lee, *Biol. Pharm. Bull.*, 2006, **29**, 2256.

50. J. H. Woo, J. W. Park, S. H. Lee, Y. H. Kim, I. K. Lee, E. Gabrielson, S. H. Lee, H. J. Lee, Y. H. Kho and T. K. Kwon, *Cancer Res.*, 2003, **63**, 3430.

51. C. M. Thompson, C. A. Quinn and P. J. Hergenrother, *J. Med. Chem.*, 2009, **52**, 117.

52. H.-J. Lee, M.-C. Chung, C.-H. Lee, B.-S. Yun, H.-K. Chun and Y.-H. Kho, *J. Antibiotics*, 1997, **50**, 357.

53. J.-H. Cho and S. Y. Ko, *Helv. Chim. Acta*, 2002, **85**, 3994.

54. E. Laborde and V. P. Manchem, *Curr. Med. Chem.*, 2002, **9**, 2231.

55. For the racemic synthesis of efaroxan, see: (a) C. R. Edwards, M. J. Readhead and N. J. J. Tweddle, *Heterocycl. Chem.*, 1987, **24**, 495; (b) P. Mayer, P. Brunel and T. Imbert, *Bioorg. Med. Chem. Lett.*, 1999, **9**, 3021; (c) K. Couture, V. Gouverneur and C. Mioskowski, *Bioorg. Med. Chem. Lett.*, 1999, **9**, 3023; (d) J. Clews, N. G. Morgan and C. A. Ramsden, *J. Heterocycl. Chem.*, 2001, **38**, 519; (e) C. B. Chapleo, P. L. Myers, R. C. M. Butler, J. A. Davis, J. C. Doxey, S. D. Higgins, M. Myers, A. G. Roach, C. F. C. Smith, M. R. Stillings and A. P. J. Welbourn, *Med. Chem.*, 1984, **27**, 570; and (f) C. Mioskowski, V. Gouverneur, K. Couture, P. Lesimple and J.-M. Autret, *WO Patent* 15624, 2000; *Chem. Abstr.*, 2000, **132**, 222438. For an example of the synthesis of an efaroxan derivative, see: (g) P. Mayer, C. Loubat and T. Imbert, *Heterocycles*, 1998, **48**, 2529

56. T. Imbert and P. Mayer, WO Patent 35682 1996; *Chem. Abstr.*, 1997, 126, 59852.

57. G. P. De, C. Silveira and F. Coelho, *Tetrahedron Lett.*, 2005, **46**, 6477.

58. K. Tsuchiya, S. Kobayashi, T. Nishikiori, T. Nakagawa and K. Tatsuta, *J. Antibiot.*, 1997, **50**, 261.

59. Y. Koguchi, J. Kohno, S. Suzuki, M. Nishio, K. Takahashi, T. Ohnuki and S. Matsubara, *J. Antibiot.*, 1999, **52**, 1069.

60. K. Sugawara, M. Hatori, Y. Nishiyama, K. Tomita, H. Kamei, M. Konishi and T. Oki, *J. Antibiot.*, 1990, **43**, 8.

61. (a) M. Groll, K. B. Kim, N. Kairies, R. Huber and C. M. Crews, *J. Am. Chem. Soc.*, 2000, **122**, 1237; (b) U. Schmidt and J. Schmidt, *J. Chem. Soc., Chem. Commun.*, 1992, 529; (c) H. Hoshi, T. Ohnuma, S. Aburaki, M. Konishi and T. Oki, *Tetrahedron Lett.*, 1993, **34**, 1047; (d) N. Sin, L. Meng, H. Auth and C. M. Crews, *Bioorg. Med. Chem. Lett.*, 1998, **6**, 1209; (e) N. Sin, K. B. Kim, M. Elofsson, L. Meng, H. Auth, B. H. K. Kwok and C. M. Crews, *Bioorg. Med. Chem. Lett.*, 1999, **9**, 2283; (f) M. Elofsson, U. Splittgerber, J. Myung, R. Mohan and C. M. Crews, *Chem. Biol.*, 1999, **6**, 811.

62. M. R. Dobler, *Tetrahedron Lett.*, 2001, **42**, 215.

63. Y. Iwabuchi, T. Sugihara, T. Esumi and S. Hatakeyama, *Tetrahedron Lett.*, 2001, **42**, 7867.

64. A. R. Carroll and W. C. Taylor, *Aust. J. Chem.*, 1991, **44**, 1615.

65. B. B. Patel, T. G. Waddell and R. M. Pagni, *Fitoterapia*, 2001, **72**, 511.

66. G. W. Kabalka and B. Venkataiah, *Tetrahedron Lett.*, 2005, **46**, 7325.

67. S. Mitra, S. R. Gurrala and R. S. Coleman, *J. Org. Chem.*, 2007, **72**, 8724.

68. G. Fráter, J. A. Bajgrowicz and P. Kraft, *Tetrahedron*, 1998, **54**, 7633.

69. G. Ohloff, *Riechstoffe und Geruchssinn*, Springer Verlag, Berlin, 1990, p. 172.

70. J. A. Bajgrowicz and G. Fráter, S.A. to Givaudan-Roure, *EP Pat.* 841,318, 1997.

71. B. Hölscher, N. A. Braun, B. Weber, C.-H. Kappey, M. Meier and W. Pickenhagen, *Helv. Chim. Acta*, 2004, **87**, 1666.

72. A. D. Patil, A. J. Freyer, L. Killmer, P. Offen, B. Carte, A. J. Jurewicz and R. K. Johnson, *Tetrahedron*, 1997, **53**, 5047.

73. Y. F. Hallock, J. H. Cardellina and M. R. Boyd, *Nat. Prod. Lett.*, 1998, **11**, 153.

74. B. M. Trost, Y. Hu and D. B. Horne, *J. Am. Chem. Soc.*, 2007, **129**, 11781.

75. (a) M. Inoue, M. W. Carson, A. J. Frontier and S. J. Danishefsky, *J. Am. Chem. Soc.*, 2001, **123**, 1878; (b) M. Inoue, A. J. Frontier and S. J. Danishefsky, *Angew. Chem. Int. Ed.*, 2000, **39**, 761; (c) C. C. Hughes and D. Trauner, *Angew. Chem. Int. Ed.*, 2002, **41**, 1569; (d) C. C. Hughes and D. Trauner, *Tetrahedron*, 2004, **60**, 9675–9686; (e) J. H. Chaplin and B. L. Flynn, *Chem. Commun.*, 2001, 1594; (f) I. Martinez, P. E. Alford and T. V. Ovaska, *Org. Lett.*, 2005, **7**, 1133; (g) X. Li, R. E. Kyne and T. V. Ovaska, *Org. Lett.*, 2006, **8**, 5153.

76. (a) S. Funayama, M. Ishibashi, Y. Anraku, K. Komiyama and S. Omura, *Tetrahedron Lett.*, 1989, **30**, 7427; (b) K. Komiyama, S. Funayama, Y. Anraku, M. Ishibashi, Y. Takahashi and S. Omura, *J. Antibiot.*, 1990, **43**, 247; (c) S. Funayama, M. Ishibashi, K. Komiyama and S. Omura, *J. Org. Chem.*, 1990, **55**, 1132; (d) M. Ishibashi, S. Funayama, Y. Anraku, K. Komiyama and S. Omura, *J. Antibiot.*, 1991, **44**, 390.

77. (a) A. B. Smith, III., J. P. Sestelo and P. G. Dormer, *J. Am. Chem. Soc.*, 1995, **117**, 10755; (b) A. B. Smith, III., J. P. Sestelo and P. G. Dormer, *Heterocycles*, 2000, **52**, 1315; (c) T. Saito, M. Morimoto, C. Akiyama, T. Matsumoto and K. Suzuki, *J. Am. Chem. Soc.*, 1995, **117**, 10757; (d) T. Saito, T. Suzuki, M. Morimoto, C. Akiyama, T. Ochiai, K. Takeuchi, T. Matsumoto and K. Suzuki, *J. Am. Chem. Soc.*, 1998, **120**, 11633.

78. B. M. Trost, H.-C. Tsui and F. D. Toste, *J. Am. Chem. Soc.*, 2000, **122**, 3534.

79. B. M. Trost, O. R. Thiel and H. C Tsui, . *J. Am. Chem. Soc.*, 2002, **124**, 11616.

80. B. M. Trost, O. R. Thiel and H. C. Tsui, *J. Am. Chem. Soc.*, 2003, **125**, 13155.

81. B. M. Trost, M. R. Machacek and H. C. Tsui, *J. Am. Chem. Soc.*, 2005, **127**, 7014.

82. (a) W. D. Celmer, K. Murai, K. V. Rao, F. W. Tanner and W. S. Marsh, *Antibiot. Ann.*, 1957-58, 484; (b) A. Tulinsky, *J. Am. Chem. Soc.*, 1964, **86**, 5368.

83. L. H. Zalkow, S. Bonetti, L. Gelbaum, M. M. Gordon, B. B. Patil and D. Van Derveer, *J. Nat. Prod.*, 1979, **42**, 603.
84. N. J. Leonard, in *The Alkaloids*, **vol. 6**, ed. R. H. F. Manske, Academic Press, New York, 1960, p. 35.
85. (a) N. Naruse, R. Kageyama-Kawase, Y. Funuhashi, T. Wakabayashi, Y. Watanabe, T. Sameshima and K. Dobashi, *J. Antibiot.*, 2000, **53**, 579; (b) T. Wakabayashi, R. Kageyama-Kawase, N. Naruse, Y. Funuhashi and K. Yoshimatsu, *J. Antibiot.*, 2000, **53**, 591.
86. N. Hata-Sugi, R. Kawase-Kageyama and T. Wakabayashi, *Biol. Pharm. Bull.*, 2002, **25**, 446.
87. R. Jogireddy and M. E. Maier, *J. Org. Chem.*, 2006, **71**, 6999.
88. (a) S. Sasaki, R. Hashimoto, M. Kikuchi, K. Inoue, T. Ikumoto, R. Hirose, K. Chiba, Y. Hoshino, T. Okamoto and T. Fujita, *J. Antibiot.*, 1994, **47**, 420; (b) T. Fujita, N. Hamamichi, M. Kikuchi, T. Matsuzaki, Y. Kitao, K. Inoue, R. Hirose, M. Yoneta, S. Sasaki and K. Chiba, *J. Antibiot.*, 1996, **49**, 846.
89. T. Fujita, N. Hamamichi, T. Matsuzaki, Y. Kitao, M. Kikuchi, M. Node and R. Hirose, *Tetrahedron Lett.*, 1995, **36**, 8599.
90. Y. Iwabuchi, M. Furukawa, T. Esumi and S. Hatakeyama, *Chem. Commun.*, 2001, 2030.
91. M. Hanaoka, H. Yamagishi, M. Marutani and C. Mukai, *Tetrahedron Lett.*, 1984, **25**, 5169.
92. H. Ishii, I.-S. Chen and S. Ueki, *Chem. Pharm. Bull.*, 1987, **35**, 2715.
93. T. H. Tsai, W. H. Chung, J. K. Chang, R. T. Hsu and N. C. Chang, *Tetrahedron*, 2007, **63**, 9825.
94. (a) S. B. Mahato, K. A. I. Siddiqui, G. Bhattacharya, T. Ghoshal, K. Miyahara, M. Sholichin and T. Kawasaki, *J. Nat. Prod.*, 1987, **50**, 245; (b) N. Petragnani, H. M. C. Ferraz and G. V. J. Silva, *Synthesis*, 1986, 157; (c) H. M. R. Hoffmann and J. Rabe, *Angew. Chem. Int. Ed. Engl.*, 1985, **24**, 94; (d) C. J. Cavallito, D. M. Fruehauf and J. H. Bailey, *J. Am. Chem. Soc.*, 1948, **70**, 3724; (e) W. Zopf, *Liebigs Ann. Chem.*, 1902, **324**, 39.
95. (a) B. K. Park, M. Nakagawa, A. Hirota and M. Nakayama, *J. Antibiot.*, 1988, **41**, 751; (b) K. Muller, *Appl. Microbiol. Biotechnol.*, 2001, **56**, 9.
96. N. Selvakumar, P. K. Kumar, K. C. S. Reddy and B. C. Chary, *Tetrahedron Lett.*, 2007, **48**, 2021.
97. (a) D. Uemura, T. Chuo, T. Haino, A. Nagatsu, S. Fukuzawa, S. Zheng and H. Chen, *J. Am. Chem. Soc.*, 1995, **117**, 1155; (b) T. Chuo, O. Kamo and D. Uemura, *Tetrahedron Lett.*, 1996, **37**, 4023; (c) T. Otofuji, A. Ogo, J. Koishi, K. Matsuo, H. Tokiwa, T. Yasumoto, K. Nishihara, E. Yamamoto, M. Saisho, Y. Kurihara and K. Hayashida, *Food Sanit. Res.*, 1981, **31**, 76; (d) S. Z. Zheng, F. L. Huang, S. C. Chen, X. F. Tan, J. B. Zuo, J. Peng and R. W. Xie, *Chin. J. Mar. Drugs*, 1990, **33**, 33.
98. (a) T. Sugimoto, J. Ishihara and A. Murai, *Tetrahedron Lett.*, 1997, **38**, 7379; (b) X. Zhang and M. C. McIntosh, 215th National Meeting of the American Chemical Society, Dallas, TX; March 29-April 2, 1998, ORGN 188.

99. J. A. McCauley, K. Nagasawa, P. A. Lander, S. G. Mischke, M. A. Semones and Y. Kishi, *J. Am. Chem. Soc.*, 1998, **120**, 7647.

100. R. H. Feling, G. O. Buchanan, T. J. Mincer, C. A. Kauffman, P. R. Jensen and W. Fenical, *Angew. Chem. Int. Ed.*, 2003, **42**, 355.

101. (a) E. J. Corey and W.-D. Z. Li, *Chem. Pharm. Bull.*, 1999, **47**, 1; (b) E. J. Corey, G. A. Reichard and R. Kania, *Tetrahedron Lett.*, 1993, **34**, 6977; (c) E. J. Corey and G. A. Reichard, *J. Am. Chem. Soc.*, 1992, **114**, 10677; (d) G. Fenteany, R. F. Standaert, G. A. Reichard, E. J. Corey and S. L. Schreiber, *Proc. Natl. Acad. Sci. USA*, 1994, **91**, 3358.

102. L. R. Reddy, P. Sarvanan and E. J. Corey, *J. Am. Chem. Soc.*, 2004, **126**, 6230.

103. L. R. Reddy, J.-F. Fournier, B. V. S. Reddy and E. J. Corey, *Org. Lett.*, 2005, **7**, 2699.

104. A. Salimbeni, R. Canevotti, F. Paleari, D. Poma, S. Caliari, F. Fici, R. Cirillo, A. R. Renzetti, A. Subissi, L. Belvisi, G. Bravi, C. Scolastico and A. Giachetti, *J. Med. Chem.*, 1995, **38**, 4806.

105. R. L. Grange, J. Ziogas, A. J. North, J. A. Angus and C. H. Schiesser, *Bioorg. Med. Chem. Lett.*, 2008, **18**, 1241.

106. J. G. Mulder, P. Diepenhorst, P. Plieger and I. E. M. Brüggemann-Rotgans, *PCT Int. Appl.* WO 93/02,083; *Chem. Abstr.*, 1993, **118**, 185844z.

107. B. T. Buu Hue, J. Dijkink, S. Kuiper, S. van Schaik, J. H. van Maarseveen and H. Hiemstra, *Eur. J. Org. Chem.*, 2006, **127**.

108. (a) D. Hauser and H. P. Sigg, *Helv. Chim. Acta*, 1971, **54**, 1178; (b) H. Liang, *Beilst. J. Org. Chem.*, 2008, **4**, 31.

109. (a) M. C. Justice, M. J. Hsu, B. Tse, T. Ku, J. M. Balkovec, D. Schmatz and J. Nielsen, *J. Biol. Chem.*, 1998, **273**, 3148; (b) L. Capa, A. Mendoza, J. L. Lavandera, F. Gomez de las Heras and J. F. Garcia-Bustos, *Antimicrob. Agents Chemother.*, 1998, **42**, 2694; (c) M. Shastry, J. Nielsen, T. Ku, M.-J. Hsu, P. Liberator, J. Anderson, D. Schmatz and M. C. Justice, *Microbiology*, 2001, **147**, 383.

110. (a) H. Liang, A. Schulé, J.-P. Vors and M. A. Ciufolini, *Org. Lett.*, 2007, **9**, 4119; (b) A. Schulé, H. Liang, J.-P. Vors and M. Ciufolini, *J. Org. Chem.*, 2009, **74**, 1587.

111. (a) H. A. Kirst, K. H. Michel, J. W. Martin, L. C. Creemer, E. H. Chio, R. C. Yao, W. M. Nakatsukasa, L. Boeck, J. L. Occolowitz, J. W. Paschal, J. B. Deeter, N. D. Jones and G. D. Thompson, *Tetrahedron Lett.*, 1991, **32**, 4839; (b) H. A. Kirst, K. H. Michel, E. H. Chio, R. C. Yao, W. M. Nakatsukasa, L. Boeck, J. L. Occolowitz, J. W. Paschal, J. B. Deeter and G. D. Thompson, in *Microbial Metabolites*, ed. C. Nash, William C Brown, Dubuque IA, 1991, **vol. 32**, 109.

112. R. Dagani, *Chem. Eng. News*, 1999, **77**, 30.

113. D. A. Evans and W. C. Black, *J. Am. Chem. Soc.*, 1993, **115**, 4497.

114. (a) L. A. Paquette, Z. L. Gao, Z. J. Ni and G. F. Smith, *J. Am. Chem. Soc.*, 1998, **120**, 2543; (b) L. A. Paquette, I. Collado and M. Purdie, *J. Am. Chem. Soc.*, 1998, **120**, 2553.

115. D. J. Mergott, S. A. Frank and W. R. Roush, *Org. Lett.*, 2002, **4**, 3157.

116. D. J. Mergott, S. A. Frank and W. R. Roush, *Proc. Natl. Acad. Sci. USA*, 2004, **101**, 11955.

117. (a) C. B. Cui, H. Kakeya and H. Osada, *J. Antibiot.*, 1996, **49**, 832; (b) C. B. Cui, H. Kakeya and H. Osada, *Tetrahedron*, 1996, **52**, 12651; (c) C. B. Cui, H. Kakeya and H. Osada, *Tetrahedron*, 1997, **53**, 59; (d) C. B. Cui, H. Kakeya, G. Okada, R. Onose, I. Ubukata, K. Takahashi, K. Isono and H. J. Osada, *Antiobiot.*, 1995, **48**, 1382; (e) C. B. Cui, H. Kakeya, G. Okada, R. Onose and H. Osada, *J. Antibiot.*, 1996, **49**, 527; (f) C. B. Cui, H. Kakeya and H. Osada, *J. Antibiot.*, 1996, **49**, 534.

118. (a) P. R. Sebahar and R. M. Williams, *J. Am. Chem. Soc.*, 2000, **122**, 5666; (b) F. von Nussbaum and S. J. Danishefsky, *Angew. Chem.*, 2000, **112**, 2259; Angew. Chem. Int. Ed., 2000, **39**, 2175; (c) H. Wang and A. Ganesan, *J. Org. Chem.*, 2000, **65**, 4685.

119. L. E. Overman and M. D. Rosen, *Angew. Chem. Int. Ed.*, 2000, **39**, 4596.

120. S. L. Midland, N. T. Keen and J. J. Sims, *J. Org. Chem.*, 1995, **60**, 1118.

121. (a) T. Honda, H. Mizutani and K. Kanai, *J. Org. Chem.*, 1996, **61**, 9374; (b) M. Carda, E. Castillo, S. Rodriguez, E. Falomir and J. A. Marco, *Tetrahedron Lett.*, 1998, **39**, 8895.

122. P. R. Krishna, M. Narsingam and V. Kannan, *Tetrahedron Lett.*, 2004, **45**, 4773.

123. (a) P. Yu, Y. Yang, Z. Y. Zhang, T. C. W. Mak and H. N. C. Wong, *J. Org. Chem.*, 1997, **62**, 6359; (b) T. J. Donohoe, J. W. Fisher and P. J. Edwards, *Org. Lett.*, 2004, **6**, 465.

124. T. A. Van Beek, R. Verpoorte and A. Baerheim Svendsen, *Tetrahedron*, 1984, **40**, 737.

125. C.-Y. Chen, M.-Y. Chang, R.-T. Hsu, S.-T. Chen and N.-C. Chang, *Tetrahedron Lett.*, 2003, **44**, 8627.

126. G. Massiot, F. Sousa Oliverira and J. Lévy, *Bull. Soc. Chim. Fr. II*, 1982, 185.

127. (a) W. H. Fenical, P. R. Jensen and T. Lindel, *US Pat. 5,473,057*, Dec. 5, 1995; (b) T. Lindel, P. R. Jensen, W. Fenical, B. H. Long, A. M. Casazza, J. Carboni and C. R. Fairchild, *J. Am. Chem. Soc.*, 1997, **119**, 8744.

128. (a) K. C. Nicoloau, F. van Delft, T. Ohshima, D. Vourloumis, J. Xu, S. Hosokawa, J. Pfefferkorn, S. Kim and T. Li, *Angew. Chem. Int. Ed. Engl.*, 1997, **36**, 2520; (b) K. C. Nicolaou, T. Ohshima, S. Hosokawa, F. L. van Delft, S. Vourloumis, J. Y. Xu, J. Pfefferkorn and S. Kim, *J. Am. Chem. Soc.*, 1998, **120**, 8674.

129. J. D. Winkler, K. J. Quinn, C. H. MacKinnon, S. D. Hiscock and E. C. McLaughlin, *Org. Lett.*, 2003, **5**, 1805.

130. (a) W. D. Celmer, K. Murai, K. V. Rao, F. W. Tanner and W. S. Marsh, *Antiobiot. Ann.*, 1957-58, 484; (b) A. Tulinsky, *J. Am. Chem. Soc.*, 1964, **86**, 5368.

131. C. M. König, K. Harms and U. Koert, *Org. Lett.*, 2007, **9**, 4777.

132. J. S. Webb, R. W. Broschard, D. B. Cosulich, J. H. Mowat and J. E. Lancaster, *J. Am. Chem. Soc.*, 1962, **84**, 3183.

133. (a) M. Igarashi, C. Hayashi, Y. Homma, S. Hattori, N. Kinoshita, M. Hamada and T. Takeuchi, *J. Antibiot.*, 2000, **53**, 1096; (b) M. Igarashi, H. Nakamura, H. Naganawa and T. Takeuchi, *J. Antibiot.*, 2000, **53**, 1102.

134. T. Takeuchi, M. Igarashi, H. Naganawa and M. Hamada, *Patent* JP 2001-55386A, 2001.

135. T. Motozaki, K. Sawamura, A. Suzuki, K. Yoshida, T. Ueki, A. Ohara, R. Munakata, K.-i. Takao and K.-i. Tadano, *Org. Lett.*, 2005, **7**, 2265.

136. K. Sawamura, K. Yoshida, A. Suzuki, T. Motozaki, I. Kozawa, T. Hayamizu, R. Munakata, K. Takao and K. Tadano, *J. Org. Chem.*, 2007, **72**, 6143.

137. T. Anzo, A. Suzuki, K. Sawamura, T. Motozaki, M. Hatta, K. Takao and K. Tadano, *Tetrahedron Lett.*, 2007, **48**, 8442.

138. (a) A. Tanaka, *Agric. Biol. Chem.*, 1980, **44**, 199; (b) C. Papagerorgiou and C. Benezra, *J. Org. Chem.*, 1985, **50**, 1144; (c) T. Ohgiya and S. Nishiyama, *Heterocycles*, 2004, **63**, 2349.

139. (a) C. R. Hutchinson, *J. Org. Chem.*, 1974, **39**, 1854; (b) J. P. Corbet and C. Benezra, *J. Org. Chem.*, 1981, **46**, 1141; (c) K. Shigetomi, T. Kishimoto, K. Shoji and M. Ubukata, *Heterocycles*, 2006, **69**, 63.

140. K. Shigetomi, T. Kishimoto, K. Shoji and M. Ubukata, *Tetrahedron: Asymmetry*, 2008, **19**, 1444.

141. S. K. Agarwal and R. P. Rastogi, *Phytochemistry*, 1978, **17**, 1663.

142. (a) D. Caine, A. S. Frobese and V. C. Ukachukwu, *J. Org. Chem.*, 1983, **48**, 740; (b) R. M. Ortuno, J. Bigorra and J. Front, *Tetrahedron*, 1987, **43**, 2199; (c) T. Sato, Y. Okumura, J. Itai and T. Fujisawa, *Chem. Lett.*, 1988, **17**, 1537; (d) C. L. Gibson and S. Handa, *Tetrahedron: Asymmetry*, 1996, **7**, 1281; (e) H. Liu, T. Zhang and Y. Li, *Chirality*, 2006, **18**, 223; (f) L. Huawei and Y. Li, *Chin. Chem. Lett.*, 2005, **16**, 716.

143. A. Kamal, T. Krishnaji and P. V. Reddy, *Tetrahedron Lett.*, 2007, **48**, 7232.

Subject Index

Page numbers in *italics* refer to entries in figures.

dienes 13–17, 18
  Diels–Alder reaction 238–42, 333, 345
  nitrodienes 26, 96, 100
  from nucleophilic addition 243, 244, 247–8
  1,4-pentadienes 88, 89
diethyl vinylphosphonate 27
dihydro-1,3-oxazoles 455, 456, 487
7-dihydro-trioxacarcinose B 536
2,3-dihydrobenzofuran carboxylic acid 505, 506
dihydropyridines 14
2,5-dihydropyrroles 398–9, 399
1,2-dihydroquinolines 414, 418, 419
diketones 41–4, 237, 243, 287
diones
  1,4-diazepane-2,5-diones 441, 443
  piperidine-2,6-diones 408, 409
dioxane 81
dioxanone 106–7
diversonol 499, 501
DMAP 92–4, 99–101, 102, 126–7
  as co-catalyst 186, 187, 188
  polyDMAP 193
  and thioureas 177, 179
DMAP-MSN 196
dolaproine 492
dolastatin 492
dolichols 502, 503
donaxaridine 392
dykellic acid 503–4
dynamic asymmetric kinetic transformation (DYKAT) 271, 370, 515, 517
Dysidea frondosa 514

(+)-efaroxan 505, 506
electrochemical catalysis 383
electrohydrocyclization 336, 339
electrophiles 28–49
  α-keto esters, lactones, lactams and diketones 41–4
  aldehydes 28–39, 40
  imines and iminium salts 44–8, 95, 96

ketones 39–41
novel 48–9
electrospray ionization mass spectrometry (ESI-MS/MS) 4, 5, *6*
eleutherobin 535, 538
emiaminal 453, 454
enones
  cyclic 23, 24, 42, 58, 98–100
    Lewis base catalysis 126–7, 129
  cyclization 336, 338, 339
  cyclohexenone 23, 121
enynes 345, 346, 437
eponemycin 507
2-*epi*-epopromycin 107, 108
epopromycin B 107, 108, 505, *507*
epoxidation 295–9, 353–5, 364
2,3-epoxy aldehydes 39
epoxy-β-aminoketone moiety 505, 507
esterification, hydroxyl group 209–16
esters, α-keto 41–4
etherification, hydroxyl group 216
ethyl acrylate 9, 37, 44, 80, 88
ethyl vinyl ketone (EVK) 20–1, 112
eupomatilones 505, 508–9, 511
Euryspongia sp. 514
Evans aldol reaction 517, 519

ferrocenyldialkylphosphines 134
five-membered-ring carbocyclics 327–32
Fleursandol® 510, 511, 513
fluoral 36, 37, 88
fluorides, allyl 216–17
2-fluoroalkylacrylates 216–17
fluoroketones 40–1, 89
fluorous phosphine 59
formaldehyde 28
formalin 28
formyl group reactivity, isoxazole ring 34, *35*
2-formylimidazole 33, 34
free radical reactions 284–90
Friedel–Crafts reaction 222–7
  ring-opening 343, 344
frondosins *511*, 512, 514–15